Frontiers in Protein Folding and Related Areas – in Memory of Professor Sir Christopher M. Dobson (1949–2019)

Frontiers in Protein Folding and Related Areas – in Memory of Professor Sir Christopher M. Dobson (1949–2019)

Editors

Kunihiro Kuwajima
Yuko Okamoto
Tuomas Knowles
Michele Vendruscolo

MDPI • Basel • Beijing • Wuhan • Barcelona • Belgrade • Manchester • Tokyo • Cluj • Tianjin

Editors
Kunihiro Kuwajima
Department of Physics
School of Science
University of Tokyo
Tokyo
Japan

Yuko Okamoto
Department of Physics
School of Science
Nagoya University
Nagoya
Japan

Tuomas Knowles
Centre for Misfolding Diseases
Yusuf Hamied Department
of Chemistry
University of Cambridge
Cambridge
United Kingdom

Michele Vendruscolo
Centre for Misfolding Diseases
Yusuf Hamied Department
of Chemistry
University of Cambridge
Cambridge
United Kingdom

Editorial Office
MDPI
St. Alban-Anlage 66
4052 Basel, Switzerland

This is a reprint of articles from the Special Issue published online in the open access journal *Molecules* (ISSN 1420-3049) (available at: www.mdpi.com/journal/molecules/special_issues/proteinfolding_areas).

For citation purposes, cite each article independently as indicated on the article page online and as indicated below:

LastName, A.A.; LastName, B.B.; LastName, C.C. Article Title. *Journal Name* **Year**, *Volume Number*, Page Range.

ISBN 978-3-0365-7321-2 (Hbk)
ISBN 978-3-0365-7320-5 (PDF)

Cover image courtesy of St John's College, University of Cambridge

Photo of Chris Dobson.

Contents

About the Editors

Kunihiro Kuwajima

Kunihiro Kuwajima is a Professor Emeritus at University of Tokyo, Institute for Molecular Science, and SOKENDAI (the Graduate University for Advanced Studies). He was a Professor at the Okazaki Institute for Integrative Bioscience and at the Institute for Molecular Science (January 2007–March 2013); served as the Dean of the Graduate School of Physical Sciences at SOKENDAI (April 2008–March 2010); was the President of the Protein Science Society of Japan (April 2010–March 2012); was the representative of the Scientific Research on Priority Areas "Water and Biomolecules", supported by MEXT (the Ministry of Education, Culture, Sports, Science, and Technology of Japan) for 5 years starting from 2003 and with Professor Chris Dobson as an International Advisor in this Priority Area; and ran a biophysics lab at the Department of Physics, University of Tokyo, for 16 years starting from 1992. His research has aimed to elucidate the molecular mechanisms of protein folding using a variety of biophysical techniques.

Yuko Okamoto

Yuko Okamoto is a Professor Emeritus at Nagoya University. He received a B.S.-M.S. in Physics from Brown University in 1979 (Grew Foundation Scholar) and a Ph.D. in Physics from Cornell University in 1984. After his postdoctoral work at Virginia Polytechnic Institute and State University, he worked as an Assistant Professor (later, an Associate Professor) at Nara Women's University from 1986 to 1995. He moved to the Institute for Molecular Science and The Graduate University for Advanced Studies (joint appointment) in 1995 as an Associate Professor. He then moved to Nagoya University as a Professor of Biophysics in 2005 and stayed in this position for 17 years until 2022. Since 2022, he has been an Invited Faculty at Information Technology Centre and a Specially Appointed Professor at Global Engagement Centre, Nagoya University, and a Senior Visiting Scientist at RIKEN. In 2011, he was an Overseas Visiting Scholar at St John's College, University of Cambridge, where Professor Chris Dobson was his host and Master of the College. His research has focused on the development of enhanced sampling methods in molecular simulations, including replica exchange molecular dynamics and other generalized ensemble algorithms, and their applications to computational physics/chemistry/biology problems (such as protein folding/misfolding, ligand binding, and the prediction of three-dimensional structures of molecules).

Tuomas Knowles

Tuomas Knowles is a Professor of Physical Chemistry and Biophysics and a Co-Director of the Centre for Misfolding Diseases at the Department of Chemistry of the University of Cambridge. His group studies the physical and chemical aspects of the behaviour of biomolecules. Much of the work of the group has been focused on the self-assembly of protein molecules, both in health and disease. The correct assembly state of protein molecules is an essential requirement for functionality, and, conversely, misassembly is commonly an event that triggers disease. Over the past ten years, Knowles has focused on elucidating the mechanisms by which proteins assemble into aberrant structures in the context of protein misfolding diseases, discovering key molecular pathways and ways to modulate them as the basis for therapeutic intervention strategies.

Michele Vendruscolo

Michele Vendruscolo is a Professor of Biophysics, the Director of the Chemistry of Health Laboratory and a Co-Director of the Centre for Misfolding Diseases at the Department of Chemistry of the University of Cambridge, where he moved over 20 years ago. His work aims to establish the fundamental principles of protein homeostasis and protein aggregation and to exploit these principles to develop methods for drug discovery in neurodegenerative diseases. He has published over 500 scientific papers and 20 patents and has given over 500 invited lectures at international meetings. He is the co-founder of Wren Therapeutics, a drug discovery company that targets protein misfolding diseases.

Preface to "Frontiers in Protein Folding and Related Areas – in Memory of Professor Sir Christopher M. Dobson (1949–2019)"

Protein folding is a fundamental theme in molecular biology. Understanding the molecular mechanisms of this process has challenged molecular biologists for over half a century. Although computational methods have now achieved remarkable success in the prediction of native structures, the underlying principles of the protein folding process have yet to be fully elucidated. In addition, we still have an incomplete understanding of the components of the protein homeostasis system, which controls protein folding in the cellular environment. This knowledge is essential, as errors in protein folding may lead to misfolding and aggregation, a phenomenon closely related to a wide range of human disorders, including Alzheimer's and Parkinson's diseases and type II diabetes.

Professor Chris Dobson was a Director of Research at the University of Cambridge and the John Humphrey Plummer Professor of Chemical and Structural Biology. He was also the 44th Master of St John's College and a Deputy Vice-Chancellor of the University of Cambridge. During his time as Master of St John's College, he directed the Quincentenary celebrations of the College in 2011. In 2012, he co-founded the Centre for Misfolding Diseases, Department of Chemistry, University of Cambridge with two of the present Editors (T.K. and M.V.) and became its first Director. He started his academic career as an undergraduate, graduate student and research fellow at the University of Oxford. He then became an Assistant Professor of Chemistry at Harvard University and a Visiting Scientist at MIT, before returning to Oxford, where he was a Professor of Chemistry until moving to Cambridge in 2001. He published more than 800 research papers and review articles, and was a Fellow of the Royal Society and a Foreign Associate of the US National Academy of Sciences. He was the recipient of a range of awards for his scientific research work, including the Davy and Royal Medals of the Royal Society, the Corday Morgan, Interdisciplinary, and Khorana Awards of the Royal Society of Chemistry, and the Hans Neurath and the Stein and Moore Awards of the Protein Society. He was also a recipient of the Heineken Prize for Biochemistry and Biophysics from the Royal Netherlands Academy of Arts and Sciences, and the Feltrinelli International Prize for Medicine from the Accademia Nazionale dei Lincei in Rome. In 2018, he was knighted in the Queen's Birthday Honours for his contributions to Science and Higher Education. He had also been a Distinguished or Endowed Lecturer at more than 40 universities across the world, as well as giving hundreds of talks and seminars at universities, research institutes and schools, across the world. Readers are also referred to the following article by two of the present Editors (T.K. and M.V., Nature Chemical Biology 16, 105 (2020)).

This book is dedicated to the memory of the late Professor Sir Christopher M. Dobson, who made outstanding contributions to the advancement of studies on protein folding, misfolding, and aggregation, as well as their links with human diseases, and played an irreplaceable role in the promotion of protein science.

Kunihiro Kuwajima, Yuko Okamoto, Tuomas Knowles, and Michele Vendruscolo
Editors

molecules

MDPI

Article

The Pathological G51D Mutation in Alpha-Synuclein Oligomers Confers Distinct Structural Attributes and Cellular Toxicity

Catherine K. Xu [1], Marta Castellana-Cruz [1], Serene W. Chen [2], Zhen Du [3], Georg Meisl [1], Aviad Levin [1], Benedetta Mannini [1], Laura S. Itzhaki [3], Tuomas P. J. Knowles [1,4], Christopher M. Dobson [1,†], Nunilo Cremades [5,*] and Janet R. Kumita [3,*]

1 Yusuf Hamied Department of Chemistry, University of Cambridge, Lensfield Road, Cambridge CB2 1EW, UK; catherine.xu@mpl.mpg.de (C.K.X.); mc2033@cam.ac.uk (M.C.-C.); gm373@cam.ac.uk (G.M.); al680@cam.ac.uk (A.L.); bm475@cam.ac.uk (B.M.); tpjk2@cam.ac.uk (T.P.J.K.)
2 Department of Life Sciences, South Kensington Campus, Imperial College London, London SW7 2AZ, UK; weiyan89@gmail.com
3 Department of Pharmacology, University of Cambridge, Tennis Court Road, Cambridge CB2 1PD, UK; zhen_du2020@163.com (Z.D.); lsi10@cam.ac.uk (L.S.I.)
4 Cavendish Laboratory, Department of Physics, University of Cambridge, JJ Thomson Avenue, Cambridge CB3 0HE, UK
5 Institute for Biocomputation and Physics of Complex Systems (BIFI), University of Zaragoza, Mariano Esquillor, Edificio I+D+I, 50018 Zaragoza, Spain
* Correspondence: ncc@unizar.es (N.C.); jrk38@cam.ac.uk (J.R.K.)
† Deceased.

Citation: Xu, C.K.; Castellana-Cruz, M.; Chen, S.W.; Du, Z.; Meisl, G.; Levin, A.; Mannini, B.; Itzhaki, L.S.; Knowles, T.P.J.; Dobson, C.M.; et al. The Pathological G51D Mutation in Alpha-Synuclein Oligomers Confers Distinct Structural Attributes and Cellular Toxicity. *Molecules* **2022**, *27*, 1293. https://doi.org/10.3390/molecules27041293

Academic Editor: Vladimir N. Uversky

Received: 18 January 2022
Accepted: 10 February 2022
Published: 15 February 2022

Abstract: A wide variety of oligomeric structures are formed during the aggregation of proteins associated with neurodegenerative diseases. Such soluble oligomers are believed to be key toxic species in the related disorders; therefore, identification of the structural determinants of toxicity is of upmost importance. Here, we analysed toxic oligomers of α-synuclein and its pathological variants in order to identify structural features that could be related to toxicity and found a novel structural polymorphism within G51D oligomers. These G51D oligomers can adopt a variety of β-sheet-rich structures with differing degrees of α-helical content, and the helical structural content of these oligomers correlates with the level of induced cellular dysfunction in SH-SY5Y cells. This structure–function relationship observed in α-synuclein oligomers thus presents the α-helical structure as another potential structural determinant that may be linked with cellular toxicity in amyloid-related proteins.

Keywords: α-synuclein; toxic oligomers; Parkinson's disease; familial mutations; α-helical structure

1. Introduction

The misfolding of proteins and their aggregation into amyloid fibrils has been implicated in numerous neurodegenerative disorders, including Parkinson's disease (PD) and Alzheimer's disease (AD) [1]. In PD, aggregates of the 14 kDa protein α-synuclein are the major component of Lewy bodies and neurites, which emerge as the pathological hallmarks of the disease. In solution, α-synuclein is intrinsically disordered; however, upon interaction with membranes, the protein has been observed to adopt an α-helical structure [2], associated with the functional role of the protein in neuronal cells [3]. In addition to the random coil to α-helix transition upon membrane binding, α-synuclein can also adopt a β-sheet structure upon self-assembly into amyloid aggregates, a process in which membranes might also play a role [4–6]. Oligomeric species with varying degrees of β-sheet structure are observable in the early stages of aggregation [7]. It is these early oligomeric species, rather than the mature amyloid fibrils, that are believed to be key toxic species in the context of disease [1,8–11]. While a range of mechanisms have been proposed by which the toxic effect could be mediated, interactions with cell membranes are likely to be a main contributor to the observed cytotoxicity [12–14].

Duplications and triplications of the WT α-synuclein gene, and a number of single-point mutations, are associated with familial cases of PD, which present with both earlier onset and faster progression of the disease [15]. The causative relationship between an increased load of the protein, which results from the duplication or triplication of the gene, and earlier onset of the disease is likely to simply be a consequence of the increased aggregation propensity of α-synuclein due to its increased concentration. By contrast, the aetiology of the familial cases associated with the pathological variants remains unknown and a variety of mechanisms and reasons for the connection of these mutations to disease have been proposed [16–24].

All familial PD-associated mutations identified thus far are located in the N-terminal region of α-synuclein (Figure S1A) and have been suggested to alter the membrane binding properties and thus the function of α-synuclein [19,25–34]. However, no clear associations with protein dysfunction have been established yet. While the A30P and G51D mutations have been reported to abrogate α-synuclein–membrane interactions, the E46K variant may enhance membrane binding [25–27,29,32,35]. Furthermore, studies on the in vitro aggregation kinetics of these α-synuclein variants have yielded conflicting results [15,18–24,36–39]. While the E46K, H50Q, and A53T variants have been found to aggregate more rapidly than WT, G51D exhibits a lower propensity for aggregation, and A30P appears to be the most variable in behaviour. Nevertheless, it is clear that aggregation kinetics and membrane binding are insufficient to explain the link between these mutations and disease.

Given that intermediate oligomeric species are proposed to be a major source of toxicity, several studies have sought to characterise the effects of familial PD-associated mutations on α-synuclein oligomers [40–42]. Paslawski et al. used hydrogen/deuterium exchange mass spectrometry to study oligomers of WT α-synuclein and the A30P, E46K, and A53T variants, finding very subtle differences between deuterium exchange profiles [43]. Furthermore, through single-molecule FRET (smFRET) experiments on the same variants, the concentrations of oligomers produced in aggregation reactions were found to be the same for all variants, indicating that structural differences between variant oligomers are likely to have a more significant pathological effect than simply levels of oligomers [44]. Moreover, this study also identified differences in the intermolecular FRET efficiencies between variant oligomers, again demonstrating that these mutations affect oligomer structure.

Detailed structural characterisation of oligomers generated in situ in aggregation reactions is highly challenging due to their heterogeneous and transient nature, whereas the use of stable kinetically trapped model systems allows us to obtain in-depth structural and biological information on the nature of these species [13,43]. Here, we characterise the effects of familial PD-associated mutations on α-synuclein oligomers, and reveal a distinct α-helical structural polymorphism within the G51D oligomers that correlates with increased cellular dysfunction in SH-SY5Y cells.

2. Results and Discussion

2.1. All α-Synuclein Variants Form Oligomers with Similar Size and Morphology

Oligomers from the familial PD-associated α-synuclein variants were successfully generated using our previously described protocols [45]. The oligomers were characterised using transmission electron microscopy (TEM), which showed that these variant oligomers have a similar size and overall morphology to the WT oligomers, being approximately spherical with a diameter of around 5–15 nm, consistent with dynamic light scattering analysis, which showed a clear size distinction between those of the monomers and oligomers (Figure 1, left panels; Figure S2). Detailed investigation of the oligomer size distributions by analytical ultracentrifugation (AUC) sedimentation velocity analysis and native-PAGE demonstrated conserved stable sizes for the oligomers, with all variant oligomers, except for the G51D variant, containing oligomer populations at both 10S and 15S, in different proportions (Figure 1, right panels; Figure S1B). While the A53T variant contained an additional population at 19S, and a lowly populated species at 24S, suggesting the presence of larger aggregate species for this variant, the distribution of G51D oligomers showed just one major population around 12S, suggesting that there may be interactions that restrict the preferred sizes of such oligomers.

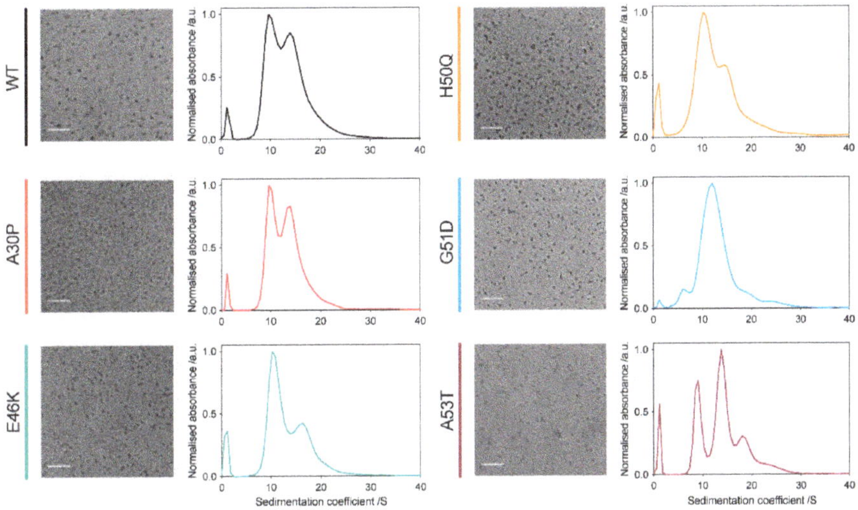

Figure 1. All α-synuclein variants form oligomers with similar size and morphology. TEM images (**left** panels) of variant oligomers, confirming that relatively homogeneous oligomer populations are produced in all cases, with a roughly spherical shape and 5–15 nm diameter (scale bar = 100 nm). AUC analysis of variant oligomers (**right** panels), demonstrating the size distributions of oligomers within each sample, with the peak at 2S arising due to the residual monomeric protein present in the oligomeric samples.

2.2. G51D Oligomers Display Marked Structural Differences, including Increased α-Helical Content and Decreased Surface Hydrophobicity

During the analysis of G51D oligomers, we found that G51D oligomers had a higher molar extinction coefficient (12,444 M^{-1} cm^{-1}) than WT oligomers (7000 M^{-1} cm^{-1}); in all variants, the oligomer extinction coefficient was higher than that of the monomeric protein (5600 M^{-1} cm^{-1}). These differences were determined by comparing the UV-vis absorbance spectra with protein quantification by the following three complementary methods: amino acid analysis, bicinchoninic acid assay, and oligomer denaturation followed by SDS-PAGE to determine the monomer concentration (Figure S3). We further analysed the spectral properties of the oligomers, finding marked differences between the intrinsic tyrosine fluorescence spectra of the variants. The monomeric proteins displayed a maximum fluorescence emission peak at 305 nm, typical of tyrosine (Figure 2A). However, both the WT and G51D oligomers displayed a maximum emission peak around 345 nm, suggesting the formation of tyrosinate in the excited state [46,47]. The intensity of this 345 nm peak is much stronger in G51D oligomers while the tyrosine emission at 305 nm is less intense compared to WT oligomers, suggesting a stronger stabilisation of the tyrosinate form in the excited state in the G51D oligomers than in the WT protein, which explains the increased extinction coefficient observed for G51D oligomers (Figure 2B) [46,47].

α-Synuclein contains 4 tyrosine residues, located at positions 39, 125, 133, and 136 (Figure 2C). The latter three, located in the C-terminus of the protein, are unlikely to be involved in structural rearrangements as this region remains unstructured in both the membrane-bound α-helical conformation and the aggregated oligomer and fibrillar forms [13]. Y39 is thus the only tyrosine residue that is likely to be in a different local environment in the monomeric and oligomeric species, so we employed the Y39F variant to investigate its role in the spectral properties of the oligomers. We compared the circular dichroism (CD) spectra of the Y39 oligomers to the WT oligomers (Figure 2D), confirming their similar secondary structure, and then measured the intrinsic fluorescence properties of the monomeric and oligomeric forms of this variant (Figure 2E,F). As predicted, Y39F oligomers display tyrosine fluorescence emission at 305 nm and they lack the strong 345 nm emission that is present in the WT oligomers and accentuated in the G51D oligomers, suggesting structural differences between the monomeric and oligomeric conformations in the N-terminal region of the protein, around position 39, which are stronger in the case of the G51D pathological variant. Overall, these data highlight the structural differences between G51D oligomers and the oligomers from the other variants.

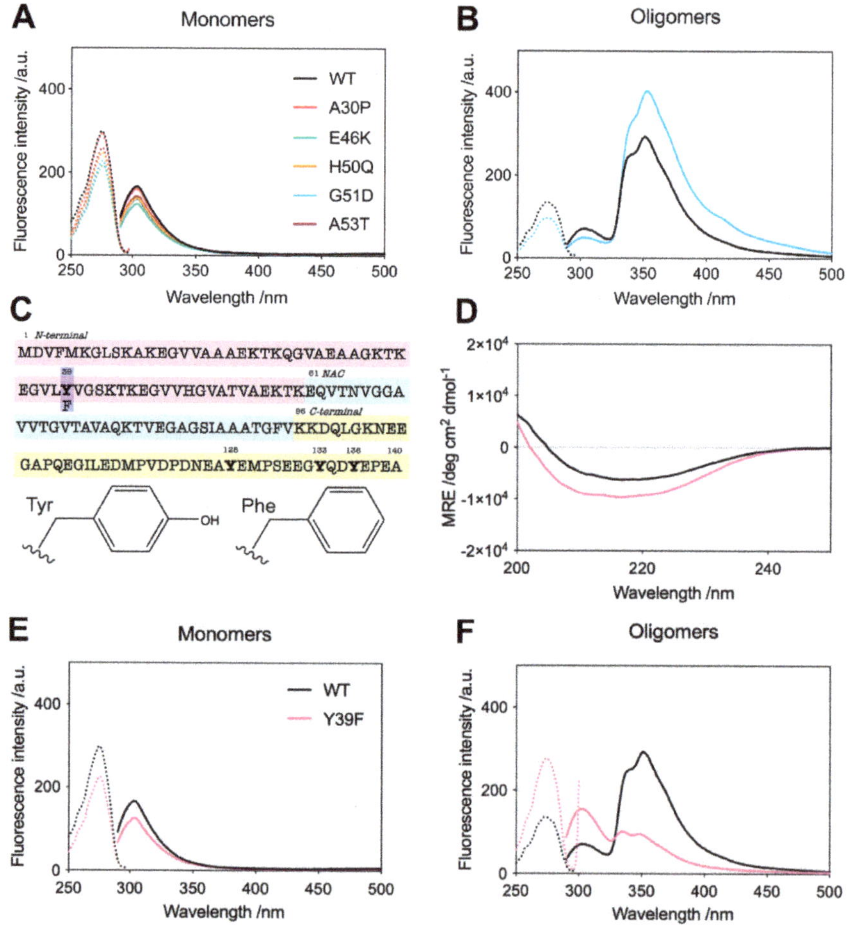

Figure 2. (**A**) Excitation (dotted lines) and fluorescence emission (solid lines) spectra for the monomeric α-synuclein variants. (**B**) Excitation (dotted lines) and emission (solid lines) spectra for the WT (black) and G51D (blue) oligomers. (**C**) Primary sequence of α-synuclein showing the location of Y39 and the change from Tyr-to-Phe. (**D**) CD spectra of the WT (black) and Y39F (pink) oligomers. (**E**) Excitation (dotted lines) and emission (solid lines) spectra for the WT (black) and Y39F (pink) monomers. (**F**) Excitation (dotted lines) and emission (solid lines) spectra for the WT (black) and Y39F (pink) oligomers. All spectra are representative of three independent experiments.

Further analysis of the oligomers by Fourier transform infrared (FTIR) and CD spectroscopies revealed that all the oligomers contain a β-sheet structure, intermediate between that of their respective disordered monomeric and β-sheet-rich fibrillar states (Figures 3 and 4). The FTIR spectra revealed that all oligomers contain an antiparallel β-sheet structure, in contrast to the dominant parallel β-sheet structure of the fibrils, as previously noted [45,48]. Remarkably, for the G51D oligomer preparations, we observed differing amounts of additional α-helical content by CD spectroscopy, despite all oligomers being prepared under identical conditions (Figure 4).

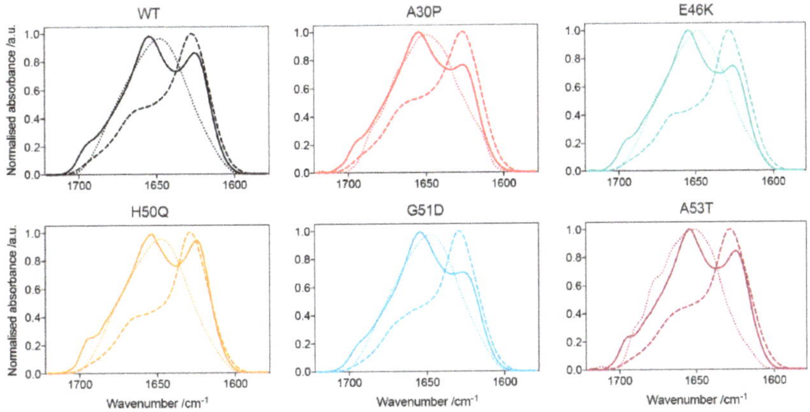

Figure 3. FTIR spectra of the variant α-synuclein species (solid-line: oligomers; dotted-line: monomers; dashed-line: fibrils). Peaks around 1695 cm^{-1} are characteristic of an antiparallel β-sheet structure, observed only in the oligomeric samples. Monomers display spectra typical of a random coil structure, whereas fibrils exhibit a significant degree of intermolecular β-sheet structure (peak around 1625 cm^{-1}), with approximately double the β-sheet content of the oligomeric species.

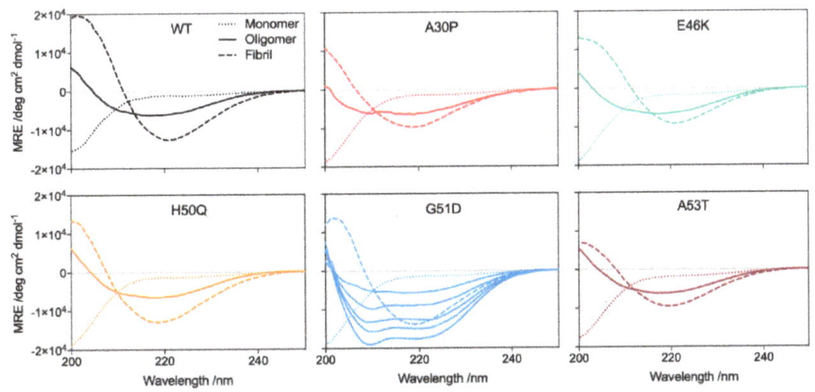

Figure 4. Representative ($n > 5$) CD spectra of the α-synuclein variants (solid line: oligomers; dotted line: monomers; dashed line: fibrils). All variant oligomers display a β-sheet content intermediate between that of the respective monomers and fibrils. Several G51D oligomer preparations are shown indicating the presence of a variable helical content.

We explored the possible experimental factors that may have influenced this heterogeneity in the G51D oligomer structure. This structural variation was not due to monomer modification, for example, by oxidation or the presence of strongly bound metal ions or detergents, since oligomers successively produced from the same monomer source yielded different structures (Figure 5A,B). We next investigated the lyophilisation process, during which oligomers are formed [48]. Lyophiliser vacuum pump pressures ranging between 0.03 and 0.7 mBar led to the formation of oligomers that all contained α-helical content, indicating that the pump pressure is not sufficient to determine oligomer structure (Figure 5C). However, the preparation of oligomers on different occasions from aliquots that were lyophilised together but stored at −20 °C for different lengths of time resulted in oligomers with identical structures, demonstrating that oligomer structure is determined during lyophilisation (Figure 5D). Despite all the other α-synuclein protein variants being treated identically, only G51D formed oligomers with significant α-helical content. The distinct structural polymorphism in G51D oligomers, therefore, appears to arise from inherent differences in the aggregation process under limited hydration conditions [48].

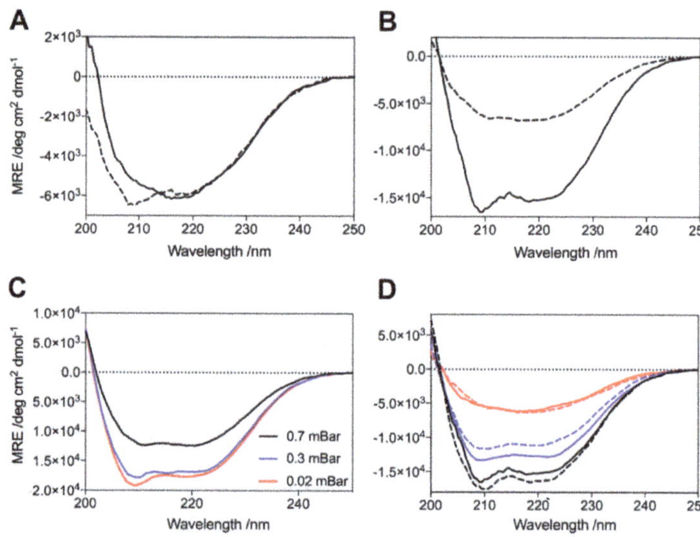

Figure 5. Investigation of the origins of the α-helical content of the G51D oligomers. (**A**,**B**) Oligomers produced from the flow-throughs of previous oligomer preparations do not present the same structure as the previous oligomers. Two examples are shown comparing the CD spectra of the first oligomers (solid lines) with the second oligomers (dashed). (**C**) Lyophiliser vacuum pump pressure does not dictate oligomer structure. The same monomer purification batch was lyophilised at three different vacuum pump pressures; all resulting oligomers contained α-helical structure, with no clear dependence on pressure. (**D**) Oligomer structure is determined during lyophilisation. Spectra are shown for three different preparations of lyophilised monomer (red, blue, black). Half of each batch was prepared immediately after lyophilisation (solid lines) and the other half stored at −20 °C prior to oligomer preparation on a later day (dashed lines). In each case, the structures of oligomers produced from the same lyophilised monomer stock were almost identical.

Although the different G51D oligomer preparations gave rise to distinctive α-helical spectra in the CD analysis, their corresponding FTIR spectra were comparable. The FTIR spectra indicate that all preparations contained a similar β-sheet content, and a similar combined content of α-helical and random coil structure. These results suggest that the variations in the α-helical content evident by CD are compensated for by concomitant changes in the random coil content such that the β-sheet content is largely unchanged, resulting in similar FTIR spectra. Furthermore, the β-sheet content of the G51D oligomers was also similar to that of the other variant oligomers (Figure 6C,D) [45]. These results suggest that the α-helical structure arises from regions that remain disordered in the WT and other variant oligomers. Despite the differences in the secondary structural content, no variations in the size and morphology of the helical G51D oligomers were observed (Figure 6A,B).

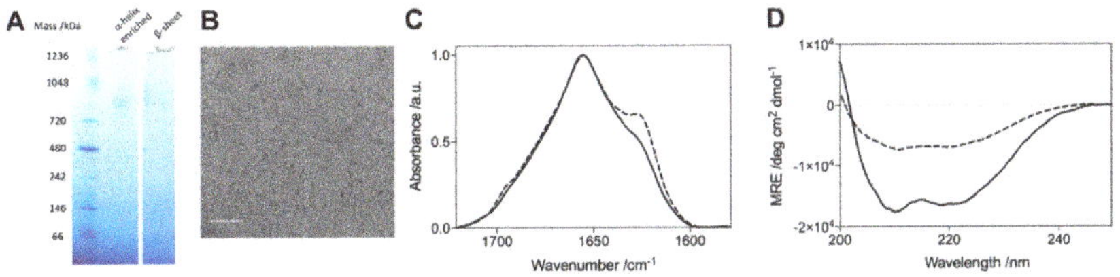

Figure 6. Comparison of G51D oligomers with a varying α-helical content by (**A**) Native-PAGE stained with Coomassie blue; (**B**) TEM image of the G51D oligomers with an increased helical content (scale bar = 100 nm); representative FTIR (**C**) and CD (**D**) spectra of G51D oligomers with increased α-helicity (solid line) and G51D oligomers with very little α-helical content (dashed line).

2.3. G51D Oligomer Polymorphs with High Helical Content Exhibit the Highest Cytotoxicity

With this set of oligomeric species, we sought to investigate their toxicity towards cells using the MTT test, an indicator of cellular stress [49]. We previously investigated the toxicity of WT oligomers to SH-SY5Y cells, and the MTT reduction to 83% (standard deviation = 8%) reported here is in good agreement with previous studies [50]. Surprisingly, unlike the A30P, E46K, H50Q, and A53T variant oligomers that showed insignificant MTT reduction effects under the experimental conditions used, G51D oligomers displayed significantly higher toxicity than the WT oligomers (Figure 7A).

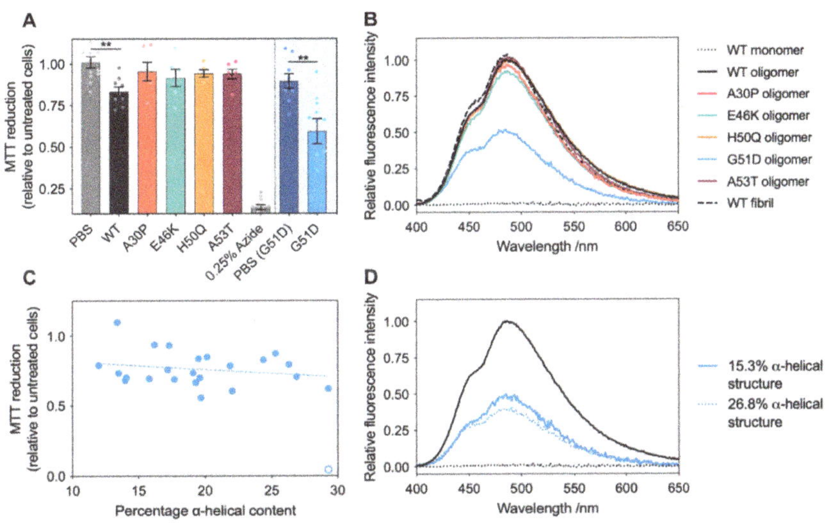

Figure 7. (**A**) Variant oligomer toxicity tested with the MTT assay on SH-SY5Y cells, with error bars showing the standard error ($n \geq 5$) and data for all individual replicates overlaid as points. Due to the very low concentration yields of G51D oligomers, an additional higher volume PBS control was required, shown as PBS (G51D). Statistical analysis was run using one-way ANOVA (all variants except for G51D) or Student's *t*-test (G51D only) ** ($p \leq 0.01$). (**B**) Representative ANS fluorescence emission spectra of variant oligomers ($n \geq 3$). (**C**) MTT reduction as a function of the α-helical content of G51D oligomers as estimated by far-UV CD spectra deconvolution [51,52]. The fitted linear relationship between the toxicity and percentage α-helical content in G51D oligomers is shown by the dotted line (the outlier data point (open circle) was not included in this analysis). (**D**) Representative ANS fluorescence emission spectra of WT (black) and G51D oligomers (blue) with different degrees of α-helical structure.

Oligomer toxicity has been shown to be largely dependent on high surface hydrophobicity, proposed to increase oligomer toxicity by increasing the oligomers' affinity for the membrane interior via non-specific hydrophobic interactions, thus facilitating membrane disruption and cellular dysfunction [53–55]. We therefore investigated if this was a key factor contributing to the increased toxicity observed for the G51D oligomers by measuring the interactions of the variant oligomers with 8-anilino-naphthalene sulphate (ANS), a dye whose fluorescence emission is enhanced upon binding to solvent-exposed hydrophobic regions in the proteins [54,56]. Interestingly, whereas A30P, E46K, H50Q, and A53T oligomers showed similar solvent-accessible hydrophobicity to the WT oligomers, the oligomers generated by the G51D variant exhibited a significantly reduced hydrophobic surface (Figure 7B). Based on these studies, it would therefore be expected that G51D would be the least toxic of all the α-synuclein variants. Furthermore, as all variant oligomers exhibited the same size range, the previously identified toxicity determinants of small size and high hydrophobicity were clearly not sufficient to explain the dramatically higher cytotoxicity of the G51D oligomers [56].

Given that the variation in cellular dysfunction between experiments for the G51D oligomers was higher than any of the other variants, and the observation that G51D oligomers can have variable degrees of α-helical content, we probed further whether the increased variance in the measured cell toxicity may be linked to this observed structural polymorphism. In order to explore whether the variation in the α-helical content of G51D oligomers correlates with changes in cellular dysfunction, we characterised structurally distinct G51D oligomers. By deconvoluting the CD spectra, we were able to estimate the relative secondary structural content of oligomer preparations [51,52]. These fits reproduced our experimental data with extremely low residuals, indicating that this is a robust method for comparatively analysing our spectra (Figure S4). Deconvolution of CD spectra of WT oligomers suggests that they contain around 11% of α-helical structure, which was not detected by solid-state NMR analysis [13], indicating that the percentage of α-helical content we report here should only be used as a relative quantification between oligomer samples.

Having already observed that the size and morphology of G51D oligomers do not vary with changes in their secondary structure (Figure 6), we additionally confirmed that preparations with different degrees of α-helical structure yielded almost identical hydrophobicity readouts by ANS (Figure 7D). Furthermore, the WT and G51D oligomers were detected by the A11 antibody (proposed to bind toxic oligomers) with similar affinities (Figure S6) [57]. We thus found that the secondary structure content is the only significant structural and morphological difference between these G51D oligomer preparations. Even considering the large variability observed in the MTT assay measurements (Figure 7A), upon testing the toxicities of these distinct G51D oligomers, we identified a clear trend between an increased α-helical content and increased cellular dysfunction (Figure 7C). However, no correlation was observed between the cell toxicity and level of β-sheet or random coil structures, suggesting that the variations in cellular dysfunction can be primarily attributed to the changes in the α-helical content (Figure S5). These results thus suggest that α-helical content constitutes an additional determinant of α-synuclein oligomer toxicity.

Indeed, α-helical content may be particularly significant in the context of the aggregation and toxicity of α-synuclein, given its propensity to form highly α-helix-rich structures upon binding to lipids, which is believed to be the first requirement for triggering oligomer-mediated cell damage [58]. Moreover, α-helical content has previously been observed during the aggregation of several α-synuclein variants [59,60].

Detailed work on the WT oligomers generated through our methods has identified the mechanistic features of α-synuclein oligomer-induced membrane disruption: first, the disordered N-terminal regions of the oligomers were found to act as anchors to the membrane by folding into α-helices, allowing the structured hydrophobic β-sheet core of the oligomer to insert into the interior of the lipid bilayer [13]. The membrane anchoring step, therefore, seems to be critical for the induction of toxicity of α-synuclein oligomers. In our study, despite the lower hydrophobic nature of the G51D oligomers, we observed an enhanced cellular toxicity, which is correlated with an increased degree of α-helical structure, relative to the WT oligomers, which, according to our tyrosinate fluorescence emission analysis, is likely to occur in the N-terminal region of the protein, close to residue Y39. This may suggest that G51D oligomers have a significant pre-formed helical structure in the N-terminal region of the protein that facilitates more efficient anchoring of the oligomers to the membranes. This may allow for a more efficient insertion of the hydrophobic β-sheet core into the lipid bilayer, thus causing membrane disruption, or alternatively, the induction of other cellular dysfunction mechanisms triggered at the plasmatic membrane level that do not involve membrane bilayer perturbation. In support of the latter toxicity mechanism, G51D oligomers have previously been reported to present a much reduced membrane disruption ability relative to other variants [33]. Regardless of the specific mechanisms involved, our results indicate that amyloid oligomer toxicity may not be solely determined by oligomer surface hydrophobicity but is likely to also be dependent on other structural features of the oligomers.

3. Materials and Methods

3.1. Preparation of Oligomers

Oligomers were prepared as previously described [45]. Briefly, α-synuclein was purified into PBS [61], and subsequently dialysed against water (4 L) (ON, 4 °C). In total, 6 mg aliquots were lyophilised (48 h), followed by resuspension in buffer (500 μL PBS). The resuspended protein was passed through 0.22 μm filters and incubated (20–24 h, 37 °C). The samples were ultracentrifuged (1 h, 288,000 rcf, 20 °C) in a TLA 120.2 rotor, using an Optima TLX Ultracentrifuge (both Beckman Coulter, High Wycombe, UK) to remove aggregates and large oligomers. Remaining monomer was removed using a 100 kDa centrifugation filter (4 × 2 min, 9300 rcf). The flow-through containing predominantly monomer from the first three passes was kept and reused up to five times. The oligomer concentration was determined using UV-vis spectroscopy, using molar extinction coefficients of 7000 M^{-1} cm^{-1} for WT, E46K, H50Q, and A53T, and 12,444 M^{-1} cm^{-1} for A30P and G51D, with molar extinction coefficients determined using amino acid analysis and BCA assays.

3.2. Bicinchoninic Acid Assay

Bicinchoninic acid (BCA) assays were performed using a kit and bovine serum albumin (BSA) (both Thermo Scientific, Rockford, IL, USA), and carried out in Corning 96 well plates (3635). In total, 200 μL of working reagent were added to 25 μL of sample, and incubated at 37 °C for 45 min. Following incubation, the absorbance at 562 nm of each sample was recorded on a FLUOstar Optima plate reader (BMG Labtech, Aylesbury, UK). A standard curve was generated using concentrations of stock BSA between 0 and 250 μg mL^{-1}, which was used to determine the protein concentrations in the sample. In order to account for potential differences in the behaviour of BSA and α-synuclein in the assay, samples were normalised to a known α-synuclein standard sample.

3.3. ANS Binding

8-Anilino-1-sulfonic acid (ANS) was added to samples (5 μM protein) to a final concentration of 250 μM and subsequently incubated (30 min, 20 °C). Fluorescence emission spectra were recorded between 400 and 650 nm with an excitation wavelength of 350 nm, using a Cary Eclipse Fluorescence spectrophotometer (Agilent, Santa Clara, CA, USA).

3.4. Circular Dichroism Spectroscopy

α-Synuclein far-UV spectra were recorded in a quartz cuvette (1 mm path length), on a JASCO J-810 equipped with a Peltier thermal-controlled cuvette holder (Jasco (UK) Ltd., Dunmow, UK) at 20 °C. In total, 15–30 spectra were averaged and recorded between 250 and 200 nm, with a data pitch of 0.5 nm, bandwidth of 1 nm, scanning speed of 50 nm/min, and response time of 4 s. Spectra were deconvoluted using the BestSel web server [51,52].

3.5. Dot Blot Analysis

In total, 1 μg of monomer or oligomer was deposited onto a 0.2 μm PVDF membrane (Millipore (UK) Ltd., West Lothian, UK) and left to dry at room temperature (RT). The membranes were then blocked (5% (*w*/*v*) BSA in PBS, 1 h, RT) and subsequently incubated with A11 or 211 antibody in 5% BSA in PBS (overnight, 4 °C) at 1:5000 and 1:2000 dilutions, respectively. Membranes were washed in PBS + 0.01% Tween-20 (PBST) (3 × 10 min, RT) then incubated with secondary antibody (Alexa Fluor-488 goat anti-mouse for A11, and Alexa Fluor-488 goat anti-rabbit for 211, both 1:5000 (Thermo Fisher Scientific, Waltham, MA, USA)) in PBST (1 h, RT). Following washing in PBST (3 × 10 min, RT), membranes were imaged on a Typhoon Trio scanner and the images analysed using ImageQuant TL v2005 (both Amersham Bioscience, Little Chalfont, UK).

3.6. FTIR Spectroscopy

FTIR measurements were performed on a Vertex 70 (Bruker, Billerica, MA, USA), fitted with a Platinum ATR (Diamond F) (oligomer and fibril measurements) or BioATRCell II (Bruker, Billerica, MA, USA). A 2–15 μM oligomer sample (2 μL) was deposited onto the detector and dried, followed by washing with milliQ water. In total, 5 spectra, each averaged over 128 scans, with atmospheric compensation and background correction, were recorded per sample. Recorded spectra were baseline corrected in the 1720–1580 cm^{-1} (amide I) region, and normalised.

3.7. Intrinsic Fluorescence Spectroscopy

Intrinsic fluorescence spectra were recorded on a Cary Eclipse fluorescence spectrophotometer. For emission spectra, samples were excited at 276 nm while for excitation spectra, emission was monitored at 305 nm.

3.8. MTT Cell Viability Assay

Human SH-SY5Y neuroblastoma cells (A.T.C.C., Manassas, VA, USA) were cultured in 1:1 Dulbecco's modified Eagle's medium (DMEM)-F12+GlutaMax supplement (Thermo Fisher Scientific, Waltham, MA, USA) supplemented with 10% foetal bovine serum. The cells were maintained in a 5.0% CO_2 humidified atmosphere at 37 °C and grown to 80% confluence for a maximum of 20 passages. SH-SY5Y cells were plated in a 96-well plate at a concentration of 10,000 cells/well and treated for 24 h at 37 °C with the different α-synuclein species. After this, the cells were incubated with 0.5 mg/mL MTT 23 (3-(4,5-dimethylthiazol-2-yl)-2,5-diphenyltetrazolium bromide) in RPMI (Thermo Fisher Scientific, Waltham, MA, USA) solution at 37 °C for 4 h and subsequently lysed with a solution of 20% SDS, 50% N,N-dimethylformamide, pH 4.7 at 37 °C for 2 h. Absorbance values of blue formazan were determined at 590 nm using a Clariostar plate reader (BMG Labtech, Aylesbury, UK). We further verified that the observed differences in the MTT reduction between the α-synuclein samples was not due to LPS contamination (Figure S7).

3.9. Native Polyacrylamide Gel Electrophoresis

Samples were mixed with 4X loading buffer and run on NativePAGE™ 4–16% Bis-Tris gels (Thermo Scientific, Rockford, IL, USA), alongside NativeMark™ Unstained Protein Standard (Invitrogen by Thermo Fisher Scientific, Carlsbad, CA, USA), for 105 min at 125 V using NativePAGE™ Cathode Buffer Additive and NativePAGE™ Running Buffer (both Novex, Carlsbad, CA, USA). Gels were destained with a mixture of water, ethanol, and acetic acid (5:4:1 volume ratio).

3.10. Transmission Electron Microscopy

Each sample (10 μM, 10 μL) was adsorbed onto carbon-coated 400 mesh, 3 mm copper grids (EM Resolutions, Saffron Walden, UK). Once dry, grids were washed (2 × 10 μL water), followed by staining with 2% (*w*/*v*) uranyl acetate, and further washes (2 × 5 μL water). The samples were imaged on a FEI Tecnai G2 transmission electron microscope operating at 80 kV (Cambridge Advanced Imaging Centre (CAIC), University of Cambridge, Cambridge, UK). Images were analysed using the SIS Megaview II Image Capture system.

4. Conclusions

In conclusion, our findings demonstrate that the structure of stable oligomers formed from a number of mutational variants (A30P, E46K, A53T, H50Q) are very similar to the WT protein; however, distinct structural polymorphs are produced from the G51D variant. Despite the decreased hydrophobicity within the G51D oligomers, they display more cytotoxicity than the other variant oligomers. Previous extensive experimental work has established size and hydrophobicity as the key molecular determinants for oligomer toxicity [7,44,56,62]; our results indicate that additional oligomer structure–toxicity relationships may exist. The ability to study the polymorphic G51D oligomers allowed us to show that the α-helical content is a good predictor of their cytotoxicity and thus constitutes an additional structural attribute that correlates with cellular dysfunction. Secondary structure polymorphism in fibrils has been studied in detail in the context of strains and toxicity [63–65]. Although the vast majority of such structures consist predominantly of the canonical amyloid β-sheet structure, recent work has identified the α-helical content as a key structural element in fibrils of the most toxic member of the phenol-soluble modulin (PSM) family, with toxicity being clearly linked to this secondary structure element [66,67]. In light of the profound effect of small structural changes on the properties of fibrils, the significant intrinsic heterogeneity of oligomer structures, in particular our findings of the G51D oligomers, may be a key determinant of their aggregation propensity and toxicity.

Supplementary Materials: The following supporting information can be downloaded online. Figure S1 (primary structure of α-synuclein and native-PAGE analysis of oligomers); Figure S2 (dynamic light scattering analysis of variant monomers and oligomers); Figure S3 (assays to confirm extinction coefficients of oligomer species); Figure S4 (BestSel fitting of experimental CD data); Figure S5 (analysis of the relationship between G51D oligomer secondary structure and MTT reduction); Figure S6 (dot blot analysis of WT and G51D oligomers with conformational specific antibody A11); Figure S7 (quantification of LPS in α-synuclein stocks).

Author Contributions: Conceptualization, C.K.X., C.M.D., N.C. and J.R.K.; methodology, C.K.X., M.C.-C., S.W.C., Z.D., G.M., A.L. and B.M.; formal analysis, C.K.X., M.C.-C., S.W.C., Z.D., N.C. and J.R.K.; writing—original draft preparation, C.K.X., M.C.-C., N.C. and J.R.K.; writing—review and editing, C.K.X., M.C.-C., S.W.C., Z.D., G.M., A.L., B.M., L.S.I., T.P.J.K., N.C. and J.R.K.; supervision, L.S.I., T.P.J.K., C.M.D., N.C. and J.R.K.; funding acquisition, C.M.D., N.C. and J.R.K. All authors have read and agreed to the published version of the manuscript.

Funding: This research was supported by the Cambridge Centre for Misfolding Diseases and funded, in part, by the Wellcome Trust (094425/Z/10/Z; J.R.K. and C.M.D.) and the Ministry of Economy and Competitiveness of Spain (MINECO RYC-2012-12068), and Ministry of Science, Innovation and Universities of Spain (MICIU/AEI/FEDER PGC2018-096335-B100) (N.C.). This work was also supported by Herchel Smith research studentships (C.K.X. and Z.D.), a Ramon Jenkins Fellowship from Sidney Sussex College Cambridge (G.M.), and an Oppenheimer Early Career Fellowship (A.L.). J.R.K. is currently supported by an MRC Career Development Award (MR/W01632X/1).

Institutional Review Board Statement: Not applicable.

Informed Consent Statement: Not applicable.

Data Availability Statement: Not applicable.

Acknowledgments: Dedicated to the memory of Christopher M. Dobson FRS, who helped and inspired us over many years and who is missed greatly by all who knew him. We thank Karin Müller and the Cambridge Advanced Imaging Centre for their support in acquiring TEM data. We also thank Patrick Flagmeier, Suman De, Francesco Simone Ruggeri, and David Klenerman for helpful discussions and input.

Conflicts of Interest: The authors declare no conflict of interest. The funders had no role in the design of the study; in the collection, analyses, or interpretation of data; in the writing of the manuscript, or in the decision to publish the results.

Sample Availability: DNA plasmids are available from the authors.

References

1. Chiti, F.; Dobson, C.M. Protein misfolding, amyloid formation, and human disease: A summary of progress over the last decade. *Annu. Rev. Biochem.* **2017**, *86*, 27–68. [CrossRef] [PubMed]
2. Fusco, G.; De Simone, A.; Gopinath, T.; Vostrikov, V.; Vendruscolo, M.; Dobson, C.M.; Veglia, G. Direct observation of the three regions in alpha-synuclein that determine its membrane-bound behaviour. *Nat. Comm.* **2014**, *5*, 3827. [CrossRef] [PubMed]
3. Snead, D.; Eliezer, D. Alpha-synuclein function and dysfunction on cellular membranes. *Exp. Neurobiol.* **2014**, *23*, 292–313. [CrossRef] [PubMed]
4. Auluck, P.K.; Caraveo, G.; Lindquist, S. α-Synuclein: Membrane interactions and toxicity in Parkinson's disease. *Annu. Rev. Cell Dev. Biol.* **2010**, *26*, 211–233. [CrossRef] [PubMed]
5. Butterfield, S.M.; Lashuel, H.A. Amyloidogenic protein-membrane interactions: Mechanistic insight from model systems. *Angew. Chem. Int. Ed.* **2010**, *49*, 5628–5654. [CrossRef]
6. Galvagnion, C.; Buell, A.K.; Meisl, G.; Michaels, T.C.T.; Vendruscolo, M.; Knowles, T.P.J.; Dobson, C.M. Lipid vesicles trigger α-synuclein aggregation by stimulating primary nucleation. *Nat. Chem. Biol.* **2015**, *11*, 229–234. [CrossRef]
7. Cremades, N.; Cohen, S.I.A.; Deas, E.; Abramov, A.Y.; Chen, A.Y.; Orte, A.; Sandal, M.; Clarke, R.W.; Dunne, P.; Aprile, F.A.; et al. Direct observation of the interconversion of normal and toxic forms of alpha-synuclein. *Cell* **2012**, *149*, 1048–1059. [CrossRef]
8. Bemporad, F.; Chiti, F. Protein misfolded oligomers: Experimental approaches, mechanism of formation, and structure-toxicity relationships. *Chem. Biol.* **2012**, *19*, 315–327. [CrossRef]
9. Glabe, C.G. Common mechanisms of amyloid oligomer pathogenesis in degenerative disease. *Neurobiol. Aging* **2006**, *27*, 570–575. [CrossRef]
10. Hartl, F.U. Protein misfolding diseases. *Annu Rev. Biochem* **2017**, *86*, 21–26. [CrossRef]
11. Giampà, M.; Amundarain, M.J.; Herrera, M.G.; Tonali, N.; Dodero, V.I. Implementing complementary approaches to shape the mechanism of α-synuclein oligomerization as a model of amyloid aggregation. *Molecules* **2021**, *27*, 88. [CrossRef] [PubMed]
12. Andreasen, M.; Lorenzen, N.; Otzen, D. Interactions between misfolded protein oligomers and membranes: A central topic in neurodegenerative diseases? *Biochim. Biophys. Acta* **2015**, *1848*, 1897–1907. [CrossRef] [PubMed]
13. Fusco, G.; Chen, S.W.; Williamson, P.T.F.; Cascella, R.; Perni, M.; Jarvis, J.A.; Cecchi, C.; Vendruscolo, M.; Chiti, F.; Cremades, N.; et al. Structural basis of membrane disruption and cellular toxicity by α-synuclein oligomers. *Science* **2017**, *358*, 1440. [CrossRef] [PubMed]
14. Musteikyte, G.; Jayaram, A.K.; Xu, C.K.; Vendruscolo, M.; Krainer, G.; Knowles, T.P.J. Interactions of alpha-synuclein oligomers with lipid membranes. *Biochim. Biophys. Acta Biomembr.* **2021**, *1863*, 183536. [CrossRef]
15. Spillantini, M.G.; Schmidt, M.L.; Lee, V.M.; Trojanowski, J.Q.; Jakes, R.; Goedert, M. Alpha-synuclein in Lewy bodies. *Nature* **1997**, *388*, 839–840. [CrossRef]
16. Conway, K.A.; Lee, S.J.; Rochet, J.C.; Ding, T.T.; Williamson, R.E.; Lansbury, P.T. Acceleration of oligomerization, not fibrillization, is a shared property of both alpha-synuclein mutations linked to early-onset Parkinson's disease: Implications for pathogenesis and therapy. *Proc. Natl. Acad. Sci. USA* **2000**, *97*, 571–576. [CrossRef]
17. Flagmeier, P.; Meisl, G.; Vendruscolo, M.; Knowles, T.P.J.; Dobson, C.M.; Buell, A.K.; Galvagnion, C. Mutations associated with familial Parkinson's disease alter the initiation and amplification steps of α-synuclein aggregation. *Proc. Natl. Acad. Sci. USA* **2016**, *113*, 10328–10333. [CrossRef]
18. Fredenburg, R.A.; Rospigliosi, C.; Meray, R.K.; Kessler, J.C.; Lashuel, H.A.; Eliezer, D.; Lansbury, P.T. The impact of the E46K mutation on the properties of α-synuclein in its monomeric and oligomeric states. *Biochemistry* **2007**, *46*, 7107–7118. [CrossRef]
19. Khalaf, O.; Fauvet, B.; Oueslati, A.; Dikiy, I.; Mahul-Mellier, A.-L.; Ruggeri, F.S.; Mbefo, M.K.; Vercruysse, F.; Dietler, G.; Lee, S.-J.; et al. The H50Q mutation enhances α-synuclein aggregation, secretion, and toxicity. *J. Biol. Chem.* **2014**, *289*, 21856–21876. [CrossRef]
20. Lashuel, H.A.; Hartley, D.; Petre, B.M.; Walz, T.; Lansbury, P.T. Neurodegenerative disease: Amyloid pores from pathogenic mutations. *Nature* **2002**, *418*, 291. [CrossRef]
21. Lemkau, L.R.; Comellas, G.; Kloepper, K.D.; Woods, W.S.; George, J.M.; Rienstra, C.M. Mutant protein A30P α-synuclein adopts wild-type fibril structure, despite slower fibrillation kinetics. *J. Biol. Chem.* **2012**, *287*, 11526–11532. [CrossRef] [PubMed]
22. Li, J.; Uversky, V.N.; Fink, A.L. Effect of familial Parkinson's disease point mutations A30P and A53T on the structural properties, aggregation, and fibrillation of human α-synuclein. *Biochemistry* **2001**, *40*, 11604–11613. [CrossRef]
23. Narhi, L.; Wood, S.J.; Steavenson, S.; Jiang, Y.; Wu, G.M.; Anafi, D.; Kaufman, S.A.; Martin, F.; Sitney, K.; Denis, P.; et al. Both familial Parkinson's disease mutations accelerate alpha-synuclein aggregation. *J. Biol. Chem.* **1999**, *274*, 9843–9846. [CrossRef] [PubMed]
24. Rutherford, N.J.; Moore, B.D.; Golde, T.E.; Giasson, B.I. Divergent effects of the H50Q and G51D SNCA mutations on the aggregation of alpha-synuclein. *J. Neurochem.* **2014**, *131*, 859–867. [CrossRef] [PubMed]

25. Bodner, C.R.; Maltsev, A.S.; Dobson, C.M.; Bax, A. Differential phospholipid binding of alpha-synuclein variants implicated in Parkinson's disease revealed by solution NMR spectroscopy. *Biochemistry* **2010**, *49*, 862–871. [CrossRef]
26. Bussell, R.; Eliezer, D. Effects of Parkinson's disease-linked mutations on the structure of lipid-associated α-synuclein. *Biochemistry* **2004**, *43*, 4810–4818. [CrossRef]
27. Choi, W.; Zibaee, S.; Jakes, R.; Serpell, L.C.; Davletov, B.; Anthony Crowther, R.; Goedert, M. Mutation E46K increases phospholipid binding and assembly into filaments of human α-synuclein. *FEBS Lett.* **2004**, *576*, 363–368. [CrossRef]
28. Fusco, G.; De Simone, A.; Arosio, P.; Vendruscolo, M.; Veglia, G.; Dobson, C.M. Structural ensembles of membrane-bound α-synuclein reveal the molecular determinants of synaptic vesicle affinity. *Sci. Rep.* **2016**, *6*, 27125. [CrossRef]
29. Jensen, P.H.; Nielsen, M.S.; Jakes, R.; Dotti, C.G.; Goedert, M. Binding of α-synuclein to brain vesicles is abolished by familial Parkinson's disease mutation. *J. Biol. Chem.* **1998**, *273*, 26292–26294. [CrossRef]
30. Jo, E.; Fuller, N.; Rand, R.P.; St George-Hyslop, P.; Fraser, P.E. Defective membrane interactions of familial Parkinson's disease mutant A30P α-synuclein. *J. Mol. Biol.* **2002**, *315*, 799–807. [CrossRef]
31. Jo, E.; McLaurin, J.; Yip, C.M.; St George-Hyslop, P.; Fraser, P.E. α-synuclein membrane interactions and lipid specificity. *J. Biol. Chem.* **2000**, *275*, 34328–34334. [CrossRef] [PubMed]
32. Perrin, R.J.; Woods, W.S.; Clayton, D.F.; George, J.M. Interaction of human α-synuclein and Parkinson's disease variants with phospholipids: Structural analysis using site-directed mutagenesis. *J. Biol. Chem.* **2000**, *275*, 34393–34398. [CrossRef] [PubMed]
33. Stefanovic, A.N.D.; Lindhoud, S.; Semerdzhiev, S.A.; Claessens, M.M.A.E.; Subramaniam, V. Oligomers of Parkinson's disease-related α-synuclein mutants have similar structures but distinctive membrane permeabilization properties. *Biochemistry* **2015**, *54*, 3142–3150. [CrossRef]
34. Ysselstein, D.; Joshi, M.; Mishra, V.; Griggs, A.M.; Asiago, J.M.; McCabe, G.P.; Stanciu, L.A.; Post, C.B.; Rochet, J.-C. Effects of impaired membrane interactions on α-synuclein aggregation and neurotoxicity. *Neurobiol. Dis.* **2015**, *79*, 150–163. [CrossRef] [PubMed]
35. Middleton, E.R.; Rhoades, E. Effects of curvature and composition on α-synuclein binding to lipid vesicles. *Biophys. J.* **2010**, *99*, 2279–2288. [CrossRef] [PubMed]
36. Fares, M.-B.; Ait-Bouziad, N.; Dikiy, I.; Mbefo, M.K.; Jovicic, A.; Kiely, A.; Holton, J.L.; Lee, S.-J.; Gitler, A.D.; Eliezer, D.; et al. The novel Parkinson's disease linked mutation G51D attenuates in vitro aggregation and membrane binding of α-synuclein, and enhances its secretion and nuclear localization in cells. *Hum. Mol. Genet.* **2014**, *23*, 4491–4509. [CrossRef] [PubMed]
37. Ghosh, D.; Mondal, M.; Mohite, G.M.; Singh, P.K.; Ranjan, P.; Anoop, A.; Ghosh, S.; Jha, N.N.; Kumar, A.; Maji, S.K. The Parkinson's disease-associated H50Q mutation accelerates α-synuclein aggregation in vitro. *Biochemistry* **2013**, *52*, 6925–6927. [CrossRef]
38. Giasson, B.I.; Uryu, K.; Trojanowski, J.Q.; Lee, V.M.Y. Mutant and wild type human α-synucleins assemble into elongated filaments with distinct morphologies in vitro. *J. Biol. Chem.* **1999**, *274*, 7619–7622. [CrossRef]
39. Li, J.; Uversky, V.N.; Fink, A.L. Conformational behavior of human α-synuclein is modulated by familial Parkinson's disease point mutations A30P and A53T. *Neurotoxicology* **2002**, *23*, 553–567. [CrossRef]
40. Bucciantini, M.; Giannoni, E.; Chiti, F.; Baroni, F.; Formigli, L.; Zurdo, J.; Taddei, N.; Ramponi, G.; Dobson, C.M.; Stefani, M. Inherent toxicity of aggregates implies a common mechanism for protein misfolding diseases. *Nature* **2002**, *416*, 507–511. [CrossRef]
41. Chiti, F.; Dobson, C.M. Protein misfolding, functional amyloid, and human disease. *Annu. Rev. Biochem.* **2006**, *75*, 333–366. [CrossRef] [PubMed]
42. Stefani, M.; Dobson, C.M. Protein aggregation and aggregate toxicity: New insights into protein folding, misfolding diseases and biological evolution. *J. Mol. Med.* **2003**, *81*, 678–699. [CrossRef] [PubMed]
43. Paslawski, W.; Mysling, S.; Thomsen, K.; Jørgensen, T.J.D.; Otzen, D.E. Co-existence of two different α-synuclein oligomers with different core structures determined by hydrogen/deuterium exchange mass spectrometry. *Angew. Chem. Int. Ed.* **2014**, *53*, 7560–7563. [CrossRef]
44. Tosatto, L.; Horrocks, M.H.; Dear, A.J.; Knowles, T.P.J.; Dalla Serra, M.; Cremades, N.; Dobson, C.M.; Klenerman, D. Single-molecule FRET studies on alpha-synuclein oligomerization of Parkinson's disease genetically related mutants. *Sci. Rep.* **2015**, *5*, 16696. [CrossRef] [PubMed]
45. Chen, S.W.; Drakulic, S.; Deas, E.; Ouberai, M.; Aprile, F.A.; Arranz, R.; Ness, S.; Roodveldt, C.; Guilliams, T.; De-Genst, E.J.; et al. Structural characterization of toxic oligomers that are kinetically trapped during α-synuclein fibril formation. *Proc. Natl. Acad. Sci. USA* **2015**, *112*, E1994–E2003. [CrossRef] [PubMed]
46. Prendergast, F.G.; Hampton, P.D.; Jones, B. Characteristics of tyrosinate fluorescence emission in α and β-purothionins. *Biochemistry* **1984**, *23*, 6690–6697. [CrossRef] [PubMed]
47. Szabo, A.G.; Lynn, K.R.; Krajcarski, D.T.; Rayner, D.M. Tyrosinate fluorescence maxima at 345 nm in proteins lacking tryptophan at pH 7. *FEBS Lett.* **1978**, *94*, 249–252. [CrossRef]
48. Camino, J.D.; Gracia, P.; Chen, S.W.; Sot, J.; de la Arada, I.; Sebastian, V.; Arrondo, J.L.R.; Goni, F.M.; Dobson, C.M.; Cremades, N. The extent of protein hydration dictates the preference for heterogeneous or homogeneous nucleation generating either parallel or antiparallel beta-sheet alpha-synuclein aggregates. *Chem. Sci.* **2020**, *11*, 11902–11914. [CrossRef]
49. Mosmann, T. Rapid colorimetric assay for cellular growth and survival: Application to proliferation and cytotoxicity assays. *J. Immunol. Methods* **1983**, *65*, 55–63. [CrossRef]

50. Perni, M.; Galvagnion, C.; Maltsev, A.; Meisl, G.; Müller, M.B.D.; Challa, P.K.; Kirkegaard, J.B.; Flagmeier, P.; Cohen, S.I.A.; Cascella, R.; et al. A natural product inhibits the initiation of alpha-synuclein aggregation and suppresses its toxicity. *Proc. Natl. Acad. Sci. USA* **2017**, *114*, E1009–E1017. [CrossRef]
51. Micsonai, A.; Wien, F.; Bulyáki, É.; Kun, J.; Moussong, É.; Lee, Y.-H.; Goto, Y.; Réfrégiers, M.; Kardos, J. BeStSel: A web server for accurate protein secondary structure prediction and fold recognition from the circular dichroism spectra. *Nucleic Acids Res.* **2018**, *46*, W315–W322. [CrossRef] [PubMed]
52. Micsonai, A.; Wien, F.; Kernya, L.; Lee, Y.-H.; Goto, Y.; Réfrégiers, M.; Kardos, J. Accurate secondary structure prediction and fold recognition for circular dichroism spectroscopy. *Proc. Natl. Acad. Sci. USA* **2015**, *112*, E3095–E3103. [CrossRef]
53. Bolognesi, B.; Kumita, J.R.; Barros, T.P.; Esbjörner, E.K.; Luheshi, L.M.; Crowther, D.C.; Wilson, M.R.; Dobson, C.M.; Favrin, G.; Yerbury, J.J. ANS binding reveals common features of cytotoxic amyloid species. *ACS Chem. Biol.* **2010**, *5*, 735–740. [CrossRef] [PubMed]
54. Campioni, S.; Mannini, B.; Zampagni, M.; Pensalfini, A.; Parrini, C.; Evangelisti, E.; Relini, A.; Stefani, M.; Dobson, C.M.; Cecchi, C.; et al. A causative link between the structure of aberrant protein oligomers and their toxicity. *Nat. Chem. Biol.* **2010**, *6*, 140–147. [CrossRef] [PubMed]
55. Oma, Y.; Kino, Y.; Toriumi, K.; Sasagawa, N.; Ishiura, S. Interactions between homopolymeric amino acids (HPAAs). *Protein Sci.* **2007**, *16*, 2195–2204. [CrossRef] [PubMed]
56. Mannini, B.; Mulvihill, E.; Sgromo, C.; Cascella, R.; Khodarahmi, R.; Ramazzotti, M.; Dobson, C.M.; Cecchi, C.; Chiti, F. Toxicity of protein oligomers is rationalized by a function combining size and surface hydrophobicity. *ACS Chem. Biol.* **2014**, *9*, 2309–2317. [CrossRef] [PubMed]
57. Kayed, R.; Head, E.; Sarsoza, F.; Saing, T.; Cotman, C.W.; Necula, M.; Margol, L.; Wu, J.; Breydo, L.; Thompson, J.L.; et al. Fibril specific, conformation dependent antibodies recognize a generic epitope common to amyloid fibrils and fibrillar oligomers that is absent in prefibrillar oligomers. *Mol. Neurodegener.* **2007**, *2*, 18. [CrossRef]
58. Stefanovic, A.N.D.; Stöckl, M.T.; Claessens, M.M.A.E.; Subramaniam, V. Alpha-synuclein oligomers distinctively permeabilize complex model membranes. *FEBS J.* **2014**, *281*, 2838–2850. [CrossRef]
59. Apetri, M.M.; Maiti, N.C.; Zagorski, M.G.; Carey, P.R.; Anderson, V.E. Secondary structure of α-synuclein oligomers: Characterization by Raman and Atomic Force Microscopy. *J. Mol. Biol.* **2006**, *355*, 63–71. [CrossRef]
60. Ghosh, D.; Singh, P.K.; Sahay, S.; Jha, N.N.; Jacob, R.S.; Sen, S.; Kumar, A.; Riek, R.; Maji, S.K. Structure based aggregation studies reveal the presence of helix-rich intermediate during α-synuclein aggregation. *Sci. Rep.* **2015**, *5*, 9228. [CrossRef]
61. Hoyer, W.; Antony, T.; Cherny, D.; Heim, G.; Jovin, T.M.; Subramaniam, V. Dependence of α-synuclein aggregate morphology on solution conditions. *J. Mol. Biol.* **2002**, *322*, 383–393. [CrossRef]
62. Winner, B.; Jappelli, R.; Maji, S.K.; Desplats, P.A.; Boyer, L.; Aigner, S.; Hetzer, C.; Loher, T.; Vilar, M.; Campioni, S.; et al. In vivo demonstration that alpha-synuclein oligomers are toxic. *Proc. Natl. Acad. Sci. USA* **2011**, *108*, 4194–4199. [CrossRef] [PubMed]
63. Boyer, D.R.; Li, B.; Sun, C.; Fan, W.; Sawaya, M.R.; Jiang, L.; Eisenberg, D.S. Structures of fibrils formed by α-synuclein hereditary disease mutant H50Q reveal new polymorphs. *Nat. Struct. Mol. Biol.* **2019**, *26*, 1044–1052. [CrossRef]
64. Peelaerts, W.; Bousset, L.; Van der Perren, A.; Moskalyuk, A.; Pulizzi, R.; Giugliano, M.; Van den Haute, C.; Melki, R.; Baekelandt, V. Alpha-synuclein strains cause distinct synucleinopathies after local and systemic administration. *Nature* **2015**, *522*, 340–344. [CrossRef] [PubMed]
65. Riek, R.; Eisenberg, D.S. The activities of amyloids from a structural perspective. *Nature* **2016**, *539*, 227–235. [CrossRef]
66. Salinas, N.; Colletier, J.-P.; Moshe, A.; Landau, M. Extreme amyloid polymorphism in Staphylococcus aureus virulent PSMα peptides. *Nat. Commun.* **2018**, *9*, 3512. [CrossRef]
67. Tayeb-Fligelman, E.; Tabachnikov, O.; Moshe, A.; Goldshmidt-Tran, O.; Sawaya, M.R.; Coquelle, N.; Colletier, J.-P.; Landau, M. The cytotoxic Staphylococcus aureus PSMα reveals a cross-a amyloid-like fibril. *Science* **2017**, *355*, 831–833. [CrossRef]

molecules

Article

The Amyloid Fibril-Forming β-Sheet Regions of Amyloid β and α-Synuclein Preferentially Interact with the Molecular Chaperone 14-3-3ζ

Danielle M. Williams [1], David C. Thorn [2], Christopher M. Dobson [3,†], Sarah Meehan [3], Sophie E. Jackson [3], Joanna M. Woodcock [4] and John A. Carver [2,*]

1 Department of Chemistry, School of Physical Sciences, The University of Adelaide, Adelaide, SA 5001, Australia; dwilliams@biocina.com
2 Research School of Chemistry, The Australian National University, Acton, ACT 2601, Australia; dr.david.thorn@gmail.com
3 Yusuf Hamied Department of Chemistry, University of Cambridge, Lensfield Road, Cambridge CB2 1EW, UK; sarah.meehan.cam@gmail.com (S.M.); sej13@cam.ac.uk (S.E.J.)
4 Centre for Cancer Biology (An Alliance between SA Pathology and the University of South Australia), Adelaide, SA 5000, Australia; Joanna.Woodcock@sa.gov.au
* Correspondence: john.carver@anu.edu.au; Tel.: +61-2-6215-9748
† Dedicated to the memory of Sir Chris Dobson FRS. Chris' fundamental and ground-breaking studies of protein folding, unfolding and aggregation, in particular amyloid fibril formation, have led to major advances in understanding the development of the many diseases associated with these processes. On the personal level, Chris was a strong supporter, mentor, colleague and friend to many of this manuscript's authors.

Citation: Williams, D.M.; Thorn, D.C.; Dobson, C.M.; Meehan, S.; Jackson, S.E.; Woodcock, J.M.; Carver, J.A. The Amyloid Fibril-Forming β-Sheet Regions of Amyloid β and α-Synuclein Preferentially Interact with the Molecular Chaperone 14-3-3ζ. *Molecules* **2021**, *26*, 6120. https://doi.org/10.3390/molecules26206120

Academic Editors: Kunihiro Kuwajima, Yuko Okamoto, Tuomas Knowles and Michele Vendruscolo

Received: 17 August 2021
Accepted: 3 October 2021
Published: 11 October 2021

Abstract: 14-3-3 proteins are abundant, intramolecular proteins that play a pivotal role in cellular signal transduction by interacting with phosphorylated ligands. In addition, they are molecular chaperones that prevent protein unfolding and aggregation under cellular stress conditions in a similar manner to the unrelated small heat-shock proteins. In vivo, amyloid β (Aβ) and α-synuclein (α-syn) form amyloid fibrils in Alzheimer's and Parkinson's diseases, respectively, a process that is intimately linked to the diseases' progression. The 14-3-3ζ isoform potently inhibited in vitro fibril formation of the 40-amino acid form of Aβ ($A\beta_{40}$) but had little effect on α-syn aggregation. Solution-phase NMR spectroscopy of ^{15}N-labeled $A\beta_{40}$ and A53T α-syn determined that unlabeled 14-3-3ζ interacted preferentially with hydrophobic regions of $A\beta_{40}$ (L11-H21 and G29-V40) and α-syn (V3-K10 and V40-K60). In both proteins, these regions adopt β-strands within the core of the amyloid fibrils prepared in vitro as well as those isolated from the inclusions of diseased individuals. The interaction with 14-3-3ζ is transient and occurs at the early stages of the fibrillar aggregation pathway to maintain the native, monomeric, and unfolded structure of $A\beta_{40}$ and α-syn. The N-terminal regions of α-syn interacting with 14-3-3ζ correspond with those that interact with other molecular chaperones as monitored by in-cell NMR spectroscopy.

Keywords: 14-3-3 proteins; molecular chaperone; amyloid β; α-synuclein; NMR spectroscopy; amyloid fibril

1. Introduction

Protein aggregation is a characteristic of many diseases, the majority of which are age-related and neurological. Common examples of protein aggregation diseases (also known as protein misfolding or protein conformational diseases) include Alzheimer's (AD), Parkinson's (PD), Huntington's, and Creutzfeldt–Jakob [1,2]. The protein aggregates or deposits associated with these diseases contain a predominant peptide or protein that, in the majority of cases, adopts an amyloid fibrillar form. Amyloid fibrils are a highly stable, aggregated proteinaceous state with the polypeptide arranged mainly in a cross β-sheet conformation that results in an extended, overall fibrillar structure up to micrometers in length [1,2]. The conversion of a protein from its native, functional state to an amyloid fibril is a multi-step process that usually involves a nucleation-dependent mechanism that has various intermediate states, including the formation of prefibrillar oligomers that act as nuclei to sequester and convert natively structured proteins into the fibrillar form. The prefibrillar oligomers are proposed to be the entities that cause cell toxicity and, hence, are intimately involved in the disease processes [1,2].

Within and outside cells, protein levels and their conformations are maintained within a narrow regime to minimize the possibility of protein unfolding, misfolding, and aggregation. The general term for this process is proteostasis, a portmanteau of protein and homeostasis [3]. The action of molecular chaperone proteins in stabilizing the conformation, stoichiometry, and interactions of other proteins is one of the major means of maintaining cellular proteostasis. Under stress conditions such as elevated temperature, the principal intracellular chaperones that function to prevent protein unfolding and aggregation are the small heat-shock proteins (sHsps) [4]. Their levels are upregulated many-fold under such conditions and with aging [4] and in diseases associated with protein aggregation [5,6]. Overall, the ATP-independent action of sHsps is crucially important in cellular well-being under normal and stress conditions.

In addition to sHsps, other proteins have an sHsp-like chaperone ability that may supplement or complement that of sHsps. In this context, intracellular 14-3-3 proteins exhibit chaperone action against a variety of unfolding proteins under stress conditions such as elevated temperature [7–10]. They are present at high levels in the brain. In humans, there are seven closely related 14-3-3 proteins. The principal role of 14-3-3 proteins is their involvement in cellular signal transduction processes via their interaction with phosphorylated substrate proteins. As such, they function as adapters and participate in a variety of cellular pathways, including apoptosis, transcription and the stress response [9]. In addition to their intracellular presence, 14-3-3 proteins are found extracellularly, for example in exosomes and in cerebrospinal fluid of people with neurodegenerative diseases such as Creutzfeldt–Jakob disease, AD, and multiple sclerosis [11]. 14-3-3 proteins are dimers composed of subunits of ~28 kDa in mass. Each subunit adopts a predominantly α-helical conformation with nine α-helices. The dimer is arranged in a double cup-like shape with two amphipathic binding grooves where phosphorylated ligands bind. The dimer interface is provided by an interaction between N-terminal helices of each subunit.

The two major neurological diseases associated with protein aggregation and deposition are AD and PD [1,2]. AD is characterized by the extracellular deposition of plaques containing mainly variants of the amyloid beta peptide (Aβ) in an amyloid fibrillar form, along with intracellular neurofibrillary tangles containing mainly the tau protein. The principal Aβ peptides are 40 and 42 amino acids in length. In PD, Lewy body deposits are the defining morphological intracellular feature; they are mainly comprised of the protein α-synuclein (α-syn), also deposited as amyloid fibrils. In AD and PD, other proteins are associated with these deposits, including sHsps and 14-3-3 proteins [9,12,13]. Their presence may arise from the utilization by cells of their chaperone ability as an attempt to prevent the aggregation of Aβ and α-syn to form amyloid fibrils.

In this study, we report on an NMR spectroscopic and biophysical analysis of the interaction of 14-3-3ζ, a major 14-3-3 isoform, with Aβ, and a similar interaction between 14-3-3ζ and A53T α-syn, a mutant of α-syn associated with familial PD that aggregates more rapidly than the wild type (WT) protein. The findings have implications for the in vivo association of these species and their involvement in AD, PD, and other related diseases of protein aggregation where amyloid fibrillar aggregation occurs.

2. Results and Discussion

2.1. Interaction of 14-3-3ζ with Amyloid β Peptides

Thioflavin T (ThT) is a dye whose fluorescence increases markedly upon binding to the β-sheet regions of amyloid fibrils, a phenomenon that is routinely utilized to monitor amyloid fibril formation in peptides and proteins [14]. Figure 1 shows the ThT fluorescence profiles for the 40- and 42-amino acid forms of Aβ ($A\beta_{40}$ and $A\beta_{42}$, respectively) with time in the absence and presence of increasing quantities of 14-3-3ζ. The $A\beta_{42}$ peptide is more hydrophobic than $A\beta_{40}$ due to the presence of the additional Ile-Ala dipeptide at its C-terminus. Consistent with the observation by others [15], $A\beta_{42}$ aggregates at a much faster rate (compare Figure 1a,b, noting the very different time scales), without the presence of a lag phase, compared to $A\beta_{40}$ which has a very long lag phase of around 33 h under the conditions used. The protein 14-3-3ζ was much more effective at inhibiting the aggregation of $A\beta_{40}$ than $A\beta_{42}$. Thus, a 1.0:0.5 molar subunit ratio of $A\beta_{40}$:14-3-3ζ completely inhibited the former's aggregation (Figure 1a) whereas a two-molar excess of 14-3-3ζ only partially reduced fibril formation of $A\beta_{42}$ (Figure 1b). In separate experiments, $A\beta_{40}$ that had been seeded with $A\beta_{40}$ fibrils aggregated much earlier, i.e., a lag phase of 12 h (Figure S1). An equimolar ratio of $A\beta_{40}$:14-3-3ζ significantly delayed the onset of $A\beta_{40}$ aggregation (lag phase of 29 h) and partially suppressed its extent of aggregation as monitored by ThT fluorescence. More pronounced effects on the lag phase and extent of $A\beta_{40}$ fibril formation were observed at a 1.0:2.0 molar ratio, with complete inhibition of aggregation occurring at a 1.0:4.0 molar ratio of $A\beta_{40}$:14-3-3ζ (Figure S1).

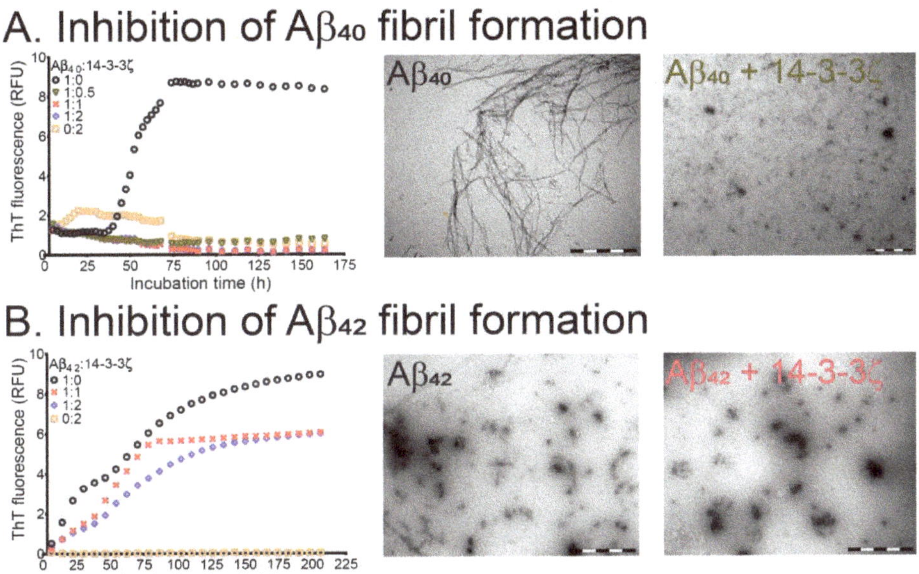

Figure 1. The molecular chaperone ability of 14-3-3ζ to inhibit amyloid fibril formation of $A\beta_{40}$ and $A\beta_{42}$. Amyloid fibril formation, as monitored by ThT fluorescence, of 15 μM $A\beta_{40}$ (**A**) and $A\beta_{42}$ (**B**) at 37 °C in the absence (black circles) and presence of 14-3-3ζ at 1.0:0.5 (brown triangles; $A\beta_{40}$ only), 1:1 (red crosses), and 1:2 (blue diamonds) molar ratios of $A\beta_{40}/A\beta_{42}$:14-3-3ζ. Also shown is 14-3-3ζ incubated alone at 30 μM (beige squares). On the right of part (**A**) are TEM images of $A\beta_{40}$ in the absence and presence of 14-3-3ζ at a 1.0:0.5 molar ratio of $A\beta_{40}$:14-3-3ζ. Samples for TEM imaging were taken upon completion of the aggregation assay after 150 h of incubation. On the right of part (**B**) are TEM images of $A\beta_{42}$ in the absence and presence of 14-3-3ζ at an equimolar ratio. Each image was acquired from the samples after 200 min. of incubation. Scale bars in all images represent 1 μm.

Transmission electron microscopy (TEM) images confirmed the conclusions from the ThT data of Figure 1. In the absence of 14-3-3ζ, both Aβ peptides formed amyloid fibrils, with $A\beta_{40}$ fibrils being longer than those of $A\beta_{42}$ (compare Figure 1a,b). TEM images of $A\beta_{42}$ showed agglomerates of fibrils alongside well-separated, distinct fibrillar species (Figure 1b). In the presence of 14-3-3ζ, the formation of amorphous-like aggregates was observed for both Aβ peptides, a phenomenon that commonly occurs when fibril-inhibiting molecules (large and small) interact with amyloid fibril-forming peptides and proteins (e.g., [16]). In summary, 14-3-3ζ inhibited the aggregation and fibril formation of Aβ$_{40}$ at stoichiometric ratios. By contrast, 14-3-3ζ only partially inhibited the aggregation of Aβ$_{42}$ at these ratios.

Other studies have examined the in vitro chaperone ability of 14-3-3ζ against amorphously aggregating target proteins [7,8,10]. 14-3-3ζ is unlike sHsps as it is not highly promiscuous in its chaperone ability, i.e., 14-3-3ζ exhibits relatively selective chaperone ability against aggregating target proteins. However, like sHsps [17–20], 14-3-3ζ is a more efficient chaperone at inhibiting target proteins that are aggregating slowly and amorphously. Thus, target protein aggregation rate (i.e., kinetics) may be the major factor in determining the greater ability of 14-3-3ζ to inhibit the aggregation of Aβ$_{40}$ more efficiently compared to the aggregation of Aβ$_{42}$. The poorer efficiency of 14-3-3ζ at inhibiting the amyloid fibril formation of seeded Aβ$_{40}$ (Figure S1), in which faster aggregation of the peptide occurred than for unseeded Aβ$_{40}$ (Figure 1a), is consistent with this conclusion. Likewise, sHsps are more efficient at inhibiting slowly aggregating amyloid fibril-forming target proteins such as α-syn [12,21]. According to Mori et al. and Kollmer et al., Aβ$_{40}$ is the major secreted form of the Aβ peptides in vivo and is the predominant species present in AD extracellular plaques [22,23]. The presence of 14-3-3 proteins extracellularly [11], where Aβ peptides are primarily located, implies that the ability of 14-3-3ζ to inhibit Aβ$_{40}$ aggregation has physiological significance. Accordingly, the residue-specific interaction between Aβ$_{40}$ and 14-3-3ζ was explored by solution-phase NMR spectroscopy.

Figure 2a shows the ^{1}H-^{15}N HSQC spectrum of ^{15}N-labeled Aβ$_{40}$ at physiological pH and 5 °C, in the absence and presence of unlabeled 14-3-3ζ, at up to a four-molar excess of the chaperone protein. The NMR experiments were conducted at a low temperature to negate (or minimize) the possibility that Aβ$_{40}$ would form amyloid fibrils during the timeframe of acquisition of the spectra. Thus, the NMR experiments monitored interactions between Aβ$_{40}$ and 14-3-3ζ at the earliest stage of Aβ$_{40}$ amyloid fibril formation.

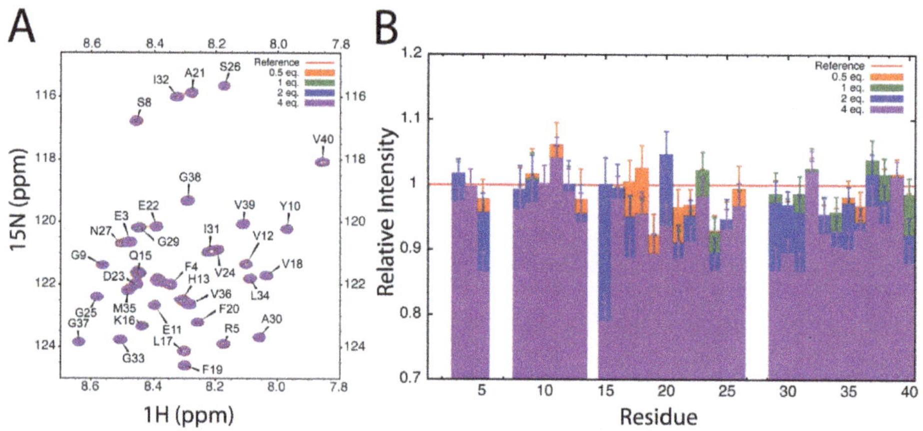

Figure 2. ^{1}H-^{15}N HSQC NMR spectrum of uniformly ^{15}N-labeled Aβ$_{40}$ in the absence and presence of 14-3-3ζ. (**A**) Amide region of the ^{1}H-^{15}N HSQC spectrum at 5 °C of ^{15}N-labeled Aβ$_{40}$ in the presence of unlabeled 14-3-3ζ at 1.0:0.0–4.0 molar ratios of Aβ$_{40}$:14-3-3ζ (red contour levels: 0.0, orange: 0.5, green: 1.0, blue: 2.0, and purple: 4.0 refer to the molar ratio of 14-3-3ζ to Aβ$_{40}$). Assignments are from Hou and Zagorski [24]. No change in the ^{1}H and ^{15}N chemical shifts of Aβ$_{40}$ occurred upon addition of 14-3-3ζ. (**B**) The change in the relative intensities of Aβ$_{40}$ ^{1}H-^{15}N cross-peaks at 1.0:0.5-4.0 molar ratios of Aβ$_{40}$:14-3-3ζ relative to Aβ$_{40}$ alone (orange line). The intensities of the contour levels were corrected for dilution effects. Error bars represent the standard deviation in background noise of the spectra.

Cross-peaks of Aβ$_{40}$ are labeled in Figure 2a [24]. Cross-peaks were not observed for six of the 40 amino acids of Aβ$_{40}$ (D1, A2, H6, D7, H14 and K28), presumably due to broadening associated with intermediate exchange. No change in chemical shifts of the observed cross-peaks occurred in the presence of 14-3-3ζ, implying weak and transient interactions and a relatively overall fast exchange between the two species. There was a non-uniform decrease in the intensity, or broadening, of the $A\beta_{40}$ cross-peaks across its amino acid sequence in the presence of increasing concentrations of 14-3-3ζ compared to the spectrum of $A\beta_{40}$ on its own (Figure 2b). The decrease in cross-peak intensity was most marked for Q15 to G25, encompassing the hydrophobic core of the peptide (L17-V18-F19-F20-A21), and most of the cross-peaks arising from the C-terminal region $A\beta_{40}$ (G29 to V40), which is also hydrophobic in character. The NMR data imply that these regions interact with 14-3-3ζ during its chaperone function to inhibit $A\beta_{40}$ amyloid fibril formation. As with the chaperone mechanism of the sHsp αB-crystallin in preventing amyloid fibril formation of $A\beta_{40}$ [25], the interaction between the 14-3-3ζ and $A\beta_{40}$ is transient in nature, which facilitates the potent stoichiometry of 14-3-3ζ in inhibiting $A\beta_{40}$ aggregation.

The recent determination of the structures of the amyloid fibrillar forms of Aβ peptides by cryo-electron microscopy (cryoEM) has provided an atomic-level description of the arrangement of the polypeptide backbone. Of most relevance is the cryoEM structure of $A\beta_{40}$ fibrils isolated from the brain tissue of AD patients, post-mortem [23]. From this study, the four β-strands in the β-sheet fibril core of $A\beta_{40}$ arise from residues A2-S8, Y10-H13, Q15-F19, and I32-L34. The latter two β-strands correspond well to the residues whose cross-peak intensities in the ^1H-^{15}N NMR spectrum of $A\beta_{40}$ are decreased in the presence of 14-3-3ζ (Figure 2b). They also correspond well to the residues that form the cross β-sheet core of Aβ, as determined by solid state NMR spectroscopy [26,27]. Furthermore, of all the $A\beta_{40}$ cross-peaks in Figure 2b, the intensity of the Q15 cross-peak was decreased the most (by over 20%) in the presence of a four-molar excess of 14-3-3ζ. Thus, 14-3-3ζ interacts preferentially with at least part of the $A\beta_{40}$ peptide that forms its fibril core and, since it does so with $A\beta_{40}$ amino acids within and nearby to Q15-F19 and I32-L34, it is surmised that these two β-strands are potentially the first to assemble during $A\beta_{40}$ fibril formation. Furthermore, 14-3-3ζ inhibits $A\beta_{40}$ fibril formation via interfering at the earliest stage of its aggregation pathway. Again, this mechanism has distinct parallels with that exhibited by sHsps during their prevention of amyloid fibril formation of a variety of proteins (summarized in [5,6,28]). Although Aβ-containing amyloid plaques are extracellular deposits, there is an intracellular component to Aβ aggregation prior to the peptide's export into the extracellular medium [29]. Thus, in vivo interaction of Aβ with 14-3-3ζ intracellularly, along with the other 14-3-3 isoforms and sHsps, may have physiological importance in modulating Aβ aggregation and maintaining intracellular proteostasis. Furthermore, the extracellular presence of 14-3-3 proteins [11] would also provide the opportunity for their interaction with Aβ in this environment.

2.2. Interaction of 14-3-3ζ with α-Synuclein

A53T α-syn is a mutant that aggregates at a faster rate compared to the WT protein and is associated with familial PD [30]. As monitored by ThT fluorescence, 14-3-3ζ had minimal ability to inhibit amyloid fibril formation of A53T α-syn at an equimolar ratio under physiological conditions of over 22 h of co-incubation (Figure S2). In agreement with this, Plotegher et al. [31] investigated the ability of all seven human 14-3-3 isoforms to inhibit α-syn fibril formation and found that two of the isoforms (η and τ (also termed θ)) were potent inhibitors of α-syn aggregation, whereas the others, including 14-3-3ζ, were ineffective. The inability of 14-3-3ζ to decrease A53T α-syn aggregation to any significant degree at a stoichiometric ratio may be due to the latter protein's relatively rapid rate of aggregation, which commences around four hours of incubation compared to 33 h for the $A\beta_{40}$ peptide (Figure 1a). The selective nature of 14-3-3ζ chaperone action against aggregating target proteins is also a factor since only the 14-3-3 η and τ/θ isoforms inhibit α-syn aggregation [31]. TEM images of A53T α-syn amyloid fibrils formed in the absence and presence of 14-3-3ζ were of very similar length and morphology, consistent with the inability of 14-3-3ζ to modify A53T α-syn fibril formation (Figure S2). Atomic force microscopic analysis of amyloid fibrils produced from incubated mixtures of α-syn and 14-3-3ζ also showed little difference in fibril morphology compared to that of mature α-syn amyloid fibrils [31].

A quartz crystal microbalance (QCM) measures the mass change of a surface upon the addition or removal of molecules as monitored by the change in frequency of a quartz crystal resonator. QCM was used to monitor the change in mass of WT α-syn preformed amyloid fibrils attached to the crystal upon flowing solutions of monomeric α-syn, an equimolar ratio of α-syn:14-3-3ζ and 14-3-3ζ itself over the crystal. The rate of change of frequency was essentially the same for the addition of α-syn in the absence or presence of 14-3-3ζ (Figure 3a), indicating that α-syn monomers bound to the α-syn fibrils on the crystal surface (thus increasing their size and mass), a process that was not affected by the presence of 14-3-3ζ. 14-3-3ζ did not bind to the α-syn fibrils. Thus, if any interaction between the two proteins occurs, it does so early along the aggregation pathway of α-syn, as with the interaction between α-syn and the sHsp, αB-crystallin [32,33]. The dynamic light scattering (DLS) profiles of WT α-syn and 14-3-3ζ were similar, with hydrodynamic radii at 37 °C and pH 7 of 2.789 ± 0.003 nm for WT α-syn and 2.818 ± 0.003 nm for 14-3-3ζ. The latter value compares well with the literature [8,34]. When both proteins were mixed together at an equimolar ratio, however, a single peak was observed at the larger hydrodynamic radius of 3.062 ± 0.001 nm (Figure 3b), implying some association between the two proteins under physiological conditions.

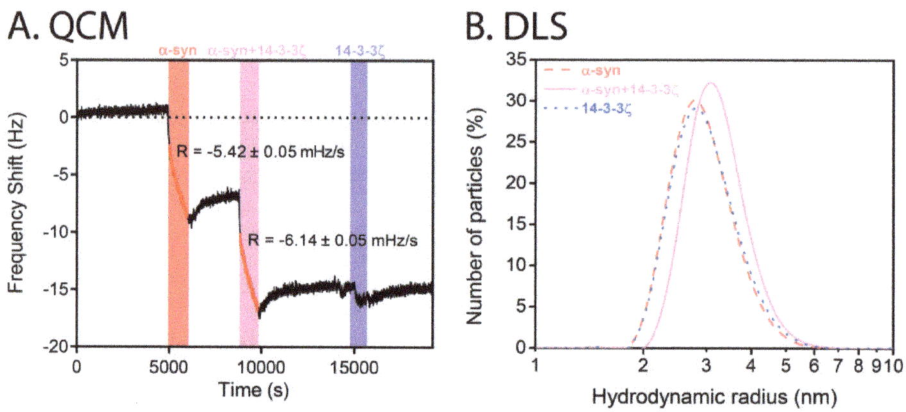

Figure 3. (**A**) QCM traces of the elongation of preformed WT α-syn amyloid fibrils in the presence and absence of 14-3-3ζ. Preformed amyloid fibrils were attached on to the chip and equilibrated with buffer for 24 h. At approximately 5000 s intervals, 10 μM WT α-syn monomer (red), 1.0:1.0 molar ratio of α-syn monomer:14-3-3ζ (pink), and 10 μM 14-3-3ζ (blue) were added. After the addition of each protein sample, the chip was washed with buffer until a plateau in signal was obtained (white). The rate of frequency change (R) is indicated for when α-syn monomer and a 1.0:1.0 molar ratio of α-syn:14-3-3ζ were washed over the QCM crystal. (**B**) DLS of WT α-syn and 14-3-3ζ. DLS profile and hydrodynamic radius of 20 μM WT α-syn at 25 °C in the absence (red dashed line) and presence (pink continuous line) of 14-3-3ζ at a 1.0:1.0 molar ratio. Also shown is the DLS profile of 20 μM 14-3-3ζ (blue dotted line).

Accordingly, the interaction of α-syn and 14-3-3ζ was investigated by NMR spectroscopy at pH 7.4 and 10 °C using ^{15}N-labeled A53T α-syn in the presence of unlabeled 14-3-3ζ at 1:1 and 1:2 molar ratios of A53T α-syn:14-3-3ζ. As with the interaction between Aβ_{40} and 14-3-3ζ, the NMR experiments were acquired at a low temperature to circumvent the possibility of fibril formation by A53T α-syn. Figure 4a shows the ^{1}H-^{15}N HSQC NMR spectrum of A53T α-syn under these conditions. The ^{1}H NH chemical shifts are all contained within the 'random coil' chemical shift regime of 7.6 to 8.6 ppm, as expected since α-syn is a 140-amino acid intrinsically disordered protein of little or no stable secondary structure. In the presence of unlabeled 14-3-3ζ, no alteration in chemical shifts occurred of the α-syn cross-peaks. This was also observed for the ^{1}H-^{15}N HSQC NMR spectrum of a mixture of α-syn and 14-3-3η [31] and for α-syn interacting with αB-crystallin [32] and an unrelated molecular chaperone, Hsp70 [35]. The intensities of the cross-peaks in the HSQC spectra of A53T α-syn in the presence of 14-3-3ζ were measured, and their intensities relative to those in the spectrum of A53T α-syn only are plotted in Figure 4b. Alteration in cross-peak intensity was localized to the N-terminus (V3 to K10) and from V40 to K60, and possibly towards the C-terminus between D121 and D135. Together, these data imply that the interaction between A53T α-syn and 14-3-3ζ is weak and transient, with fast exchange. In their NMR studies of the interaction of α-syn with 14-3-3η, Plotegher et al. [31] did not examine the effects on the intensity of α-syn cross-peaks due to the presence of 14-3-3η. In contrast to the results presented herein with 14-3-3ζ, 14-3-3η was an effective inhibitor of α-syn amyloid fibril formation [31]. Overall, the results from both studies are consistent with a transient and weak interaction between the two proteins which, owing to some subtle structural and mechanistic differences between isoforms, leads to inhibition of α-syn oligomerization in the presence of 14-3-3η but not 14-3-3ζ.

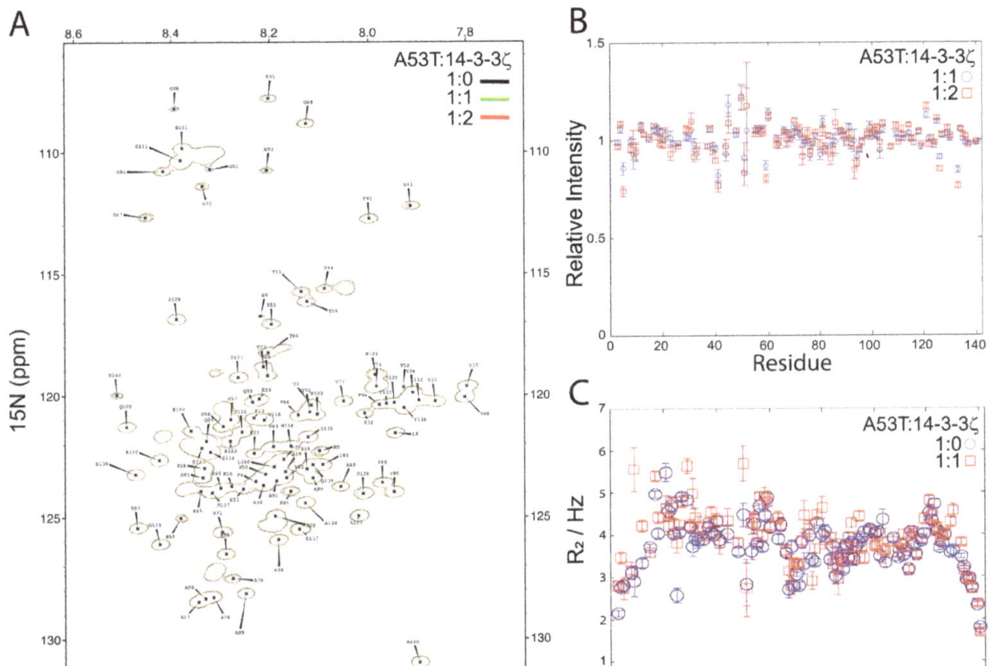

Figure 4. ^{1}H-^{15}N HSQC NMR spectrum of uniformly ^{15}N-labeled A53T α-syn in the absence and presence of 14-3-3ζ. (**A**) Amide region of the ^{1}H-^{15}N HSQC spectrum at 10 °C of ^{15}N-labeled A53T α-syn (black) following the addition of one (green) and two (red) molar equivalents of unlabeled 14-3-3ζ. Assignments are from Dedmon et al. [35]. No change in the ^{1}H and ^{15}N chemical shifts of A53T α-syn occurred upon addition of 14-3-3ζ. (**B**) The change in the relative intensity of ^{15}N-labeled A53T α-syn cross-peaks in the presence of 14-3-3ζ at 1.0:1.0 and 1.0:2.0 molar ratios, relative to the absence of 14-3-3ζ. The intensities of the contour levels were corrected for dilution effects. Three HSQC spectra were acquired for each sample and the data are plotted as the average ± SEM. (**C**) The spin-spin relaxation rate (R_2) for individual backbone nitrogens of ^{15}N-labeled A53T α-syn in the absence (blue) and presence (red), at a 1.0:1.0 molar ratio, of unlabeled 14-3-3ζ.

The transverse (spin-spin) ^{15}N relaxation rate (R_2) of ^{15}N-labeled A53T α-syn cross-peaks in the absence and presence of a molar equivalent of 14-3-3ζ was determined (Figure 4c). Interaction between the two proteins, even if transient, will alter (most likely increase) the R_2 values of the residues involved. On average, there was a slight increase in R_2 values for A53T α-syn in the presence of 14-3-3ζ which was concentrated in the first 50 or so amino acids, in agreement with the cross-peak intensity data in Figure 4b, i.e., comparison of R_2 values in the absence and presence of 14-3-3ζ was consistent with an interaction between the two proteins that mainly involved the N-terminal region of A53T.

The preferential interaction of the N-terminal region of α-syn with 14-3-3ζ is consistent with the in vitro and in-cell NMR data of Burmann et al. [36]. ^{1}H-^{15}N HSQC spectra of ^{15}N-labeled α-syn revealed that six molecular chaperones (they did not examine 14-3-3 proteins or sHsps) 'commonly recognize a canonical motif in α-synuclein, consisting of the N-terminus (12 residues) and a segment (six residues) around Tyr39'. The interaction is transient and weak and, inside cells, maintains α-syn in its monomeric, unfolded, and functional (i.e., non-amyloidogenic) state. The two interacting N-terminal regions of α-syn are hydrophobic in nature, consistent with the primary role of hydrophobic interactions in the interaction of molecular chaperones. Mass spectrometric determination of the interactome of α-syn in mammalian cells revealed that many molecular chaperones are involved in interacting with α-syn via its N-terminal region, including 14-3-3ζ, along with three other 14-3-3 isoforms (ε, γ and θ/τ) [36]. Thus, during intracellular proteostasis, a diversity of molecular chaperones interacts with a common N-terminal interface of α-syn to prevent its association and amyloid fibril formation.

The association of α-syn and 14-3-3 proteins intra- and extracellularly has been demonstrated in other studies. Immunoprecipitation and immunoblotting of rat brain homogenates revealed co-association of α-syn and 14-3-3 proteins [37]. They also noted that α-syn and 14-3-3 proteins have significant sequence similarity in their N-terminal regions, i.e., of L8-E61 in α-syn and L44-S99 in 14-3-3ζ which would facilitate their mutual association. Wang et al. [38] determined from a cell biological study that the θ/τ isoform of 14-3-3 complexes to α-syn in a chaperone interaction, thereby preventing α-syn oligomerization and regulating the cell-to-cell transfer of cytotoxic α-syn. As a result, 14-3-3θ/τ reduces α-syn cell toxicity and, hence, may play an important role in the pathology associated with PD.

Recently, Doherty et al. [39] identified that the N-terminal regions G36 to S42 and K45 to E57, particularly the former, are critical for regulating the aggregation of α-syn in vitro and in the nematode, *C. elegans*. These regions align well with those that interact with 14-3-3ζ and are obvious targets for the development of α-syn aggregation inhibitors. Furthermore, in the cryoEM-derived structure of the amyloid fibril form of a variant (residues 1 to 121) of α-syn without the last 19 amino acids of its acidic, proline-rich and unstructured C-terminal region, K43 to K58 is β-strand 3 that forms the interface between the two protofibrils in the overall fibrillar structure [40]. From cryoEM and solid-state NMR experiments on full-length fibrillar α-syn, a similar interface is present, but two different overall morphologies are formed compared to that for 1-121 α-syn, highlighting the polymorph nature of the protein in its amyloid fibrillar state [41]. Most of the first 37 residues of α-syn are not observed, presumably because they are disordered and mobile and therefore not part of the fibril core [40]. Thus, 14-3-3 proteins interacting transiently, in a chaperone manner, with K43 to K58 of A53T α-syn may preclude the formation of β-strand 3 (and its dimer association) in the earliest stages of α-syn aggregation.

The cryoEM structures of α-syn fibrils (filaments) extracted from inclusions in the brains of individuals with multiple system atrophy (MSA) are also polymorphic. They comprise two types of fibrils that each contain two different protofibrils [42]. The MSA α-syn fibrils are different from those formed in vitro from recombinant α-syn in terms of the number and arrangement of β-strands. However, the MSA and recombinant α-syn structures contain a common interface between the protofibrils involving the N-terminal region of α-syn. For the MSA α-syn protofibrils, this encompasses Q24 to A56 (twice), G36 to V63 and L38 to T64. Again, these data are consistent with the role of the N-terminal region in regulating the self-association of α-syn and its interaction with molecular chaperones such as 14-3-3ζ.

3. Materials and Methods

Reagents. All reagents were of analytical grade and purchased from Sigma-Aldrich (Australia). Aβ peptides (1–40 and 1–42) were purchased from Bachem Ltd. (Weil am Rhein, Germany). In situ and ex situ Thioflavin T (ThT) fluorescence measurements were conducted in black, clear bottom 96 microwell plates (Greiner Bio-One, Baden-Württemberg, Germany) using SealPlate MiniStrips (Astral Scientific, Australia) to prevent evaporation. Uranyl acetate, used for negative staining of samples for TEM, was obtained from Agar Scientific (Essex, UK). All solutions were prepared using deionized water purified to a resistivity of 18.2 MΩ·cm and subsequently filtered through a 0.22 μm membrane (Millipore, Australia).

Protein expression and purification. The 14-3-3ζ and Tobacco Etch Virus (TEV) protease plasmid constructs were a kind gift from Prof. James Murphy (Walter and Eliza Hall Institute of Medical Research, Australia) and Prof. Michael Parker (University of Melbourne, Australia), respectively. The expression plasmids for WT and A53T α-syn were a gift from Dr Tim Guilliams (University of Cambridge, UK). Recombinant 14-3-3ζ-His6 fusion proteins were expressed and purified as described previously [8,43,44]. Following purification, cleavage of the His6 tag was achieved using the TEV protease, and the cleavage products were purified using Ni-NTA column chromatography (Qiagen). WT and A53T α-syn were expressed and purified using the protocol of Narhi et al. [45]. ^{15}N-labeled α-syn was prepared as outlined in Dedmon et al. [35]. Recombinant ^{15}N-labeled Aβ peptides (1–40 and 1–42) were prepared by co-expression with an affibody [46]. The purified proteins were stored at −20 °C.

All protein and peptide concentrations were determined via absorbance measurements at either 276 nm or 280 nm using a Cary 5000 UV-vis spectrophotometer (Varian Ltd., Australia). A molar extinction coefficient (ε) of 5600 M^{-1} cm^{-1} was used for WT and A53T α-syn, measured at 276 nm. An ε of 23,860 M^{-1} cm^{-1} was used for 14-3-3ζ, measured at 280 nm.

Dynamic Light Scattering of 14-3-3ζ and α-synuclein. For 14-3-3ζ and α-syn alone or at a 1.0:1.0 molar ratio (7.2 μM) in 50 mM phosphate containing 100 mM NaCl and 2 mM EDTA, pH 7.4, time-resolved DLS analysis was performed at 37 °C using a Zetasizer Nano-ZS (Malvern Instruments, Worcestershire, UK). The particle diameter-intensity distribution and mean hydrodynamic diameter were determined from 13 acquired correllograms using the program CONTIN [47] and the method of cumulants [48], respectively, via Dispersion Technology Software (Malvern Instruments Ltd., Worcestershire, UK).

Transmission Electron Microscopy imaging of amyloid fibrils formed in vitro. An aliquot of the protein solutions from in situ ThT assays (6 μL) was transferred onto a carbon-coated nickel transmission electron microscopy (TEM) grid (SPI Supplies, West Chester, PA, USA). The grid was then washed using filtered MilliQ water (2 × 10 μL) before negative staining with uranyl acetate solution (8 μL, 2% *w*/*v*, in MilliQ). Between each step and after staining, excess solvent was removed by filter paper. After staining, the grids were left to air dry. Grids were viewed on a Philips CM 100 Transmission Electron Microscope (Eindhoven, Netherlands) between 13,500 and 64,000 times magnification operating at 120 kV.

Chaperone assays to monitor the effect of 14-3-3ζ on amyloid fibril formation of amyloid β and α-synuclein. All in vitro experiments in which amyloid fibrils were formed from $Aβ_{40}$ and $Aβ_{42}$ or A53T α-syn were undertaken at 37 °C in 50 mM phosphate, 100 mM NaCl, and pH 7.4. The formation of amyloid fibrils, in the absence or presence of 14-3-3ζ was assessed by ThT fluorescence (20 μM, excitation 440 nm, emission 490 nm) using a Fluorostar Optima plate reader (BMG Labtechnologies, Australia).

$Aβ_{40}$ and $Aβ_{42}$ were dissolved in ammonium hydroxide (final concentration 3.8 mM) and then diluted to 500 μM in water and stored at −80 °C. Further dilutions were made in

50 mM phosphate buffer containing 100 mM NaCl, pH 7.4, to achieve final concentrations for plate reader aggregation assays. The kinetics of Aβ_{40} and Aβ_{42} (15 µM, 100 µL) amyloid fibril formation, in the presence and absence of 14-3-3ζ, were monitored by the change in ThT fluorescence.

A53T α-syn and 14-3-3ζ solutions were prepared in 50 mM phosphate buffer containing 100 mM NaCl, pH 7.4. Each sample of either A53T α-syn (70 µM), 14-3-3ζ (70 µM), or A53T α-syn and 14-3-3ζ (70 µM of each protein) was separated into four Eppendorf tubes, each containing 500 µL. All samples were then wrapped in aluminium foil, with air-holes to aid in air circulation and temperature equilibration. They were incubated at 37 °C and shaken at 200 rpm for 24 h. For each protein solution, samples (2 × 5 µL) were taken every 2 h and added to 96 well plates (Greiner Bio-One, Baden-Wurttemberg, Germany) containing ThT (20 µM, 60 µL). The ThT fluorescence of each plate was read at 37 °C. Samples (8 µL) were also taken at 0, 4, 10, and 21 h for TEM imaging.

Two-dimensional ^{1}H-^{15}N HSQC NMR spectroscopy of ^{15}N-labeled Aβ_{40} and unlabeled 14-3-3ζ. Samples of ^{15}N-labeled Aβ_{40} and 14-3-3ζ were prepared separately in 20 mM phosphate containing 0.5 mM EDTA and 0.02 % *w*/*v* NaN_3, pH 7.4. All spectra were recorded at 5 °C on a Bruker Avance 500 NMR spectrometer (Bruker, UK) operating at a magnetic field strength of 11.7 T and a ^{1}H frequency of 500.1 MHz and a ^{15}N frequency of 50.7 MHz equipped with ^{1}H-^{15}N, ^{13}C, ^{2}H z-gradient TCI cryoprobe. ^{1}H chemical shifts were referenced to water as per Cavanagh et al. [49]. Data were processed with NMRPipe [50] and analyzed with Sparky v. 3.112 [51] software.

Gradient-enhanced two-dimensional ^{1}H-^{15}N heteronuclear single quantum coherence (HSQC) correlation spectra were acquired using water suppression via a Watergate pulse sequence. Spectra were acquired with 640 and 64 complex points and spectral widths of 5001.324 Hz and 532.180 Hz for the ^{1}H and ^{15}N dimensions, respectively. The carrier was set on-resonance with water in the ^{1}H dimension and 16 scans were recorded per increment for 70 µM Aβ_{40}. Up to four molar equivalences of 14-3-3ζ were then added in identical buffer, and subsequent NMR spectra were acquired with identical parameters. No chemical shift changes were observed in the HSQC spectra, and intensity changes were calculated (following correction for dilution), with error bars indicating the standard deviation in background noise of the spectra.

Two-dimensional ^{1}H-^{15}N HSQC NMR spectroscopy of ^{15}N-labeled A53T α-synuclein and unlabeled 14-3-3ζ. All NMR spectra were acquired at 10 °C on a Bruker Avance III 700 NMR spectrometer operating at a magnetic field strength of 16.4 T and an ^{1}H frequency of 700.1 MHz and a ^{15}N frequency of 71.0 MHz equipped with a TXI cryoprobe. Data were processed with NMRPipe [50] and analyzed with Sparky v. 3.112 [51] software using previously reported assignments [35,52]. ^{15}N A53T α-syn (100 µM) was dissolved in 50 mM sodium phosphate containing 100 µM NaCl (10% D_2O, pH 7.4). DSS (4,4-dimethyl-4-silapentane-1-sulfonic acid, 0.1%) was used as a chemical shift reference [53]. Three sensitivity-enhanced ^{1}H-^{15}N HSQC spectra were acquired, with 256 increments, a sweep width of 25 ppm in the indirect dimension, and 4 scans per increment.

^{15}N spin-spin relaxation rates (R_2) were determined for relaxation times of 15.8 and 126.7 ms with a Carr-Purcell-Meiboom-Gill (CPMG) frequency of 1 kHz, encoded as a ^{1}H-^{15}N HSQC with 128 increments and 16 scans per increment, with a recycle delay of 2 s. One equivalent of 14-3-3ζ was then added in identical buffer, resulting in final α-syn and 14-3-3ζ concentrations of 84 µM, and a second set of NMR spectra were acquired with identical parameters. No chemical shift changes were observed in the HSQC spectra, and intensity changes were calculated (following correction for dilution) from the mean of the three spectra, with error bars indicating the standard error of the mean. All A53T α-syn cross-peak positions were obtained from an independent HSQC experiment to avoid the introduction of systematic bias that results from coupled measurements of peak position and intensity.

Quartz Crystal Microbalance of α-synuclein and 14-3-3ζ interaction. QCM experiments were performed as outlined in Shammas et al. [54].

4. Conclusions

Solution-phase NMR spectroscopy has provided detailed characterization of the regions or interfaces of two disease-related, amyloid fibril-forming proteins, $A\beta_{40}$ and α-syn, that interact with the molecular chaperone, 14-3-3ζ. For both $A\beta_{40}$ and α-syn, the hydrophobic regions that interact with 14-3-3ζ are integral components of the β-sheet core within the final amyloid fibrillar structures of $A\beta_{40}$ and α-syn prepared in vitro, and in vivo samples isolated from diseased individuals. The implication is that 14-3-3ζ interacts preferentially with these β-strand-forming regions of $A\beta_{40}$ and α-syn early along their aggregation pathway and thereby interferes with their propensity to associate and form a β-sheet as the first step in amyloid fibril formation. In the case of α-syn, the N-terminal V40 to K60 region that interacts with 14-3-3ζ mainly encompasses the regions that regulate the protein's aggregation [39].

As with sHsp chaperone action, intimate details of the mechanism by which 14-3-3 proteins elicit their chaperone action are not known, and neither are the reasons for the variation in their ability to inhibit the amorphous and fibrillar aggregation of destabilized peptides and proteins [7,8,10,31]. The chaperone action of 14-3-3ζ probably arises from at least partial dissociation of the dimer and exposure of the chaperone interaction site(s) at the dimer interface [44]. The significant variation in 14-3-3ζ chaperone ability (in this case, fibril formation of the Aβ peptides and A53T α-syn) is due to a combination of factors that have parallels with the chaperone action of the ten human sHsps [5,6]: (i) slowly aggregating target peptides or proteins provide the opportunity for 14-3-3ζ to interact efficiently, for example, via dissociation and exposure of its chaperone binding site(s); (ii) the nature of the partially folded intermediate state of the target peptide or protein prior to it forming the prefibrillar aggregate; (iii) the size/dimensions of the target peptide or protein. The seven 14-3-3 isoforms have different functional roles in vivo, most of which have not been elucidated. For example, the θ/τ isoform inhibits α-syn oligomerization and fibril formation [31,38], whereas the other isoforms, apart from η, are ineffective [31]. Furthermore, 14-3-3θ/τ reduces cell-to-cell transfer of α-syn, as occurs in the pathology of PD [38]. Similar diverse functionality, and some redundancy, occur within sHsps [5,6].

In conclusion, this study has provided insights into the means by which 14-3-3ζ (and, presumably, other 14-3-3 isoforms) exerts its chaperone action to inhibit amyloid fibril formation. The conclusions are generally applicable as there are strong parallels between the results reported herein and the in vitro and in-cell NMR studies of the interaction of α-syn with other molecular chaperones [36].

Supplementary Materials: The following are available online. Figure S1: Amyloid fibril formation, as monitored by ThT fluorescence, of seeded $A\beta_{40}$ (15 μM) in the absence and presence of 14-3-3ζ at 1.0:1.0–4.0 molar ratios of $A\beta_{40}$:14-3-3ζ, Figure S2: Amyloid fibril formation of A53T α-syn (70 μM) in the absence of 14-3-3ζ and at a 1.0:1.0 molar ratio of A53T α-syn:14-3-3ζ, as monitored by ThT fluorescence and TEM.

Author Contributions: Conceptualization, D.M.W., J.A.C., S.E.J., S.M., J.M.W. and C.M.D.; Methodology, D.M.W., D.C.T., S.M., C.M.D., S.E.J., J.M.W. and J.A.C.; Validation, D.M.W., D.C.T., S.M. and J.A.C.; Formal Analysis, D.M.W., S.M., D.C.T. and J.A.C.; Investigation, D.M.W. and S.M.; Resources, C.M.D., S.E.J. and J.A.C.; Data Curation, D.M.W. and D.C.T.; Writing–Original Draft Preparation, D.M.W. and J.A.C.; Writing–Review & Editing, D.M.W., J.A.C., S.M., D.C.T., S.E.J. and J.M.W.; Visualization, D.M.W. and D.C.T.; Supervision, J.A.C., S.M. and J.M.W.; Project Administration, J.A.C., S.M. and J.M.W.; Funding Acquisition, D.M.W., J.A.C., S.E.J. and C.M.D. All authors have read and agreed to the published version of the manuscript.

Funding: This research was supported by grants from the Australian Research Council and National Health and Medical Research Council of Australia to J.A.C. Most of this work was conducted in the laboratories of C.M.D. and S.E.J. during a visit by D.M.W. D.M.W. is grateful for support from an Australian Postgraduate Award, the Adelaide University Graduate Union-R.C. Heddle Award, the Australian Federation of University Women, the University of Adelaide Faculty of Health Sciences Travelling Abroad Fellowship, and the George Murray Scholarship.

Institutional Review Board Statement: Not applicable.

Informed Consent Statement: Not applicable.

Data Availability Statement: The data presented in this study are available on request from the corresponding author.

Conflicts of Interest: The authors declare no conflict of interest. The funders had no role in the design of the study; in the collection, analyses, or interpretation of data; in the writing of the manuscript, or in the decision to publish the result.

Sample Availability: Samples of the proteins are not available from the authors.

References

1. Chiti, F.; Dobson, C.M. Protein misfolding, functional amyloid, and human disease. *Annu. Rev. Biochem.* **2006**, *75*, 333–366. [CrossRef]
2. Knowles, T.P.; Vendruscolo, M.; Dobson, C.M. The amyloid state and its association with protein misfolding diseases. *Nat. Rev. Mol. Cell Biol.* **2014**, *15*, 384–396. [CrossRef]
3. Powers, E.T.; Morimoto, R.I.; Dillin, A.; Kelly, J.W.; Balch, W.E. Biological and chemical approaches to diseases of proteostasis deficiency. *Annu. Rev. Biochem.* **2009**, *78*, 959–991. [CrossRef]
4. Walther, D.M.; Kasturi, P.; Zheng, M.; Pinkert, S.; Vecchi, G.; Ciryam, P.; Morimoto, R.I.; Dobson, C.M.; Vendruscolo, M.; Mann, M.; et al. Widespread proteome remodeling and aggregation in aging *C. elegans*. *Cell* **2015**, *161*, 919–932. [CrossRef]
5. Treweek, T.M.; Meehan, S.; Ecroyd, H.; Carver, J.A. Small heat-shock proteins: Important players in regulating cellular proteostasis. *Cell. Mol. Life Sci.* **2015**, *72*, 429–451. [CrossRef] [PubMed]
6. Hayashi, J.; Carver, J.A. The multifaceted nature of alphaB-crystallin. *Cell Stress Chaperones* **2020**, *25*, 639–654. [CrossRef] [PubMed]
7. Yano, M.; Nakamuta, S.; Wu, X.; Okumura, Y.; Kido, H. A novel function of 14-3-3 protein: 14-3-3zeta is a heat-shock-related molecular chaperone that dissolves thermal-aggregated proteins. *Mol. Biol. Cell* **2006**, *17*, 4769–4779. [CrossRef] [PubMed]
8. Williams, D.M.; Ecroyd, H.; Goodwin, K.L.; Dai, H.; Fu, H.; Woodcock, J.M.; Zhang, L.; Carver, J.A. NMR spectroscopy of 14-3-3zeta reveals a flexible C-terminal extension: Differentiation of the chaperone and phosphoserine-binding activities of 14-3-3zeta. *Biochem. J.* **2011**, *437*, 493–503. [CrossRef] [PubMed]
9. Pair, F.S.; Yacoubian, T.A. 14-3-3 proteins: Novel pharmacological targets in neurodegenerative diseases. *Trends Pharmacol. Sci.* **2021**, *42*, 226–238. [CrossRef]
10. Sluchanko, N.N.; Gusev, N.B. Moonlighting chaperone-like activity of the universal regulatory 14-3-3 proteins. *FEBS J.* **2017**, *284*, 1279–1295. [CrossRef]
11. Kaplan, A.; Bueno, M.; Fournier, A.E. Extracellular functions of 14-3-3 adaptor proteins. *Cell Signal.* **2017**, *31*, 26–30. [CrossRef]
12. Cox, D.; Carver, J.A.; Ecroyd, H. Preventing alpha-synuclein aggregation: The role of the small heat-shock molecular chaperone proteins. *Biochim. Biophys. Acta* **2014**, *1842*, 1830–1843. [CrossRef]
13. Foote, M.; Zhou, Y. 14-3-3 proteins in neurological disorders. *Int. J. Biochem. Mol. Biol.* **2012**, *3*, 152–164. [PubMed]
14. LeVine, H., 3rd. Quantification of beta-sheet amyloid fibril structures with thioflavin T. *Methods Enzymool.* **1999**, *309*, 274–284.
15. Jarrett, J.T.; Berger, E.P.; Lansbury, P.T., Jr. The carboxy terminus of the beta amyloid protein is critical for the seeding of amyloid formation: Implications for the pathogenesis of Alzheimer's disease. *Biochemistry* **1993**, *32*, 4693–4697. [CrossRef] [PubMed]
16. Hudson, S.A.; Ecroyd, H.; Dehle, F.C.; Musgrave, I.F.; Carver, J.A. (−)-Epigallocatechin-3-gallate (EGCG) maintains kappa-casein in its pre-fibrillar state without redirecting its aggregation pathway. *J. Mol. Biol.* **2009**, *392*, 689–700. [CrossRef]
17. Lindner, R.A.; Treweek, T.M.; Carver, J.A. The molecular chaperone alpha-crystallin is in kinetic competition with aggregation to stabilize a monomeric molten-globule form of alpha-lactalbumin. *Biochem. J.* **2001**, *354*, 79–87. [CrossRef]
18. Carver, J.A.; Lindner, R.A.; Lyon, C.; Canet, D.; Hernandez, H.; Dobson, C.M.; Redfield, C. The interaction of the molecular chaperone alpha-crystallin with unfolding alpha-lactalbumin: A structural and kinetic spectroscopic study. *J. Mol. Biol.* **2002**, *318*, 815–827. [CrossRef]
19. Treweek, T.M.; Rekas, A.; Lindner, R.A.; Walker, M.J.; Aquilina, J.A.; Robinson, C.V.; Horwitz, J.; Perng, M.D.; Quinlan, R.A.; Carver, J.A. R120G alphaB-crystallin promotes the unfolding of reduced alpha-lactalbumin and is inherently unstable. *FEBS J.* **2005**, *272*, 711–724. [CrossRef]
20. Kulig, M.; Ecroyd, H. The small heat-shock protein alphaB-crystallin uses different mechanisms of chaperone action to prevent the amorphous versus fibrillar aggregation of alpha-lactalbumin. *Biochem. J.* **2012**, *448*, 343–352. [CrossRef]
21. Cox, D.; Selig, E.; Griffin, M.D.; Carver, J.A.; Ecroyd, H. Small heat-shock proteins prevent alpha-synuclein aggregation via transient interactions and their efficacy is affected by the rate of aggregation. *J. Biol. Chem.* **2016**, *291*, 22618–22629. [CrossRef]
22. Mori, H.; Takio, K.; Ogawara, M.; Selkoe, D.J. Mass spectrometry of purified amyloid beta protein in Alzheimer's disease. *J. Biol. Chem.* **1992**, *267*, 17082–17086. [CrossRef]
23. Kollmer, M.; Close, W.; Funk, L.; Rasmussen, J.; Bsoul, A.; Schierhorn, A.; Schmidt, M.; Sigurdson, C.J.; Jucker, M.; Fandrich, M. Cryo-EM structure and polymorphism of Abeta amyloid fibrils purified from Alzheimer's brain tissue. *Nat. Commun.* **2019**, *10*, 4760. [CrossRef]

24. Hou, L.; Zagorski, M.G. NMR reveals anomalous copper(II) binding to the amyloid Abeta peptide of Alzheimer's disease. *J. Am. Chem. Soc.* **2006**, *128*, 9260–9261. [CrossRef] [PubMed]
25. Narayan, P.; Meehan, S.; Carver, J.A.; Wilson, M.R.; Dobson, C.M.; Klenerman, D. Amyloid-beta oligomers are sequestered by both intracellular and extracellular chaperones. *Biochemistry* **2012**, *51*, 9270–9276. [CrossRef]
26. Petkova, A.T.; Ishii, Y.; Balbach, J.J.; Antzutkin, O.N.; Leapman, R.D.; Delaglio, F.; Tycko, R. A structural model for Alzheimer's beta-amyloid fibrils based on experimental constraints from solid state NMR. *Proc. Natl. Acad. Sci. USA* **2002**, *99*, 16742–16747. [CrossRef] [PubMed]
27. Paravastu, A.K.; Qahwash, I.; Leapman, R.D.; Meredith, S.C.; Tycko, R. Seeded growth of beta-amyloid fibrils from Alzheimer's brain-derived fibrils produces a distinct fibril structure. *Proc. Natl. Acad. Sci. USA* **2009**, *106*, 7443–7448. [CrossRef] [PubMed]
28. Carver, J.A.; Ecroyd, H.; Truscott, R.J.W.; Thorn, D.C.; Holt, C. Proteostasis and the regulation of intra- and extracellular protein aggregation by ATP-independent molecular chaperones: Lens alpha-crystallins and milk caseins. *Acc. Chem. Res.* **2018**, *51*, 745–752. [CrossRef] [PubMed]
29. Friedrich, R.P.; Tepper, K.; Ronicke, R.; Soom, M.; Westermann, M.; Reymann, K.; Kaether, C.; Fandrich, M. Mechanism of amyloid plaque formation suggests an intracellular basis of Abeta pathogenicity. *Proc. Natl. Acad. Sci. USA* **2010**, *107*, 1942–1947. [CrossRef] [PubMed]
30. Conway, K.A.; Harper, J.D.; Lansbury, P.T. Accelerated in vitro fibril formation by a mutant alpha-synuclein linked to early-onset Parkinson disease. *Nat. Med.* **1998**, *4*, 1318–1320. [CrossRef]
31. Plotegher, N.; Kumar, D.; Tessari, I.; Brucale, M.; Munari, F.; Tosatto, L.; Belluzzi, E.; Greggio, E.; Bisaglia, M.; Capaldi, S.; et al. The chaperone-like protein 14-3-3eta interacts with human alpha-synuclein aggregation intermediates rerouting the amyloidogenic pathway and reducing alpha-synuclein cellular toxicity. *Hum. Mol. Genet.* **2014**, *23*, 5615–5629. [CrossRef] [PubMed]
32. Rekas, A.; Adda, C.G.; Aquilina, J.A.; Barnham, K.J.; Sunde, M.; Galatis, D.; Williamson, N.A.; Masters, C.L.; Anders, R.F.; Robinson, C.V.; et al. Interaction of the molecular chaperone alphaB-crystallin with alpha-synuclein: Effects on amyloid fibril formation and chaperone activity. *J. Mol. Biol.* **2004**, *340*, 1167–1183. [CrossRef]
33. Rekas, A.; Jankova, L.; Thorn, D.C.; Cappai, R.; Carver, J.A. Monitoring the prevention of amyloid fibril formation by alpha-crystallin. Temperature dependence and the nature of the aggregating species. *FEBS J.* **2007**, *274*, 6290–6304. [CrossRef]
34. Sluchanko, N.N.; Tugaeva, K.V.; Greive, S.J.; Antson, A.A. Chimeric 14-3-3 proteins for unraveling interactions with intrinsically disordered partners. *Sci. Rep.* **2017**, *7*, 12014. [CrossRef] [PubMed]
35. Dedmon, M.M.; Christodoulou, J.; Wilson, M.R.; Dobson, C.M. Heat shock protein 70 inhibits alpha-synuclein fibril formation via preferential binding to prefibrillar species. *J. Biol. Chem.* **2005**, *280*, 14733–14740. [CrossRef] [PubMed]
36. Burmann, B.M.; Gerez, J.A.; Matecko-Burmann, I.; Campioni, S.; Kumari, P.; Ghosh, D.; Mazur, A.; Aspholm, E.E.; Sulskis, D.; Wawrzyniuk, M.; et al. Regulation of alpha-synuclein by chaperones in mammalian cells. *Nature* **2020**, *577*, 127–132. [CrossRef] [PubMed]
37. Ostrerova, N.; Petrucelli, L.; Farrer, M.; Mehta, N.; Choi, P.; Hardy, J.; Wolozin, B. Alpha-synuclein shares physical and functional homology with 14-3-3 proteins. *J. Neurosci.* **1999**, *19*, 5782–5791. [CrossRef]
38. Wang, B.; Underwood, R.; Kamath, A.; Britain, C.; McFerrin, M.B.; McLean, P.J.; Volpicelli-Daley, L.A.; Whitaker, R.H.; Placzek, W.J.; Becker, K.; et al. 14-3-3 proteins reduce cell-to-cell transfer and propagation of pathogenic alpha-synuclein. *J. Neurosci.* **2018**, *38*, 8211–8232. [CrossRef]
39. Doherty, C.P.A.; Ulamec, S.M.; Maya-Martinez, R.; Good, S.C.; Makepeace, J.; Khan, G.N.; van Oosten-Hawle, P.; Radford, S.E.; Brockwell, D.J. A short motif in the N-terminal region of alpha-synuclein is critical for both aggregation and function. *Nat. Struct. Mol. Biol.* **2020**, *27*, 249–259. [CrossRef]
40. Guerrero-Ferreira, R.; Taylor, N.M.; Mona, D.; Ringler, P.; Lauer, M.E.; Riek, R.; Britschgi, M.; Stahlberg, H. Cryo-EM structure of alpha-synuclein fibrils. *Elife* **2018**, *7*, e36402. [CrossRef]
41. Guerrero-Ferreira, R.; Taylor, N.M.; Arteni, A.A.; Kumari, P.; Mona, D.; Ringler, P.; Britschgi, M.; Lauer, M.E.; Makky, A.; Verasdonck, J.; et al. Two new polymorphic structures of human full-length alpha-synuclein fibrils solved by cryo-electron microscopy. *Elife* **2019**, *8*, e48907. [CrossRef]
42. Schweighauser, M.; Shi, Y.; Tarutani, A.; Kametani, F.; Murzin, A.G.; Ghetti, B.; Matsubara, T.; Tomita, T.; Ando, T.; Hasegawa, K.; et al. Structures of alpha-synuclein filaments from multiple system atrophy. *Nature* **2020**, *585*, 464–469. [CrossRef]
43. Woodcock, J.M.; Coolen, C.; Goodwin, K.L.; Baek, D.J.; Bittman, R.; Samuel, M.S.; Pitson, S.M.; Lopez, A.F. Destabilisation of dimeric 14-3-3 proteins as a novel approach to anti-cancer therapeutics. *Oncotarget* **2015**, *6*, 14522–14536. [CrossRef] [PubMed]
44. Woodcock, J.M.; Goodwin, K.L.; Sandow, J.J.; Coolen, C.; Perugini, M.A.; Webb, A.I.; Pitson, S.M.; Lopez, A.F.; Carver, J.A. Role of salt bridges in the dimer interface of 14-3-3zeta in dimer dynamics, N-terminal alpha-helical order, and molecular chaperone activity. *J. Biol. Chem.* **2018**, *293*, 89–99. [CrossRef] [PubMed]
45. Narhi, L.; Wood, S.J.; Steavenson, S.; Jiang, Y.; Wu, G.M.; Anafi, D.; Kaufman, S.A.; Martin, F.; Sitney, K.; Denis, P.; et al. Both familial Parkinson's disease mutations accelerate alpha-synuclein aggregation. *J. Biol. Chem.* **1999**, *274*, 9843–9846. [CrossRef] [PubMed]
46. Macao, B.; Hoyer, W.; Sandberg, A.; Brorsson, A.C.; Dobson, C.M.; Hard, T. Recombinant amyloid beta-peptide production by coexpression with an affibody ligand. *BMC Biotechnol.* **2008**, *8*, 82. [CrossRef] [PubMed]
47. Provencher, S.W. CONTIN: A general purpose constrained regularization program for inverting noisy linear algebraic and integral equations. *Comput. Phys. Commun.* **1982**, *27*, 229–242. [CrossRef]

48. Koppel, D.E. Analysis of macromolecular polydispersity in intensity correlation spectroscopy: The method of cumulants. *J. Chem. Phys.* **1972**, *57*, 4814–4820. [CrossRef]
49. Cavanagh, J.; Fairbrother, W.J.; Palmer, A.G.; Rance, M.; Skelton, N.J. Heteronuclear NMR experiments. In *Protein NMR Spectroscopy*; Cavanagh, J., Fairbrother, W.J., Palmer, A.G., Rance, M., Skelton, N.J., Eds.; Academic Press: Burlington, MA, USA, 2007; pp. 533–678.
50. Delaglio, F.; Grzesiek, S.; Vuister, G.W.; Zhu, G.; Pfeifer, J.; Bax, A. NMRPipe: A multidimensional spectral processing system based on UNIX pipes. *J. Biomol. NMR* **1995**, *6*, 277–293. [CrossRef]
51. Goddard, T.D.; Kneller, D.G. *SPARKY 3*; University of California: San Francisco, CA, USA, 2008.
52. Bussell, R., Jr.; Eliezer, D. Residual structure and dynamics in Parkinson's disease-associated mutants of alpha-synuclein. *J. Biol. Chem.* **2001**, *276*, 45996–46003. [CrossRef]
53. Wishart, D.S.; Bigam, C.G.; Holm, A.; Hodges, R.S.; Sykes, B.D. 1H, 13C and 15N random coil NMR chemical shifts of the common amino acids. I. Investigations of nearest-neighbor effects. *J. Biomol. NMR* **1995**, *5*, 67–81. [CrossRef] [PubMed]
54. Shammas, S.L.; Waudby, C.A.; Wang, S.; Buell, A.K.; Knowles, T.P.; Ecroyd, H.; Welland, M.E.; Carver, J.A.; Dobson, C.M.; Meehan, S. Binding of the molecular chaperone alphaB-crystallin to Abeta amyloid fibrils inhibits fibril elongation. *Biophys. J.* **2011**, *101*, 1681–1689. [CrossRef] [PubMed]

molecules

MDPI

Article

High-Efficiency Expression and Purification of DNAJB6b Based on the pH-Modulation of Solubility and Denaturant-Modulation of Size

Sara Linse

Department of Biochemistry and Structural Biology, Lund University, P.O. Box 124, 221 00 Lund, Sweden; sara.linse@biochemistry.lu.se

Abstract: The chaperone DNAJB6b delays amyloid formation by suppressing the nucleation of amyloid fibrils and increases the solubility of amyloid-prone proteins. These dual effects on kinetics and equilibrium are related to the unusually high chemical potential of DNAJB6b in solution. As a consequence, the chaperone alone forms highly polydisperse oligomers, whereas in a mixture with an amyloid-forming protein or peptide it may form co-aggregates to gain a reduced chemical potential, thus enabling the amyloid peptide to increase its chemical potential leading to enhanced solubility of the peptide. Understanding such action at the level of molecular driving forces and detailed structures requires access to highly pure and sequence homogeneous DNAJB6b with no sequence extension. We therefore outline here an expression and purification protocol of the protein "as is" with no tags leading to very high levels of pure protein based on its physicochemical properties, including size and charge. The versatility of the protocol is demonstrated through the expression of an isotope labelled protein and seven variants, and the purification of three of these. The activity of the protein is bench-marked using aggregation assays. Two of the variants are used to produce a palette of fluorescent DNAJB6b labelled at an engineered N- or C-terminal cysteine.

Keywords: self-assembly; extraction; solubilization

Citation: Linse, S. High-Efficiency Expression and Purification of DNAJB6b Based on the pH-Modulation of Solubility and Denaturant-Modulation of Size. *Molecules* **2022**, *27*, 418. https://doi.org/10.3390/molecules27020418

Academic Editors: Kunihiro Kuwajima, Yuko Okamoto, Tuomas Knowles and Michele Vendruscolo

Received: 30 November 2021
Accepted: 2 January 2022
Published: 10 January 2022

1. Introduction

DNAJB6 belongs to a group of chaperons, also called J-domain proteins or HSP40, characterized by the presence of a J-domain [1–3]. Two member proteins, DNAJB6b and DNAJB8, are recognized as potent inhibitors of amyloid formation in vitro and in vivo by a range of proteins and peptides including IAPP from diabetes type II [4], poly-Q peptides from Huntington's disease [1,5–9], α-synuclein from Parkinson's disease [10–12] and amyloid β peptide from Alzheimer's disease [13–15]. While amyloid is recognized as a generic state that most peptides/proteins or parts thereof may attain under some solution conditions [16], DNAJB6b and DNAJB8 are thus promiscuous amyloid formation inhibitors acting on a wide range of proteins.

The current study concerns the 241-residue DNAJB6b isoform, which is found in the cytosol and nucleus of cells. The DNAJB6a isoform is only found in the nucleus, differs at 10 residue positions in the C-terminal domain, and contains a nuclear localization signal in a C-terminal extension. Familial mutations in the DNAJB6 gene are connected to limb-girdle muscular dystrophy [17–19]. As reviewed [20], most of the familial mutations in DNAJB6 are found in the region including amino acids 89–100.

In a solution, depending on the total protein concentration, DNAJB6b may form large and highly polydisperse oligomers [6]. The three-dimensional structure of the protein has been studied using nuclear magnetic resonance spectroscopy yielding high-resolution structural models of the N-terminal J-domain rich in α-helices as well as the C-terminal domain rich in β-sheet [21]. However, the relative orientation of the two globular domains remains elusive and is most likely highly variable because the domains are connected by a very long linker whose composition is of low complexity. A model for a dimeric form of DNAJB6b based on covalent cross-linking data has been presented [22].

The current study is motivated by the very potent roles of DNAJB6b as an inhibitor of amyloid formation and an enhancer of amyloid solubility [14]. The retardation of amyloid formation is understood as an interaction with pre-fibrillar species hindering their nucleation into a fibrillar structure [13–15], a mode of action shared with several anti-amyloid chaperones including, for example, clusterin [23] and HSP70 [24].

The increase in solubility, i.e., the monomer concentration in equilibrium with aggregates of Aβ42, amounts to an impressive factor of ca. 500 in the presence of a 0.1:1 molar ratio of DNAJB6b:Aβ42. This can be reconciled with the formation of co-aggregates between Aβ42 and DNAJB6b with altered structure compared to pure Aβ42 fibrils [25]. This is a consequence of the second law of thermodynamics; if a molecule is present in co-existing phases at equilibrium, the chemical potential of the molecule must be the same in all these phases [26,27]. Thus, if the same peptide, in this case Aβ42, displays a difference in solubility in samples with a different composition, i.e., the chemical potential of the free monomer is not the same in these samples, then it follows that the structure of the aggregates in these samples must be different enough that the chemical potential of the monomers in the aggregates is altered. The unfavorable increase in the chemical potential of Aβ42 in the co-aggregates must be compensated by a reduction in the chemical potential of the chaperone in the co-aggregates relative to pure chaperone assemblies, such that the free energy of the system as a whole is lowered upon formation of the co-aggregates. This further implies that the chaperone on its own has a high chemical potential, the molecular origin of which remains elusive.

Biophysical and structural investigations of DNAJB6b alone and in co-aggregates with amyloid peptides may enhance our understanding of the molecular driving forces and structural features behind both the co-assembly process and the unusually high chemical potential of DNAJB6b. This will require highly pure and homogeneous protein devoid of tags or other sequence extensions to ensure that the physicochemical properties are not altered. The most straightforward way to achieve this is to express the proteins "as is" with their native sequence and to develop a purification protocol based on their physico-chemical properties, thus eliminating the need for expensive proteases and post-cleavage steps.

Here we develop facile and scalable protocols for the expression and purification of DNAJB6b "as is", i.e., devoid of tags or other sequence additions. For the initial isolation from an *E. coli* cell pellet, we utilize the large difference in DNAJB6b solubility depending on the solution pH. For the subsequent purification of the protein, we utilize the large difference in size depending on the presence of a moderate concentration of denaturant (1.5–2.0 M GuHCl). Consecutive size-exclusion steps in the presence and absence of denaturant are thus used to remove any remaining large and small contaminants, respectively. The robustness of the protocol is validated through activity assays and through the expression and purification of sequence variants, two of which facilitate the production of a palette of fluorescent DNAJB6b with different excitation and emission wavelengths for spectroscopic and microscopic studies.

2. Results

2.1. Expression

The protein was expressed in *Escherichia coli* BL21 DE3 pLysS star "as is", i.e., with no tags or sequence extensions, from a synthetic gene with *E. coli*-preferred codons to yield the human DNAJB6b protein with the amino acid sequence given in Figure 1A, and the predicted 3D-structure in Figure 1B. As shown in Figure 1C, the expression level of wt DNAJB6b is around 1.5 g/L in rich auto-induction medium and only slightly lower in M9 minimal medium used for ^{13}C and ^{15}N labelling.

```
A wt     M-VDYYEVLGVQRHASPEDIKKAYRKLALKWHPDKNPENKEEAERKFKQVAEAYEVLSDAKK-61
  NCys    C

  wt     KRDIYDKYGKEGLNGGGGGGSHFDSPFEFGFTFRNPDDVFREFFGGRDPFSFDFFEDPFEDF-123
  P96R                                      R

  wt     FGNRRGPRGSRSRGTGSFFSAFSGFPSFGSGFSSFDTGFTSFGSLGHGGLTSFSSTSFGGS-185
  ST18A                                   AA  A  AA  A      AA AAAA   A
  ΔST             ----------------------------------------------------

  wt     GMGNFKSISTSTKMVNGRKITTKRIVENGQERVEVEEDGQLKSLTINGKEQLLRLDNK-241
  C-Cys                                                           C
  T193A           A
  ST5A         A AAAA
  ST18A        A AAAA
```

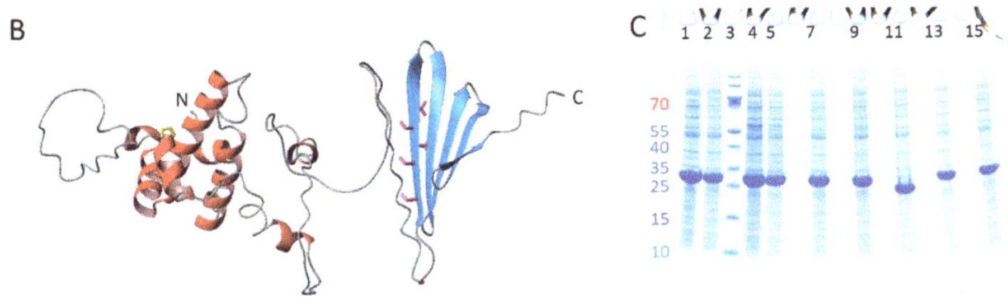

Figure 1. Expression of DNAJB6b wt and mutants. (**A**) Sequence of DNAJB6b wt and the substitutions made in the mutants of this study. (**B**) Prediction of DNAJB6b wt using Alphafold2 [28] with the J-domain shown in red, the C-terminal domain in blue, Pro96 in yellow, Ser190, Ser192, Thr193, Ser194 and Thr195 in pink. (**C**) Whole *E. coli* extract after expression of DNAJB6b wt and mutants in rich medium (lane 1, 4, 5, 7, 9, 11, 13 and 15) or M9 minimal medium (lane 2). Lane 1, 2 = wt DNAJB6b. Lane 3 = M_w standard with the M_w of the 7 smallest proteins given to the left of lane 1. Lane 4 = NCys- DNAJB6b. Lane 5 = CCys- DNAJB6b. Lane 7 = DNAJB6b-ST5A. Lane 9 = DNAJB6b -ST18A. Lane 11 = DNAJB6b-ΔST. Lane 13 = DNAJB6b -P96R. Lane 13 = DNAJB6b -T193A. Cell pellets from 1 mL were dissolved in 800 μL of 8 M urea, pH 8.0, mixed 1:1 with SDS loading buffer and 3 μL loaded per lane. The quantity loaded in each lane thus corresponds to cells from 2 μL culture.

2.2. Isolation

The purification protocol starts with isolation of the protein from *E. coli* using repeated sonication and centrifugation at pH 6.0, conditions under which the protein forms a white precipitate and *E. coli* proteins are removed in the supernatant (Figure 2). At each sonication step, the pellet, after centrifugation, has a white-grey body covered with brown matter. The brown matter is scraped off using a spatula and discarded before the next round of sonication. After five rounds of sonication at pH 6.0, centrifugation and scraping, the pellet is sonicated at pH 8.0, conditions under which DNAJB6 rapidly goes into solution (Figure 2). Charged contaminants are then removed by passing this final sonicate through an anion exchange resin (Q sepharose big beads) and then through a cation exchange resin (SP sepharose HP) (Figure 2).

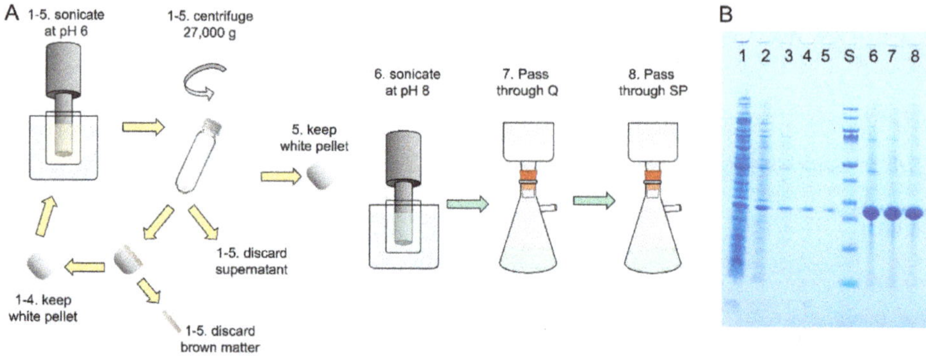

Figure 2. Initial isolation of DNAJB6b from *E. coli* cell pellet. (**A**) Outline of the isolation steps. 1–5. Sonication in 20 mM MES, pH 6.0 (supernatant after centrifugation). 6. Sonication in 10 mM Tris/HCL, 1 mM EDTA, pH 8.0. 7. Passage through Q-sepharose big beads. 8. Passage through SP sepharose HP. (**B**) SDS PAGE on a 10–20% polyacrylamide gel with lane 1–5 loaded with sonicate 1–5, lane S with M_W standard with sizes given to the right of the gel, lane 6 with sonicate 6, lane 7 with the Q flow-through, and lane 8 with the SP sepharose flow-through. The total time required for step 1–8 is ca. 2 h. The flow-through of SP sepharose (lane 8) was used for purification of the protein using ammonium sulphate precipitation and size-exclusion chromatography (see Figure 3).

The total isolation procedure takes ca. 2 h from the removal of the cell pellet from the freezer to aliquoting the flow-through from the SP sepharose. Using the cell pellet from 0.25 L, the procedure results in a 200 mL solution with a total of ca. 300 mg semi-pure DNAJB6b, thus corresponding to ca. 1.2 g protein per liter of culture (Table 1).

Table 1. Expression and purification procedure and yield estimation.

Step	Amount of DNAJB6b per Litre Culture	Estimation Based on
Expression in *E. coli*	1500 mg/L	Gel band intensity
Sonication at pH 6.0, removal of soluble *E. coli* proteins and lightly precipitated ones.	1200 mg/L	Gel band intensity
Sonication at pH 8.0 followed by passage through anion and cation exchange resins	1200 mg/L	Gel band intensity
Ammonium sulphate precipitation to remove non-protein contaminants and some contaminating proteins	900 mg/L	Gel band intensity
Size exclusion chromatography to remove large and small contaminants (two steps)	600 mg/L	Absorbance in 2 M GuHCl

2.3. Ammonium Sulphate Precipitation

Further purification is achieved using ammonium sulfate (AMS) precipitation to remove non-protein contaminants as well as some of the protein contaminants. After finding the relevant AMS concentration range using small volume samples and SDS PAGE (Supplementary Figure S1), bulk precipitation was performed for 25 mL aliquots and brought forward to the next step.

The time consumption for this step is ca. 30 min, and the losses are ca. 25% meaning that after this step, there remains ca. 900 mg protein per liter of culture.

2.4. Size Exclusion Chromatography

Size-exclusion chromatography (SEC) was used to remove low and high Mw contaminants and to exchange the buffer to the one used in biophysical experiments (Figure 3). The first SEC step is performed with 2 M GuHCl to favor dissociation of DNAJB6 oligomers and using a Superdex200 resin (separation range 10–600 kDa) to remove large protein contaminants from the AMS precipitate, which was dissolved in 6 M GuHCl prior to injection. The second SEC step is performed under native conditions and using a Superos6 resin (separation range 5–5000 kDa) to remove lower Mw contaminants. Based on SDS PAGE analysis, this combination of SEC steps results in a highly pure protein.

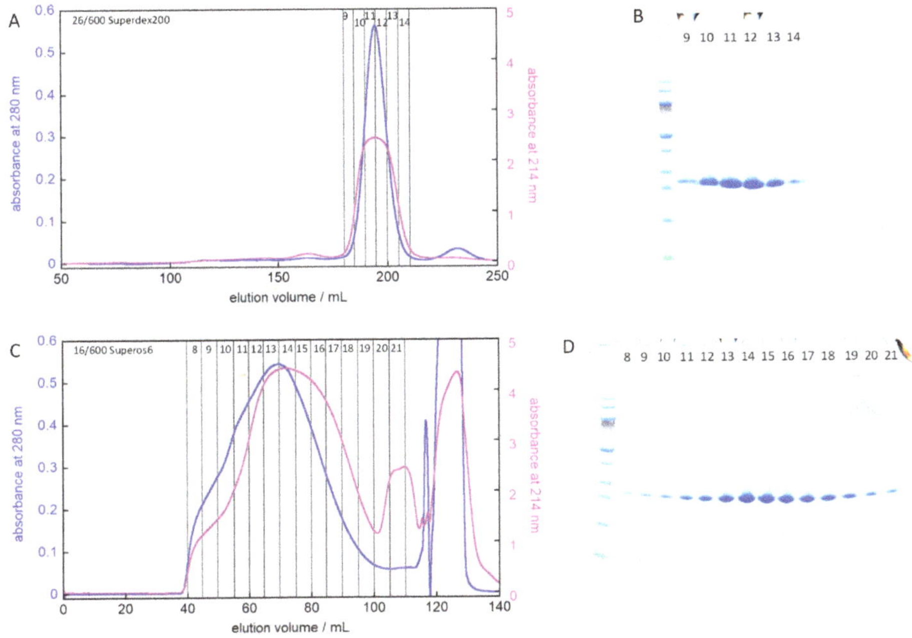

Figure 3. Purification of DNAJB6b using size exclusion chromatograph. (**A**) An aliquot from the flow-through of SP sepharose (see lane 8 in Figure 2) was precipitated by AMS and the 10–21% fraction dissolved in 10 mL 6 M GuHCl, 20 mM sodium phosphate, 0.2 mM EDTA, pH 8.0 and injected on a 26/600 Superdex200 column operated in 2 M GuHCl, 20 mM sodium phosphate, 0.2 mM EDTA, pH 8.0. (**B**) SDS PAGE of fractions 9–14 on a 10–20% polyacrylamide gel. Fraction 10–13 were concentrated and injected on a 16/600 Superose6 column. (**C**) The elution of the 16/600 Superos6 column operated in 20 mM sodium phosphate, 0.2 mM EDTA, pH 8.0. The injected sample was 5 mL of fractions 10–13 from panels A, B lyophilized down to 1/3 of the original volume, i.e., in 6 M GuHCl, 60 mM sodium phosphate, 0.6 mM EDTA, pH 8.0. (**D**) SDS PAGE of fractions 8–21 on a 10–20% polyacrylamide gel. In panels A and C, the absorbance at 280 and 214 nm are shown in blue and purple, respectively. Fractions 11–18 are kept for use in biophysical experiments.

The run time for the two SEC steps is ca. 100 and 80 min, respectively, but with the intervening lyophilization preferably run overnight. The loss per SEC step is ca. 25% meaning that after these final two steps, the total protocol results in ca. 600 mg ultra-pure protein per liter of culture (Table 1). The total operation time is ca. 5.5 h, plus the time required to run SDS PAGE.

2.5. Concentration Determination

DNAJB6b in solution exists as polydisperse oligomers that scatter light, which compromises concentration determination using absorbance. This is solved by diluting the samples 3:1 with 8 M GuHCl, yielding a 75% solution in 2 M GuHCl (Figure 4). Back-calculation to 100% gives the sample concentration of each fraction. Based on this information, the total yield after the second SEC step is 600 mg purified protein per liter of culture.

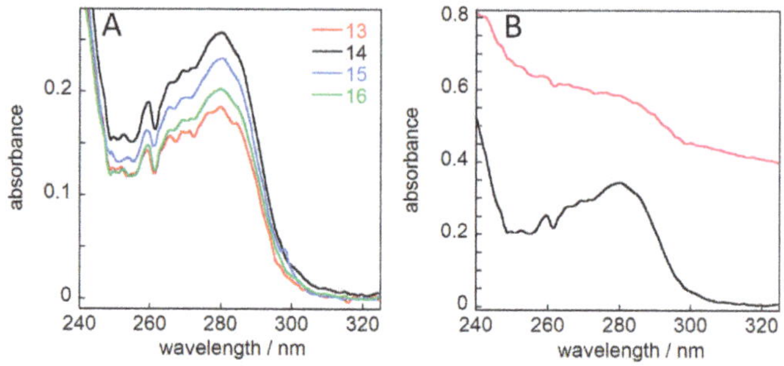

Figure 4. Concentration determination. (**A**) Absorbance spectra of fractions 13–16 from the elution of Superos6 as shown in Figure 3C,D. A small sample from each fraction was mixed 3:1 with 8 M GuHCl to bring the samples to 2 M GuHCl, which leads to the dissociation of the large polydisperse oligomers. (**B**) Absorbance spectrum of fraction 14 in buffer (pink), in which case the large polydisperse oligomers lead to significant light-scattering prohibiting concentration determination, and in 2 M GuHCl, which permits absorbance to be used for concentration determination.

2.6. Robustness of Expression Protocol

The robustness of the tag-free expression system was evaluated for seven variants referred to as NCys with a cysteine residue added at the N-terminus, CCys, with a cysteine residue added at the C-terminus, STA5 with 5 alanine residues replacing Ser190, Ser192, Thr193, Ser194, and Thr195, ST18A with 18 different S -> A and T -> A substitutions, ΔST with a large part of the Ser/Thr-rich region deleted, P96R with an Arg residue replacing Pro96, and T193A with Thr193 changed to Ala. The amino acid substitutions in the variants are given in Figure 1. We find that all seven variants express to very high levels, similar to wt, and clearly dominate over all the *E. coli* proteins in the whole-cell extracts (Figure 1C).

2.7. Robustness of Purification Protocol

The robustness of purification was evaluated for three variants, NCys, CCys, and STA5. STA5 was purified using the same protocol as for wt (Figure S1). NCys and CCys (Figure 5) were purified using the same protocol as for wt, except that 1 mM DTT was included in all buffers.

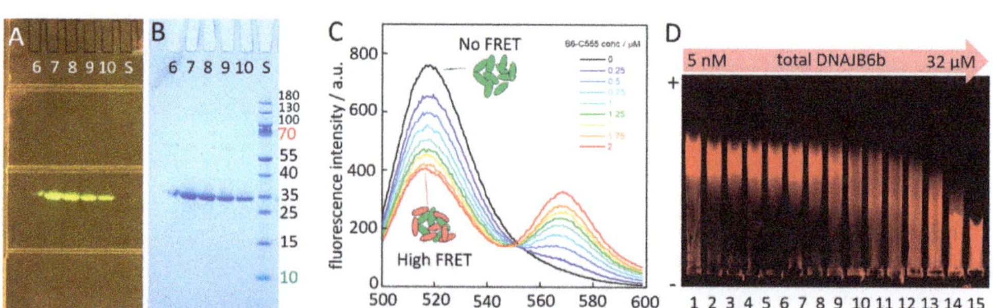

Figure 5. Fluorescent DNAJB6b. (**A**,**B**) Examples of SDS PAGE after labelling and SEC to remove free

dye for DNAJB6b-CCys labelled with Alexa-488. (**A**)The gel imaged on a "dark reader" with blue excitation filter and orange emission filters. (**B**) The same gel photographed after staining with coomassie (quick stain). Panel (**C**) shows fluorescence emission spectra recorded for samples with 1 μM constant concentration of DNAJB6b-CCys-Alexa-488 and varying concentrations of DNAJB6b-CCys-Alexa-555: 0 (black), 0.25 (marine), 0.5 (blue), 0.75 (light blue), 1.0 (cyan), 1.25 (green), 1.5 (yellow), 1.75 (orange) and 2.0 μM (red). (**D**) Native gel electrophoresis. Agarose gel imaged on an IR fluorescence scanner. In each well is loaded 5 nM DNAJB6b-CCys-IR680 mixed with different concentrations of unlabeled DNAJB6b-wt to yield the following total concentrations of DNAJB6b: 5 nM (lane 1), 9 nM (2), 13 nM (3), 21 nM (4), 37 nM (5), 69 nM (6), 133 nM (7), 255 nM (8), 505 nM (9), 1.0 μM (10), 2.0 μM (11), 4.0 μM (12), 8.0 μM (13), 16.0 μM (14), 32 μM (15). The gel is oriented with the positive pole at the top of the image and the negative pole at the bottom.

2.8. *Activity of the Purified Protein*

The activity of the DNAJB6b protein was investigated by monitoring the fibril formation of Aβ42 by ThT fluorescence in the absence and presence of the purified DNAJB6b wt or STA5 at a series of concentrations (Figure 6). The wt protein is a very potent inhibitor of Aβ42 aggregation, and we observe a substantial delay of the sigmoidal transition from 0.0005:1 molar ratio of DNAJB6b:Aβ42 and upwards. The data are well fitted by a model that includes primary nucleation, elongation, and secondary nucleation, keeping the rate constants for elongation and secondary nucleation as global parameters and the rate constants for primary nucleation as a local parameter specific to each DNAJB6b concentration (see solid lines in Figure 6). Inhibition of primary nucleation can thus explain all the data obtained in the presence of DNAJB6b. The data for the DNAJB6b-STA5 mutant are also well fitted assuming inhibition of primary nucleation, but the variant appears less effective than the wt (Figure 6). These findings are in agreement with those of earlier reports [13,14].

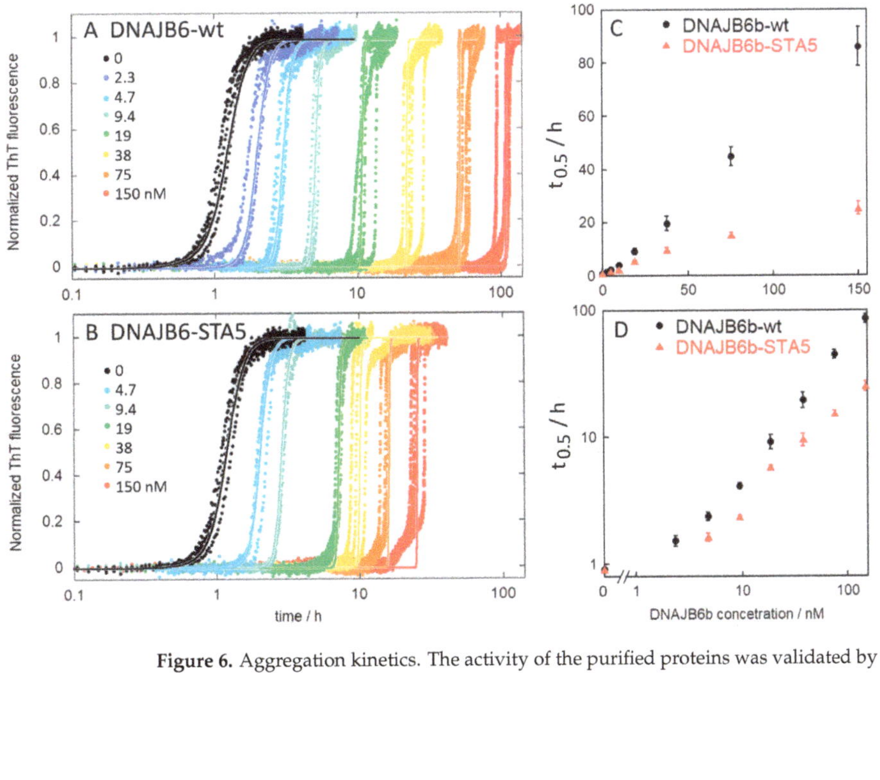

Figure 6. Aggregation kinetics. The activity of the purified proteins was validated by monitoring the

fibril formation of 4 μM Aβ42 by ThT fluorescence in the absence (black) and presence (colors) of DNAJB6b-wt (**A**) or DNAJB6b-STA5 (**B**) at concentrations ranging from 2 to 150 nM, i.e., at molar ratios ranging from 0.0005 to 0.0375. The solid lines show fitted curves assuming inhibition of primary nucleation. The half times of aggregation as extracted from the data in panels A and B are shown as averages and standard deviations over 4 replicates for DNAJB6b-wt (black) and DNAJB6b-STA5 (red) with linear (**C**) and logarithmic (**D**) axes.

2.9. Fluorophore Labelling

Each of the cysteine containing constructs was labelled with a palette of fluorophores including Alexa488, Alexa555, Alexa647, and IRdye-680. DTT was first removed using SEC and maleimide dyes were added from stocks in DMSO, followed by a second SEC step to isolate the labelled protein and to remove the excess free dye (Figure 5).

2.10. Polydispersity

Depending on the total protein concentration, DNAJB6b may form large and highly polydisperse aggregates in solution [6]. This is evident from the SEC under non-denaturing conditions, in which case DNAJB6b elutes as a very broad peak from a superos6. This is also evident from native electrophoresis in an agarose gel of IR680-labelled DBAJB6b at a constant concentration of 5 nM with a variable concentration of unlabelled DBAJB6b (Figure 5).

3. Discussion

The results of the current study show that *E. coli* cells are highly tolerant to DNAJB6b expression. In the whole-cell extracts (Figure 1), the overexpressed chaperone clearly dominates over all the *E. coli* proteins, and this is observed for DNAJB6b wt as well as all seven mutants tested. This provides a very good starting state for isolation and purification of the proteins and relies on the use of synthetic genes with *E. coli* optimized codons in the expression plasmids on the DNAJB6b protein causing no impairment of *E. coli* growth.

Studies of protein biophysical properties require highly pure protein devoid of tags or other sequence extensions. The main advantage of the current expression protocol is that the production of tag-free proteins makes the following isolation and purification straightforward. No expensive proteases or post-cleavage steps are needed. Instead, the protocol is built on the physico-chemical properties of the protein in terms of charge and size and the modulation of these properties by pH and denaturant, respectively. We have expressed the protein "as is" with its native sequence and developed a purification protocol based on its physico-chemical properties.

In the initial isolation of the protein from *E. coli* cell pellet (Figure 2), we utilize the large difference in DNAJB6b solubility depending on pH. Intriguingly, DNAJB6b is highly soluble in the form of polydisperse oligomers at pH 8.0 and displays a minimum solubility around pH 6.0, although the isoelectric point is 7.3 if calculated from its amino acid sequence using model pKa values ([29]; Figure 7A). Precipitation at pH 6.0 likely results from titration of the N-terminal domain and the linker, because the C-terminal domain (residues 187-241) displays constant positive net charge at around +2 between pH 8.0 and 6.0 (Figure 7B). The net charge of the N-terminal domain changes from ca. +1.2 at pH 8.0 to ca. +3.6 at pH 6.0, while that of the linker changes from ca. −4.0 to −2.5 (Figure 7B). The low solubility at pH 6.0 allows the initial isolation protocol to use repeated sonication at pH 6.0 and removal of soluble *E. coli* proteins by decanting the supernatant over the precipitate and by removal of the less dense precipitate by scraping away brown matter from the top of the white DNAJB6b pellet. This scraping procedure sacrifices yield for purity, but the final yield is very high anyway given the extreme expression level and for the subsequent biophysical studies it is much more important to focus on achieving as high purity as possible. The isolation procedure ends with solubilization of the protein through sonication at pH 8.0 and removal of charged contaminants through stepwise passage of the sonicate through anion and cation exchange resins. Here, it is fortuitous that DNAJB6b carries such a low net negative charge of ca. −0.8 that it is not retained by anion nor cation exchange resins.

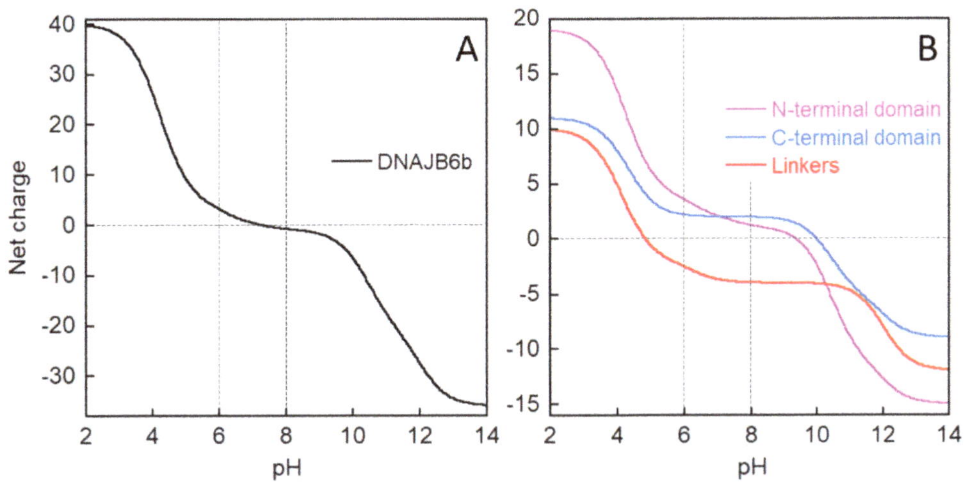

Figure 7. Net charge of DNAJB6b (**A**) and its parts (**B**) as a function of pH titration as calculated based on model compound pKa values [29].

Because DNAJB6 forms highly polydisperse oligomers, it elutes very broadly during size exclusion chromatography and is difficult to separate from impurities of all sizes except possibly those smaller than the DNAJB6b monomer. To circumvent this problem, we used a first SEC step in the presence of enough denaturant that the oligomers dissociate and DNAJB6b elutes as a relatively narrow peak on a Superdex200 column relatively late in the chromatogram meaning high Mw protein contaminants are removed by this step. This first SEC step was preceded by ammonium sulphate (AMS) precipitation to remove non-protein contaminants as well as some protein contaminants. The second SEC step on a superos6 column serves to exchange the buffer to the one used in the following biophysical experiments and to remove any remaining small proteins. In this step, the protein is injected denatured in 6 M GuHCl but folds and oligomerizes as it elutes ahead of the denaturant.

The result is highly pure DNAJB6b amenable for detailed studies of its biophysical properties, such as structure and dynamics, stability, polydispersity, exchange rates, etc. aiming at an understanding of the molecular origin of its unusually high chemical potential [26] and the driving force for its co-assembly with amyloid peptides. The ease of adaption to mutational and sequence length variants is another advantage of the protocol. NCys and CCys were produced to enable site-specific covalent attachment of fluorophores or nanogod labels to either terminus using maleimide chemistry. Although fluorophores should be used with care as an appendix the size of 7–10 residues with mixed non-polar and charged character is added to the protein under study, the possibility of including a small fraction of labelled DNAJB6b will enable the study of the biophysical properties of mostly unlabelled oligomers. STA5 and STA18 were chosen based on earlier work showing reduced inhibitory capacity [14]. ΔST was chosen based on earlier work showing reduced polydispersity and a larger fraction of monomeric protein for this variant [21]. P96R and T193A were based on the disease association of these mutants [20].

4. Methods

4.1. Gene Synthesis and Cloning

The protein was expressed in *Escherichia coli* BL21 DE3 PlyS star "as is", i.e., with no tags or sequence extensions, from a synthetic gene with *E. coli*-preferred codons to yield the human DNAJB6b wt protein with the amino acid sequence as shown in Figure 1, as well as seven variants. The genes were cloned between NdeI and BamHI restriction sites in a Pet3a vector. The gene synthesis and cloning were purchased from Genscipt (Piscataway, NJ, USA).

4.2. Expression in Rich Medium

The plasmid was transformed into Ca^{2+} competent *E. coli* BL21 DE3 PLysS Star and spread on LB-agar plates with 50 mg/L ampicillin and 30 mg/L chloramphenicol. Single small and well-isolated colonies were used to inoculate 50 mL LB cultures with 50 mg/L ampicillin and 30 mg/L chloramphenicol, which were grown at 37 °C in 250 mL baffled flasks with 125 rpm orbital shaking for 8 h. The OD at 600 nm was measured and the equivalent of 0.5 mL at OD600 = 0.8 was transferred to 500 mL pre-warmed (37 °C) overnight express medium. One liter of medium was prepared mixing the following separately autoclaved solutions: (a) 1 mL 1 M $MgSO_4$, (b) 50 mL"20 × 5052": 5 g glycerol, 0.5 g glucose, 2 g α-lactose, 44.6 mL H_2O, (c) 50 mL "20 × NPS": 3.3 g $(NH_4)_2SO_4$, 6.8 g KH_2PO_4, 7.1 g Na_2HPO_4, 45 mL H_2O, and (d) 900 mL "1.1 × LB": 10 g Bacto^{TM} tryptone, 5 g Bacto^{TM} yeast extract, 10 g, H_2O up to 900 mL, and adding 50 mg/L ampicillin and 30 mg/L chloramphenicol from concentrated stocks. The overnight express cultures were grown in 2500 mL baffled flasks with 125 rpm orbital shaking for 15 h, after which the cells were harvested by centrifugation at 6000 rpm for 10 min in a JLA 8.1000 rotor.

4.3. Expression in M9 Minimal Medium with Isotope Labels

The plasmid was transformed into Ca^{2+} competent *E. coli* BL21 DE3 PLysS Star and spread on LB-agar plates with 50 mg/L ampicillin and 30 mg/L chloramphenicol. Single small and well-isolated colonies were used to inoculate 50 mL LB cultures with 50 mg/L ampicillin and 30 mg/L chloramphenicol, which were grown at 37 °C in 250 mL baffled flasks with 125 rpm orbital shaking for 14 h. 3 mL was transferred to 500 mL pre-warmed (37 °C) M9 medium. One liter of medium was prepared by mixing the following separately autoclaved solutions: (a) 980 mL M9 salt solution: 4.65 g $Na_2HPO_4{\cdot}2H_2O$, 3 g KH_2PO_4, 0.5 g NaCl, tap water up to 980 mL. (b) 10 mL of 200 g/L 13C-glucose, (c) 10 mL of 100 g/L $^{15}NH_4Cl$, (d) 1 mL 1 M $MgSO_4$, (e) 0.4 mL of 250 mM $CaCl_2$ and the following sterile filtered solutions 0.6 mL of 30 mM $FeCl_3$, (f) 1 mL of 1 mg/mL vitamin B1 and adding 50 mg/L ampicillin and 30 mg/L chloramphenicol from concentrated stocks. The cultures were grown in 2500 mL baffled flasks with 125 rpm orbital shaking until they reached an OD at 600 mm of 0.8, at which point 0.2 mM IPTG was added from a freshly prepared stock and the cells were harvested 5 h later by centrifugation at 6000 rpm for 10 min in a JLA 8.1000 rotor.

4.4. Purification Procedure

Step 1—sonication at pH 6.0. The pellet from 0.5 L culture was sonicated in 80 mL 20 mM MES, pH 6.0, in a glass beaker immersed in an ice-water slurry using a Soni-Prep 150 sonicator, large horn, 50% duty cycle, for three minutes. The slurry was centrifuged for 6 min at 4 °C at 18,000 rpm in a Beckman 25:50 rotor. The supernatant (lane 1 in Figure 2) was discarded and the top of the pellet was scraped with a spatula to remove brown matter. The pellet was re-sonicated four times in 80 mL 20 mM MES, pH 6.0, then centrifuged for 6 min at 4 °C at 18,000 rpm as above. After each centrifugation, the supernatant (lane 2–5 in Figure 2) was discarded and the top of the pellet was scraped with a spatula to remove brown matter, the amount of which significantly reduced after each sonication and centrifugation round.

Step 2—sonication at pH 8.0. The pellet after the fifth sonication above was sonicated and dispersed using a magnetic stir bar in a total of 200 mL 10 mM tris, 1 mM EDTA, pH 8.0, resulting in a clear solution (lane 6 in Figure 1).

Step 3—ion exchange removal of charged contaminants. The DNJAB6b solution after step 2 was first passed through 50 mL Q Sepharose big beads (anion exchange resin, lane 7 in Figure 2 shows the flow-through) and then through 25 mL SP HP resin (cation exchange resin, lane 7 in Figure 1 shows the flow-through).

Step 4—ammonium sulphate precipitation. The flow-through of the SP HP resin was mixed with 100% ammonium sulphate (AMS), pH 8.0 (equilibrated over ammonium sulphate salt at room temperature) at a 100:17.6 volume ratio resulting in 15% AMS, incubated for 10 min in an ice box, and centrifuged for 10 min at 8000 rpm in a JLA 8.1000 rotor. The supernatant was mixed with 100% AMS pH 8.0 at a 117.6:10.6 volume ratio, resulting in 22% AMS (the 15–22% fraction). The sample was incubated for 10 min in an ice box and centrifuged for 10 min at 8000 rpm in a JLA 8.1000 rotor.

Step 5—size exclusion chromatography. Each pellet from step 4 was dissolved in 6 M GuHCl and separated by size exclusion chromatography (SEC) on a 26/600 mm Superdex200 column (GE Healthcare) operated in 2 M GuHCl, 20 mM sodium phosphate, 0.2 mM EDTA, pH 8.0 (Figure 3A,B). Under these conditions, DNAJB6b elutes as a monomer or dimer. Fractions were analyzed using the absorbance at 214, 260, 280 nm and by SDS PAGE (10–20% gel from Novex in a Tris/Tricine buffer system). Fractions containing DNAJB6b were lyophilized to concentrate them three times, followed by SEC on a 16/600 mm Superose6 column (GE Healthcare) operated in 20 mM sodium phosphate, 0.2 mM EDTA, pH 8.0 (Figure 3C,D). Under these conditions, DNAJB6b elutes as highly polydisperse oligomers. Fractions were analyzed using the absorbance at 214, 260, 280 nm and by SDS PAGE (10–20% gel from Novex in a Tris/Tricine buffer system). Fractions containing DNAJB6b were stored on ice or frozen at −20 °C until used in biophysical experiments. The concentration of DNAJB6b was determined by mixing 150 µL from each fraction with 50 µL 8M GuHCl, recording the absorbance between 360 and 232 nm, multiplying the absorbance at 280 nm by 4/3, and dividing by the extinction coefficient 14,400 $M^{-1}cm^{-1}$.

4.5. Adaption of the Purification Protocol for Mutants

Variants with retained charge distribution, e.g., STA5, STA18, and T193A, can be purified using the same protocol as the wt. The purification of charge substitution mutants may require adjustment to the pH values of the buffers used. In such a case, a quick pH scouting with crude sonicates may be helpful to find out under which pH the variant is minimally soluble. When purifying the NCys and CCys mutants, it is possible to use the same protocol as for wt with the only modification being the inclusion of 1 mM DTT in all buffers to prevent the formation of covalent dimers.

4.6. Aggregation Kinetics

Aβ(M1-42), was expressed and purified as described [30] and monomers were isolated in 20 mM sodium phosphate, 0.2 mM EDTA, pH 8.0. For DNAJB6b wt, we used an aliquot from fraction 14 from the SEC run on Superos6 shown in Figure 3C,D. For DNAJB6b STA5 we used an aliquot from fraction 14 in the SEC run on Superos6 (Figure S1). Samples were kept and mixed on ice to contain 4 µM Aβ42 alone or 4 µM Aβ42 with DNAJB6b wt or STA5 at concentrations ranging from 2 to 150 nM. All samples contained 6 uM thioflavin T (ThT) and were distributed as multiple replicates in wells of a Corning 3881 96-well plate (a half area PEGylated black polystyrene plate with a transparent bottom). The plate was inserted in a plate reader (BMG omega) equilibrated at 37 °C. The fluorescence was read through the bottom of the plate using an excitation filter at 440 nM and an emission filter at 480 nm. The data were normalized using Amylofit [31] and could only be fitted using models assuming inhibition of primary nucleation.

4.7. Fluorophore Labelling

Purified DNAJB6-CCys or DNAJB6-NCys was subjected to SEC on a G25 column operated in 20 mM sodium phosphate, pH 8.0, to remove DTT. Alexa488, Alexa555, Alexa647, or IRdye680 was added from a 5 mM stock in DMSO, 1.5 molar equivalents. The solution was gently mixed and incubated at room temperature for 2 h. Excess dye was removed by

SEC on a 10 × 300 mm Superdex75 column operated in 20 mM sodium phosphate, 0.2 mM EDTA, and pH 8.0.

5. Conclusions

We have presented here a versatile protocol for producing highly pure DNAJB6b wt and variants. The protocol relies on inexpensive tools as it is based on the physicochemical properties of the protein, which is expressed as is with no sequence additions. The purified protein will thus be amenable to more detailed studies of these properties and their molecular origin using a range of biophysical tools including NMR and optical spectroscopy, as well as a multitude of surface and scattering techniques. The ease of isotope labeling will be an advantage both for NMR spectroscopy and neutron scattering with contrast variation to distinguish different components in co-aggregates. The palette of fluorescently labelled variants opens for additional experiments using fluorescence correlation spectroscopy, microscopy, and fluorescence resonance energy transfer. The plasmids described here will be shared with the research community and the protocols we describe for expression, purification, and labelling are easy to set up in any lab at low cost.

Supplementary Materials: The following supporting information can be downloaded online. Figure S1: Purification of DNAJB6b-STA5.

Funding: This work was supported by the Swedish Research Council (SL 2015-00143).

Data Availability Statement: All data will be shared with readers upon reasonable request.

Acknowledgments: We thank Harm Kampinga, Groningen and Cecilia Emanuelsson, Lund, for stimulating scientific discussions regarding relevant DNAJb6b mutants.

Conflicts of Interest: The author declares no conflict of interest.

Sample Availability: All plasmids will be shared with readers upon reasonable request.

References

1. Hageman, J.; Rujano, M.A.; van Waarde, M.A.; Kakkar, V.; Dirks, R.P.; Govorukhina, N.; Oosterveld-Hut, H.M.; Lubsen, N.H.; Kampinga, H.H. A DNAJB chaperone subfamily with HDAC-dependent activities suppresses toxic protein aggregation. *Mol. Cell* **2010**, *37*, 355–369. [CrossRef] [PubMed]
2. Zarouchioti, C. DNAJ protein in neurodegeneration: Essential and protective factors. *Phil. Trans. R. Soc. B* **2017**, *373*, 20160534. [CrossRef] [PubMed]
3. Kampinga, H.H.; Andreasson, C.; Barducci, A.; Cheetham, M.E.; Cyr, D.; Emanuelsson, C.; Genevaux, P.; Gestwicki, J.E.; Goloubinoff, P.; Huerta-Cepas, J.; et al. Function, evolution, and structure of J-domain proteins. *Cell Stress Chaperones* **2019**, *24*, 7–15. [CrossRef]
4. Chien, V.; Aitken, J.F.; Zhang, S.; Buchanan, C.M.; Hickey, A.; Brittain, T.; Cooper, G.J.; Loomes, K.M. The chaperone proteins HSP70, HSP40/DnaJ and GRP78/BiP suppress misfolding and formation of β-sheet-containing aggregates by human amylin: A potential role for defective chaperone biology in Type 2 diabetes. *Biochem. J.* **2010**, *432*, 113–121. [CrossRef] [PubMed]
5. Gillis, J.; Schipper-Krom, S.; Juenemann, K.; Gruber, A.; Coolen, S.; van den Nieuwendijk, R.; van Veen, H.; Overkleeft, H.; Goedhart, J.; Kampinga, H.H.; et al. The DNAJB6 and DNAJB8 protein chaperones prevent intracellular aggregation of polyglutamine repeats. *J. Biol. Chem.* **2013**, *288*, 17225–17237. [CrossRef]
6. Månsson, C.; Kakkar, V.; Monsellier, E.; Sourigues, Y.; Härmark, J.; Kampinga, H.H.; Melki, R.; Emanuelsson, C. DNAJB6 is a peptide-binding chaperone which can suppress amyloid fibrillation of polyglutamine peptides at substoichiometric molar ratios. *Cell Stress Chaperones* **2014**, *19*, 227–239. [CrossRef] [PubMed]
7. Kakkar, V.; Månsson, C.; de Mattos, E.P.; Bergink, S.; van der Zwaag, M.; van Waarde, M.A.W.H.; Kloosterhuis, N.J.; Melki, R.; van Cruchten, R.T.P.; Al-Karadaghi, S.; et al. The S/T-Rich Motif in the DNAJB6 Chaperone Delays Poly-glutamine Aggregation and the Onset of Disease in a Mouse Model. *Mol. Cell* **2016**, *62*, 272–283. [CrossRef] [PubMed]
8. Rodríguez-González, C.; Lin, S.; Arkan, S.; Hansen, C. Co-chaperones DNAJA1and DNAJB6 are critical for regulation of polyglutamine aggregation. *Sci. Rep.* **2020**, *10*, 8130. [CrossRef] [PubMed]
9. Thiruvalluvan, A.; de Mattos, E.P.; Brunsting, J.F.; Bakels, R.; Serlidaki, D.; Barazzuol, L.; Conforti, P.; Fatima, A.; Koyuncu, S.; Cattaneo, E.; et al. DNAJB6, a Key Factor in Neuronal Sensitivity to Amyloidogenesis. *Mol. Cell* **2020**, *78*, 346–358.e9. [CrossRef] [PubMed]
10. Aprile, F.A.; Källstig, E.; Limorenko, G.; Vendruscolo, M.; Ron, D.; Hansen, C. The molecular chaperones DNAJB6 and Hsp70 cooperate to suppress alpha-synuclein aggregation. *Sci. Rep.* **2017**, *7*, 9039. [CrossRef] [PubMed]

11. Deshayes, N.; Arkan, S.; Hansen, C. The Molecular Chaperone DNAJB6, but Not DNAJB1, Suppresses the Seeded Aggregation of Alpha-Synuclein in Cells. *Int. J. Mol. Sci.* **2019**, *20*, 4495. [CrossRef]
12. Arkan, S.; Ljungberg, M.; Kirik, D.; Hansen, C. DNAJB6 suppresses alpha-synuclein induced pathology in an animal model of Parkinson's disease. *Neurobiol. Dis.* **2021**, *158*, 105477. [CrossRef]
13. Månsson, C.; Arosio, P.; Hussein, R.; Kampinga, H.H.; Hashem, R.M.; Boelens, W.C.; Dobson, C.M.; Knowles, T.P.J.; Linse, S.; Emanuelsson, E. Interaction of the molecular chaperone DNAJB6 with growing amyloid-beta 42 (Aβ42) aggregates leads to sub-stoichiometric inhibition of amyloid formation. *J. Biol Chem.* **2014**, *289*, 31066–31076. [CrossRef] [PubMed]
14. Månsson, C.; van Cruchten, R.T.P.; Weininger, U.; Yang, X.; Cukalevski, R.; Arosio, P.; Dobson, C.M.; Knowles, T.; Akke, M.; Linse, S.; et al. Conserved S/T Residues of the Human Chaperone DNAJB6 Are Required for Effective Inhibition of Aβ42 Amyloid Fibril Formation. *Biochemistry* **2018**, *57*, 4891–4902. [CrossRef] [PubMed]
15. Österlund, N.; Lundqvist, M.; Ilag, L.L.; Gräslund, A.; Emanuelsson, C. Amyloid-β oligomers are captured by the DNAJB6 chaperone: Direct detection of interactions that can prevent primary nucleation. *J. Biol. Chem.* **2020**, *295*, 8135–8144. [CrossRef] [PubMed]
16. Dobson, C.M. Protein misfolding, evolution and disease. *Trends Biochem. Sci.* **1999**, *24*, 329–332. [CrossRef]
17. Sarparanta, J.; Jonson, P.H.; Golzio, C.; Sandell, S.; Luque, H.; Screen, M.; McDonald, K.; Stajich, J.M.; Mahjneh, I.; Vihola, A.; et al. Mutations affecting the cytoplasmic functions of the co-chaperone DNAJB6 cause limb-girdle muscular dystrophy. *Nat. Genet.* **2012**, *44*, 450–455. [CrossRef]
18. Nam, T.S.; Li, W.; Heo, S.H.; Lee, K.H.; Cho, A.; Shin, J.H.; Kim, Y.O.; Chae, J.H.; Kim, D.S.; Kim, M.K.; et al. A novel mutation in DNAJB6, p.(Phe91Leu), in childhood-onset LGMD1D with a severe phenotype. *Neuromuscul. Disord.* **2015**, *25*, 843–851. [CrossRef]
19. Palmio, J.; Jonson, P.H.; Evilä, A.; Auranen, M.; Straub, V.; Bushby, K.; Sarkozy, A.; Kiuru-Enari, S.; Sandell, S.; Pihko, H.; et al. Novel mutations in DNAJB6 gene cause a very severe early-onset limb-girdle muscular dystrophy 1D disease. *Neuromuscul. Disord.* **2015**, *25*, 835–842. [CrossRef]
20. Finsterer, J. Genotypic and Phenotypic Heterogeneity of LGMD1D due to DNAJB6 Mutations. *Yonsei Med. J.* **2018**, *59*, 1008–1009. [CrossRef]
21. Karamanos, T.K.; Tugarinov, V.; Clore, G.M. Unraveling the structure and dynamics of the human DNAJB6b chaperone by NMR reveals insights into Hsp40-mediated proteostasis. *Proc. Natl. Acad. Sci. USA* **2019**, *116*, 21529–21538. [CrossRef]
22. Söderberg, C.A.G.; Månsson, C.; Bernfur, K.; Rutsdottir, G.; Härmark, J.; Rajan, S.; Al-Karadaghi, S.; Rasmussen, M.; Höjrup, P.; Hebert, H.; et al. Structural modelling of the DNAJB6 oligomeric chaperone shows a peptide-binding cleft lined with conserved S/T-residues at the dimer interface. *Sci. Rep.* **2018**, *26*, 5199. [CrossRef]
23. Kumita, J.R.; Poon, S.; Caddy, G.L.; Hagan, C.L.; Dumoulin, M.; Yerbury, J.J.; Stewart, E.M.; Robinson, C.V.; Wilson, M.R.; Dobson, C.M. The extracellular chaperone clusterin potently inhibits human lysozyme amyloid formation by interacting with prefibrillar species. *J. Mol. Biol.* **2007**, *369*, 157–167. [CrossRef] [PubMed]
24. Dedmon, M.M.; Christodoulou, J.; Wilson, M.R.; Dobson, C.M. Heat shock protein 70 inhibits alpha-synuclein fibril formation via preferential binding to prefibrillar species. *J. Biol. Chem.* **2005**, *280*, 14733–14740. [CrossRef] [PubMed]
25. Linse, S.; Thalberg, K.; Knowles, T.P.J. The unhappy chaperone. *QRB Discov.* **2021**, *2*, e7. [CrossRef]
26. Atkins, P.; De Paula, J. *Physical Chemistry*, 7th ed.; Oxford University Press: Oxford, UK, 2002; p. 141.
27. Evans, D.F.; Wennerström, H. *The Colloidal Domain: Where Physics, Chemistry, Biology, and Technology Meet*, 2nd ed.; Whiley: Hoboken, NJ, USA, 1999.
28. Jumper, J.; Evans, R.; Pritzel, A.; Green, T.; Figurnov, M.; Ronneberger, O.; Tunyasuvunakool, K.; Bates, R.; Žídek, A.; Potapenko, A.; et al. Highly accurate protein structure prediction with AlphaFold. *Nature* **2021**, *596*, 583–589. [CrossRef] [PubMed]
29. Nozaki, Y.; Tanford, C. Examination of titration behavior. *Methods Enzymol.* **1967**, *11*, 715–734.
30. Linse, S. Expression and Purification of Intrinsically Disordered Aβ Peptide and Setup of Reproducible Aggregation Kinetics Experiment. *Methods Mol. Biol.* **2020**, *2141*, 731–754.
31. Meisl, G.; Kirkegaard, J.B.; Arosio, P.; Michaels, T.C.; Vendruscolo, M.; Dobson, C.M.; Linse, S.; Knowles, T.P. Molecular mechanisms of protein aggregation from global fitting of kinetic models. *Nat. Protoc.* **2016**, *11*, 252–272. [CrossRef]

molecules

MDPI

Article

The RHIM of the Immune Adaptor Protein TRIF Forms Hybrid Amyloids with Other Necroptosis-Associated Proteins

Max O. D. G. Baker [1], Nirukshan Shanmugam [1], Chi L. L. Pham [1], Sarah R. Ball [1], Emma Sierecki [2], Yann Gambin [2], Megan Steain [3] and Margaret Sunde [4,*]

1 School of Medical Sciences, The University of Sydney, Sydney, NSW 2006, Australia; max.baker@pasteur.fr (M.O.D.G.B.); nirukshan@uow.edu.au (N.S.); chilelan.pham@mq.edu.au (C.L.L.P.); sball@wrentherapeutics.com (S.R.B.)

2 EMBL Australia Node in Single Molecule Science, and School of Medical Sciences, University of New South Wales, Sydney, NSW 2052, Australia; e.sierecki@unsw.edu.au (E.S.); y.gambin@unsw.edu.au (Y.G.)

3 School of Medical Sciences, Sydney Institute for Infectious Diseases, The University of Sydney, Sydney, NSW 2006, Australia; megan.steain@sydney.edu.au

4 School of Medical Sciences, Sydney Institute for Infectious Diseases, The University of Sydney Nano Institute, The University of Sydney, Sydney, NSW 2006, Australia

* Correspondence: margaret.sunde@sydney.edu.au; Tel.: +61-2-9351-6955

Abstract: TIR-domain-containing adapter-inducing interferon-β (TRIF) is an innate immune protein that serves as an adaptor for multiple cellular signalling outcomes in the context of infection. TRIF is activated via ligation of Toll-like receptors 3 and 4. One outcome of TRIF-directed signalling is the activation of the programmed cell death pathway necroptosis, which is governed by interactions between proteins that contain a RIP Homotypic Interaction Motif (RHIM). TRIF contains a RHIM sequence and can interact with receptor interacting protein kinases 1 (RIPK1) and 3 (RIPK3) to initiate necroptosis. Here, we demonstrate that the RHIM of TRIF is amyloidogenic and supports the formation of homomeric TRIF-containing fibrils. We show that the core tetrad sequence within the RHIM governs the supramolecular organisation of TRIF amyloid assemblies, although the stable amyloid core of TRIF amyloid fibrils comprises a much larger region than the conserved RHIM only. We provide evidence that RHIMs of TRIF, RIPK1 and RIPK3 interact directly to form heteromeric structures and that these TRIF-containing hetero-assemblies display altered and emergent properties that likely underlie necroptosis signalling in response to Toll-like receptor activation.

Keywords: RHIM; TRIF; necroptosis; functional amyloid; fibrils; RIPK

Citation: Bąker, M.O.D.G.; Shanmugam, N.; Pham, C.L.L.; Ball, S.R.; Sierecki, E.; Gambin, Y.; Steain, M.; Sunde, M. The RHIM of the Immune Adaptor Protein TRIF Forms Hybrid Amyloids with Other Necroptosis-Associated Proteins. *Molecules* **2022**, *27*, 3382. https://doi.org/10.3390/molecules27113382

Academic Editors: Kunihiro Kuwajima, Yuko Okamoto, Tuomas Knowles and Michele Vendruscolo

Received: 9 April 2022
Accepted: 23 May 2022
Published: 24 May 2022

1. Introduction

TIR-domain-containing adapter-inducing interferon-β (TRIF) is a key innate immune adaptor protein that forms signalling complexes downstream of the activation of Toll-like receptors 3 and 4 (TLR3/4) [1]. TLRs are membrane-associated receptors that are capable of detecting damage-associated and pathogen-associated molecular patterns (DAMPs and PAMPs) [2]. TLR3 detects double-stranded RNA [3], triggering a response to a range of viruses. TLR4 is best known for responding to lipopolysaccharide in gram-negative bacteria [4] and can also respond to gram-positive bacterial and viral PAMPs [5–8]. As TRIF acts downstream of both TLR3 and TLR4, it plays an important role in inducing immune defence to a wide range of pathogenic stress.

TRIF is a multidomain protein, with each domain capable of signalling for specific cellular outcomes. It contains an N-terminal domain, a TIR (Toll/interleukin1 (IL-1) receptor) domain and a C-terminal Receptor Homotypic Interaction Motif (RHIM) sequence (Figure 1A). The RHIM sequence in the C-terminal region of TRIF has been shown to mediate interactions that can result in the activation of the lytic inflammatory cell death programme known as necroptosis, which is initiated during innate immune stress caused by pathogenic infection [9,10].

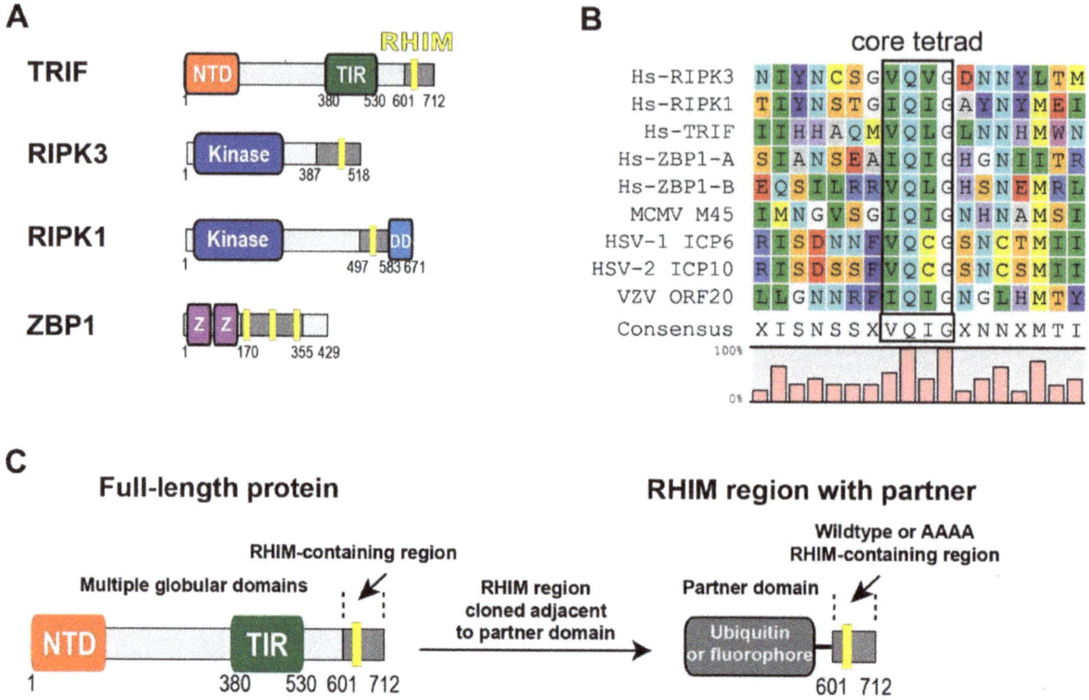

Figure 1. (**A**) Domain architecture of human RHIM-containing proteins, which contain multiple globular domains that serve diverse functional roles. RHIMs are present within disordered protein regions. Z: Zα domain; DD: death domain; NTD: N-terminal domain; TIR: Toll/interleukin-1 (IL-1) receptor domain. (**B**) Alignment of RHIM sequences from human and virus proteins. ZBP1 contains at least two RHIMs, presented as RHIM-A and RHIM-B. The conserved core tetrad is indicated in the boxed region. (**C**) Design of recombinant RHIM-containing fusion proteins with RHIM-containing regions required for necroptosis (light grey) and ubiquitin or fluorescent partner domain.

Necroptotic death was initially observed downstream of tumour necrosis factor (TNF) receptor ligation, resulting from the interaction of RHIM-containing receptor interacting protein kinase 1 (RIPK1) with the RHIM-containing protein RIPK3 [11,12]. The protein complex formed by the hetero-assembly of RIPK1 and RIPK3 is termed the necrosome [13,14] and is an ultra-high molecular weight functional amyloid signalling complex [15]. The necrosome formed by interactions between RIPK1 and RIPK3 is a heteromeric amyloid assembly driven by interactions between RHIM sequences in these proteins [15].

RHIMs are conserved protein sequences located within disordered regions in four multi-domain immunity-associated proteins (Figure 1A). RHIMs contain a highly conserved core tetrad of residues with the consensus sequence (V/I)Q(V/I/L/C)G (Figure 1B). This core tetrad is important for function, as mutagenesis of this sequence to AAAA in many different experimental workflows ablates RHIM-driven protein function [12,15–18]. Additionally, the RHIMs are essential for necrosome assembly and amyloid formation: mutation of key amyloid-forming residues in RHIMs results in a loss of necroptosis capability in cells [15]. Recent structural determination of necrosome assemblies has revealed that the RHIM sequence forms the amyloid core of these fibrils, which explains the importance of this sequence for both amyloid assembly and signalling function [19,20].

In addition to RIPK1 and RIPK3, RHIMs that drive cell death have also been identified in the cytoplasmic nucleic acid sensor ZBP1 [17,21,22] and in TRIF. The RHIM from TRIF was first identified as a mediator of apoptosis resulting from TLR activation [23].

Subsequent studies revealed that the RHIM from TRIF could also induce necroptosis by RHIM-driven interaction with RIPK3 in multiple cell types, including macrophages, endothelial cells and fibroblasts [16,24]. The formation of the necrosome and RIPK3 phosphorylation leads to the phosphorylation, conformational change and oligomerisation of the pseudokinase mixed lineage kinase-like protein (MLKL) [25–27]. MLKL is the executioner protein of necroptosis [25] and induces cell membrane lysis when activated by RIPK3 [26–28].

TRIF is well-established as a pleiotropic adaptor protein involved in Toll-like receptor signalling [1,16], but little is known about the structure of the hetero-assemblies that define its role in programmed cell death. TRIF is capable of signalling for necroptosis in the absence of RIPK1 in murine fibroblasts and endothelial cells; however, it is required for necroptosis in macrophages [16]. The requirement of RIPK1 to modulate TRIF-RIPK3 signalling may be cell-context dependent, and potentially reliant on the presence of other cellular co-factors. The ability of TRIF to form either homo- or heteromeric amyloid assemblies has not been determined, although cell-based experiments have indicated its ability to form insoluble and fibrillar structures in cells [29,30].

Here we have used a range of biophysical approaches to characterise the amyloidogenic capacity of the TRIF RHIM. We have delineated the region of TRIF that forms a protected amyloid-structured core and our results demonstrate that this region extends beyond the 18-residues of the RHIM that are identified by homology. Our results demonstrate the capacity of the TRIF RHIM to form heteromeric assemblies with RIPK1 and RIPK3 through direct interaction.

2. Results

2.1. The RHIM from TRIF Self-Assembles into Amyloid Structures

A protein construct encoding the RHIM and surrounding residues of wildtype TRIF was cloned with His-tagged ubiquitin or fluorescent partner domains to assist in the expression, purification and detection of amyloid-based assembly and protein:protein interactions (Figure 1C). Protein constructs encoding for RIPK1, RIPK3 and a TRIF core tetrad mutant (VQLG to AAAA) were also cloned (Supplementary Figure S1). The AAAA mutation was chosen because this change to the core residues of the motif has been shown to reduce the ability of RHIM-containing proteins to signal for cell death by necroptosis.

These protein constructs are named for the partner domain and the stretch of residues incorporated from the original RHIM protein (for instance, Ub-TRIF$_{601–712}$ represents the residues 601–712 of TRIF with an N-terminal ubiquitin partner domain). Proteins were produced by overexpression in *E. coli* and purified by affinity chromatography. Previous experience has shown that RHIM-containing proteins are prone to rapid self-assembly into insoluble aggregates in buffers of physiological salt and pH [31]. Therefore, proteins were purified and maintained as monomers in chaotropic 8 M urea-containing buffer before removal of denaturant, triggering oligomerisation. This methodology has been previously utilised for the characterization of other RHIM proteins [18,20,31,32].

The capability of the TRIF RHIM to drive the formation of homomeric amyloid assemblies was assessed first with a suite of hallmark assays. The fluorescent dye thioflavin T (ThT) and the aniline dye Congo red show characteristic spectral changes when they bind to the cross-β structure in amyloid fibrils and both are widely used to report on the formation of amyloid protein assemblies [33,34]. Both the wildtype and the mutant TRIF-RHIM containing proteins spontaneously form structures that generate enhanced ThT fluorescence emission at 485 nm and induce a red shift in λ_{max} in the absorbance spectrum of Congo red to 540 nm. These features are consistent with the formation of cross-β structure (Figure 2A,B).

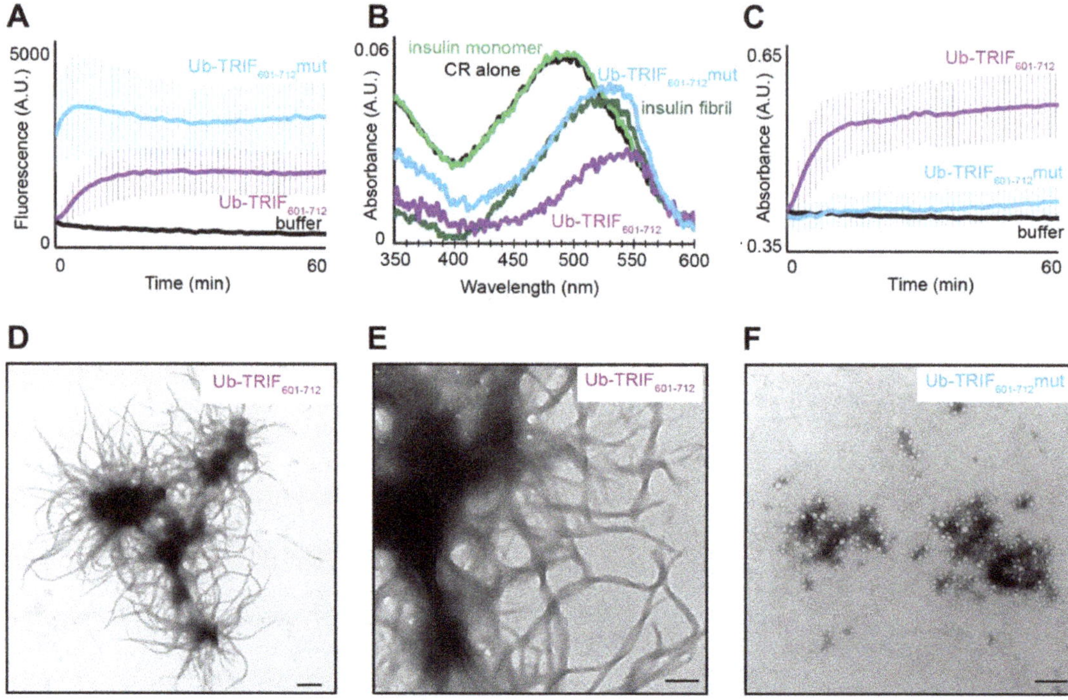

Figure 2. (**A**) Time course of amyloid assembly by Ub-TRIF$_{601-712}$ wild type and mut, initiated by dilution from denaturant into assembly-permissive buffer at 2.5 μM, and assessed by ThT fluorescence. Curves indicate average of three independent replicates, and error bars indicate one standard deviation. (**B**) Congo red absorbance spectra from Ub-TRIF$_{601-712}$ WT and mut macromolecular assemblies, compared to insulin monomer, insulin amyloid fibril and Congo red and buffer samples. (**C**) Time course of static light scattering from Ub-TRIF$_{601-712}$ RHIM assemblies, following dilution from denaturant into assembly-permissive buffer at 2.5 μM. Curves indicate average from three independent replicates, and error bars indicate one standard deviation. (**D**) Representative transmission electron micrograph of 1 μM Ub-TRIF$_{601-712}$ assemblies depicting 'sea anemone' morphology. Scale bar 500 nm (**E**) Representative transmission electron micrograph of 1 μM Ub-TRIF$_{601-712}$ at higher magnification. Scale bar 200 nm. (**F**) Representative transmission electron micrograph of amorphous 1 μM Ub-TRIF$_{601-712}$mut aggregates. Scale bar 500 nm.

The increase in ThT fluorescence in the presence of wildtype TRIF RHIM follows a sigmoidal curve, which is reminiscent of other well-characterised amyloid-forming proteins. The mutant version of the protein displays a modest increase in intensity from a high initial measured point, indicating that rapid assembly into the ThT-binding conformation starts before the first measurement is recorded. There is a difference in the final intensity of the ThT fluorescence between the two proteins but since ThT intensity is not a quantitative measure of amyloid structure and the mode of ThT binding can affect quantum yield, no conclusion can be drawn from these data about the amount of amyloid formed in the two samples.

This experiment was repeated for a range of protein concentrations of wildtype and mutant versions of the TRIF RHIM (Supplementary Figure S2A). For the wildtype, the extent of ThT fluorescence correlated with the concentration of protein and a sigmoidal increase in fluorescence intensity was apparent, consistent with other well-characterised amyloid-forming systems. For the mutant, the ThT fluorescence increased correspondingly with protein concentration, but even the early timepoints showed an intensity above buffer-only, indicative of rapid formation of ThT-binding species.

Ub-TRIF$_{601–712}$ and Ub-TRIF$_{601–712}$mut assemblies both induced a shift in the absorbance maximum exhibited by Congo red, towards 540 nm and similar to that seen with insulin fibrils (Figure 2B), with the mutant form of TRIF yielding a higher absorbance at 540 nm than the wildtype protein. The ThT fluorescence and Congo red data indicate that both wildtype and AAAA TRIF RHIM constructs do form structures with elements of cross-β architecture, albeit with different kinetics of assembly and possibly with a slightly different environment for ThT binding, reflected by the different final ThT intensity.

Static light scattering was performed in parallel with the ThT assays and showed a large increase in scattering signal for Ub-TRIF$_{601–712}$ but little signal for Ub-TRIF$_{601–712}$mut over the course of 60 min at 2.5 μM (Figure 2C). We also examined the concentration-dependence of light scattering for both protein constructs (Supplementary Figure S2B). For Ub-TRIF$_{601–712}$, light scattering increased with concentration, typical of amyloid fibrils which are large insoluble structures. When studying Ub-TRIF$_{601–712}$mut, no scattering was visible for the 60 min time period at concentrations between 0.625 μM and 2.5 μM, and was severely attenuated at 5 μM compared to Ub-TRIF$_{601–712}$ at the same concentration. These data demonstrate that the mutant form of the TRIF RHIM does not form large structures and suggest that the VQLG core tetrad is required for the assembly of the TRIF RHIM into large amyloid fibrils.

Transmission electron microscopy (TEM) revealed the difference in the size and morphology of the assemblies formed by the proteins that underly the difference in light scattering signal. Electron micrographs in Figure 2 demonstrate that Ub-TRIF$_{601–712}$ assembles into fibrils that show an unusual higher-level assembly. The long Ub-TRIF$_{601–712}$ fibrils appear to radiate from a central dense core, reminiscent of a sea anemone with tendrils (Figure 2D). The dense cores have diameters 0.5–1 μm, and the tendrils range in length from 0.5 to >3 μm, resulting in an overall diameter of >5 μm. The dense core is composed of very closely intertwined fibrils, while the emanating fibrils are individual fibrils that display a clear twist (Figure 2E). In contrast to the long fibrils formed by the wildtype TRIF sequence, the Ub-TRIF$_{601–712}$mut was observed to assemble into small aggregates 0.5–2 μm in diameter (Figure 2F).

The ThT, Congo red and TEM results demonstrate that the wildtype TRIF RHIM drives the formation of amyloid fibrils which have a propensity to form large clusters. The AAAA mutation abolishes the ability of the TRIF RHIM to form long amyloid fibrils, as evidenced by TEM and light scattering. The Ub-TRIF$_{601–712}$mut forms only small irregular aggregates. Some residual elements of cross-β structure may remain in these aggregates, as evidenced by the concentration-dependent increase in ThT signal at 485 nm and increase in absorbance at 540 nm in the Congo red absorbance spectrum. However, the data show that the intact VQLG core tetrad of TRIF is required for this RHIM to support the formation of long, regular amyloid fibrils, suggesting that amyloid fibril formation is important for the function of the RHIM within the necroptosis signalling pathway.

2.2. *Identifying the Protected Core of TRIF Amyloid Assemblies*

Recent structural elucidation of functional amyloids involved in cellular signalling processes, including the RIPK1-RIPK3 necrosome and Orb2 assemblies that are associated with memory formation in *Drosophila*, has revealed that these fibrils are composed of an amyloid structured core scaffold and peripheral partner domains that remain active in solution [19,35]. We wished to identify the region of TRIF that forms the core amyloid scaffold. Other investigations of amyloid-forming proteins have used protease digestion experiments to identify the stable, hydrogen-bonded scaffold [36]. The stable secondary structure and tight interdigitation of side chains in this scaffold gives relative protection against proteolysis compared to distal flexible regions.

Here we have exposed TRIF-containing amyloid fibrils to digestion by subtilisin, a non-specific peptidase, and then have used mass spectrometry to identify the residual sequences, which represent the amyloid core. ThT-positive fibrils formed by YPet-TRIF$_{601–712}$ were utilised for these experiments, as the partner protein was resistant to subtilisin degradation and hence did not contribute to the insoluble, proteolysis-resistant fraction. Fibrils formed by YPet-TRIF$_{601–712}$ were treated with subtilisin, which led to the rapid release of the YPet domain into the soluble fraction, while the insoluble fraction contained multiple protease resistant fragments, with the most dominant ~6 kDa (Figure 3A). This fragment was subjected to formic acid treatment to achieve depolymerisation (Figure 3B) and then analysed by LC-MS.

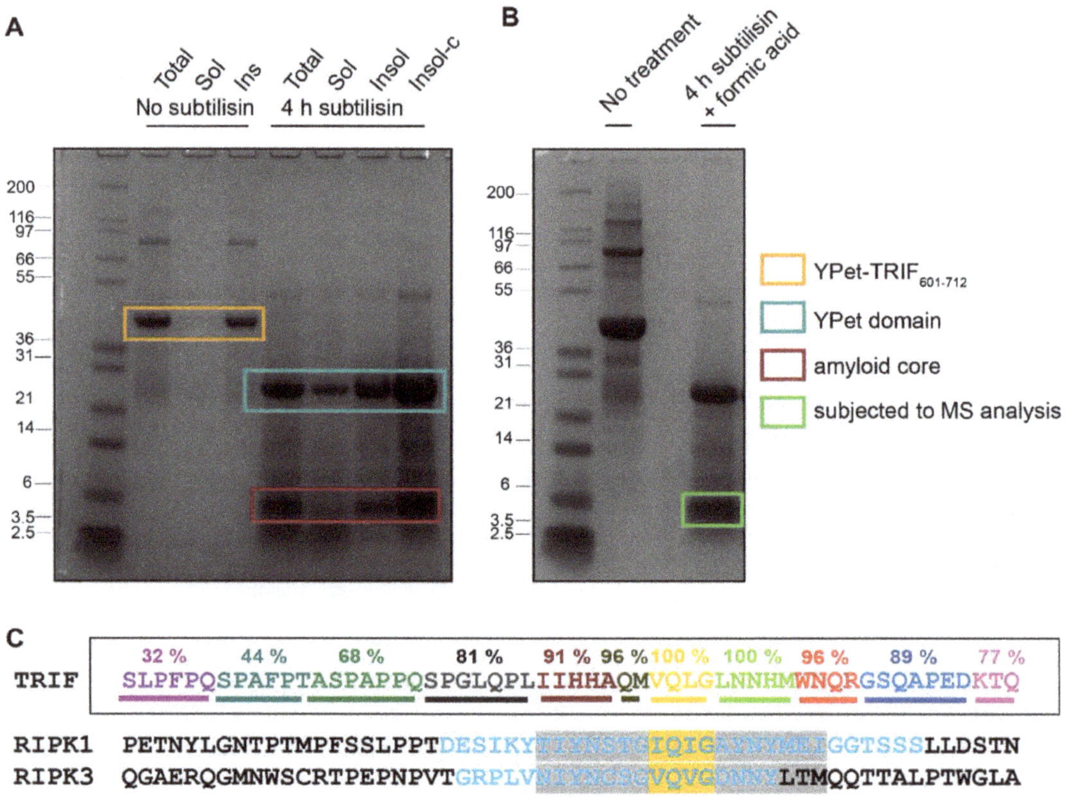

Figure 3. (**A**) SDS PAGE analysis of products of subtilisin digestion of YPet-TRIF$_{601–712}$. Samples are total (Tot), soluble (Sol), insoluble (Insol) or 10X concentrated insoluble (Insol-c). (**B**) Digestion products of YPet-TRIF$_{601–712}$ after lyophilisation, resuspension in 90% formic acid and additional lyophilisation, indicating band excised from gel and subjected to mass spectrometry. (**C**) The sequences of the RHIM-containing regions of human TRIF, RIPK1 and RIPK3 aligned by the core tetrad, coloured yellow. % Prevalence of detected TRIF peptides indicated above sequence. All sequences identified by mass spectrometry available in SI Figure S2. Regions within RIPK1 and RIPK3 previously identified as important for amyloid interactions [15] and functional RHIM-dependent signalling [13] indicated in blue and grey, respectively.

The stable amyloid-forming region of TRIF was successfully identified in this way (Figure 3C). The region of the wildtype TRIF protein used as the starting material in these experiments is 112-residues long. From this region, up to 44 residues flanking the core tetrad were conserved in close to 70% of peptide reads, spanning from S654 to Q709. The core tetrad of VQLG was present in every sequence detected. Regions immediately adjacent to the core tetrad are highly represented, with slight decreases in protection from proteolysis in regions more distal to the tetrad. The full list of peptide sequences successfully identified, including any posttranslational modifications, is provided in Supplementary Figure S2. This extent of residues potentially involved in amyloid formation by TRIF was surprising, as Waltz and Tango algorithms do not predict any amyloidogenic sequences within this TRIF C-terminal region. AmylPred identifies amyloidogenic propensity in the nine residues on the N-terminal side of the tetrad (PLIIHHAQM). All post-translational modifications corresponded to oxidation of methionine residues, which likely occurred as part of the electrospray ionisation process [37].

These data indicate that the VQLG sequence and its immediately flanking regions form the most protected, and hence structured, component of TRIF amyloid fibrils. Alignment of these protected regions in TRIF with the RHIM region previously identified by other groups, delineated with mutagenesis and functional assays, pulldown interaction studies, measures of heteromeric amyloid formation and solid state NMR [13,15,23], indicates that the amyloid-structured region of TRIF is relatively large and likely extends beyond the ~18-residue sequence that controls interactions between the different RHIM-containing proteins (Figure 3C). While the full extent of this structured region may not be necessary for interactions with other proteins, this first unbiased analysis of the amyloid-forming region of the TRIF protein suggests that a large segment of the C-terminal region of TRIF becomes structured when the protein self-assembles.

2.3. *TRIF Directly Interacts with RIPK1 and RIPK3 to Form Heteromeric Assemblies*

Experiments conducted in cells have indicated that assembly of TRIF into large macromolecular structures is vital for the engagement of necroptosis [16,30]. Confocal coincidence spectroscopy (CCS) has previously been used successfully to probe the formation of large amyloid-based homomeric and heteromeric assemblies [31,38]. In CCS, initially monomeric RHIM-containing fluorescent fusion proteins (Figure 1C; Supplementary Figure S1) are mixed together and then analysed in a ~1 fL confocal volume, allowing for characterisation and quantification of individual molecules and any complexes formed. Fluorescence emission intensity is recorded from this volume over time. Baseline fluorescence arises from monomeric protein in solution. Deflections from baseline in one or both fluorescence channels indicate the presence of higher-order protein assemblies. Co-incidence of signals from the two channels reflects heteromeric assembly [38].

CCS was performed on YPet-$TRIF_{601-712}$, YPet-$TRIF_{601-712}$mut, mCherry-$RIPK1_{497-583}$ and mCherry-$RIPK3_{387-518}$ alone and in pairs. Fluorescence data were recorded each millisecond for three-minute intervals. A representative 250 ms trace for each experimental condition is shown in Figure 4A.

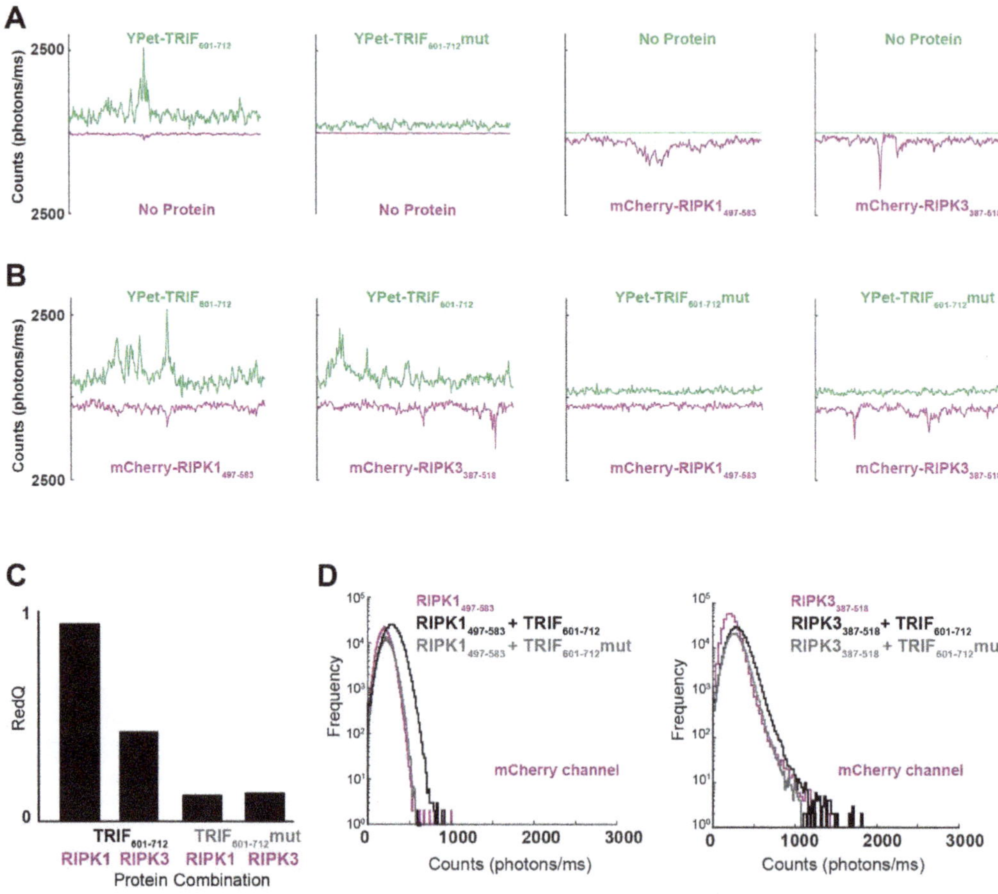

Figure 4. (**A**) Representative 250 ms time traces from individual RHIM-containing proteins at 0.175 μM under assembly-permissive conditions. YPet and mCherry fluorescence emission was measured simultaneously, allowing for detection of coincidence of different RHIM-containing proteins in individual assemblies diffusing through the confocal volume. Proteins in each sample are indicated on each trace. (**B**) Time traces from mixtures of RHIM-containing proteins: TRIF-RHIM$_{601–712}$wt with RIPK1$_{497–583}$ and RIPK3$_{387–518}$ RHIMs, and TRIF-RHIM$_{601–712}$mut with RIPK1$_{497–583}$ and RIPK3$_{387–518}$ RHIMs. (**C**) Comparison of RedQ scores of TRIF-RHIM$_{601–712}$wt, TRIF-RHIM$_{601–712}$mut, RIPK1$_{497–583}$ and RIPK3$_{387–518}$ RHIM combination pairs. (**D**) Photon count histograms from protein assemblies.

For YPet-TRIF$_{601–712}$, many large homomeric assemblies were apparent. No such large homomeric assemblies were detected when YPet-TRIF$_{601–712}$mut was incubated under amyloid assembly-permissive conditions. mCherry-RIPK1$_{497–583}$, and mCherry-RIPK3$_{387–518}$ also formed homomeric assemblies under these conditions (Figure 4A).

When YPet-TRIF$_{601–712}$ and mCherry-RIPK1$_{497–583}$ were incubated together under assembly-permissive conditions, many large peaks were detected in the YPet channel, and multiple smaller peaks were observed in the mCherry channel (Figure 4B). This interaction appears to be dependent on the core tetrad of TRIF, as no large peaks were detected when -YPet-TRIF$_{601–712}$mut and mCherry-RIPK1$_{497–583}$ were co-incubated (Figure 4B). For co-incubations of -YPet-TRIF$_{601–712}$ and mCherry-RIPK3$_{387–518}$, many peaks were detected in both the YPet and mCherry emission channels, reflecting the high propensity of both proteins to assemble (Figure 4B). In the mixture of YPet-TRIF$_{601–712}$mut and mCherry-

RIPK3$_{387–518}$, many peaks were detected in the mCherry channel, but few were detected in the YPet channel (Figure 4B).

It is possible to utilise quantitative analysis to measure the interaction capability of YPet and mCherry proteins in this CCS system. One such measure of interaction propensity is termed '*colourQ*' and defined as follows:

$$colourQ = \frac{number\ of\ Colour\ A\ peaks\ aligned\ with\ Colour\ B\ peaks}{total\ number\ of\ Colour\ A\ peaks} \quad (1)$$

RedQ measurements (which describe the proportion of green peaks aligned with red peaks) were performed on the two-protein pairs described above (Figure 4C). For co-incubations of mCherry-RIPK1$_{497–583}$ with YPet-TRIF$_{601–712,}$ the peaks were highly co-localised (RedQ = 0.9). For mixtures of YPet-TRIF$_{601–712}$mut and mCherry-RIPK1$_{497–583}$, a RedQ score of 0.11 was reported, indicating a low level of interaction. These data indicate that the TRIF VQLG core tetrad motif is important for co-assembly with RIPK1. For mixtures of YPet-TRIF$_{601–712}$ and mCherry-RIPK3$_{387–518}$, these co-incubated proteins showed a RedQ score of 0.40. The interaction of TRIF with RIPK3 appears dependent on the VQLG core tetrad of TRIF, as coincidence of detection between YPet-TRIF$_{601–712}$mut and mCherry-RIPK3$_{387–518}$ was low (RedQ = 0.13; Figure 4C). The RHIM from TRIF appears more amenable to interaction with RIPK1 than RIPK3 (RedQ = 0.9 vs. RedQ = 0.4). This may be due to the higher propensity of RIPK3 to self-assemble, which may reduce its likelihood for heteromeric interaction.

The number and size of complexes detected in each sample was also analysed by binning the detected fluorescent signals according to intensity and plotting the number of peaks of each intensity range, in a photon counting histogram (PCH; Figure 4D). Comparison of the PCHs for YPet-TRIF$_{601–712}$ and YPet-TRIF$_{601–712}$mut with mCherry-RIPK1$_{497–583}$ shows that the average size of RIPK1-containing complexes increases in the presence of YPet-TRIF$_{601–712}$ (black) compared to RIPK1 alone (magenta), indicating that hetero-assembly with TRIF drives RIPK1 into larger complexes. This change in peak size distribution is not detected when mCherry-RIPK1$_{497–518}$ is co-incubated with the mutant form of TRIF (grey). This confirms the crucial role of the core tetrad in RHIM-based higher order assembly. Analysis of mixtures of TRIFwt and RIPK3 shows that on average these heteromeric complexes (black) are slightly larger than those formed by the RIPK3 RHIM construct on its own (magenta). Additionally, a small number of larger heteromeric complexes (1500–2000 photons/ms; Figure 4D right panel) are observed. This is consistent with a reduced interaction propensity and lower RedQ for this complex compared to the TRIF:RIPK1 complex. No change in average size of oligomer was detected when the mutant form of the TRIF RHIM was co-incubated with mCherry-RIPK3$_{387–518}$, confirming the dependence on an intact wildtype tetrad sequence for interaction between the RHIMs of TRIF and RIPK3. Taken together, these data indicate that the RHIM of TRIF supports complex formation with RIPK1 and RIPK3 and co-assembly with TRIF recruits RIPK1 into macromolecular complexes that are larger than homomeric RIPK1 assemblies.

2.4. Hetero-Assemblies Formed between TRIF and RIPK1 or RIPK3 Are Larger and More Stable Than Homo-Assemblies Formed by These Proteins

Sodium dodecyl sulphate agarose gel electrophoresis (SDS-AGE) experiments were used to probe the native size and stability of TRIF homomeric and heteromeric assemblies containing TRIF and one of either RIPK1 or RIPK3. Previous work has shown that the differential stability of RHIM homo- and heteromeric assemblies can be revealed by treatment with 2% SDS [18,31,32].

YPet-TRIF$_{601–712}$ when maintained in 8 M urea remains monomeric (Figure 5A, left panel, lane 1). TRIF assemblies formed following removal of denaturant were retained in the well (Figure 5A, left panel, lane 2). These were confirmed by TEM to be >5 μM in length. These homomeric amyloids formed by TRIF are susceptible to SDS depolymerisation and are depolymerised to oligomers of a range of sizes and monomer (Figure 5A, left panel, lane 3).

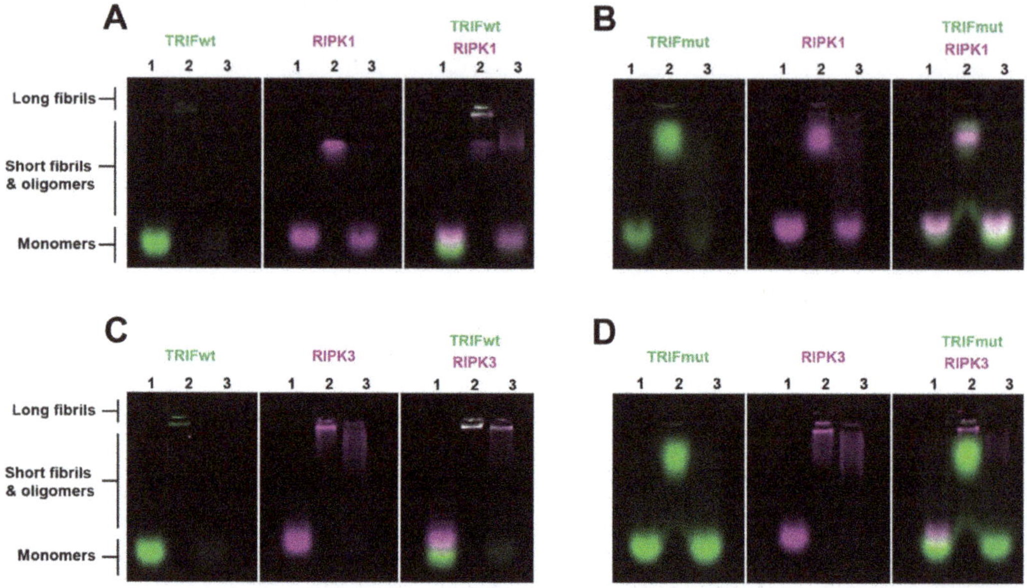

Figure 5. SDS-agarose gel electrophoresis reveals structural properties arising from heteromeric assembly of TRIF and other RHIM-containing proteins. Single or equimolar at 5 μM two-protein combinations carrying YPet or mCherry fluorophores were prepared in (1) monomeric or (2) co-assembled forms and (3) treated with SDS, then electrophoresed on an agarose gel and imaged. Protein combinations used: (**A**) YPet-TRIF$_{601–712}$ and mCherry-RIPK1$_{497–583}$, (**B**) YPet-TRIFmut$_{601–712}$ and mCherry-RIPK1$_{497–583}$, (**C**) YPet-TRIF$_{601–712}$ and mCherry-RIPK3$_{387–518}$, (**D**) YPet-TRIFmut$_{601–712}$ and mCherry-RIPK3$_{387–518}$. All conditions were imaged with both GFP and mCherry appropriate filter sets, and then overlaid using FIJI ImageJ. All experiments were conducted multiple times, and a representative gel is shown. Migration of long fibrils, short fibrils/oligomers and monomeric forms are indicated.

mCherry-RIPK1$_{497–583}$ monomer completely traversed to the gel front (Figure 5A, middle panel, lane 1). After dialysis, mCherry-RIPK1$_{497–583}$ migrated partway down the gel, indicating the formation of oligomeric species or small fibrils, which is consistent with the small fibril assemblies made by this protein that we have previously observed by TEM in prior studies (Figure 5A, middle panel, lane 2) [31]. SDS treatment caused the reversion of dialysed mCherry-RIPK1$_{497–583}$ samples to monomer, indicating low stability of RIPK1-only assemblies (Figure 5A, middle panel, lane 3).

When mCherry-RIPK1$_{497–583}$ and YPet-TRIF$_{601–712}$ were co-dialysed, both proteins were retained in the well, and the oligomer band for mCherry-RIPK1$_{497–583}$ was weaker, indicating that TRIF:RIPK1 interaction results in assembly of structures that are larger than the RIPK1-alone structures (Figure 5, right panel, lane 2). These findings are consistent with the observation of larger RIPK1 assemblies by CCS described earlier (Figure 4C). When this mixture was incubated with 2% SDS, a small amount of protein remains visible in the well, along with a streak near the expected oligomer band size (Figure 5, right panel, lane 3), indicating that the hetero-assembly complex containing of RIPK1 and TRIF is partly resistant to SDS, and is more stable than the homomeric assemblies of these two proteins.

A similar set of experiments was performed with YPet-TRIF$_{601–712}$mut and mCherry-RIPK1$_{497–583}$ (Figure 5B). Unlike the wildtype version of YPet-TRIF$_{601–712}$, the mutant formed oligomers (as previously observed by TEM in Figure 2F) instead of fibrils and these were also not SDS resistant (Figure 5B, first panel, lane 3). mCherry-RIPK1$_{497–583}$ alone formed the same oligomer species as previously described (Figure 5B, middle panel) [31]. Co-dialysis of mCherry-RIPK1$_{497–583}$ and YPet-TRIF$_{601–712}$mut did not see them colocalise (no white overlay) nor was any specific change in translocation or stability characteristics visible for either protein (Figure 5B, third panel, lanes 2 and 3). These findings were consistent with confocal CCS, where wildtype TRIF and RIPK1 RHIMs were shown to interact, and it was observed that the wildtype core tetrad of TRIF was required for TRIF to impart changes on RIPK1 (Figure 4B,C).

The homo- and heteromerization of TRIF and RIPK3 were investigated using the same protocols. YPet-TRIF$_{601–712}$ formed large assemblies that were SDS-soluble (Figure 5C, left panel), as previously described. Homomeric mCherry-RIPK3$_{387–518}$ formed large assemblies that were predominantly retained in the well, again consistent with prior studies (Figure 5C, middle panel) [15,31]. Notably, mCherry-RIPK3$_{387–518}$ homo-assemblies were partially resistant to SDS dissolution, with SDS treatment generating large fibrils that migrated only a short distance from the well (Figure 5C, middle panel, lane 3). Analysis of mixtures of TRIF and RIPK3 revealed novel properties induced by heteromerization (Figure 5C, right panel). When dialysed together, YPet-TRIF$_{601–712}$ and mCherry-RIPK3$_{387–518}$ both remained colocalised within the wells and the RIPK3 in these large hetero-assemblies was completely resistant to SDS treatment. The large fibrils observed in the wells could be a mixture of homomeric RIPK3 fibrils, homomeric TRIF fibrils and heteromeric RIPK3:TRIF fibrils. We observe that the stability of a proportion of the RIPK3 and TRIF material is increased, indicating that TRIF and RIPK3 hetero-interaction serves to stabilise both proteins within amyloid assemblies.

These emergent features of TRIF:RIPK3 hetero-assembly are driven by the wildtype core tetrad of TRIF, as evidenced by the lack of co-localisation when these experiments were repeated with YPet-TRIF$_{601–712}$mut and mCherry-RIPK3$_{387–518}$ (Figure 5D, right panel). Oligomers of mutant TRIF and large RIPK3 fibrils were observed at different locations and the mutant TRIF assemblies were depolymerised by SDS.

2.5. Microscopy of Hetero-Assemblies Containing TRIF and Partner Proteins Elucidates Extensive Cointegrated Protein Architectures

The hetero-assemblies formed by TRIF and RIPK1 and RIPK3 were imaged by both TEM and fluorescence confocal microscopy to characterise their size, morphology and integration and localisation of the constituent proteins.

Two prominent morphologies were observed by TEM when Ub-TRIF$_{601–712}$ and mCherry-RIPK1$_{497–583}$ were allowed to co-assemble (Figure 6A): a network of fibrils (Figure 6A, Morphology A) and a dense fibril bundle with some smaller fibrils emanating from the core (Figure 6A, Morphology B). The relationship between Morphology A and Morphology B is unclear—it is possible that Morphology A is on a pathway to Morphology B, or that the different morphologies arise from different clusters of each of the constituent proteins in a single superstructure. Assemblies containing YPet-TRIF$_{601–712}$ and mCherry-RIPK1$_{497–583}$ were assessed by confocal microscopy (Figure 6B) and very large clusters of fibrils were observed and appeared to contain both proteins, with YPet and mCherry colocalised (Pearson correlation value = 0.95). This co-assembly was dependent on the presence of a WT tetrad sequence, since samples prepared with both YPet-TRIF$_{601–712}$mut and mCherry-RIPK1$_{497–583}$ present were observed to contain structures composed almost entirely of RIPK1 (Figure 6C), as indicated by the Pearson correlation score of 0.08. No large structures containing YPet-TRIF$_{601–712}$mut were observed.

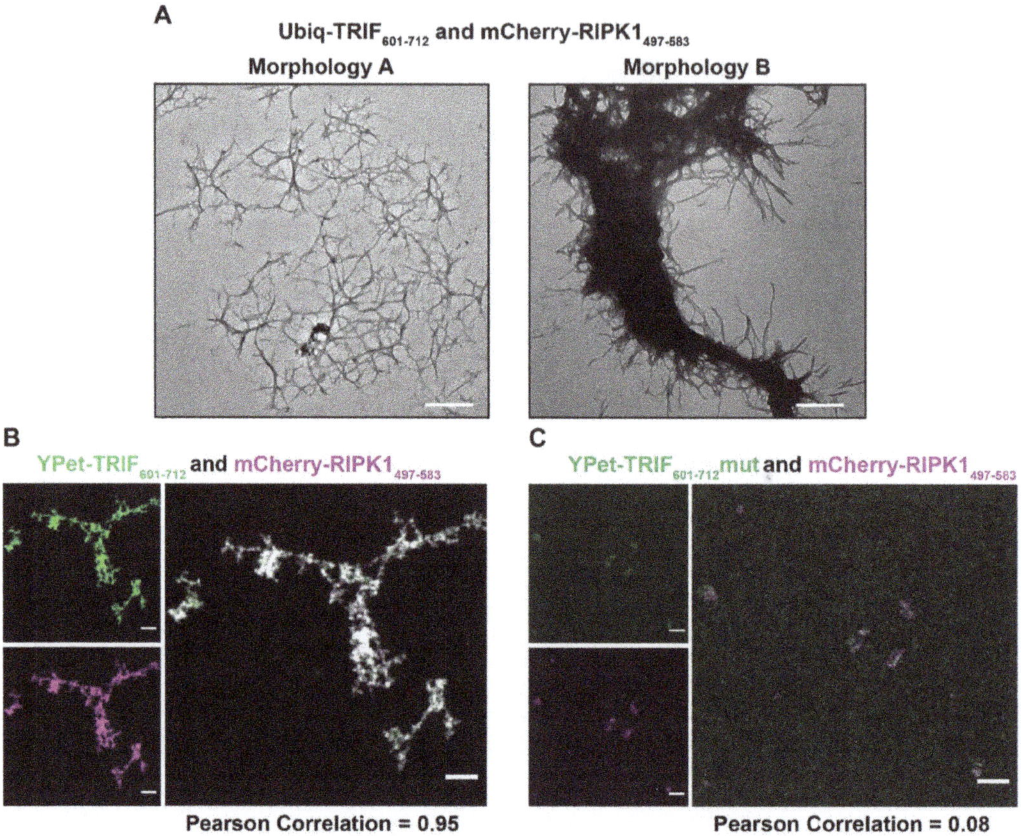

Figure 6. Confocal and transmission electron microscopy of mixtures of TRIF and RIPK1. (**A**) Transmission electron microscopy of mixtures of 1 μM Ub-TRIF$_{601–712}$ and mCherry-RIPK1$_{497–583}$. There were two common morphologies for these mixtures, Morphology A and Morphology B. Scale bars represent 500 nm. (**B**,**C**) Confocal images of mixtures of 2.5 μM TRIFwt$_{601–712}$ and TRIFmut$_{601–712}$ and RIPK1$_{497–583}$ RHIM-containing proteins. Images are displayed as recorded in GFP channel, mCherry channel and merged. Scale bar indicates 10 μm. Image analysis performed in Fiji. Pearson Correlation determined by JaCoP plugin.

Co-assembled structures of Ub-TRIF$_{601–712}$ and YPet-RIPK3$_{387-518}$ imaged by electron microscopy had a single prominent morphological identity (Figure 7A) comprised of dense aggregated fibrillar material. Higher-magnification of these TRIF-RIPK3 co-assemblies reveals individual fibrils emanating from the surface and visible on the periphery of the dense bundles (Supplementary Figure S4). Coincubation of YPet-TRIF$_{601–712}$ and mCherry-RIPK3$_{387–518}$ and subsequent observation by confocal microscopy confirmed both proteins were contained within these assemblies (Figure 7B). The observed assemblies had extensive fibrillar morphology and were highly colocalised (Pearson correlation = 0.95). This co-assembly requires a WT tetrad sequence: in mixtures of YPet-TRIF$_{601–712}$mut and mCherry-RIPK3$_{387–518}$ the proteins were poorly colocalised (Pearson correlation = 0.18). The observed assemblies displayed low levels of YPet-TRIF$_{601–712}$mut incorporation and appeared to be mostly composed of mCherry-RIPK3$_{387–518}$ (Figure 7C). In order to definitively ascertain whether RIPK3-TRIF co-assemblies were indeed hetero-amyloids, we imaged these assemblies in the presence of ThT, using widefield fluorescence microscopy (Supplementary Figure S5). The large assemblies composed of wildtype YPet-TRIF$_{601–712}$

and mCherry-RIPK3$_{387–518}$, and corresponding to those observed by confocal microscopy (Figure 7B), were highly co-localised with signal for ThT confirming the cross-β amyloid architecture of the assemblies. Mixtures of YPet-TRIF$_{601–712}$mut and mCherry-RIPK3$_{387–518}$, generated smaller assemblies with a weaker YPet signal but strong mCherry signal, corresponding to low levels of the TRIF mut relative to RIPK3 RHIM and in line with the structures observed using confocal imaging (Figure 7C). These structures were highly colocalised with ThT signal. However, RIPK3 homomeric fibrils would also be expected to bind ThT so the extent of heteromeric TRIF$_{601–712}$mut:RIPK3 heterofibril assembly could not be distinguished.

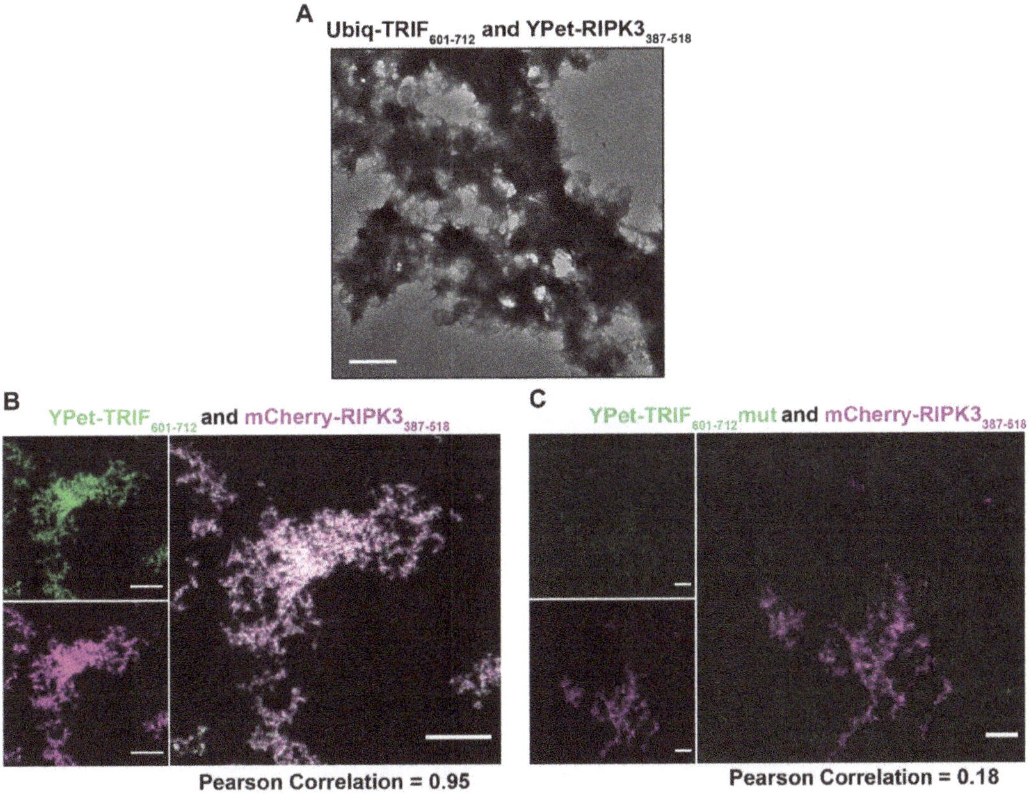

Figure 7. Confocal and transmission electron microscopy of mixtures of TRIF and RIPK3. (**A**) Transmission electron microscopy of mixtures of 1 μM Ub-TRIF$_{601–712}$ and mCherry-RIPK3$_{387–518}$. Scale bars represent 500 nm. (**B**,**C**) Confocal images of mixtures of 2.5 μM TRIF$_{601–712}$wt and TRIF$_{601–712}$mut and RIPK3$_{387–518}$ RHIM-containing proteins. Images are displayed as recorded in GFP channel, mCherry channel and merged. Scale bar indicates 10 μm. Image analysis performed in Fiji. Pearson correlation determined by JaCoP plugin.

3. Discussion

Previous studies reporting on TRIF have indicated its ability to form large insoluble filamentous aggregates in cells and to interact with RIPK1 and RIPK3 [29,30] and the results presented here provide the first structural characterisation of those RHIM:RHIM interactions between TRIF and key necroptosis-associated proteins. We have demonstrated that the RHIM from TRIF is capable of spontaneously forming fibrillar amyloid assemblies, like other RHIM-containing proteins [15,31,32]. The assembly of the TRIF RHIM-containing domain into organised amyloid fibrils is crucially dependant on the core tetrad VQLG residues of the motif. Although the AAAA mutant forms ThT and Congo red positive structures, the loss of the key interacting residues abrogates the domain's ability to form ordered and stable fibrils, as observed by SDS treatment followed by agarose electrophoresis, coincidence confocal spectroscopy and confocal and electron microscopy.

The mass spectrometry data presented here reveal that the protected amyloid core of TRIF assemblies is larger than the sequence motif defined by homology to other RHIM proteins and by pulldown experiments [16]. The structures of heteromeric RIPK1:RIPK3 and homomeric RIPK3 RHIM cores, determined by solid state NMR, show that the residues involved in amyloid formation by RHIM proteins can vary to some extent [19,20]. Additionally, both homomeric and heteromeric forms of RHIM-containing proteins may be important in the determination of cell fate [20,39–41]. RIPK3 homo-assembles are comprised of a single protofilament composed of three β-strands [20]. RIPK3 hetero-assemblies with RIPK1 form a serpentine fold comprised of two protofilaments [19,41]. The number, identity and importance of the specific amino acid residues that comprise the amyloid core in these two different conformers vary. It is possible that the architecture of amyloid formed by RHIM-containing proteins differs depending on homo-assembly or specific hetero-assembly status. This would be consistent with the characterisation of mosaic RIPK1:RIPK3 necrosomes by Chen et al. (2022) where the supramolecular architecture of the necrosome reflects its functional impact in cells [41]. Elucidation of TRIF amyloid homo-assemblies or hetero-assemblies with RIPK1 and/or RIPK3 by solid state NMR or cryo-EM is an important avenue for future study. These structures would allow for comparison with the elucidated RIPK1:RIPK3 structure [19], and reveal differences in assemblies formed by TRIF compared to RIPK1 or RIPK3 that may underlie physiological function.

The suite of experiments used in this study has revealed that the RHIM of TRIF binds directly to both RIPK1 and RIPK3 RHIMs, with distinct effects on the two interacting partners. When TRIF binds to RIPK1, it drives RIPK1 into heteromeric assemblies that are larger and more stable than RIPK1 homomeric assemblies. When TRIF forms two-protein hetero-assemblies with RIPK3, these are larger assemblies than formed by either protein alone and they are more stable than TRIF homomeric assemblies. The experiments reported here describe the two-protein assemblies formed by either human TRIF and RIPK1 or TRIF and RIPK3 and they form part of the extensive suite of protein interactions that drive programmed cell death [42]. However, in-cell experiments indicate that TRIF signalling for necroptosis may also involve hetero-interactions between all three of TRIF, RIPK1 and RIPK3 in mouse cells [16].

In murine fibroblasts and endothelial cells, TRIF and RIPK3 are capable of inducing necroptosis without the activity of RIPK1, whereas in macrophages RIPK1 is required for signalling [16]. The data reported here indicate that human TRIF is capable of direct interactions with both RIPK3 and RIPK1 but do not provide an explanation for the cell-type- or host-dependence of TRIF signalling outcomes. Therefore, it appears likely that other cellular co-factors are involved in induction of necroptosis cascades in these different cell types. A recent report studying co-factor driven modulation of ZBP1-RIPK3 necroptosis described that caspase 6 binds directly to RIPK3 to facilitate its interactions with ZBP1 [43]. It is possible that caspase 6 could also facilitate interaction between RIPK3 and TRIF, or modulate the amyloidogenicity or stoichiometry of amyloid complexes, rendering them more or less amenable to activation of the downstream pseudokinase MLKL. Little information is known about the cellular concentrations of RHIM proteins during microbial infection, their specific cellular localisation, or any important post-translational modifications, including ubiquitination [44–46]. These characteristics may also modulate or control the structural properties of immunity-associated structural assemblies beyond the intrinsic amyloidogenicity and interaction-propensity of the proteins themselves.

Overall, this work has demonstrated that TRIF is capable of forming heteromeric amyloid structures with unique biophysical and biochemical characteristics that may underlie their signalling function. The identification of TRIF as another protein that exerts signalling function by oligomeric assembly adds to the growing number of recognised important supramolecular signalling complexes in immunity, known as signalosomes [47]. These higher-order protein assemblies have specific architectural features that drive the activation of downstream effector proteins [47,48]. They include Fas/FADD complexes for caspase 8 activation [49]; MyD88 and partner proteins in the Myddosome [50]; ubiquitination complexes including TRAF 6 y [51], multiple variants of the inflammasome [52–54], CARD9/CARD11 filaments [55] and MAVS filaments [56]. This work, utilising truncated RHIM-containing domains, provides the first step towards understanding of the size, stoichiometry and stability of assemblies generated in cells by full-length RHIM-containing proteins during genuine infection. Future work will focus on the characterisation of multi-protein RHIM-based complexes to understand how these assemblies form, what effects co-factors may have and what role they play in signalling.

4. Materials and Methods

4.1. Production of RHIM Protein Fragments

Synthetic gene sequences encoding RHIM-containing regions of human RIPK1, RIPK3, TRIF and ZBP1 were purchased from Genscript and cloned into modified pET15b vectors to generate genes for His-tagged fusion proteins with YPet, mCherry or ubiquitin as N-terminal fusion partners for the RHIM domains (as illustrated in Figure 1). Successful cloning was confirmed by sequencing at the Australian Genome Research Facility. Recombinant expression was achieved in *E. coli* BL21(DE3) cells, with induction of gene expression by IPTG (0.5–1 mM) and 2–4 h incubation at 37 °C following addition of IPTG. Harvested cell pellets were stored at −20 °C until purification.

4.2. Purification of Proteins by Nickel Nitriloacetic Acid Affinity Chromatography

Cell pellets were resuspended in 20 mM Tris, 150 mM NaCl, 2 mM EDTA pH 8.0 at 5 mL/by shaking at 180 rpm at 37 °C for 30 min, with intermittent vortexing. Cells were lysed on ice by sonication and then centrifuged at 16,000× *g* and supernatant removed. The remaining pellet was washed and solubilised in 8 M urea, 20 mM Tris, pH 8.0 with mixing at 4 °C overnight. Remaining insoluble material was removed by centrifugation at 16,000× *g* for 30 min. β-mercaptoethanol was added to a final concentration of 4.8 mM and the solubilised protein was incubated with Ni NTA bead (agarose, supplier) with gentle agitation for 2 h. Resin was washed with 10 column volumes of 8 M urea, 20 mM Tris, 25 mM imidazole, pH 8 containing 4.8 mM β-mercaptoethanol and proteins were eluted with 8 M urea, 20 mM Tris, 300 mM imidazole, pH 8.0 containing 4.8 mM β-mercaptoethanol. Fractions containing the desired protein were pooled, concentrated to approximately 1 mL using Amicon ultra15 MWCO 10,000 centrifugal units and then buffer-exchanged into 8 M urea, 20 mM Tris, pH 8.0. Protein concentration was determined using the Pierce BCA Protein Assay Kit (ThermoFisher-Scientific, Scoresby, VIC, Australia), using a fresh standard curve prepared with bovine serum albumin in relevant buffer for each set of protein concentration measurements. Where protein stocks contained urea, the equivalent urea concentration was included in the samples used to create the standard curve. In all samples, the urea concentration was adjusted to less than 100 mM before protein determination using this kit.

4.3. Thioflavin T Formation and Light Scattering Assays

Buffer containing 25 mM NaH_2PO_4, 150 mM NaCl, pH 7.4, 40 μM ThT and 0.5 mM DTT was pre-loaded into black Costar 96 well fluorescent plates (#3631), the top sealed with Corning Plastic Sealing Tape (Product #6575) and incubated to a temperature of 37 °C in a POLARstar Omega microplate reader (BMG Labtech). Amyloid assembly was initiated by diluting desired proteins out of urea-containing buffer to required concentrations in the wells (specific concentrations varied between experiments and are indicated in each case), with final urea concentration below 300 mM. Fluorescence intensity to indicate amyloid formation was recorded with excitation filter 440 nm (+/− 10 nm) and emission filter 480 +/− 10 nm. Aggregate size was determined by simultaneous measurement of scattering of light at 350 nm. Assays were performed in triplicate. Data analysis was performed in Microsoft Excel and GraphPad Prism.

4.4. Congo Red Assays

RHIM-containing proteins were dialysed overnight against 25 mM NaH_2PO_4, 150 mM NaCl, 0.5 mM DTT, pH 7.4 at 0.3 mg/mL to allow assembly into amyloid. Positive control insulin amyloid fibrils were prepared by heating a solution of insulin at 2.5 mg/mL in 20 mM glycine, pH 2.0 with shaking at 700 rpm for 6 h, followed by 10-fold dilution. Negative control monomeric insulin was prepared at 0.25 mg/mL in 20 mM glycine, pH 2.0 at room temperature. Congo red from a filtered stock solution of 100 μM in 90% 25 mM NaH_2PO_4, 150 mM NaCl, 10% ethanol was added to a final concentration of 20 μM and the absorbance spectrum collected over 350–850 nm using cuvettes with 1cm pathlength in a Nanodrop™ 2000C Spectrophotometer (Thermo Fisher Scientific, Scoresby, VIC, Australia), relative to buffer only. The absorbance from a protein-only sample was subtracted from the spectrum arising from Congo red in the presence of each protein.

4.5. Confocal Microscopy

Proteins of interest were diluted to 2.5 μM into sodium phosphate buffer into black Costar 96-well fluorescent plates (Corning, New York, NY, USA) and incubated at 37 °C. Amyloid formation in one sample was recorded by addition of ThT and when amyloid formation was complete, as judged by plateau phase of ThT fluorescence, samples without ThT were prepared for confocal imaging. A protein sample measuring 30 μL was dispensed onto a microscope slide and a coverslip carefully lowered to exclude air bubbles before sealing and storage in the dark overnight at 4 °C before imaging on a Leica SP5 Confocal Microscope.

4.6. Widefield Fluorescence Microscopy

Proteins of interest were diluted to 2.5 μM into assembly-permissive sodium phosphate buffer (25 mM NaH_2PO_4, 150 mM NaCl, 0.5 mM DTT, pH 7.4) into black Costar 96-well fluorescent plates (Corning) and incubated with 40 μM ThT at 37 °C. After two hours, 30 μL of protein sample was dispensed onto a microscope slide and a coverslip carefully lowered to exclude air bubbles before sealing and storage in the dark overnight at 4 °C before imaging on a Cytation 3 Imager (Biotek, Winusky, VT, USA). CFP, GFP and Texas Red filter sets were used to detect ThT fluorescence, YPet and mCherry respectively. Image analysis was performed in ImageJ (Fiji).

4.7. Transmission Electron Microscopy

Samples of individual proteins or desired combinations of RHIM-containing proteins prepared in buffers containing 8 M urea were dialysed overnight against 25 mM NaH_2PO_4, 150 mM NaCl, 0.5 mM DTT, pH 7.4 at 1 μM concentration. Copper transmission electron microscopy grids with formvar film support and carbon coating (ProSciTech, Kirwan, QLD, Australia) were floated on 20 μL of dialysed protein suspensions for 10 min. Excess solution was discarded, and the grid washed with three droplets of MilliQ water before staining with 2% uranyl acetate. Grids were imaged using a Tecnai T12 microscope operating at 120 kV and images captured with a side-mounted CCD camera and RADIUS 2.0 imaging software (EMSIS GmbH, Münster, Germany).

4.8. Subtilisin Digestion and Mass Spectrometry Analysis

YPet-$TRIF_{601-712}$ (50 μM) was dialysed out of 8 M urea buffer into sodium phosphate buffer overnight. Subtilisin was added at 200:1 YPet-TRIF:subtilisin molar ratio for 4 h in order to completely digest proteins. Samples were then separated by centrifugation into soluble and insoluble fractions. The insoluble fraction was resuspended at 1× and 10× concentration in 8 M urea to identify potential insoluble bands comprising the amyloid core. These experiments were repeated in a workflow suitable for mass spectrometry, where insoluble pellets were resuspended in 90% formic acid, freeze dried, and then resuspended in sodium phosphate buffer. The samples were then electrophoresed by SDS-PAGE and bands of interest were excised. The excised samples were trypsin digested and underwent liquid chromatography mass spectrometry to identify peptide sequence.

4.9. Confocal Coincidence Spectroscopy

All of the confocal coincidence spectroscopy data were collected on a home built NanoBright Zeiss Axio Observer microscope system, as described in [57]. Proteins, either alone or in pre-mixed combinations, were diluted 100 fold out of 8 M urea, 100 mM NaH_2PO_4, 20 mM Tris, pH 8.0 into 100 mM NaH_2PO_4, 20 mM Tris, 0.5 mM DTT, pH 8.0, resulting in final protein concentration of 0.175 µM in 80 µM urea. Two lasers (emitting at 488 nm and 561 nm) were focused within the solution and fluorescence emission collected with 525/20 nm bandpass filter and 580 nm longpass filter separated by 565 nm dichroic, for YPet and mCherry signals respectively. Signals were detected simultaneously for both channels and binned at 1 ms intervals. Data analysis was performed using custom routines in MatLab, adapted from workflows described in [38,57].

4.10. Sodium Dodecyl Sulfate Agarose Gel Electrophoresis

Proteins of interest were diluted to 5 µM in 8 M urea-containing buffer, either alone or in mixtures. A sample of each was retained for later use as a monomeric, unassembled control. The remaining sample was dialysed against 100 mM NaH_2PO_4, 20 mM Tris, 0.5 mM DTT, pH 8.0 overnight. The next day, glycerol was added to samples to a final concentration of 4%, and bromophenol blue to a final concentration of 0.0008%. SDS (from 20% stock solution) or equivalent volume of water, was added to some samples to 2% and then samples incubated for 10 min before electrophoresis through a 1% agarose gel in 40 mM Tris, 20 mM acetic acid, 1 mM EDTA, 0.1% SDS, pH 8.3, at 50 V for 120 min. Gels were imaged on a ChemiDoc (Bio Rad Laboratories, Hercules, CA, USA), where YPet signal was detected using 605/50 nm emission filters, and mCherry signal was detected using 695/55 nm filters.

Supplementary Materials: The following supporting information can be downloaded at: https://www.mdpi.com/article/10.3390/molecules27113382/s1, Figure S1: Schematic representation of fusion proteins used in this study. Figure S2: Concentration-dependence of assembly formation by Ub-TRIF RHIM constructs. In all cases, error bars indicate one standard deviation. (A). Concentration dependence of ThT fluorescence over time for Ub-$TRIF_{601-712}$ and Ub-$TRIF_{601-712}$mut. Concentrations used are 0.625–5 µM, as indicated. (B). Concentration dependence of assembly size over time for Ub-$TRIF_{601-712}$ and Ub-$TRIF_{601-712}$mut as determined by static light scattering. Concentrations used are 0.625–5 µM, as indicated. Figure S3: Complete list of sequences identified by mass spectrometry from the YPet-$TRIF_{601-712}$ RHIM core. Alignment of peptides identified by mass spectrometry, including posttranslational modifications. The oxidation of methionine likely resulted from the process of electrospray ionisation during mass spectrometry experiments. Residue series are coloured according to their prevalence. Figure S4: Morphological detail of TRIF-RIPK3 hetero-assemblies analysed by negative stain transmission electron microscopy. TRIF and RIPK3 form dense fibrillar bundles or clusters when co-assembled at 1 µM and individual fibrils emanate from the surface of the clusters. Boxed areas indicate zoom regions of the larger image. Black arrowheads indicate fibrillar material visible within or extending from the dense clusters. All scale bars represent 500 nm. Figure S5: Widefield fluorescence microscopy of TRIF-RIPK3 hetero-assemblies co-stained with ThT. RHIM-containing proteins (as indicated above) from urea-containing monomeric stocks were diluted to 2.5 µM in assembly-permissive buffer containing 40 µM ThT. Protein assemblies were imaged using a Cytation 3 Imager (BioTek). CFP, GFP and Texas Red filter sets were used to detect ThT fluorescence, YPet and mCherry respectively. Scale bars indicate 20 µm.

Author Contributions: Conceptualization, M.S. (Megan Steain) and M.S. (Margaret Sunde); methodology, M.O.D.G.B., C.L.L.P., N.S., Y.G. and E.S.; validation, M.O.D.G.B., C.L.L.P. and N.S.; investigation, M.O.D.G.B., C.L.L.P., N.S., M.S. (Margaret Sunde) and S.R.B.; resources, M.S. (Margaret Sunde), Y.G. and E.S.; data curation, M.O.D.G.B. and N.S.; writing—original draft preparation, M.O.D.G.B.; writing—review and editing, M.O.D.G.B., C.L.L.P., N.S., M.S. (Megan Steain) and M.S. (Margaret Sunde); visualization, M.O.D.G.B. and M.S. (Margaret Sunde); supervision, M.S. (Megan Steain) and M.S. (Margaret Sunde); project administration, Margaret Sunde.; funding acquisition, Margaret Sunde. All authors have read and agreed to the published version of the manuscript.

Funding: This research was funded by AUSTRALIAN RESEARCH COUNCIL Discovery Project DP180101275 awarded to M.S. and E.S., M.O.D.G.B., N.S. and S.B. were supported by Research Training Program Scholarships from the Australian Government (Department of Education, Skills and Employment). MODGB received a Paulette Isabel Jones PhD Completion Scholarship award from The University of Sydney.

Institutional Review Board Statement: Not applicable.

Informed Consent Statement: Not applicable.

Data Availability Statement: Not applicable.

Acknowledgments: This work celebrates the work of Chris Dobson, who inspired amyloid researchers worldwide. The authors acknowledge the facilities and the scientific and technical assistance of staff from Sydney Microscopy and Microanalysis Core Research Facility at the University of Sydney. They thank Ben Crossett from Sydney Mass Spectrometry for performing the mass spectrometry analysis and James WP Brown for sharing scripts for analysis of confocal coincidence spectroscopy data.

Conflicts of Interest: The authors declare no conflict of interest. The funders had no role in the design of the study; in the collection, analyses, or interpretation of data; in the writing of the manuscript, or in the decision to publish the results.

Sample Availability: Samples of the compounds are available from the authors.

References

1. Ullah, M.O.; Sweet, M.; Mansell, A.; Kellie, S.; Kobe, B. TRIF-dependent TLR signaling, its functions in host defense and inflammation, and its potential as a therapeutic target. *J. Leukoc. Biol.* **2016**, *100*, 27–45. [CrossRef] [PubMed]
2. Kawasaki, T.; Kawai, T. Toll-Like Receptor Signaling Pathways. *Front. Immunol.* **2014**, *5*, 461. [CrossRef] [PubMed]
3. Vercammen, E.; Staal, J.; Beyaert, R. Sensing of Viral Infection and Activation of Innate Immunity by Toll-Like Receptor 3. *Clin. Microbiol. Rev.* **2008**, *21*, 13–25. [CrossRef] [PubMed]
4. Park, B.S.; Lee, J.-O. Recognition of lipopolysaccharide pattern by TLR4 complexes. *Exp. Mol. Med.* **2013**, *45*, e66. [CrossRef]
5. Flo, T.H.; Ryan, L.; Latz, E.; Takeuchi, O.; Monks, B.G.; Lien, E.; Halaas, Ø.; Akira, S.; Skjåk-Bræk, G.; Golenbock, D.T.; et al. Involvement of Toll-like Receptor (TLR) 2 and TLR4 in Cell Activation by Mannuronic Acid Polymers. *J. Biol. Chem.* **2002**, *277*, 35489–35495. [CrossRef]
6. Yang, S.; Sugawara, S.; Monodane, T.; Nishijima, M.; Adachi, Y.; Akashi, S.; Miyake, K.; Hase, S.; Takada, H. *Micrococcus luteus* Teichuronic Acids Activate Human and Murine Monocytic Cells in a CD14- and Toll-Like Receptor 4-Dependent Manner. *Infect. Immun.* **2001**, *69*, 2025–2030. [CrossRef]
7. Haynes, L.M.; Moore, D.D.; Kurt-Jones, E.A.; Finberg, R.W.; Anderson, L.J.; Tripp, R.A. Involvement of Toll-Like Receptor 4 in Innate Immunity to Respiratory Syncytial Virus. *J. Virol.* **2001**, *75*, 10730–10737. [CrossRef]
8. Vaure, C.; Liu, Y. A Comparative Review of Toll-Like Receptor 4 Expression and Functionality in Different Animal Species. *Front. Immunol.* **2014**, *5*, 316. [CrossRef]
9. Baker, M.O.; Shanmugam, N.; Pham, C.L.; Strange, M.; Steain, M.; Sunde, M. RHIM-based protein:protein interactions in microbial defence against programmed cell death by necroptosis. *Semin. Cell Dev. Biol.* **2020**, *99*, 86–95. [CrossRef]
10. Muendlein, H.I.; Connolly, W.M.; Magri, Z.; Smirnova, I.; Ilyukha, V.; Gautam, A.; Degterev, A.; Poltorak, A. ZBP1 promotes LPS-induced cell death and IL-1β release via RHIM-mediated interactions with RIPK1. *Nat. Commun.* **2021**, *12*, 86. [CrossRef]
11. Micheau, O.; Tschopp, J. Induction of TNF Receptor I-Mediated Apoptosis via Two Sequential Signaling Complexes. *Cell* **2003**, *114*, 181–190. [CrossRef]
12. Upton, J.; Kaiser, W.J.; Mocarski, E.S. Cytomegalovirus M45 Cell Death Suppression Requires Receptor-interacting Protein (RIP) Homotypic Interaction Motif (RHIM)-dependent Interaction with RIP1. *J. Biol. Chem.* **2008**, *283*, 16966–16970. [CrossRef] [PubMed]
13. Sun, X.; Yin, J.; Starovasnik, M.A.; Fairbrother, W.J.; Dixit, V.M. Identification of a Novel Homotypic Interaction Motif Required for the Phosphorylation of Receptor-interacting Protein (RIP) by RIP3. *J. Biol. Chem.* **2002**, *277*, 9505–9511. [CrossRef] [PubMed]

14. Cho, Y.S.; Challa, S.; Moquin, D.; Genga, R.; Ray, T.D.; Guildford, M.; Chan, F.K.-M. Phosphorylation-Driven Assembly of the RIP1-RIP3 Complex Regulates Programmed Necrosis and Virus-Induced Inflammation. *Cell* **2009**, *137*, 1112–1123. [CrossRef]
15. Li, J.; McQuade, T.; Siemer, A.B.; Napetschnig, J.; Moriwaki, K.; Hsiao, Y.-S.; Damko, E.; Moquin, D.; Walz, T.; McDermott, A.; et al. The RIP1/RIP3 Necrosome Forms a Functional Amyloid Signaling Complex Required for Programmed Necrosis. *Cell* **2012**, *150*, 339–350. [CrossRef] [PubMed]
16. Kaiser, W.J.; Sridharan, H.; Huang, C.; Mandal, P.; Upton, J.; Gough, P.J.; Sehon, C.A.; Marquis, R.W.; Bertin, J.; Mocarski, E.S. Toll-like Receptor 3-mediated Necrosis via TRIF, RIP3, and MLKL. *J. Biol. Chem.* **2013**, *288*, 31268–31279. [CrossRef]
17. Guo, H.; Omoto, S.; Harris, P.A.; Finger, J.N.; Bertin, J.; Gough, P.J.; Kaiser, W.J.; Mocarski, E.S. Herpes Simplex Virus Suppresses Necroptosis in Human Cells. *Cell Host Microbe* **2015**, *17*, 243–251. [CrossRef]
18. Steain, M.; Baker, M.O.D.G.; Pham, C.L.L.; Shanmugam, N.; Gambin, Y.; Sierecki, E.; McSharry, B.P.; Avdic, S.; Slobedman, B.; Sunde, M.; et al. Varicella zoster virus encodes a viral decoy RHIM to inhibit cell death. *PLoS Pathog.* **2020**, *16*, e1008473. [CrossRef]
19. Mompeán, M.; Li, W.; Li, J.; Laage, S.; Siemer, A.B.; Bozkurt, G.; Wu, H.; McDermott, A.E. The Structure of the Necrosome RIPK1-RIPK3 Core, a Human Hetero-Amyloid Signaling Complex. *Cell* **2018**, *173*, 1244–1253.e10. [CrossRef]
20. Wu, X.; Ma, Y.; Zhao, K.; Zhang, J.; Sun, Y.; Li, Y.; Dong, X.; Hu, H.; Liu, J.; Wang, J.; et al. The structure of a minimum amyloid fibril core formed by necroptosis-mediating RHIM of human RIPK3. *Proc. Natl. Acad. Sci. USA* **2021**, *118*, e2022933118. [CrossRef]
21. Upton, J.; Kaiser, W.J.; Mocarski, E.S. DAI/ZBP1/DLM-1 Complexes with RIP3 to Mediate Virus-Induced Programmed Necrosis that Is Targeted by Murine Cytomegalovirus vIRA. *Cell Host Microbe* **2012**, *11*, 290–297. [CrossRef] [PubMed]
22. Guo, H.; Gilley, R.P.; Fisher, A.; Lane, R.; Landsteiner, V.J.; Ragan, K.B.; Dovey, C.M.; Carette, J.; Upton, J.; Mocarski, E.S.; et al. Species-independent contribution of ZBP1/DAI/DLM-1-triggered necroptosis in host defense against HSV1. *Cell Death Dis.* **2018**, *9*, 816. [CrossRef] [PubMed]
23. Kaiser, W.J.; Offermann, M.K. Apoptosis Induced by the Toll-Like Receptor Adaptor TRIF Is Dependent on Its Receptor Interacting Protein Homotypic Interaction Motif. *J. Immunol.* **2005**, *174*, 4942–4952. [CrossRef] [PubMed]
24. He, S.; Liang, Y.; Shao, F.; Wang, X. Toll-like receptors activate programmed necrosis in macrophages through a receptor-interacting kinase-3–mediated pathway. *Proc. Natl. Acad. Sci. USA* **2011**, *108*, 20054–20059. [CrossRef] [PubMed]
25. Murphy, J.M.; Czabotar, P.E.; Hildebrand, J.M.; Lucet, I.S.; Zhang, J.-G.; Alvarez-Diaz, S.; Lewis, R.; Lalaoui, N.; Metcalf, D.; Webb, A.I.; et al. The Pseudokinase MLKL Mediates Necroptosis via a Molecular Switch Mechanism. *Immunity* **2013**, *39*, 443–453. [CrossRef]
26. Meng, Y.; Davies, K.A.; Fitzgibbon, C.; Young, S.N.; Garnish, S.E.; Horne, C.R.; Luo, C.; Garnier, J.-M.; Liang, L.-Y.; Cowan, A.D.; et al. Human RIPK3 maintains MLKL in an inactive conformation prior to cell death by necroptosis. *Nat. Commun.* **2021**, *12*, 6783. [CrossRef]
27. Murphy, J.M. The Killer Pseudokinase Mixed Lineage Kinase Domain-Like Protein (MLKL). *Cold Spring Harb. Perspect. Biol.* **2019**, *12*, a036376. [CrossRef]
28. Samson, A.L.; Zhang, Y.; Geoghegan, N.D.; Gavin, X.J.; Davies, K.A.; Mlodzianoski, M.J.; Whitehead, L.W.; Frank, D.; Garnish, S.E.; FitzGibbon, C.; et al. MLKL trafficking and accumulation at the plasma membrane control the kinetics and threshold for necroptosis. *Nat. Commun.* **2020**, *11*, 3151. [CrossRef]
29. Lim, J.; Park, H.; Heisler, J.; Maculins, T.; Roose-Girma, M.; Xu, M.; McKenzie, B.; Campagne, M.V.L.; Newton, K.; Murthy, A. Autophagy regulates inflammatory programmed cell death via turnover of RHIM-domain proteins. *eLife* **2019**, *8*, e44452. [CrossRef]
30. Gentle, I.E.; McHenry, K.T.; Weber, A.; Metz, A.; Kretz, O.; Porter, D.; Häcker, G. TIR-domain-containing adapter-inducing interferon-β (TRIF) forms filamentous structures, whose pro-apoptotic signalling is terminated by autophagy. *FEBS J.* **2017**, *284*, 1987–2003. [CrossRef]
31. Pham, C.L.; Shanmugam, N.; Strange, M.; O'Carroll, A.; Brown, J.W.; Sierecki, E.; Gambin, Y.; Steain, M.; Sunde, M. Viral M45 and necroptosis-associated proteins form heteromeric amyloid assemblies. *EMBO Rep.* **2018**, *20*, e46518. [CrossRef] [PubMed]
32. Shanmugam, N.; Baker, M.O.; Sanz-Hernandez, M.; Sierecki, E.; Gambin, Y.; Steain, M.; Pham, C.L.; Sunde, M. Herpes simplex virus encoded ICP6 protein forms functional amyloid assemblies with necroptosis-associated host proteins. *Biophys. Chem.* **2020**, *269*, 106524. [CrossRef] [PubMed]
33. Levine, H. Thioflavine T interaction with synthetic Alzheimer's disease*β*-amyloid peptides: Detection of amyloid aggregation in solution. *Protein Sci.* **1993**, *2*, 404–410. [CrossRef] [PubMed]
34. Klunk, W.; Pettegrew, J.W.; Abraham, D.J. Quantitative evaluation of congo red binding to amyloid-like proteins with a beta-pleated sheet conformation. *J. Histochem. Cytochem.* **1989**, *37*, 1273–1281. [CrossRef]
35. Hervas, R.; Rau, M.J.; Park, Y.; Zhang, W.; Murzin, A.G.; Fitzpatrick, J.A.J.; Scheres, S.H.W.; Si, K. Cryo-EM structure of a neuronal functional amyloid implicated in memory persistence in *Drosophila*. *Science* **2020**, *367*, 1230–1234. [CrossRef]
36. Lassé, M.; Ulluwishewa, D.; Healy, J.; Thompson, D.; Miller, A.; Roy, N.C.; Chitcholtan, K.; Gerrard, J.A. Evaluation of protease resistance and toxicity of amyloid-like food fibrils from whey, soy, kidney bean, and egg white. *Food Chem.* **2016**, *192*, 491–498. [CrossRef]
37. Chen, M.; Cook, K.D. Oxidation Artifacts in the Electrospray Mass Spectrometry of Aβ Peptide. *Anal. Chem.* **2007**, *79*, 2031–2036. [CrossRef]
38. Brown, J.W.P.; Bauer, A.; Polinkovsky, M.E.; Bhumkar, A.; Hunter, D.J.B.; Gaus, K.; Sierecki, E.; Gambin, Y. Single-molecule detection on a portable 3D-printed microscope. *Nat. Commun.* **2019**, *10*, 5662. [CrossRef]

39. Cook, W.D.; Moujalled, D.M.; Ralph, T.J.; Lock, P.; Young, S.N.; Murphy, J.; Vaux, D.L. RIPK1- and RIPK3-induced cell death mode is determined by target availability. *Cell Death Differ.* **2014**, *21*, 1600–1612. [CrossRef]
40. Hu, H.; Wu, X.; Wu, G.; Nan, N.; Zhang, J.; Zhu, X.; Zhang, Y.; Shu, Z.; Liu, J.; Liu, X.; et al. RIP3-mediated necroptosis is regulated by inter-filament assembly of RIP homotypic interaction motif. *Cell Death Differ.* **2020**, *28*, 251–266. [CrossRef]
41. Chen, X.; Zhu, R.; Zhong, J.; Ying, Y.; Wang, W.; Cao, Y.; Cai, H.; Li, X.; Shuai, J.; Han, J. Mosaic composition of RIP1–RIP3 signalling hub and its role in regulating cell death. *Nature* **2022**, *24*, 471–482. [CrossRef] [PubMed]
42. Verdonck, S.; Nemegeer, J.; Vandenabeele, P.; Maelfait, J. Viral manipulation of host cell necroptosis and pyroptosis. *Trends Microbiol.* **2021**, *30*, 593–605. [CrossRef] [PubMed]
43. Zheng, M.; Karki, R.; Vogel, P.; Kanneganti, T.-D. Caspase-6 Is a Key Regulator of Innate Immunity, Inflammasome Activation, and Host Defense. *Cell* **2020**, *181*, 674–687.e13. [CrossRef] [PubMed]
44. Garcia, L.R.; Tenev, T.; Newman, R.; Haich, R.O.; Liccardi, G.; John, S.W.; Annibaldi, A.; Yu, L.; Pardo, M.; Young, S.N.; et al. Ubiquitylation of MLKL at lysine 219 positively regulates necroptosis-induced tissue injury and pathogen clearance. *Nat. Commun.* **2021**, *12*, 3364. [CrossRef]
45. Liu, Z.; Dagley, L.F.; Shield-Artin, K.; Young, S.N.; Bankovacki, A.; Wang, X.; Tang, M.; Howitt, J.; Stafford, C.A.; Nachbur, U.; et al. Oligomerization-driven MLKL ubiquitylation antagonizes necroptosis. *EMBO J.* **2021**, *40*, e103718. [CrossRef]
46. Onizawa, M.; Oshima, S.; Schulze-Topphoff, U.; Oses-Prieto, J.A.; Lu, T.; Tavares, R.; Prodhomme, T.; Duong, B.; Whang, M.I.; Advincula, R.; et al. The ubiquitin-modifying enzyme A20 restricts ubiquitination of the kinase RIPK3 and protects cells from necroptosis. *Nat. Immunol.* **2015**, *16*, 618–627. [CrossRef]
47. Wu, H. Higher-Order Assemblies in a New Paradigm of Signal Transduction. *Cell* **2013**, *153*, 287–292. [CrossRef]
48. Wu, H.; Fuxreiter, M. The Structure and Dynamics of Higher-Order Assemblies: Amyloids, Signalosomes, and Granules. *Cell* **2016**, *165*, 1055–1066. [CrossRef]
49. Wang, L.; Yang, J.K.; Kabaleeswaran, V.; Rice, A.J.; Cruz, A.C.; Park, A.Y.; Yin, Q.; Damko, E.; Jang, S.B.; Raunser, S.; et al. The Fas–FADD death domain complex structure reveals the basis of DISC assembly and disease mutations. *Nat. Struct. Mol. Biol.* **2010**, *17*, 1324–1329. [CrossRef]
50. Lin, S.-C.; Lo, Y.-C.; Wu, H. Helical assembly in the MyD88–IRAK4–IRAK2 complex in TLR/IL-1R signalling. *Nature* **2010**, *465*, 885–890. [CrossRef]
51. Yin, Q.; Lin, S.-C.; Lamothe, B.; Lu, M.; Lo, Y.-C.; Hura, G.; Zheng, L.; Rich, R.L.; Campos, A.D.; Myszka, D.G.; et al. E2 interaction and dimerization in the crystal structure of TRAF6. *Nat. Struct. Mol. Biol.* **2009**, *16*, 658–666. [CrossRef] [PubMed]
52. Cai, X.; Chen, J.; Xu, H.; Liu, S.; Jiang, Q.-X.; Halfmann, R.; Chen, Z.J. Prion-like Polymerization Underlies Signal Transduction in Antiviral Immune Defense and Inflammasome Activation. *Cell* **2014**, *156*, 1207–1222. [CrossRef] [PubMed]
53. Sborgi, L.; Ravotti, F.; Dandey, V.P.; Dick, M.S.; Mazur, A.; Reckel, S.; Chami, M.; Scherer, S.; Huber, M.; Böckmann, A.; et al. Structure and assembly of the mouse ASC inflammasome by combined NMR spectroscopy and cryo-electron microscopy. *Proc. Natl. Acad. Sci. USA* **2015**, *112*, 13237–13242. [CrossRef]
54. Sharif, H.; Hollingsworth, L.R.; Griswold, A.R.; Hsiao, J.C.; Wang, Q.; Bachovchin, D.A.; Wu, H. Dipeptidyl peptidase 9 sets a threshold for CARD8 inflammasome formation by sequestering its active C-terminal fragment. *Immunity* **2021**, *54*, 1392–1404.e10. [CrossRef] [PubMed]
55. Holliday, M.J.; Witt, A.; Gama, A.R.; Walters, B.T.; Arthur, C.P.; Halfmann, R.; Rohou, A.; Dueber, E.C.; Fairbrother, W.J. Structures of autoinhibited and polymerized forms of CARD9 reveal mechanisms of CARD9 and CARD11 activation. *Nat. Commun.* **2019**, *10*, 3070. [CrossRef]
56. Wu, B.; Peisley, A.; Tetrault, D.; Li, Z.; Egelman, E.; Magor, K.; Walz, T.; Penczek, P.A.; Hur, S. Molecular Imprinting as a Signal-Activation Mechanism of the Viral RNA Sensor RIG-I. *Mol. Cell* **2014**, *55*, 511–523. [CrossRef] [PubMed]
57. Gambin, Y.; Ariotti, N.; McMahon, K.A.; Bastiani, M.; Sierecki, E.; Kovtun, O.; Polinkovsky, M.E.; Magenau, A.; Jung, W.; Okano, S.; et al. Single-molecule analysis reveals self assembly and nanoscale segregation of two distinct cavin subcomplexes on caveolae. *eLife* **2014**, *3*, e01434. [CrossRef] [PubMed]

molecules

MDPI

Article

Conversion of the Native N-Terminal Domain of TDP-43 into a Monomeric Alternative Fold with Lower Aggregation Propensity

Matteo Moretti [1,†], Isabella Marzi [1,†], Cristina Cantarutti [2], Mirella Vivoli Vega [3], Walter Mandaliti [2], Maria Chiara Mimmi [4], Francesco Bemporad [1], Alessandra Corazza [2,5] and Fabrizio Chiti [1,*]

1 Department of Experimental and Clinical Biomedical Sciences "Mario Serio", University of Florence, 50134 Florence, Italy; matteomoretti109@gmail.com (M.M.); isabella.marzi@unifi.it (I.M.); francesco.bemporad@unifi.it (F.B.)
2 Department of Medical Area, University of Udine, 33100 Udine, Italy; cristina.cantarutti@uniud.it (C.C.); walter.mandaliti@gmail.com (W.M.); alessandra.corazza@uniud.it (A.C.)
3 School of Biochemistry, University of Bristol, Bristol BS8 1TD, UK; mirella.vivolivega@bristol.ac.uk
4 Department of Molecular Medicine, University of Pavia, 27100 Pavia, Italy; chiara.mimmi@unipv.it
5 INBB, Viale Medaglie d'Oro 305, 00136 Roma, Italy
* Correspondence: fabrizio.chiti@unifi.it; Tel.: +39-055-275-1220
† These authors contributed equally to the work.

Abstract: TAR DNA-binding protein 43 (TDP-43) forms intraneuronal cytoplasmic inclusions associated with amyotrophic lateral sclerosis and ubiquitin-positive frontotemporal lobar degeneration. Its N-terminal domain (NTD) can dimerise/oligomerise with the head-to-tail arrangement, which is essential for function but also favours liquid-liquid phase separation and inclusion formation of full-length TDP-43. Using various biophysical approaches, we identified an alternative conformational state of NTD in the presence of Sulfobetaine 3-10 (SB3-10), with higher content of α-helical structure and tryptophan solvent exposure. NMR shows a highly mobile structure, with partially folded regions and β-sheet content decrease, with a concomitant increase of α-helical structure. It is monomeric and reverts to native oligomeric NTD upon SB3-10 dilution. The equilibrium GdnHCl-induced denaturation shows a cooperative folding and a somewhat lower conformational stability. When the aggregation processes were compared with and without pre-incubation with SB3-10, but at the identical final SB3-10 concentration, a slower aggregation was found in the former case, despite the reversible attainment of the native conformation in both cases. This was attributed to protein monomerization and oligomeric seeds disruption by the conditions promoting the alternative conformation. Overall, the results show a high plasticity of TDP-43 NTD and identify strategies to monomerise TDP-43 NTD for methodological and biomedical applications.

Keywords: neurodegeneration; self-assembly; oligomerisation; native-like; micelle

Citation: Moretti, M.; Marzi, I.; Cantarutti, C.; Vivoli Vega, M.; Mandaliti, W.; Mimmi, M.C.; Bemporad, F.; Corazza, A.; Chiti, F. Conversion of the Native N-Terminal Domain of TDP-43 into a Monomeric Alternative Fold with Lower Aggregation Propensity. *Molecules* **2022**, *27*, 4309. https://doi.org/10.3390/molecules27134309

Academic Editors: Kunihiro Kuwajima, Yuko Okamoto, Tuomas Knowles and Michele Vendruscolo

Received: 27 May 2022
Accepted: 2 July 2022
Published: 5 July 2022

1. Introduction

Amyotrophic lateral sclerosis (ALS) is a fatal disorder characterized by a progressive degeneration of the upper and lower motor neurons of the brain, brainstem, and spinal cord, but also in the frontotemporal cortex and hippocampus in a fraction of patients [1–3], leading to muscle weakness, wasting and spasticity [4]. In 2006, the transactive response (TAR) DNA-binding protein (TDP-43) was identified as the main component of intraneuronal cytosolic ubiquitinated protein aggregates found in ALS and tau-negative, ubiquitin-positive frontotemporal lobar degeneration (FTLD) patients [1,2]. Numerous post-translational modifications of TDP-43 have been described in the aggregates, such as abnormal polyubiquitination and hyperphosphorylation as well as partial proteolysis to form C-terminal fragments [1,2,5,6].

The TDP-43 protein contains 414 amino acid residues and the encoding gene *TARDBP* is located on chromosome 1. It comprises an N-terminal domain (NTD, aa 1–76), a nuclear localization signal (NLS, aa 82–98), two RNA recognition motifs (RRM1, aa 106–176, and RRM2, aa 191–259), a nuclear export signal (NES, aa 239–250), and a C-terminal domain (CTD, aa 274–414), which encompasses a prion-like glutamine/asparagine-rich (Q/N) region (aa 345–366) and a glycine-rich region (aa 366–414) [7–12]. TDP-43 was first proposed to be natively dimeric or at least to exist in a monomer-dimer equilibrium under normal physiological conditions [13–16]. The dimerisation of TDP-43 was shown to occur through interactions of the N-terminal domains, and these interactions were proposed to be responsible for the dimerisation of the entire full-length protein [14–16]. It was also found that the deletion of the N terminus, or even the first nine residues, is sufficient to abolish NTD folding, dimerisation and the TDP-43–regulated RNA splicing [15]. With the elucidation of the monomeric structure of NTD with NMR spectroscopy, models of the dimer were proposed [12,17,18]. Later on, the NTD and full-length TDP-43 were found to form oligomers beyond the dimer and that TDP-43 oligomerisation was necessary for its function as mutants destabilising the NTD-NTD interface without affecting the NTD folded structure inhibit the splicing activity of full-length TDP-43 expressed in cells [19,20].

Although NTD-mediated TDP-43 dimerisation/oligomerisation appears to be essential for TDP-43 function [14,15,19,20], extensive NTD-mediated TDP-43 oligomerisation was also found to enhance the liquid-liquid phase separation (LLPS) at physiological concentrations [20–22], as well as formation of solid phase cytotoxic inclusions in the cytoplasm [15,23].

The NTD monomer consists of seven-eight β-strands and a single α-helix arranged in an axin-1 DIX domain fold [12,17–20], which facilitates dimer formation and further oligomerisation by a head-to-tail disposition of the individual subunits [19,20,24]. The conformational stability of the domain, as measured by the hydrogen/deuterium exchange at pH 3.8 and 25 °C, was found to be 15.9 $\pm$ 0.5 kJ mol^{-1} [12]. Using equilibrium urea denaturation curves monitored with intrinsic fluorescence and far-UV circular dichroism at pH 7.4 and 25 °C, higher conformational stability values of 20.1 $\pm$ 1.5 kJ mol^{-1} and 21.8 $\pm$ 1.5 kJ mol^{-1} were found, respectively [24], most probably due to the neutral pH of the measurements. The β-strands, α-helix, and three of four turns are rigid, although the loop formed by residues 47–53 appeared mobile [17]. All X-Pro peptide bonds adopt a trans configuration and the two cysteine residues at positions 39 and 50 are reduced and distantly separated on the surface of the protein [12,17,19,20].

The NTD of TDP-43 follows a folding process characterized by a series of steps [24]. The NTD unfolded state (U) consists of a pre-equilibrium between molecules having all the X-Pro peptide bonds in a native trans configuration and others with one or more X-Pro peptide bonds in a non-native cis configuration. The unfolded state U converts into a collapsed state (CS), which then converts into the fully folded monomer (F) following two parallel pathways: in the first CS converts directly to F, whereas in the second, CS converts into an intermediate state (I) that is on-pathway to F. All these U, CS and I states maintain the partition equilibrium between cis and trans-X-Pro peptide bonds, but folding can occur only from molecules having all-trans X-Pro bonds. Once the fully folded state F is attained, the NTD is able to dimerise into a head-to-tail homodimer that presents the further ability to form higher molecular weight assemblies [24]. Following urea denaturation at equilibrium, a native-like dimeric state was also detected at low urea concentrations (F*). Moreover, the various NMR studies of TDP-43 NTD show spectral differences in the various experimental conditions in which the protein domain is folded [10,12,14,17,18,20]. Overall, these structural and folding studies have revealed that TDP-43 NTD is a highly plastic protein able to adopt different conformational states.

In this work, we purified the NTD of TDP-43 and identified conditions to obtain two conformational states of the protein—one that is fully folded, native and dimeric and another, obtained through the use of a zwitterionic detergent Sulfobetaine 3-10 (SB3-10), that presents a monomeric, alternative, non-native, partially folded structure. We will show that the SB3-10-stabilised conformational state of TDP-43 NTD retains a cooperative fold and high conformational stability in spite of a remarkably different three-dimensional structure but does not retain the ability to dimerise and has a remarkably lower propensity to oligomerise and aggregate.

2. Results

2.1. The Alternative NTD Conformation Is Enriched with α-Helical Structure and Less Densely Packed

The NTD of TDP-43, characterised by a 6x His-tag followed by TEV protease cleavage site and the first 77 residues of TDP-43, was purified and found to be electrophoretically pure. Initially, the structure and oligomerisation of the protein domain were studied in 5 mM sodium phosphate buffer, 50 mM NaCl, 1 mM dithiothreitol (DTT), pH 7.4, in the absence or presence of 3.0% (*w*/*v*) SB3-10, at 25 °C. This zwitterionic detergent was initially used to stabilize the protein domain during purification, as it is normally used in protein purification protocols. However, we soon realized that it led to a different conformational state of the protein domain and we, therefore, decided to use it more systematically to investigate the structural plasticity of the TDP-43 NTD.

The far UV circular dichroism (CD) spectrum for the protein domain without SB3-10 featured two negative peaks at ca. 195 and 209 nm with a distinct small positive band at 233 nm (Figure 1A). The obtained CD spectrum was very similar to the other TDP-43 NTD spectra reported in the literature [14,18,24]. It was also found to be stable within 24 h, as ten CD spectra recorded at regular time intervals within this timeframe were superimposable and led to similar estimates of secondary structure types (Figure 1A, inset) when analysed with the BeStSel algorithm, which takes into account sheet twisting as a deconvolution parameter [25]. The secondary structure distribution is also in agreement with the values deduced from the X-ray and NMR structures of this protein domain [19,20]. The CD spectrum of the protein domain with 3.0% SB3-10, on the other hand, exhibits two negative peaks at ca. 207 nm and 224 nm and a positive peak at ca. 192 nm, thus showing a higher α-helical content (Figure 1A). Similar, to the spectrum in 0.0% SB3-10, it was found to be stable within 24 h but led to a different distribution of secondary structure types when analysed with the BeStSel algorithm, with a higher content of α-helical structure at the expense of the β-sheet structure, and a higher quantity of turns ($p < 0.001$, Figure 1A, inset).

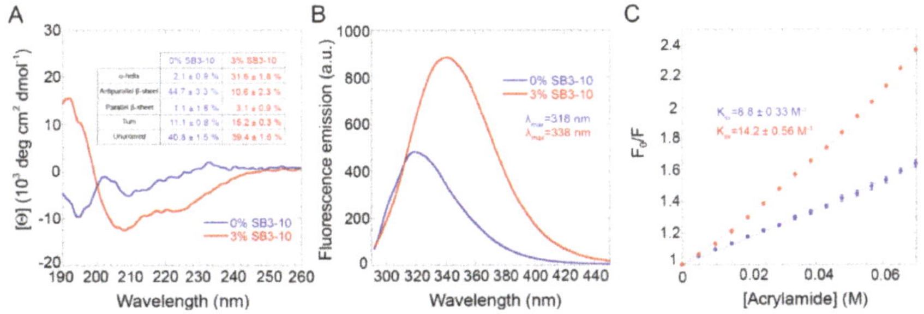

Figure 1. Structural characterization of TDP-43 NTD. Far-UV CD spectra (**A**), Intrinsic fluorescence spectra (**B**) and Stern-Volmer assays (**C**) of TDP-43 NTD in 0.0% (blue) and 3.0% (red) SB3-10. All SB3-10 percentages are *w*/*v*. The writings inside the panels indicate the values obtained for the analysed parameters.

The intrinsic fluorescence spectrum of TDP-43 NTD in the absence of SB3-10 showed a single peak at 319 nm, indicating a fully folded structure, with the Trp68 residue well buried in the hydrophobic core of the protein (Figure 1B). In 3.0% SB3-10, the intrinsic fluorescence increased in intensity and underwent a red-shift to 339 nm, showing a Trp68 residue more exposed to the solvent and a partially folded structure (Figure 1B). The degree of solvent exposure of the Trp68 residue of TDP-43 NTD was also studied using a Stern–Volmer assay with acrylamide as a quencher (Figure 1C). TDP-43 NTD treated with 0.0% SB3-10 showed a linear plot of F_0/F versus acrylamide concentration with a K_{SV} value of 8.8 ± 0.3 M^{-1}, while TDP-43 NTD treated with 3.0% SB3-10 showed a non-linear plot, with a K_{SV} value of 14.3 ± 0.6 M^{-1} and a K_{ST} value of 4.5 ± 0.2 M^{-1} (Figure 1C). The higher K_{SV} value ($p < 0.01$) and the presence of a static quenching component (K_{ST}) indicate a greater exposure of the Trp68 residue to the solvent for the protein domain in 3.0% SB3-10.

2.2. The Alternative NTD Conformation Studied with NMR Is Highly Dynamic

We then performed an NMR analysis on the ^{15}N and ^{13}C labelled protein domains in both conditions. The ^{1}H-^{15}N HSQC spectrum in 3% (w/v) SB3-10 shows a low degree of resonance spreading, typical of ⟨-helix rich and partially folded proteins (Figure 2A), in contrast to the HSQC spectrum of the native NTD in 0% SB3-10 previously shown [17,18,20] and here analysed as a control (Figure S1). At 25 °C, 84 backbone amide peaks can be observed in the HSQC spectrum in 3% SB3-10 and at 17 °C, this number raises to 99 (Figure S2). Considering that the NTD has 72 non-proline residues, we infer that some residues exhibit multiple forms. Seven backbone peaks could not be unambiguously assigned (E21, R44, N45, C50, M51, R55 and L56). Thirteen alternative forms were assigned. Four of them were found to be related to C39 oxidation (A33, A38, L41, R42) because DTT addition led to a decrease in the peak intensities of the second form (Figure S3A). Cα/β chemical shift values of C39 are however compatible with the absence of a disulphide bond, indicating a different type of thiol oxidation. The other nine residues were not affected by the addition of DTT and are residues 24–29, 59–60 and 63 that form a well-defined cluster in the native structure (Figure S3B), indicating structural mobility at these sites. In particular, three forms were detected for A63. Spectra acquired at low temperature (17 °C) show that one of the two conformations (hereinafter referred to as form A) is poorly populated relative to form B, whereas it becomes more populated as the temperature increases suggesting an equilibrium between the two species. ^{1}H, ^{13}C and ^{15}N chemical shifts have been deposited in the BioMagResBank (http://www.bmrb.wisc.edu, accessed on 14 June 2022) under the BMRB accession number 51492.

Based on chemical shift values, we assessed the secondary structure composition by Talos+ [26], chemical shift index (CSI) [27], and Secondary Structure Propensity (SSP) [28]. All methods confirmed, in agreement with the CD analysis, the increase in α-helical structure and the concomitant decrease in β-sheet structure. In addition to the 28–33 helix, which adopts a helical structure in native NTD as well, other two non-native helices were found in the N-terminal and C-terminal domains (Table S1, Figures 2B and S4). Moreover, form A was found to have a longer helix 26–33, starting from residue 26 rather than residue 28. A consensus prediction of extended conformation for residues 16–18, the central portion of ®-strand 2 in native NTD, was found for the three predictive methods. In native NTD β-strand 2 forms an antiparallel β-sheet with ®-strand 1 that, according to CSI (Table S1) and SSP analysis (Figure S4), also shows a propensity for ®-strand conformation, although with a low score, in agreement with the little antiparallel β-sheet structure observed with far-UV CD (10.6 ± 2.3%). Talos+ analysis indicates that those strands of the protein are highly dynamic with random coil indexes (RCI) below or equal to 0.7 (Figure 2C).

We also recorded ^{1}H-^{15}N HSQC spectra at different temperature values from 17 to 39 °C (Figure S6). The slope of the variation of the peak chemical shift as a function of temperature reveals whether or not the amide group is involved in a hydrogen bond; in particular, a slope < -4.5 ppb/K reveals a hydrogen bond [29]. Because of the overlap of some peaks, it was not possible to extract this information for all the assigned residues. The hydrogen bond pattern is largely consistent with the secondary structure obtained from CSI and Talos+ analysis (Table S1 and Figure S5).

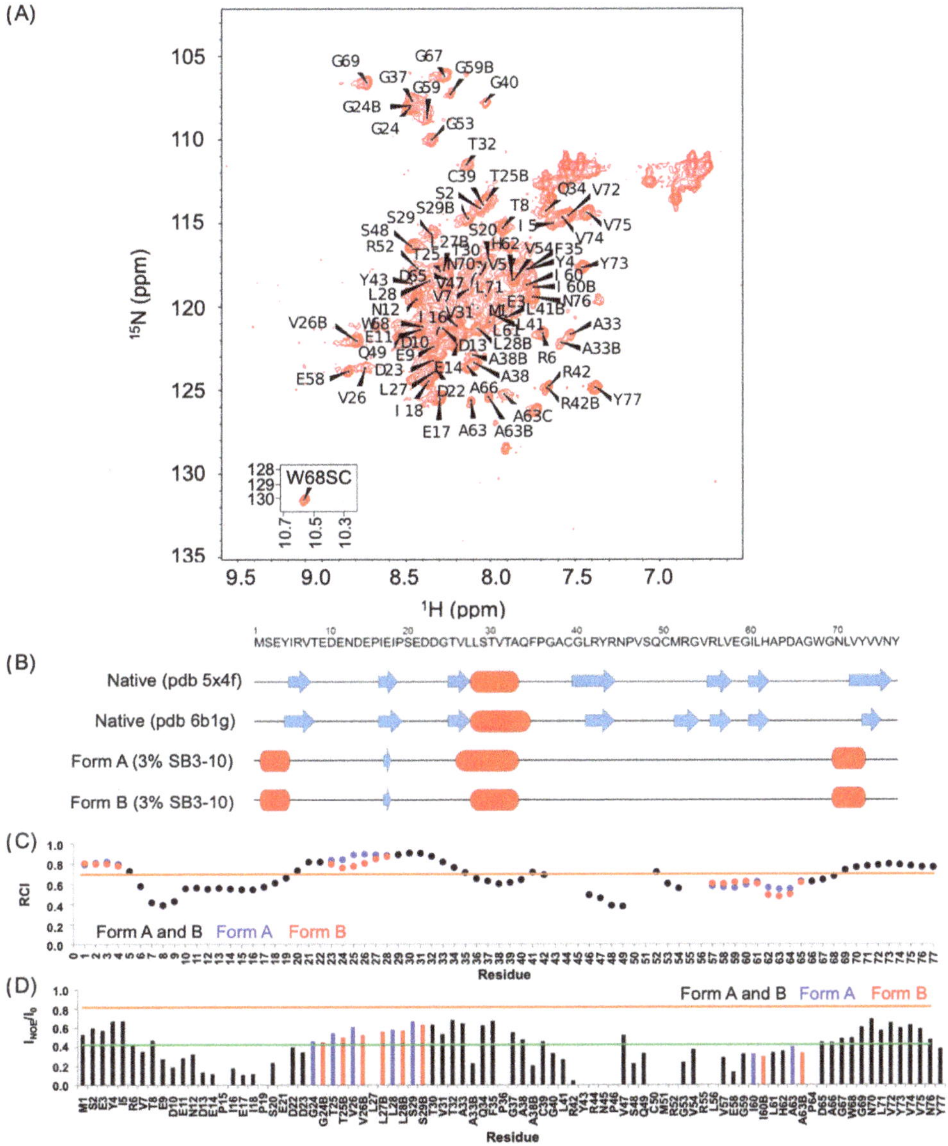

Figure 2. NMR characterization of TDP-43 NTD in 3% SB3-10. (**A**) ^{1}H-^{15}N HSQC spectrum of TDP-43 NTD in 3% (*w*/*v*) SB3-10 at 25 °C, recorded at 700 MHz. (**B**) Comparison of the secondary structure of native NTD without SB3-10 (pdb id: 5x4f and 6b1g) and of the alternative conformation in 3% SB3-10 reporting the consensus analysis of Talos+, CSI and SSP for forms (**A**) and (**B**). Blue arrows indicate extended structures (β-strands) and red cylinders stand for α-helices. (**C**) Random coil index (RCI) values as predicted by Talos+. The black, blue and red bars correspond to both forms, form (**A**) and form (**B**), respectively. RCI values correspond to order parameters (S^2). Residues with $S^2 \leq 0.7$ (orange line) are considered too flexible to produce reliable torsion angles [26]. (**D**) ^{15}N{^{1}H} NOE ratios at 25 °C. Colour codes as in panel (**C**). The green and orange lines indicate the average NOE value and the NOE value expected for a rigid structure (0.82) [30], respectively.

The nature of this alternative TDP-43 NTD conformation in 3% (*w*/*v*) SB3-10 was also studied by $^{15}N\{^1H\}$ NOE experiments [30]. The average NOE value calculated over the sequence is quite low (0.42 ± 0.18, maximum value 0.69) and the predicted structured regions also show NOE values lower than expected (Figure 2D), confirming the overall dynamic nature of this conformation. Furthermore, form B presents slightly lower NOE values. Experimental NOEs and predicted RCIs are in agreement and report reduced flexibility in the structured regions (at the termini and the central helix) and high mobility in the regions encompassing residues 6–19 and 46–67. By decreasing the temperature from 25 to 17 °C, we observed a remarkable increase in the NOE values in the 39–42 region (Figure S6). This region lacks a secondary structure, as inferred from NMR chemical shifts, but in the native conformation, it has an extended secondary structure (Table S1 and Figure 2B). Residues R6 and V57 also decrease significantly their flexibility at 17 °C and both are involved in the formation of β-strands in the native conformation.

2.3. *The Alternative NTD Conformation Is a Monomer*

The hydrodynamic diameters of both TDP-43 NTD conformations were determined using dynamic light scattering (DLS). The distribution of light scattering intensity vs. the apparent hydrodynamic diameter revealed that TDP-43 NTD in the absence of SB3-10 has a monodispersed distribution with a hydrodynamic diameter of 5.1 ± 0.3 nm, while in the presence of 3.0% SB3-10, the hydrodynamic diameter was 4.7 ± 0.2 nm (Figure 3A). The hydrodynamic diameters expected for the TDP-43 NTD devoid of any tag are 3.23 and 4.28 nm for folded monomer and dimer, respectively [24]. Since our TDP-43 NTD has an additional unfolded tag, it can be concluded that native TDP-43 NTD in 0.0% SB3-10 occurs as a folded dimer, possibly adopting a higher-order oligomer. By contrast, the decrease of the hydrodynamic diameter of the protein domain from 5.1 ± 0.3 nm to 4.7 ± 0.2 nm upon SB3-10 addition, concomitantly with its partial unfolding suggested by the CD, fluorescence and NMR analyses reported above, suggests that the protein domain is a partially folded monomer in 3.0% SB3-10.

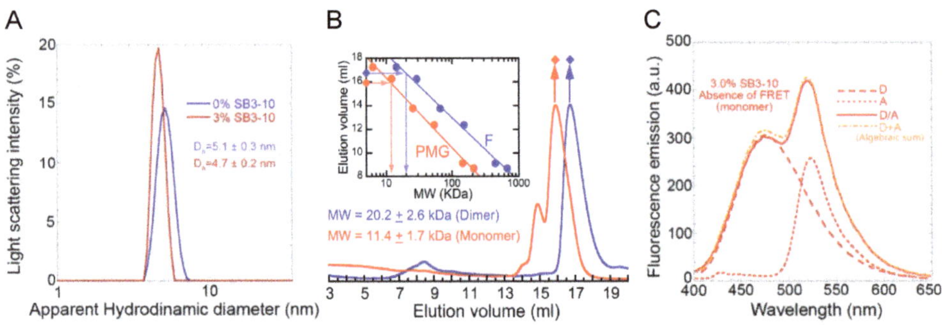

Figure 3. Oligomeric state characterization of TDP-43 NTD. DLS distributions (**A**), analytical size exclusion chromatography (SEC) profiles (**B**) and FRET analyses (**C**) of TDP-43 NTD in 0.0% (blue) and 3.0% (red) SB3-10. All SB3-10 percentages are *w*/*v*. In panel (**B**) the inset shows the calibration curve obtained with proteins of known masses (blue) and re-determined mathematically for pre-molten globule states, as described in Methods (red). In panel (**C**) the spectra refer to NTD-D (dashed line), NTD-A (dotted line), NTD-D/NTD-A (continuous line) and the algebraic sum of the first two spectra (mixed orange dotted/dashed line). The writings inside the panels indicate the values obtained for the analysed parameters.

The two different conformations were then studied using analytical SEC. The elution volume (V_e) peak was 16.7 ± 0.3 mL in 0.0% SB3-10 (Figure 3B, blue). Through the use of the calibration curve obtained with folded proteins of known mass (Figure 3B, inset, blue), this V_e was found to correspond to a molecular mass of 20.2 ± 2.6 kDa, closer to that of a dimer rather than a monomer (expected molecular weights of 22.2 and 11.1 kDa, respectively). This is in agreement with previous SEC studies [24]. The alternative form in 3.0% SB3-10 eluted as a major peak at 16.0 ± 0.3 mL preceded by a minor one at 14.9 ± 0.3 mL (Figure 3B, red). Using the same analysis with a calibration curve corrected for partially unfolded (pre-molten globule) standard proteins (Figure 3B, inset, red), a molecular weight of 11.4 ± 1.7 kDa was obtained for the major peak, confirming its monomeric state.

Lastly, Förster Resonance Energy Transfer (FRET) was used to further investigate the conformational state of the TDP-43 NTD in 3.0% SB3-10. Since the NTD has two cysteine residues at position 39 and 50, a mutant that maintained only Cys39 was purified, namely C50S. This C50S mutant was labelled with the thiol-reactive probes 5-((((2-iodoacetyl)amino)ethyl)amino)naphthalene-1-sulfonic acid (1,5-IAEDANS) as a donor (D) and 6-iodoacetamidofluorescein (6-IAF) as an acceptor (A). FRET analysis was then performed acquiring the fluorescence spectra of the samples containing only NTD-D, only NTD-A and both NTD-D and NTD-A at a 1:1 molar ratio (Figure 3C, dashed, dotted and continuous red lines, respectively). The spectrum obtained from the algebraic sum of the first two spectra (Figure 3C, mixed dotted/dashed orange line) was found to be superimposable to the third spectrum (Figure 3C, continuous red line). Hence, unlike the TDP-43 NTD head-to-tail dimer previously studied in 0.0% SB3-10, in which significant FRET was observed [24], in the presence of 3.0% SB3-10 FRET was not observed, further confirming the monomeric state of the alternative TDP-43 NTD conformation.

2.4. *Alternative NTD Exhibits a Lower Cooperativity Than Native NTD*

The conformational stabilities of the two forms of TDP-43 NTD were analysed by means of intrinsic tryptophan fluorescence and using GdnHCl as a chemical denaturant (Figure 4). 30 samples containing TDP-43 NTD and GdnHCl concentrations ranging from 0.0 to 5.0 M were prepared and their tryptophan emission spectra were recorded. The analysis was then repeated adding 3.0% SB3-10 in all samples. Afterwards, the ratio between the fluorescence values in two distinct regions of wavelength was measured for each spectrum and the obtained ratio was plotted as a function of the molar concentration of GdnHCl (Figure 4). In both cases, the GdnHCl-induced denaturation of TDP-43 NTD follows a two-state equilibrium and can be fitted to the model edited by Santoro and Bolen [31]. The results of the analysis with 0.0% SB3-10 yielded values of 22.3±1.4 kJ/mol, 8.7 ± 0.6 kJ/(mol M) and 2.6 ± 0.2 M for ΔG_{U-F}^{H20}, m_{eq} and C_m, respectively (Figure 4A). The ΔG_{U-F}^{H20} value is identical, within experimental error ($p > 0.05$), to that obtained previously for the same protein domain using urea as a chemical denaturant, i.e., 21.8 ± 1.5 kJ/mol [24]. For TDP-43 NTD treated with 3.0% SB3-10, values of 19.2 ± 3.3 kJ/mol, 5.7 ± 1.2 kJ/(mol M) and 3.4 ± 0.2 M were obtained for the same parameters, respectively (Figure 4B). The two sets of values indicate a cooperative transition in both cases, with significantly lower cooperativity for the alternative conformation in 3.0% SB3-10, as indicated by a m_{eq} value decreased by 35 ± 15% ($p < 0.05$) relative to that measured in its absence. This reduction indicates an increase in the solvent-exposed surface area. The overall stability is also lower with the ΔG_{U-F}^{H20} value decreasing by ca. 3 kJ/mol, although the difference does not reach statistical significance in this case.

2.5. *The Native-to-Alternative NTD Transition Is Reversible and Associated with SB3-10 Micelle Formation*

In order to investigate the existence and reversibility of a possible transition between the two conformations in 0.0% and 3.0% SB3-10, intrinsic fluorescence spectra were recorded treating TDP-43 NTD with increasing concentrations of SB3-10 from 0.0% to 3.0%. The data show that starting from a concentration of ca. 0.6% SB3-10 there is a sharp increase in the fluorescence intensity up to ca. 1.4% (Figure 5A). It is also possible to see a red-shift of the wavelength of maximum fluorescence from 319 nm to 339 nm, within the same range of SB3-10 concentration (Figure 5A). The centre of spectral mass (COM) for each spectrum was calculated and the COM was plotted vs. SB3-10 percent concentration (Figure 5B).

The SB3-10-induced conformational change of TDP-43 NTD follows an apparent two-state equilibrium and can be fitted to the model edited by Santoro and Bolen [31]. The results of this analysis yielded values of 16.0 ± 1.6 kJ/mol, 16.2 ± 1.4 kJ/(mol M) and 0.98% (w/v) for ΔG_{U-F}^{H20}, m_{eq} and C_m, respectively. Spectra recorded at 0.6% SB3-10 at the same protein concentration, before and after the protein was incubated at 1.8% SB31-0 for 1 h, were essentially superimposable indicating the reversibility of the transition.

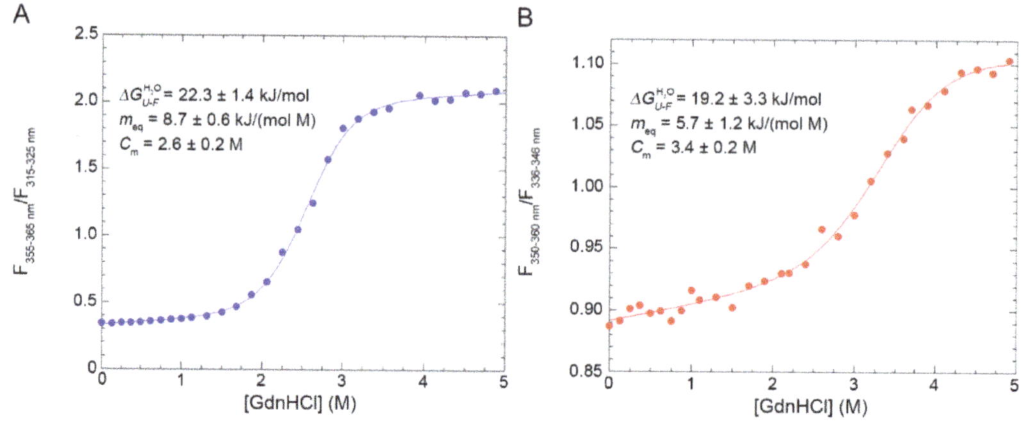

Figure 4. GdnHCl-induced denaturation curves of TDP-43 NTD. The curves refer to TDP-43 NTD in 0.0% (**A**) and 3.0% (**B**) SB3-10. All SB3-10 percentages are w/v. The ratio between the fluorescence values observed at the indicated wavelength windows, in which fluorescence emission was marked at high and low GdnHCl concentrations, respectively, was obtained at each GdnHCl concentration and plotted as a function of the GdnHCl concentration. Data were analysed with a best-fitting procedure using the two-state model equation [31]. The writings inside the panels indicate the values obtained for the analysed parameters.

In order to explore if the conformational transition observed for TDP-43 NTD upon addition of SB3-10 was due to the formation of micelles by this compound, the solution containing 3.0% SB3-10 without TDP-43 NTD was analysed using DLS (Figure 5C). The distribution of light scattering intensity vs. the apparent hydrodynamic diameter revealed that at 3.0% SB3-10 the formation of micelles was evident, with a monodispersed distribution and a hydrodynamic diameter of 3.87 ± 0.05 nm. The overall intensity of scattered light was one order of magnitude lower than that observed when the protein was present under the same solution conditions, ruling out that the size distribution reported for the protein in Figure 3A is dominated by SB3-10. The SB3-10 concentration-dependent formation of the micelles was then studied by monitoring the light scattering intensity of the solution with an increasing concentration of SB3-10 in the absence of protein (Figure 5D). The distribution of light scattering intensity vs. SB3-10 percent concentration revealed that micelles begin to form starting from a concentration of 0.6% and then increase stabilizing at 1.4%, with a midpoint around 1.0%. It therefore appears that the SB3-10-induced conformational transition of TDP-43 NTD monitored with intrinsic fluorescence and the formation of SB3-10 micelles monitored with light scattering are fully superimposable, indicating that the conformational transition of this protein domain is induced by the formation of SB3-10 micelles.

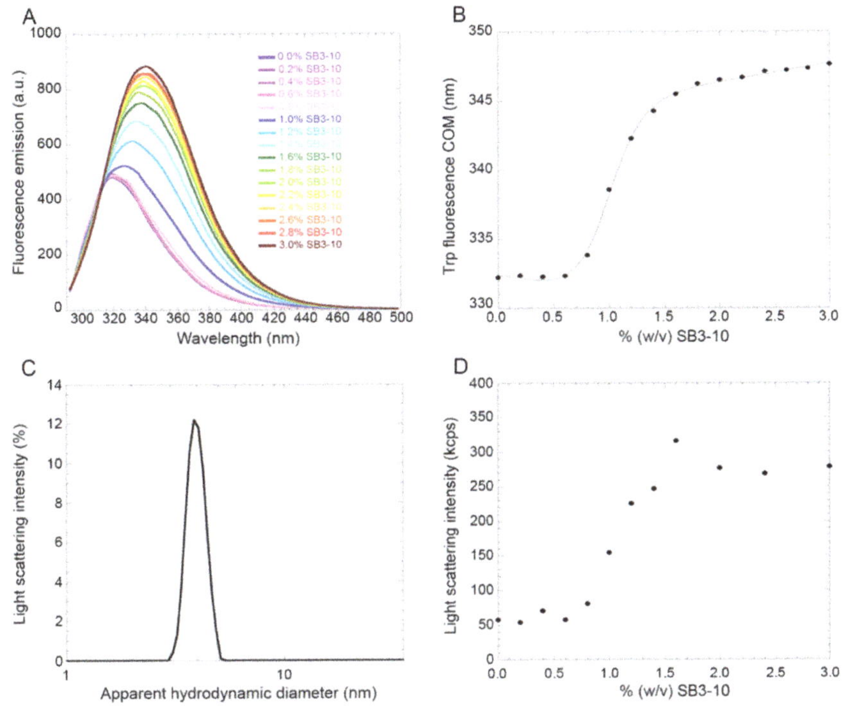

Figure 5. SB3-10-induced denaturation curve of TDP-43 NTD. (**A**) Fluorescence spectra of TDP-43 NTD (excitation 280 nm) at the indicated SB3-10 percentages (*w*/*v*). (**B**) Plot of spectral COM versus SB3-10 percentage (*w*/*v*). Data were analysed with a best-fitting procedure using the two-state model equation [31]. (**C**) Representative DLS distribution of a solution containing 3.0% (*w*/*v*) SB3-10 in the absence of protein. Similar distributions were obtained at all SB3-10 percentages analysed here, using a fixed attenuator and measurement position on the instrument. (**D**) Plot of light scattering intensity as a function of the SB3-10 percentage (*w*/*v*) in the absence of protein, using a fixed attenuator and measurement position on the instrument.

2.6. Alternative NTD Has a Lower Propensity to Aggregate

Native NTD is known to oligomerise in a head-to-tail fashion at sufficiently high concentrations, well beyond the dimer [19,20]. Such oligomers represent a functional state of the NTD [19,20], but extensive and uncontrolled oligomerisation leads liquid–liquid phase separation and possibly inclusion formation of the full-length protein [20,22]. Using analytical ultracentrifugation (AUC), an isodesmic dissociation constant of 40 μM was found for the native NTD oligomers [20]. In order to investigate the oligomerisation process of TDP-43 NTD in the absence of SB3-10, NTD was incubated at a concentration of 0.5 mg/mL (45 μM), in 5 mM sodium phosphate buffer, 50 mM NaCl, 1 mM DTT, pH 7.4, 25 °C, which are solution conditions known to promote oligomerisation [20,32].

Under these conditions, the size distribution by light scattering intensity showed initially (0 h) a significant amount of dimeric or low molecular weight oligomeric state but also high molecular weight species (Figure 6A,E). Since light scattering intensity scales with the square of the mass, the latter species are initially quantitatively negligible. However, prolonged incubation leads to the slow decay of the Gaussian curve associated with the dimer/oligomer, until it disappears after 3 h (Figure 6A,E). A similar decay was observed in the presence of small concentrations of SB3-10, up to 0.6%, that is in the pre-transition region of the conversion between the two conformational states studied here, when the protein domain maintains initially its native folded state (Figure 6B,F). The apparent rate constant of decay of the DLS peak did not change significantly with SB3-10 concentration (Figure 6I, r = 0.433, $p > 0.05$), nor was the light scattering intensity associated with this species at time 0 h found to correlate significantly with SB3-10 concentration (Figure 6I, r = 0.303, $p > 0.05$).

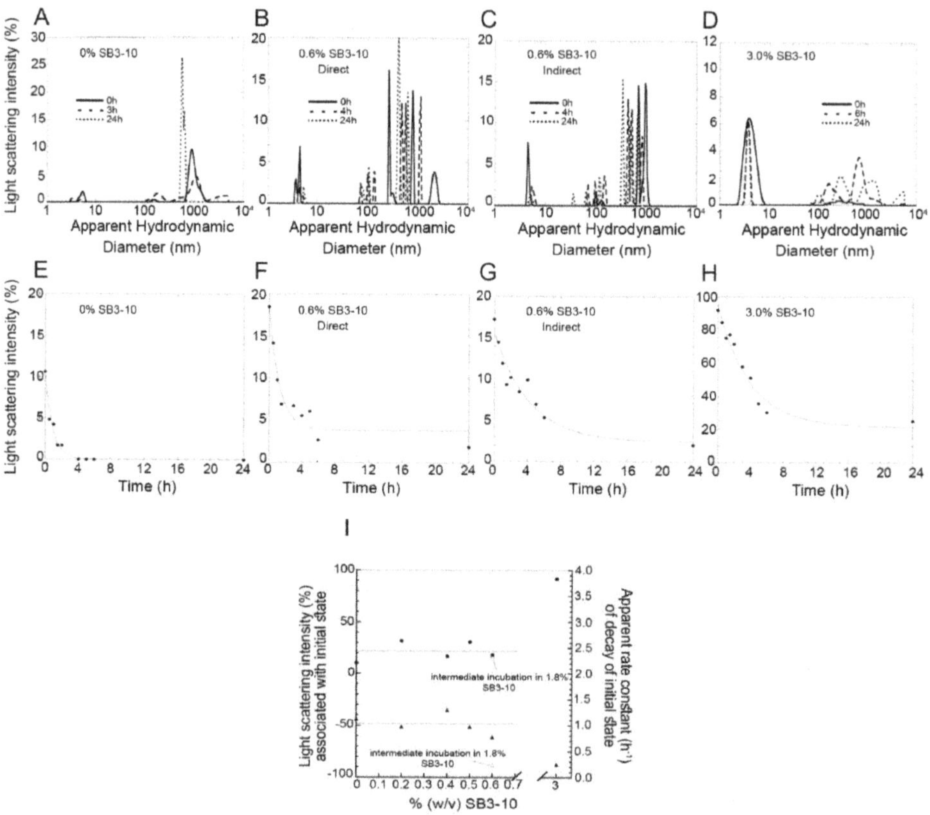

Figure 6. Aggregation kinetics of TDP-43 NTD at different SB3-10 percent concentrations monitored with DLS. (**A–D**) Size distributions of TDP-43 NTD (0.5 mg/mL, 45 μM) in 5 mM sodium phosphate buffer, 50 mM NaCl, 1 mM DTT, pH 7.4, 25 °C, in the presence of 0.0% (*w*/*v*) SB3-10 (**A**), 0.6% (*w*/*v*) SB3-10 (**B**,**C**), 3.0% (*w*/*v*) SB3-10 (**D**), after 0 h (continuous lines), 3–6 h (dashed lines) and 24 h (dotted lines). (**E–H**) Light scattering intensity associated with the initial state (low-molecular weight monomer/dimer/oligomer) of TDP-43 NTD versus time. Conditions as in the corresponding top panels. In panels (**C**), (**G**) TDP-43 NTD was pre-incubated in 1.8% (*w*/*v*) SB3-10 for 1 h and then diluted to 0.6% (*w*/*v*) SB3-10 at the same final conditions as in panels (**B**,**F**). (**I**) Light scattering intensity associated with initial state (circles), and apparent rate constant of decay of initial state (triangles), versus SB3-10 percentage (*w*/*v*). The empty symbols refer to the sample pre-incubated in 1.8% (*w*/*v*) SB3-10 for 1 h and then diluted to 0.6% (*w*/*v*) SB3-10. Horizontal lines indicate mean values. All data were acquired using a fixed attenuator and measurement position on the instrument.

In another experiment, NTD was treated differently: starting from the native conformation in 0.0% SB3-10, the protein domain was brought to 1.8% SB3-10 for 1 h, i.e., in the post-transition region in which it adopts the alternative monomeric conformational state, and then diluted down to 0.6% SB3-10, under final protein concentration and solution conditions identical to those of the previous experiments (Figure 6C,G). In this case, the decay of light scattering associated with the low molecular weight peak was slower, with a rate constant 4-fold lower than that observed under identical final conditions but without the intermediate incubation in 1.8% SB3-10 for 1 h (Figure 6I). The rate constants values were indeed $1.03 \pm 0.19\ h^{-1}$ and $0.24 \pm 0.04\ h^{-1}$, without and with preincubation in 1.8% SB3-10, respectively ($p < 0.01$). In other words, the NTD that adopted transiently the alternative monomeric conformation in 1.8% SB3-10 was found to aggregate, in 0.6% SB3-10, more slowly than the NTD brought to 0.6% SB3-10 directly, in spite of otherwise identical final conditions.

In the presence of 3.0% SB3-10, i.e., in the post-transition region when the protein adopts the alternative monomeric conformational state, the size distribution was dominated at time 0 h by a single monomeric peak, accounting for ca. 92% of light scattering intensity, which then disappeared slowly (Figure 6D,H). The decay occurred in this case with an apparent rate constant 4-fold lower ($0.25 \pm 0.04\ h^{-1}$) than that observed in 0.0–0.6% SB3-10 ($1.03 \pm 0.19\ h^{-1}$, $p < 0.01$), but apparently identical to that observed in 0.6% SB3-10 with intermediate incubation in 1.8% SB3-10 ($0.24 \pm 0.04\ h^{-1}$, $p > 0.05$) (Figure 6I).

2.7. *Aggregation of Either NTD Conformation Is Not Associated with Detectable Structural Variation*

In order to monitor the occurrence of possible structural changes of NTD during the time-dependent self-assembly of the dimer or low molecular weight oligomers monitored with DLS at low SB3-10 concentrations (0.0–0.6%) or of the monomeric alternative conformational state at higher SB3-10 concentrations (>1.6%) we recorded far-UV CD and intrinsic fluorescence spectra of the same protein samples analysed with DLS (same protein concentrations, solution conditions, time points).

The far-UV CD spectra obtained in 0.0%, 0.2% and 0.6% SB3-10 did not show significant variations with time; in agreement, intrinsic fluorescence did not change with time within this range of concentrations, maintaining the characteristic peaks of the native fully folded state (Figure S7A–D), thus indicating a native-like oligomerisation. Similarly, the far-UV CD spectra obtained in 1.8% and 3.0% SB3-10 did not show detectable changes with time, maintaining the characteristic peaks of the alternative partially folded state (Figure S7E,F). In 0.6% SB3-10 the spectra were similar without (Figure S7C) or with (Figure S7D) the intermediate 1 h incubation in 1.8% SB3-10, indicating the reversibility of secondary structure changes undergone by the protein.

The intrinsic fluorescence spectra in 0.0%, 0.2% and 0.6% SB3-10 maintained a λ_{max} value at ca. 319 nm during self-assembly, indicating the maintenance of the overall exposure of Trp68, typical of the fully folded state (Figure S8A–D). Similarly, the spectra in 1.8% and 3.0% SB3-10 maintained the λ_{max} value at ca. 339 nm during aggregation, which is typical of the alternative conformational state, indicating again a process of self-assembly in which the protein molecules maintained a similar exposure of Trp68 (Figure S8E,F). Similar to the CD analysis, in 0.6% SB3-10 the spectra were similar without (Figure S8C) or with (Figure S8D) the intermediate 1-h incubation in 1.8% SB3-10. A slight progressive decrease in intrinsic fluorescence was observed under all the conditions studied, attributable to the increasing light scattering as larger assemblies accumulate.

2.8. *NTD Aggregation Does Not Lead to Formation of Cross-β Structure*

We then assayed the ability of the TDP-43 NTD samples, obtained under the various conditions tested, to bind the ThT dye and increase its fluorescence at 485 nm, which is typical of amyloid protein aggregates. The sample in 0.0% SB3-10 was found to have a fluorescence value F identical to that of free ThT in the absence of protein F_0 (Figure 7). A very small increase in ThT fluorescence F relative to the F_0 value was observed in all the remaining conditions after 6 h, but the increase (F_0/F) was generally lower than 1.25 (Figure 7). The samples in 3.0% SB3-10 aged beyond 6 h were found to increase the ThT fluorescence to higher extents, typically between 1.25 and 1.4 (Figure 7). This is in sharp contrast with the over 5-fold fluorescence increase expected for amyloid [33,34], ruling out the formation of amyloid-like species for TDP-43 NTD under these experimental conditions.

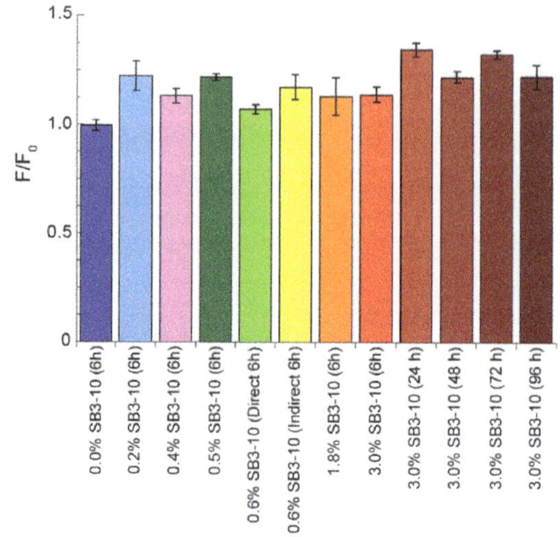

Figure 7. Thioflavin T (ThT) assay of TDP-43 NTD at different SB3-10 percent concentrations. TDP-43 NTD (0.5 mg/mL, 45 μM) in 5 mM sodium phosphate buffer, 50 mM NaCl, 1 mM DTT, pH 7.4, 25 °C, in the presence of the indicated SB3-10 percent concentrations (w/v) and time points. ThT assay was performed with excitation at 440 nm and fluorescence emission at 450–600 nm and reported as the ratio between ThT fluorescence at 485 nm in the presence (F) and absence (F_0) of TDP-43 NTD (n = 3, mean ± SEM). Differences were in some cases statistically significant relative to the first sample in 0.0% SB3-10 at 6 h but considered to be non-relevant because an increase of the F/F_0 value is expected to be at least 5-fold for amyloid (see text for details).

3. Discussion

The NTD of TDP-43 is able to dimerise with a head-to-tail arrangement into a dimeric structure that has been solved by both NMR and X-ray crystallography [19,20]. Following the head-to-tail interaction, this process does not terminate at the dimeric level but proceeds to form larger oligomers [19,20]. The role of NTD dimerisation and oligomerisation in the aggregation process of full-length TDP-43 is still discussed. It has been proposed to be responsible for the oligomerisation of the entire full-length protein [12,14–20]. Oligomerisation of full-length TDP-43 mediated by the NTD is essential for TDP-43 function [14,15,19,20], but also favours liquid–liquid phase separation and solid-phase inclusion formation of the full-length protein [20–22]. The structural plasticity of the TDP-43 NTD resulting in its in dimerisation/oligomerisation is also observed in its folding process from a fully unfolded state, in which the protein domain forms a number of partially folded states before achieving the fully folded dimeric structure, and even populates, at low denaturant concentrations, a native-like dimeric state distinct from the fully native dimeric conformation [24]. Such conformational heterogeneity is also witnessed by NMR spectral differences in the various experimental conditions in which the protein domain is folded [10,12,14,17,18,20].

The use of very small SB3-10 percent concentrations, typically in the range of 1.4–3.0% (w/v), has allowed us to isolate a stable and well-defined alternative conformation. While the native conformation in the absence of SB3-10 presents a far-UV CD spectrum comparable to those reported in the literature [14,18,24], the alternative one has a different spectrum characterized by a higher content of α-helical structure, a lower content of β-sheet structure and a slightly higher content of turns. NMR spectroscopy confirmed these observations (discussed below). The intrinsic fluorescence spectrum and Stern–Volmer assay also indicate that the Trp68 residue is more exposed to the solvent in this new conformational state relative to the native structure. The DLS, SEC and FRET analyses all indicate that the NTD in 3.0% SB3-10 is a monomer, unlike the native dimer populated in the absence of this detergent but under identical conditions in terms of protein concentration and solution conditions.

The NMR spectral properties of the SB3-10 dependent conformational state differ from those previously observed for the native NTD, as it clearly appears from the low resonance spreading of its HSQC spectrum. In particular, the NMR characterisation indicates the presence of a partially unfolded state together with the conservation of the native α-helix and of the native antiparallel β-sheet contributed by strands 1-2, although the latter is remarkably more dynamic than the native one (Figure 8A,B). On the basis of chemical shift values, two other non-native α-helices located in the N- and C-termini are detected encompassing residues 1–4 and 68–77, respectively (Figure 8A,B). Moreover, the first part of the central native helix along with the preceding β-strand (residues 24–29) and other spatially close residues that in the native structure faces this portion (residues 59,60,63) form a cluster that is present in two different forms in equilibrium, named here A and B (Figure 8A,B). The intensity of the two forms is temperature dependent, suggesting that the two forms might be related to two conformers with different compactness. For some residues, the two forms show an NH combined chemical shift difference quite marked (between 0.25 and 0.94 ppm) indicating a different chemical environment that reflects distinct conformations and also indicates structure formation under these conditions for at least one form.

The alternative conformation adopted in 3.0% (w/v) SB3-10 is able to return to its initial native conformation if diluted to a smaller concentration of detergent, conditions in which the native conformation is stable, showing how this process is fully reversible. Its conformational stability (ΔG^{H20}_{U-F}) in 3.0% SB3-10 is only approximately 3 kJ/mol lower than that of the native conformation in 0.0% SB3-10, as determined with GdnHCl-induced equilibrium denaturation. The m_{eq} value, which reports on the cooperativity of the GdnHCl-induced unfolding transition and on the change of solvent-exposed surface area (ΔASA) upon unfolding, is 35 ± 15% lower than that of the native state but is sufficiently high to indicate a cooperative transition. The lower m_{eq} value suggests, however, significantly lower cooperativity, possibly due to the increase of the surface area exposed to the solvent in the folded alternative conformation compared to the folded native conformation.

In our time-dependent aggregation assays, we observed a slow, yet detectable, aggregation in the sample left untreated with SB3-10 and in those samples in which the concentration of the detergent was kept low. In both cases, the disappearance of the low molecular weight species (either monomeric, dimeric or oligomeric) was not accompanied by detectable changes in the far-UV CD and intrinsic fluorescence spectra, indicating the maintenance of the initial conformational states in both cases and further supporting the notion that TDP-43 NTD self-assembly consists in a native-like aggregation process similar to that observed for other proteins [35–39]. Aggregation is slower in 3.0% (w/v) SB3-10, but this effect cannot be attributed necessarily to the SB3-10-mediated conformational change of the TDP-43 NTD, as it may well arise from the effect of the detergent in the formation of intermolecular interactions. However, when the aggregation processes were compared with and without pre-incubation with 1.8% (w/v) SB3-10, but with an identical final SB3-10 concentration of 0.6% (w/v) promoting the native state, a slower aggregation was observed in the NTD sample pre-incubated in SB3-10, in spite of the rapid reversibility of the conformational change and the native conformation adopted by the protein domain in both cases. This can be attributed to the ability of the condition promoting temporarily the alternative conformation to monomerise the protein and disrupt any intermolecular interaction (Figure 8C). Under physiological conditions, the native monomers, dimers and oligomers exist initially in a pre-equilibrium where the mutual conversions are slow (Figure 8, left). This condition is favourable for aggregation as dimers and oligomers are present already. The alternative state has a very low propensity to oligomerise (Figure 8C, centre), but the lower propensity of the native state to oligomerise and self-assemble is maintained when the alternative state is re-located under native conditions to form the native state (Figure 8C, right) because the conversion to dimers and oligomers is slow.

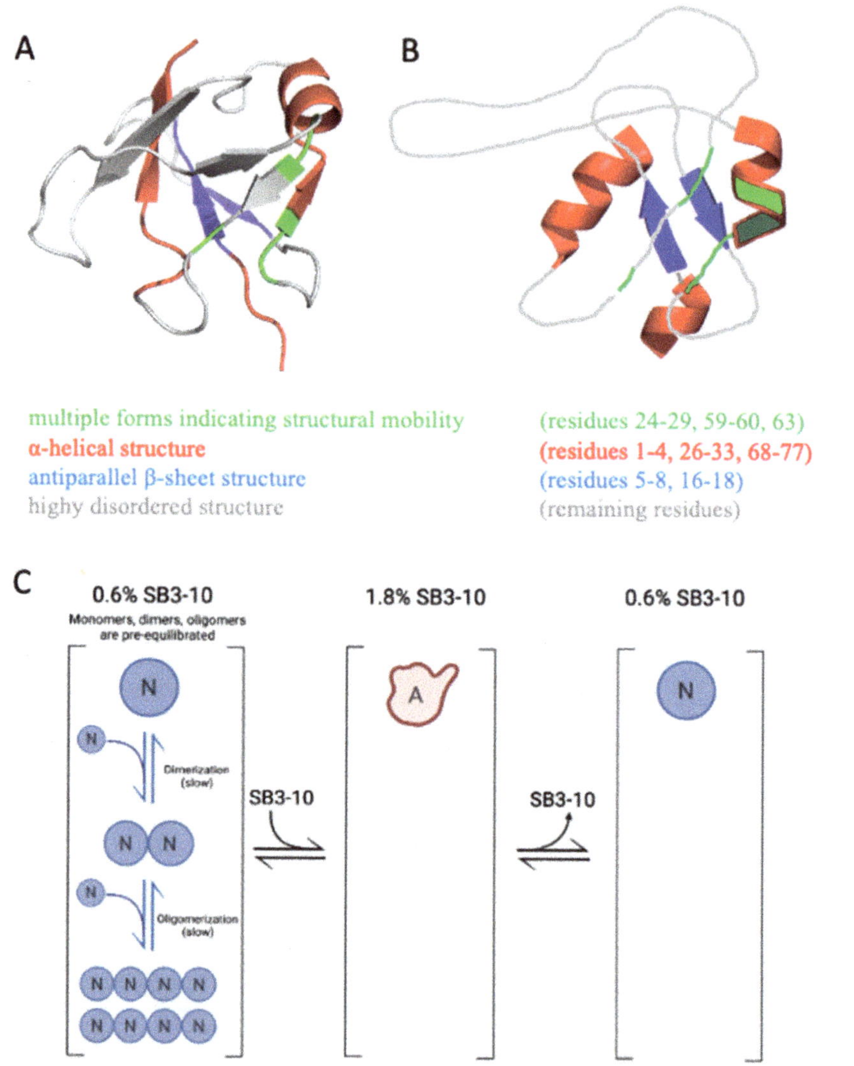

Figure 8. (**A**) NMR structure of monomeric native TDP-43 NTD (PDB ID 5x4f) colour coded to illustrate residues adopting multiple forms (green), α-helical structure (red), antiparallel β-sheet structure (blue) and highly disordered structure or loops (grey) in the alternative conformation studied here in 3% (w/v) SB3-10. (**B**) Proposed model of the same alternative conformation of TDP-43 NTD in 3% (w/v) SB3-10, colour coded as in panel (**A**). Orientation of helices, β-sheet and loops in the images is arbitrary as the relative orientation of secondary structure elements has not been attempted. (**C**) Scheme showing that TDP-43 NTD has a low propensity to dimerise/oligomerise in the alternative conformation. N and A indicate the native and alternative conformational states, respectively. N monomers, dimers and oligomers are initially pre-equilibrated (slow conversions). A is a monomer. The lower propensity of TDP-43 NTD to oligomerise and self-assemble is maintained when A is re-located in native conditions to form N (right) because the conversion to dimers and oligomers is slow.

Under conditions in which the alternative conformational state was observed (3.0% SB3-10), the critical micellar concentration (CMC) of the detergent was largely exceeded, leading to hypothesize that the NTD conformational change was induced by the newly formed SB3-10 micelles. The SB3-10 titration of the TDP-43 NTD structural change monitored spectroscopically, on the one hand, and of the SB3-10 micelle formation monitored with light scattering intensity, on the other hand, indicate that the two transitions coincide as they start and end at identical SB3-10 concentrations and also have similar transition midpoints. This indicates that the structural change of TDP-43 NTD is driven by SB3-10 micelle formation. Therefore, the role of biological micelles formed intraneuronally can be decisive in this regard and promote structural conversions into alternative conformations of TDP-43 NTD that may resist oligomerisation, LLPS and inclusion formation, also affecting the full-length protein.

4. Conclusions

The results presented here show that small concentrations of SB3-10 are able to promote an alternative conformational state of the NTD of TDP-43, highlighting its high structural plasticity. This new conformational state exhibits a different secondary structure, hydrophobic packing, size and oligomerisation state from the fully native state, but presents a fold with cooperativity and conformational stability only slightly lower than those of the native state. The solution conditions explored here represent a valid strategy to stabilize the NTD of TDP-43 as a monomer and this may be helpful to gain a better understanding of TDP-43 biological functions and its role in ALS and FTLD pathology. Furthermore, we speculate that the alternative NTD conformation might also form in vivo since biological micelles are normally found in neurons. This may be of physiological relevance, as this distinct conformational state might act differently in terms of functional oligomerisation and pathological aggregation also in vivo, although this awaits experimental demonstration. It could be interesting to further investigate the role of micelles in the monomerization and conformational change of the TDP-43 NTD and to explore if they are able to affect also the full-length protein. Our ability to stabilize TDP-43 in the monomer state is achievable and its employment for studies and approaches in vivo could be useful for future scientific applications.

5. Materials and Methods

5.1. Chemicals

SB3-10, acrylamide and ThT were from Sigma-Aldrich (St. Louis, MO, USA). DTT was from Thermo Fisher Scientific (Waltham, MA, USA).

5.2. Gene Cloning, Expression and Purification

Gene cloning, expression and purification of TDP-43 NTD and the Cys50Ser (C50S) single-point mutant were performed as previously described [24]. The purified protein contained 77 residues and the MHHHHHHSSGVDLGTENLYFQS sequence fused to the N-terminus for a total of 99 residues. It was maintained at 1.6–3.0 mg/mL (150–270 μM) in 5 mM sodium phosphate buffer, 50 mM NaCl, 1 mM DTT, pH 7.4, −20 °C. Protein purity was checked with SDS page. Protein concentration was measured using a SHIMADZU UV-1900 UV-Vis spectrophotometer at a wavelength of 280 nm with an extinction coefficient (ε_{280}) of 12,950 M^{-1} cm^{-1}.

5.3. Far-UV Circular Dichroism Spectroscopy

The TDP-43 NTD sample was centrifuged at 18,000× *g* for 15 min, 4 °C, and diluted to prepare two samples containing 0.5 mg/mL (45 μM) NTD, in 5 mM sodium phosphate buffer, 50 mM NaCl, 1 mM DTT, pH 7.4, one in the absence and the other in the presence of 3.0% (*w*/*v*) SB3-10. Spectra were acquired at 25 °C in the far-UV between 190 and 260 nm using a 0.1 mm path length cell on a Jasco J-810 spectropolarimeter (Tokyo, Japan) equipped with a thermostated cell holder attached to a Thermo Haake C25P water bath (Karlsruhe,

Germany). Spectra were then blank subtracted, truncated when the high tension (HT) signal was higher than 700 V and normalized to mean residue ellipticity using

$$[\theta] = \frac{\theta}{\left(\frac{10 \times N\ residues \times optical\ path \times concentration}{molecular\ weight}\right)}$$

where $[\theta]$ is the mean residue ellipticity in deg cm^2 dmol^{-1}, θ is the ellipticity in mdeg, the optical path is in cm, concentration is in g/L, and the molecular weight is in g/mol.

5.4. Fluorescence Spectroscopy

The TDP-43 NTD sample was centrifuged at 18,000× *g* for 15 min, at 4 °C, and two samples containing 0.5 mg/mL (45µM) NTD, in 5 mM sodium phosphate buffer, 50 mM NaCl and 1 mM DTT, pH 7.4, were prepared, one in the absence and the other in the presence of 3.0% SB3-10. Fluorescence spectra were acquired at 25 °C from 290 to 500 nm (excitation at 280 nm) using a 3 × 3 mm black wall quartz cell cuvette on an Agilent Cary Eclipse spectrofluorimeter (Agilent Technologies, Santa Clara, CA, USA) equipped with a thermostated cell holder attached to an Agilent PCB 1500 water Peltier system (Agilent Technologies, Santa Clara, CA, USA). Excitation and emission slits were 5 nm. Spectra were then blank subtracted.

5.5. Acrylamide Quenching Experiment

The TDP-43 NTD sample was centrifuged at 18,000× *g* for 15 min, at 4 °C, and two 1 mL samples containing 0.05 mg/mL (4.5 µM) NTD, in 5 mM sodium phosphate buffer, 50 mM NaCl and 1 mM DTT, pH 7.4, were prepared, one in the absence and the other in the presence of 3.0% SB3-10. For each sample, a first fluorescence spectrum was acquired at 25 °C from 300 to 420 nm (excitation at 280 nm) using a 10 × 4 mm quartz cuvette under magnetic stirring on an Agilent Cary Eclipse spectrofluorimeter (Agilent Technologies, Santa Clara, CA, USA) equipped with a thermostated cell holder attached to an Agilent PCB 1500 water Peltier system. Excitation and emission slits were 5 nm. Then, 5 µL of 1 M acrylamide was added directly to the cuvette and another spectrum was recorded. This step was repeated 15 times. For each recorded spectrum, the total fluorescence was corrected to take account of the dilution with the acrylamide solution. The quenching of acrylamide in the 0.0% SB3-10 sample was analyzed with the Stern-Volmer equation

$$\frac{F_0}{F} = 1 + K_{SV}[Q]$$

where F_0 and F are the integrated fluorescence intensity areas at 300–400 nm in the absence and presence of acrylamide, respectively, K_{SV} is the Stern-Volmer constant and $[Q]$ is the concentration of the quencher (acrylamide) in the cuvette. The quenching of acrylamide in the 3.0% SB3-10 sample was instead analysed using

$$\frac{F_0}{F} = 1 + K_{SV}[Q]\ e^{K_{ST} \times [Q]}$$

where K_{ST} is a constant that considers the static quenching caused by the binding of acrylamide to tryptophan residues.

5.6. Nuclear Magnetic Resonance Spectroscopy

NMR spectra were acquired with a Bruker Avance 500 MHz and a Bruker Avance Neo 700 MHz NMR spectrometers on ^{13}C, ^{15}N uniformly labelled TDP-43 NTD samples (250 µM) in 5 mM sodium phosphate, 50 mM NaCl, 3% (*w*/*v*) SB3-10, 1 mM DTT, 5% (*v*/*v*) D_2O. For reference purposes, trimethylsilylpropanoic acid (TSP) was added [40]. For backbone assignment a series of 2D HSQC and 3D HNCA, HN(CO)CA, HNCACB, HNCO, and HN(CA)CO spectra were acquired at 25 °C and 17 °C with 2048 × 256, 2048 × 50 × 96, 2048 × 50 × 96, 2048 × 50 × 128, 2048 × 50 × 72, 2048 × 50 × 72 complex points, respectively. The spectral widths were 13.8 ppm (^{1}H), 35 ppm (^{15}N) and 30 ppm (^{13}C in HNCA and HN(CO)CA), 16 ppm (^{13}C in HNCO and HN(CA)CO), 80 ppm (^{13}C in HNCACB). Secondary structures of individual residues were assessed by chemical shift index (CSI) [27], Talos+ [26] and Secondary Structure Propensity (SSP) [28]. According to Wishart [27], helix (H) or ®-strand (E) were indicated in Table S1 only if present a consensus for CSI for Cα, Cβ, C®, and CO. A series of 2D HSQC spectra were acquired at different temperature values (290–312K every 2 K). 2D ^{15}N{^{1}H} NOE measurements were also performed using a 700 MHz spectrometer at 25 °C and 17 °C, with a 5 s relaxation or saturation delay, 256 t_1 increments of 2048 complex data points, and 48 scans/t_1. The spectra were processed with Topspin 3.5 (Bruker Biospin) and analysed in Sparky [41]. NOE values were calculated as the ratio of peak height with and without saturation.

5.7. Dynamic Light Scattering

The TDP-43 NTD sample was centrifuged at 18,000× *g* for 15 min, at 4 °C, and filtered with Whatman Anotop 0.02 µm cut-off filters. Two samples containing 1.35 mg/mL (121 µM) NTD, in 5 mM sodium phosphate buffer, 50 mM NaCl and 1 mM DTT, pH 7.4, were prepared, one in the absence and the other in the presence of 3.0% SB3-10. Their size distributions (distribution of apparent hydrodynamic diameter by light scattering intensity) were then acquired on a Malvern Panalytical Zetasizer Nano S DLS device (Malvern Panalytical, Malvern, UK), thermostated at 25 °C with a Peltier temperature controller using a 3 × 3 mm black wall quartz cell cuvette. The refractive index and viscosity, acquired using a 2WAJ ABBE bench refractometer from Optika Microscopes (Bergamo, Italy) and a Viscoball viscometer (Fungilab, Barcelona, Spain), were 1.331 and 0.8998 cp for the 0.0% SB3-10 sample and 1.338 and 1.0530 cp for the 3.0% SB3-10 sample, respectively. The measurements were acquired with the cell position 4.20 and attenuator index 10. The light scattering intensity was also measured for a blank sample containing 3.0% SB3-10 without protein and was found to be negligible relative to the corresponding sample with protein, ruling out that the apparent hydrodynamic diameter measured for the protein sample under these conditions is affected by SB3-10 micelles. In another experiment, the same buffer solutions containing SB3-10 concentrations ranging from 0.0 to 3.0%, in the absence of protein were also analysed. Their light scattering intensities were then plotted as a function of SB3-10 concentration to monitor micelle formation.

5.8. Analytical SEC

The mother solution containing TDP-43 NTD was centrifuged at 18,000× *g* for 15 min, at 4 °C, and two samples containing 0.5 mg/mL (45µM) NTD, in 5 mM sodium phosphate buffer, 50 mM NaCl and 1 mM DTT, pH 7.4, were prepared, one in the absence and the other in the presence of 3.0% SB3-10. 100 µL of NTD sample were loaded in a Superdex® 200 Increase 10/300 GL column (GE Healthcare, Chicago, IL, USA) pre-equilibrated at 4 °C with the corresponding protein buffer and run using an Akta Pure 25L System (GE Healthcare, Chicago, IL, USA). The experimental V_e was determined as the volume of the buffer passed through the column between sample injection and the point of highest absorbance at 280 nm. A calibration curve was determined by loading protein standards (100 µL) of known molecular weight (MW) separately, such as thyroglobulin (669 kDa), apoferritin (443 kDa), alcohol dehydrogenase (150 kDa), albumin (66 kDa), carbonic anhydrase (29 kDa) and α-lactalbumin (14.2 kDa), and obtaining their experimental V_e. A plot of V_e versus log[MW] was obtained with the 6 standard data points and fitted to a linear regression curve. This calibration curve was then used to interpolate the experimental V_e (y axis) of NTD samples and to obtain the corresponding MW (x axis) and, thus, their oligomeric states. The MW values determined with this approach are valid only for globular proteins. For the NTD sample in 3.0% SB3-10, the calibration curve was re-determined mathematically to account for the differences between the hydrodynamic radii of folded and pre-molten globule states [42]. Indeed, the relationships between the hydrodynamic radii of a folded protein (R_h^{N}), or a pre-molten globule (R_h^{PMG}), and the MW in Da, can be obtained using previously published equations [41]. The experimental V_e obtained for the NTD sample in 3.0% SB3-10 was interpolated into the re-determined calibration curve to obtain its MW.

5.9. Förster Resonance Energy Transfer (FRET)

The C50S mutant of TDP-43 NTD (containing only Cys39) was centrifuged at 18,000× g for 15 min, 4 °C, and diluted to prepare two samples containing 0.5 mg/mL (45 μM) C50S NTD, in 5 mM sodium phosphate buffer, 50 mM NaCl, 1 mM DTT, pH 7.4. The C50S NTD samples were labelled using the thiol-reactive probes 1,5-IAEDANS as a donor (D) and 6-IAF as an acceptor (A), respectively. Both fluorescent probes (Thermo Fisher Scientific, Waltham, MA, USA) were dissolved in dimethylformamide (DMF) at 30 mM concentration and added separately to the two C50S samples to a final probe:protein molar ratio of 10:1, for 22 h at 20 °C, under gentle and constant shaking. Then, gravity chromatography was performed to remove the excess of the unreacted probes using 6 mL of G-15 resin (Pharmacia, Uppsala, Sweden), previously equilibrated with 5 mM sodium phosphate buffer, 50 mM NaCl, pH 7.4.Labelledd protein fractions were collected and the concentration of the probe in each fraction was measured on a SHIMADZU UV-1900 UV-Vis spectrophotometer using ε_{336} = 5700 $M^{-1}cm^{-1}$ and ε_{491} = 8200 $M^{-1}cm^{-1}$ for 1,5-IAEDANS and 6-IAF, respectively. The C50S NTD concentration was determined using ε_{280} = 12,950 $M^{-1}cm^{-1}$ after subtraction of the absorbance contribution of the probes at 280 nm. Fractions where the probe:protein ratio was 1:1 were pooled and concentrated using centrifugal filter devices with a 3 kDa molecular weight cut-off (MWCO) cellulose membrane (Millipore, Burlington, MA, USA).

Two samples containing 0.1 mg/mL (9 μM) C50S NTD labelled with D and A, respectively, were prepared in 5 mM sodium phosphate buffer, 50 mM NaCl, 1 mM DTT and 3% SB3-10, pH 7.4, and mixed at 1:1 molar ratio (D:A). Two samples containing only C50S NTD labelled with D and C50S NTD labelled with A at the concentration of 0.05 mg/mL (4.5 μM) were also prepared. Fluorescence spectra of the various samples were acquired at 25 °C from 400 to 650 nm (excitation at 336 nm) using a 10 × 2 mm quartz cuvette on an Agilent Cary Eclipse spectrofluorimeter (Agilent Technologies, Santa Clara, CA, USA) equipped with a thermostated cell holder attached to an Agilent PCB 1500 water Peltier system. Excitation and emission slits were 5 nm. Spectra were then blank-subtracted.

5.10. SB3-10 Titration on NTD

The TDP-43 NTD sample was centrifuged at 18,000× g for 15 min, at 4 °C. The experiment was performed by preparing 16 samples containing NTD at the concentration of 0.5 mg/mL (45μM), in 5 mM sodium phosphate buffer, 50 mM NaCl and 1 mM DTT, pH 7.4, and SB3-10 concentrations ranging from 0.0 to 3.0%. Fluorescence spectra were acquired at 25 °C from 290 to 500 nm (excitation at 280 nm) using a 3 × 3 mm black wall quartz cell cuvette on an Agilent Cary Eclipse spectrofluorimeter (Agilent Technologies, Santa Clara, CA, USA) equipped with a thermostated cell holder attached to an Agilent PCB 1500 water Peltier system. Excitation and emission slits were 5 nm. Spectra were then subtracted from blanks containing only buffers. For each spectrum, COM was calculated according to

$$\mathrm{COM} = \frac{\sum_i F_i}{\sum_i v_i \cdot F_i}$$

where F_i is the fluorescence emission at a wavenumber of ν_i. The resulting COM values were then plotted as a function of SB3-10 concentration and fitted using the model edited by Santoro and Bolen [31].

5.11. Guanidine Hydrochloride (GdnHCl)-Induced Denaturation

The TDP-43 NTD sample was centrifuged at 18,000× g for 15 min, at 4 °C. The experiment was performed by preparing 30 samples containing NTD at the concentration of 0.05 mg/mL (4.5 μL), in 5 mM sodium phosphate buffer, 50 mM NaCl and 1 mM, pH 7.4, and GdnHCl concentrations ranging from 0.0 to 4.9 M. In a second experiment,

30 additional samples were prepared in the presence of 3.0% SB3-10. Fluorescence spectra were acquired at 25 °C from 290 to 500 nm (excitation at 280 nm) using a 10 × 2 mm quartz cuvette on an Agilent Cary Eclipse spectrofluorimeter (Agilent Technologies, Santa Clara, CA, USA), equipped with a thermostated cell holder attached to an Agilent PCB 1500 water Peltier system. Excitation and emission slits were 5 nm. Spectra were then subtracted from blanks containing only buffers. For each spectrum, the ratio between the sum of the fluorescence in bands of 10 nm in the post-transition region and the pre-transition region was calculated. These specific bands were chosen to contain the NTD fluorescence peak observed at 0.0 M GdnHCl and 4.9 M GdnHCl for the pre-transition and the post-transition, respectively. The resulting fluorescence ratios were then plotted as a function of the GdnHCl concentration and the traces obtained were then fitted to the model edited by Santoro and Bolen [31].

5.12. *Aggregation Kinetics Using DLS, Far-UV CD and Intrinsic Fluorescence Spectroscopies*

The TDP-43 NTD sample was centrifuged at 18,000× *g* for 15 min, at 4 °C. Samples containing NTD at the concentration of 0.5 mg/mL (45 μM), in 5 mM sodium phosphate buffer, 50 mM NaCl, 1 mM DTT, pH 7.4, 25 °C were prepared using different SB3-10 concentrations (0.0, 0.2, 0.4, 0.6, 1.8 and 3.0%). One sample was obtained by treating the protein, initially in 0.0% SB3-10, with 1.8% SB3-10 for 1 h and then diluting it down to 0.6% SB3-10. Samples were analyzed every 30 min for the first 2 h, then hourly up to 6 h and the next day at 24 h. Their size distributions were acquired at 25 °C using a 3 × 3 mm black wall quartz cell cuvette on a Malvern Panalytical Zetasizer Nano S DLS device (Malvern, Worcestershire, UK), thermostated at 25 °C with a Peltier temperature controller. The refractive index and viscosity set on the instrument were changed according to the SB3-10 content of the sample. The measurements were acquired with the cell position 4.20 and attenuator index 10. The light scattering intensity percentage of the soluble protein peak was then calculated for each sample and plotted as a function of time. Their far-UV CD spectra were acquired at 25 °C between 190 and 260 nm using a 0.1 mm path length cell on the same Jasco J-810 spectropolarimeter described above. Spectra were then blank subtracted, masked when the high tension (HT) signal was higher than 700 V and normalized to mean residue ellipticity ([θ]). Their fluorescence spectra were acquired at 25 °C from 290 to 500 nm (excitation at 280 nm) using a 10 × 2 mm quartz cuvette on an Agilent Cary Eclipse spectrofluorimeter (Agilent Technologies, Santa Clara, CA, USA), equipped with a thermostated cell holder attached to an Agilent PCB 1500 water Peltier system. Excitation and emission slits were 5 nm. Spectra were then blank subtracted.

5.13. *Thioflavin T (ThT) Assay*

TDP-43 NTD samples were prepared at various SB3-10 concentrations, as described in the previous subsection and then incubated for 6 h at 25 °C, while the sample in 3.0% SB3-10 was incubated also for 24, 48, 72 and 96 h. 100 μL of buffer or protein sample were then mixed with 400 μL of 25 μM ThT. Fluorescence spectra were acquired at 25 °C from 450 to 600 nm (excitation at 440 nm) using a 10 × 2 mm quartz cuvette on an Agilent Cary Eclipse spectrofluorimeter (Agilent Technologies, Santa Clara, CA, USA) equipped with a thermostated cell holder attached to an Agilent PCB 1500 water Peltier system. Excitation and emissions slits were 5 and 10 nm, respectively. All spectra were blank subtracted (using only PBS as a blank). The ratio F_0/F was then calculated for each sample, where F and F_0 are the blank-subtracted fluorescence values at 485 nm of ThT+protein and free ThT, respectively. An over 5-fold ThT fluorescence increase was considered to be diagnostic for amyloid [33,34,43,44].

5.14. *Statistical Analysis*

Data were expressed as means ± SEM. Data pairs were compared using the Student's *t*-test and the resulting *p* values were indicated in the text for each comparison. A *p* value < 0.05 was considered to be statistically significant.

Supplementary Materials: The following supporting information can be downloaded at: https://www.mdpi.com/article/10.3390/molecules27134309/s1, Table S1: Comparison between secondary structures of native TDP-43 NTD and the alternative conformation. Comparison performed according to CSI and Talos+ analysis and NH involvement into H bonds as inferred from temperature coefficients. H stands for α-helix, E for extended β-strand and x indicates uncertainty; Figure S1: Change of 1H-15N HSQC spectrum upon addition of 3% SB3-10. Overlay of 1H-15N HSQC spectra of TDP-43 NTD recorded at 500 MHz and 25 °C in absence (green) and presence (red) of 3% (*w*/*v*) SB3-10. The superimposition of the two spectra clearly shows the reduced spectral dispersion in the presence of 3% SB3-10; Figure S2: Change of 1H-15N HSQC spectrum upon temperature decrease. Overlay of 1H-15N HSQC spectra of TDP-43 NTD recorded at 700 MHz at 25 °C (red) and 17 °C (blue) in the presence of 3% (*w*/*v*) SB3-10. The superimposition of the two spectra shows the increase in the number of backbone amide peaks at lower temperature; Figure S3: Change of 1H-15N HSQC spectrum upon DTT addition (A) Superimposition of 1H-15N HSQC spectra of the TDP-43 NTD alternative conformation recorded at 700 MHz and 25 °C before (black) and after (red) the addition of freshly prepared DTT. The blue circles highlight peaks where the second form is highly decreased after DTT addition. (B) Native TDP-43 NTD protein cartoon (pdb 5x4f) showing the residues presenting double forms in the HSQC spectrum that are not affected by DTT addition coloured in red; Figure S4: SSP analysis of alternative TDP-43 NTD conformation for both form A (A) and form B (B). SSP indicates the tendency to adopt secondary structure (SSP > 0 corresponds to α-helix, SSP < 0 corresponds to β-strand). A SSP score of ± 1 corresponds to fully formed secondary structure, while an intermediate SSP value indicate a propensity for those residues to adopt the corresponding secondary structure; Figure S5: Difference between NOE values found at different temperature values. The differences were obtained at 17 °C (290 K) and 25 °C (298 K) for the TDP-43 NTD alternative form in 3% (*w*/*v*) SB3-10. The black, blue and red bars correspond to both forms, form A and form B, respectively; Figure S6: Overlay of 1H-15N HSQC spectra of TDP-43 NTD. Spectra are in 3% (*w*/*v*) SB3-10 recorded at 700 MHz at different temperatures: 17 °C (blue), 23 °C (green), 29 °C (yellow), 35 °C (orange) and 39 °C (red); Figure S7: Aggregation kinetics of TDP-43 NTD at different SB3-10 percent concentrations monitored with far-UV CD. Far-UV CD spectra of TDP-43 NTD (0.5 mg/mL, 45 μM) in 5 mM sodium phosphate buffer, 50 mM NaCl, 1 mM DTT, pH 7.4, 25 °C, in the presence of 0.0% (*w*/*v*) SB3-10 (A), 0.2% (*w*/*v*) SB3-10 (B), 0.6% (*w*/*v*) SB3-10 (C,D), 1.8% (*w*/*v*) SB3-10 (E), 3.0% (*w*/*v*) SB3-10 (F), at the indicated time points. In panel D TDP-43 NTD was pre-incubated in 1.8% (*w*/*v*) SB3-10 for 1 h and then diluted to 0.6% (w/v) SB3-10 at the same final conditions as in panel (C); Figure S8: Aggregation kinetics of TDP-43 NTD at different SB3-10 percent concentrations monitored with intrinsic fluorescence. Intrinsic Fluorescence spectra of TDP-43 NTD (0.5 mg/mL, 45 μM) in 5 mM sodium phosphate buffer, 50 mM NaCl, 1 mM DTT, pH 7.4, 25 °C, in the presence of 0.0% (*w*/*v*) SB3-10 (A), 0.2% (*w*/*v*) SB3-10 (B), 0.6% (*w*/*v*) SB3-10 (C,D), 1.8% (*w*/*v*) SB3-10 (E), 3.0% (*w*/*v*) SB3-10 (F), at the indicated time points. In panel D TDP-43 NTD was pre-incubated in 1.8% (*w*/*v*) SB3-10 for 1 h and then diluted to 0.6% (*w*/*v*) SB3-10 at the same final conditions as in panel (C).

Author Contributions: Conceptualization, F.B; Formal analysis, M.M., I.M., F.B., C.C., W.M., M.C.M., A.C. and F.C.; Funding acquisition, A.C. and F.C.; Investigation, M.M. (NTD Purification; Development of the alternative conformation; Fluorescence spectroscopy; Far-UV CD; Acrylamide Quenching Experiment; DLS; Analytical SEC; Guanidine hydrochloride (GdnHCl)-induced denaturation; SB3-10 titration on NTD; Titration of SB3-10), I.M. (GdnHCl-induced denaturation; Aggregation kinetics using DLS, far-UV CD and intrinsic fluorescence spectroscopies; ThT assay; FRET), C.C. (*NMR*), W.M. (NMR), M.C.M. (NMR) and A.C. (NMR); Methodology, M.M., I.M., M.V.V., C.C., W.M., M.C.M., A.C. and F.C.; Project administration, F.C.; Resources, M.V.V. (NTD Purification protocol, Bacterial glycerolates); Supervision, F.C.; Visualization, M.M. (Figures 1 and 3–6), I.M. (Figures 3 and 6–8, S7 and S8), C.C. (Figures 2, S1 and S6), A.C. (Figures 2, 8, S1 and S6) and F.C. (Figure 8); Writing—original draft, M.M., I.M., C.C., A.C. and F.C; Writing—review and editing, F.B. and F.C. All authors have read and agreed to the published version of the manuscript.

Funding: This work was supported by a full grant from the *Fondazione Italiana di Ricerca per la Sclerosi Laterale Amiotrofica* (AriSLA, project TDP-43-STRUCT), by a grant from Università di Firenze-Fondazione Cassa di Risparmio di Firenze (Projects TDP43SLA), and by the University of Florence (Fondi Ateneo to F.C. and F.B.). We acknowledge the use of the NMR at the Centro Grandi Strumenti of the University of Pavia (Italy) and, consequently, the Italian Ministry of Research and University, for a "Dipartimenti di Eccellenza 2018–2022 grant" to the Molecular Medicine Department (University of Pavia).

Institutional Review Board Statement: Not applicable.

Informed Consent Statement: Not applicable.

Data Availability Statement: Not applicable.

Acknowledgments: The authors thank Silvia Errico and Tommaso Staderini (University of Florence, Italy) and Teresa Recca (University of Florence, Italy) for technical support.

Conflicts of Interest: The authors declare no conflict of interest.

Sample Availability: Samples of the compounds are not available from the authors.

References

1. Arai, T.; Hasegawa, M.; Akiyama, H.; Ikeda, K.; Nonaka, T.; Mori, H.; Mann, D.; Tsuchiya, K.; Yoshida, M.; Hashizume, Y.; et al. TDP-43 is a component of ubiquitin-positive tau-negative inclusions in frontotemporal lobar degeneration and amyotrophic lateral sclerosis. *Biochem. Biophys. Res. Commun.* **2006**, *351*, 602–611. [CrossRef] [PubMed]
2. Neumann, M.; Sampathu, D.M.; Kwong, L.K.; Truax, A.C.; Micsenyi, M.C.; Chou, T.T.; Bruce, J.; Schuck, T.; Grossman, M.; Clark, C.M.; et al. Ubiquitinated TDP-43 in frontotemporal lobar degeneration and amyotrophic lateral sclerosis. *Science* **2006**, *314*, 130–133. [CrossRef] [PubMed]
3. Mackenzie, I.R.; Neumannn, M.; Bigio, E.H.; Cairns, N.J.; Alafuzoff, I.; Kril, J.; Kovacs, G.G.; Ghetti, B.; Halliday, G.; Holm, I.E.; et al. Nomenclature forneuropathologic subtypesof frontotemporal lobar degeneration: Consensus recommendations. *Acta Neuropathol.* **2009**, *117*, 15–18. [CrossRef] [PubMed]
4. Robberecht, W.; Philips, T. The changing scene of amyotrophic lateral sclerosis. *Nat. Rev. Neurosci.* **2013**, *14*, 248–264. [CrossRef]
5. Polymenidou, M.; Cleveland, D.W. The seeds of neurodegeneration: Prion-like spreading in ALS. *Cell* **2011**, *147*, 498–508. [CrossRef]
6. Ravits, J.M.; La Spada, A.R. ALS motor phenotype heterogeneity, focality, and spread: Deconstructing motor neuron degeneration. *Neurology* **2009**, *73*, 805–811. [CrossRef]
7. Cohen, T.J.; Lee, V.M.Y.; Trojanowski, J.Q. TDP-43 functions and pathogenic mechanisms implicated in TDP-43 proteinopathies. *Trends Mol. Med.* **2011**, *17*, 659–667. [CrossRef]
8. Lukavsky, P.J.; Daujotyte, D.; Tollervey, J.R.; Ule, J.; Stuani, C.; Buratti, E.; Baralle, F.E.; Damberger, F.F.; Allain, F.H. Molecular basis of UG-rich RNA recognition by the human splicing factor TDP-43. *Nat. Struct. Mol. Biol.* **2013**, *20*, 1443–1449. [CrossRef]
9. Kuo, P.H.; Chiang, C.H.; Wang, Y.T.; Doudeva, L.G.; Yuan, H.S. The crystal structure of TDP-43 RRM1-DNA complex reveals the specific recognition for UG- and TG-rich nucleic acids. *Nucleic Acids Res.* **2014**, *42*, 4712–4722. [CrossRef]
10. Qin, H.; Lim, L.Z.; Wei, Y.; Song, J. TDP-43N terminus encodes a novel ubiquitin-like fold and its unfolded form in equilibrium that can be shifted by binding to ssDNA. *Proc. Natl. Acad. Sci. USA* **2014**, *111*, 18619–18624. [CrossRef]
11. Jiang, L.L.; Zhao, J.; Yin, X.F.; He, W.T.; Yang, H.; Che, M.X.; Hu, H.Y. Two mutations G335D and Q343R within the amyloidogenic core region of TDP-43 influence its aggregation and inclusion formation. *Sci. Rep.* **2016**, *6*, 23928. [CrossRef] [PubMed]
12. Mompean, M.; Romano, V.; Pantoja-Uceda, D.; Stuani, C.; Baralle, F.E.; Buratti, E.; Laurents, D.V. The TDP-43 N-terminal domain structure at high resolution. *FEBS J.* **2006**, *283*, 1242–1260. [CrossRef] [PubMed]
13. Shiina, Y.; Arima, K.; Tabunoki, H.; Satoh, J. TDP-43 dimerizes in human cells in culture. *Cell. Mol. Neurobiol.* **2010**, *30*, 641–652. [CrossRef] [PubMed]
14. Chang, C.K.; Wu, T.H.; Wu, C.Y.; Chiang, M.H.; Toh, E.K.; Hsu, Y.C.; Lin, K.F.; Liao, Y.H.; Huang, T.H.; Huang, J.J. The N-terminus of TDP-43 promotes its oligomerization and enhances DNA binding affinity. *Biochem. Biophys. Res. Commun.* **2012**, *425*, 219–224. [CrossRef] [PubMed]
15. Zhang, Y.J.; Caulfield, T.; Xu, Y.F.; Gendron, T.F.; Hubbard, J.; Stetler, C.; Sasaguri, H.; Whitelaw, E.C.; Cai, S.; Lee, W.C.; et al. The dual functions of the extreme N-terminus of TDP-43 in regulating its biological activity and inclusion formation. *Hum. Mol. Genet.* **2013**, *22*, 3112–3122. [CrossRef]
16. Wang, Y.T.; Kuo, P.H.; Chiang, C.H.; Liang, J.R.; Chen, Y.R.; Wang, S.; Shen, J.C.; Yuan, H.S. The truncated C-terminal RNA recognition motif of TDP-43 protein plays a key role in forming proteinaceous aggregates. *J. Biol. Chem.* **2013**, *288*, 9049–9057. [CrossRef]
17. Mompean, M.; Romano, V.; Pantoja-Uceda, D.; Stuani, C.; Baralle, F.E.; Buratti, E.; Laurents, D.V. Point mutations in the N-terminal domain of transactive response DNA-binding protein 43 kDa (TDP-43) compromise its stability, dimerization, and functions. *J. Biol. Chem.* **2017**, *292*, 11992–12006. [CrossRef]
18. Jiang, L.L.; Xue, W.; Hong, J.Y.; Zhang, J.T.; Li, M.J.; Yu, S.N.; He, J.H.; Hu, H.Y. The N-terminal dimerization is required for TDP-43 splicing activity. *Sci. Rep.* **2017**, *7*, 6196. [CrossRef]
19. Afroz, T.; Hock, E.M.; Ernst, P.; Foglieni, C.; Jambeau, M.; Gilhespy, L.A.B.; Laferriere, F.; Maniecka, Z.; Plückthun, A.; Mittl, P.; et al. Functional and dynamic polymerization of the ALS-linked protein TDP-43 antagonizes its pathologic aggregation. *Nat. Commun.* **2017**, *8*, 45. [CrossRef]

20. Wang, A.; Conicella, A.E.; Schmidt, H.B.; Martin, E.W.; Rhoads, S.N.; Reeb, A.N.; Nourse, A.; Ramirez Montero, D.; Ryan, V.H.; Rohatgi, R.; et al. A single N-terminal phosphomimic disrupts TDP-43 polymerization, phase separation, and RNA splicing. *EMBO J.* **2018**, *37*, e97452. [CrossRef]
21. Carter, G.C.; Hsiung, C.H.; Simpson, L.; Yang, H.; Zhang, X. N-terminal Domain of TDP43 Enhances Liquid-Liquid Phase Separation of Globular Proteins. *J. Mol. Biol.* **2013**, *433*, 166948. [CrossRef] [PubMed]
22. Rengifo-Gonzalez, J.C.; El Hage, K.; Clément, M.J.; Steiner, E.; Joshi, V.; Craveur, P.; Durand, D.; Pastré, D.; Bouhss, A. The cooperative binding of TDP-43 to GU-rich RNA repeats antagonizes TDP-43 aggregation. *eLife* **2021**, *10*, e67605. [CrossRef] [PubMed]
23. Sasaguri, H.; Chew, J.; Xu, Y.F.; Gendron, T.F.; Garrett, A.; Lee, C.W.; Jansen-West, K.; Bauer, P.O.; Perkerson, E.A.; Tong, J.; et al. The extreme N-terminus of TDP-43 mediates the cytoplasmic aggregation of TDP-43 and associated toxicity in vivo. *Brain Res.* **2016**, *1647*, 57–64. [CrossRef] [PubMed]
24. Vivoli-Vega, M.; Guri, P.; Chiti, F.; Bemporad, F. Insight into the Folding and Dimerization Mechanisms of the N-Terminal Domain from Human TDP-43. *Int. J. Mol. Sci.* **2020**, *21*, 6259. [CrossRef] [PubMed]
25. Micsonai, A.; Wien, F.; Kernya, L.; Lee, Y.H.; Goto, Y.; Réfrégiers, M.; Kardos, J. Accurate secondary structure prediction and fold recognition for circular dichroism spectroscopy. *Proc. Natl. Acad. Sci. USA* **2015**, *112*, 3095–3103. [CrossRef]
26. Shen, Y.; Delaglio, F.; Cornilescu, G.; Bax, A. TALOS+: A Hybrid Method for Predicting Protein Backbone Torsion Angles from NMR Chemical Shifts. *J. Biomol. NMR* **2009**, *44*, 213–223. [CrossRef]
27. Wishart, D.S.; Sykes, B.D. The 13C Chemical-Shift Index: A Simple Method for the Identification of Protein Secondary Structure Using 13C Chemical-Shift Data. *J. Biomol. NMR* **1994**, *4*, 171–180. [CrossRef]
28. Marsh, J.A.; Singh, V.K.; Jia, Z.; Forman-Kay, J.D. Sensitivity of secondary structure propensities to sequence differences between alpha- and gamma-synuclein: Implications for fibrillation. *Protein Sci.* **2006**, *15*, 2795–2804. [CrossRef]
29. Baxter, N.J.; Williamson, M.P. Temperature Dependence of 1H Chemical Shifts in Proteins. *J. Biomol. NMR* **1997**, *9*, 359–369. [CrossRef]
30. Kay, L.E.; Torchia, D.A.; Bax, A. Backbone Dynamics of Proteins as Studied by Nitrogen-15 Inverse Detected Heteronuclear NMR Spectroscopy: Application to Staphylococcal Nuclease. *Biochemistry* **1989**, *28*, 8972–8979. [CrossRef]
31. Santoro, M.M.; Bolen, D.W. Unfolding free energy changes determined by the linear extrapolation method. 1. Unfolding of phenylmethanesulfonyl alpha-chymotrypsin using different denaturants. *Biochemistry* **1988**, *27*, 8063–8068. [CrossRef] [PubMed]
32. Wang, L.; Lim, L.; Dang, M.; Song, J. A novel mechanism for ATP to enhance the functional oligomerization of TDP-43 by specific binding. *Biochem. Biophys. Res. Commun.* **2019**, *514*, 809–814. [CrossRef] [PubMed]
33. LeVine, H., 3rd. Thioflavine T interaction with synthetic Alzheimer's disease beta-amyloid peptides: Detection of amyloid aggregation in solution. *Protein Sci.* **1993**, *2*, 404–410. [CrossRef]
34. Xue, C.; Lin, T.Y.; Chang, D.; Guo, Z. Thioflavin T as an amyloid dye: Fibril quantification, optimal concentration and effect on aggregation. *R. Soc. Open Sci.* **2017**, *4*, 160696. [CrossRef] [PubMed]
35. Canet, D.; Last, A.M.; Tito, P.; Sunde, M.; Spencer, A.; Archer, D.B.; Redfield, C.; Robinson, C.V.; Dobson, C.M. Local cooperativity in the unfolding of an amyloidogenic variant of human lysozyme. *Nat. Struct. Biol.* **2002**, *9*, 308–315. [CrossRef]
36. Soldi, G.; Bemporad, F.; Torrassa, S.; Relini, A.; Ramazzotti, M.; Taddei, N.; Chiti, F. Amyloid formation of a protein in the absence of initial unfolding and destabilization of the native state. *Biophys. J.* **2005**, *89*, 4234–4244. [CrossRef]
37. Plakoutsi, G.; Bemporad, F.; Calamai, M.; Taddei, N.; Dobson, C.M.; Chiti, F. Evidence for a mechanism of amyloid formation involving molecular reorganisation within native-like precursor aggregates. *J. Mol. Biol.* **2005**, *351*, 910–922. [CrossRef]
38. Eakin, C.M.; Berman, A.J.; Miranker, A.D. A native to amyloidogenic transition regulated by a backbone trigger. *Nat. Struct. Mol. Biol.* **2006**, *13*, 202–208. [CrossRef]
39. Garcia-Pardo, J.; Graña-Montes, R.; Fernandez-Mendez, M.; Ruyra, A.; Roher, N.; Aviles, F.X.; Lorenzo, J.; Ventura, S. Amyloid formation by human carboxypeptidase D transthyretin-like domain under physiological conditions. *J. Biol. Chem.* **2014**, *289*, 33783–33796. [CrossRef]
40. Wishart, D.S.; Bigam, C.G.; Yao, J.; Abildgaard, F.; Dyson, H.J.; Oldfield, E.; Markley, J.L.; Sykes, B.D. 1H, 13C and 15N Chemical Shift Referencing in Biomolecular NMR. *J. Biomol. NMR* **1995**, *6*, 135–140. [CrossRef]
41. Lee, W.; Tonelli, M.; Markley, J.L. NMRFAM-SPARKY: Enhanced Software for Biomolecular NMR Spectroscopy. *Bioinformatics* **2015**, *31*, 1325–1327. [CrossRef] [PubMed]
42. Uversky, V.N. Size-exclusion chromatography in structural analysis of intrinsically disordered proteins. *Methods Mol. Biol.* **2012**, *896*, 179–194. [PubMed]
43. Picotti, P.; De Franceschi, G.; Frare, E.; Spolaore, B.; Zambonin, M.; Chiti, F.; de Laureto, P.P.; Fontana, A. Amyloid fibril formation and disaggregation of fragment 1-29 of apomyoglobin: Insights into the effect of pH on protein fibrillogenesis. *J. Mol. Biol.* **2007**, *367*, 1237–1245. [CrossRef] [PubMed]
44. Motamedi-Shad, N.; Monsellier, E.; Chiti, F. Amyloid formation by the model protein muscle acylphosphatase is accelerated by heparin and heparan sulphate through a scaffolding-based mechanism. *J. Biochem.* **2009**, *146*, 805–814. [CrossRef] [PubMed]

molecules

MDPI

Article

Interdiction in the Early Folding of the p53 DNA-Binding Domain Leads to Its Amyloid-like Misfolding

Fernando Bergasa-Caceres * and **Herschel A. Rabitz**

Department of Chemistry, Princeton University, Princeton, NJ 08544, USA; hrabitz@princeton.edu
* Correspondence: bergasa@princeton.edu

Abstract: In this article, we investigate two issues: (a) the initial contact formation events along the folding pathway of the DNA-binding domain of the tumor suppressor protein p53 (core p53); and (b) the intermolecular events leading to its conversion into a prion-like form upon incubation with peptide P8(250-257). In the case of (a), the calculations employ the sequential collapse model (SCM) to identify the segments involved in the initial contact formation events that nucleate the folding pathway. The model predicts that there are several possible initial non-local contacts of comparative stability. The most stable of these possible initial contacts involve the protein segments ^{159}AMAIY^{163} and ^{251}ILTII^{255,} and it is the only native-like contact. Thus, it is predicted to constitute "Nature's shortcut" to the native structure of the core domain of p53. In the case of issue (b), these findings are then combined with experimental evidence showing that the incubation of the core domain of p53 with peptide P8(250-257), which is equivalent to the native protein segment ^{250}PILTIITL^{257}, leads to an amyloid conformational transition. It is explained how the SCM predicts that P8(250-257) effectively interdicts in the formation of the most stable possible initial contact and, thereby, disrupts the subsequent normal folding. Interdiction by polymeric P8(250-257) seeds is also studied. It is then hypothesized that enhanced folding through one or several of the less stable contacts could play a role in P8(250-257)-promoted core p53 amyloid misfolding. These findings are compared to previous results obtained for the prion protein. Experiments are proposed to test the hypothesis presented regarding core p53 amyloid misfolding.

Keywords: cancer; prion; folding; pathway; interdiction; peptide

Citation: Bergasa-Caceres, F.; Rabitz, H.A. Interdiction in the Early Folding of the p53 DNA-Binding Domain Leads to Its Amyloid-like Misfolding. *Molecules* **2022**, *27*, 4810. https://doi.org/10.3390/molecules27154810

Academic Editors: Kunihiro Kuwajima, Yuko Okamoto, Tuomas Knowles and Michele Vendruscolo

Received: 21 June 2022
Accepted: 23 July 2022
Published: 27 July 2022

1. Introduction

Since its discovery in 1979 [1–4], considerable evidence has accumulated on the critical importance of the tumor suppressor factor p53 in the natural protection mechanisms against cancer development [5,6]. Around 50% of all cancers show altered functionality of p53, frequently through mutation [7–9]. Thus, considerable efforts have focused on understanding the natural roles of p53 in protecting against cancer [10–12], and also in finding ways to restore its natural activity in cancer patients [13,14]. The most important role of p53 seems to be tumor suppression through the induction of apoptosis cellular programs in response to stress signals [15,16]. Additional tumor suppression-related activities have been discovered, such as metabolic regulation, autophagy and changes in the oxidative state of the cell [17–19]. Recently, it has also been found that p53 has further roles in normal physiology and the etiology of other diseases [20]. Thus, the complete biophysical characterization of p53 and its functionalities is an issue of prime importance.

Similar to other transcription factors, the 393 amino acids of p53 make it a multi-domain protein [21,22]. Its five domains include an N-terminal activation domain (residues 1-63), a proline-rich domain (residues 64-93), a DNA-binding domain, which is known as the core domain (i.e., core p53, residues 94-297), a tetramerization domain (residues 298-355), and a C-terminal domain (residues 356-393). The DNA binding of p53 to cellular DNA is carried out by the core domain of p53, and more than 95% of the tumor-inducing mutations of p53 occur in the core p53 [9,23,24]. Furthermore, p53 is a very flexible protein that can attain different closely related native conformations [25,26].

A striking recent discovery has been that under amyloidogenic experimental conditions, p53 and its individual domains undergo amyloid misfolding and aggregation [27–33]. These promoting factors include: (a) several point mutations that destabilize core p53; (b) denaturing conditions; (c) incubation with specific aggregation-seeding external agents, such as the peptide P8(250-257) resembling the native segment ^{250}PILTIITL^{257} [34,35]. The amyloid aggregates of p53 have been shown to display prion-like properties in vivo that could have a direct bearing on the evolution of several tumor types. These properties include: (a) Several types of cancer tissues show abnormal amyloid-like aggregates of misfolded p53 [32]; (b) p53 amyloid formation leads to cellular pro-metastatic gain-of-function [36]; (c) p53 amyloid-fibrils seed misfolding and aggregation when internalized in cells [37], and (d) p53 amyloid formation in cells impairs its transcriptional regulation function [34].

The amyloid misfolding and aggregation of the prion protein is understood to play a key role in triggering several neurodegenerative diseases [38], including Creutzfeldt-Jakob disease in humans [38] and mad cow and scrapie in cows and sheep, respectively [38]. It is likely that similar misfolding and aggregation mechanisms underlie other more widespread human pathologies, such as Parkinson's disease [39,40], Lewy-body dementia [39,40] and, possibly, Alzheimer's disease [41]. The existence of a prion-like amyloid misfolding-aggregation behavior for p53 suggests the possibility that there might be substantial molecular commonalities between at least some types of cancer and neurodegenerative diseases. There is, however, still some uncertainty concerning whether the p53 aggregates are infectious in the same way as prion aggregates [42]. The firm establishment of such a common molecular basis for apparently unrelated and widespread diseases could be of major importance, leading to opportunities for the development of related therapeutic strategies [33]. Here, prion or amyloid-like refer only to the misfolding of p53 into amyloid species that can aggregate into higher-order oligomers and fibrils and propagate inside an individual, not to the possible infectious properties of such amyloids.

The purpose of this paper is to: (a) predict the possible initial contact formation events along the folding mechanism of the core domain of P53; and (b) study the effects on the dominant native folding pathway of incubation with peptide P8(250-257). This peptide has been experimentally shown to promote the prion-like transformation of the core domain of P53 upon incubation with peptide P8(250-257), which resembles the native sequence segment 250-257 [34,35]. The calculations will employ the sequential collapse model (SCM) for the protein folding pathways [43,44].

Our calculations predict that there are several possible initial contact formation events that potentially lead to multiple folding pathways. It is also predicted that the more stable possible initial contact (i.e., the dominant contact) involves the protein segment 251-255. This prediction leads naturally within the model to the hypothesis that incubation of the core domain of p53 under denaturing conditions, with P8(250-257), may interdict in the dominant folding pathway, which is a process previously studied for several viral proteins with the SCM [45,46]. The possible interdiction effect arises because P8(250-257) competes with the key interactions defining the dominant early contact formation event. The experimental observation that such incubation produces the transformation of the core domain into a prion-like species is then explained to be naturally understood within the model, as evidence that such a misfolding is likely linked to enhanced folding (i.e., actually protein misfolding) through one or several of the secondary pathways initiated by the less stable contacts. In this paper, these findings will be compared to similar previously obtained results for the misfolding of the prion protein [47]. Possible experiments to test the hypothesis for p53 are discussed.

2. Results

2.1. The Physical Basis of Non-Local Early Contact Formation in the SCM

The physical basis of the SCM and its most up-to-date formulation has been recently explained in detail [44], and the associated calculation methodology is summarized in the Methods section of this paper. Here, a brief introduction to the main concepts is presented that are relevant to the issues investigated in the present paper. The SCM considers early non-local contacts based on the entropy of formation of the resultant protein loops in the unfolded state and the hydrophobic stabilization energy of the protein segments that define the contacts. The SCM has successfully predicted many of the observed features of the protein folding pathways at a low resolution [44]. Within the SCM, the folding of proteins with more than ~100 amino acids is nucleated by the formation of a specific early non-local contact, called the primary contact, which defines the earliest folding phase. The primary contacts between two protein segments centered at the residues i and j, separated by a distance along the protein sequence n_{ij}, form references to an optimal distance n_{op} such that $n_{ij} \geq n_{op} \approx 65$ amino acids, where the actual contact at n_{ij} is determined by the excluded volume-related entropic consequences of forming early protein loops [44]. The physical basis for the formation of the early non-local contacts in the SCM is illustrated in Figure 1.

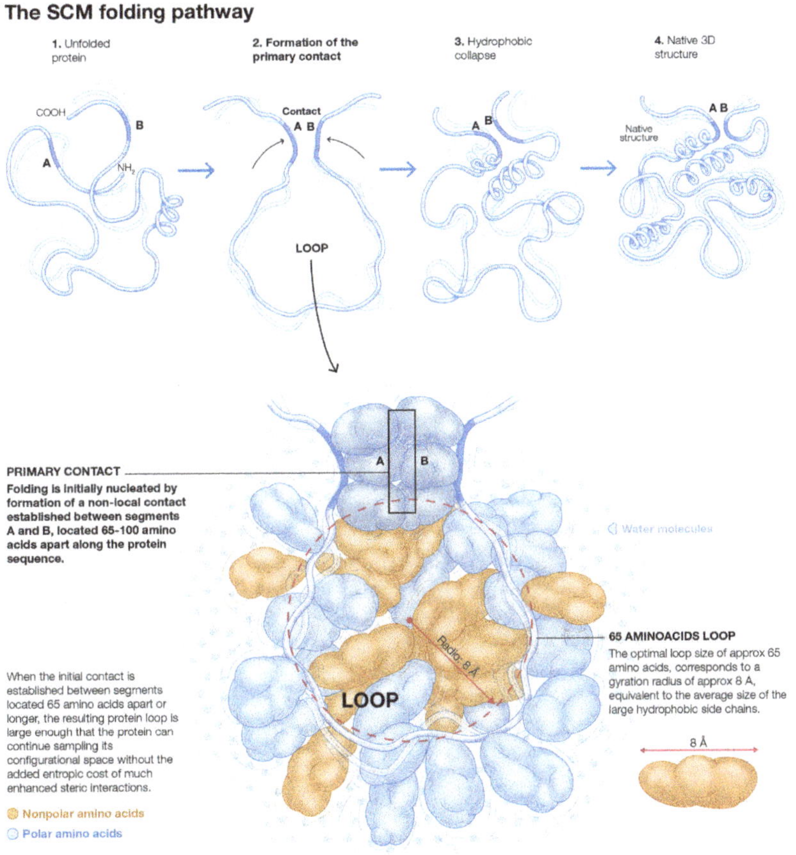

Figure 1. The physical basis for the formation of early non-local contacts in the SCM. The segments forming the contact are labeled A and B.

For a given protein sequence, there might be several viable primary contacts that nucleate parallel folding pathways [44]. As at most, only a few simultaneous primary contacts can be established in proteins of length $n \geq n_{op}$, while most of the tertiary structure contacts will still be defined by contacts at a shorter range established in later folding phases [44]. The formation of the primary contact in the SCM defines the primary loop, which subsequently collapses through two-state kinetics [44]. The nucleation by an early primary contact has been referred to within the model as "Nature's shortcut to protein folding" [44]. The short-range contacts established in later folding phases are defined by fluctuating shorter loops called minimal loops in the model, which are expected to be of length n_{min} ~15 amino acids [43]. Thus, within the model, the observed persistence length of the loops in a native protein is expected to be significantly lower than n_{op} and closer to n_{min}, in agreement with the experimental observations [48]. It is important to bear in mind, however, that the SCM is concerned with the optimal sizes of loops in the fluctuating unfolded chain rather than with the topology of the fully folded protein. The final length of the topological elements of the 3D structure can vary from their open-chain seeding loops, as contacts in the folded structure are refined by optimal packing, secondary structure formation and the establishment of all the relevant interactions [49]. Because proteins longer than ~100 amino acids do not generally undergo a complete two-state collapse [44,50] but rather fold through multi-step pathways, consistent and simple physical reasoning implies that there is a limit to the size of the primary loop (i.e., ~100 amino acids) that can successfully lead to the native SCM folding pathway.

The concept of folding nucleation by non-local contacts is not exclusive to the SCM, having arisen earlier in the context of the diffusion-collision model [51], the loop hypothesis [52] and the energy landscape picture [53]. Furthermore, it has appeared in simulations of the transition state of two-state folding proteins [54]. Protein topology has been considered an essential element of folding mechanisms in a number of theoretical efforts [48,55–58]. The particular feature of the SCM is that the early non-local contacts are highly specific, as in the loop hypothesis [52], and the SCM provides a general method to determine their location at specific distances along the primary sequence [43,44].

2.2. Primary Contacts of Core p53

Following the methodology employed in the SCM before [43,44], which is described in detail in the Methods section, our calculations searched for the most stable possible hydrophobic contacts between pairs of 5-amino acid segments centered at amino acids i and j, located at a distance n_{ij} along the sequence such that $65 \leq n_{ij} \leq 100$ amino acids. The calculations employ the primary sequence and experimental hydrophobicity values to determine the location of the possible primary contacts. The results were seen to be robust when extending the segment lengths up to seven amino acids. Here, we will keep the analysis focused on the 5-amino acid segment results for the purpose of consistency and the ability to compare with the previous work within the SCM. The predicted primary contacts and their stabilities are listed in Table 1. The stabilities in the table correspond to the contacts as formed in the earliest folding phase, not on the fully folded structure.

Table 1. Possible primary contacts for the core domain of core p53, their stabilization energy (*k*T), and their location in the native 3D structure.

	Contact	Stability	3D Structure
C1	^{159}AMAIY^{163} on ^{251}ILTII^{255}	8.9 ± 0.3	Native
C2	^{143}VQLWV^{147} on ^{216}VVVPY^{220}	8.2 ± 0.4	Non-Native
C3	^{143}VQLWV^{147} on ^{234}YNYMC^{238}	7.8 ± 0.4	Non-Native
C4	^{133}MFCQL^{137} on ^{216}VVVPY^{220}	7.8 ± 0.4	Non-native

The most stable primary contact, C1, is predicted to form between the segments ^{159}AMAIV^{163} and ^{251}ILTII^{255}, with a predicted stability of $\Delta G_{cont}(C1) \approx -8.9\ kT$. It is a native contact on the 3D structure [59], as shown in Figure 2.

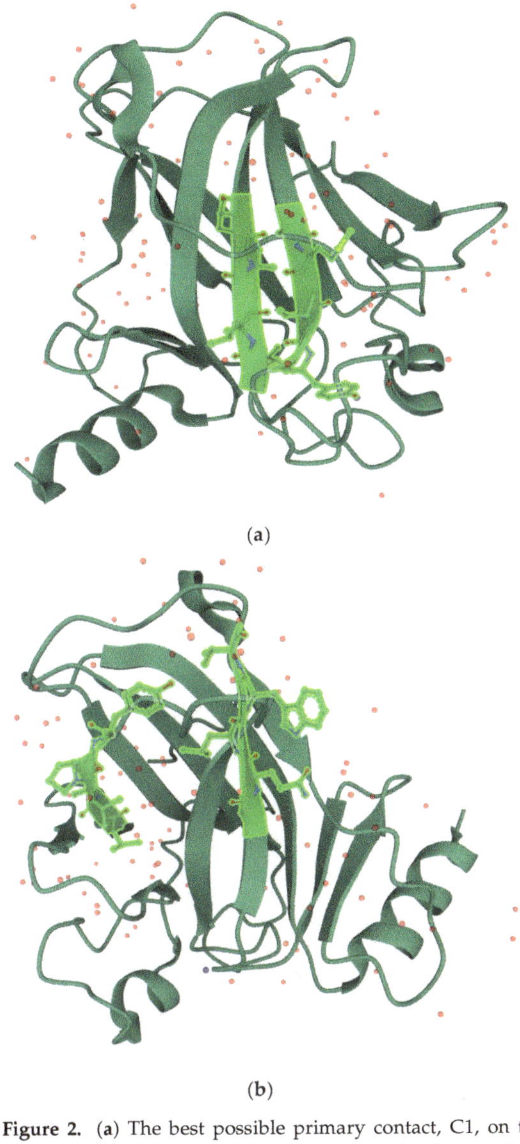

(**a**)

(**b**)

Figure 2. (**a**) The best possible primary contact, C1, on the structure of core p53 (PDB 2OCJ); (**b**) contact C2 is on the same structure. As can be observed in the figures, C1 is a good contact on the 3D structure, while C2 does not, as the side chains of the two segments that define it are not close in the folded structure. The figures were prepared using Mol* [60].

The second best possible primary contact, C2, is established between the segments ^{143}VQLWV^{147} and ^{216}VVVPY^{220}, with a contact stability of $\Delta G_{cont}(C2) \approx -8.2\ kT$. Contact C2 is not a good match for the 3D structure, as shown in Figure 2. There are two additional possible non-native contacts, C3 and C4, both with a lower stability of $\Delta G_{cont}(C3) = \Delta G_{cont}(C4) \approx -7.8\ kT$. All the other possible primary contacts are more than

~3 kT less stable than the dominant one and non-native on the tertiary structure, with corresponding populations of more than ~one order of magnitude smaller than that of the best contact, and we have not included them in the analysis. The issue of the multiplicity of the possible primary contacts in the SCM has been considered in previous work [44]. Because no major rearrangements of the protein core are expected post-collapse, it is generally assumed within the model that the primary contacts which are non-native in the 3D folded structure likely do not correspond to the native pathways leading to the functional folded structure [44]. In all the naturally folding proteins studied to date within the SCM, it has been observed that the most stable primary contact is native-like [44].

On the basis of the above results, it is reasonable within the SCM to expect that the contact C1 nucleates the native folding pathways for core p53. Thus, C1 constitutes "Nature's shortcut" to the folding of core p53 [44]. On the other hand, contacts C2, C3 and C4 probably represent non-native-initial contacts that must break up in order for core p53 to enter the native folding pathway nucleated by C1. The nucleation of native folding by the non-native primary contacts would imply a considerable later rearrangement of the protein core, which is entropically unfavorable [61].

2.3. Comparison of the Predicted Primary Contact Populations with Experimental Data on Core p53 Folding

If the folding of core p53 proceeds through the four predicted possible nucleation events, it is reasonable to expect that each of the four corresponding folding channels must involve a fraction of the folding proteins that are equivalent to the initial relative population of the corresponding primary contact. Based on their relative stabilities, the predicted relative populations of contacts C1, C2, C3 and C4 are, respectively, 46% for C1, 23% for C2 and 15% for both C3 and C4. From a kinetic point of view, as $\Delta G_{cont}(C3) = \Delta G_{cont}(C4)$, the break-up of both contacts should take place within similar time scales, and thus, in a low-resolution experiment, they would probably appear as a single channel traversed by the combined populations of C3 and C4, with an apparent population of ~31%. The uncertainty in the energy values $\Delta G_{cont}(C)$ accounts for the single standard deviation confidence intervals of ~[+7%,−7%] for a variation in the population associated with C1, [−6%,+8%] for C2, and [−4%,+6%] for each of C3 and C4.

Experimental evidence exists for core p53 that supports the existence of three kinetically distinct folding channels a, b, and a less well-resolved channel c, with ~50% of the native population folding through a and ~25% through each of b and c [62,63]. The predicted population of each native contact does not necessarily precisely reflect the overall population of molecules fully folding through each folding pathway, as there might be differences in the folding kinetics arising, for example, from the break-up of the non-native primary contacts [63]. Within the SCM, however, it is reasonable to assume that the intermediates nucleated by the four primary contacts will undergo a hydrophobic collapse with broadly similar kinetics once the native primary contact is established [64]. Thus, taking into account their equivalent stabilization energy, C3 and C4 should be difficult to resolve in separate channels in standard kinetic experiments. However, it is fair to say that the populations of the four early native-like intermediates predicted here correlate well with the experimental observations concerning the populations of molecules traversing each of the three folding channels detected experimentally. C1 would nucleate channel a with a population of ~46%, C2 represents channel 2 with a population of ~23%, and C3 and C4 add up to a less resolved channel c with a population of ~31%. Because of the lack of single-residue resolution in the available experimental data, this correlation should be considered as just a consistency test of the SCM predictions.

2.4. Interdiction of the Dominant Folding Pathway through Competition of Peptide P8(250-257) with the Formation of Primary Contact C1

In recent work, we proposed that the SCM's primary contact predictions provide natural targets for folding interdicting peptide drugs, which are aimed to treat viral diseases, such as SARS-CoV-2, Ebola and influenza A [45,46] by blocking the initial folding steps of specific viral proteins. Such peptide drugs could be designed by employing, as templates, the segments naturally involved in the primary contact, where one of the segments, S_1, would be the basis of the folding interdicting peptide (FIP), and the other segment, S_2, would constitute the folding interdiction target region (FITR), as described in Figure 3. In such an interdiction mechanism, the FIP would compete with S_1 to bind to S_2, leading to a decrease in the number of proteins folding through the pathway initiated by the contact S_1-S_2.

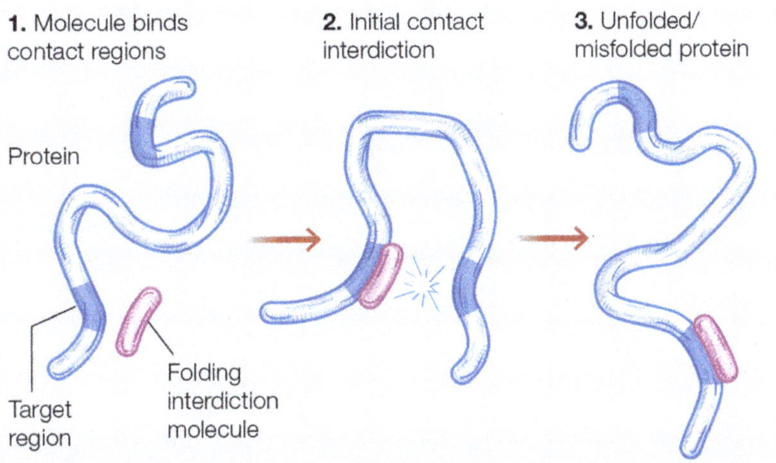

Figure 3. Protein folding interdiction, based on the identification of a FITR target.

It is interesting to investigate whether the FITR concept could be applied to core p53, such that specific peptide drugs can be designed to modulate its folding dynamics and, potentially, its physiological activity. In particular, it would be of considerable importance to determine, through selective interdiction of the possible folding pathways, whether any or several of them are involved in the prion-like transition of p53. If that was the case, it might be possible to design specific peptide drugs aiming to interdict the misfolding pathways, thus maximizing the population of natively folded p53. Below, we will explain that such a folding interdiction experiment has already been partially carried out for the core domain of p53 [34,35].

Experiments have shown that the incubation of the core domain of p53 with peptide P8(250-257), defined by the native sequence ^{250}PILTIITL^{257}, induces amyloid-like aggregation [34,35]. Peptide P8(250-257) is just the segment ^{251}ILTII^{255} predicted to form the native primary contact C1 with segment ^{159}AMAIY^{163}, plus residues P250, T256 and L257, amino acids that are non-polar or hydrophobic and, thus, will tend to promote further contact formation. Thus, within the current model, the prediction is clear that the intermolecular effect of P8(250-257) on the folding of core p53 is the formation of the contact between P8(250-257) and the segment ^{159}AMAIY^{163}, leading to interdiction in the native folding pathway of core p53.

Experimental evidence also shows that P8(250-257) tends to form β-sheet-rich aggregates by itself [35], suggesting that these aggregates may act as a seed for the prion-like misfolding of p53 [35]. Most of segment ^{250}PILTIITL^{257} defines a strand of a β-sheet in the native structure of core p53, in close interaction with the strands ^{264}LLGRNSFEVRV^{274} and ^{156}RVRAMAIY^{163} that includes the segment ^{159}AMAIY^{163} which also nucleates C1 [59]. Thus, it is reasonable to hypothesize that the ß-sheet-rich aggregates of P8(250-257) can also interdict in the formation of the primary contact C1 by interacting with the segments that define C1. This result also suggests that P8(250-257) alone might be able to interdict in the formation of C1, not just by its attachment to segment ^{159}AMAIY^{163}, but also to segment ^{251}ILTII^{255} itself, thus enhancing the interdiction effect. Only more detailed experimental and theoretical studies can determine which of the possible interdiction modes described here, if any, is dominant in triggering the experimentally observed misfolding effects.

2.5. Possible Coincidence of the p53 and Prion Protein Misfolding Mechanisms

If interdiction in the formation of the dominant native-like primary contact C1 triggers the misfolding and subsequent aggregation of core p53 when incubated with peptide P8(250-257), a relevant question is by what molecular mechanisms interdiction might lead to misfolding. It is important to keep in mind that there might be more than one mechanism leading to misfolding [65]. For the neuropathogenic protein, for example, different mutations lead to distinct amyloid formation and aggregation dynamics [65–67]. In the case of core p53, several mutations with no obvious connection to the interdiction mechanism described here (i.e., involving the mutations outside the predicted primary contacts), such as R175H, R249S, R273H, C242S and R248Q, lead to amyloid formation under the appropriate conditions [68,69].

In recent work, we proposed that a few specific mutation-related misfolding events of the murine prion protein *m*PrP(90-231) were related to the protein traversing through a folding pathway nucleated by a primary contact of lesser stability than the dominant contact [47]. Furthermore, we showed that several known pathogenic mutations have the effect of increasing the relative population of prion proteins entering the secondary pathways defined by the less stable contacts [47]. It then becomes an interesting question of whether a similar mechanism could be at play in the interdiction-triggered prion-like conversion of core p53.

Within the current model, the experimental observation that the incubation of core p53 with peptide P8(250-257) leads to its conversion into a prion-like species, combined with our previous results for the prion protein, suggests that one or several of the less stable contacts might nucleate a folding pathway, thus leading to misfolded core p53. Then, interdiction in the dominant native pathway, with a concomitant reduction in the population of molecules folding through the native pathway, could lead to increased folding through the less stable primary contacts, including the pathogenic pathway/s leading to abnormal levels of misfolded p53 that aggregate into amyloid-like inclusions. This hypothetical mechanism is shown in Figure 4.

At the current level of the analysis, it is impossible to predict unequivocally whether such a mechanism is in play, and, in the absence of detailed experimental evidence, it should be considered as a hypothesis. Another possibility is that interdiction in the formation of the dominant primary contact by either monomeric or aggregated P8(250-257) leads to a direct amyloid transition. A possible experiment to test the secondary folding channel hypothesis could be carried out by interdicting in the formation of the non-native contacts under amyloidogenic conditions. If the prion-like transition of core p53 proceeds through the initial formation of any of the non-native contacts C2–C4, interdiction in the formation of that contact should inhibit it. Thus, for example, if the amyloid transition was nucleated by the formation of C2, the incubation of core p53 with the peptides derived from either ^{143}VQLWV^{147} or ^{216}VVVPY^{220} should diminish the formation of prion-like structures under amyloidogenic conditions. There is considerable interest in the investigation of therapeutic drugs that might retrieve the active conformation of p53 [70] and, more specifically, that interfere with the formation and effects of the p53 aggregates [71]. The results presented here could potentially open a new avenue towards this goal. However, as discussed above, there might be more than one mechanism leading to an amyloidogenic transition, and the set-up for an experiment such as the one described above, leading to a strong clear-cut result, is likely to be a complex undertaking.

Finally, it is important to point out that the aggregation mechanisms in vivo can be decisively influenced by additional factors, such as pH and metal ion concentrations [72] and other proteins [73]. Thus, the complete validation of the biological relevance of the mechanism for p53 aggregation proposed here would require in vivo testing, probably employing the full range of existing testing techniques [74].

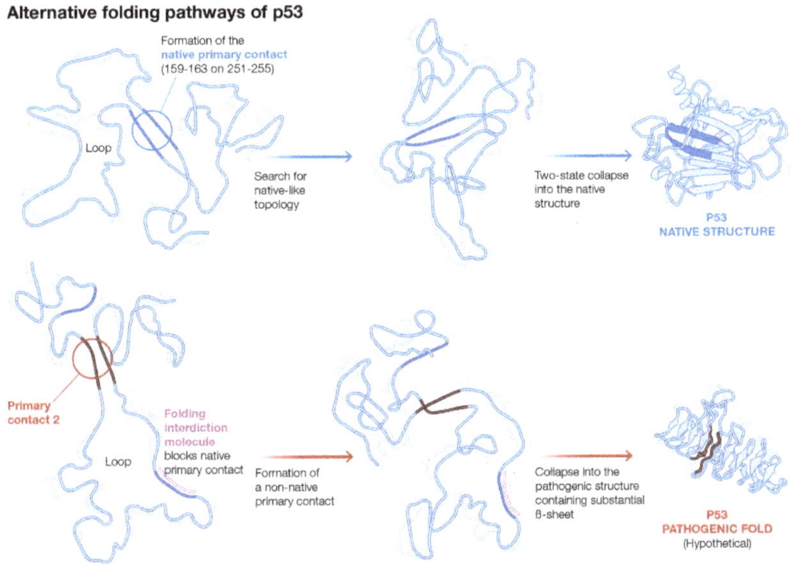

Figure 4. Hypothetical misfolding mechanism of core p53 triggered by interdiction with peptide P8(250-257).

3. Conclusions

In this paper, we presented the theoretical predictions for the earliest folding events of the core domain of p53. Several possible initial contact-forming events were identified, potentially leading to a multiplicity of folding pathways. It was also explained that the experimentally observed prion-like transition of the core domain of p53 upon incubation in the denaturing conditions with peptide P8(250-257) could be understood within the SCM through a folding interdiction mechanism, as proposed in earlier work. It was explained how additional experiments could possibly confirm this hypothesis and open a new path to the design of peptide drugs that are able to modulate p53 folding dynamics.

4. Methods: Determination of the Primary Contact in the SCM Model

The physical basis of the SCM and its most up-to-date formulation has been recently explained in full detail [44]. Here, only a brief summary of the methodology to determine the primary contact is presented.

Based on the model presented in the previous sections, whether there is a non-local contact in an otherwise unfolded state is dependent upon the stability of the potential contact candidates at the loop lengths of $n \geq n_{op}$ amino acids. In the SCM, the stability of a contact formed by the number n_{cont} of amino acids, $\Delta G_{contact}(n_{cont}, n_{loop})$, can be written as:

$$\Delta G_{contact}(n_{cont}, n_{loop}) \approx \Delta G_{int,H}(n_{cont}) + \Delta G_{loop}(n_{loop}) + \Delta G_{cont,S}(n_{cont}) \quad (1)$$

Here, ΔG_{loop} represents the entropic free energy cost of the loop, as discussed in Section 2.1. The term $\Delta G_{int,H}$ denotes all the enthalpic interactions that help stabilize the contact, which possibly includes the hydrophobic interactions, van der Waals interactions, hydrogen bonds, disulfide bonds and salt bridges [49], and its value satisfies $\Delta Gint < 0$. The term $\Delta G_{cont,S} > 0$ represents the entropic cost of constraining the side chains of the amino acids, defining the contact as such that the contact is stable and it opposes contact formation. A segment-specific determination of the value $\Delta G_{cont,S}(n_{cont})$ for a given contact would require detailed molecular dynamics techniques. However, a heuristic estimate can be made from earlier work within the SCM, which showed that the average entropic cost of folding per amino acid for a sample of thirteen proteins was $\Delta G_{folding/residue,S} \approx 0.85$ kT/residue [75], and the maximum was $\Delta G_{folding/residue,S} \approx 1.09$ kT/residue. As these are estimates for the entropic cost for folding per residue of the complete proteins that include highly buried as well as flexible exposed regions, it is then reasonable to expect that the entropic cost of a contact-forming region must be closer to the highest calculated values or $\Delta G_{folding/residue,S}$. Here, we will assume that $\Delta G_{contact,S}(n_{contact})$ for a contact including the n_{cont} amino acids is approximately $\Delta G_{folding/residue,S}$, determined by the number of residues defining the contact, such that $\Delta G_{cont,S}(n_{cont}) \approx 1.09\ n_{cont}$. This result is clearly an approximation, but it suffices for establishing a cutoff in the number of possible contacts, which is consistent with the available structural data.

Hydrophobic interactions are well understood to constitute the main driving force of the folding process [49]. Other interactions, such as hydrogen bonds, are weaker [76], or like disulfide bonds and salt bridges, form later along the folding pathway [49]. Thus, for an early contact that forms from the unfolded state, we can take $\Delta G_{int}(n_{op}) \approx \Delta G_{hyd}(n_{op})$, where $\Delta G_{hyd}(n_{op})$ represents the stabilizing effect of hydrophobicity in the early contacts, and Equation (1) can be written as

$$\Delta G_{contact}(n_{cont}, n_{loop}) \approx \Delta G_{hyd}(n_{cont}) + \Delta G_{loop}(n_{loop}) + \Delta G_{contact,S}(n_{contact}) \quad (2)$$

Since the hydrophobic stabilization energy of the contact ΔG_{hyd} is determined by the hydrophobicity of the segments involved, the hydrophobicity values h_k are obtained from the Fauchere–Pliska scale [77] and assigned to each residue in accordance with previous calculations within the SCM.

Because the amino acid side chains are significantly larger than the typical peptide bond length, early contacts between two hydrophobic amino acids will inherently involve segments, including several amino acids, adjacent to the initial contact. The stability of this early hydrophobic contact will determine where the folding process is initiated. This picture is not unlike the zapping model of Dill and collaborators [78] and includes a well-defined nucleation step, as expected in most protein folding models [79]. Here, the typical early contact segment size will be taken to be ~5 amino acids, in line with previous calculations within the SCM. The 5-amino acid window size is based on the geometric considerations underlying the SCM: With an average effective fluctuating width of the unfolded protein chain of w ~2 $\int(n) \approx 15.8$ Å and a peptide bond length of 3.5 Å, the minimum number n_{cont} of amino acids that can define a contact in the open fluctuating chain should be $n_{cont} \sim \mathrm{int}[2\int(n)/3.5] = 5$ amino acids. The results for the location of the most stable primary contact were seen to be robust to the employment of up to seven amino acid windows, while some deviations were observed when the window was reduced to four amino acids. In practice, within the SCM, the hydrophobicity h_k of each residue is added over a segment contact window of five amino acids centered at residue i, resulting in a segment hydrophobicity $h_{i,5}$ (a value of ~0.45 is equivalent to a change in the energy of kT, with the margin of error being ~0.1 kT [77]).

In order to determine the best contact, the $h_{i,5}$ values of a segment centered at residue i are added to the h_j value of a segment centered at residue j, located at a distance n_{ij} at least n_{op} amino acids apart along the sequence, and no longer than the maximum primary loop length of ~100 amino acids, to give a contact stability of:

$$\Delta G_{cont}(n_{cont}, n_{loop}) \approx kT\,[-(h_{i,5} + h_{j,5})/0.45 + 3/2 \ln n_{ij} + 10.9] \quad 100 \geq n_{ij} \geq 65 \quad (3)$$

Finally, the relatively simple algorithm presented here has sufficed to successfully predict primary contacts within the SCM so far. However, specific issues critical for drug design, such as identifying the optimal interdiction molecules and fully characterizing their interactions, will probably require a more sophisticated approach, including state-of-the-art molecular dynamics [80].

Author Contributions: F.B.-C. identified the key experimental observations to be explained and carried out the analysis and preparation of the paper; H.A.R. was involved at all stages, including the conceptualization phase. All authors have read and agreed to the published version of the manuscript.

Funding: This research received no external funding.

Institutional Review Board Statement: Not applicable.

Informed Consent Statement: Not applicable.

Data Availability Statement: Not applicable.

Conflicts of Interest: The authors declare no conflict of interest.

Abbreviations

SCM, Sequential Collapse Model.

References

1. Kress, M.; May, E.; Cassingena, R.; May, P. Simian Virus 40-Transformed Cells Express New Species of Proteins Precipitable by Anti-Simian Virus 40 Tumor Serum. *Virol. J.* **1979**, *31*, 472–483. [CrossRef] [PubMed]
2. Lane, D.P.; Crawford, L.V. T Antigen is Bound to a Host Protein in SV40-Transformed Cells. *Nature* **1979**, *278*, 261–263. [CrossRef] [PubMed]
3. Linzer, D.I.; Levine, A.J. Characterization of a 54K Dalton Cellular SV40 Tumor Antigen Present in SV40-Transformed Cells and Uninfected Embryonal Carcinoma Cells. *Cell* **1979**, *17*, 43–52. [CrossRef]
4. DeLeo, A.B.; Jay, G.; Appella, E.; Dubois, G.C.; Law, L.W.; Old, L.J. Detection of a Transformation-Related Antigen in Chemically Induced Sarcomas and Other Transformed Cells of the Mouse. *Proc. Natl. Acad. Sci. USA* **1979**, *76*, 2420–2424. [CrossRef] [PubMed]
5. Baker, S.J.; Fearon, E.R.; Nigro, J.M.; Hamilton, S.R.; Preisinger, A.C.; Jessup, J.M.; van Tuinen, P.; Ledbetter, D.H.; Barker, D.F.; Nakamura, Y.; et al. Chromosome 17 Deletions and p53 Mutations in Colorectal Carcinomas. *Science* **1989**, *244*, 217–221. [CrossRef]
6. Takahashi, T.; Nau, N.M.; Chiba, I.; Birrer, M.J.; Rosenberg, R.K.; Vinocour, M.; Levitt, M.; Pass, H.; Gazdar, A.F.; Minna, J.D. p53: A Frequent Target for Genetic Abnormalities in Lung Cancer. *Science* **1989**, *246*, 491–494. [CrossRef] [PubMed]
7. Freed-Pastor, W.A.; Prives, C. Mutant p53: One Name, Many Proteins. *Genes Dev.* **2012**, *26*, 1268–1286. [CrossRef]
8. Muller, P.A.J.; Vousden, K.H. Mutant p53 in Cancer: New Functions and Therapeutic Opportunities. *Cancer Cell* **2014**, *25*, 304–317. [CrossRef]
9. Muller, P.A.J.; Vousden, K.H. p53 mutations in cancer. *Nat. Cell Biol.* **2013**, *15*, 2–8. [CrossRef]
10. Wang, X.; Simpson, E.R.; Brown, K.A. p53: Protection against tumor growth beyond effects on cell cycle and apoptosis. *Cancer Res.* **2015**, *75*, 5001–5007. [CrossRef]
11. Mantovani, F.; Collavin, L.; Del Sal, G. Mutant p53 as a guardian of the cancer cell. *Cell Death Differ.* **2019**, *26*, 199–212. [CrossRef] [PubMed]
12. Lees, A.; Sessler, T.; McDade, S. Dying to survive—The p53 paradox. *Cancers* **2021**, *13*, 3257. [CrossRef]
13. Ventura, A.; Kirsch, D.G.; McLaughlin, M.E.; Tuveson, D.A.; Grimm, J.; Lintault, L.; Newman, J.; Reczek, E.E.; Weissleder, R.; Jacks, T. Restoration of p53 function leads to tumour regression in vivo. *Nature* **2007**, *445*, 661–665. [CrossRef] [PubMed]
14. Lozano, G. Restoring p53 in cancer: The promises and the challenges. *JMCB* **2019**, *11*, 615–619. [CrossRef] [PubMed]
15. Aubrey, B.J.; Kelly, G.L.; Janic, A.; Herold, M.J.; Strasser, A. How does p53 induce apoptosis and how does this relate to p53-mediated tumour suppression? *Cell Death Differ.* **2018**, *25*, 104–113. [CrossRef] [PubMed]
16. Fridman, J.S.; Lowe, S.W. Control of apoptosis by p53. *Oncogene* **2003**, *22*, 9030–9040. [CrossRef] [PubMed]
17. Lahalle, A.; Lacroix, M.; De Blasio, C.; Cissé, M.Y.; Linares, L.K.; Le Cam, L. The p53 pathway and metabolism: The tree that hides the forest. *Cancers* **2021**, *13*, 133. [CrossRef] [PubMed]
18. Shi, Y.; Norberg, E.; Vakifahmetoglu-Norberg, H. Mutant p53 as a regulator and target of autophagy. *Front. Oncol.* **2021**, *10*, 607149. [CrossRef]
19. Shi, T.; van Soest, D.M.K.; Polderman, P.E.; Burgering, B.M.T.; Dansen, T.B. DNA damage and oxidant stress activate p53 through differential upstream signaling pathways. *Free Radic. Biol. Med.* **2021**, *172*, 298–311. [CrossRef]
20. Tsai, Y.-Y.; Su, C.-H.; Tarn, W.-Y. p53 activation in genetic disorders: Different routes to the same destination. *Int. J. Mol. Sci.* **2021**, *22*, 9307. [CrossRef]
21. May, P.; May, E. Twenty years of p53 research: Structural and functional aspects of the p53 protein. *Oncogene* **1999**, *18*, 7621–7636. [CrossRef]
22. Joerger, A.C.; Fersht, A. The tumor suppressor p53: From structures to drug discovery. *Cold Spring Harb. Perspect. Biol.* **2010**, *2*, a000919. [CrossRef]

23. Pavletich, N.P.; Chambers, K.A.; Pabo, C.O. The DNA-binding domain of p53 contains the four conserved regions and the major mutation hot spots. *Genes Dev.* **1993**, *7*, 2556–2564. [CrossRef] [PubMed]
24. Baugh, E.H.; Ke, H.; Levine, A.J.; Bonneau, R.A.; Chan, C.S. Why are there hotspot mutations in the *TP53* gene in human cancers? *Cell Death Differ.* **2018**, *25*, 154–160. [CrossRef] [PubMed]
25. Ory, K.; Legros, Y.; Auguin, C.; Soussi, T. Analysis of the most representative tumour-derived p53 mutants reveals that changes in protein conformation are not correlated with loss of transactivation or inhibition of cell proliferation. *EMBO J.* **1994**, *13*, 3496–3504. [CrossRef]
26. Gamble, J.; Milner, J. Evidence that immunological variants of p53 represent alternative protein conformation. *Virology* **1988**, *162*, 452–458. [CrossRef]
27. Bosari, S.; Viale, G.; Roncalli, M.; Graziani, D.; Borsani, G.; Lee, A.K.; Coggi, G. p53 Gene Mutations, p53 Protein Accumulation and Compartmentalization in Colorectal Adenocarcinoma. *Am. J. Pathol.* **1995**, *147*, 790–798. [PubMed]
28. Ishimaru, D.; Andrade, L.R.; Teixeira, L.S.; Quesado, P.A.; Maiolino, L.M.; Lopez, P.M.; Cordeiro, Y.; Costa, L.T.; Heckl, W.M.; Weissmuller, G.; et al. Fibrillar Aggregates of the Tumor Suppresor p53 Core Domain. *Biochemistry* **2003**, *42*, 9022–9027. [CrossRef] [PubMed]
29. Rigacci, S.; Bucciantini, M.; Relini, A.; Pesce, A.; Gliozzi, A.; Berti, A.; Stefani, M. The (1-63) Region of the p53 Transactivation Domain Aggregates in Vitro into Cytotoxic Amyloid Assemblies. *Biophys. J.* **2008**, *94*, 3635–3646. [CrossRef] [PubMed]
30. Bom, A.P.D.; Rangel, L.P.; Costa, D.C.F.; de Oliveira, G.A.P.; Sanchez, D. Mutant p53 aggregates into prion-like amyloid oligomers and fibrils. *J. Biol. Chem.* **2012**, *287*, 28152–28162. [CrossRef] [PubMed]
31. Wang, G.; Fersht, A.R. Mechanism of initiation of aggregation of p53 revealed by Φ analysis. *Proc. Natl. Acad. Sci. USA* **2015**, *112*, 2437–2442. [CrossRef] [PubMed]
32. Navalkar, A.; Ghosh, S.; Pandey, S.; Paul, A.; Datta, D.; Maji, S.K. Prion-like amyloids in cancer. *Biochemistry* **2020**, *59*, 146–155. [CrossRef] [PubMed]
33. Lee, A.S.; Galea, C.; DiGiammarino, E.L.; Jun, B.; Murti, G.; Ribeiro, R.C.; Zambetti, G.; Schultz, C.P.; Kriwacki, R.W. Reversible amyloid formation by the p53 tetramerization domain and a cancer-associated mutant. *J. Mol. Biol.* **2003**, *327*, 699–709. [CrossRef]
34. Ghosh, S.; Salot, S.; Sengupta, S.; Navalkar, A.; Ghosh, D.; Jacob, R.; Das, S.; Kumar, R.; Jha, N.N.; Sahay, S.; et al. p53 amyloid formation leading to its loss of function: Implications in cancer pathogenesis. *Cell Death Differ.* **2017**, *24*, 1784–1798. [CrossRef] [PubMed]
35. Ghosh, S.; Ghosh, D.; Ranganathan, S.; Anoop, A.; Kumar, P.S.; Jha, N.N.; Padinhateeri, R.; Maji, S.K. Investigating the intrinsic aggregation potential of evolutionarily conserved segments in p53. *Biochemistry* **2014**, *53*, 5995–6010. [CrossRef]
36. Zhang, C.; Liu, J.; Xu, D.; Zhang, T.; Hu, W.; Feng, Z. Gain-of-function mutant p53 in cancer progression and therapy. *J. Mol. Cell. Biol.* **2020**, *12*, 674–687. [CrossRef]
37. Forget, K.J.; Tremblay, G.; Roucou, X. p53 aggregates penetrate cells and induce the co-aggregation of intracellular p53. *PLoS ONE* **2013**, *8*, e69242. [CrossRef]
38. Prusiner, S.B. *Prion Biology and Diseases;* Cold Spring Harbor Laboratory Press: Cold Spring Harbor, NY, USA, 2014.
39. Uversky, V.N. α-Synuclein Misfolding and Neurodegenerative Diseases. *Curr. Protein Pept. Sci.* **2009**, *9*, 507–540. [CrossRef]
40. Angot, E.; Steiner, J.A.; Hansen, C.; Li, J.-Y.; Brundin, P. Are synucleinopathies prion-like diseases? *Lancet Neurol.* **2010**, *9*, 1128–1138. [CrossRef]
41. Kondo, J.; Ihara, Y.; Sairoh, T. Molecular Cloning of cDNA Encoding an Unrecognized Component of Amyloid in Alzheimer Disease. *Proc. Natl. Acad. Sci. USA* **1993**, *90*, 11282–11287.
42. Billant, O.; Friocourt, G.; Roux, P.; Voisset, C. p53, a victim of the prion fashion. *Cancers* **2021**, *13*, 269. [CrossRef]
43. Bergasa-Caceres, F.; Ronneberg, T.A.; Rabitz, H.A. Sequential Collapse Model for Protein Folding Pathways. *J. Phys. Chem. B* **1999**, *103*, 9749–9758. [CrossRef]
44. Bergasa-Caceres, F.; Haas, E.; Rabitz, H.A. Nature's Shortcut to Protein Folding. *J. Phys. Chem. B* **2019**, *123*, 4463–4476. [CrossRef] [PubMed]
45. Bergasa-Caceres, F.; Rabitz, H.A. Interdiction of Protein Folding for Therapeutic Drug Development in SARS-CoV-2. *J. Phys. Chem. B* **2020**, *124*, 8201–8208. [CrossRef]
46. Bergasa-Caceres, F.; Rabitz, H.A. Protein folding interdiction strategy for therapeutic drug development in viral diseases: Ebola VP40 and Influenza A M1. *Int. J. Mol. Sci.* **2022**, *23*, 3906. [CrossRef] [PubMed]
47. Bergasa-Caceres, F.; Rabitz, H.A. Identification of Two Early Folding Stage Prion Non-Local Contacts Suggested to Serve as Key Steps in Directing the Final Fold to Be Either Native or Pathogenic. *Int. J. Mol. Sci.* **2021**, *22*, 8619. [CrossRef] [PubMed]
48. Berezovsky, I.N.; Trifonov, E.N. Van der Waals Locks: Loop-n-Lock Structure of Globular Proteins. *J. Mol. Biol.* **2001**, *307*, 1419–1426. [CrossRef]
49. Dill, K.A. Dominant Forces in Protein Folding. *Biochemistry* **1990**, *29*, 7133–7155. [CrossRef] [PubMed]
50. Kuwajima, K. The molten globule, and two-state vs. non-two-state folding of globular proteins. *Biomolecules* **2020**, *10*, 407. [CrossRef] [PubMed]
51. Karplus, M.; Weaver, D.L. Diffusion–collision Model for Protein Folding. *Biopolymers* **1979**, *18*, 1421–1437. [CrossRef]
52. Ittah, V.; Haas, E. Nonlocal Interactions Stabilize Long Range Loops in the Initial Folding Intermediates of Reduced Bovine Pancreatic Trypsin Inhibitor. *Biochemistry* **1995**, *34*, 4493–4506. [CrossRef] [PubMed]
53. Sali, A.; Shakhnovich, E.; Karplus, M. How Does a Protein Fold? *Nature* **1994**, *369*, 248–251. [PubMed]

54. Vendruscolo, M.; Paci, E.; Dobson, C.M.; Karplus, M. Three Key Residues Form a Critical Contact Network in a Protein Folding Transition State. *Nature* **2001**, *409*, 641–645. [CrossRef] [PubMed]
55. Alm, E.; Baker, D. Prediction of Protein-Folding Mechanisms from Free-energy Landscapes Derived from Native Structures. *Proc. Natl. Acad. Sci. USA* **1999**, *96*, 11305–11310. [CrossRef] [PubMed]
56. Munoz, V.; Eaton, W.A. A Simple Model for Calculating the Kinetics of Protein Folding from Three-dimensional Structures. *Proc. Natl. Acad. Sci. USA* **1999**, *96*, 11311–11316. [CrossRef] [PubMed]
57. Clementi, C.; Jennings, P.A.; Onuchic, J.N. How Native-state Topology Affects the Folding of Dihydrofolate Reductase and Interleukin-1ß. *Proc. Natl. Acad. Sci. USA* **2000**, *97*, 5871–5876. [CrossRef] [PubMed]
58. Makarov, D.E.; Plaxco, K.W. The Topomer Search Model: A Simple, Quantitative Theory of Two-State Protein Folding Kinetics. *Protein Sci.* **2003**, *12*, 17–26. [CrossRef]
59. Wang, Y.; Rosengarth, A.; Luecke, H. Structure of the human p53 core domain in the absence of DNA. *Acta Crystallogr. D Biol. Crystallogr.* **2007**, *63*, 276–281. [CrossRef]
60. Sehnal, D.; Bittrich, S.; Deshpande, M.; Svobodová, R.; Berka, K.; Bazgier, V.; Velankar, S.; Burley, S.K.; Koča, J.; Rose, A.S.; et al. Mol* viewer: Modern web app for 3D visualization and analysis of large biomolecular structures. *Nucleic Acids Res.* **2021**, *49*, W431–W437. [CrossRef]
61. Bergasa-Caceres, F.; Rabitz, H.A. Role of topology in the cooperative collapse of the protein core in the sequential collapse model. Folding pathway of α-lactalbumin and hen lysozyme. *J. Phys. Chem. B* **2001**, *105*, 2874–2880. [CrossRef]
62. Butler, J.S.; Loh, S.N. Kinetic partitioning during folding of the p53 DNA binding domain. *J. Mol. Biol.* **2005**, *350*, 906–918. [CrossRef] [PubMed]
63. Butler, J.S.; Loh, S.N. Folding and misfolding mechanism of the p53 DNA binding domain at physiological temperature. *Prot. Sci.* **2006**, *15*, 2457–2465. [CrossRef] [PubMed]
64. Bergasa-Caceres, F.; Rabitz, H.A. Relating contact order to the rate of cooperative collapse in the sequential collapse model for protein folding pathways. *Chem. Phys. Lett.* **2003**, *376*, 612–617. [CrossRef]
65. Bullock, A.N.; Henckel, J.; De Decker, B.S.; Johnson, C.M.; Nikolova, P.V.; Proctor, M.R.; Lane, D.P.; Fersht, A.R. Thermodynamic stability of wild-type and mutant p53 core domain. *Proc. Natl. Acad. Sci. USA* **1997**, *94*, 14338–14342. [CrossRef] [PubMed]
66. Ang, H.C.; Joerger, A.C.; Mayer, S.; Fersht, A.R. Effects of common cancer mutations on stability and DNA binding of full length p53 compared with isolated core domains. *J. Biol. Chem.* **2006**, *281*, 21934–21941. [CrossRef] [PubMed]
67. Dea, A.J.; Michaels, T.C.T.; Meisl, G.; Klenerman, D.; Wu, S.; Perrett, S.; Linse, S.; Dobson, C.M.; Knowles, T.P.J. Kinetic diversity of amyloid oligomers. *Proc. Natl. Acad. Sci. USA* **2020**, *22*, 12087–12094.
68. Singh, J.; Udgaonkar, J.B. Molecular Mechanism of the Misfolding and Oligomerization of the Prion Protein: Current Understanding and Its Implication. *J. Phys. Chem. B* **2015**, *54*, 4431–4442. [CrossRef] [PubMed]
69. Singh, J.; Udgaonkar, J.B. The Pathogenic Mutation T182A Converts the Prion Protein into a Molten Globule-like Conformation Whose Misfolding to Oligomers but Not to Fibrils is Drastically Accelerated. *J. Phys. Chem. B* **2015**, *55*, 459–469. [CrossRef] [PubMed]
70. Foster, B.A.; Coffey, H.A.; Moprin, M.J.; Rastinejad, F. Pharmacological rescue of mutant p53 conformation and function. *Science* **1999**, *286*, 2507–2510. [CrossRef] [PubMed]
71. Silva, J.L.; Cino, E.A.; Soares, I.N.; Ferreira, V.F.; de Oliveira, G.A.P. Targeting the prion-like aggregation of mutant p53 to combat cancer. *Acc. Chem. Res.* **2018**, *51*, 151–160. [CrossRef] [PubMed]
72. Poulson, B.G.; Szczepski, K.; Lachowicz, J.I.; Jaremko, L.; Emwas, A.-H.; Jaremko, M. Aggregation of biologically important peptides and proteins: Inhibition or acceleration depending on protein and metal ion concentrations. *RSC Adv.* **2020**, *10*, 215–227. [CrossRef]
73. Emwas, A.H.; Alghrably, M.; Dhahri, M.; Sharfalddin, A.; Alsiary, R.; Jaremko, M.; Faa, G.; Campagna, M.; Congiu, T.; Piras, M.; et al. Living with the enemy: From protein-misfolding pathologies we know, to those we want to know. *Ageing Res. Rev.* **2021**, *70*, 101391. [CrossRef]
74. Jaremko, M.; Jaremko, L.; Kim, H.-Y.; Cho, M.-K.; Schwieters, C.D.; Giller, K.; Becker, S.; Zweckstetter, M. Cold denaturation of a protein dimer monitored at atomic resolution. *Nat. Chem. Biol.* **2013**, *9*, 264–272. [CrossRef]
75. Bergasa-Caceres, F.; Rabitz, H.A. Low Entropic Barrier to the Hydrophobic Collapse of the Prion Protein: Effects of Intermediate States and Conformational Flexibility. *J. Phys. Chem. A* **2010**, *114*, 6978–6982. [CrossRef] [PubMed]
76. Pace, C.N.; Fu, H.; Fryar, K.L.; Landua, J.; Trevino, S.R.; Schell, D.; Thurlkill, R.L.; Imura, S.; Scholtz, J.M.; Gajiwala, K.; et al. Contribution of Hydrogen Bonds to Protein Stability. *Protein Sci.* **2014**, *23*, 652–661. [CrossRef] [PubMed]
77. Fauchere, J.L.; Pliska, V. Hydrophobic Parameters II of Amino-Acid Side Cains from the Partitioning of *N*-Acetyl-Amino-Acid Amides. *Eur. J. Med. Chem.* **1983**, *18*, 369–375.
78. Ozkan, S.B.; Wu, G.A.; Chodera, J.D.; Dill, K.A. Protein Folding by Zipping and Assembly. *Proc. Natl. Acad. Sci. USA* **2007**, *104*, 11987–11992. [CrossRef] [PubMed]
79. Fersht, A.R. Nucleation mechanisms in protein folding. *Curr. Op. Struct. Biol.* **1997**, *7*, 3–9. [CrossRef]
80. Nishikawa, N.; Sakae, Y.; Gouda, T.; Tsujimura, Y.; Okamoto, Y. Structural analysis of a trimer of ß$_2$-microglobulin fragment by molecular dynamics simulations. *Biophys. J.* **2019**, *116*, 781–790. [CrossRef] [PubMed]

molecules

Article

Effect of Cholesterol Molecules on Aβ1-42 Wild-Type and Mutants Trimers

Trung Hai Nguyen [1,2], Phuong H. Nguyen [3,4], Son Tung Ngo [1,2] and Philippe Derreumaux [3,4,5,*]

1 Laboratory of Theoretical and Computational Biophysics, Ton Duc Thang University, Ho Chi Minh City, Vietnam; nguyentrunghai@tdtu.edu.vn (T.H.N.); ngosontung@tdtu.edu.vn (S.T.N.)
2 Faculty of Applied Sciences, Ton Duc Thang University, Ho Chi Minh City, Vietnam
3 Laboratoire de Biochimie Théorique, UPR 9080 CNRS, Université de Paris, 13 rue Pierre et Marie Curie, 75005 Paris, France; phuong.nguyen@ibpc.fr
4 Institut de Biologie Physico-Chimique, Fondation Edmond de Rothschild, PSL Research University, 75005 Paris, France
5 Institut Universitaire de France (IUF), 75005 Paris, France
* Correspondence: philippe.derreumaux@ibpc.fr

Abstract: Alzheimer's disease displays aggregates of the amyloid-beta (Aβ) peptide in the brain, and there is increasing evidence that cholesterol may contribute to the pathogenesis of the disease. Though many experimental and theoretical studies have focused on the interactions of Aβ oligomers with membrane models containing cholesterol, an understanding of the effect of free cholesterol on small Aβ42 oligomers is not fully established. To address this question, we report on replica exchange with a solute tempering simulation of an Aβ42 trimer with cholesterol and compare it with a previous replica exchange molecular dynamics simulation. We show that the binding hot spots of cholesterol are rather complex, involving hydrophobic residues L17–F20 and L30–M35 with a non-negligible contribution of loop residues D22–K28 and N-terminus residues. We also examine the effects of cholesterol on the trimers of the disease-causing A21G and disease-protective A2T mutations by molecular dynamics simulations. We show that these two mutations moderately impact cholesterol-binding modes. In our REST2 simulations, we find that cholesterol is rarely inserted into aggregates but rather attached as dimers and trimers at the surface of Aβ42 oligomers. We propose that cholesterol acts as a glue to speed up the formation of larger aggregates; this provides a mechanistic link between cholesterol and Alzheimer's disease.

Keywords: aggregation; amyloid-beta; mutants; cholesterol; simulations

Citation: Nguyen, T.H.; Nguyen, P.H.; Ngo, S.T.; Derreumaux, P. Effect of Cholesterol Molecules on Aβ1-42 Wild-Type and Mutants Trimers. *Molecules* **2022**, *27*, 1395. https://doi.org/10.3390/molecules27041395

Academic Editors: Yuko Okamoto, Kunihiro Kuwajima, Tuomas Knowles and Michele Vendruscolo

Received: 4 January 2022
Accepted: 15 February 2022
Published: 18 February 2022

1. Introduction

The two hallmarks of Alzheimer's disease (AD) are extracellular amyloid-beta (Aβ) plaques of Aβ42 and Aβ40 peptides, the 42-residue species being the most toxic and intracellular neurofibrillary tangles built from hyperphosphorylated tau protein [1]. Despite extensive research, all drugs targeting Aβ and tau oligomers have failed in AD [2,3]. Among many cellular factors contributing to AD development, cholesterol plays a critical role via different actions.

First, cholesterol is present in micro-dissected AD senile plaques with a molar ratio of 1:1 [4] and purified AD paired helical fragments of tangles [5]. The level of plasma cholesterol is 10% higher in AD patients than in normal individuals [6]. Cholesterol levels in the brain positively correlate with the severity of dementia in AD patients [7] in contrast to Aβ plaque burden, which correlates weakly with disease severity [8].

Second, cholesterol impacts Aβ production through the cleavage of the amyloid protein precursor (APP) [9]. Based on nuclear magnetic resonance and electron paramagnetic resonance, E22-N27 residues in the bulk solution and membrane-buried residues G29-G33 (using Aβ42 amino acid numbering) play a key role in cholesterol binding to

APP [10]. Recently, it was proposed from atomistic molecular dynamics (MD) simulations that cholesterol modulates the conformation and activity of the C-terminal domain of APP directly through hydrogen bonding and indirectly through induction of the liquid-ordered phase [11]. Furthermore, Aβ accumulation in neurons is tightly regulated by cholesterol production in astrocytes [12], and elevated membrane cholesterol alters lysosomal degradation to induce Aβ degradation [13]. Finally, the apolipoprotein E gene, a major transporter of cholesterol in the brain, and in particular its e4 allele, the most genetic risk factor for AD, is associated with higher total cholesterol [14].

Third, numerous in vitro experiments have explored Aβ aggregation in a water–phospholipid–cholesterol membrane environment. They showed that cholesterol favors the formation of Aβ pores in the membrane of brain cells [15]. Cholesterol inserted into membranes facilitates aggregation of Aβ at the surface of the membranes at physiological concentrations, as reported by atomic force microscopy images [16]. Lipid membranes containing cholesterol also enhance the primary nucleation rate of Aβ42 aggregation by up to 20-fold, as reported by aggregation kinetics experiments [17].

At the theoretical level, many all-atom MD simulations were performed to study the interactions between Aβ42 peptides and membranes with various contents of cholesterol, revealing the mechanisms by which cholesterol changes the mechanisms of Aβ monomers and dimers binding and affinity to the lipid bilayer [16,18–20]. The binding of Aβ42 amyloid fibrils to the lipid bilayer was also investigated using coarse-grained MD simulations and showed that the addition of cholesterol increases the binding frequency and alters the binding interface and contacts [21].

Little is known about the interactions of Aβ42 peptides with free cholesterol in the bulk solution. The effect of unmodified cholesterol and charged cholesterol derivatives on Aβ40 fibril formation was explored using a Thioflavin T kinetic aggregation assay, atomic force microscopy and dynamic light scattering. At a concentration lower than the critical micelle concentration, and therefore assuming that cholesterol exists mainly as monomeric molecules, unmodified cholesterol and positively and negatively charged cholesterol derivatives accelerate the aggregation rate of Aβ40, the aggregation half time being reduced by 15 to 25% compared to Aβ40 peptide in phosphate buffer [22].

Using transmission electron microscopy, Harris et al. demonstrated the binding of soluble 10 nm diameter cholesterol-PEG 600 micelles to Aβ42 fibrils and proposed that binding involves the central hydrophobic core CHC spanning residues L17–A21 [23]. Atomistic 20 ns MD simulation of a system of four Aβ42-12 cholesterol molecules in a box of water showed stable contact between the benzyl group of F19 and the cholesterol micelle, which forms a flat surface and, notably, the steroid group of cholesterol [24]. It was suggested that the carboxyl terminus spanning residues have a higher tendency to interact with cholesterol and form β-sheet conformation [24].

It is well-established that lipids can be removed from the membrane because of oligomers–membrane interactions [25–27]. Recent experimental studies indicate the critical role of free phospholipids at a nanomolar to micromolar concentration in equilibrium with the membrane (large unilamellar vesicles) in forming Aβ–lipids complexes that favor amyloid–membrane poration [28,29]. We recently performed atomistic replica exchange molecular dynamics (REMD) simulations of an Aβ42 dimer and trimer with cholesterol using the protein AMBER99sb-ildn force field (also called AMBER ff99sb-ildn) with the TIP3P water force field. We reported on the drastic effect of cholesterol on the conformational ensemble of both Aβ42 species and showed multiple transient binding modes involving the residues L17–A21 and L30–M35 [30].

In this study, we further studied the impact of cholesterol with a ratio of 1:1 on Aβ42 trimers in the bulk solution. First, replica exchange with solute tempering (REST2) was chosen for sampling, as this procedure converges faster to equilibrium than REMD while requiring a smaller number of replicas [31–33]. Second, we used CHARMM36m, as this force field is more relevant for intrinsically disordered proteins [34] and small oligomers of Aβ [35–39]. This allows comparison with our previous REMD study using

AMBER99sb-ildn TIP3P [30]. Third, we examined the impact of the A2T mutation, known to be AD protective, and the disease-causing A21G mutation [3] on the interactions of the Aβ trimer–cholesterol by MD simulations. It is to be noted that the A21G and A2T Aβ42 oligomers were extensively studied by experimental and theoretical means in the bulk solution [40–46]; however, no study reports on their interactions with free cholesterol.

2. Materials and Methods

As in our previous REMD study [30], the cholesterol molecule was geometrically optimized by quantum mechanics with the B3LYP functional and the 6-31G(d,p) basis set, and the cholesterol force field was parameterized using the general amber force field (GAFF) [42], the restrained electrostatic potential (RESP) method [47] and the B3LYP functional and the 6-31G(d,p) basis set [48].

In all simulations, Aβ42 peptides at pH 7 have $NH3^+$ and $CO2^-$ termini, deprotonated Glu and Asp, protonated Arg and Lys and neutral His with a protonated N_ε atom. The Aβ42/cholesterol system was neutralized by sodium ions. Prior to MD and REST2 simulations, all systems were minimized with harmonic restraints on the positions of the peptide Cα atoms and then equilibrated by MD at 310 K for 1 ns in the NVT ensemble, followed by 2 ns in the NPT ensemble.

In REST2, we used the same initial structure of the Aβ42 trimer–cholesterol system as in our REMD simulation inserted into a cubic water box of 7.26 nm size and 382.66 nm^3 volume (10,869 water molecules). This structure with a high β-hairpin propensity of each chain was already discussed in many simulations of Aβ monomers [3,22] and oligomers [28,37,38,49,50] and in exploring amyloid oligomers with a peptide model system [51]. The equilibrated structure shown in Figure 1 started from three separated β-hairpins with one hairpin perpendicular to the other two peptides, and three cholesterol molecules randomly positioned and orientated with respect to Aβ42 peptides with a minimal distance of 1 nm from all Aβ42 atoms.

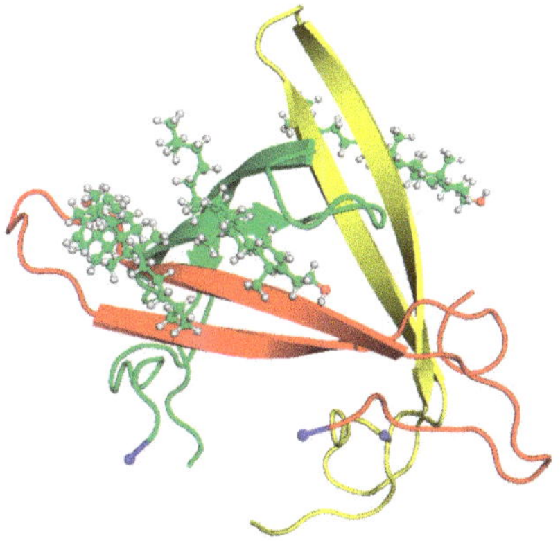

Figure 1. Initial REST2 conformation of Aβ42 trimer—three cholesterols with Aβ42 forming β-sheets at residues 13–21 and 29–35 of chain A (red), residues 16–22 and 31–37 of chain B (green) and residues 13–21 and 27–35 of chain C (yellow). Each N-terminus is shown by a blue ball. The three all-atom cholesterol molecules are shown in green.

REST2 simulation with the CHARMM36m-TIP3P force field was performed using NAMD [52]. We used a time step of 2 fs, a cutoff of 1.2 nm for Van der Waals interactions and a cutoff of 1.1 nm for electrostatic interactions using the particle mesh Ewald (PME) method [53]. REST2 scales the solute interactions by λ with the solute consisting of Aβ42 and cholesterol molecules, scales the solute–water interactions by $\lambda^{1/2}$ and leaves the water–water interactions unaltered. Using T_{min} = 310 K and T_{max} = 500 K with the number of replicas set to 16, we performed REST2 simulation in the NPT ensemble at the temperature of 310 K for 250 ns on 16 replicas exchanging the following solute–solute corresponding temperatures of 310, 320, 330.4, 341.1, 352.1, 363.5, 375.3, 387.5, 400, 413, 426, 440.1, 454.4, 469.1, 484.3 and 500 K, i.e., for λ varying between 1 and 0.75. Exchanges of configurations between neighboring replicas were attempted every 1 ps. In the REST2 simulation, we used Nose–Hoover Langevin pressure control and Langevin temperature control as described in NAMD.

Note that MD simulations of the selected WT, A2T and A21G systems were performed at 310 K using GROMACS [54] with the velocity-rescaling thermostat [55], a cutoff of 1.2 nm for Van der Waals interactions and a cutoff of 1.1 nm for electrostatic interactions using the PME method.

To determine the interface between Aβ42 chains and cholesterols, we calculated the distances between heavy atoms of cholesterols and non-hydrogen atoms of the side chain residues of each Aβ42 chain. We considered that a contact formed between a side chain residue and cholesterol if there was at least one distance below 0.45 nm. For clarity, we report on the probability of side-chain contacts per Aβ42 chain and cholesterol molecule. We also calculated the percentage of cholesterol monomers, dimers+monomer and trimers. Monomers of cholesterol exist if there are no contacts between any two cholesterols, and trimers of cholesterol are formed if there is at least one contact between all cholesterols, a contact being defined if the intermolecular distance between any two heavy atoms is less than 0.45 nm. The DSSP protocol was used to determine the secondary structure of Aβ42 peptides [56].

The full energy landscape was approximated by a 2D energy landscape. First, the trajectory was projected on the first two principal components obtained after the diagonalization of the positional fluctuation covariance matrix of the backbone peptides and cholesterol atoms. Then, the free energy landscape was constructed from the previous data using the formula $-RT \times \log(H(x,y))$, where H(x,y) is the histogram of the two selected order parameters x and y [57]. The population of each minimum was determined by counting all conformations around each minimum. The representative structure or center of each cluster was obtained by the Daura clustering method [58].

In what follows, we used for comparison the time intervals 50–250 ns of the present REST2 simulation at 310 K and our previous REMD simulation at 315 K. Note that if death is imminent for humans at 315 K, atomistic protein simulations at 310 and 315 K led to very similar thermodynamic and structural properties. Additionally, a pure cholesterol bilayer was already explored at 310 K by all-atom MD simulations [59], and REST2 was successfully shown to simulate the weak binding of Aβ40 peptides on a lipid bilayer [32] and the lateral equilibration in mixed cholesterol-DPPC bilayers [60] very well.

3. Results and Discussion

The sampling of the REST2 simulation using 16 replicas is first illustrated by the good overlap of the distributions of the potential energy between neighboring replicas (Figure 2A). The exchanges of coordinates as a function of simulation time for replicas 1 (Figure 2B) and 16 (Figure 2C) indicate exploration of the full replica space in the nanosecond time scale. Overall, the average acceptance probability at different replicas is 0.3. The convergence of the REST2 simulation at 310 K is also assessed by the very high similarity of the two FELs using the single REST2 trajectory at the time intervals 50–220 ns and 50–250 ns (Figure 3A,B).

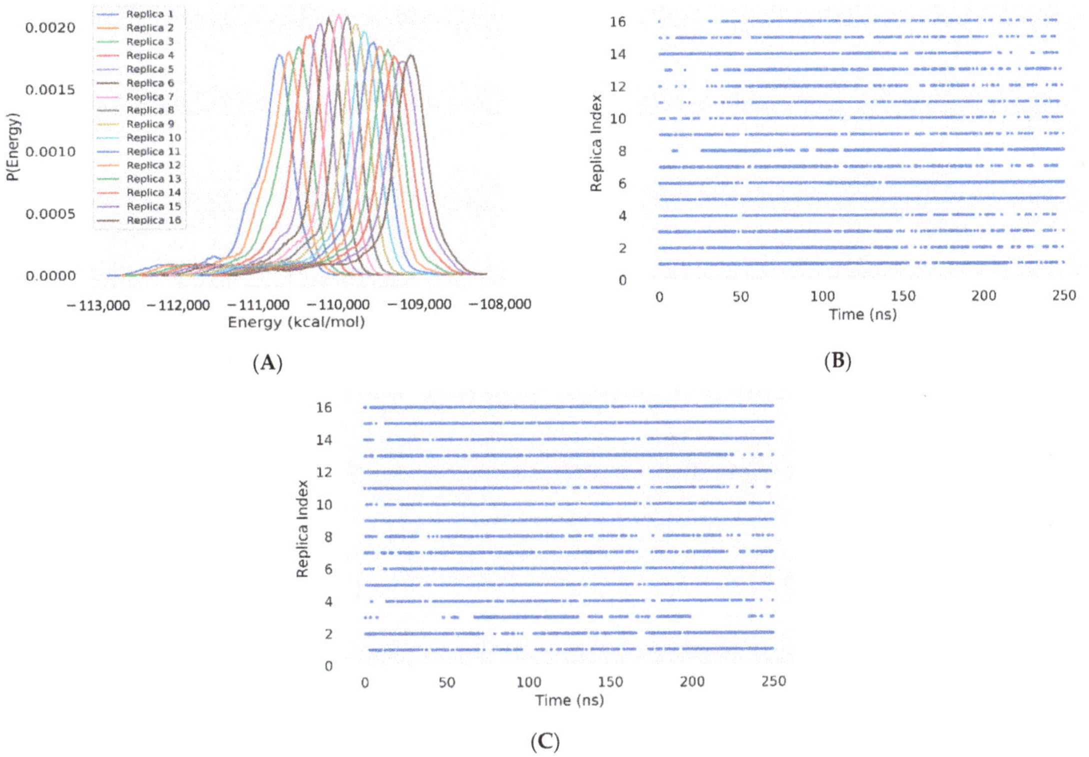

Figure 2. REST2 Sampling. (**A**) Overlap of the total potential energies for replicas 1 to 16. (**B**,**C**) Exchanges of coordinates as a function of simulation time of replica 1 and replica 16, respectively.

The secondary structure compositions of Aβ42 slightly vary between the two simulations, reaching β-sheet and coil contents of 37% and 36% in REMD vs. 36% and 43% in REST2. There are, however, differences along the amino acid sequence. Both methods give the same β-strand content for the residues 15–21 and 28–37 (Figure 4A) and the same turn probability for residues 22–27 (Figure 4B). The β-strand probability of residues 3–5 and the turn character of residues 6–10 are, however, reduced by a factor of two from REMD to REST2 (18% β-strand and 30% turn by REST2), leading to an enhancement of coil character of the N-terminus (residues 1–10) in REST2 compared to REMD (Figure 4C). Additionally, REST2 explores more β-strands at positions 36–41 (Figure 4A).

Differences in the conformational space explored by the REST2 and REMD simulations are further analyzed by comparing the two FEL's using the combined REMD and REST2 trajectories to compute the first two principal components. Though the direct comparison is not possible because we use two distinct force fields, there is some overlap, but overall, the free energy landscapes are different. In contrast to the REMD FEL (Figure 3C), the REST2 FEL is divided into well-separated regions (Figure 3D). Additionally, the amplitudes of fluctuation along the PC1 and PC2 are also much larger in REST2, and the REST2 FEL is dominated by two states, S1 and S2, representing 40% and 22% of the full ensemble. In contrast, the REMD FEL is dominated by two states representing 12% and 28% of the conformational ensemble.

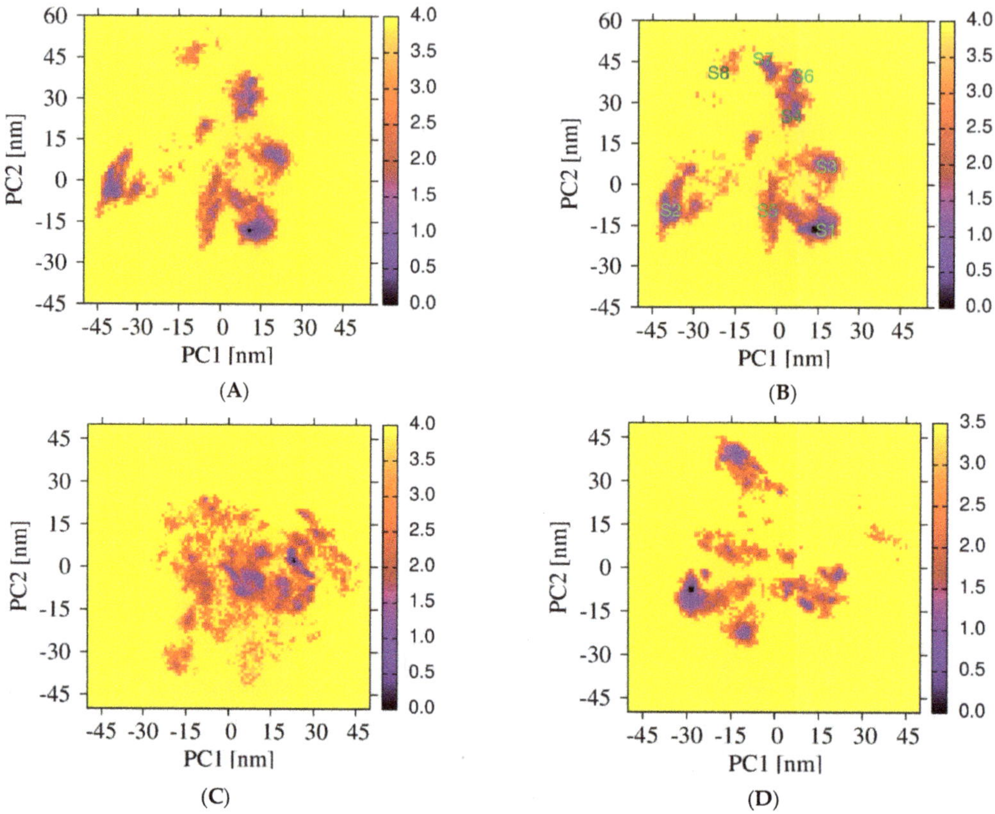

Figure 3. Free energy landscapes of Aβ42 trimer + 3 cholesterols from PCA. REST2 FEL by using the single REST2 trajectory at 310 K and time intervals 50–220 ns (**A**) and 50–250 ns (**B**). REMD FEL (**C**) and REST2 FEL (**D**) by using the combined REST2 (310 K) and REMD (315 K) trajectories and the time interval 50–250 ns. Note that the PC1 and PC2 differ from REMD to REST2.

Deviation between the REST2 and REMD conformational ensembles is observed for the probability of side-chain contacts per cholesterol molecule and Aβ42 chain (Figure 5). Using REST2, the highest binding spot with cholesterol molecules involves the side-chains of the CHC region (residues 17–21) with a probability of 33% for the aromatic interaction with F19 and 17% for the interaction with L17, and then the side-chains of residues L30, I31 and I32 and residues L34 and M35 (average probability of 12%). The same residues are identified with REMD, but the probability decreases notably for F19 to 25% and increases moderately for the residues H6, Y10, V12, H13 and N27, with most probabilities remaining < 10%, however. There is almost no difference between the REMD and REST2 populations of (free monomers of cholesterol, dimers + monomer of cholesterol and trimers of cholesterol), as they reach (0.5%, 43% and 56%) in REST2 vs. (0.6%, 47% and 52%) in REMD. It is found that the network of interactions between the side-chains of Aβ42 varies with the aggregated forms of cholesterol. Using REST2, there are 15 contacts per Aβ42 chain and per cholesterol with a probability > 7.5% when cholesterol is in dimers + monomer form, while there are 8 contacts formed when cholesterol is in trimer form, and this is accompanied by a substantial probability reduction with the CHC region and residues 30–41 (Figure 6).

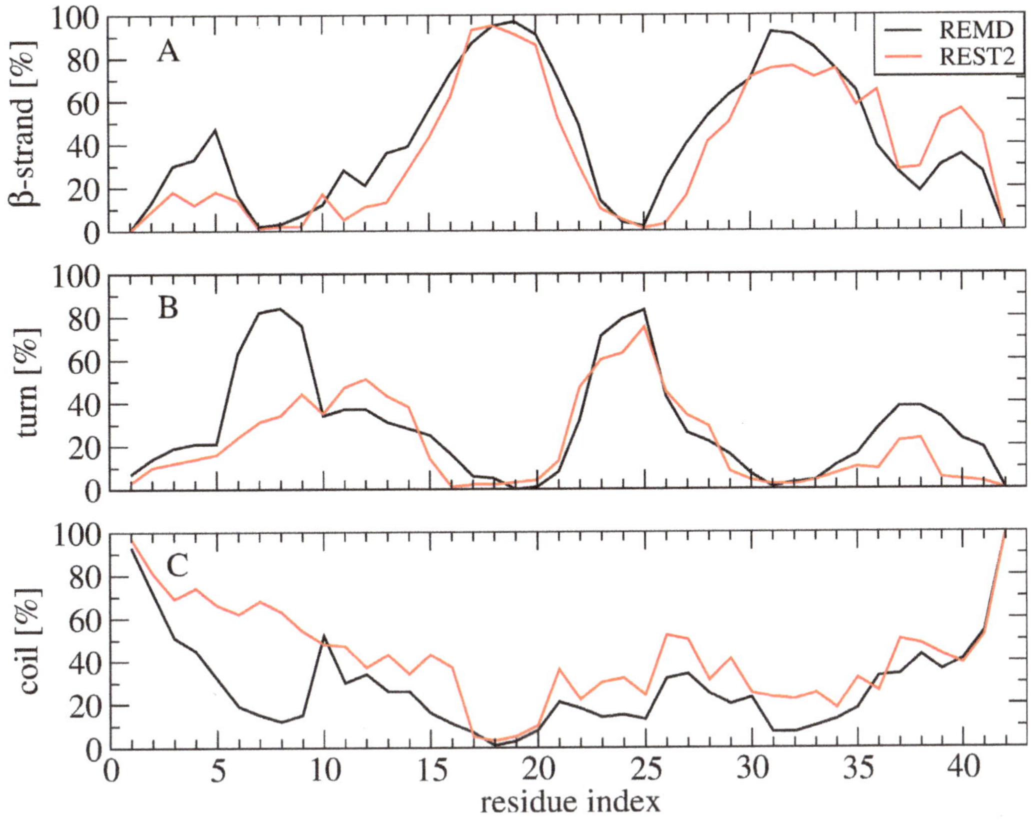

Figure 4. Secondary structure composition along the amino acid sequence using the time interval 50–250 ns of the REMD simulation at 315 K and the time interval 50–250 ns of the REST2 simulation at 310 K. (**A**) beta-strand, (**B**) turn and (**C**) coil content.

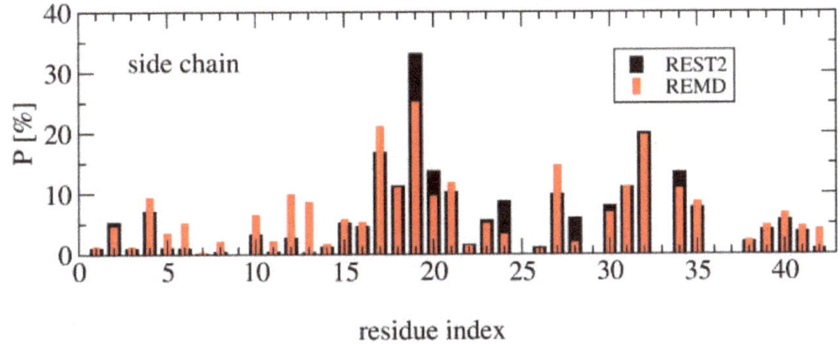

Figure 5. Probability of side-chain contacts between each cholesterol molecule and each amino acid of each Aβ42 peptide using the time interval 50–250 ns of REMD at 315 K, and REST2 at 310 K.

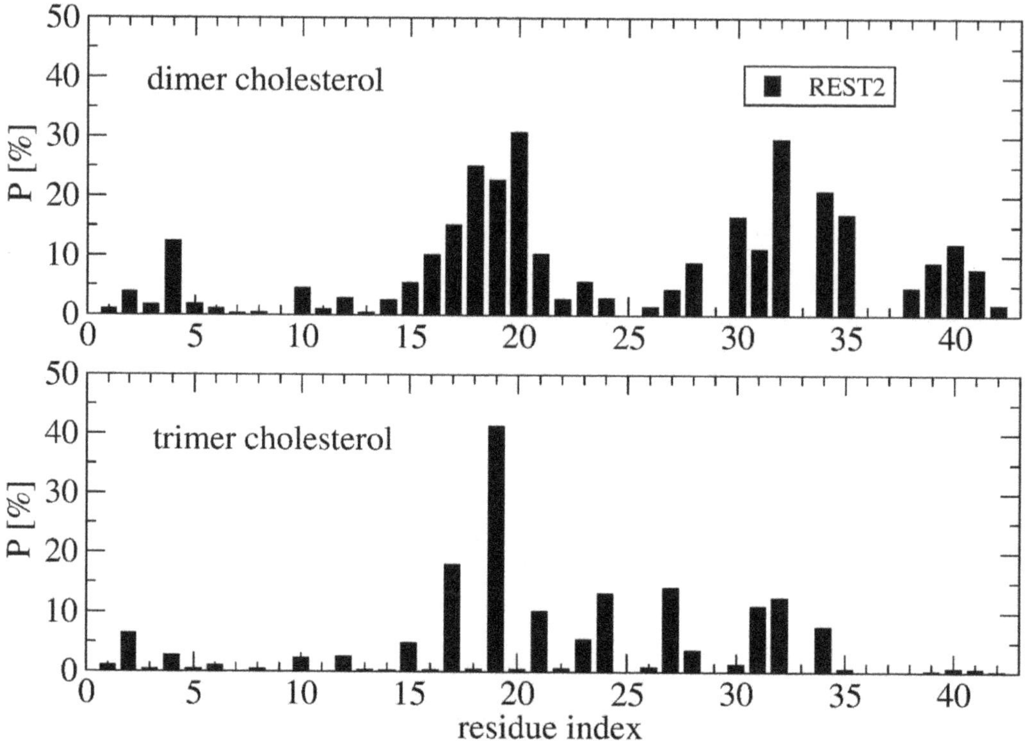

Figure 6. Probability of contact between the cholesterol molecules in dimer+monomer form and trimer form and the side-chains of Aβ42 using the time intervals 50–250 ns of REST2 at 310 K.

Overall, there are non-negligible differences between the REST2 and REMD results. REST2 results emphasize the role of the CHC region L17–A21 and the hydrophobic residues L30–M35 and V39–I41 in the binding with the dimer form of cholesterol, and the residues L17, F19, A21, V24, N27, I31 and I32 in the binding with the trimer forms of cholesterol. This binding mechanistic view between the hydrophobic region of Aβ42 and cholesterol explains why the nonvesicle forms of the negatively charged cholesterol sulfate and the cationic cholesterol derivative, 3β(N-dimethylaminoethane)carbaloyl)-cholesterol, moderately change the aggregation kinetics of Aβ40 [22]. Our cholesterol-binding mechanism with an average probability of contacts of 16% averaged over the residues 17–21 is clearly more complex than that previously described by short MD simulations, which only emphasized the critical role of the side-chain of F19 in the potentiation effect of cholesterol on Aβ40 fibril formation [24].

The representative structures of the eight free energy minima designated as S1–S8 on the FEL with decreasing populations are shown in Figure 7. The first minimum, S1, with a population of 40%, is characterized by three chains forming well-defined β-hairpins and chains A and B forming a short-twisted four-stranded antiparallel β-sheet. The β-strands cover residues L17–A21 and N27–I31 in chain A, residues L17–A21 and I32–V36 in chain B and residues H14–A21 and K28–M35 in chain C, indicating that the β-hairpins are distinct and extend beyond the CHC region by including some residues in the loop region (residues E22–K28). The second minimum (S2, population of 22%) displays two highly bent and twisted β-hairpins with strands covering residues L17–F20 and A30–I41 in chain A, residues L17–F20, K28–L34 and V39–I41 in chain C and chain B adopting a five stranded β-sheet at R5–S8, Y10–Q15, L17–E22, I31–V36 and V39–I41. The orientation of the three chains is complex as the C-terminus residues (30–41) of chain A are antiparallel with the K28–L34 residues of chain C, and the CHC of chain C is parallel to residues V39–I41 of chain C.

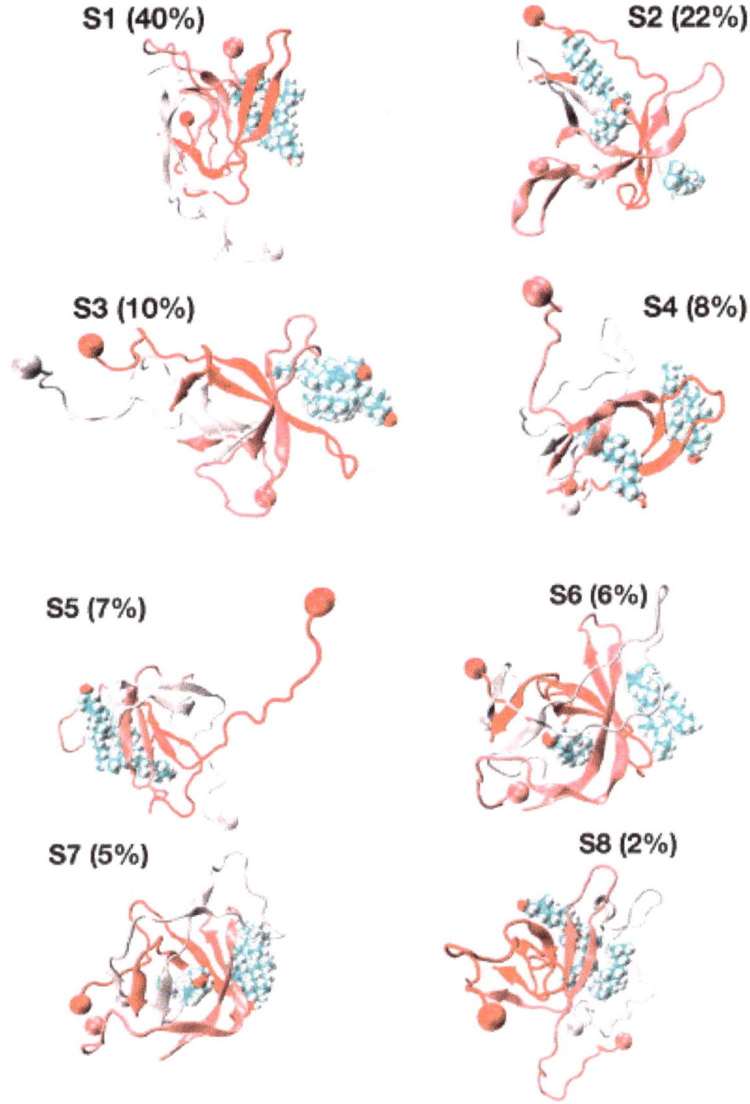

Figure 7. Representative structures and populations of eight free energy minima from REST2 simulation at 310 K. Chain A is in red, chain C is in white and chain B is in pastel pink. The ball shows the N-terminus of each chain. The cholesterols are visualized in all-atom and Van der Waals representations.

The S3, S4, S6 and S7 minima, with a total population of 29%, have a β-strand content of 36% and share the same topological features: two peptides with highly twisted β-hairpins (chains A and B) covering distinct amino acids perpendicular to each other, and a third disordered and compact peptide (chain C) with no preferred interface with the other two chains. For instance, there is a small intermolecular β-sheet between residues V40–I41 and K28–G38 in S3 and between residues L34–V36 and V39–I41 in S4. The S5 state with a population of 7%, dominated by a coil (49%) with low β-strand content (29%), forms a twisted four-stranded antiparallel β-sheet (chains A and B) with chain C having very deformed β-strands. Finally, the S8 state with a population of 2% is amorphous and has high turn and coil contents of 36% and 38%. S8 has a low β-strand content of 24% with strands at residues Q15–V18, L30–G33 and V40–I41 in chain A and residues V18–E22, I31–M35 and V39–V40 in chain B, forming a short intermolecular parallel β-sheet between the two termini of chains A and B. In this state, chain C is almost devoid of any secondary structure with the exception of a small helix covering residues A30–G33.

The various globular shapes of β-strand contents varying between 24% (S8 state) and 42% (S2) are all characterized by different orientations and packings of the chains, but also different binding interfaces and hot spots with cholesterol (Figure 8). In the S1 state, the residue hot spots with a trimer of cholesterol involve residues H14, K16, N27, K28 and I31 of chain A, and residues A2, F19, A21, D23, L30, I32 and L34 of chain B. In this state, cholesterol is located at the surface of the Aβ trimer (Figure 7, S1 state). The S3 state has almost the same residues for binding and a trimer of cholesterol located at the surface. The S5 state is also characterized by a trimer of cholesterol at the surface, but binding involves many residues in the F4–L34 region of chain B, the residues V24, K28, I31, I41 and A42 of chain A, and residues E22 and A42 of chain C (Figure 8). In total, 57% of Aβ states have a trimer of cholesterol at the surface.

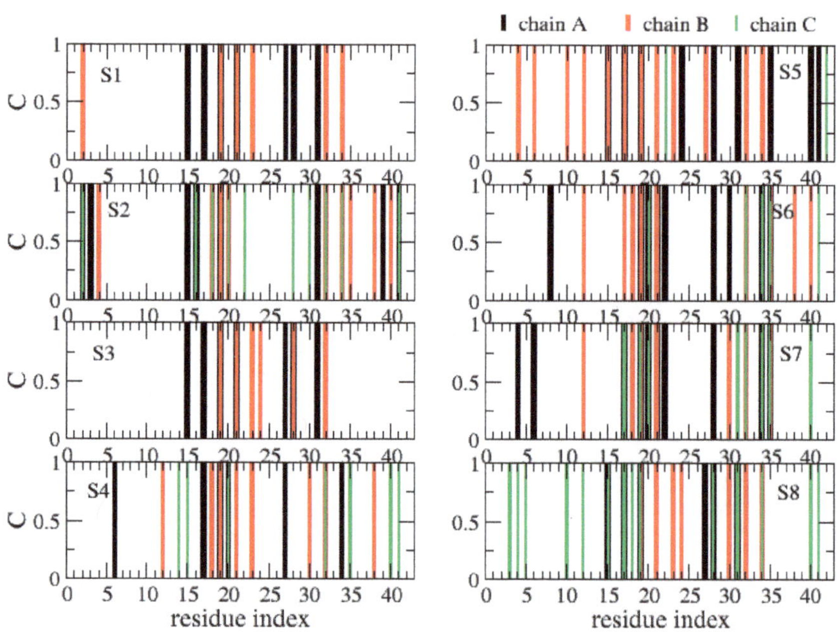

Figure 8. Contact between each Aβ42 chain (A, B, C) and cholesterols. Shown are results for each free energy minimum Si (i = 1 . . . 8). One Aβ42 side-chain and one cholesterol molecule are considered in contact if there is at least one intermolecular distance below 0.45 nm.

The S4, S6 and S7 states representing 22% of the ensemble are characterized by a dimer of cholesterol at the surface and one cholesterol inserted into the complex (Figure 7). In these states, we find that the residues 16–22 are essential for binding, but many residues from the region K28 to A42 and a few residues in the N-terminus (residues F4, H6, S8 and V12) participate as well (Figure 8). Finally, the S2 state reveals a dimer of cholesterol inserted in Aβ trimer and one cholesterol at the surface with binding residues D1–F4, Q15–V24 and K28–I41. In contrast, the S8 state reveals a trimer of cholesterol in the interior of oligomers with binding residues D1–Q15 and K16–V36 (Figure 7 panel S8 and Figure 8).

Overall, these results are very different from a previous MD simulation which only reported on interactions of Aβ42 peptides bound to the surface cholesterol micelle [22]. Our cholesterol-binding sites of Aβ42 also differ from combined docking modelling and surface pressure studies of Langmuir monolayers that suggested the region E22–M35 as the minimal cholesterol-binding site. It must be stressed that the docking procedure used the NMR structure of an Aβ monomer mixed with detergent micelles, characterized by an alpha-helix at residues Q15–V36 with a hinge at residues G25–N27 [61]. However, it is interesting that our simulations report on the contribution of the loop residues E22–K28 in the binding process.

To further understand the binding mechanism of cholesterol, Figure 9 report the time-averaged probability of contacts between the cholesterol molecules and the side chains of wild-type (WT) Aβ42, A21G Aβ42 and A2T Aβ42 obtained from a total of 3 microseconds per system (namely 1 microsecond MD simulation starting from the S1, S2 and S4 structures at 310 K shown on Figure 7). Though the S1, S2 and S4 structures of WT represent 70% of the full conformation space of the wild-type sequence, they were selected because they display hot spots involving the N-terminus, the CHC, the loop region and the C-terminus. It is important to stress that these two mutations were also selected because A2T introduces a hydrophilic residue and thus potentially reduces the hydrophobic surface with cholesterol, and A21G because it reduces the total hydrophobic character of the CHC region. Our results show that both mutations do not change the profile of interactions per Aβ42 residue and per cholesterol in all regions of Aβ42, including the N-terminus and the CHC region (Figure 9). These results suggest that cholesterol should moderately alter the aggregation kinetics of Aβ42 A2T and Aβ42 A21G in the bulk solution, with an enhancement that should be comparable to that observed for WT Aβ42 [22].

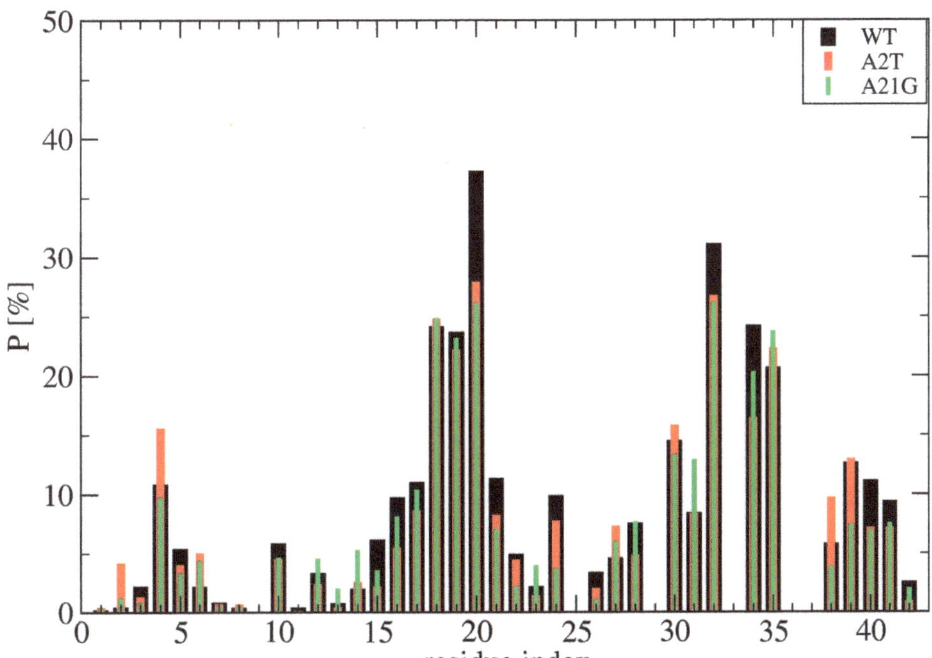

Figure 9. Time-averaged probability of contacts between the cholesterol molecules and the side-chain of Aβ42 obtained from 1 microsecond MD trajectory at 310 K of the wild-type (WT, black), A2T mutation (red) and A21G mutation (green) sequence. Shown are the averaged values over the three simulations starting from the S1, S2 and S4 structures displayed in Figure 7.

4. Conclusions

In summary, we determined the Aβ42/cholesterol trimeric states by means of REST2 simulations and a force field designed for intrinsically disordered proteins. Consistent with our previous REMD simulation using a force field for well-structured proteins [30], we found that the conformational space does not contain the aggregation-prone state of the parallel U- and S-shape Aβ42 fibrils [2] but displays some antiparallel dimers with short intermolecular β-sheets built on strands located at different regions that go beyond the CHC region. Based on our previous REMD study of Aβ42/cholesterol dimers showing an increase of the population of β-hairpin and β-sheet contents upon cholesterol addition compared to a pure bulk solution [30], it is likely that all Aβ42 oligomers mixed with free cholesterol will show more β-structural and antiparallel β-sheet features of the peptides than in pure bulk solution as the oligomer size augments [50].

Our simulations also show that the formation of Aβ42 trimers in the presence of cholesterol involves two binding interfaces. A first highly populated one, where cholesterols are mainly located at the surface of Aβ oligomers (either as a trimer or a dimer of cholesterol), and a second much less populated where cholesterol is fully inserted in the interior of the Aβ oligomers. The predominance of the first binding interface implies that cholesterol should act as a glue to speed up the formation of larger aggregates and therefore catalyzes primary nucleation. This explains the modest but non-negligible experimentally observed acceleration of the aggregation rate of Aβ40 and Aβ42 in the presence of cholesterol [22].

Finally, we found that the binding hot spots of Aβ are very complex and go much beyond the CHC region (residues L17–A21) and the hydrophobic residues L30–M35. Our mechanism cannot be generalized to all amyloid proteins. Indeed, it was shown that free cholesterol has an inhibitory effect on the aggregation of the 37-residue amylin protein, which has two hydrophobic regions, LANFLV and FGAIL, separated by HSSNN, in both solutions and on model membranes [62]. Clearly, a better understanding of the interactions of free cholesterols or free phospholipids on amyloid aggregates in the brain either by computational [63] or experimental [25,64] means must be further explored. To this end, we are coupling a coarse-grained protein force field in an aqueous solution [65,66] with coarse-grained cholesterols and phospholipids [67] to explore larger aggregates and the impact of other disease-causing and disease-protecting mutations [68–71].

Author Contributions: Conceptualization, P.D.; Data curation, T.H.N. and P.H.N.; Formal analysis, T.H.N., P.H.N., S.T.N. and P.D.; Investigation, P.H.N.; Methodology, T.H.N., P.H.N. and S.T.N.; Writing—review & editing, P.D. All authors have read and agreed to the published version of the manuscript.

Funding: This research was funded by Initiative d'Excellence program from the French state grant numbers DYNAMO ANR11-LBX-0011 and CACSICE ANR-11-EQPC-0008.

Institutional Review Board Statement: Not applicable.

Informed Consent Statement: Not applicable.

Data Availability Statement: Not applicable.

Acknowledgments: This work was supported by grants from the French IDRIS and CINES computer centers.

Conflicts of Interest: The authors declare no conflict of interest.

References

1. Benson, M.D.; Buxbaum, J.N.; Eisenberg, D.S.; Merlini, G.; Saraiva, M.J.M.; Sekijima, Y.; Sipe, J.D.; Westermark, P. Amyloid Nomenclature 2018: Recommendations by the International Society of Amyloidosis (ISA) Nomenclature Committee. *Amyloid* **2018**, *25*, 215–219. [CrossRef] [PubMed]
2. Doig, A.J.; Del Castillo-Frias, M.P.; Berthoumieu, O.; Tarus, B.; Nasica-Labouze, J.; Sterpone, F.; Nguyen, P.H.; Hooper, N.M.; Faller, P.; Derreumaux, P. Why Is Research on Amyloid-β Failing to Give New Drugs for Alzheimer's Disease? *ACS Chem. Neurosci.* **2017**, *8*, 1435–1437. [CrossRef] [PubMed]

3. Nasica-Labouze, J.; Nguyen, P.H.; Sterpone, F.; Berthoumieu, O.; Buchete, N.V.; Coté, S.; De Simone, A.; Doig, A.J.; Faller, P.; Garcia, A.; et al. Amyloid β Protein and Alzheimer's Disease: When Computer Simulations Complement Experimental Studies. *Chem. Rev.* **2015**, *115*, 3518–3563. [CrossRef] [PubMed]
4. Panchal, M.; Loeper, J.; Cossec, J.C.; Perruchini, C.; Lazar, A.; Pompon, D.; Duyckaerts, C. Enrichment of Cholesterol in Microdissected Alzheimer's Disease Senile Plaques as Assessed by Mass Spectrometry. *J. Lipid Res.* **2010**, *51*, 598–605. [CrossRef] [PubMed]
5. Gellermann, G.P.; Appel, T.R.; Davies, P.; Diekmann, S. Paired helical filaments contain small amounts of cholesterol, phosphatidylcholine and sphingolipids. *Biol. Chem.* **2006**, *387*, 1267–1274. [CrossRef] [PubMed]
6. Popp, J.; Meichsner, S.; Kölsch, H.; Lewczuk, P.; Maier, W.; Kornhuber, J.; Jessen, F.; Lütjohann, D. Cerebral and Extracerebral Cholesterol Metabolism and CSF Markers of Alzheimer's Disease. *Biochem. Pharmacol.* **2013**, *86*, 37–42. [CrossRef]
7. Cutler, R.G.; Kelly, J.; Storie, K.; Pedersen, W.A.; Tammara, A.; Hatanpaa, K.; Troncoso, J.C.; Mattson, M.P. Involvement of oxidative stress-induced abnormalities in ceramide and cholesterol metabolism in brain aging and Alzheimer's disease. *Proc. Natl. Acad. Sci. USA* **2004**, *101*, 2070–2075. [CrossRef]
8. Makin, S. The amyloid hypothesis on trial. *Nature* **2018**, *559*, S4–S7. [CrossRef]
9. Xiong, H.; Callaghan, D.; Jones, A.; Walker, D.G.; Lue, L.F.; Beach, T.G.; Sue, L.I.; Woulfe, J.; Xu, H.; Stanimirovic, D.B.; et al. Cholesterol retention in Alzheimer's brain is responsible for high beta- and gamma-secretase activities and Abeta production. *Neurobiol. Dis.* **2008**, *29*, 422–437. [CrossRef]
10. Barrett, P.J.; Song, Y.; Van Horn, W.D.; Hustedt, E.J.; Schafer, J.M.; Hadziselimovic, A.; Beel, A.J.; Sanders, C.R. The amyloid precursor protein has a flexible transmembrane domain and binds cholesterol. *Science* **2012**, *336*, 1168–1171. [CrossRef]
11. Pantelopulos, G.A.; Panahi, A.; Straub, J.E. Impact of Cholesterol Concentration and Lipid Phase on Structure and Fluctuation of Amyloid Precursor Protein. *J. Phys. Chem. B* **2020**, *124*, 10173–10185. [CrossRef] [PubMed]
12. Wang, H.; Kulas, J.A.; Wang, C.; Holtzman, D.M.; Ferris, H.A.; Hansen, S.B. Regulation of beta-amyloid production in neurons by astrocyte-derived cholesterol. *Proc. Natl. Acad. Sci. USA* **2021**, *118*, e2102191118. [CrossRef] [PubMed]
13. Takeuchi, S.; Ueda, N.; Suzuki, K.; Shimozawa, N.; Yasutomi, Y.; Kimura, N. Elevated Membrane Cholesterol Disrupts Lysosomal Degradation to Induce β-Amyloid Accumulation: The Potential Mechanism Underlying Augmentation of β-Amyloid Pathology by Type 2 Diabetes Mellitus. *Am. J. Pathol.* **2019**, *189*, 391–404. [CrossRef] [PubMed]
14. Corder, E.H.; Saunders, A.M.; Strittmatter, W.J.; Schmechel, D.E.; Gaskell, P.C.; Small, G.W.; Roses, A.D.; Haines, J.L.; Pericak-Vance, M.A. Gene dose of apolipoprotein E type 4 allele and the risk of Alzheimer's disease in late onset families. *Science* **1993**, *261*, 921–923. [CrossRef]
15. Di Scala, C.; Yahi, N.; Boutemeur, S.; Flores, A.; Rodriguez, L.; Chahinian, H.; Fantini, J. Common Molecular Mechanism of Amyloid Pore Formation by Alzheimer's β-amyloid Peptide and α-synuclein. *Sci. Rep.* **2016**, *6*, 28781. [CrossRef]
16. Banerjee, S.; Hashemi, M.; Zagorski, K.; Lyubchenko, Y.L. Cholesterol in Membranes Facilitates Aggregation of Amyloid β Protein at Physiologically Relevant Concentrations. *ACS Chem. Neurosci.* **2021**, *12*, 506–516. [CrossRef]
17. Habchi, J.; Chia, S.; Galvagnion, C.; Michaels, T.C.T.; Bellaiche, M.M.J.; Ruggeri, F.S.; Sanguanini, M.; Idini, I.; Kumita, J.R.; Sparr, E.; et al. Cholesterol Catalyses Aβ42 Aggregation through a Heterogeneous Nucleation Pathway in the Presence of Lipid Membranes. *Nat. Chem.* **2018**, *10*, 673–683. [CrossRef]
18. Lockhart, C.; Klimov, D.K. Cholesterol Changes the Mechanisms of Aβ Peptide Binding to the DMPC Bilayer. *J. Chem. Inf. Model.* **2017**, *57*, 2554–2565. [CrossRef]
19. Grouleff, J.; Irudayam, S.J.; Skeby, K.K.; Schiøtt, B. The Influence of Cholesterol on Membrane Protein Structure, Function, and Dynamics Studied by Molecular Dynamics Simulations. *Biochim. Biophys. Acta* **2015**, *1848*, 1783–1795. [CrossRef]
20. Yu, X.; Zheng, J. Cholesterol Promotes the Interaction of Alzheimer β-amyloid Monomer with Lipid Bilayer. *J. Mol. Biol.* **2012**, *421*, 561–571. [CrossRef]
21. Dias, C.L.; Jalali, S.; Yang, Y.; Cruz, L. Role of Cholesterol on Binding of Amyloid Fibrils to Lipid Bilayers. *J. Phys. Chem. B* **2020**, *124*, 3036–3042. [CrossRef] [PubMed]
22. Elbassal, E.A.; Liu, H.; Morris, C.; Wojcikiewicz, E.P.; Du, D. Effects of Charged Cholesterol Derivatives on Aβ40 Amyloid Formation. *J. Phys. Chem. B* **2016**, *120*, 59–68. [CrossRef]
23. Harris, J.R.; Milton, N.G. *Cholesterol in Alzheimer's Disease and Other Amyloidogenic Disorders*; Subcellular Biochemistry; Springer: Berlin/Heidelberg, Germany, 2010; Volume 51, pp. 47–75.
24. Zhou, X.; Xu, J. Free Cholesterol Induces Higher β-sheet Content in Aβ Peptide Oligomers by Aromatic Interaction with Phe19. *PLoS ONE* **2012**, *7*, e46245.
25. Ait-Bouziad, N.; Lv, G.; Mahul-Mellier, A.L.; Xiao, S.; Zorludemir, G.; Eliezer, D.; Walz, T.; Lashuel, H.A. Discovery and Characterization of Stable and Toxic Tau/phospholipid Oligomeric Complexes. *Nat. Commun.* **2017**, *8*, 1678. [CrossRef] [PubMed]
26. Sciacca, M.F.; Kotler, S.A.; Brender, J.R.; Chen, J.; Lee, D.K.; Ramamoorthy, A. Two-step mechanism of membrane disruption by Aβ through membrane fragmentation and pore formation. *Biophys. J.* **2012**, *103*, 702–710. [CrossRef] [PubMed]
27. Nguyen, P.H.; Ramamoorthy, A.; Sahoo, B.R.; Zheng, J.; Faller, P.; Straub, J.E.; Dominguez, L.; Shea, J.E.; Dokholyan, N.V.; De Simone, A.; et al. Amyloid Oligomers: A Joint Experimental/Computational Perspective on Alzheimer's Disease, Parkinson's Disease, Type II Diabetes, and Amyotrophic Lateral Sclerosis. *Chem. Rev.* **2021**, *121*, 2545–2647. [CrossRef] [PubMed]

28. Scollo, F.; Tempra, C.; Lolicato, F.; Sciacca, M.F.M.; Raudino, A.; Milardi, D.; La Rosa, C. Phospholipids Critical Micellar Concentrations Trigger Different Mechanisms of Intrinsically Disordered Proteins Interaction with Model Membranes. *J. Phys. Chem. Lett.* **2018**, *9*, 5125–5129. [CrossRef]
29. Sciacca, M.F.; Lolicato, F.; Tempra, C.; Scollo, F.; Sahoo, B.R.; Watson, M.D.; García-Viñuales, S.; Milardi, D.; Raudino, A.; Lee, J.C.; et al. Lipid-Chaperone Hypothesis: A Common Molecular Mechanism of Membrane Disruption by Intrinsically Disordered Proteins. *ACS Chem. Neurosci.* **2020**, *11*, 4336–4350. [CrossRef]
30. Ngo, S.T.; Nguyen, P.H.; Derreumaux, P. Cholesterol Molecules Alter the Energy Landscape of Small Aβ1-42 Oligomers. *J. Phys. Chem. B* **2021**, *125*, 2299–2307. [CrossRef]
31. Wang, L.; Friesner, R.A.; Berne, B.J. Replica exchange with solute scaling: A more efficient version of replica exchange with solute tempering (REST2). *J. Phys. Chem. B* **2011**, *115*, 9431–9438. [CrossRef]
32. Smith, A.K.; Lockhart, C.; Klimov, D.K. Does Replica Exchange with Solute Tempering Efficiently Sample Aβ Peptide Conformational Ensembles? *J. Chem. Theory Comput.* **2016**, *12*, 5201–5214. [CrossRef] [PubMed]
33. Noda, K.; Tachi, Y.; Okamoto, Y. Structural Characteristics of Monomeric Aβ42 on Fibril in the Early Stage of Secondary Nucleation Process. *ACS Chem. Neurosci.* **2020**, *11*, 2989–2998. [CrossRef]
34. Huang, J.; Rauscher, S.; Nawrocki, G.; Ran, T.; Feig, M.; de Groot, B.L.; Grubmüller, H.; MacKerell, A.D., Jr. CHARMM36m: An Improved Force Field for Folded and Intrinsically Disordered Proteins. *Nat. Methods* **2017**, *14*, 71–73. [CrossRef] [PubMed]
35. Man, V.H.; He, X.; Derreumaux, P.; Ji, B.; Xie, X.Q.; Nguyen, P.H.; Wang, J. Effects of All-Atom Molecular Mechanics Force Fields on Amyloid Peptide Assembly: The Case of $A\beta_{16-22}$Dimer. *J. Chem. Theory Comput.* **2019**, *15*, 1440–1452. [CrossRef] [PubMed]
36. Samantray, S.; Yin, F.; Kav, B.; Strodel, B. Different Force Fields Give Rise to Different Amyloid Formation Pathways in Molecular Dynamics Simulations. *J. Chem. Inf. Model.* **2020**, *60*, 6462–6475. [CrossRef] [PubMed]
37. Nguyen, P.H.; Derreumaux, P. Structures of the intrinsically disordered Aβ, tau and α-synuclein proteins in aqueous solution from computer simulations. *Biophys. Chem.* **2020**, *264*, 106421. [CrossRef] [PubMed]
38. Nguyen, P.H.; Campanera, J.M.; Ngo, S.T.; Loquet, A.; Derreumaux, P. Tetrameric Aβ40 and Aβ42 β-Barrel Structures by Extensive Atomistic Simulations. II. In Aqueous Solution. *J. Phys. Chem. B* **2019**, *123*, 6750–6756. [CrossRef]
39. Strodel, B. Amyloid aggregation simulations: Challenges, advances and perspectives. *Curr. Opin. Struct. Biol.* **2021**, *67*, 145–152. [CrossRef]
40. Huet, A.; Derreumaux, P. Impact of the mutation A21G (Flemish variant) on Alzheimer's beta-amyloid dimers by molecular dynamics simulations. *Biophys. J.* **2006**, *91*, 3829–3840. [CrossRef]
41. Benilova, I.; Gallardo, R.; Ungureanu, A.A.; Castillo Cano, V.; Snellinx, A.; Ramakers, M.; Bartic, C.; Rousseau, F.; Schymkowitz, J.; De Strooper, B. The Alzheimer disease protective mutation A2T modulates kinetic and thermodynamic properties of amyloid-β (Aβ) aggregation. *J. Biol. Chem.* **2014**, *289*, 30977–30989. [CrossRef]
42. Zheng, X.; Liu, D.; Roychaudhuri, R.; Teplow, D.B.; Bowers, M.T. Amyloid β-Protein Assembly: Differential Effects of the Protective A2T Mutation and Recessive A2V Familial Alzheimer's Disease Mutation. *ACS Chem. Neurosci.* **2015**, *6*, 1732–1740. [CrossRef]
43. Nguyen, P.H.; Sterpone, F.; Pouplana, R.; Derreumaux, P.; Campanera, J.M. Dimerization Mechanism of Alzheimer $A\beta_{40}$ Peptides: The High Content of Intrapeptide-Stabilized Conformations in A2V and A2T Heterozygous Dimers Retards Amyloid Fibril Formation. *J. Phys. Chem. B* **2016**, *120*, 12111–12126. [CrossRef]
44. Sharma, B.; Ranganathan, S.V.; Belfort, G. Weaker N-Terminal Interactions for the Protective over the Causative Aβ Peptide Dimer Mutants. *ACS Chem. Neurosci.* **2018**, *9*, 1247–1253. [CrossRef]
45. Aggarwal, L.; Biswas, P. Hydration Thermodynamics of the N-Terminal FAD Mutants of Amyloid-β. *J. Chem. Inf. Model.* **2021**, *61*, 298–310. [CrossRef]
46. Yang, X.; Meisl, G.; Frohm, B.; Thulin, E.; Knowles, T.P.J.; Linse, S. On the role of sidechain size and charge in the aggregation of Aβ42 with familial mutations. *Proc. Natl. Acad. Sci. USA* **2018**, *115*, E5849–E5858. [CrossRef] [PubMed]
47. Wang, J.; Wang, W.; Kollman, P.A.; Case, D.A. Automatic Atom Type and Bond Type Perception in Molecular Mechanical Calculations. *J. Mol. Graph. Model.* **2006**, *25*, 247–260. [CrossRef] [PubMed]
48. Wang, J.; Wolf, R.M.; Caldwell, J.W.; Kollman, P.A.; Case, D.A. Development and Testing of a General Amber Force Field. *J. Comput. Chem.* **2004**, *25*, 1157–1174. [CrossRef] [PubMed]
49. Cao, Y.; Jiang, X.; Han, W. Self-Assembly Pathways of β-Sheet-Rich Amyloid-β(1–40) Dimers: Markov State Model Analysis on Millisecond Hybrid-Resolution Simulations. *J. Chem. Theory Comput.* **2017**, *13*, 5731–5744. [CrossRef] [PubMed]
50. Zheng, W.; Tsai, M.Y.; Wolynes, P.G. Comparing the Aggregation Free Energy Landscapes of Amyloid Beta(1–42) and Amyloid Beta(1–40). *J. Am. Chem. Soc.* **2017**, *139*, 16666–16676. [CrossRef]
51. Kreutzer, A.G.; Nowick, J.S. Elucidating the Structures of Amyloid Oligomers with Macrocyclic β-Hairpin Peptides: Insights into Alzheimer's Disease and Other Amyloid Diseases. *Acc. Chem. Res.* **2018**, *51*, 706–718. [CrossRef] [PubMed]
52. Phillips, J.C.; Braun, R.; Wang, W.; Gumbart, J.; Tajkhorshid, E.; Villa, E.; Chipot, C.; Skeel, R.D.; Kalé, L.; Schulten, K. Scalable molecular dynamics with NAMD. *J. Comput. Chem.* **2005**, *26*, 1781–1802. [CrossRef] [PubMed]
53. Man, V.H.; Nguyen, P.H.; Derreumaux, P. High-Resolution Structures of the Amyloid-β 1–42 Dimers from the Comparison of Four Atomistic Force Fields. *J. Phys. Chem. B* **2017**, *121*, 5977–5987. [CrossRef] [PubMed]
54. van der Spoel, D.; Lindahl, E.; Hess, B.; Groenhof, G.; Mark, A.E.; Berendsen, H.J. GROMACS: Fast, Flexible, and Free. *J. Comput. Chem.* **2005**, *26*, 1701–1718. [CrossRef] [PubMed]

55. Bussi, G.; Donadio, D.; Parrinello, M. Canonical sampling through velocity rescaling. *J. Chem. Phys.* **2007**, *126*, 014101. [CrossRef]
56. Kabsch, W.; Sander, C. Dictionary of Protein Secondary Structure: Pattern Recognition of Hydrogen-bonded and Geometrical Features. *Biopolymers* **1983**, *22*, 2577–2637. [CrossRef]
57. Lu, Y.; Derreumaux, P.; Guo, Z.; Mousseau, N.; Wei, G. Thermodynamics and dynamics of amyloid peptide oligomerization are sequence dependent. *Proteins* **2009**, *75*, 954–963. [CrossRef]
58. Smith, L.J.; Daura, X.; van Gunsteren, W.F. Assessing Equilibration and Convergence in Biomolecular Simulations. *Proteins* **2002**, *48*, 487–496. [CrossRef]
59. Plesnar, E.; Subczynski, W.K.; Pasenkiewicz-Gierula, M. Comparative computer simulation study of cholesterol in hydrated unary and binary lipid bilayers and in an anhydrous crystal. *J. Phys. Chem. B.* **2013**, *117*, 8758–8769. [CrossRef]
60. Huang, K.; García, A.E. Acceleration of Lateral Equilibration in Mixed Lipid Bilayers Using Replica Exchange with Solute Tempering. *J. Chem. Theory Comput.* **2014**, *10*, 4264–4272. [CrossRef]
61. Di Scala, C.; Yahi, N.; Lelièvre, C.; Garmy, N.; Chahinian, H.; Fantini, J. Biochemical identification of a linear cholesterol-binding domain within Alzheimer's β amyloid peptide. *ACS Chem. Neurosci.* **2013**, *4*, 509–517. [CrossRef]
62. Cho, W.J.; Trikha, S.; Jeremic, A.M. Cholesterol regulates assembly of human islet amyloid polypeptide on model membranes. *J. Mol. Biol.* **2009**, *393*, 765–775. [CrossRef] [PubMed]
63. Fatafta, H.; Kav, B.; Bundschuh, B.F.; Loschwitz, J.; Strodel, B. Disorder-to-order transition of the amyloid-β peptide upon lipid binding. *Biophys Chem.* **2022**, *280*, 106700. [CrossRef] [PubMed]
64. Confer, M.P.; Holcombe, B.M.; Foes, A.G.; Holmquist, J.M.; Walker, S.C.; Deb, S.; Ghosh, A. Label-Free Infrared Spectroscopic Imaging Reveals Heterogeneity of β-Sheet Aggregates in Alzheimer's Disease. *J. Phys. Chem. Lett.* **2021**, *12*, 9662–9671. [CrossRef] [PubMed]
65. Maupetit, J.; Tuffery, P.; Derreumaux, P. A coarse-grained protein force field for folding and structure prediction. *Proteins* **2007**, *69*, 394–408. [CrossRef] [PubMed]
66. Sterpone, F.; Melchionna, S.; Tuffery, P.; Pasquali, S.; Mousseau, N.; Cragnolini, T.; Chebaro, Y.; St-Pierre, J.F.; Kalimeri, M.; Barducci, A.; et al. The OPEP protein model: From single molecules, amyloid formation, crowding and hydrodynamics to DNA/RNA systems. *Chem. Soc. Rev.* **2014**, *43*, 4871–4893. [CrossRef] [PubMed]
67. Brandner, A.F.; Timr, S.; Melchionna, S.; Derreumaux, P.; Baaden, M.; Sterpone, F. Modelling lipid systems in fluid with Lattice Boltzmann Molecular Dynamics simulations and hydrodynamics. *Sci. Rep.* **2019**, *9*, 16450. [CrossRef]
68. Viet, M.H.; Nguyen, P.H.; Derreumaux, P.; Li, M.S. Effect of the English familial disease mutation (H6R) on the monomers and dimers of Aβ40 and Aβ42. *ACS Chem. Neurosci.* **2014**, *5*, 646–657. [CrossRef]
69. Nguyen, P.H.; Tarus, B.; Derreumaux, P. Familial Alzheimer A2 V mutation reduces the intrinsic disorder and completely changes the free energy landscape of the Aβ1–28 monomer. *J. Phys. Chem. B* **2014**, *118*, 501–510. [CrossRef]
70. Viet, M.H.; Nguyen, P.H.; Ngo, S.T.; Li, M.S.; Derreumaux, P. Effect of the Tottori familial disease mutation (D7N) on the monomers and dimers of Aβ40 and Aβ42. *ACS Chem. Neurosci.* **2013**, *4*, 1446–1457. [CrossRef]
71. Côté, S.; Laghaei, R.; Derreumaux, P.; Mousseau, N. Distinct dimerization for various alloforms of the amyloid-beta protein: Aβ(1–40), Aβ(1–42), and Aβ(1–40)(D23N). *J. Phys. Chem. B* **2012**, *116*, 4043–4055. [CrossRef]

molecules

Article

Pathway Dependence of the Formation and Development of Prefibrillar Aggregates in Insulin B Chain

Yuki Yoshikawa [1], Keisuke Yuzu [1], Naoki Yamamoto [2], Ken Morishima [3], Rintaro Inoue [3], Masaaki Sugiyama [3], Tetsushi Iwasaki [1,4], Masatomo So [5,†], Yuji Goto [6], Atsuo Tamura [1] and Eri Chatani [1,*]

1 Graduate School of Science, Kobe University, 1-1 Rokkodai, Nada, Kobe 657-8501, Hyogo, Japan; yoshikawa.yuki@kao.com (Y.Y.); 192s227s@stu.kobe-u.ac.jp (K.Y.); tiwasaki@kobe-u.ac.jp (T.I.); tamuatsu@kobe-u.ac.jp (A.T.)
2 Division of Biophysics, Physiology, School of Medicine, Jichi Medical University, 3311-1 Yakushiji, Shimotsuke 329-0498, Tochigi, Japan; nyamamoto@jichi.ac.jp
3 Institute for Integrated Radiation and Nuclear Science, Kyoto University, 2 Asashiro-Nishi, Kumatori, Sennan-gun 590-0494, Osaka, Japan; morishima.ken.8e@kyoto-u.ac.jp (K.M.); inoue.rintaro.5w@kyoto-u.ac.jp (R.I.); sugiyama.masaaki.5n@kyoto-u.ac.jp (M.S.)
4 Biosignal Research Center, Kobe University, 1-1 Rokkodai, Nada, Kobe 657-8501, Hyogo, Japan
5 Institute for Protein Research, Osaka University, 3-2 Yamadaoka, Suita 565-0871, Osaka, Japan; m.so@leeds.ac.uk
6 Global Center for Medical Engineering and Informatics, Osaka University, 2-1 Yamadaoka, Suita 565-0871, Osaka, Japan; gtyj8126@protein.osaka-u.ac.jp
* Correspondence: chatani@crystal.kobe-u.ac.jp
† Current address: Astbury Centre for Structural Molecular Biology, University of Leeds, Leeds LS2 9JT, UK.

Citation: Yoshikawa, Y.; Yuzu, K.; Yamamoto, N.; Morishima, K.; Inoue, R.; Sugiyama, M.; Iwasaki, T.; So, M.; Goto, Y.; Tamura, A.; et al. Pathway Dependence of the Formation and Development of Prefibrillar Aggregates in Insulin B Chain. *Molecules* **2022**, 27, 3964. https://doi.org/10.3390/molecules27133964

Academic Editors: Kunihiro Kuwajima, Yuko Okamoto, Tuomas Knowles and Michele Vendruscolo

Received: 30 May 2022
Accepted: 19 June 2022
Published: 21 June 2022

Abstract: Amyloid fibrils have been an important subject as they are involved in the development of many amyloidoses and neurodegenerative diseases. The formation of amyloid fibrils is typically initiated by nucleation, whereas its exact mechanisms are largely unknown. With this situation, we have previously identified prefibrillar aggregates in the formation of insulin B chain amyloid fibrils, which have provided an insight into the mechanisms of protein assembly involved in nucleation. Here, we have investigated the formation of insulin B chain amyloid fibrils under different pH conditions to better understand amyloid nucleation mediated by prefibrillar aggregates. The B chain showed strong propensity to form amyloid fibrils over a wide pH range, and prefibrillar aggregates were formed under all examined conditions. In particular, different structures of amyloid fibrils were found at pH 5.2 and pH 8.7, making it possible to compare different pathways. Detailed investigations at pH 5.2 in comparison with those at pH 8.7 have suggested that the evolution of protofibril-like aggregates is a common mechanism. In addition, different processes of evolution of the prefibrillar aggregates have also been identified, suggesting that the nucleation processes diversify depending on the polymorphism of amyloid fibrils.

Keywords: amyloid; insulin B chain; nucleation; prefibrillar aggregates; protofibrils

1. Introduction

Amyloid fibrils are protein aggregates that are associated with many serious human diseases [1]. They typically show needle-like morphology, which is formed by the intertwining of protofilaments with a characteristic cross-β structure [2–4]. Recently, detailed structural investigations using cryo-electron microscopy and solid-state NMR techniques have determined many structures of amyloid fibrils at higher spatial resolution [5–7]. The molecular structures have demonstrated a common characteristic that polypeptide chains are folded into a planar structure, and then stack along the fibril axis. This is suggestive of highly ordered and periodic architecture of amyloid fibrils, and, probably because of this structure property, the formation of amyloid fibrils typically exhibits a nucleation step similar to the crystallization of a wide range of substances [8,9]. Nucleation is the first step in the emergence of amyloid structures in a reaction system, and primarily determines the progression of amyloid formation as well as the length of the lag phase as a rate-limiting step. Although recent modeling analyses have suggested that secondary nucleation and fragmentation are also involved in the late lag phase [10–12], investigating early aggregation of protein or peptide molecules is important for elucidating the initiation of amyloid formation and for developing strategies for preventing and treating amyloid-related diseases at early stages. However, detailed mechanisms of protein assembly that progresses during the formation of amyloid nuclei remain largely unclear.

To elucidate the mechanism of amyloid nucleation, studies have been carried out to propose appropriate reaction schemes as well as kinetic parameters that can reproduce experimental observations. The most fundamental and simplest mechanism of nucleation is assumed to progress in one step, and in this case, the process is accomplished without passing through any thermodynamically stable intermediates [13]. On the other hand, prefibrillar aggregates have often been identified in early stages of amyloid formation reactions, and some of them have been suggested to play an important role for amyloid nucleation [14–17]. The nucleus conformational conversion (NCC) proposed in the Sup35 study is the first model describing multistep nucleation mediated by prefibrillar aggregates [14], and since then, the involvement of oligomers in nucleation has often been argued. Nevertheless, previous studies have identified aggregates involved as off-pathway intermediates in the formation of amyloid fibrils [18,19], and the role of prefibrillar aggregates often seems controversial. Trapping and characterizing structures and formation processes of prefibrillar aggregates in a wide range of proteins and discussing their detailed roles in nucleation are, therefore, important for facilitating our understanding of protein assembly mechanisms in the initial stage of amyloid formation.

Given this research background, we previously found that prefibrillar aggregates function as a nucleation intermediate in the amyloid formation of an insulin-derived peptide B chain. Insulin is composed of two polypeptide chains, and the B chain is one of them, composed of 30 amino acid residues. High amyloid propensity of the B chain was previously revealed [20], and in our analysis, large amounts of prefibrillar aggregates were clarified to accumulate before the appearance of amyloid fibrils [21]. The formed prefibrillar aggregates were metastable and the application of agitation significantly promoted the formation of amyloid fibrils. It was also suggested that targeting the prefibrillar aggregates is an efficient strategy to inhibit amyloid formation [22]. These observations indicate the deep involvement of the B chain prefibrillar aggregates in the nucleation process. It is therefore expected that the B chain will serve as a useful model system for examining the detailed mechanism of the nucleation process mediated by prefibrillar aggregates.

In this study, we have found that similar prefibrillar aggregate-mediated amyloid formation proceeds over a wide pH range between 5.2 and 9.1. With a main focus on the amyloid formation at pH 5.2 showing a different pathway from that previously investigated at pH 8.7 [21,22], we have analyzed structural properties of prefibrillar aggregates and their formation processes through the combination of various analytical techniques, i.e., thioflavin T (ThT) fluorescence, attenuated total reflectance Fourier transform infrared (ATR-FTIR) spectroscopy, circular dichroism (CD) spectroscopy, atomic force microscopy (AFM), small-angle X-ray scattering (SAXS), and dynamic light scattering (DLS). The structural properties of the prefibrillar aggregates have suggested that they are protofibril-like, exhibiting partially organized β-structure and rod-like shape. It has also been clarified that both of their structure and size gradually developed during the formation of prefibrillar aggregates. By measuring the time course of prefibrillar aggregation of the B chain at pH 5.2, we discuss the formation processes of prefibrillar aggregates, and furthermore, their pathway dependence in comparison with that at pH 8.7.

2. Results

2.1. Amyloid Formation at Various pH Values

We monitored aggregation reactions at different pH values ranging from 3.0 to 9.1. When reactions were performed under agitated conditions (i.e., shaking at 1200 rpm), an increase in ThT fluorescence intensity was observed at almost all pH conditions except pH 3.0 (Figure 1A). In addition, the AFM images of the reaction products with positive ThT fluorescence intensity showed characteristic morphologies typical of amyloid fibrils (Figure 1B), suggesting that the B chain has a strong propensity to form amyloid fibrils over a wide pH range. At pH 5.2, many amyloid fibrils were observed in the AFM image despite the fact that the ThT fluorescence intensity was lower than the others (Figure 1A, inset, and Figure 1B).

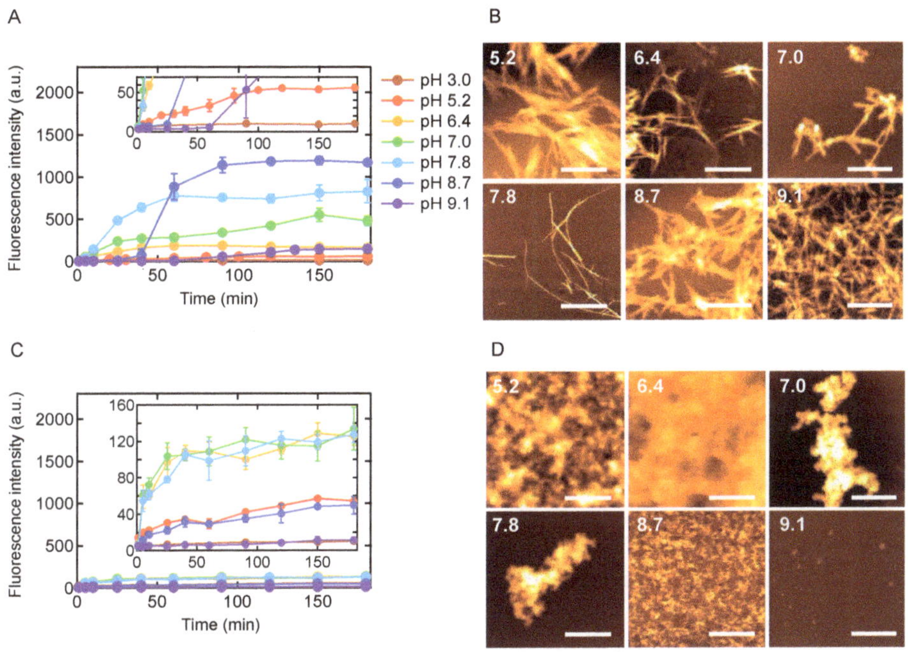

Figure 1. Aggregation reactions of insulin B chain at various pHs under (**A**,**B**) agitated and (**C**,**D**) quiescent conditions. The B chain at a concentration of 1.40 mg/mL was incubated at 25 °C, and in agitated conditions, the sample was shaken continuously at 1200 rpm. (**A**,**C**) Time courses of ThT fluorescence intensity under (**A**) agitated or (**C**) quiescent conditions. The mean values of triple data are plotted with standard deviations as error bars. (**B**,**D**) AFM images of aggregates after incubation for 2.5 h under (**B**) agitated or (**D**) quiescent conditions. Scale bars represent 1 μm.

In our previous investigation at pH 8.7, prefibrillar aggregates were transiently accumulated early in the amyloid formation reaction [21]. They were characterized as a metastable species being trapped under quiescent conditions. To examine whether similar prefibrillar aggregates are involved at different pH values other than pH 8.7, aggregation reactions were monitored without agitation. As a result, a slight increase in ThT fluorescence was observed at pH range between 5.2 and 8.7 (Figure 1C). When the morphology of the reaction products was analyzed by AFM, nonfibrillar aggregates were observed in all conditions showing positive ThT fluorescence (Figure 1D). At pH 9.1, a small number of particle-like aggregates was observed even though ThT fluorescence was almost under detection limits (Figure 1C, inset, and Figure 1D), suggesting that nonfibrillar aggregation progresses to some extent. Taken together, amyloid formation accompanying nonfibrillar aggregates has been suggested to progress under a wide range of pH between 5.2 and 9.1.

2.2. Structure of Nonfibrillar Aggregates Formed under Quiescent Conditions

To characterize the structure of nonfibrillar aggregates observed under quiescent conditions, we measured ATR-FTIR spectra. Although the amount of nonfibrillar aggregates formed at pH 9.1 was small, the measurement was realized by increasing sample volume. At all pH values, the aggregates were richer in random coil, as represented by a larger peak around 1650 cm^{-1}, than in amyloid fibrils, although the β-sheet component at around 1625 cm^{-1} was contained to some extent (Figure 2). This suggests that nonfibrillar aggregates are not fully disordered but contain β structure. The β-sheet peak of nonfibrillar aggregates tended to be located slightly at a lower wavenumber than that of amyloid fibrils. Furthermore, a characteristic peak was commonly observed at around 1695 cm^{-1}, presumably corresponding to antiparallel β-sheet [23]. These spectral properties suggest that the β structure contained in nonfibrillar aggregates has different properties from that of amyloid fibrils.

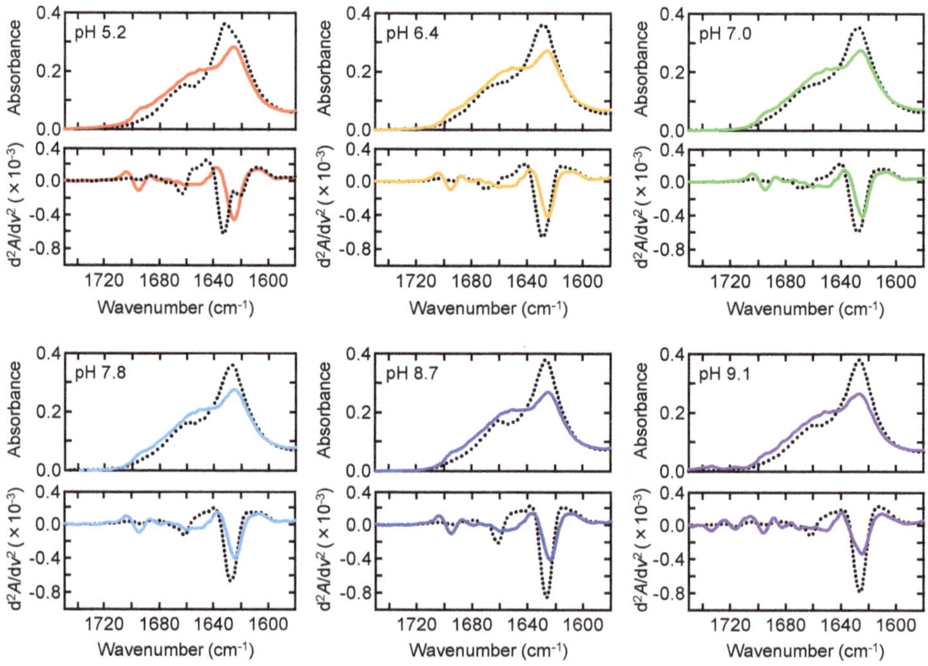

Figure 2. Structural properties of nonfibrillar aggregates formed under quiescent conditions at various pHs. ATR-FTIR absorption spectra and their second derivatives are shown in upper and lower panels, respectively. Solid lines indicate the spectra of nonfibrillar aggregates formed under quiescent conditions, and the spectra of amyloid fibrils formed under agitated conditions are also represented by black dotted lines. Spectra were normalized so that the integrated intensity of the amide I band ranging from 1580 to 1750 cm^{-1} was set to be equal.

We next examined whether the nonfibrillar aggregates play a role as intermediates of amyloid formation. In the previous work conducted at pH 8.7, nonfibrillar aggregates formed under quiescent conditions were transformed into amyloid fibrils by supplying a mechanical stimulus with shaking or ultrasonic wave [21]. To test whether similar agitation-induced conversion to amyloid fibrils occurs at other pH conditions, nonfibrillar aggregates formed by incubating for 2.5 h under quiescent conditions were shaken for 2.5 h, and then FTIR spectra were measured. The resulting spectra showed a similar shape to that of amyloid fibrils in all conditions (Figure 3A), and the formation of amyloid fibrils was confirmed by AFM at pH 5.2 as a representative (Figure 3B), raising a possibility that these nonfibrillar aggregates function as precursors of amyloid fibrils.

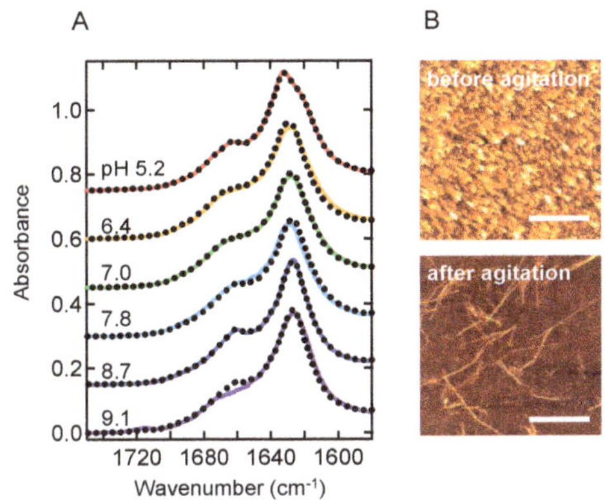

Figure 3. Agitation-induced conversion of nonfibrillar aggregates to amyloid fibrils. (**A**) ATR-FTIR absorption spectra of nonfibrillar aggregates measured after subjected to agitation. The spectra of amyloid fibrils formed under agitated conditions (the same spectra shown in Figure 2) are overlaid with dotted lines for references. (**B**) AFM images at pH 5.2 before and after agitation. Scale bars represent 1 μm.

2.3. Formation Process of Prefibrillar Aggregates at pH 5.2

When the FTIR spectra of the formed amyloid fibrils were compared at different pHs, the position and intensity of main absorption peaks showed pH dependence. In particular, the spectrum at pH 5.2 had two β-sheet peaks at 1621 cm^{-1} and 1632 cm^{-1}, which were clearly identified by the second derivative, suggesting that the pH 5.2 amyloid fibrils contain two types of β-sheet structures with different hydrogen bond arrangements. On the other hand, the spectra at the other pH values showed no clear separation of the β-sheet peak (Figure 2). The difference in amyloid structure between pH 5.2 and pH 8.7 was also supported by a comparison of proteinase K digestion (Figure S1) and cytotoxicity (Figure S2). Given the polymorphism in the final product, it was expected that the formation of B chain amyloid fibrils proceeds at pH 5.2 in a different pathway from other pH conditions. We therefore selected pH 5.2 as a target of investigation in comparison with pH 8.7 analyzed in the previous study [21,22].

To characterize the detailed process of aggregation at pH 5.2, a time-dependent change in secondary structure was tracked by using far-UV CD spectra. The spectrum changed in two steps under agitated conditions, supporting that the formation of amyloid fibrils progresses via prefibrillar aggregates (Figure S3). The formation process of prefibrillar aggregates was successfully observed under quiescent conditions, although it was much faster than at pH 8.7, showing a large burst phase (Figure 4A,B, blue plots). When additional measurements were performed at lower B chain concentrations, earlier structural changes could be captured because aggregation rate slowed down. The time-dependent changes in spectra were reasonably explained by an exponential or a biexponential function at each B chain concentration (Figure 4A). Combining all the convergent spectra estimated by the fitting of the spectral time courses, we found that the change in CD spectrum converged to a total of four different spectra, i.e., the first and second spectra at 0.22 mg/mL; the second spectrum at 0.25 mg/mL; the third spectrum at 0.30 mg/mL; the third spectrum at 0.50 mg/mL; the third and fourth spectra at 0.70 mg/mL; and the fourth spectrum at 1.40 mg/mL (Figure 4C). Given that each of them was observed for at least two different concentrations, the appearance of these spectra was considered to be sequential over the increase of peptide concentration. This result suggests that there are four steps in secondary structural change during the formation of prefibrillar aggregates at pH 5.2. At pH 8.7, on the other hand, the spectral change at 1.40 mg/mL did not show a significant burst phase, and two steps were identified (Figure 4D). The difference in the number of steps suggests that the process of prefibrillar aggregation diversifies in a pathway-dependent manner. In addition, although difficult to detect in FTIR spectroscopy, the final shape of the CD spectrum showed a slightly different minimum wavelength between pH 5.2 and 8.7 (Figure 4E), implying that these prefibrillar aggregates have structural differences to some extent.

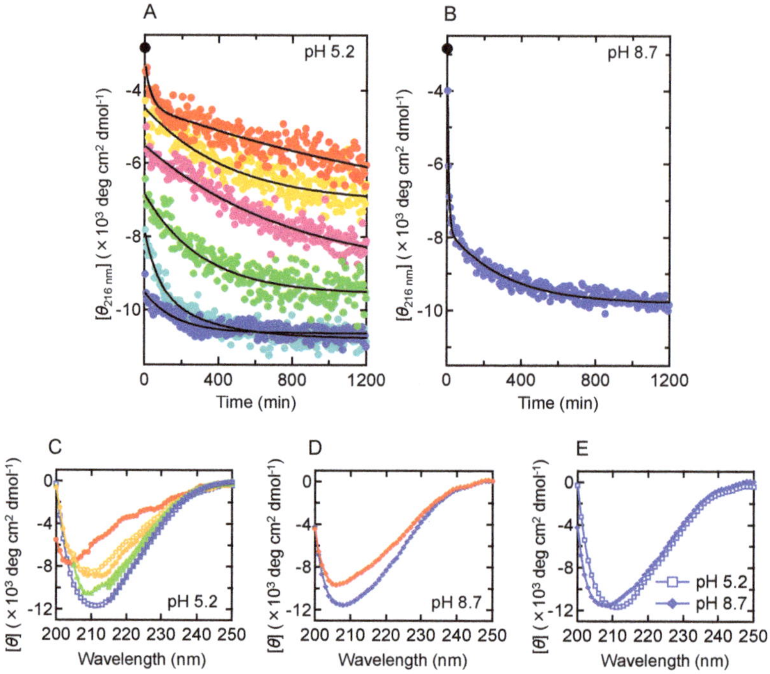

Figure 4. Difference in the formation process of prefibrillar aggregates between pH 5.2 and pH 8.7 as revealed by CD spectral changes. (**A**) Time-dependent changes in the values of mean residue molar ellipticity at 216 nm $[\theta_{216\,nm}]$ at pH 5.2. Plots at six different peptide concentrations (0.22 mg/mL, red; 0.25 mg/mL, orange; 0.30 mg/mL, magenta; 0.50 mg/mL, green; 0.70 mg/mL, cyan; 1.40 mg/mL, blue) are shown. Black lines represent regression curves obtained by fitting analysis with an exponential or a biexponential function, using Equation (1) (for 0.25 mg/mL, 0.30 mg/mL, 0.50 mg/mL, and 1.40 mg/mL) or Equation (2) (at 0.22 mg/mL and 0.70 mg/mL). A black circle indicates the value of the monomeric B chain measured in NaOH. (**B**) Time-dependent changes in the values of $[\theta_{216\,nm}]$ at pH 8.7. Plots at a peptide concentration of 1.40 mg/mL are shown. A black line represents a regression curve obtained by spectral fitting using a biexponential function, Equation (2). A black circle indicates the value of the monomeric B chain. (**C**) CD spectra of intermediate states at pH 5.2 reproduced by spectral extrapolation at each phase in the spectral fitting. Overlaying spectra obtained at different peptide concentrations (closed circles, 0.22 mg/mL; open circles, 0.25 mg/mL; closed triangles, 0.30 mg/mL, open triangles, 0.50 mg/mL; closed squares, 0.70 mg/mL; open squares, 1.40 mg/mL) suggested four different states, which are colored by red, orange, green, and blue in order of appearance, respectively. (**D**) CD spectra of intermediate states at pH 8.7 reproduced by spectral extrapolation at each phase in the spectral fitting. The result suggested two different states, as shown by red and blue diamonds in order of appearance. (**E**) Comparison of final convergent spectra between the pH 5.2 and the pH 8.7 prefibrillar aggregates.

To track the progress of aggregation during the organization of secondary structure, the time course measurement of ^{1}H-NMR spectra was also performed. It was conducted at a peptide concentration of 0.22 mg/mL, the lowest concentration used for the CD measurement to track the earliest conformational changes (see Figure 4A). The results showed that proton signals almost disappeared immediately after the start of the reaction, suggesting rapid progression of aggregation (Figure 5A). The decay of the NMR peaks demonstrated that the monomer concentration dropped to 4% within 7 min from the start of the reaction, and by comparing with the changes in CD spectra, it was revealed that disordered aggregation occurred before the formation of secondary structure (Figure 5B). Considering that the proton signal typically disappears when the molecular diameter becomes 30–40 nm [21], the initial aggregates were estimated to have a significant number of associations of the B chain peptides. However, their accurate size could not be measured by DLS or SAXS in this work due to the experimental limitation that their formation was only allowed at low peptide concentrations. This reaction process is obviously different when compared to the reaction at pH 8.7, where aggregation and conformational changes occur synchronously [21].

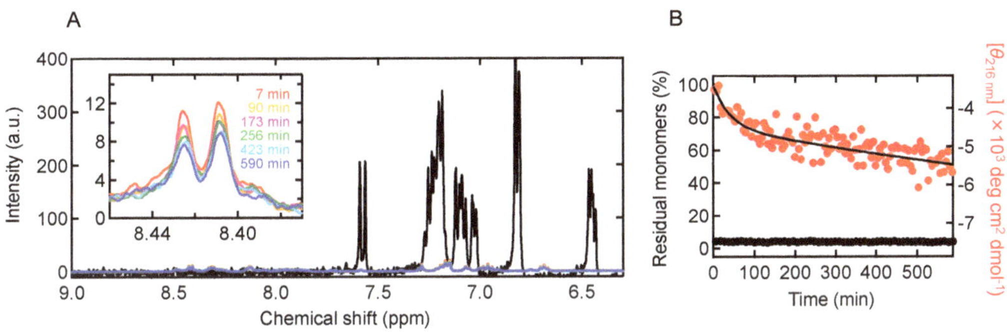

Figure 5. Quantification of residual monomers in the process of the formation of prefibrillar aggregates at pH 5.2 using ^{1}H-NMR spectra. (**A**) Time-dependent change in ^{1}H-NMR spectra in a low magnetic field region. The measurement was performed at a B chain concentration of 0.22 mg/mL. Inset shows a magnified view of peaks of two histidine ε protons, which were used for quantifying the fraction of residual monomers. The black line is the spectrum of a monomer state obtained with the B chain dissolved in a NaOH solution, and it should be noted that the peaks of two histidine ε protons are observed at around 7.6 ppm due to difference in pH. (**B**) Time course of the fraction of the residual monomers estimated from the area of the histidine peaks. For comparison, the time-dependent change in [$\theta_{216\ nm}$] at 0.22 mg/mL shown in Figure 4A is overlaid.

2.4. Size and Shape of the pH 5.2 Aggregates

To analyze the size and shape of prefibrillar aggregates formed at pH 5.2, SAXS measurements were performed. The B chain solution at pH 5.2 was loaded into the cell at a concentration of 1.40 mg/mL, and SAXS profiles were obtained continuously until the reaction completed. Figure 6A shows SAXS profiles monitored at several time points in the process of the formation of prefibrillar aggregates. The slope of the $I(q)$ against q exhibited a value close to −1, suggesting that prefibrillar aggregates have a rod-like shape. Although this result apparently contradicts the AFM measurements where granular particles were observed (Figure 1D), this is presumably because prefibrillar aggregates are fragile and unstable against washing of the sample plate when preparing the AFM sample [22].

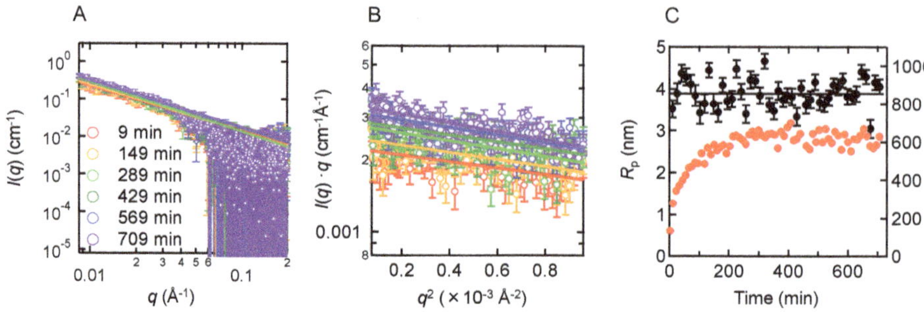

Figure 6. Characterization of the size and shape of prefibrillar aggregates at pH 5.2 monitored by SAXS. The measurement was performed at 1.40 mg/mL. (**A**) Double logarithm plots of the SAXS profiles at different time points. Solid lines indicate fitted lines using Equation (3) in Porod region. (**B**) Cross-section plots. Solid lines indicate fitted curves obtained using Equation (4). (**C**) Time changes of R_p (black) and L (red), which were obtained from R_c in panel B and the DLS data in Figure S4 using Equations (5) and (6). A solid line indicates the mean of the R_p values.

Under an assumption that the shape of the peptide aggregates can be approximately regarded as a rod, we could evaluate the cross-section inertia radius (R_c) from the slope of a cross-section plot. The slope of the cross-section plot was almost constant (Figure 6B) and the time dependence of R_c suggested that the intermediates maintain the same radius during the measurement time period. The radius of the rod (R_p), which is equal to $\sqrt{2}R_c$ (see Equation (5)), was calculated to be 3.9 ± 0.3 nm (Figure 6C), and no significant difference was found from that at pH 8.7 estimated in our previous study (3.7 ± 0.1 nm) within experimental errors [24]. By using diffusion coefficient D_T obtained from DLS and R_p, we further attempted to estimate the length (L) of the rod through Equation (6). The L values (Figure 6C) showed time-dependent elongation of the rod towards the length of 610 nm, which was longer than the value at pH 8.7 (480 nm) [24]; however, the lengths have relatively large errors and distribution as suggested by DLS (Figure S4), and the approximate shape appeared to be similar, between pH 5.2 and pH 8.7.

3. Discussion

The insulin B chain has been shown to have an ability to form amyloid fibrils under a wide range of pH conditions with the help of agitation. In all of the amyloid formations investigated, prefibrillar aggregates were observed immediately after the reaction started, and had a common tendency to accumulate under quiescent conditions as a metastable state. The structure of prefibrillar aggregates was not completely disordered but contained some amount of β-sheet, which is validated by the fact that ThT fluorescence intensity of these aggregates showed slightly positive values. Furthermore, they transformed into amyloid fibrils when subjected to agitation, which would provide mechanical stimuli to prefibrillar aggregates. This behavior is similar to that of prefibrillar aggregates at pH 8.7, which have been suggested to function as a nucleation precursor. It has therefore been suggested that the B chain has a strong propensity for nucleation in a multistep way under a wide range of reaction conditions.

From the FTIR spectra along with proteinase K digestibility and cytotoxicity, structural differences have been revealed between amyloid fibrils formed at pH 5.2 and at pH 8.7. The differences in the final product indicate that the formation pathway of amyloid fibrils varies with pH, which has allowed us to examine prefibrillar aggregates in different pathways. Figure 7 summarizes the reaction schemes proposed at pH 5.2 and pH 8.7. As a basic property of prefibrillar aggregates, rod-like shape and immature β-sheet structure analogous to protofibrils have been revealed irrespective of pathways. In addition, the elongation of the protofibril-like aggregates along the long axis was observed similar to pH

8.7, although the observation was limited at pH 5.2 because the reaction rate was too rapid to be fully tracked.

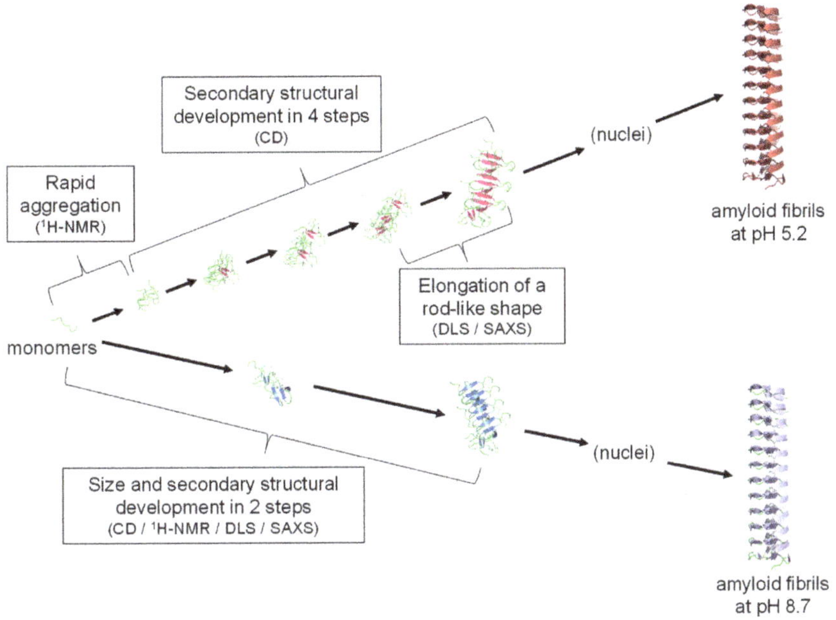

Figure 7. Schematic illustration representing prefibrillar aggregate-mediated amyloid formation of B chain. Two different pathways observed at pH 5.2 and pH 8.7 are shown at the top and bottom of this figure, respectively.

While the basic property of prefibrillar aggregates was similar between pH 5.2 and pH 8.7, there were differences in detailed processes. The most notable was the number of steps, with a larger number of intermediates observed at pH 5.2 than at pH 8.7. It was additionally demonstrated that the clustering of peptide molecules preceded their structural development at pH 5.2, unlike pH 8.7, where size and structural development occurred synchronously. The factor that contributes to the difference in the formation process between pH 5.2 and pH 8.7 is considered to be the charged state of the B chain peptide. Given that the pK_a values of histidine residues at the 5th and the 10th positions, and cysteine residues at the 7th and 19th positions are between these two pHs, these residues are deduced to influence the modes of aggregation. Although only slight differences in size and structure of prefibrillar aggregates could be detected from the analytical techniques used in this work, the pH-dependent change in charge distribution is predicted to modify regions where intermolecular interactions are likely to occur and then to guide to different pathways reaching distinct prefibrillar aggregates.

The tracking of prefibrillar aggregation in the B chain prior to amyloid formation in a comparative way of the two different pathways has shed light on multistep nucleation mechanism of amyloid fibrils. The simplest scheme describing amyloid nucleation is a one-step without intermediates, and in this case, a classical nucleation theory established in crystallography provides an important framework for explaining it reasonably from an energetic view [25,26]. According to this theory, energy change upon crystal formation is described as the sum of the bulk free energy and the interfacial free energy as positive and negative driving forces, respectively. The change in balance of these two energy terms as a function of crystal size and shape produces energy cost until the crystal reaches a critical size, which corresponds to an energy required for nucleation.

In the case of the B chain, however, the nucleation does not fit the classical nucleation theory, and instead, prefibrillar aggregation progresses transiently at an initial stage. The formation of prefibrillar aggregates prior to the appearance of amyloid fibrils is presumed to be due to their kinetic stability. The less organized structure of prefibrillar aggregates tends to exhibit low interfacial energy cost, and therefore the initial energy required for their formation is expected to be low. However, since prefibrillar aggregates are less thermodynamically stable, they need to grow into a more energetically stable state, and eventually, into the most stable amyloid fibril state. Given that the formation of oligomers and protofibrils have been widely observed in many proteins [27], the gradual structural development of prefibrillar aggregates towards amyloid fibrils as observed in the B chain may be one of the representative schemes of amyloid nucleation.

In the multistep nucleation scheme, the process of transforming prefibrillar aggregates to amyloid fibrils is considered the rate-limiting step. It is estimated that direct structural conversion, as proposed in the NCC model, is a candidate mechanism for the structural transformation. Indeed, although not very numerous, some experiments have shown that prefibrillar aggregates convert directly into amyloid fibrils without dissociating into monomers [15,17,28]. Furthermore, computational simulation studies have identified a conversion pathway between prefibrillar and fibrillar forms [29,30]. In addition to this, secondary nucleation, which has recently attracted much attention as an important reaction to proliferate amyloid fibrils [11,12,31], is predicted an alternate mechanism. In this case, the surface of prefibrillar aggregates is conceived to function as a reaction field for secondary nucleation, and prefibrillar aggregates themselves then eventually dissociate to supply peptides for the growth of more stable amyloid fibrils. Although it is not possible to conclude exactly the mechanism actually adopted, direct conversion might be a strong candidate considering that amyloid formation occurred via prefibrillar aggregates even at pH 5.2, where almost all fraction of the monomers was depleted after the formation of prefibrillar aggregates.

The previous observation at pH 8.7 has suggested that specific structural features of prefibrillar aggregates play an important role when they mediate nucleation. Prefibrillar aggregates are predicted to link to amyloid fibrils on the energy landscape when their structure allows the transition energy to amyloid fibrils to be reduced enough to be overcome. Interestingly, FTIR spectra commonly showed that prefibrillar aggregates of the B chain contained a small amount of antiparallel β-sheet structures, as seen in many other oligomers and protofibrils [29,32–34], although the significance of such a structural property for the transition to the fibril state is still unknown. Given that similar structural properties are observed regardless of pH conditions, the nonfibrillar aggregates at other pH conditions are also expected to convert to amyloid fibrils, as at pH 8.7 and 5.2. However, nucleation from coexisting monomers may also need to be considered, especially at pH 9.1 where only a small amount of aggregates is formed. Clarifying detailed structures of various prefibrillar aggregates, as well as capturing the moment when they form nuclei, will reveal the exact role of prefibrillar aggregates.

Furthermore, the identified variety in evolution of prefibrillar aggregates has provided insights into amyloid polymorphism. Diverse amyloid structures formed from the same protein or polypeptide have recently attracted attention since they have been proposed to be involved in different pathologies [35]. The present study has suggested that different patterns of prefibrillar aggregation accompany polymorphic pathways. Prefibrillar aggregation is estimated to proceed under conditions that exceed the peptide solubility limit, and where the peptides form a variety of assembly structures reflecting the state of the peptides. The resulting diversity of prefibrillar aggregates is expected to guide various pathways for amyloid formation. Investigating amyloid formation pathways in terms of prefibrillar aggregates will provide deeper understanding not only of nucleation itself, but also of pathogenesis associated with polymorphism.

4. Materials and Methods

4.1. Purification of Insulin B Chain

Insulin B chain was isolated from human insulin (FUJIFILM Wako Pure Chemical, Osaka, Japan) as in our previous works [21,22]. The B chain was dissolved in 10 mM NaOH and stocked at −80 °C before use. The purity of the B chain was assessed by the ^{1}H signals of ε protons in tyrosine residues obtained using the NMR spectrometer, AVANCEIII HD (Bruker, Billerica, MA, USA). The concentration of B chain was determined by using the absorption coefficient of 0.90 $(mg/mL)^{-1}cm^{-1}$ at 280 nm in 10 mM NaOH solution.

4.2. Formation of B Chain Amyloid Fibrils

The formation of amyloid fibrils of B chain was carried out with peptide concentration ranging from 0.22 to 1.40 mg/mL (i.e., 64 to 408 μM) and at pH ranging from 3.0 to 9.1. The B chain in 10 mM NaOH was diluted by an approximately half volume of an appropriate buffer, i.e., glycine for pH 3.0, acetate for pH 5.2, phosphate for pH 6.4, 7.0, and 7.8, and Tris for pH 8.7 and 9.1, to reach a final peptide concentration. The final buffer concentration was 50 mM, and the sample contained 5 mM NaCl generated by neutralizing NaOH in the B chain solution. All experiments were performed at 25 °C under agitated or quiescent conditions. In experiments conducted under agitated conditions, the samples in microtubes were incubated under continuous shaking at 1200 rpm using a ThermoMixer C (Eppendorf, Hamburg, Germany). In experiments conducted under quiescent conditions, samples prepared in microtubes were placed in an air incubator.

4.3. ThT Assay

The formation of amyloid fibrils was monitored by ThT fluorescence. In the assay, 5 μL of a sample solution was mixed with 1.5 mL of a ThT assay solution, which was composed of 5 μM ThT and 50 mM glycine buffer (pH 8.5). One minute after the incubation, fluorescent intensity was recorded at 485 nm with an excitation at 445 nm using the spectrofluorometer, RF-5300pc (Shimadzu, Kyoto, Japan).

4.4. AFM

AFM images were obtained using micro cantilever (OLYMPUS, Tokyo, Japan) and the dynamic force mode with Probestation NanoNavi II/IIe (Hitachi High-Tech Science, Tokyo, Japan). Five to twenty microliters of a sample were loaded to a mica plate, left for one minute, and then rinsed using 200 μL of water. The sweep rate was set to 0.5 or 1.0 Hz.

4.5. ATR-FTIR Spectroscopy

ATR-FTIR spectra were measured with a Nicolet iS5 FT-IR equipped with an iD5 ATR accessory (Thermo Fisher Scientific, Waltham, MA, USA). Samples of all pH conditions were precipitated by centrifugation in the presence of 0.5 M NaCl and resuspended in water. This process was repeated three times to replace the buffer solution in the sample with water. The suspension was placed on a diamond crystal prism and then dried. FTIR measurement was performed by collecting 128 interferograms at a resolution of 4 cm^{-1}.

4.6. CD Spectroscopy

Far-UV CD spectra were measured to investigate the secondary structure of the product, as well as to track the time-dependent structural changes. CD spectra were obtained using a CD spectrometer, J-1100 (JASCO, Tokyo, Japan). A sample was placed in a quartz cell with a path length of 0.2 mm or 1.0 mm. Each scan was performed at 200 nm/min, and eight individual scans were averaged to obtain one spectrum. A time-course plot obtained under quiescent conditions was fitted using an exponential or a biexponential equation below:

$$[\theta](t) = [\theta]_1 - ([\theta]_1 - [\theta]_0)\exp\left(-\frac{t}{\tau_1}\right) \quad (1)$$

$$[\theta](t) = [\theta]_2 - ([\theta]_1 - [\theta]_0)\exp\left(-\frac{t}{\tau_1}\right) - ([\theta]_2 - [\theta]_1)\exp\left(-\frac{t}{\tau_2}\right) \quad (2)$$

where τ_i and $[\theta]_i$ represent the apparent time constants of the *i*th phase and the asymptotic value of mean residue molar ellipticity after the completion of the *i*th phase, respectively. For fitting, the time dependence of the $[\theta]$ values at intervals of 1 nm over the wavelength range of 200–250 nm at 0.22 and 1.40 mg/mL, 205–250 nm at 0.25 and 0.30 mg/mL, or 215–250 nm at 0.50 and 0.70 mg/mL was used.

4.7. ^{1}H-NMR Spectroscopy

The monomer concentration was estimated from the decrease of the NMR spectrum signal. A sample was placed in a glass tube with an inner diameter of 5 mm and incubated at 25 °C. NMR spectra were measured using AVANCE III HD with a superconducting magnet with a Larmor frequency of 400.13 MHz (Bruker, Germany). The spectrometer was controlled using the programs of Topspin 1.5 and IconNMR (Bruker). High homogeneity of the magnetic field was achieved by Topspim, a routine tool built by Topspin 1.5, and a pulse program, zgesgp, was used for spectral measurements.

4.8. SAXS

SAXS measurements were performed by NANOPIX equipped with a HyPix-6000 (Rigaku, Tokyo, Japan). A Cu K-α line (MicroMAX-007HF; Rigaku) was used as a beam source and the camera length was set to 1.33 m. A sample liquid was loaded into a made-to-order sample cell consisting of a pair of quartz windows having an optical path length of 1 mm, incubated under quiescent conditions at 25 °C, and time-lapse measurements were performed by continuous data acquisition with an exposure time of 15 min for each profile. The magnitude of scattering vector q ($q = 4\pi\sin(\theta/2)/\lambda$, where λ and θ indicate X-ray wavelength and the scatter angle, respectively) ranged from 0.0085 to 0.2 Å^{-1}. The Porod region of the double logarithm plots of the SAXS profiles was analyzed to evaluate the approximate shape of aggregates using the equation below [36]:

$$\log I(q) = \log I(0) + a \cdot \log(q) \quad (3)$$

where $I(0)$ and a represent the intensity at $q = 0$ and the slope of the double logarithm plot, respectively. In the analysis of aggregates with a rod-like shape, cross-section plots were constructed and fitted using the equation below [37]:

$$\log\{I(q) \cdot q\} = A - \frac{R_c^2}{2}q^2 \quad (4)$$

where A represents the constant, and R_c represents the radius of the cross-section. The maximum q value for the fitting region is restricted to $R_c\, q < 1.3$. The radius of a rod, i.e., R_p, is described as

$$R_p = \sqrt{2}R_c \quad (5)$$

The length of prefibrillar aggregate, L, is obtained by Broersma's relationship [38]:

$$D_T = \frac{k_B T}{3\pi\eta_0 L}\ln\left(\frac{L}{R_p}\right) \quad (6)$$

where D_T is the diffusion coefficient obtained by DLS analyses (see Supplementary Materials), and k_B, T, and η_0 represent the Boltzmann constant, temperature, and viscosity, respectively.

Supplementary Materials: The following supporting information can be downloaded at: https://www.mdpi.com/article/10.3390/molecules27133964/s1, Supplementary Methods: Proteinase K digestion of amyloid fibrils; Cytotoxicity assay; DLS. Figure S1: Proteinase K digestion of amyloid fibrils formed at pH 5.2 and pH 8.7.; Figure S2: Cell viability against the B chain amyloid fibrils formed at pH 5.2 and pH 8.7.; Figure S3: CD spectra monitored in the formation of amyloid fibrils under agitated conditions at pH 5.2.; Figure S4: Time-dependent changes in hydrodynamic diameter during the formation of prefibrillar aggregates at pH 5.2. Ref. [39] cited in Supplementary Materials.

Author Contributions: Conceptualization, Y.Y. and E.C.; investigation, Y.Y., K.Y., N.Y., K.M., R.I., M.S. (Masaaki Sugiyama), T.I., M.S. (Masatomo So), Y.G., A.T. and E.C; data curation, Y.Y., K.Y., N.Y., K.M., R.I., T.I. and M.S. (Masatomo So); writing—original draft preparation, Y.Y., K.Y., N.Y. and E.C.; writing—review and editing, T.I., K.M., R.I., M.S. (Masaaki Sugiyama), M.S. (Masatomo So), A.T. and Y.G.; supervision, E.C.; project administration, E.C.; funding acquisition, E.C., K.M., R.I., M.S. (Masaaki Sugiyama) and Y.G. All authors have read and agreed to the published version of the manuscript.

Funding: This research was funded by JSPS KAKENHI, grant numbers JP17H06352, JP20H03224, JP20K21396, JP19KK0071, JP20K06579, JP19K16088, JP21K15051, JP18H05229, JP18H05534, and JP18H03681.

Institutional Review Board Statement: Not applicable.

Informed Consent Statement: Not applicable.

Data Availability Statement: Data are contained within the article or Supplementary Materials.

Acknowledgments: SAXS experiment at Institute for Integrated Radiation and Nuclear Science, Kyoto University (KURNS) were performed under Proposal Numbers, 30028, 31028, and R2043 to E.C. This study was also supported partially by the project for Construction of the basis for the advanced materials science and analytical study by the innovative use of quantum beam and nuclear sciences in KURNS, and by the Japan Society for the Promotion of Science, Core-to-Core Program A: Advanced Research Networks.

Conflicts of Interest: The authors declare no conflict of interest.

Sample Availability: There are no samples of the compounds that are available from the authors.

References

1. Benson, M.D.; Buxbaum, J.N.; Eisenberg, D.S.; Merlini, G.; Saraiva, M.J.M.; Sekijima, Y.; Sipe, J.D.; Westermark, P. Amyloid nomenclature 2020: Update and recommendations by the international society of amyloidosis (ISA) nomenclature committee. *Amyloid* **2020**, *27*, 217–222. [CrossRef] [PubMed]
2. Dobson, C.M. Protein folding and misfolding. *Nature* **2003**, *426*, 884–890. [CrossRef]
3. Riek, R.; Eisenberg, D.S. The activities of amyloids from a structural perspective. *Nature* **2016**, *539*, 227–235. [CrossRef] [PubMed]
4. Chiti, F.; Dobson, C.M. Protein misfolding, amyloid formation, and human disease: A summary of progress over the last decade. *Annu. Rev. Biochem.* **2017**, *86*, 27–68. [CrossRef]
5. Tycko, R. Amyloid polymorphism: Structural basis and neurobiological relevance. *Neuron* **2015**, *86*, 632–645. [CrossRef]
6. Fitzpatrick, A.W.P.; Falcon, B.; He, S.; Murzin, A.G.; Murshudov, G.; Garringer, H.J.; Crowther, R.A.; Ghetti, B.; Goedert, M.; Scheres, S.H.W. Cryo-EM structures of tau filaments from Alzheimer's disease. *Nature* **2017**, *547*, 185–190. [CrossRef] [PubMed]
7. Iadanza, M.G.; Jackson, M.P.; Hewitt, E.W.; Ranson, N.A.; Radford, S.E. A new era for understanding amyloid structures and disease. *Nat. Rev. Mol. Cell Biol.* **2018**, *19*, 755–773. [CrossRef] [PubMed]
8. So, M.; Hall, D.; Goto, Y. Revisiting supersaturation as a factor determining amyloid fibrillation. *Curr. Opin. Struct. Biol.* **2016**, *36*, 32–39. [CrossRef]
9. Chatani, E.; Yamamoto, N. Recent progress on understanding the mechanisms of amyloid nucleation. *Biophys. Rev.* **2018**, *10*, 527–534. [CrossRef]
10. Arosio, P.; Knowles, T.P.; Linse, S. On the lag phase in amyloid fibril formation. *Phys. Chem. Chem. Phys.* **2015**, *17*, 7606–7618. [CrossRef] [PubMed]
11. Tornquist, M.; Michaels, T.C.T.; Sanagavarapu, K.; Yang, X.; Meisl, G.; Cohen, S.I.A.; Knowles, T.P.J.; Linse, S. Secondary nucleation in amyloid formation. *Chem. Commun.* **2018**, *54*, 8667–8684. [CrossRef]
12. Michaels, T.C.T.; Saric, A.; Habchi, J.; Chia, S.; Meisl, G.; Vendruscolo, M.; Dobson, C.M.; Knowles, T.P.J. Chemical kinetics for bridging molecular mechanisms and macroscopic measurements of amyloid fibril formation. *Annu. Rev. Phys. Chem.* **2018**, *69*, 273–298. [CrossRef]
13. Ferrone, F. Analysis of protein aggregation kinetics. *Methods Enzymol.* **1999**, *309*, 256–274.
14. Serio, T.R.; Cashikar, A.G.; Kowal, A.S.; Sawicki, G.J.; Moslehi, J.J.; Serpell, L.; Arnsdorf, M.F.; Lindquist, S.L. Nucleated conformational conversion and the replication of conformational information by a prion determinant. *Science* **2000**, *289*, 1317–1321. [CrossRef]

15. Bleiholder, C.; Dupuis, N.F.; Wyttenbach, T.; Bowers, M.T. Ion mobility-mass spectrometry reveals a conformational conversion from random assembly to β-sheet in amyloid fibril formation. *Nat. Chem.* **2011**, *3*, 172–177. [CrossRef] [PubMed]
16. Chatani, E.; Imamura, H.; Yamamoto, N.; Kato, M. Stepwise organization of the β-structure identifies key regions essential for the propagation and cytotoxicity of insulin amyloid fibrils. *J. Biol. Chem.* **2014**, *289*, 10399–10410. [CrossRef] [PubMed]
17. Pavlova, A.; Cheng, C.Y.; Kinnebrew, M.; Lew, J.; Dahlquist, F.W.; Han, S. Protein structural and surface water rearrangement constitute major events in the earliest aggregation stages of tau. *Proc. Natl. Acad. Sci. USA* **2016**, *113*, E127–E136. [CrossRef] [PubMed]
18. Miti, T.; Mulaj, M.; Schmit, J.D.; Muschol, M. Stable, metastable, and kinetically trapped amyloid aggregate phases. *Biomacromolecules* **2015**, *16*, 326–335. [CrossRef]
19. Cao, Y.; Adamcik, J.; Diener, M.; Kumita, J.R.; Mezzenga, R. Different folding states from the same protein sequence determine reversible vs irreversible amyloid fate. *J. Am. Chem. Soc.* **2021**, *143*, 11473–11481. [CrossRef]
20. Hong, D.P.; Ahmad, A.; Fink, A.L. Fibrillation of human insulin a and B chains. *Biochemistry* **2006**, *45*, 9342–9353. [CrossRef] [PubMed]
21. Yamamoto, N.; Tsuhara, S.; Tamura, A.; Chatani, E. A specific form of prefibrillar aggregates that functions as a precursor of amyloid nucleation. *Sci. Rep.* **2018**, *8*, 62. [CrossRef]
22. Yamamoto, N.; Akai, T.; Inoue, R.; Sugiyama, M.; Tamura, A.; Chatani, E. Structural insights into the inhibition of amyloid fibril formation by fibrinogen via interaction with prefibrillar intermediates. *Biochemistry* **2019**, *58*, 2769–2781. [CrossRef] [PubMed]
23. Krimm, S.; Bandekar, J. Vibrational spectroscopy and conformation of peptides, polypeptides, and proteins. *Adv. Protein Chem.* **1986**, *38*, 181–364. [PubMed]
24. Yamamoto, N.; Inoue, R.; Makino, Y.; Shibayama, N.; Naito, A.; Sugiyama, M.; Chatani, E. Structural development of amyloid precursors in insulin B chain and the inhibition effect by fibrinogen. *BioRxiv* **2021**. [CrossRef]
25. Erdemir, D.; Lee, A.Y.; Myerson, A.S. Nucleation of crystals from solution: Classical and two-step models. *Acc. Chem. Res.* **2009**, *42*, 621–629. [CrossRef] [PubMed]
26. Karthika, S.; Radhakrishnan, T.K.; Kalaichelvi, P. A review of classical and nonclassical nucleation theories. *Cryst. Growth Des.* **2016**, *16*, 6663–6681. [CrossRef]
27. Langkilde, A.E.; Vestergaard, B. Methods for structural characterization of prefibrillar intermediates and amyloid fibrils. *FEBS Lett.* **2009**, *583*, 2600–2609. [CrossRef]
28. Liang, C.; Ni, R.; Smith, J.E.; Childers, W.S.; Mehta, A.K.; Lynn, D.G. Kinetic intermediates in amyloid assembly. *J. Am. Chem. Soc.* **2014**, *136*, 15146–15149. [CrossRef]
29. Zheng, W.; Tsai, M.Y.; Chen, M.; Wolynes, P.G. Exploring the aggregation free energy landscape of the amyloid-β protein (1-40). *Proc. Natl. Acad. Sci. USA* **2016**, *113*, 11835–11840. [CrossRef]
30. Chiricotto, M.; Melchionna, S.; Derreumaux, P.; Sterpone, F. Hydrodynamic effects on β-amyloid (16–22) peptide aggregation. *J. Chem. Phys.* **2016**, *145*, 035102. [CrossRef]
31. Hasecke, F.; Niyangoda, C.; Borjas, G.; Pan, J.; Matthews, G.; Muschol, M.; Hoyer, W. Protofibril-fibril interactions inhibit amyloid fibril assembly by obstructing secondary nucleation. *Angew. Chem. Int. Ed. Engl.* **2021**, *60*, 3016–3021. [CrossRef] [PubMed]
32. Laganowsky, A.; Liu, C.; Sawaya, M.R.; Whitelegge, J.P.; Park, J.; Zhao, M.; Pensalfini, A.; Soriaga, A.B.; Landau, M.; Teng, P.K.; et al. Atomic view of a toxic amyloid small oligomer. *Science* **2012**, *335*, 1228–1231. [CrossRef] [PubMed]
33. Zou, Y.; Li, Y.; Hao, W.; Hu, X.; Ma, G. Parallel β-sheet fibril and antiparallel β-sheet oligomer: New insights into amyloid formation of hen egg white lysozyme under heat and acidic condition from ftir spectroscopy. *J. Phys. Chem. B* **2013**, *117*, 4003–4013. [CrossRef] [PubMed]
34. Samdin, T.D.; Kreutzer, A.G.; Nowick, J.S. Exploring amyloid oligomers with peptide model systems. *Curr. Opin. Chem. Biol.* **2021**, *64*, 106–115. [CrossRef]
35. Chatani, E.; Yuzu, K.; Ohhashi, Y.; Goto, Y. Current understanding of the structure, stability and dynamic properties of amyloid fibrils. *Int. J. Mol. Sci.* **2021**, *22*, 4349. [CrossRef]
36. Glatter, O.; Kratky, O. *Small Angle X-ray Scattering*; Academic Press: London, UK; New York, NY, USA, 1982.
37. Holasek, A.; Wawra, H.; Kratky, O.; Mittelbach, P. Small-angle scattering of Bence-Jones protein. *Biochim. Biophys. Acta* **1964**, *79*, 76–87. [CrossRef]
38. Broersma, S. Viscous force constant for a closed cylinder. *J. Chem. Phys.* **1960**, *32*, 1632–1635. [CrossRef]
39. Nagano, T.; Iwasaki, T.; Onishi, K.; Awai, Y.; Terachi, A.; Kuwaba, S.; Asano, S.; Katasho, R.; Nagai, K.; Nakashima, A.; et al. Ly6d-induced macropinocytosis as a survival mechanism of senescent cells. *J. Biol. Chem.* **2021**, *296*, 100049. [CrossRef]

molecules

MDPI

Article

Shotgun Proteomics Revealed Preferential Degradation of Misfolded In Vivo Obligate GroE Substrates by Lon Protease in *Escherichia coli*

Tatsuya Niwa, Yuhei Chadani and Hideki Taguchi *

Cell Biology Center, Institute of Innovative Research, Tokyo Institute of Technology, Yokohama 226-8503, Japan; tniwa@bio.titech.ac.jp (T.N.); chadani.y.aa@m.titech.ac.jp (Y.C.)
* Correspondence: taguchi@bio.titech.ac.jp

Abstract: The *Escherichia coli* chaperonin GroEL/ES (GroE) is one of the most extensively studied molecular chaperones. So far, ~80 proteins in *E. coli* are identified as GroE substrates that obligately require GroE for folding in vivo. In GroE-depleted cells, these substrates, when overexpressed, tend to form aggregates, whereas the GroE substrates expressed at low or endogenous levels are degraded, probably due to misfolded states. However, the protease(s) involved in the degradation process has not been identified. We conducted a mass-spectrometry-based proteomics approach to investigate the effects of three ATP-dependent proteases, Lon, ClpXP, and HslUV, on the *E. coli* proteomes under GroE-depleted conditions. A label-free quantitative proteomic method revealed that Lon protease is the dominant protease that degrades the obligate GroE substrates in the GroE-depleted cells. The deletion of DnaK/DnaJ, the other major *E. coli* chaperones, in the Δ*lon* strain did not cause major alterations in the expression or folding of the obligate GroE substrates, supporting the idea that the folding of these substrates is predominantly dependent on GroE.

Keywords: molecular chaperone; chaperonin; GroEL; protease; Lon protease; proteomics; proteostasis

Citation: Niwa, T.; Chadani, Y.; Taguchi, H. Shotgun Proteomics Revealed Preferential Degradation of Misfolded In Vivo Obligate GroE Substrates by Lon Protease in *Escherichia coli*. *Molecules* **2022**, *27*, 3772. https://doi.org/10.3390/molecules27123772

Academic Editors: Kunihiro Kuwajima, Yuko Okamoto, Tuomas Knowles and Michele Vendruscolo

Received: 26 April 2022
Accepted: 8 June 2022
Published: 11 June 2022

1. Introduction

Most proteins must fold into their native structures to gain their functions [1]. However, protein folding often competes with the formation of aggregates, which not only impair protein function but also cause cytotoxicity in some cases [2–4]. In cells, molecular chaperones facilitate the proper folding of various proteins by preventing aggregate formation [2,3]. Indeed, previous studies have revealed that a significant fraction of the proteins in any cell requires at least one or several chaperones. In *Escherichia coli*, GroEL/ES (GroE) and DnaK/DnaJ (DnaKJ)-GrpE are known to assist in the proper folding of a large subset of proteins [2,3].

GroE is the only essential chaperone for the growth of *E. coli* [5,6]. One of the decades-long efforts in chaperonin biology is to identify the substrate proteins that are obligately dependent on GroE for proper folding in the cell (in vivo obligate GroE substrates). Kerner et al. [6] identified hundreds of proteins that interact with GroEL in *E. coli*, using a mass spectrometry (MS)-based proteomics approach. The GroEL-interactors are classified into three categories (Classes I, II, and III) according to the relative amount of the proteins bound to GroEL against the total cellular amount of each protein. In their study, the Class III substrates, which are most enriched in the GroE complex, were classified as potential obligate GroE substrates. Subsequently, Fujiwara et al. [7] conducted a further systematic assessment by using a conditional GroE expression strain, *E. coli* MGM100 [8], and found that only a subset of the Class III substrates has obligate GroE dependences for their folding in vivo. These substrates are regarded as in vivo obligate GroE substrates, and termed Class IV substrates. Further analyses based on data from a cell-free proteomics approach identified several in vivo obligate GroE substrates [9].

Many of these in vivo obligate GroE substrates formed aggregates when overexpressed in the GroE-depleted cells [7]. In addition to the aggregate formation, a subset of the GroE substrates expressed under leaky conditions are degraded in the GroE-depleted cells [7]. This degradation is thought to be a consequence of the failure of the substrates to fold under the GroE-depleted conditions, resulting in targets for proteases. In the *E. coli* cytosol, three ATP-dependent proteases, Lon, ClpXP, and HslUV, are involved in the degradation of misfolded proteins [10–12]. However, the physiological differences of these proteases in the clearance of misfolded proteins derived from chaperone deficiencies at the proteome level are largely unknown.

To evaluate the abundance of a wide range of proteins in cells, comprehensive and deep quantitative proteomic analysis technologies are needed. Over the past two decades, shotgun proteomics using liquid chromatography coupled to tandem MS (LC-MS/MS) has achieved remarkable progress, including precise quantitative analyses such as SILAC [13]. SILAC requires stable isotopic labeling, which restricts the available media for cultivation. Recent advances in label-free quantification technologies, such as a method based on the signal intensity of the MS1 ion chromatogram (so-called LFQ intensity) [14] and DIA/SWATH [15], have expanded the options for medium selection and culture conditions. Using the SWATH-MS acquisition method, we aim to investigate the "fate" of the in vivo obligate GroE substrates under GroE-depleted conditions. Specifically, the degradation properties of these substrates when misfolded was evaluated by deletions of three major cytosolic proteases (Figure 1A). The results showed that most of the obligate GroE substrates were degraded by Lon protease in the GroE-depleted cells. Further analysis using a DnaKJ-deleted strain and its derivative again revealed that GroE is predominantly involved in the folding of most obligate GroE substrates.

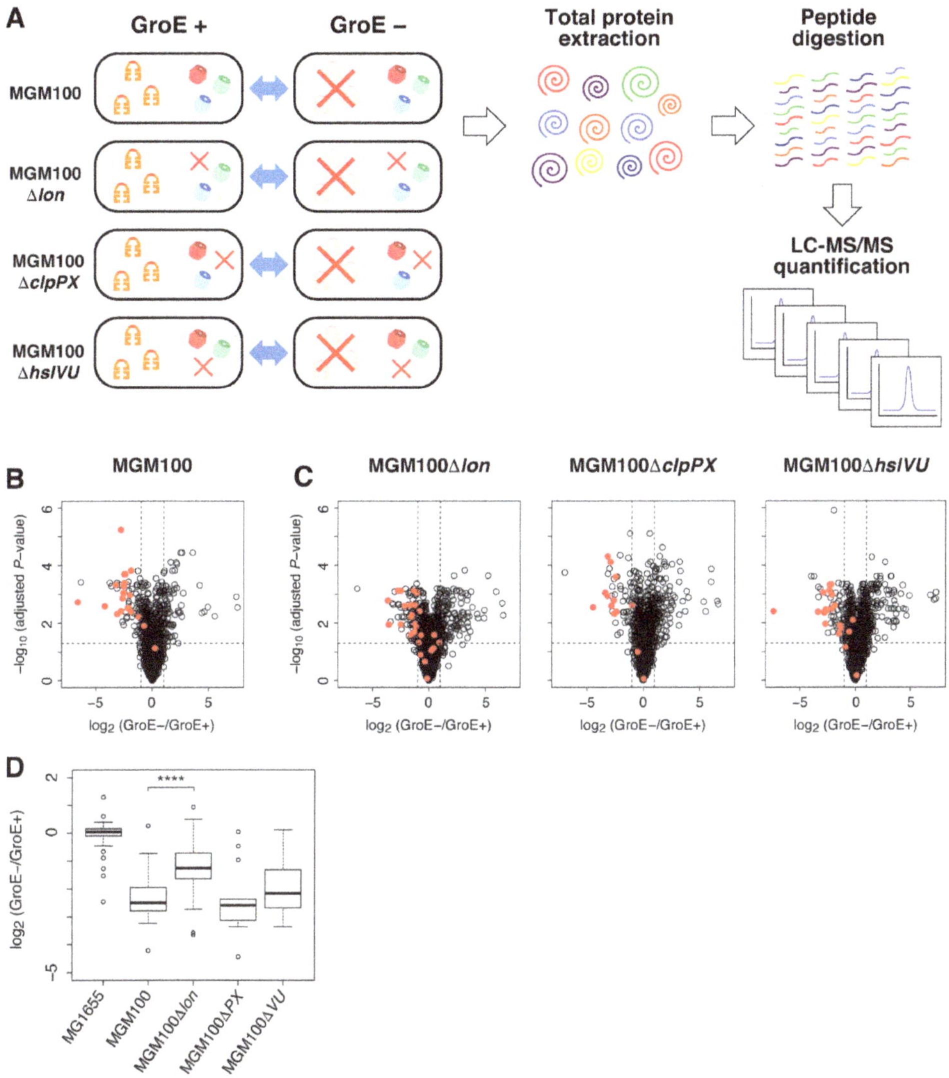

Figure 1. Proteome changes of the in vivo obligate GroE substrates by GroE depletion. (**A**) Schematic illustration of the experiment. The wildtype cells and the protease deletion variants were grown under the GroE-normal and GroE-depleted conditions. The total proteins were then extracted and digested into peptides by the Lys-C/Trypsin proteases. The digested peptides were measured and quantified by LC-MS/MS. (**B**) Proteomic changes in MGM100 by GroE depletion represented as a volcano plot. The horizontal axis indicates the value of the fold changes by the GroE depletion taken as $\log_2$, and the vertical axis indicates p-values by Welch's t-test (two-sided) with three technical replicates in each sample taken as $-\log_{10}$. For multiplicity correction, the p-values are adjusted by the Benjamini-Hochberg method. Red dots indicate the in vivo obligate GroE substrates. (**C**) Proteomic changes in the MGM100-derived protease-deletion strains by the GroE depletion, depicted as volcano plots. Red dots indicate the in vivo obligate GroE substrates. (**D**) Distribution of the fold changes of the in vivo obligate GroE substrates in each strain depicted as a boxplot. The box portions and the central bands are described according to the 25th percentile and the median, respectively. **** p-value < 0.001 (Wilcoxon's rank-sum test).

2. Results

2.1. Obligate GroE Substrates Tend to Be Degraded by Lon under GroE-Depleted Conditions

To investigate the degradation properties of the in vivo obligate GroE substrates under GroE-depleted conditions, we employed the SWATH-MS acquisition method. SWATH-MS can evaluate the relative abundances of proteins in a label-free manner at the proteome level in nutrient-rich media. To deplete the expression of GroE in cells, we used the MGM100 strain, in which the *groESL* gene is controlled by the arabinose-inducible *BAD* promoter to regulate the expression by sugar, as used in previous studies [7,9] (Figure 1A). First, we compared the expression amounts of the whole-cell proteins under GroE-depleted conditions with those under GroE-normal conditions. Based on this analysis, we evaluated the fold changes of ~1300 proteins (Supplementary Dataset S1). Among these evaluated proteins, ~20 in vivo obligate GroE substrates were included (Table 1). As reported previously, the MS analysis confirmed the GroE depletion-induced large proteome alteration, including the over expression of MetE and several heat shock proteins such as ClpB and DnaK [7] (Supplementary Dataset S1). Importantly, the expression of the obligate GroE substrates showed a strong tendency to decrease when GroE was depleted (Figure 1B). This result suggests that many obligate GroE substrates tend to be degraded, rather than forming aggregates in the GroE-depleted cells.

Table 1. Fold change values of the in vivo obligate GroE substrates in MGM100 and MGM100-derived protease-deletion strains.

Gene Name	Kerner 2005	Fujiwara 2010	Niwa 2016	MGM100 Number of Detection Ara	MGM100 Number of Detection Glc	MGM100 FC	MGM100Δ*lon* Number of Detection Ara	MGM100Δ*lon* Number of Detection Glc	MGM100Δ*lon* FC	MGM100Δ*clpPX* Number of Detection Ara	MGM100Δ*clpPX* Number of Detection Glc	MGM100Δ*clpPX* FC	MGM100Δ*hslVU* Number of Detection Ara	MGM100Δ*hslVU* Number of Detection Glc	MGM100Δ*hslVU* FC
argP	3	4		3	3	1.191	3	3	1.407	3	3	1.033	3	3	1.089
ltaE	3	4		3	3	0.604	3	3	1.170	3	3	0.723	3	3	0.857
metK	3	4		3	3	0.433	3	3	1.902	3	3	0.513	3	3	0.683
add	3	4		3	3	0.271	3	3	0.316	3	3	0.194	3	3	0.280
dapA	3	4		3	3	0.269	3	3	0.483	3	3	0.187	3	3	0.234
asd	3	4		3	3	0.249	3	3	0.569	3	3	0.171	3	3	0.419
rfbC	3	4		3	3	0.235	3	3	0.450	3	1		3	3	0.386
serC	2	4		3	3	0.221	3	3	0.428	3	3	0.167	3	3	0.363
hemB	3	4		3	3	0.188	3	3	0.395	3	3	0.112	3	3	0.202
pmbA	3	4		3	3	0.178	3	3	0.225	3	0		3	3	0.176
lipA	3	4		3	3	0.177	3	3	0.410	3	3	0.136	3	3	0.201
nuoC	2	4		3	3	0.170	3	3	0.080	3	3	0.115	3	3	0.221
pepQ	3	4		3	3	0.169	3	3	0.177	3	3	0.143	3	3	0.155
fabF	3	4		3	3	0.158	3	3	0.336	3	2		3	3	0.228
kdsA	2	4		3	3	0.145	3	3	0.328	3	3	0.098	3	3	0.157
pyrD		4		3	3	0.145	3	3	0.176	3	2		3	2	
gatY	3	4		3	3	0.114	3	3	0.150	3	3	0.046	3	3	0.098
deoA	3	4		3	3	0.106	3	1		3	0		3	3	0.145
sdhA	3	4		3	3	0.054	3	3	0.084	3	0		3	3	0.135
araA	3	4		3	3	0.010	3	1		3	2		3	3	0.006
fbaB	3	4		3	0		3	3	1.323	3	1		3	0	
ftsE	3	4		3	0		3	3	0.620	3	0		3	2	
nagZ	3	4		3	0		3	3	0.539	3	0		3	0	
ybhA			○	3	0		3	3	0.332	3	1		3	0	
tas	2		○	3	0		3	3	0.788	3	0		3	0	
nagD			○	3	2		3	3	0.598	3	0		3	2	
alaA	3	4		3	1		3	3	0.418	3	0		3	3	0.535
dadA	3	4		3	1		3	3	0.350	3	0		3	0	

Next, we investigated the relevance of the Lon, ClpXP, and HslUV cytosolic proteases on the degradation of misfolded proteins in the GroE-depleted cells. We made new deletion strains of each protease in MGM100 as the background strain (MGM100Δ*lon*, MGM100Δ*clpPX*, and MGM100Δ*hslVU*). Then, we investigated their proteome changes elicited by the GroE-depletion. Strikingly, the volcano plot, which indicates the variation of expression amounts on the horizontal axis and statistical certainty on the vertical axis, showed that the decrease in the obligate GroE substrates under GroE-depleted conditions was largely recovered in the MGM100Δ*lon* strain (Figure 1C and Supplementary Dataset S1). This recovery trend was statistically significant from the fold change distributions of the obligate GroE substrates (p = 0.000942, by Wilcoxson's rank-sum test), represented as a boxplot, which depicted the distribution of the values in each sample (Figure 1D). In addition, some of the previously known Lon substrates including LipA, one of the obligate GroE substrates, were increased in the MGM100Δ*lon* strain (Supplementary Table S1), corroborating the assumption based on the previous findings. In contrast, the volcano plots of the ClpXP and HslUV deletion strains did not show this trend (Figure 1C and Supplementary Dataset S1). The boxplot revealed a weak recovery of the GroE substrates in the MGM100Δ*hslVU* cells, but the difference was not statistically significant ($p = 0.3273$, by Wilcoxson's rank-sum test) (Figure 1D). No significant changes by GroE depletion were observed for the GroE substrates belonging to other classes (Class I, II, and III minus) in any of the *E. coli* strains (Supplementary Figure S1). Furthermore, the MS analysis revealed that five GroE substrates (FbaB, FtsE, NagZ, YbhA, and Tas), which were identified in the GroE-normal cells but not in the GroE-depleted cells were reproducibly identified in the MGM100Δ*lon* strain under GroE-depleted conditions (Table 1 and Supplementary Figure S1B). This result suggests that the proteolysis of the five proteins by Lon protease under GroE-depleted conditions was circumvented in the Lon-deleted cells.

2.2. Deletion of DnaKJ Barely Affects the Folding of Most In Vivo Obligate GroE Substrates

Our previous reconstituted cell-free translation (PURE system) analysis revealed that almost all of the obligate GroE substrates have a strong tendency to form aggregates without chaperones [7,16], but for many of them the aggregation-formation is rescued by the DnaKJ system [17]. Therefore, we assumed that the folding of the in vivo obligate GroE substrates might depend on not only GroE but also DnaKJ. If this is the case, then the Lon deletion in DnaKJ-deleted cells would affect the abundance of the in vivo obligate GroE substrates, even in the presence of GroE. Accordingly, we compared the proteome changes between the wildtype, *dnaKJ*-deleted (Δ*dnaKJ*), and *dnaKJ*&*lon*-deleted (Δ*dnaKJ*Δ*lon*) strains. As reported previously, the deletion of *dnaKJ* caused drastic proteome changes [18] (Figure 2A and Supplementary Dataset S2). Among ~1300 evaluated proteins, 60~70 and 60~80 proteins were specifically up- and down-regulated (fold change > 2 or <0.5 and adjusted p-value < 0.05) by the deletion of *dnaKJ* or *dnaKJ* and *lon*, respectively (Figure 2A and Supplementary Dataset S2). In contrast, the proteome change by the deletion of *lon* in addition to *dnaKJ* was small (Figure 2A and Supplementary Dataset S2). Although 30~40 proteins were specifically up-regulated (fold change > 2 and adjusted p-value < 0.05) by the deletion of *lon*, their fold change values were not large compared to the results in wildtype vs. Δ*dnaKJ* or wildtype vs. Δ*dnaKJ*Δ*lon* (Figure 2A and Supplementary Dataset S2). In addition, only about five to nine proteins were down-regulated in the Δ*dnaKJ*Δ*lon* strain against Δ*dnaKJ* (Figure 2A and Supplementary Dataset S2). Notably, the obligate GroE substrates did not show any remarkable changes in both deletion strains (Figure 2A, Table 2 and Supplementary Figure S2A). This result suggests that DnaKJ is not an additional factor associated with the folding of the obligate GroE substrates in cells.

However, the possibility that these substrates form aggregates before degradation remained, and hence the amounts of the proteins are not changed. To assess this possibility, we prepared the pellet fraction from the lysates of each strain by centrifugation and conducted the same proteomic analysis. Although the reproducibility of the fold change values was not as good as that of the total proteome analysis (Supplementary Figure S2B), the results clearly demonstrated that only a small subset of the obligate GroE substrates accumulated in the Δ*dnaKJ* cells and the Δ*dnaKJ*Δ*lon* cells (Figure 2B and Table 2). In other words, the absence of DnaKJ does not induce the aggregate formation of many obligate GroE substrates. In summary, the results suggest that there is no additional benefit to having DnaKJ for the folding of many obligate GroE substrates when GroE is present.

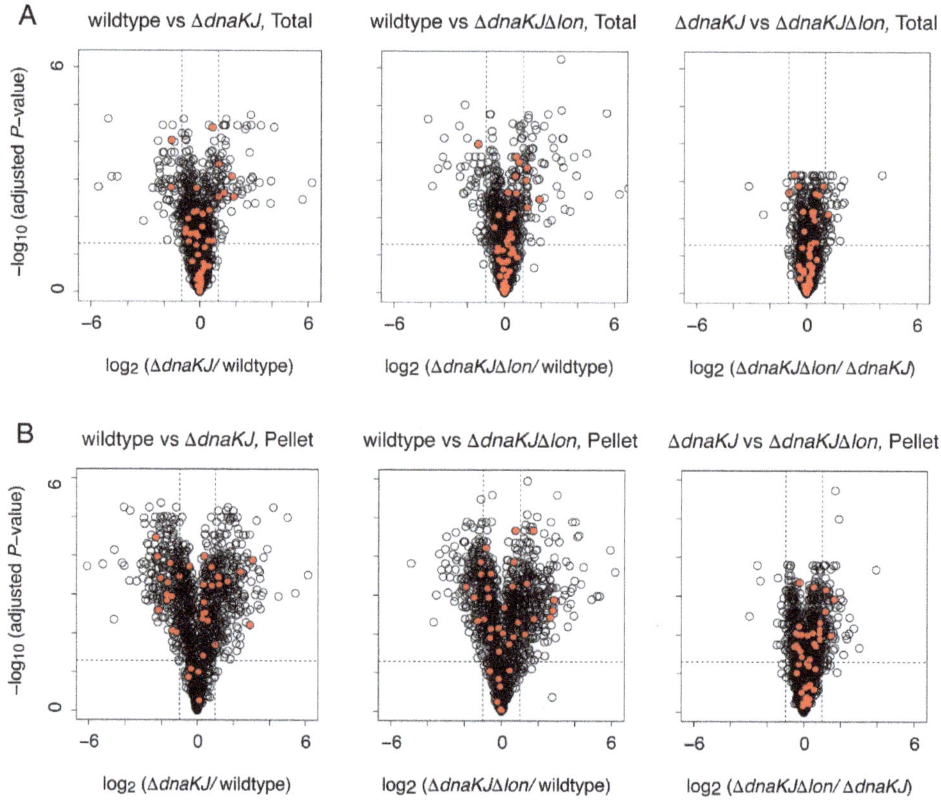

Figure 2. Proteome changes of the in vivo obligate GroE substrates by DnaK/DnaJ deletion. (**A**) Proteome changes of Δ*dnaKJ* and Δ*dnaKJ*Δ*lon* in the total fraction depicted as volcano plots. The horizontal axis indicates the value of the fold changes taken as $\log_2$, and the vertical axis indicates *p*-values by Welch's *t*-test (two-sided) with three technical replicates in each sample taken as $-\log_{10}$. For multiplicity correction, the *p*-values are adjusted by the Benjamini-Hochberg method. Red dots indicate the in vivo obligate GroE substrates. (**B**) Proteome changes of Δ*dnaKJ* and Δ*dnaKJ*Δ*lon* in the pellet fraction depicted as volcano plots. Red dots indicate the in vivo obligate GroE substrates.

Table 2. Fold change values of the in vivo obligate GroE substrates for Δ*dnaKJ* and Δ*dnaKJ*Δ*lon* strain.

Gene Name	Kerner 2005	Fujiwara 2010	Niwa 2016	Number of Detection			Fold Change			Number of Detection			Fold Change		
				wt Total	Δ*KJ* Total	Δ*KJ* Δ*lon* Total	Δ*KJ*/wt Total	Δ*KJlon*/ wt Total	Δ*KJlon*/ Δ*KJ* Total	wt ppt	Δ*KJ* ppt	Δ*KJ* Δ*lon* ppt	Δ*KJ*/wt ppt	Δ*KJlon*/ wt ppt	Δ*KJlon*/ Δ*KJ* ppt
dadX	3	4		3	3	3	3.611	3.691	1.022	3	3	3	5.136	7.144	1.391
nanA	3	4		3	3	3	3.351	1.711	0.511	3	3	3	1.047	0.904	0.864
sdhA	3	4		3	3	3	2.466	2.230	0.905	3	3	3	1.725	1.578	0.915
lldD	3	4		3	3	3	2.035	1.477	0.726	3	3	3	1.218	0.890	0.731
dadA	3	4		3	3	3	2.022	2.320	1.147	3	3	3	1.754	1.847	1.053
gdhA		4		3	3	3	1.632	2.291	1.404	3	3	3	2.147	3.361	1.566
metK	3	4		3	3	3	1.612	1.008	0.625	3	3	3	1.264	0.996	0.788
nuoC	2	4		3	3	3	1.423	1.232	0.866	3	3	3	1.311	1.102	0.840
ybjS	3	4		3	3	3	1.395	1.592	1.141	3	3	3	2.299	2.393	1.041
rfbC	3	4		3	3	3	1.378	1.079	0.783	3	3	3	8.265	6.135	0.742
yqaB	3	4		3	3	3	1.209	1.323	1.094	2	3	3			0.820
ycfH	3	4		3	3	3	1.206	0.935	0.775	3	0	2			
lsrF	3	4		3	3	3	1.154	1.381	1.196	3	3	3	0.713	0.855	1.200
ybhA			○	3	3	3	1.135	1.197	1.055	3	3	3	7.462	6.802	0.912
gatY	3	4		3	3	3	1.114	1.541	1.384	3	3	3	2.262	3.266	1.443

Table 2. *Cont.*

Gene Name	Kerner 2005	Fujiwara 2010	Niwa 2016	Number of Detection			Fold Change			Number of Detection			Fold Change		
				wt Total	ΔKJ Total	ΔKJ Δlon Total	ΔKJ/wt Total	ΔKJlon/ wt Total	ΔKJlon/ ΔKJ Total	wt ppt	ΔKJ ppt	ΔKJ Δlon ppt	ΔKJ/wt ppt	ΔKJlon/ wt ppt	ΔKJlon/ ΔKJ ppt
fabF	3	4		3	3	3	1.088	0.910	0.836	3	3	3	0.496	0.592	1.194
nagD			○	3	3	3	1.077	1.073	0.996	3	3	3	1.293	0.892	0.690
deoA	3	4		3	3	3	1.047	0.931	0.888	3	3	3	0.443	0.644	1.454
nagZ	3	4		3	3	3	1.045	0.954	0.914	3	0	2			
add	3	4		3	3	3	1.039	1.135	1.092	3	3	3	3.164	2.571	0.813
ftsE	3	4		3	3	3	1.038	1.127	1.086	3	3	3	1.270	1.051	0.827
lipA	3	4		3	3	3	1.032	1.343	1.302	3	3	3	1.360	1.327	0.976
kdsA	2	4		3	3	3	1.004	1.006	1.002	3	3	3	0.303	0.555	1.829
alaA	3	4		3	3	3	0.989	1.054	1.066	3	3	3	1.987	1.499	0.754
asd	3	4		3	3	3	0.962	1.086	1.130	3	3	3	0.368	0.666	1.809
pmbA	3	4		3	3	3	0.943	1.051	1.114	3	3	3	1.500	1.662	1.109
serC	2	4		3	3	3	0.927	1.160	1.251	3	3	3	0.223	0.409	1.837
argP	3	4		3	3	3	0.920	1.339	1.456	3	3	3	1.252	1.548	1.236
pepQ	3	4		3	3	3	0.916	1.127	1.230	3	3	3	0.199	0.458	2.297
ltaE	3	4		3	3	3	0.890	1.192	1.339	3	3	3	0.344	0.615	1.787
pyrD		4		3	3	3	0.886	0.838	0.947	3	3	3	0.728	0.667	0.916
dapA	3	4		3	3	3	0.879	0.949	1.080	3	3	3	0.210	0.466	2.217
uxaC	3	4		3	3	3	0.870	0.772	0.887	3	2	3		0.263	
tldD	3	4		3	3	3	0.859	0.944	1.100	3	3	3	0.784	0.966	1.232
hemB	3	4		3	3	3	0.812	1.531	1.887	3	3	3	0.303	0.949	3.135
pyrC	2	4		3	3	3	0.785	0.797	1.015	3	2	3		0.520	
argE	3	4		3	3	2	0.745			3	3	3	1.076	1.346	1.251
nfo	3	4		3	3	3	0.686	0.676	0.984	3	0	2			
yajO	3	4		3	3	3	0.648	0.802	1.239	3	0	0			
tas	2		○	3	3	3	0.633	1.002	1.583	3	1	0			
gpr	2		○	3	3	3	0.581	0.756	1.300	3	3	3	0.236	0.651	2.762
cysH			○	3	3	3	0.566	1.257	2.221	3	3	2	0.380		
fbaB	3	4		3	3	3	0.332	0.372	1.120	3	0	0			
frdA	3	4		3	3	3	0.326	0.373	1.147	3	3	3	0.315	0.359	1.141
yafD	3	4		0	3	3			1.206	1	3	3			1.461
fadA		4		2	3	3			0.740	1	3	3			1.068
yigB			○	2	2	2				2	3	3			1.542
yjhH		4								0	3	3			1.219
dusB	3	4								1	3	3			1.119
eutB	3	4		1	0	1				1	3	3			0.962

2.3. Metabolic Perturbations by Protease Deletions under GroE-Depleted Conditions Revealed by Clustering Analysis

In the above analyses, we only focused on the changes in the amounts of the obligate GroE substrates. However, although the GroE depletion alone causes drastic changes in protein expression and metabolism [7], the additional deletion of the proteases may elicit further perturbations of the proteome or metabolome. Therefore, we assessed the proteome changes caused by deleting each protease under GroE-depleted conditions. For this purpose, we performed a clustering approach with the fold change values, defined as the ratio of protein abundances under GroE-normal and GroE-depleted conditions in each strain. We chose ~1000 proteins, with fold change values quantified in all four strains, for clustering by the k-means method. The number of clusters was set to six, and the fold change values were converted to logarithmic values before the clustering. The clustering analysis returned four clusters containing small numbers of proteins (Clusters 1~4) and two clusters with larger numbers of proteins (Clusters 5~6) (Figure 3A and Supplementary Dataset S3). Clusters 1~4 exhibited relatively more significant differences in their fold-changes than the other two, suggesting that these four clusters could provide some information about the specific changes caused by the deletion of the proteases. We then applied an enrichment analysis with annotation by the KEGG BRITE hierarchy [19] to characterize these four clusters (Table 3). As shown in Table 3, many metabolic pathways were enriched in each cluster. Especially, Cluster 3, in which the fold change pattern in MGM100Δlon was only decreased, had the highest number of metabolic pathways with fluctuations, including amino acid metabolism and nucleic acid metabolism. This result suggests that the deletion of Lon protease in the GroE-depleted cells causes a further metabolic perturbation in addition to the GroE depletion.

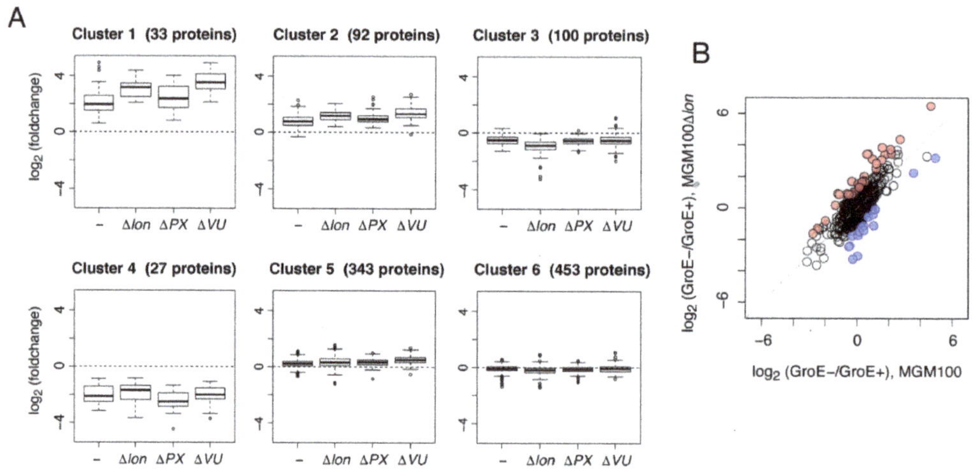

Figure 3. Clustering analysis to investigate the proteome perturbations induced by the protease deletions. (**A**) Distribution of the fold changes of the MGM100 and MGM100-derived protease-deletion strains in each cluster. Fold changes were defined as the protein abundance ratios under GroE-normal and GroE-depleted conditions in each strain. The box portions and the central bands are described according to the 25th percentile and the median, respectively. (**B**) Fold change comparison between MGM100 and MGM100Δ*lon*. Red dots indicate the proteins specifically up-regulated by the Lon deletion, and blue dots indicate the proteins specifically down-regulated by the Lon deletion.

Table 3. Enrichment analysis with the KEGG BRITE hierarchy for each cluster.

Cluster	Annotation (KEGG BRITE Hierarchy3)	Odds Ratio	p-Value *	Number of Proteins in Population	Number of Proteins in Subgroup
Cluster 1	Selenocompound metabolism	24.95	0.0009	7	3
	Monobactam biosynthesis	16.15	0.0133	6	2
	Folate biosynthesis	9.23	0.0302	9	2
	Cysteine and methionine metabolism	8.83	0.0008	25	5
	Lysine biosynthesis	8.08	0.0370	10	2
	Glycine, serine and threonine metabolism	5.93	0.0086	27	4
Cluster 2	Histidine metabolism	31.93	0.0025	4	3
	Nitrogen metabolism	10.55	0.0407	4	2
	Chaperones and folding catalysts	5.41	0.0002	31	10
	Starch and sucrose metabolism	4.77	0.0215	13	4
	Glutathione metabolism	3.90	0.0357	15	4
Cluster 3	C5-Branched dibasic acid metabolism	Inf	0.0090	2	2
	Lysine degradation	Inf	0.0090	2	2
	Photosynthesis proteins	16.48	0.0003	8	5
	Oxidative phosphorylation	11.54	5.5×10^{-7}	21	11
	Tetracycline biosynthesis	9.61	0.0476	4	2
	Citrate cycle (TCA cycle)	6.64	0.0016	15	6
	Purine metabolism	5.33	1.2×10^{-5}	42	14
	Pyrimidine metabolism	4.89	0.0003	31	10
	One carbon pool by folate	4.84	0.0463	9	3
	Propanoate metabolism	3.90	0.0369	14	4
	Enzymes	1.96	0.0015	561	68

Table 3. *Cont.*

Cluster	Annotation (KEGG BRITE Hierarchy3)	Odds Ratio	*p*-Value *	Number of Proteins in Population	Number of Proteins in Subgroup
Cluster 4	Bacterial chemotaxis	Inf	0.0006	2	2
	Galactose metabolism	43.52	1.8×10^{-8}	15	7
	Monobactam biosynthesis	20.12	0.0090	6	2
	Lysine biosynthesis	10.07	0.0253	10	2
	Glycine, serine and threonine metabolism	5.18	0.0295	27	3
	Enzymes	3.12	0.0076	561	21

* *p*-value was calculated by Fisher's exact test. Only the annotations whose *p*-values were less than 0.05 are listed.

To investigate the perturbations from the deletion of Lon more directly, we defined the proteins with specific changes between MGM100 and MGM100Δ*lon* from the distribution of the fold changes (Figure 3B). The results of the enrichment analysis for the up-regulated and down-regulated proteins in MGM100Δ*lon* showed that the deletion of Lon caused the up-regulation of some metabolic enzymes related to amino acid synthesis under GroE-depleted conditions (Table 4 and Supplementary Dataset S4).

Table 4. Enrichment analysis for the proteins with specific changes by the deletion of Lon protease.

Group	Annotation (KEGG BRITE Hierarchy3)	Odds Ratio	*p*-Value *	Number of Proteins in Population	Number of Proteins in Subgroup
Up-regulated in MGM100Δ*lon*	Bacterial motility proteins	Inf	0.0344	1	1
	Monobactam biosynthesis	14.69	0.0158	6	2
	Selenocompound metabolism	11.76	0.0216	7	2
	Cysteine and methionine metabolism	7.96	0.0012	25	5
	Lysine biosynthesis	7.35	0.0434	10	2
	Glyoxylate and dicarboxylate metabolism	6.02	0.0213	18	3
	Glycine, serine and threonine metabolism	5.36	0.0117	27	4
Down-regulated in MGM100Δ*lon*	Glycosylphosphatidylinositol (GPI)-anchored proteins	Inf	0.0162	1	1
	Aminobenzoate degradation	Inf	0.0162	1	1
	Glyoxylate and dicarboxylate metabolism	14.38	0.0025	18	3

* *p*-value was calculated by Fisher's exact test. Only the annotations whose *p*-values were less than 0.05 are listed.

Another minor change was observed in Cluster 1, as its fold change pattern revealed a large increase in both MGM100Δ*lon* and MGM100Δ*hslVU* as compared to the other two strains. This cluster included some proteins induced in the stationary growth phase, such as Sra, Dps, WrbA, ElaB, and OsmC.

3. Discussion

In this analysis, we have shown that many in vivo obligate GroE substrates are degraded by cytosolic proteases under GroE-depleted conditions, and Lon protease is mainly responsible for this degradation (Figure 1). Conversely, DnaKJ does not act as a dominant factor for the folding of these substrates, although a few obligate GroE substrates tended to form aggregates by the deletion of DnaKJ (Figure 2). Based on these results, a plausible scheme of the behavior of the obligate GroE substrates under GroE-depleted conditions is depicted in Figure 4. When GroE is absent, these substrates cannot complete their folding and are degraded by proteases such as Lon. In contrast, when DnaKJ was absent, most of the obligate GroE substrates can complete their folding with the aid of GroE, although a small fraction of the obligate GroE substrates form aggregates in cells. However, since various chaperones such as GroE are highly up-regulated in Δ*dnaKJ* cells [18], our observation might be affected by these up-regulated chaperones' effects. Considering this point, our results do not exclude the possibility that DnaKJ is also involved in the folding of the obligate GroE substrates under some conditions.

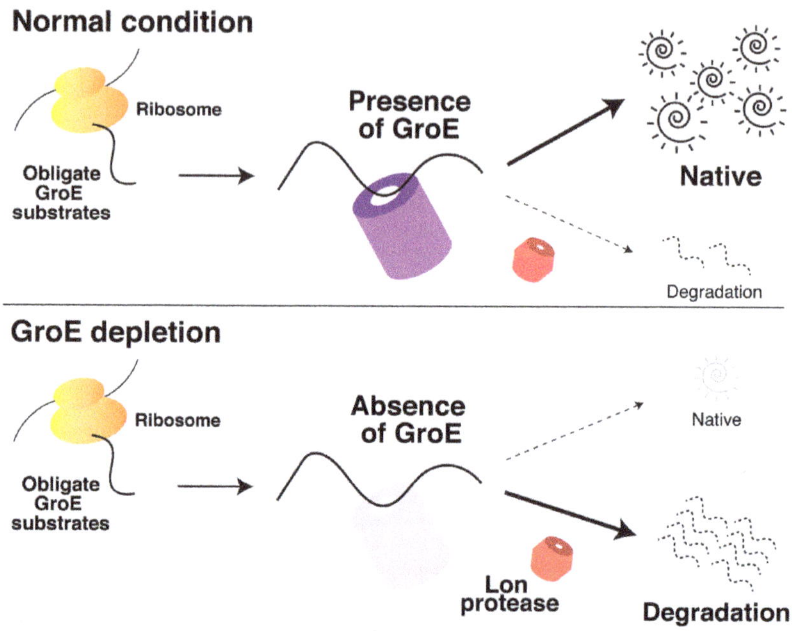

Figure 4. Schematic illustration of the fate of in vivo obligate GroE substrates in GroE-depleted cells. Under normal conditions, the in vivo obligate GroE substrates can fold into their native structures with the aid of GroE. Upon GroE depletion, these GroE substrates tend to be degraded by proteases, preferentially by the Lon protease, since they cannot complete their folding without GroE.

The statistical analysis of the proteome changes in our experiments suggested that the deletion of Lon may cause additional metabolic perturbations, such as in amino acid synthesis. Note that the *lon* deletion does not show large proteome changes under nutrient-rich medium conditions (Niwa T. et al., in preparation). Of course, since the GroE depletion itself reportedly induces large metabolic changes, including the depletion of several amino acids, S-adenosylmethionine, and NADPH [7], this phenomenon may be significant only under exceptional metabolic conditions. However, this observation might reflect the possibility that protein degradation affects the metabolism in an unappreciated manner, although it may be significant only under extreme conditions such as severe energy deficiency.

In addition, HslUV may also be involved in the degradation of misfolded proteins and additional metabolic perturbations, as shown in Figure 1B,C and Figure 3A and Table 3. Although HslUV is expressed abundantly in cells and its expression is induced by heat stress, knowledge about the physiological role of HslUV is limited. Our observations suggest that HslUV may have certain specific functions in the cell, with some overlapping with Lon. However, the evidence for this overlapped role is not strong, and the details are still unclear.

The up-regulation of some stationary-phase-induced proteins in both MGM100Δ*lon* and MGM100Δ*hslVU* suggests that the deletion of these two proteases may cause additional changes related to starvation or another specific factor could invoke the transition of the growth phase under GroE-depleted conditions. However, the connection between them remains unclear. We confirmed that RpoS, a factor responsible for various stress responses and the growth phase transition [20,21], does not appear to be involved in this change since its expression was up-regulated only in MGM100Δ*clpPX* (Supplementary Dataset S1). Accordingly, further investigations are needed to clarify the physiological roles and the overlapped manners of these cytosolic proteases in detail.

In summary, our study has partially uncovered the fate of the in vivo obligate GroE substrates under GroE-depleted conditions. Furthermore, the inability to degrade the misfolded protein perturbs proper intracellular metabolism. The intimate link between chaperones and proteases in cellular proteostasis is important but not well understood; hence, the approach conducted here would be valuable for analyses of other organisms, including eukaryotes.

4. Materials and Methods

4.1. Bacterial Strains

E. coli strains used in this study are listed in Supplementary Table S2. The DNA fragment amplified from JW0013-KC (Δ*dnaK*::FRT-Km^R-FRT) [22], using the primers PT0071 (AAATTGGGCAGTTGAAACCAGAC) and PM0195 (GATGTTTCGCTTGGTGGTCGAATGGGCAGG), and that from JW0014-KC (Δ*dnaJ*::FRT-Km^R-FRT), using the primers PT0072 (TACAGGTGCTCGCATATCTTCAACG) and PM0196 (CCTGCCCATTCGACCACCAAGCGAAACATC), were mutually annealed and amplified using PT0071 and PT0072. The purified DNA was electroporated into the *E. coli* strain BW25113 harboring pKD46 [23], and the transformants resistant to 40 μg/mL kanamycin were stored as ECY0262 (BW25113Δ*dnaKJ*, Km^R).

The DNA fragment amplified from JW3903-KC (Δ*hslV*::FRT-Km^R-FRT), using the primers PT0223 (CCATCTATAATTGCATTATG) and PM0195, and that from JW3902-KC (Δ*hslU*::FRT-Km^R-FRT), using the primers PT0224 (CTGAGTTCGGCTAATTTGTTG) and PM0196, were mutually annealed and amplified using PT0223 and PT0224. The purified DNA was electroporated into the *E. coli* strain BW25113 harboring pKD46, and the transformants resistant to 40 μg/mL kanamycin were stored as ECY0289 (BW25113ΔhslVU, Km^R).

The DNA fragment amplified from JW0427-KC (Δ*clpP*::FRT-Km^R-FRT), using the primers PT0065 (ATGGTGATGCCGTACCCATAACAC) and PM0195, and that from JW0428-KC (Δ*clpX*::FRT-Km^R-FRT), using the primers PT0066 (AGCCCGATCCGCCATCTAACTTAGC) and PM0196, were mutually annealed and amplified using PT0065 and PT0066. The purified DNA was electroporated into the *E. coli* strain BW25113 harboring pKD46, and the transformants resistant to 40 μg/mL kanamycin were stored as ECY0290 (BW25113ΔclpPX, Km^R).

Amplified pKD3 [23] plasmid DNA was electroporated into ECY0289, ECY0290, and JW0429 (Δ*lon*::FRT-Km^R-FRT) harboring pCP20A (modification of pCP20 [23], with an inactivated cat selection marker), and the transformants resistant to 20 μg/mL chloramphenicol were stored as ECY0292 (Δ*hslVU*, Cm^R), ECY0293 (Δ*clpPX*, Cm^R), and ECY0210 (Δ*lon*, Cm^R), respectively.

Phage P1-mediated transduction was used to introduce the Δ*dnaKJ*, Δ*hslVU*, Δ*clpPX* and Δ*lon* mutations from ECY0262, ECY0292, ECY0293 and ECY0210, respectively.

4.2. Cell Culture and Sample Preparation for the LC-MS/MS Analysis

For the analysis of the MGM100 and MGM100-derived mutants, cells were grown in LB medium supplemented with 0.2% arabinose at 37 °C and harvested at an early logarithmic growth phase (0.2~0.3 OD_{660}). After washing with LB medium, the cells were inoculated into LB medium supplemented with 1 mM diaminopimelate and either 0.2% arabinose or 0.2% glucose. At this time, the OD_{660} of the culture solution was set to 0.04. After 3 h of cultivation at 37 °C, the cells were harvested. The OD_{660} at the time of collection was around 0.8~1.0. For the analyses of Δ*dnaKJ* and Δ*dnaKJ*Δ*lon*, cells were grown in LB medium at 37 °C and harvested at a logarithmic growth phase (~1.0 OD_{660}).

The harvested cells were resuspended in the PTS solution [24] (100 mM Tris-HCl (pH 9.0), 12 mM sodium deoxycholate, 12 mM sodium N-lauroylsarcosinate) and boiled at

95 °C for 5 min. The solution was then frozen at −80 °C for 10 min. Next, the solution was sonicated in an ultrasonic bath for 20 min at room temperature for further cell disruption. After the cell disruption, the protein concentration was measured using the BCA protein assay kit (Thermo Fisher Scientific, Waltham, MA, USA) and fixed at 50 μg in 100 μL or 25 μg in 50 μL by dilution with the PTS solution.

The obtained total proteins were reduced by 10 mM dithiothreitol and incubated for 30 min at room temperature. Afterwards, 50 mM iodoacetamide was added, and the solution was incubated for 20 min at room temperature in the dark for alkylation. After the reduction and alkylation, the solution was diluted 5-fold with 50 mM ammonium bicarbonate and digested by adding Lys-C protease (FUJIFILM-Wako, Osaka, Japan) at 1/100 of the total protein weight and incubating for 3 h at room temperature. The fragmented peptides were further digested by adding Trypsin Gold (Promega, Madison, Wisconsin, USA) at 1/50 of the total protein weight, and incubated at 37 °C overnight. After the digestion, an equal volume of ethyl acetate and 1/20 volume of 10% trifluoroacetic acid (TFA) were added to the peptide solution and mixed vigorously. The mixture was centrifuged at 15,700× *g* for 2 min, and the upper ethyl acetate layer containing the surfactants was withdrawn. The resulting lower water layer was dried with a centrifugal evaporator. The peptides were then re-dissolved in 0.1% TFA and 2% acetonitrile and desalted with a handmade Stage Tip [25] composed of an SDB-XC Empore Disk (3 M, Maplewood, MN, USA). The peptides bound to the Stage Tip were eluted with 0.1% TFA and 80% acetonitrile. After the desalting, the peptides were dried by a centrifugal evaporator again and re-dissolved in 0.1% TFA and 2% acetonitrile for LC-MS/MS measurements.

For the proteome analysis of pellet fractions, cells were resuspended in lysis buffer (50 mM Tris-HCl (pH 7.5), 100 mM NaCl, 1 mM EDTA) supplemented with a protease inhibitor cocktail (cOmplete™ mini, EDTA free, Roche, Basel, Switzerland) and disrupted by sonication. The resulting lysate was centrifuged at 20,000× *g* for 10 min, and the supernatant was discarded. After washing twice with the lysis buffer, the pellet was dissolved in the PTS solution. The protein concentration was measured with a BCA protein assay kit and fixed at 10 μg in 25 μL by dilution with the PTS solution. The subsequent processes were performed as described above.

4.3. LC-MS/MS Measurement and Data Analysis

The LC-MS/MS measurements were conducted with an Eksigent NanoLC Ultra and TripleTOF 4600 tandem-mass spectrometer or an Eksigent NanoLC 415 and TripleTOF 6600 mass spectrometer (AB Sciex, Framingham, MA, USA). The trap column used for nanoLC was a 5.0 mm × 0.3 mm ODS column with a particle size of 5 μm (L-column2, Chemical Evaluation and Research Institute, Tokyo, Japan). The separation column was a 12.5 cm × 75 μm capillary column packed with 3 μm C18-silica particles (Nikkyo Technos, Tokyo, Japan). The detailed settings for the LC-MS/MS measurements are summarized in Supplementary Table S3. The measurement was conducted three times for each sample. One biological replicate set was used for the analysis of the MGM100 and its derivative strains, and two biological replicate sets were used for Δ*dnaKJ* and Δ*dnaKJ*Δ*lon* strains.

Data analysis was performed by the DIA-NN software (version 1.7.16, https://github.com/vdemichev/diann, accessed on 30 April 2021) [26]. The library for SWATH-MS was obtained from the SWATH atlas (http://www.swathatlas.org/, accessed on 30 April 2021); the original data were acquired by Midha et al. [27]. The fold changes between mean intensities and *p*-values by Welch's *t*-test were calculated by an in-house R script (R.app for Mac, version 3.6.2). For the correction of multiple testing, *p*-values were adjusted by the Benjamini-Hochberg method (using the "p.adjust" function). Only the proteins with intensities obtained in all three measurements in both samples were used to calculate fold changes. The enrichment analysis was performed with an in-house R script (using Fisher's exact test). The KEGG BRITE hierarchy information (*E. coli* MG1655 strain) was downloaded from the website (https://www.genome.jp/brite/eco00001) on 23 March 2016.

Supplementary Materials: The following supporting information can be downloaded at: https://www.mdpi.com/article/10.3390/molecules27123772/s1, Figure S1: Additional data of proteome analysis of MGM100 and MGM100-derived strains. Figure S2: Additional data of proteome analysis of Δ*dnaKJ* and Δ*dnaKJ*Δ*lon* strains. Table S1: Abundance changes of known Lon substrates in MGM100 and MGM100Δlon strains. Table S2: *E. coli* strains used in this study. Table S3: Parameters for the SWATH acquisition. Dataset S1: Proteome data for the MGM100 and MGM100-derived strains. Dataset S2: Proteome data for ΔdnaKJ and ΔdnaKJΔlon. Dataset S3: List of proteins with cluster numbers. Dataset S4: List of proteins up- or down-regulated by the Lon deletion. References [8,22,28–37] are cited in the supplementary materials.

Author Contributions: T.N. and Y.C. performed experiments; T.N., Y.C. and H.T. conceived the study, designed experiments, and analyzed the results; H.T. supervised the entire project; T.N., Y.C. and H.T. wrote the manuscript. All authors have read and agreed to the published version of the manuscript.

Funding: This research was funded by MEXT Grants-in-Aid for Scientific Research (Grant Numbers JP26116002, JP18H03984, JP20H05925 to H.T., and 17K15073 to T.N.).

Data Availability Statement: The mass spectrometry proteomics data have been deposited in the jPOST repository [38] (https://repository.jpostdb.org/, accessed on 30 April 2021), with reference number JPST001558.

Acknowledgments: We thank Kazuhiro Takemoto for advice on statistical analyses, Eri Uemura for technical support, the Bio-Support Center at Tokyo Tech for DNA sequencing, and the Cell Biology Center Research Core Facility at Tokyo Tech for the TripleTOF 6600 mass spectrometer measurements.

Conflicts of Interest: The authors declare no conflict of interest.

Sample Availability: The *E. coli* strains used in this study are available from the authors.

Abbreviations

GroE: GroEL/ES, DnaKJ: DnaK/DnaJ, MS: Mass spectrometry, LC-MS/MS: Liquid chromatography coupled to tandem mass spectrometry, SILAC: Stable isotope labeling using amino acids in cell culture, LFQ intensity: Label-free quantification intensity, DIA: Data-independent acquisition, SWATH: Sequential window acquisition of all theoretical fragment ion spectra, PURE system: Protein synthesis using recombinant elements system, KEGG: Kyoto Encyclopedia of Genes and Genomes, Lys-C: Lysyl Endopeptidase.

References

1. Anfinsen, B.C. Principles That Govern the Folding of Protein Chains. *Science* **1973**, *181*, 223–230. [CrossRef] [PubMed]
2. Tyedmers, J.; Mogk, A.; Bukau, B. Cellular Strategies for Controlling Protein Aggregation. *Nat. Rev. Mol. Cell Biol.* **2010**, *11*, 777–788. [CrossRef]
3. Balchin, D.; Hayer-Hartl, M.; Hartl, F.U. In Vivo Aspects of Protein Folding and Quality Control. *Science* **2016**, *353*, aac4354. [CrossRef]
4. Chiti, F.; Dobson, C.M. Protein Misfolding, Amyloid Formation, and Human Disease: A Summary of Progress Over the Last Decade. *Annu. Rev. Biochem.* **2017**, *86*, 27–68. [CrossRef]
5. Fayet, O.; Ziegelhofer, T.; Georgopoulos, C. The GroES and GroEL Heat Shock Gene Products of *Escherichia coli* Are Essential for Bacterial Growth at All Temperatures. *J. Bacteriol.* **1989**, *171*, 1379–1385. [CrossRef]
6. Kerner, M.J.; Naylor, D.J.; Ishihama, Y.; Maier, T.; Chang, H.C.; Stines, A.P.; Georgopoulos, C.; Frishman, D.; Hayer-Hartl, M.; Mann, M.; et al. Proteome-Wide Analysis of Chaperonin-Dependent Protein Folding in *Escherichia coli*. *Cell* **2005**, *122*, 209–220. [CrossRef]
7. Fujiwara, K.; Ishihama, Y.; Nakahigashi, K.; Soga, T.; Taguchi, H. A Systematic Survey of In Vivo Obligate Chaperonin-Dependent Substrates. *EMBO J.* **2010**, *29*, 1552–1564. [CrossRef] [PubMed]
8. McLennan, N.; Masters, M. GroE Is Vital for Cell-Wall Synthesis. *Nature* **1998**, *392*, 139. [CrossRef] [PubMed]
9. Niwa, T.; Fujiwara, K.; Taguchi, H. Identification of Novel in Vivo Obligate GroEL/ES Substrates Based on Data from a Cell-Free Proteomics Approach. *FEBS Lett.* **2016**, *590*, 251–257. [CrossRef]
10. Chandu, D.; Nandi, D. Comparative Genomics and Functional Roles of the ATP-Dependent Proteases Lon and Clp during Cytosolic Protein Degradation. *Res. Microbiol.* **2004**, *155*, 710–719. [CrossRef]
11. Mahmoud, S.A.; Chien, P. Regulated Proteolysis in Bacteria. *Annu. Rev. Biochem.* **2018**, *87*, 677–696. [CrossRef] [PubMed]

12. Wettstadt, S.; Llamas, M.A. Role of Regulated Proteolysis in the Communication of Bacteria with the Environment. *Front. Mol. Biosci.* **2020**, *7*, 294. [CrossRef]
13. Ong, S.-E.; Blagoev, B.; Kratchmarova, I.; Kristensen, D.B.; Steen, H.; Pandey, A.; Mann, M. Stable Isotope Labeling by Amino Acids in Cell Culture, SILAC, as a Simple and Accurate Approach to Expression Proteomics. *Mol. Cell. Proteom.* **2002**, *1*, 376–386. [CrossRef] [PubMed]
14. Rozanova, S.; Barkovits, K.; Nikolov, M.; Schmidt, C.; Urlaub, H.; Marcus, K. Quantitative Mass Spectrometry-Based Proteomics: An Overview. *Methods Mol. Biol.* **2021**, *2228*, 85–116. [CrossRef] [PubMed]
15. Gillet, L.C.; Navarro, P.; Tate, S.; Röst, H.; Selevsek, N.; Reiter, L.; Bonner, R.; Aebersold, R. Targeted Data Extraction of the MS/MS Spectra Generated by Data-Independent Acquisition: A New Concept for Consistent and Accurate Proteome Analysis. *Mol. Cell. Proteom.* **2012**, *11*, O111.016717. [CrossRef] [PubMed]
16. Niwa, T.; Ying, B.W.; Saito, K.; Jin, W.; Takada, S.; Ueda, T.; Taguchi, H. Bimodal Protein Solubility Distribution Revealed by an Aggregation Analysis of the Entire Ensemble of *Escherichia coli* Proteins. *Proc. Natl. Acad. Sci. USA* **2009**, *106*, 4201–4206. [CrossRef]
17. Niwa, T.; Kanamori, T.; Ueda, T.; Taguchi, H. Global Analysis of Chaperone Effects Using a Reconstituted Cell-Free Translation System. *Proc. Natl. Acad. Sci. USA* **2012**, *109*, 8937–8942. [CrossRef]
18. Calloni, G.; Chen, T.; Schermann, S.M.; Chang, H.C.; Genevaux, P.; Agostini, F.; Tartaglia, G.G.; Hayer-Hartl, M.; Hartl, F.U. DnaK Functions as a Central Hub in the E. Coli Chaperone Network. *Cell Rep.* **2012**, *1*, 251–264. [CrossRef]
19. Kanehisa, M.; Furumichi, M.; Tanabe, M.; Sato, Y.; Morishima, K. KEGG: New Perspectives on Genomes, Pathways, Diseases and Drugs. *Nucleic Acids Res.* **2017**, *45*, D353–D361. [CrossRef]
20. Weber, H.; Polen, T.; Heuveling, J.; Wendisch, V.F.; Hengge, R. Genome-Wide Analysis of the General Stress Response Network in *Escherichia coli*: SigmaS-Dependent Genes, Promoters, and Sigma Factor Selectivity. *J. Bacteriol.* **2005**, *187*, 1591–1603. [CrossRef]
21. Hengge-Aronis, R. Signal Transduction and Regulatory Mechanisms Involved in Control of the Sigma(S) (RpoS) Subunit of RNA Polymerase. *Microbiol. Mol. Biol. Rev.* **2002**, *66*, 373–395. [CrossRef] [PubMed]
22. Baba, T.; Ara, T.; Hasegawa, M.; Takai, Y.; Okumura, Y.; Baba, M.; Datsenko, K.A.; Tomita, M.; Wanner, B.L.; Mori, H. Construction of *Escherichia coli* K-12 in-Frame, Single-Gene Knockout Mutants: The Keio Collection. *Mol. Syst. Biol.* **2006**, *2*, 2006.0008. [CrossRef]
23. Datsenko, K.A.; Wanner, B.L. One-Step Inactivation of Chromosomal Genes in *Escherichia coli* K-12 Using PCR Products. *Proc. Natl. Acad. Sci. USA* **2000**, *97*, 6640–6645. [CrossRef]
24. Masuda, T.; Tomita, M.; Ishihama, Y. Phase Transfer Surfactant-Aided Trypsin Digestion for Membrane Proteome Analysis. *J. Proteome Res.* **2008**, *7*, 731–740. [CrossRef] [PubMed]
25. Rappsilber, J.; Ishihama, Y.; Mann, M. Stop And Go Extraction Tips for Matrix-Assisted Laser Desorption/Ionization, Nanoelectrospray, and LC/MS Sample Pretreatment in Proteomics. *Anal. Chem.* **2003**, *75*, 663–670. [CrossRef] [PubMed]
26. Demichev, V.; Messner, C.B.; Vernardis, S.I.; Lilley, K.S.; Ralser, M. DIA-NN: Neural Networks and Interference Correction Enable Deep Proteome Coverage in High Throughput. *Nat. Methods* **2020**, *17*, 41–44. [CrossRef]
27. Midha, M.K.; Kusebauch, U.; Shteynberg, D.; Kapil, C.; Bader, S.L.; Reddy, P.J.; Campbell, D.S.; Baliga, N.S.; Moritz, R.L. A Comprehensive Spectral Assay Library to Quantify the *Escherichia coli* Proteome by DIA/SWATH-MS. *Sci. Data* **2020**, *7*, 389. [CrossRef]
28. Mizusawa, S.; Gottesman, S. Protein Degradation in *Escherichia coli*: The Lon Gene Controls the Stability of SulA Protein. *Proc. Natl. Acad. Sci. USA* **1983**, *80*, 358–362. [CrossRef]
29. Stout, V.; Torres-Cabassa, A.; Maurizi, M.R.; Gutnick, D.; Gottesman, S. RcsA, an Unstable Positive Regulator of Capsular Polysaccharide Synthesis. *J. Bacteriol.* **1991**, *173*, 1738–1747. [CrossRef]
30. Leffers, J.; Gottesman, S. Lambda Xis Degradation In Vivo by Lon and FtsH. *J. Bacteriol.* **1998**, *180*, 1573–1577. [CrossRef]
31. Biran, D.; Gur, E.; Gollan, L.; Ron, E.Z. Control of Methionine Biosynthesis in *Escherichia coli* by Proteolysis. *Mol. Microbiol.* **2000**, *37*, 1436–1443. [CrossRef]
32. Griffith, K.L.; Shah, I.M.; Wolf, R.E. Proteolytic Degradation of *Escherichia coli* Transcription Activators SoxS and MarA as the Mechanism for Reversing the Induction of the Superoxide (SoxRS) and Multiple Antibiotic Resistance (Mar) Regulons. *Mol. Microbiol.* **2004**, *51*, 1801–1816. [CrossRef] [PubMed]
33. Neher, S.B.; Villen, J.; Oakes, E.C.; Bakalarski, C.E.; Sauer, R.T.; Gygi, S.P.; Baker, T.A. Proteomic Profiling of ClpXP Substrates after DNA Damage Reveals Extensive Instability within SOS Regulon. *Mol. Cell* **2006**, *22*, 193–204. [CrossRef] [PubMed]
34. Bissonnette, S.A.; Rivera-Rivera, I.; Sauer, R.T.; Baker, T.A. The IbpA and IbpB Small Heat-Shock Proteins Are Substrates of the AAA+ Lon Protease. *Mol. Microbiol.* **2010**, *75*, 1539–1549. [CrossRef]
35. Langklotz, S.; Narberhaus, F. The *Escherichia coli* Replication Inhibitor CspD Is Subject to Growth-Regulated Degradation by the Lon Protease. *Mol. Microbiol.* **2011**, *80*, 1313–1325. [CrossRef] [PubMed]
36. Westphal, K.; Langklotz, S.; Thomanek, N.; Narberhaus, F. A Trapping Approach Reveals Novel Substrates and Physiological Functions of the Essential Protease FtsH in *Escherichia coli*. *J. Biol. Chem.* **2012**, *287*, 42962–42971. [CrossRef]

37. Arends, J.; Griego, M.; Thomanek, N.; Lindemann, C.; Kutscher, B.; Meyer, H.E.; Narberhaus, F. An Integrated Proteomic Approach Uncovers Novel Substrates and Functions of the Lon Protease in *Escherichia coli*. *Proteomics* **2018**, *18*, e1800080. [CrossRef]
38. Okuda, S.; Watanabe, Y.; Moriya, Y.; Kawano, S.; Yamamoto, T.; Matsumoto, M.; Takami, T.; Kobayashi, D.; Araki, N.; Yoshizawa, A.C.; et al. JPOSTrepo: An International Standard Data Repository for Proteomes. *Nucleic Acids Res.* **2017**, *45*, D1107–D1111. [CrossRef]

molecules

MDPI

Article

The Denaturant- and Mutation-Induced Disassembly of *Pseudomonas aeruginosa* Hexameric Hfq Y55W Mutant

Victor Marchenkov, Natalia Lekontseva, Natalia Marchenko, Ivan Kashparov, Victoriia Murina †, Alexey Nikulin, Vladimir Filimonov * and Gennady Semisotnov *

Institute of Protein Research, Russian Academy of Sciences, Institutskaya Street 4, 142290 Pushchino, Russia; march@phys.protres.ru (V.M.); lekontseva@vega.protres.ru (N.L.); lita@phys.protres.ru (N.M.); ivkashp@vega.protres.ru (I.K.); victoriia.murina@gmail.com (V.M.); nikulin@vega.protres.ru (A.N.)
* Correspondence: vladimir@vega.protres.ru (V.F.); nina@vega.protres.ru (G.S.); Tel.: +7-(496)-731-8461 (V.F.); +7-(496)-731-8409 (G.S.)
† Present address: SARomics Biostructures AB, Medicon Village, Scheelevägen 2, SE-223 63 Lund, Sweden.

Abstract: Although oligomeric proteins are predominant in cells, their folding is poorly studied at present. This work is focused on the denaturant- and mutation-induced disassembly of the hexameric mutant Y55W of the Qβ host factor (Hfq) from mesophilic *Pseudomonas aeruginosa* (*Pae*). Using intrinsic tryptophan fluorescence, dynamic light scattering (DLS), and high-performance liquid chromatography (HPLC), we show that the dissociation of Hfq Y55W occurs either under the effect of GuHCl or during the pre-denaturing transition, when the protein concentration is decreased, with both events proceeding through the accumulation of stable intermediate states. With an extremely low pH of 1.4, a low ionic strength, and decreasing protein concentration, the accumulated trimers and dimers turn into monomers. Also, we report on the structural features of monomeric Hfq resulting from a triple mutation (D9A/V43R/Y55W) within the inter-subunit surface of the protein. This globular and rigidly packed monomer displays a high thermostability and an oligomer-like content of the secondary structure, although its urea resistance is much lower.

Keywords: Hfq hexamer; mutations; unfolding intermediates; fluorescence; thermodynamics

Citation: Marchenkov, V.; Lekontseva, N.; Marchenko, N.; Kashparov, I.; Murina, V.; Nikulin, A.; Filimonov, V.; Semisotnov, G. The Denaturant- and Mutation-Induced Disassembly of *Pseudomonas aeruginosa* Hexameric Hfq Y55W Mutant. *Molecules* **2022**, *27*, 3821. https://doi.org/10.3390/molecules27123821

Academic Editors: Kunihiro Kuwajima, Yuko Okamoto, Tuomas Knowles, Michele Vendruscolo and Carmelo Corsaro

Received: 20 April 2022
Accepted: 11 June 2022
Published: 14 June 2022

1. Introduction

The mysterious puzzle of how protein chains, initially disordered, form their spatial structure has not yet been solved. While the main folding steps of relatively small globular proteins are generally known [1–4], those of large and oligomeric proteins remain unclear [4–7]. One of these issues is the relationship between the protein's quaternary structure, its stability, and folding.

This study is focused on the hexameric protein Hfq from *Pseudomonas aeruginosa* that acts as the main mediator in the regulatory network of gene expression by small RNAs [8,9]. It belongs to the Sm/LSm superfamily of proteins with a ring-like quaternary structure containing six identical subunits of about 10 kDa each [10]. There are some common properties of Hfqs from various organisms. One of the properties is the increased thermostability of Hfq, even from mesophilic bacteria [11,12]. This property, for example, greatly simplifies the purification process of these proteins from cellular lysates, i.e., it requires only heating to 80 °C and then separating the aggregates of contaminating proteins by centrifugation [12–14]. The second common property of Hfq proteins is their ring-like quaternary structure that provides the binding of single-stranded RNA [12] via multiple sites on the protein surface. Thus, the study of the stability of the Hfq quaternary structure is important to elucidate the protein function [15]. The Hfq protein from mesophilic *Pseudomonas aeruginosa* displays super-thermostability, with a half-transition temperature at neutral pH and about 116 °C [11,16]. Previously, we have shown that the substitution of Ala for the conserved residues Gln8, Asp40, and Tyr55 results in a decrease of Hfq thermostability, while its hexameric quaternary structure remains unchanged [16]. A two-fold decrease of the protein concentration does not affect the main differential scanning calorimetry (DSC) peak but visibly influences its left shoulder [16]. This observation allows us to propose a three-state model of Hfq thermal unfolding.

Here, we verify this suggestion by equilibrium denaturation studies of the protein, which allow us to obtain information about its conformational stability [17,18]. In thermodynamic experiments on oligomeric proteins, their concentration was used as an additional regulator of equilibrium [15,19–22]. These experiments require techniques other than differential scanning calorimetry (DSC) or circular dichroism (CD) which are efficient within a wide protein concentration range. In DSC, the lower limit of protein concentration is about 10 μM, while its ten-fold increase can provoke protein aggregation during thermal unfolding. The CD usage is often hampered by a low signal amplitude at low protein concentrations. Unlike these methods, fluorescence works within a sufficiently broad range of protein concentrations [18,21]. Because the wild type (WT) Hfq lacks tryptophan (Trp) [14], which is the most effective natural fluorescence probe, we replaced a tyrosine residue within the inter-subunit contact (Y55) with tryptophan. A preliminary test of the Hfq structure by the Swiss-Model program [23,24] showed no serious distortions resulting from this substitution. The expression of the plasmid encoding this mutant in *Escherichia coli* (*E. coli*) gave a product with a hexameric conformation close to that reported for the WT protein [14], as shown by standard biochemical tests and crystallographic studies.

In addition to substitutions in the Hfq sequence at position H57 and some other positions (e.g., Q8A) analyzed previously [11,16], here, we report on the single (Y55W) and triple (Y55W/D9A/V43R) Hfq mutants. The aim of introducing a tryptophan residue into the inter-subunit area is to use fluorescence in examining the denaturant-induced Hfq Y55W disassembly. We replaced D9 with Ala for two reasons: firstly, to avoid the spontaneous hydrolysis of the peptide bond, D9-P10 [25], and secondly, to remove a negative charge compensating, to some extent, for the positive charge of the conserved K56 protected from solvent within the inter-subunit contacts [14]. It turned out that single substitutions do not noticeably change the hexameric Hfq structure at neutral pH, but rather, decrease its stability. However, an additional bulky positive charge introduced into the inter-subunit area through the V43R substitution resulted in both inhibition of the hexamer formation and biosynthesis of the folded monomer. Note that V43R mutation of *Escherichia coli* Hfq results in a very negative effect on the protein function [26].

2. Results

In proteins, the intrinsic probe most sensitive to its surrounding is a tryptophan residue (Trp) [27,28]. Because the WT Hfq does not contain Trp, here, Y55 belonging to the protein inter-subunit surface was substituted for Trp. To obtain the monomeric Hfq with Trp fluorescence, two additional mutations, D9A and V43R, were introduced.

2.1. Structural Properties of the Hexameric Single (Y55W) and Monomeric Triple (Y55W/D9A/V43R) Hfq Mutants

The three-dimensional crystal structure of the Hfq Y55W mutant, deposited in the PDB under the code 5I21, is presented in Figure 1; this confirms that the mutation Y55W does not change the overall hexameric structure of the protein (see also [14]). Yet, some changes must be mentioned. Figure 1 presents the superposition of the main chain conformations for one subunit of WT [14] and Y55W variants of Hfq. Small deviations are observed in the loop positions, while all elements of the mutant secondary structure remain intact. Moreover, the position of the tryptophan indole ring is slightly different from that of the benzene one for intact tyrosine 55, possibly due to the bulky side group being included. As Y55 engages in stacking-like interactions with conserved H57 [14], this local distortion may influence the protein stability.

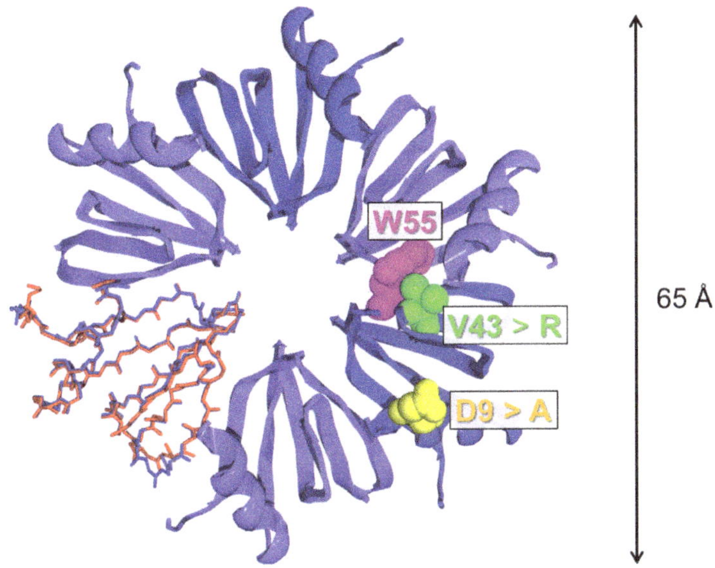

Figure 1. The crystal structure of the Hfq Y55W mutant: the tryptophan residue (W55) inserted into the hexameric structure and additional mutations (V43R, D9A) serving to obtain the protein monomeric form are shown in rectangles. One subunit represents the superposition of the subunit main chain conformations for the WT (red, PDB code 1U1T) and Y55W (blue, PDB code 5I21) Hfq variants. The double arrow to the right of the figure shows the external diameter of the protein ring.

Figure 2 presents some structural properties of the Hfq hexameric Y55W and monomeric Y55W/D9A/V43R mutants. First, the electrophoretic mobility of the monomer is much higher than that of the hexamer (Figure 2a), while its presence as a monomer in solution was confirmed by cross-linking with a glutaric aldehyde (not shown). Second, despite the similarity in the secondary structures of spatial crystal structures of WT Hfq and its Y55W mutant (Figure 1), there is a distinct difference in their far-UV CD spectra (Figure 2b). This difference may be attributed to either some destabilization of the Hfq Y55W secondary structure in solution or the optical properties of the tryptophan residue inserted into its current surrounding [29], apparently reflecting its interaction with the nearest neighbors within the monomer (subunit). At the same time, the deconvolution of the CD spectra of WT Hfq, hexameric, and monomeric mutants (using one of the latest programs available on the internet [30]) shows the similarity of the secondary structure content of these forms (Table 1).

Table 1. The secondary structure content (%) of Hfq variants as shown by far-UV CD spectra deconvolution [30].

Protein	α-Helix	β-Structure	Others
WT Hfq (hexamer)	10.8	41.0	48.2
Hfq Y55W (hexamer)	8.4	39.4	52.2
Hfq Y55W/D9A/V43R (monomer)	10.5	30.9	58.6

Third, intrinsic Trp fluorescence spectra (Figure 2c) indicate that in monomeric Hfq variants, Trp is more exposed to the solvent than in the hexamer (the spectrum is shifted to the long-wavelength region), but less exposed than in the unfolded state. Thus, the Trp fluorescence spectrum position is a good indicator of the disassembly of the protein's oligomeric structure.

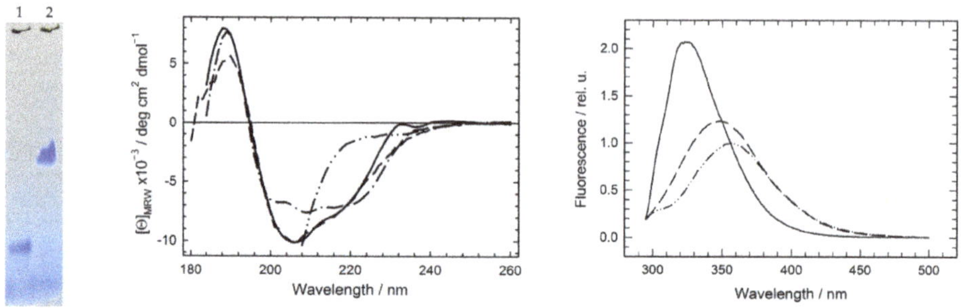

(**a**) (**b**) (**c**)

Figure 2. Structural properties of the hexameric Hfq Y55W and monomeric Hfq Y55W/D9A/V43R mutants: (**a**) Native polyacrylamide gel electrophoresis (PAGE) of the monomeric (track 1) and hexameric (track 2) forms; (**b**) Far-UV circular dichroism spectra for WT (dash-dot line), Y55W (solid line), Y55W/D9A/V43R (dashed line) Hfq variants, and any variant in the presence of 5 M GuHCl (dash-dot-dot line); (**c**) The fluorescence emission spectra for Hfq Y55W in three states (Equation (2)): the native hexameric (solid line, spectrum center of mass CM = 337 nm), the monomeric Y55W/D9A/V43R mutant(dashed line, CM = 355 nm), and unfolded monomeric (in the presence of 5 M GuHCl; dash-dot-dot line, CM = 360 nm); Protein concentration (monomer): 65 μM; Buffer: 20 mM Tris-HCl, pH 8.5.

Figure 3 shows the resistance of hexameric and monomeric Hfq mutants against urea, GuHCl, and temperature. Transverse urea-gradient electrophoresis (TUGE) allowed us to visualize the urea-induced denaturation of globular proteins caused by a change in the hydrodynamic size of the protein chain [31]. As seen in Figure 3a, hexameric Hfq Y55W retains its hydrodynamic size (i.e., does not unfold) under the effect of urea. The hexameric WT Hfq displays the same behavior toward urea (not shown). In contrast, the Hfq monomeric triple mutant undergoes an S-like urea-induced transition that results in an increased hydrodynamic size (decreased electrophoretic mobility) during the unfolding (Figure 3b). This transition is completely reversible, as confirmed by monitoring the protein refolding with transverse urea-gradient gel electrophoresis (not shown). This observation proves that the triple mutant has a rigidly packed tertiary structure with the thermodynamic parameters of its unfolding (shown in Table 2), which are typical for globular proteins of similar size [32]. Interestingly, despite the monomeric triple Hfq mutant showing low urea resistance in comparison with the monomer within the hexameric HfqY55W (Figure 3a,b), it retained relatively high thermostability. The CD spectrum of the triple mutant recorded at 90 °C (Figure 3d) showed a high content of secondary structure elements (α = 10.2%, β = 28.1%, other = 61.7%), i.e., similar to that at 25 °C (data in Table 1).

A stronger denaturant, GuHCl, causes the unfolding of both hexameric WT Hfq and its Y55W mutant (Figures 3c and S1). The GuHCl-induced changes in CD spectra for WT Hfq occur within a much wider range of denaturant concentrations with much less parameter m than that expected for the cooperative two-state unfolding of globular proteins with a molecular weight of about 55 kDa [32] (see also [11]), thus unequivocally demonstrating an alternative non-two-state type of the WT Hfq unfolding transition. The substitution Y55W essentially decreases the resistance of the Hfq hexamer against GuHCl, and its unfolding transition is apparently more cooperative (Figure 3c) than in the case of WT Hfq, resembling a simple two-state transition. However, as seen in Figure 3c, an eleven-fold decrease of Hfq Y55W concentration resulted in a shift to the left of the part of the unfolding curve, as would be expected from a process involving the dissociation of an oligomer. Therefore, the hexameric protein unfolding data were analyzed using the three-state model (see Materials and Methods). Initially, the deconvolution procedure based on the three-state model (Equations (2) and (17)) was applied to the protein denaturing transition monitored by fluorescence spectra decomposition, as described below, with parameters as shown in Table 2. Afterward, the three-state model (with the fixed values of the m and K parameters for the dissociation and unfolding processes) was applied to the fitting of the protein unfolding transitions monitored by Far-UV CD (Figure 3c).

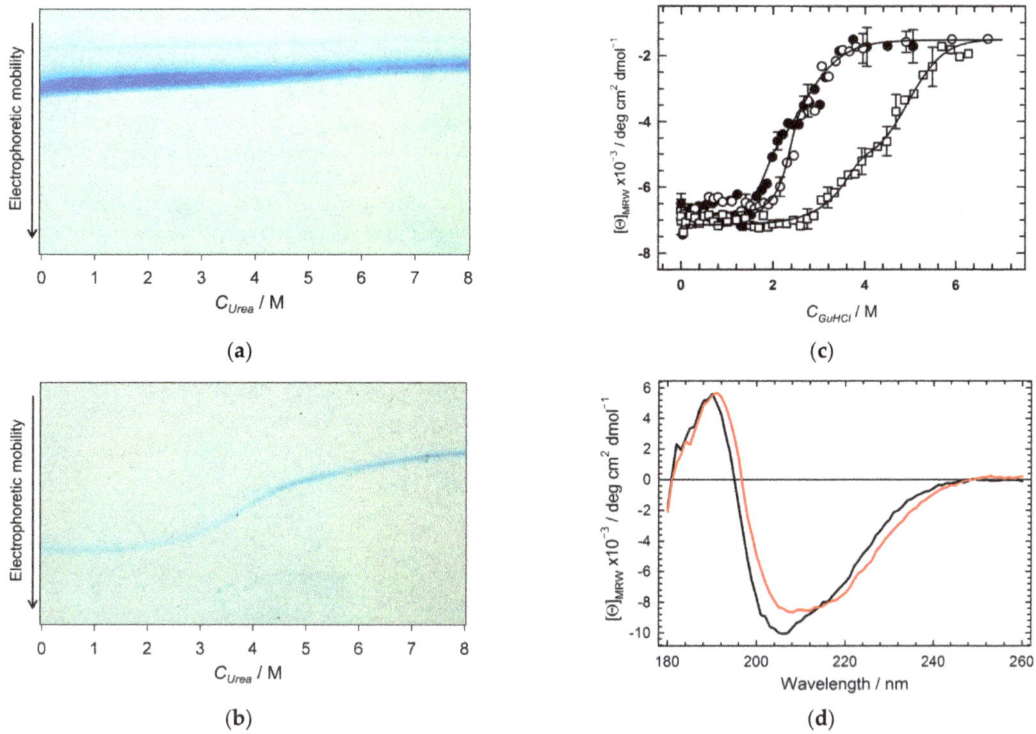

Figure 3. The effect of urea on Hfq structure (**a**,**b**), GuHCl (**c**), and temperature (**d**): (**a**) Transverse urea gradient gel electrophoresis of the Hfq Y55W hexamer; (**b**) Transverse urea gradient gel electrophoresis of the Hfq Y55W/D9A/V43R monomer; (**c**) The CD-monitored GuHCl-induced unfolding of the Hfq Y55W (circles) at 66 μM (open symbols) and 6 μM (filled symbols), and Hfq WT (squares) at 66 μM. The error bars correspond to the deviations from average values as a result of five measurements. The solid lines stand for the best fittings of the data with the parameters shown in Table 2; (**d**) The CD spectra of monomeric triple Hfq mutant at 20 °C (black line) and at 90 °C (red line) at a protein concentration of 1.24 mg/mL (124 μM of monomers).

Table 2. The thermodynamic parameters of Hfq equilibrium unfolding transitions (as shown by the linear extrapolation method (LEM) [20,21]).

HFQ Variant	Denaturant	Method	Model	m_{NI} (kJ/mol)	$K_{NI,w}$ (mol$^{5/6}$)	m_{IU} (kJ/mol)	$K_{IU,w}$	$\Delta_I{}^U G_w$ (kJ/mol)
WT	GuHCl	CD	Equations (2) and (11)	5.9	1.5×10^{-7}	5.5	2.8×10^{-5}	−26.0
Y55W (hexamer)	GuHCl	Flu.	Equations (2) and (11)	6.2	3.5×10^{-7}	5.5	3.5×10^{-3}	−14.0
	GuHCl	CD	Equations (2) and (11)	6.2	3.5×10^{-7}	5.5	3.5×10^{-3}	−14.0
	dilution at 1.6 M GuHCl	Flu.	Equations (12) and (13)	6.2	4.0×10^{-7}	-	-	-
Triple mutant	Urea	TUGE *	Equations (15) and (18)	-		3.9	4.1×10^{-3}	−13.3

* The data were derived from Figure 3b.

2.2. The Intermediate Species Accumulated during GuHCl-Induced Denaturation of the Hexameric Hfq Y55W

As seen from Figure 3c, an eleven-fold decrease in Hfq Y55W concentration resulted in a change of the GuHCl-induced unfolding, as monitored by far-UV CD within the 1.6 M–2.2 M denaturant concentration range. With this in mind, we selected a GuHCl concentration, providing a maximally populated native state of the protein with a minimally populated unfolded state. Varying protein concentrations, we analyzed the unfolding transition depending on the protein concentration. For this purpose, we used the Trp fluorescence spectra position (center of mass (CM), Equation (1)) which is mainly sensitive to hexamer dissociation (see Figure 2c) and can be registered with a reasonable noise level at protein (monomer) concentrations as low as to 0.1 μM.

Figure 4a demonstrates a set of Hfq Y55W Trp fluorescence spectra recorded within the 0–5 M GuHCl range at three protein (monomer) concentrations: 0.8 (top), 6.5 (middle), and 65 μM (bottom). These spectra within intermediate concentrations of GuHCl apparently are at least two-component.

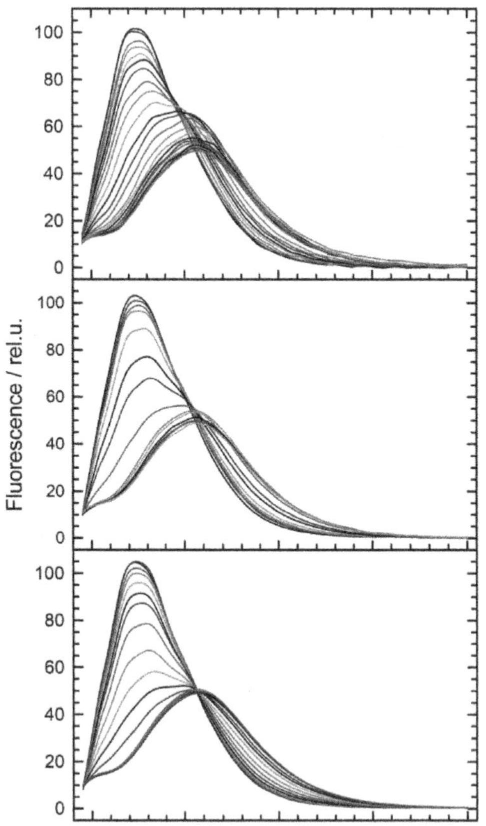

(a)

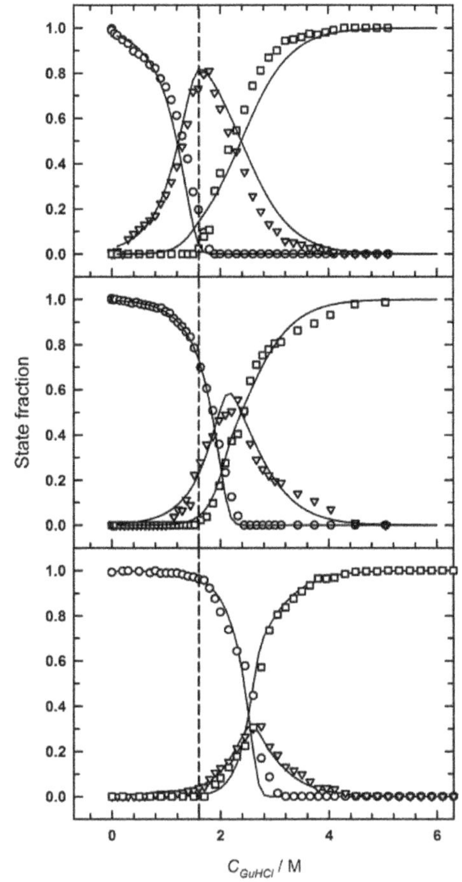

(c)

Figure 4. *Cont.*

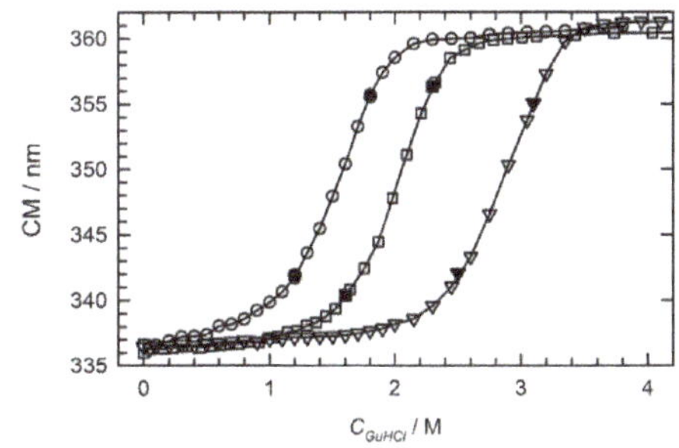

(b)

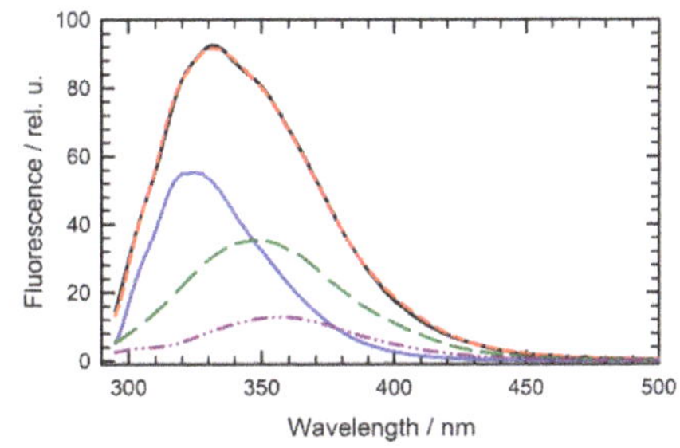

(d)

Figure 4. Analysis of the GuHCl-induced unfolding at various Hfq Y55W concentrations: (**a**) A set of selected Y55W fluorescence spectra recorded within the 0 ÷ 5 M GuHCl range (from top to bottom, grayscale) at three protein (monomer) concentrations: 0.8 (top), 6.5 (middle), and 65 μM (bottom); (**b**) The GuHCl dependence of the spectrum mass center values (Equation (1)) for three Hfq Y55W concentrations. Circles stand for 0.8 μM, squares for 6.5 μM, and triangles for 65 μM. The black symbols represent results of refolding of Hfq Y55W; (**c**) The GuHCl concentration dependences of the state fractions, as derived from the deconvolution of Y55W fluorescence spectra at three protein concentrations: 0.8 (top), 6.5 (middle), and 65 μM (bottom). Symbols show: hexamer, circles; intermediate, triangles; and unfolded monomer, squares; while solid lines represent simulations based on the three-state unfolding model (3), Equations (3)–(11), and the parameters from Table 2. The vertical dashed line indicates a GuHCl concentration of 1.6 M; (**d**) The result of the best fit of the fluorescence spectrum of Hfq Y55W (6.5 μM) at 1.98 M GuHCl (black solid line, experiment; red dashed line, the best fit to the three-state model). State contributions: hexamer, blue solid line; intermediate monomer, dark green dashed line; unfolded monomer, dark pink dash-dot-dot line. Best fit fractions: $f_n = 0.36$, $f_i = 0.47$, $f_u = 0.17$.

Figure 4b represents the dependence of the spectrum center of mass on the GuHCl concentration at different protein concentrations. The standard fluorescence spectra (Figure 2c) were used for the deconvolution of Hfq Y55W fluorescence spectra during GuHCl-induced protein unfolding at different protein concentrations. An example of such deconvolution is presented in Figure 4d. The result of such a deconvolution for the protein unfolding transitions at three protein concentrations is shown in Figure 4c. An important conclusion derived from these data is that the lower the protein concentration, the higher the accumulation level of the intermediate state fraction. Thus, at a high protein concentration, the denaturation mode is apparently close to the two-state one. This signifies that the stability of the protein inter-subunit interactions approaches that of the monomer (subunit) at high protein concentrations.

2.3. Intermediate Species Accumulated during Protein Concentration Decrease

The concentration-dependent dissociation of the oligomeric protein structure and the nature of the accumulated intermediate state were determined from the protein concentration dependence (at fixed GuHCl concentration) of the protein fluorescence spectra position and the hydrodynamic size. The GuHCl concentration was 1.6 M, providing the maximally populated intermediate at low protein concentrations (see Figure 4c).

Figure 5a presents the set of Hfq Y55W fluorescence spectra at various protein concentrations at 1.6 M GuHCl. The position of the maximum of the protein spectra recorded below 1 μM protein concentration (inset of the Figure 5a) is close to that for the monomeric (Y55W/D10A/V43R) spectrum (Figure 2c).

Figure 5b shows the dependence of state fractions (N and I) on the molar Hfq Y55W (monomer) concentration at 1.6 M GuHCl, as determined by fluorescence spectra deconvolution using the standard spectra of hexamer and monomer presented in Figure 2c.

Size exclusion chromatography was used to determine the nature of the intermediate state accumulated during a decrease in protein concentration. Figure 6a presents several standardized elution profiles of Hfq Y55W at 1.6 M GuHCl with decreasing concentrations of loaded protein, as monitored by tryptophan fluorescence.

As seen, there are two major and one minor elution peaks. The left minor peak evidently corresponds to some associates and disappears with decreasing protein concentration. The other left major peak shows hexamer elution (Figure 6a,b) and does not change its position with decreasing protein concentration, although its amplitude is diminished (Figure 6a,c). The right peak shows dimer elution (Figure 6a,b) at high protein concentrations, while at low ones, it is close to a monomer (Figure 6b). Similar behavior was reported for the dimeric factor for inversion stimulation [22,33] and may be interpreted as a protein concentration-dependent exchange between the dimeric and monomeric forms of the protein.

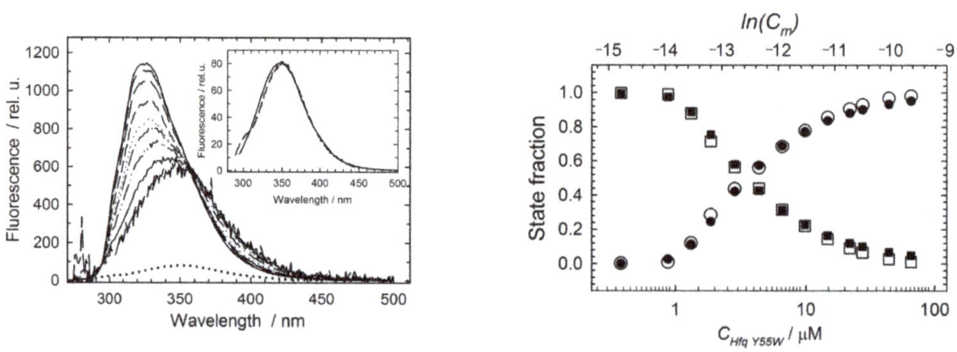

(a) (b)

Figure 5. Disassembly of Hfq Y55W in the presence of 1.6 M GuHCl by decreasing protein concentration: (**a**) The set of selected fluorescence spectra of Hfq Y55W at various protein (monomer) concentrations (from the top line to the bottom line—65 µM, 22 µM, 10 µM, 6.5 µM, 4.4 µM, 2.8 µM, 1.9 µM, 1.3 µM, 0.87 µM, 0.38 µM). The spectra are standardized to the same concentration (65 µM). At the inset: the basic spectrum of the triple mutant (solid line) assumed in fluorescence spectra deconvolution, and the spectrum of the intermediate, obtained by the dilution down to 0.38 µM at 1.6 M GuHCl (dashed line); (**b**) The dependence of the state fractions on the molar protein (monomer) concentration determined by deconvolution of the fluorescence spectra. The fractions of the native (hexameric) and intermediate states are shown by filled circles and filled squares, respectively. The open symbols are the prediction of the state distribution obtained from simulations based on the parameters from Table 2. The buffer contains 20 mM Tris-HCl, pH 7.5, and 1.6 M GuHCl.

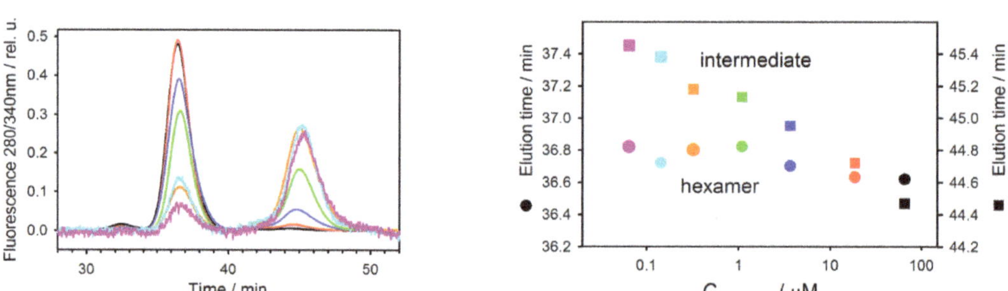

(a) (c)

Figure 6. *Cont.*

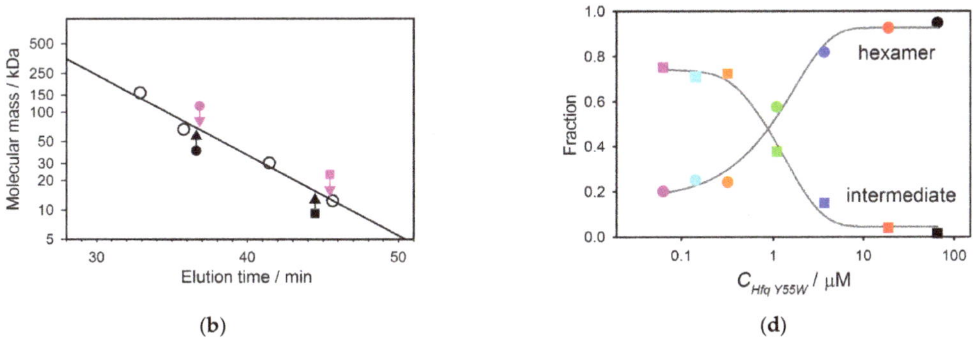

Figure 6. The size-exclusion chromatography of Hfq Y55W at 1.6 M GuHCl: (**a**) The set of standardized elution profiles of Hfq Y55W at various concentrations of the protein loaded (67 μM is shown in black, 18.9 μM—in red, 3.7 μM—in dark blue, 1.1 μM—in green, 0.33 μM—in yellow, 0.15 μM—in cyan, and 0.064 μM—in pink); (**b**) The dependence of the elution time on the protein molecular mass (the chromatographic column calibration); proteins used for calibration are aldolase (158 kDa), bovine serum albumin (66 kDa), bovine carbonic anhydrase B (29 kDa), cytochrome *c* (12.3 kDa). The arrows indicate the elution times of the respective peaks at the highest (black) and the lowest (pink) concentration of the protein loaded; (**c**) The dependence of the elution peak time on the protein concentration loaded (○—initial peak, □—peak of the dissociated intermediate; the color coding is the same); (**d**) The dependence of the protein state populations (determined as the areas of corresponding elution peaks) on the protein concentration loaded (○—hexamer, □—dissociated intermediate; color coding is the same).

2.4. Intermediate Species Accumulated at Low pH

The dissociation of hexameric Hfq Y55W occurs at low pH. Figure 7a presents a selected set of Trp fluorescence spectra at various pH, from 6.0 up to 1.4, and a protein (monomer) concentration of 67 μM. With a pH as low as 4.0, the fluorescence intensity was strongly quenched without a peak shift. By further decreasing pH, the fluorescence intensity increased, and the spectrum exhibited a red shift, achieving the position between the peaks for hexameric and monomeric forms (Figure 7b). Moreover, at a pH below 4.0, the protein fluorescence spectra seem to be at least two-component, probably due to the presence of more than one species with different spectrum positions (Figure 7a). The red shift of the protein fluorescence spectrum mainly indicates the dissociation of the protein's hexameric structure (see Figure 5). At a pH value between 4.0 and 3.0 (Figure 7a), the change in fluorescence intensity may have resulted from Trp fluorescence quenching by the nearest discharged carboxyl groups [27,28], while with a further pH decrease, the Trp fluorescence intensity increase may have been caused by diminishing the quenching due to the dissociation of the oligomeric structure. To evaluate the size of the species presented at pH 1.4, we performed DLS measurements (Figure 8a). The hexameric WT Hfq at pH 1.4 and Hfq Y55W mutant at pH 7.6 showed a symmetric peak corresponding to ~62 Å, which agrees with crystallographic data (Figure 1). At pH 1.4, the asymmetric peak for Hfq Y55W shifted toward lesser sizes. The decomposition of this peak hints at the presence of two main species with hydrodynamic sizes of ~35 Å (trimer) and ~23 Å (dimer).

To clarify the situation with the intermediates at pH 1.4, we performed size-exclusion chromatography of Hfq Y55W at pH 1.4 and various concentrations of the loaded protein. The results are presented in Figure 8b. As seen, there were three distinct peaks corresponding to at least three species.

Unfortunately, it is very difficult to estimate the exact size of these species because of ambiguous column calibration at low pH. Nevertheless, the first (left) small peak evidently corresponds to the hexamer, as was established by chromatography of WT Hfq, which does not dissociate at pH 1.4. Moreover, the elution volume of two less mobile species showed the tendency of depending on protein concentration, i.e., it shifted toward a lower hydrodynamic volume with decreasing protein concentration. Based on the results presented in Figures 7 and 8, we propose that at pH 1.4 and a low ionic strength, the hexameric Hfq Y55W mutant dissociates into trimers, dimers, and monomers, which exchange with each other. The fractions of hexamers, trimers, and dimers decrease, and the population of monomers increases with decreasing protein concentration.

Thus, like the GuHCl effect (Figure 4), the repulsion of positive charges at decreasing pH destabilizes the oligomeric structure of Hfq Y55W, while the lowering protein concentration results in its dissociation into monomers.

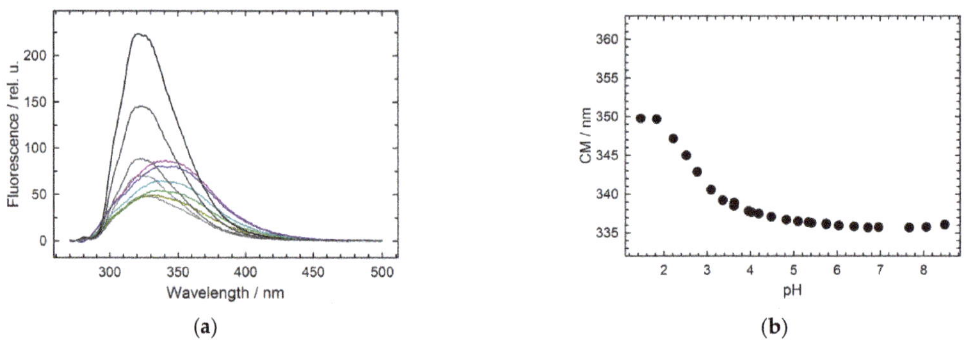

Figure 7. The pH dependence of Hfq Y55W fluorescence spectrum. (**a**) Fluorescence spectra measured at pH 6.0, 5.07, 4.15, 3.53, 3.03 (lines range from black to light grey); 2.72, 2.16, 1.76, 1.6, 1.4 (dark yellow, green, light blue, dark blue, and purple lines, respectively). Protein (monomer) concentration is 67 μM. (**b**) The pH dependence of the spectrum center of mass (CM). For different pH values, the following buffer solutions were used: 20 mM Tris-HCl, 30 mM NaCl for pH 7.6–8.5; 20 mM Na-Cacodylate-HCl, 30 mM NaCl for pH 5.35–6.93; 20 mM Na-Acetate, 30 mM NaCl for pH 3.55–5.65; 20 mM Glycine-HCl, 30 mM NaCl for pH 2.16–3.83; 20 mM HCl, 30 mM NaCl for pH 1.76; 50 mM HCl for pH 1.4.

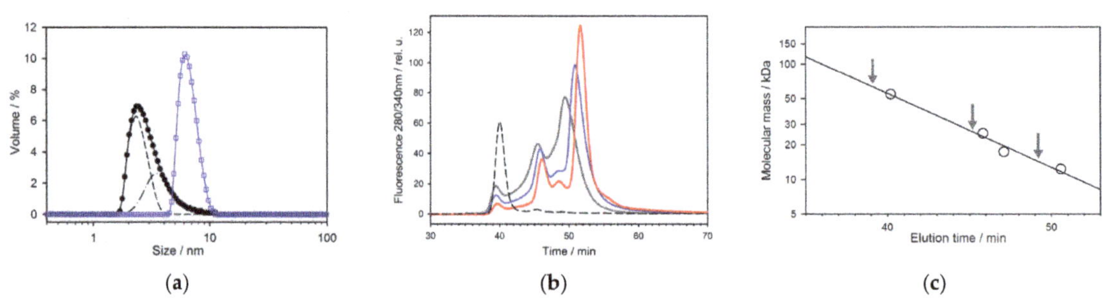

Figure 8. The hydrodynamic properties of pH-induced intermediates of Hfq Y55W. (**a**) Size distribution of Hfq Y55W at pH 7.6 or WT Hfq at pH 1.4 (blue squares) and that of Hfq Y55W at pH 1.4 (black circles) revealed by dynamic light scattering at a protein (monomer) concentration of 100 μM. The deconvolution of the mutant peak at pH 1.4 into peaks corresponding to 35 Å (dashed-dot line) and 23 Å (dashed line); (**b**) The size-exclusion chromatography of Hfq Y55W on Superdex 200 10/300 GL at pH 1.4 and various applied protein concentrations: the grey line stands for 3.0 mg/mL (300 μM monomer); the blue line for 0.6 mg/mL (60 μM monomer); the red line for 0.12 mg/mL (12 μM monomer). The dashed line shows the hexameric WT Hfq elution profile at pH 1.4; (**c**) The chromatographic column calibration. Proteins used for calibration are WT Hfq (54.6 kDa), chymotrypsinogen (25 kDa), myoglobin (17.3 kDa), cytochrome *c* (12.3 kDa). The arrows indicate the elution times of the peaks at 300 μM protein concentration loaded.

3. Discussion

A major result of this study is that the high stability of the Hfq subunit leads to a non-two-state mode of hexamer unfolding. Inter-subunit interactions play an important role in maintaining the oligomer structure of the proteins, especially in cases when isolated subunits are unable to maintain their folded conformation. In these cases, the inter-subunit interactions result in the formation of a cooperative structure which unfolds in a two-state process (for example, in the case of GroES heptamer of similar to the Hfq size [34]). As we suggested previously [11,16] and confirmed here, at neutral pH, the three-state model adequately describes the GuHCl-induced unfolding of Hfq hexamer (Figures 3 and 4). The relatively high population of the monomeric intermediate during hexamer unfolding indicates that the isolated subunits are stable, and makes it possible to independently analyze the hexamer dissociation by varying protein concentration at a fixed GuHCl concentration within the pre-denaturing range (Figure 5). At selected GuHCl activity, the fraction of the intermediate monomer strongly depends on protein concentration, reaching a level of 80% at the lower limit of protein concentration suitable for fluorescence measurements (Figures 4–6). In contrast, at the highest protein concentration used in this study, the intermediate state population was only about 30%, and at further increase in protein concentration, the unfolding would tend to the two-state model simply because something similar to an isosbestic point at 356 nm is observed (Figure 4a). In any case, with GuHCl > 3 M, the hexamer population became negligible, and only monomer unfolding was observed (Figure 4c).

Here, we demonstrated that the isolated subunit of Hfq can not only fold independently, but can also form an extremely stable structure. *Pseudomonas aeruginosa* is not a thermophile, and there is no natural need to produce extremely stable proteins. Among possible reasons, we can mention the following: (a) Intrinsically stable monomers could facilitate protein oligomerization; and (b) The folded monomers could have some unknown function in the cell at small concentrations.

Another important issue is the sensitivity of the subunit stability to the point mutations used here and elsewhere. All mutants with substitutions within the subunit interface have lower stabilities than the wild-type protein [16] (Figure 3c), while these substitutions mostly destabilize the structure of subunits. For example, the replacement of Y55 by alanine resulted in a drastic decrease in the stability of hexameric Hfq [16]. Here, we show that at neutral pH, the hexameric structure of WT Hfq and its Y55W mutant cannot be unfolded by urea (Figure 3a). The triple mutation Y55W/D9A/V43R, resulting in the inhibition of protein oligomerization, also decreased the subunit resistance to urea (Figure 3b), while the subunit thermostability remained high (Figure 3d). Moreover, the residues D9, V43, and Y55 are apparently located within the region of inter-subunit contacts for the majority of bacterial Hfqs [35]. Thus, the corresponding mutations of these residues may have been the cause of the inhibition of Hfqs assembly and, hence, their function [15]. Besides, as we demonstrated in this work, the insertion of Trp residue into an inter-subunit region of oligomeric proteins may be useful to obtain information about their dissociation.

4. Materials and Methods

4.1. Reagents

To prepare buffer solutions, commercial reagents (from Sigma-Aldrich (St. Louis, MO, USA), SERVA Electrophoresis GmbH (Heidelberg, Germany), Boehringer Mannheim GmbH (Mannheim, Germany), and Invitrogen (Waltham, MA, USA)) and bidistilled water were used.

4.2. Proteins

The QuikChange Site-Directed Mutagenesis Kit (Stratagene, La Jolla, CA, USA) was used to introduce the mutations in the Pae Hfq protein. PCR was carried out using the pET22b(t)/Hfq plasmid [14], and the primers containing the necessary substitutions:

D9A For 5′-CGCTACAAGCTCCTTACCTCAATACCCTG-3′

D9A Rev 5′-CAGGGTATTGAGGTAAGGAGCTTGTAGCG-3′
V43R For 5′-GAGTCTTTCGACCAGTTTCGCATCCTG-3′
V43R Rev 5′-CAGGATGCGAAACTGGTCGAAAGACTC-3′
Y55W For 5′-GTCAGCCAGATGGTTTGGAAGCACGCGAT-3′
Y55W Rev 5′-TCGCGTGCTTCCAAACCATCTGGCTGAC-3′

The Pae Hfq Y55W mutant was purified according to the protocol developed for the purification of a wild-type protein [14]. As nothing was known preliminarily about the physical-chemical properties of Hfq triple mutant, we added to its C-end a hexa histidine-tag. This facilitates the purification of the monomeric Hfq mutant. The resulting plasmid constructs were checked by sequencing. The plasmid containing the Hfq triple D9A/V43R/Y55W mutant gene was expressed in the *E. coli* strain BL21 (DE3). The cells were grown at 37 °C in LB medium until the absorption at 600 nm reached 0.6 optical units. The plasmid expression was induced with 0.5 mM Isopropyl β-D-1-thiogalactopyranoside (IPTG). After overnight incubation at 20 °C, the cells were harvested by centrifugation at 8000× *g* for 20 min at 4 °C, resuspended in 30 mL lysis buffer (1 M NaCl, 20 mM sodium phosphate buffer, pH 6.0), and disrupted by sonication (Thermo Fisher Scientific, Waltham, MA, USA). Cell debris and ribosomes were precipitated by stepwise (successive) centrifugation at 12,000× *g* for 30 min and at 20,800× *g* for 50 min. After the addition of 20 mM imidazole, the supernatant was loaded onto an Ni-NTA agarose column (Qiagen, Düsseldorf, Germany) equilibrated with 20 mM sodium phosphate buffer, pH 6.0, 0.2 M NaCl, 20 mM imidazole. The protein was eluted with a linear gradient of imidazole from 20 mM to 250 mM. The fractions containing the target protein were concentrated and dialyzed overnight against buffer containing 50 mM Tris-HCl, pH 8.0, 200 mM NaCl.

Protein concentration was measured spectrophotometrically using molar extinction coefficients calculated according to [36], i.e., 4470 $M^{-1}cm^{-1}$ for WT Hfq and 8480 $M^{-1}cm^{-1}$ for tryptophan-containing mutants. The values of 9140 Da and 9150 Da were taken for the molecular masses of WT Hfq and its Y55W mutant subunits.

4.3. Crystallographic Studies

Hfq Y55W crystallization was performed at 25 °C using the hanging drop vapor diffusion method on siliconized glass in Libro plates. Drops were prepared by mixing 1–2 μL of protein solution with an equal volume of precipitant containing 7% Poly(ethylene) glycol monomethyl ether (PEG 2000 MME), 2% 2-Methyl-2,4-pentanediol (MPD), 50 mM Tris-HCl, pH 6.5. The crystals were flash cooled by liquid nitrogen after the addition of 30% glycerol as a cryoprotectant.

The X-ray diffraction data were collected at the MX beam-line 14.1 with the Pilatus 6 M detector of BESSY II (Berlin, Germany). The data were processed in XDS and scaled in AIMLESS (CCP4). The data collection statistics is presented in Table 3.

Table 3. X-ray data parameters and refinement statistics for the Pae Hfq Y55W mutant. Statistics for the highest-resolution shell is shown in parentheses.

PDB ID	5I21
Wavelength (Å)	0.86
Resolution range (Å)	37.43–1.55 (1.605–1.55)
Space group	*P*6
Unit cell	114.4 Å, 114.4 Å, 28.3 Å, 90°, 90°, 120°
Total reflections	242,835 (23,797)
Unique reflections	31,293 (3038)
Multiplicity	7.8 (7.8)
Completeness (%)	1.00 (0.99)
Mean *I*/sigma (*I*)	25.89 (4.09)
Wilson *B*-factor	16.57
R-merge	0.04633 (0.5224)
R-meas	0.04967 (0.5593)

Table 3. *Cont.*

PDB ID	5I21
CC1/2	1 (0.908)
CC*	1 (0.976)
R-work	0.1877 (0.2138)
R-free	0.2135 (0.2418)
CC(work)	0.965 (0.892)
CC(free)	0.943 (0.880)
Number of non-hydrogen atoms	1791
macromolecules	1592
water molecules	198
ions (Cl-)	1
RMS (bonds)	0.006
RMS (angles)	0.84
Ramachandran favored (%)	97
Ramachandran allowed (%)	3.1
Ramachandran outliers (%)	0
Rotamer outliers (%)	1.1
Clashscore	3.96
Average B-factor (Å^2)	23.60
macromolecules	22.22
ligands	40.51
solvent	34.66

The determination of the protein structure was done by the molecular replacement technique using the Phaser program of the Phenix set [37,38]. As an initial model, the subunit of the WT Pae Hfq (PDB ID 1U1S [14]) was accepted. The asymmetric part of the crystal cell contains three protein subunits that are half of the whole hexamer, which was obtained by applying the crystallographic symmetry to this trimer. Then, the hexamer protein structure was refined using the Phenix refine program [37]. The first step included successive applications of the refinement protocols for isolated monomers (subunits) as rigid bodies, simulation of molecular annealing, and standard refinement. Then, 198 water molecules and one chloride ion were induced in the model for non-bound electron density description. The final model had good stereochemical parameters, as shown in Table 3, and was deposited in the PDB (ID 5I21).

4.4. Physical-Chemical Techniques

The protein absorption spectra were recorded using a Cary 100 Bio spectrophotometer (Varian Medical Systems, Palo Alto, CA, USA). The Trp fluorescence spectra were recorded at 293 nm excitation using a Cary Eclipse spectrofluorimeter (Varian Medical Systems, Palo Alto, CA, USA) with a $1 \times 1 \times 4$ cm^3 quartz cell. Far-UV circular dichroism spectra were measured using a Chirascan spectropolarimeter (Applied Photophysics Ltd., London, UK) with a quartz cell of 0.1 mm pathway length.

Transverse urea-gradient electrophoresis (TUGE) was performed according to a published protocol [31] with gels prepared in 20 mM Tris-HCl, pH 8.9, buffer at the 0 M–8 M urea gradient and protein run for 4 h.

The dynamic light scattering experiments were done with Zetasizer Nano ZSP (Malvern Panalytical, Malvern, UK) at 25 °C. Particle sizes were evaluated by the instrument's software.

The size-exclusion chromatography was performed with HPLC chromatograph ProStar (Varian Medical Systems, Palo Alto, CA, USA), Superdex 200 10/300 GL column, and a flow rate of 0.4 mL/min. The calibration of the column at neutral pH was performed using various globular proteins with known molecular mass in 20 mM Tris-HCl pH 7.5, 150 mM NaCl buffer. The calibration of the column at pH 1.4 (50 mM HCl) was performed using the proteins the globularity and size of which at pH 1.4 are not essentially different from the ones at pH 7.5 accordingly to the independent data.

Non-linear fitting and smoothing of experimental data, as well as their simulations, were performed using Sigma Plot software (Systat Software Inc., Chicago, IL, USA).

4.5. Theoretical Basis

To reveal the changes of the protein fluorescence spectrum position caused by the denaturants within a wide range of protein concentrations, a parameter called center of mass (*CM*) was used. This parameter was calculated from each fluorescence spectrum as:

$$CM(\text{nm}) = \frac{\sum_{i=1}^{n} \lambda_i \cdot A_{fl,i}}{\sum_{i=1}^{n} A_{fl,i}}, \tag{1}$$

where A_{fl} is the fluorescence intensity amplitude in arbitrary units at the wavelength λ varying between 295 and 500 nm.

This parameter seems to be less sensitive to the noise of the spectra registration, especially at low protein concentrations (in comparison, for example, with the ratio of the fluorescence intensities at the left and the right of the spectrum maximum).

To selectively reduce the high-frequency noise, the spectra were smoothed using the appropriate options of the SigmaPlot software (Systat Software Inc., Chicago, IL, USA). The smoothed spectra were deconvoluted into two or three components (depending on the unfolding model assumed) by least-square fitting using Equation (18) (see below). As suggested previously [16], the main unfolding model for HFQ should include the equilibrium transitions between the three basic states:

$$\frac{1}{n} N_n \overset{K_{NI}}{\longleftrightarrow} I \overset{K_{IU}}{\longleftrightarrow} U, \tag{2}$$

where *N*, *I*, *U* are the native, intermediate, and unfolded state of the protein subunits, respectively, while *n* is the number of subunits and

$$K_{NI}^{n} = \frac{[I]^n}{[N_n]}, \tag{3}$$

$$K_{IU} = \frac{[U]}{[I]}. \tag{4}$$

With the overall polypeptide concentration (C_m) introduced, and the state fractions (f_N, f_I, f_U) as

$$f_N = \frac{n[N_n]}{C_m}, \tag{5}$$

$$f_I = \frac{[I]}{C_m}, \tag{6}$$

$$f_U = \frac{[U]}{C_m}, \tag{7}$$

$$f_N + f_I + f_U = 1, \tag{8}$$

the Equations (3) and (4) can be rewritten as

$$K_{NI}^{n} = \frac{n f_I^n C_m^{n-1}}{f_N}, \tag{9}$$

$$K_{IU} = \frac{f_U}{f_I}, \tag{10}$$

Combining the above equations, we get for the fraction of the intermediate state in the scheme (2) (equations for the two other state fractions can be written similarly):

$$nC_m^{n-1}\left(\frac{f_I}{K_{NI}}\right)^n + (1+K_{IU})f_I - 1 = 0. \tag{11}$$

Since in the case of Hfq n = 6, these equations can be solved only by numerical procedures. We used Newton method for finding a single real root satisfying to $0 < f_I < 1$.

Depending on the solvent composition and protein concentration, the maximal population of the intermediate state can approach 1 (the two transitions are "well separated"). For example, with $K_{IU} \approx 0$, the first transition in the scheme (2) can be considered separately, with the two-state dissociation equation valid:

$$\frac{1}{n}N_n \overset{K_{NI}}{\longleftrightarrow} I, \tag{12}$$

$$nC_m^{n-1}\left(\frac{f_I}{K_{NI}}\right)^n + f_I - 1 = 0, \tag{13}$$

where $f_N + f_I \approx 1$.

At any selected denaturant concentration, where Equation (13) holds and $C_m = C_{m,0.5}$ (transition midpoint, $f_N = f_I = 0.5$) we get from Equation (9):

$$\ln K_{NI,0.5} = {}^{1}/{}_{n}\left[\ln(n) + (n-1)\ln\left(\frac{C_{m,0.5}}{2}\right)\right]. \tag{14}$$

This equation was used in data analysis with fixed GuHCl concentrations (Figure 5).

To describe the unfolding of the monomeric triple mutant (see Figure 3b), a typical two-state model together with the linear extrapolation method LEM [20] was used:

$$I \overset{K_{IU}}{\longleftrightarrow} U. \tag{15}$$

In this model, however, the state I displays close, but not necessarily identical, thermodynamic parameters to the intermediate state of the three-state model (3).

For the analysis of the denaturant-induced unfolding at a constant temperature, T_0 = 298.15 K, the simplest LEM was used:

$$\Delta_X^Y G = -RT_0\ln(K_{Y,w}) = \Delta_X^Y G_w - m_Y C_{den}, \tag{16}$$

where $\Delta_X^Y G_w$ and $\Delta_X^Y G$ are the Gibbs energy changes in water and in the presence of the denaturant, m_Y is the proportionality parameter (kJ/M mol), C_{den} is the molar denaturant concentration, and R is the gas constant (kJ/K mol).

Hence, to simulate the state fraction dependence on the denaturant concentration at any fixed C_m, one should set n, C_m, two proportionality coefficients, and two Gibbs energy changes in water. The dependence of state populations, and hence, fluorescence spectra, on the protein concentration can be used for estimating the spectrum of the intermediate state at a denaturant concentration where the population of the unfolded state is negligible, and the fraction of the intermediate state can be raised to 1 by protein dilution. In this case, tryptophan fluorescence could be the most useful, as spectra could be reliably measured at very low protein concentrations (in our case, down to 0.4 μM).

To change the population of various conformational states during the equilibrium protein unfolding, we did the following.

For high-frequency noise reduction, the spectra, particularly those recorded at low protein concentrations (<0.05 mg/mL), were smoothed using the appropriate options in the SigmaPlot software (Systat Software Inc., Chicago, IL, USA). In some cases, the spectra were standardized by dividing them by the intensity at some selected wavelength. Such standardization minimizes the error caused by protein concentration uncertainty. The spectra were deconvoluted into two or three components (two- or three-state models) using the least-square fitting based on the following equations:

$$Fl_{sym}(\lambda) = f_N fl_N(\lambda) + f_I fl_I(\lambda) + f_U fl_U(\lambda), \quad (17)$$

where f_X and fl_X are the populations and basic fluorescence spectra of the corresponding states. Three basic spectra (for the native, monomeric, and unfolded state) were estimated as described above (Figure 2c) and assumed to be independent of buffer composition. In the case of a negligible unfolded state population, this equation transforms into:

$$Fl_{sym}(\lambda) = f_N fl_N(\lambda) + f_I fl_I(\lambda). \quad (18)$$

Supplementary Materials: The following supporting information can be downloaded at: https://www.mdpi.com/article/10.3390/molecules27123821/s1, Figure S1: The selected far-UV CD spectra of Hfq variants for the crucial points of GuHCl-induced unfolding of the proteins.

Author Contributions: Conceptualization, V.F. and G.S.; Data curation, G.S.; Formal analysis, V.F.; Funding acquisition, N.L.; Investigation, V.M. (Victor Marchenkov), N.L. and V.M. (Victoriia Murina); Methodology, V.M. (Victor Marchenkov); Project administration, A.N.; Resources, N.L. and V.M. (Victoriia Murina); Software, V.F.; Supervision, V.F. and G.S.; Validation, N.L., N.M. and I.K.; Visualization, N.M.; Writing—original draft, V.F.; Writing—review & editing, G.S. All authors have read and agreed to the published version of the manuscript.

Funding: This research was funded by the Russian Science Foundation, grant number 22-24-00934.

Institutional Review Board Statement: Not applicable.

Informed Consent Statement: Not applicable.

Data Availability Statement: Structural data are available in the Protein Data Bank (PDB) database under the accession numbers 5I21 (http://doi.org/10.2210/pdb5I21/pdb, accessed on 13 April 2022).

Conflicts of Interest: The authors declare no conflict of interest. The funders had no role in the design of the study; in the collection, analyses, or interpretation of data; in the writing of the manuscript, or in the decision to publish the results.

Sample Availability: Samples of the Hfq mutant plasmids (Y55W and D9A/V43R/Y55W) are available from the authors.

References

1. Tanford, C. Protein Denaturation. *Adv. Protein Chem.* **1968**, *23*, 121–282. [PubMed]
2. Pace, C.N. Conformational stability of globular proteins. *Trends Biochem. Sci.* **1990**, *15*, 14–17. [CrossRef]
3. Timasheff, S.N. Water as Ligand: Preferential Binding and Exclusion of Denaturants in Protein Unfolding. *Biochemistry* **1992**, *31*, 9857–9864. [CrossRef] [PubMed]
4. Braselmann, E.; Chaney, J.L.; Clark, P.L. Folding the proteome. *Trends Biochem. Sci.* **2013**, *38*, 337–344. [CrossRef]
5. Jaenicke, R. Folding and association of proteins. *Prog. Biophys. Mol. Biol.* **1987**, *49*, 117–237. [CrossRef]
6. Garel, J.-R. Folding of large proteins: Multidomain and multisubunit proteins. In *Protein Folding*; Creighton, T.E., Ed.; W.H. Freeman and Company: New York, NY, USA, 1992; pp. 405–454.
7. Price, N. Folding and assembly of multi-subunit proteins. In *Receptor Subunits and Complexes*; Burgen, A., Barnard, E.A., Eds.; Cambridge University Press: Cambridge, UK, 1992; pp. 9–38.
8. Wagner, E.G.H. Cycling of RNAs on Hfq. *RNA Biol.* **2013**, *10*, 619–626. [CrossRef]
9. Beisel, C.L.; Storz, G. Base pairing small RNAs and their roles in global regulatory networks. *FEMS Microbiol. Rev.* **2010**, *34*, 866–882. [CrossRef]
10. Wilusz, C.J.; Wilusz, J. Eukaryotic Lsm proteins: Lessons from bacteria. *Nat. Struct. Mol. Biol.* **2005**, *12*, 1031–1036. [CrossRef]
11. Moskaleva, O.; Melnik, B.; Gabdulkhakov, A.; Garber, M.; Nikonov, S.; Stolboushkina, E.; Nikulin, A. The structures of mutant forms of Hfq from Pseudomonas aeruginosa reveal the importance of the conserved His57 for the protein hexamer organization. *Acta Crystallogr. Sect. F Struct. Biol. Cryst. Commun.* **2010**, *66*, 760–764. [CrossRef]
12. de Fernandez, M.T.F.; Hayward, W.S.; August, J.T. Bacterial Proteins Required for Replication of Phage Qβ Ribonucleic Acid: Purification and Properties of Host Factor I, a Ribonucleic Acid-Binding Protein. *J. Biol. Chem.* **1972**, *247*, 824–831. [CrossRef]
13. Mura, C.; Kozhukhovsky, A.; Gingery, M.; Phillips, M.; Eisenberg, D. The oligomerization and ligand-binding properties of Sm-like archaeal proteins (SmAPs). *Protein Sci.* **2003**, *12*, 832–847. [CrossRef] [PubMed]

14. Nikulin, A.; Stolboushkina, E.; Perederina, A.; Vassilieva, I.; Blaesi, U.; Moll, I.; Kachalova, G.; Yokoyama, S.; Vassylyev, D.; Garber, M.; et al. Structure of Pseudomonas aeruginosa Hfq protein. *Acta Crystallogr. Sect. D Biol. Crystallogr.* **2005**, *61*, 141–146. [CrossRef] [PubMed]
15. Panja, S.; Woodson, S.A. Hexamer to monomer equilibrium of E. coli Hfq in solution and its impact on RNA annealing. *J. Mol. Biol.* **2012**, *417*, 406–412. [CrossRef] [PubMed]
16. Murina, V.N.; Melnik, B.S.; Filimonov, V.V.; Ühlein, M.; Weiss, M.S.; Müller, U.; Nikulin, A.D. Effect of conserved intersubunit amino acid substitutions on Hfq protein structure and stability. *Biochemistry* **2014**, *79*, 469–477. [CrossRef]
17. Pace, C.N. Determination and Analysis of Urea and Guanidine Hydrochloride Denaturation Curves. *Methods Enzymol.* **1986**, *131*, 266–280.
18. Shirdel, S.A.; Khalifeh, K. Thermodynamics of protein folding: Methodology, data analysis and interpretation of data. *Eur. Biophys. J.* **2019**, *48*, 305–316. [CrossRef]
19. Silva, J.L.; Silveira, C.F.; Correia, A.; Pontes, L. Dissociation of a native dimer to a molten globule monomer. Effects of pressure and dilution on the association equilibrium of arc repressor. *J. Mol. Biol.* **1992**, *223*, 545–555. [CrossRef]
20. Doyle, C.M.; Rumfeldt, J.A.; Broom, H.R.; Broom, A.; Stathopulos, P.B.; Vassall, K.A.; Almey, J.J.; Meiering, E.M. Energetics of oligomeric protein folding and association. *Arch. Biochem. Biophys.* **2013**, *531*, 44–64. [CrossRef]
21. Gloss, L.M. Equilibrium and kinetic approaches for studying oligomeric protein folding. *Methods Enzymol.* **2009**, *466*, 325–357.
22. Hobart, S.A.; Ilin, S.; Moriarty, D.F.; Osuna, R.; Colón, W. Equilibrium denaturation studies of the Escherichia coli factor for inversion stimulation: Implications for in vivo function. *Protein Sci.* **2002**, *11*, 1671–1680. [CrossRef]
23. Bertoni, M.; Kiefer, F.; Biasini, M.; Bordoli, L.; Schwede, T. Modeling protein quaternary structure of homo- and hetero-oligomers beyond binary interactions by homology. *Sci. Rep.* **2017**, *7*, 10480. [CrossRef] [PubMed]
24. Waterhouse, A.; Bertoni, M.; Bienert, S.; Studer, G.; Tauriello, G.; Gumienny, R.; Heer, F.T.; de Beer, T.A.P.; Rempfer, C.; Bordoli, L.; et al. SWISS-MODEL: Homology modelling of protein structures and complexes. *Nucleic Acids Res.* **2018**, *46*, W296–W303. [CrossRef] [PubMed]
25. Piszkiewicz, D.; Landon, M.; Smith, E.L. Anomalous cleavage of aspartyl-proline peptide bonds during amino acid sequence determinations. *Biochem. Biophys. Res. Commun.* **1970**, *40*, 1173–1178. [CrossRef]
26. Ziolkowska, K.; Derreumaux, P.; Folichon, M.; Pellegrini, O.; Régnier, P.; Boni, I.V.; Hajnsdorf, E. Hfq variant with altered RNA binding functions. *Nucleic Acids Res.* **2006**, *34*, 709–720. [CrossRef]
27. Lakowicz, J.R. *Principles of Fluorescence Spectroscopy*, 3rd ed.; Springer: New York, NY, USA, 2006; ISBN 0387312781.
28. Chen, Y.; Barkley, M.D. Toward understanding tryptophan fluorescence in proteins. *Biochemistry* **1998**, *37*, 9976–9982. [CrossRef]
29. Woody, R.W. Contributions of tryptophan side chains to the far-ultraviolet circular dichroism of proteins. *Eur. Biophys. J.* **1994**, *23*, 253–262. [CrossRef]
30. Micsonai, A.; Wien, F.; Kernya, L.; Lee, Y.H.; Goto, Y.; Réfrégiers, M.; Kardos, J. Accurate secondary structure prediction and fold recognition for circular dichroism spectroscopy. *Proc. Natl. Acad. Sci. USA* **2015**, *112*, E3095–E3103. [CrossRef]
31. Creighton, T.E. Electrophoretic analysis of the unfolding of proteins by urea. *J. Mol. Biol.* **1979**, *129*, 235–264. [CrossRef]
32. Myers, J.K.; Nick Pace, C.; Martin Scholtz, J. Denaturant m values and heat capacity changes: Relation to changes in accessible surface areas of protein unfolding. *Protein Sci.* **1995**, *4*, 2138–2148. [CrossRef]
33. Topping, T.B.; Hoch, D.A.; Gloss, L.M. Folding Mechanism of FIS, the Intertwined, Dimeric Factor for Inversion Stimulation. *J. Mol. Biol.* **2004**, *335*, 1065–1081. [CrossRef]
34. Boudker, O.; Todd, M.J.; Freire, E. The structural stability of the co-chaperonin GroES. *J. Mol. Biol.* **1997**, *272*, 770–779. [CrossRef] [PubMed]
35. Arluison, V.; Folichon, M.; Marco, S.; Derreumaux, P.; Pellegrini, O.; Seguin, J.; Hajnsdorf, E.; Regnier, P. The C-terminal domain of Escherichia coli Hfq increases the stability of the hexamer. *Eur. J. Biochem.* **2004**, *271*, 1258–1265. [CrossRef] [PubMed]
36. Pace, C.N.; Vajdos, F.; Fee, L.; Grimsley, G.; Gray, T. How to measure and predict the molar absorption coefficient of a protein. *Protein Sci.* **1995**, *4*, 2411–2423. [CrossRef]
37. Liebschner, D.; Afonine, P.V.; Baker, M.L.; Bunkoczi, G.; Chen, V.B.; Croll, T.I.; Hintze, B.; Hung, L.W.; Jain, S.; McCoy, A.J.; et al. Macromolecular structure determination using X-rays, neutrons and electrons: Recent developments in Phenix. *Acta Crystallogr. Sect. D Struct. Biol.* **2019**, *75*, 861–877. [CrossRef] [PubMed]
38. McCoy, A.J.; Grosse-Kunstleve, R.W.; Adams, P.D.; Winn, M.D.; Storoni, L.C.; Read, R.J. Phaser crystallographic software. *J. Appl. Crystallogr.* **2007**, *40*, 658–674. [CrossRef]

molecules MDPI

Article

Reverse Engineering Analysis of the High-Temperature Reversible Oligomerization and Amyloidogenicity of PSD95-PDZ3

Sawaros Onchaiya [1,†], Tomonori Saotome [2,3,†], Kenji Mizutani [4], Jose C. Martinez [5], Jeremy R. H. Tame [4], Shun-ichi Kidokoro [3] and Yutaka Kuroda [1,*]

1 Department of Biotechnology and Life Science, Tokyo University of Agriculture and Technology, 2-24-16, Naka-cho, Koganei-shi 184-8588, Tokyo, Japan; s227417t@st.go.tuat.ac.jp
2 Institute of Global Innovation Research, Tokyo University of Agriculture and Technology, 3-8-1, Harumi-cho, Fuchu-shi 183-8538, Tokyo, Japan; t_saotome@vos.nagaokaut.ac.jp
3 Department of Bioengineering, Nagaoka University of Technology, 1603-1, Kamitomioka-cho, Nagaoka-shi 940-2188, Niigata, Japan; kidokoro@vos.nagaokaut.ac.jp
4 Graduate School of Medical Life Science, Yokohama City University, 1-7-29 Suehiro, Yokohama 230-0045, Kanagawa, Japan; mizutani@yokohama-cu.ac.jp (K.M.); jtame@yokohama-cu.ac.jp (J.R.H.T.)
5 Department of Physical Chemistry, Institute of Biotechnology, Faculty of Sciences, University of Granada, 18071 Granada, Spain; jcmh@ugr.es
* Correspondence: ykuroda@cc.tuat.ac.jp; Tel.: +81-42-388-7794
† These authors contributed equally to this work.

Abstract: PSD95-PDZ3, the third PDZ domain of the post-synaptic density-95 protein (MW 11 kDa), undergoes a peculiar three-state thermal denaturation (N $\leftrightarrow$ I_n $\leftrightarrow$ D) and is amyloidogenic. PSD95-PDZ3 in the intermediate state (I) is reversibly oligomerized (RO: Reversible oligomerization). We previously reported a point mutation (F340A) that inhibits both ROs and amyloidogenesis and constructed the PDZ3-F340A variant. Here, we "reverse engineered" PDZ3-F340A for inducing high-temperature RO and amyloidogenesis. We produced three variants (R309L, E310L, and N326L), where we individually mutated hydrophilic residues exposed at the surface of the monomeric PDZ3-F340A but buried in the tetrameric crystal structure to a hydrophobic leucine. Differential scanning calorimetry indicated that two of the designed variants (PDZ3-F340A/R309L and E310L) denatured according to the two-state model. On the other hand, PDZ3-F340A/N326L denatured according to a three-state model and produced high-temperature ROs. The secondary structures of PDZ3-F340A/N326L and PDZ3-wt in the RO state were unfolded according to circular dichroism and differential scanning calorimetry. Furthermore, PDZ3-F340A/N326L was amyloidogenic as assessed by Thioflavin T fluorescence. Altogether, these results demonstrate that a single amino acid mutation can trigger the formation of high-temperature RO and concurrent amyloidogenesis.

Keywords: high-temperature reversible oligomerization; amyloidogenicity; oligomeric interface residues; thermal denaturation; mutational analysis

Citation: Onchaiya, S.; Saotome, T.; Mizutani, K.; Martinez, J.C.; Tame, J.R.H.; Kidokoro, S.-i.; Kuroda, Y. Reverse Engineering Analysis of the High-Temperature Reversible Oligomerization and Amyloidogenicity of PSD95-PDZ3. *Molecules* **2022**, *27*, 2813. https://doi.org/10.3390/molecules27092813

Academic Editors: Tuomas Knowles and René Csuk

Received: 14 March 2022
Accepted: 18 April 2022
Published: 28 April 2022

1. Introduction

The two-state thermal denaturation process is a biophysical hallmark for a natively folded single-domain globular protein. A two-state thermal denaturation process exhibits a sharp endothermic peak as observed by micro-calorimetry [1–4], and the two-state unfolding can be formally confirmed by thermodynamically analyzing the heat capacity from differential scanning calorimetry (DSC) [5]. Exceptions to the two-state thermal unfolding are observed when a molten globule (MG) state forms upon thermal denaturation [6–11], usually under non-physiological conditions (acidic/high salt conditions) [8]. The equilibrium MG is a state where the secondary structure is retained, but the tertiary structure is loosely packed, similar to the kinetic intermediates observed during protein folding [7].

Post-synaptic density-95 protein (PSD-95) is a member of the membrane-associated guanylate kinase (MAGUK) family. Like other MAGUK proteins, PSD-95 consists of three PDZ domains, one SH3 domain, and one guanylate kinase [12–15]. PDZ3 is the third PDZ domain of PSD-95, containing three α-helices and six β-strands. It is small, globular, and it has a molecular weight of 11 kDa. PDZ3 undergoes a three-state denaturation process. The intermediate state is not a MG state but is an oligomer formed reversibly at temperatures as high as 60 to 70 °C, which we coined high-temperature reversible oligomer (RO) [16–19].

Recently, we found that the single mutations (F340A and L342A), which do not affect the physicochemical properties of PDZ3 at ambient temperatures, could inhibit the formation of RO and amyloids [16]. Both variants (PDZ3-F340A and PDZ3-L342A) undergoing a two-state unfolding process were designed by replacing hydrophobic residues at the interface of the tetrameric crystal structure of PDZ3 to alanine. Namely, the residues were identified by their large buried surface area (BSA) and accessible surface area (ASA). In particular, F340A mutation inhibited not only high-temperature RO but also amyloidogenesis; however, it would be of interest to understand how a single mutation could induce the formation of an RO state at a high-temperature.

This study applies a reverse engineering strategy to design point mutations to reintroduce high-temperature ROs in PDZ3-F340A. Reverse engineering test our hypothesis on the mechanisms underlying the formation of high-temperature RO in natural sequences. The PDZ3-F340A, which is RO-free, was used as a template protein. We hypothesized RO is produced by a hydrophobic residue on the surface of the monomeric protein and buried at the interface of the tetrameric structure. We thus selected three such hydrophilic residues and replaced them with Leucine (Leu) to enhance the hydrophobic interaction between the monomeric proteins. DSC analysis indicated that two variants unfolded according to the two-state model and did not form high-temperature ROs; however, RO was successfully reintroduced by the N326L mutation, and PDZ3-F340A/N326L was strongly amyloidogenic, confirming a correlation between the appearance of RO and amyloidogenesis.

2. Results and Discussion

2.1. Reverse Engineering of PDZ3

This study used PDZ3-F340A, which undergoes a two-state thermal denaturation as a template for examining whether one can find point mutations yielding RO at high temperatures. PDZ3-F340A itself was designed by mutating F340A in the wild-type of PSD95-PDZ3, which undergoes a three-state thermal denaturation with the concurrent production of high-temperature RO. The RO-producing mutations are assumed to be hydrophilic residues located on the surface of PDZ3 but at the interface of the tetrameric structural unit. The mutation site was determined by calculating the ASA and RSA using DSSP and by applying the following rules (i) high total monomeric accessible surface area (ASA) and tetrameric buried surface area (BSA), (ii) high total tetrameric relative solvent accessible area (RSA), (iii) non-hydrophobic residue, and, (iv) hydrophobic interaction between a side-chain without a steric clash (Table 1 and Table S1). We thus computed the ASA and BSA of PDZ3-F340A, modeled from the X-ray structure of PSD95-PDZ3-wt (PDB ID: 3I4W) using COOT. A large ASA indicates that the residue is located on the surface of the monomeric protein, while a large BSA indicates that the residue is in the interface of the tetrameric unit cell. We selected three hydrophilic residues based on their RSA and BSA (Figure 1a), namely R309, E310, N326, and visually confirmed their location using PyMol (Figure S1, see Supplementary Materials). Note that the calculation from DSSP was in line with those from PDBePISA [20], which we used in our previous reports [16,21]. Three variants, where the residues mentioned above were individually substituted to a Leucine (Leu) with the aim of inducing high-temperature RO by increasing hydrophobic interactions, were produced in *E. coli*, purified, and characterized as described in the following sections.

2.2. *Biophysical Characterization of the PDZ3 Variants*

The substitution of the candidate residues to Leu did not affect PDZ3 variants' native structure and physiochemical properties at ambient temperatures. Sedimentation velocity analysis indicated that all PDZ3 variants were monomeric at 25 °C (Table S2, Figure S2). CD spectra of PDZ3 variants showed that the secondary structure contents of PDZ3 variants were identical to that of PDZ3-wt at temperatures up to 60 °C (Figure S3), and the denaturation was reversible as assessed by measuring the spectra after cooling the heated sample to 25 °C. The CD spectra of PDZ3 variants contained similar fractions of antiparallel and parallel β-strands more than α-helix structures in line with the secondary structure content calculated from the X-ray structure of PDZ3-wt (Table S3). The thermal denaturation curves measured by CD at a concentration of 0.5 mg/mL and pH 7.5 were sigmoidal (Figure 2), indicating an apparent two-state denaturation of the secondary structures (N↔D). For discussion, we estimated the apparent thermodynamic parameters, and we calculated the melting temperature (T_m) and van't Hoff enthalpy ($\Delta H_{\text{van't Hoff}}$ (T_m)) from the fitting of the CD denaturation curves (Table 2). The apparent Tm of PDZ3-F340A/N326L was the highest at 72.40 °C, whereas R309L and E310L slightly decreased the apparent T_m.

Table 1. ASA, BSA, and total RSA of candidate residues of PDZ3-F340A.

Residue	Monomeric ASA (Å^2)				Tetrameric ASA (Å^2)				Tetrameric BSA (Å^2)				Total RSA (%)
	A	B	C	D	A	B	C	D	A	B	C	D	
GLU 310	142	145	140	143	28	145	140	46	114	0	0	97	95
ASN 326	109	107	110	107	109	23	21	107	0	84	89	0	89
ARG 309	216	218	220	218	27	218	220	202	189	0	0	16	75
ASN 369	154	160	157	160	154	17	157	160	0	143	0	0	73
GLU 331	175	175	173	174	175	12	173	174	0	163	0	0	73
THR 321	76	77	76	80	76	20	8	80	0	57	68	0	73
ARG 368	177	150	174	153	84	131	174	74	93	19	0	79	70
LEU 342	85	87	84	85	85	17	17	85	0	70	67	0	68
ALA 343	88	86	91	88	88	47	48	88	0	39	43	0	64
GLY 333	96	100	96	98	96	34	96	98	0	66	0	0	63
SER 320	131	132	132	131	131	100	66	131	0	32	66	0	63
ARG 313	169	169	164	172	90	169	164	81	79	0	0	91	62
PRO 311	67	69	65	67	20	69	65	17	47	0	0	50	61
GLU 395	150	146	148	149	38	146	148	149	112	0	0	0	50
LYS 380	138	138	138	140	138	113	50	140	0	25	88	0	48
GLY 322	25	24	22	25	25	2	0	25	0	22	22	0	42
ILE 389	50	48	49	48	10	48	49	11	40	0	0	37	39
VAL 328	34	35	33	33	34	2	2	33	0	33	31	0	37
ASP 366	108	91	107	88	75	68	107	77	33	23	0	11	35
GLN 391	43	44	43	46	2	44	43	9	41	0	0	37	35

Residues with ASA > 0 Å^2 of at least one of the chains are listed. The monomeric and tetrameric ASA values of four polypeptide chains (A, B, C, D) of PDZ3-F340A were calculated by DSSP. PDZ3-F340A and other variants were modeled using X-ray crystallographic data of PSD95-PDZ3 (PDB ID: 3I4W) consisting of four monomeric chains in each asymmetric unit cell using COOT (crystallographic object-oriented toolkit) [22] The modeled structures were used for calculating accessible surface area (ASA) by DSSP. Buried surface area (BSA) was calculated by subtracting the ASA in the tetrameric structure from the calculated ASA of the monomeric structure. Relative solvent accessibility (RSA) is the total tetrameric ASA of each residue divided by the maximum amino acid solvent accessibility from theoretical normalization values [23]. The selected residues are underlined.

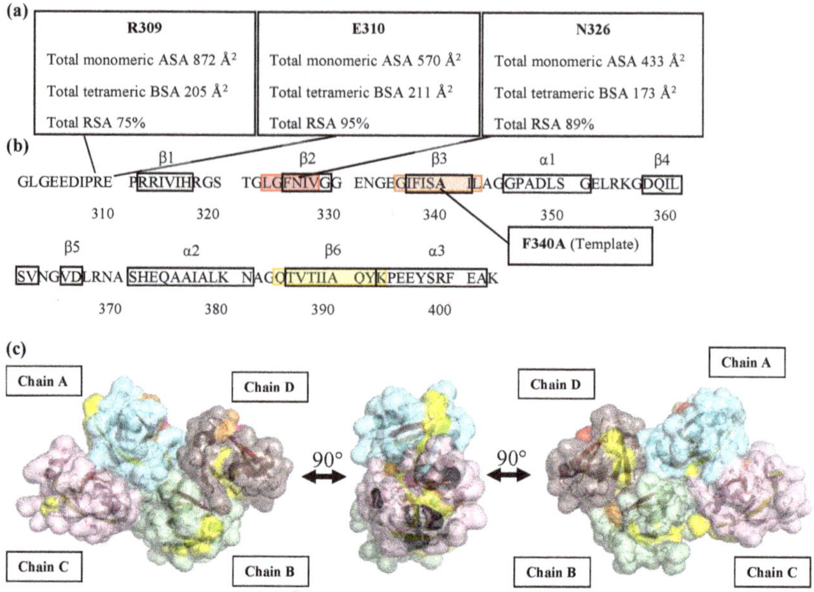

Figure 1. Secondary structure and three-dimensional accessible surface structures of PDZ3-F340A (**a**) Total monomeric accessible surface area (ASA), total tetrameric buried surface area (BSA), and total relative solvent accessible surface area (RSA) values of R309, E310, and N326 residues are shown in square boxes; (**b**) TANGO analysis of β-aggregation (amyloid) prone regions in PDZ3-F340A. The 323–328 residue region crossing the β2 strand is represented in red. The 335–342 residue region crossing the β3 strand is represented in orange. The 384–393 residue region crossing the β6 strand and α-helix 3 at C-terminus is represented in yellow; (**c**) Three images of 90° counter-clockwise views of the PDZ3-F340A tetrameric structure represent the aggregation-prone region on the interface of a tetramer unit cell using Pymol. The color code is the same as in the amino acid sequence.

Table 2. Apparent T_m and van't Hoff enthalpy ($\Delta H_{van't\ Hoff}$ (T_m)).

Name	T_m (°C)	$\Delta H_{van't\ Hoff}$ (T_m) (kJ/mol)
PDZ3-F340A/E310L	69.5 ± 0.0	375.6 ± 3.3
PDZ3-F340A/N326L	72.4 ± 0.0	309.5 ± 2.6
PDZ3-F340A/R309L	67.2 ± 0.0	363.1 ± 2.7
PDZ3-F340A	71.3 ± 0.0	367.4 ± 3.6
PDZ3-wt	71.1 ± 0.1	255.3 ± 3.2

The parameters were calculated by fitting the CD denaturation curves using a two-state model in Origin 2020b software. CD denaturation curves were measured at 0.5 mg/mL, pH 7.5, 25–90 °C and +1.0 °C/min scan rate.

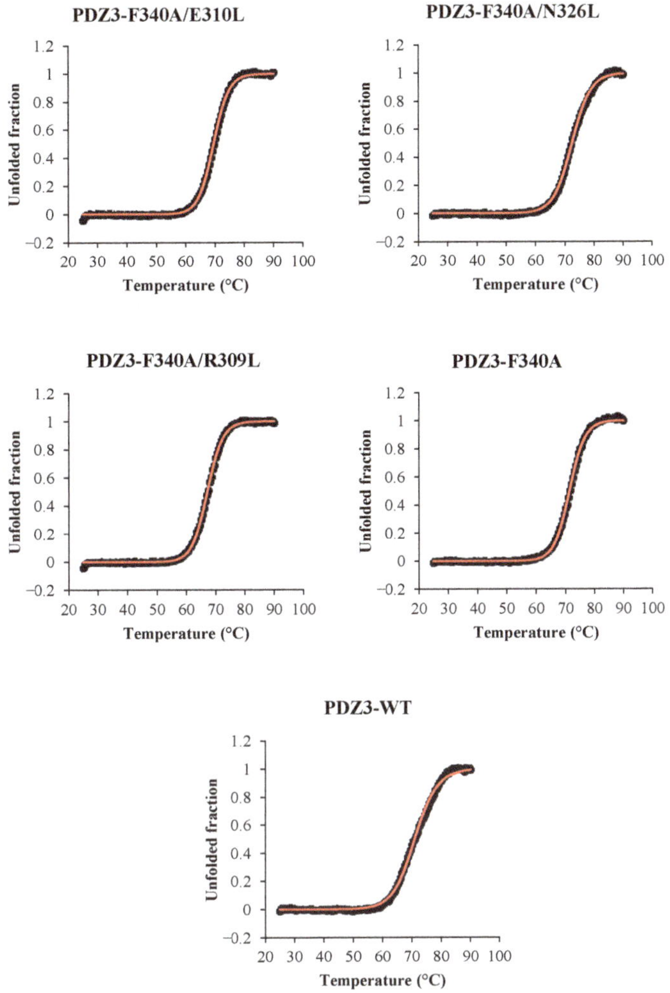

Figure 2. CD thermal denaturation curves of PDZ3 variants at 0.5 mg/mL and pH 7.5 at a scan rate of +1.0 °C/min. The CD values were monitored at 220 nm. Black dots are the experimental data, and red lines are the fitting curves.

2.3. DSC Analysis and Thermodynamic Parameters

The DSC thermograms of reversely engineered PDZ3 variants were measured at a 0.5–1 mg/mL concentration in pH 7.5 with a +1 °C/min scan rate. A single endothermic peak was observed in DSC thermograms of PDZ3-F340A, as reported earlier [16], and for PDZ3-F340A/R309L and PDZ3-F340A/E310L. On the other hand, PDZ3-F340A/N326L exhibited two distinct endothermic peaks similar to those observed for PDZ3-wt (Figure 3).

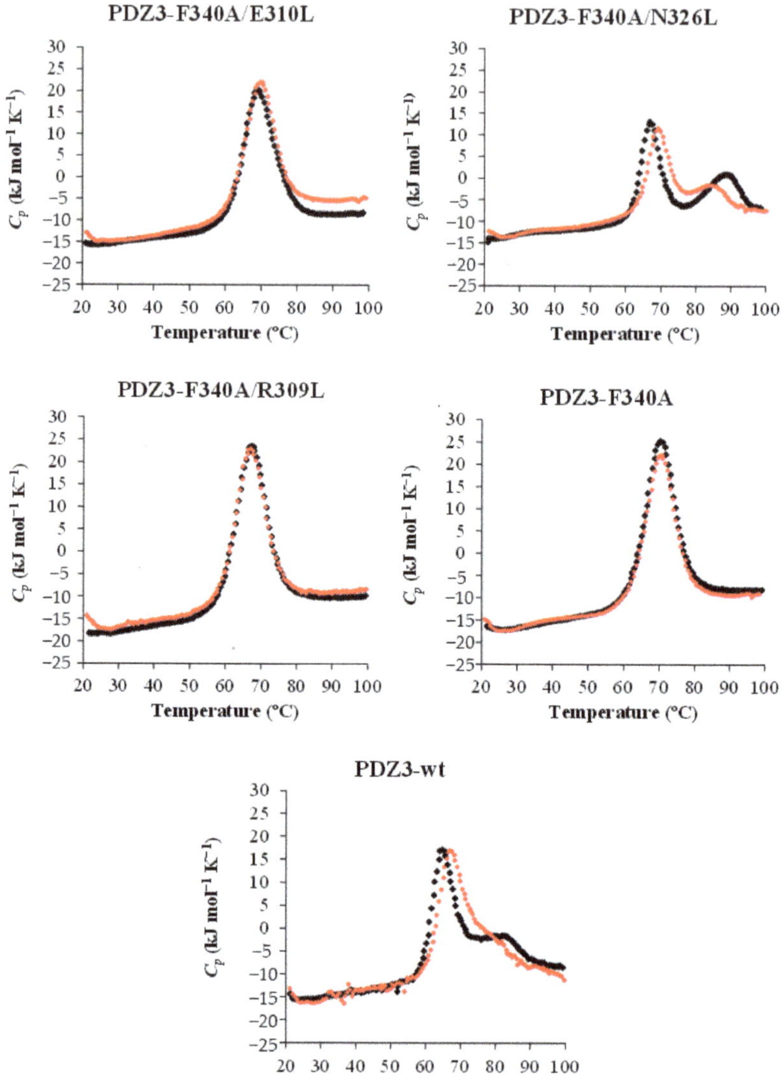

Figure 3. Concentration dependence of DSC thermograms of PDZ3 variants at 0.5–1 mg/mL, pH 7.5, and 1 °C/min scan rate. Black and red dots show DSC thermograms at a protein concentration of 1 and 0.5 mg/mL, respectively.

A detailed analysis of the DSC curves was performed using DDCL3 with a two- and a three-state model (Table 3, Figure S4). The template PDZ3-F340A, PDZ3-F340A/E310L, and PDZ3-F340A/R309L were well fitted with a two-state model (N↔D). Global fitting of curves with protein concentrations of 0.5–1.0 mg/mL showed that PDZ3-F340A/N326L and PDZ3-wt formed tetrameric and pentameric ROs (N↔1/4(I_4)↔D and N↔1/5(I_5)↔D), respectively (Figures 4 and S5). In addition, we observed a strong correlation between T_{mid} (N↔ I_n+D) and the apparent T_m determined by CD (Figures 5 and S6), but not between T_{mid} (N + I_n↔D) and T_m. This result suggests that the secondary structures of PDZ3-F340A/N326L and PDZ3-wt in the intermediate state is unfolded.

Table 3. Thermodynamics parameters were calculated by DSC measurements at 0.5–1 mg/mL of protein concentration.

Name	Concentration (mg/mL)	Transition	T_{mid} (°C)	ΔH_{cal} (T_{mid}) (kJ/mol)
PDZ3-F340A/E310L	1	N-D	68.9 ± 0.1	341.8 ± 2.7
	0.5	N-D	69.3 ± 0.1	343.1 ± 1.9
PDZ3-F340A/N326L	1	N-I_4	68.1 ± 0.2	233.4 ± 10.1
		N-D	73.4 ± 1.2	250.8 ± 16.8
	0.5	N-I_4	70.5 ± 0.1	239.0 ± 7.3
		N-D	72.8 ± 0.7	242.7 ± 4.4
PDZ3-F340A/R309L	1	N-D	67.0 ± 0.1	364.6 ± 2.6
	0.5	N-D	66.8 ± 0.1	359.9 ± 2.7
PDZ3-F340A	1	N-D	70.0 ± 0.1	369.0 ± 2.6
	0.5	N-D	70.2 ± 0.2	358.7 ± 2.7
PDZ3-wt	1	N-I_5	64.9 ± 0.1	251.4 ± 3.7
		N-D	67.9 ± 0.3	296.0 ± 4.4
	0.5	N-I_5	67.8 ± 0.1	250.4 ± 8.2
		N-D	69.5 ± 0.5	316.2 ± 8.5

Midpoint temperature (T_{mid}), calorimetric enthalpy (ΔH_{cal} (T_{mid})) of PDZ3 variants at pH 7.5 determined by fitting the DSC thermogram using DDCL3. PDZ3-F340A/E310L (N-D model); PDZ3-F340A/N326L (N-I_4-D model); PDZ3-F340A/R309L (N-D model); PDZ3-F340A (N-D model); PDZ3-wt (N-I_5-D model).

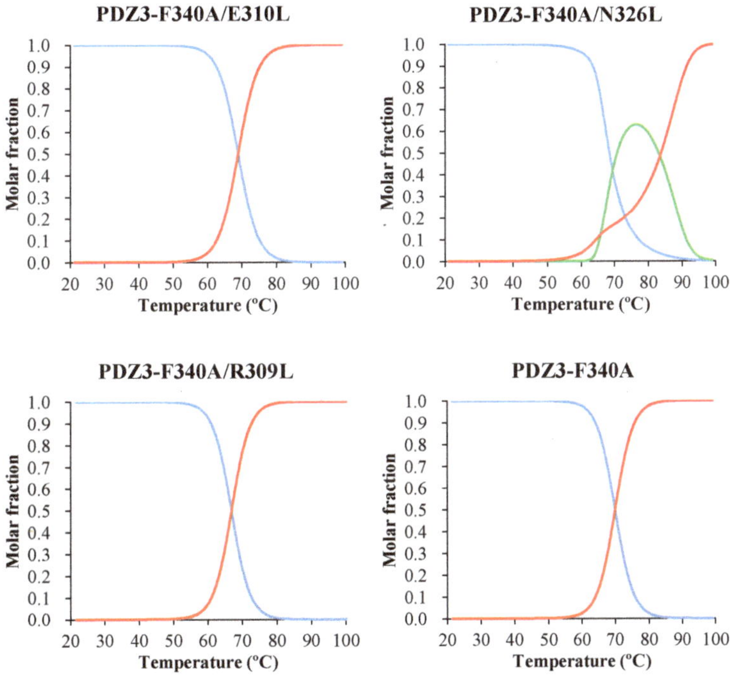

Figure 4. *Cont.*

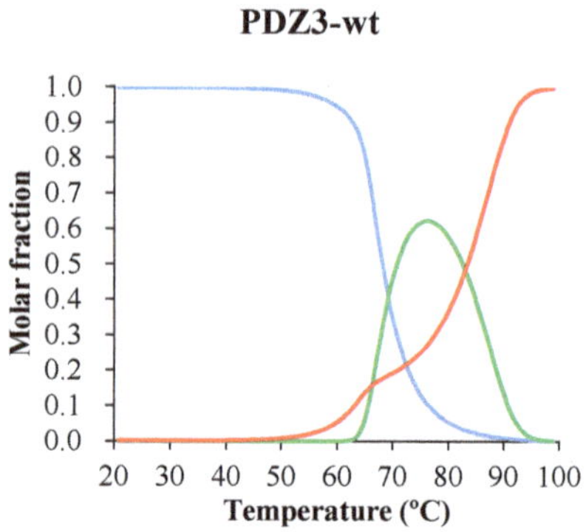

Figure 4. Molar fraction of PDZ3-F340A/E310L (N-D), F340A/N326L (N-I_4-D), F340A/R309L (N-D), PDZ3-F340A (N-D model) and PDZ3-wt (N-I_5-D model) by DDCL3 analysis of DSC thermograms at 1 mg/mL, pH 7.5, and 1 °C/min scan rate. The lines represent natively folded monomers (blue), intermediate oligomers (green), and unfolded monomers (red).

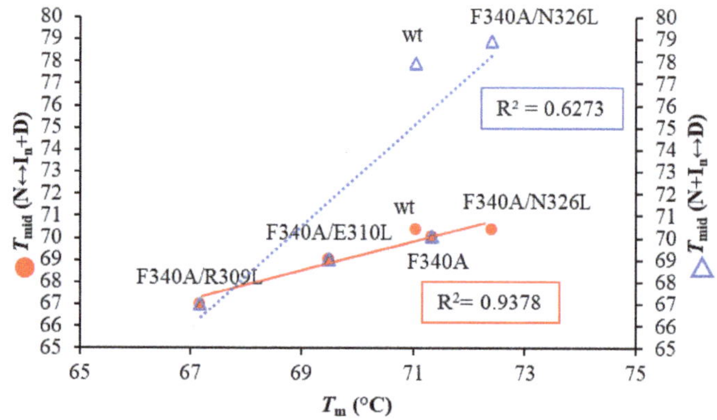

Figure 5. Correlation plots between T_m calculated by CD denaturation curve and T_{mid} (N↔I_n + D), T_{mid} (N + I_n↔D) calculated from the molar fractions as determined by DSC analysis. The CD denaturation curve and DSC measurement were measured at 0.5 mg/mL and pH 7.5 at a scan rate of +1.0 °C/min. The red circles (•) represent T_{mid} (N↔I_n+D) and the blue open triangles (Δ) present T_{mid} (N + I_n↔D).

2.4. High-Temperature ROs and Amyloidogenesis of the Variants

To gain insight into the concurrent formation of high-temperature ROs with amyloidogenicity in PDZ3, we monitored the ThT and ANS fluorescence (Figures 6 and S7). We used ThT to monitor amyloidogenesis because there is a strong relationship between beta-cross structures' formation in fiber and oligomer forms [24–26]. ANS indicates molten globule-like properties and binds to partially exposed hydrophobic surfaces and cavities but also to aggregates with molten globule-like properties [24,27,28].

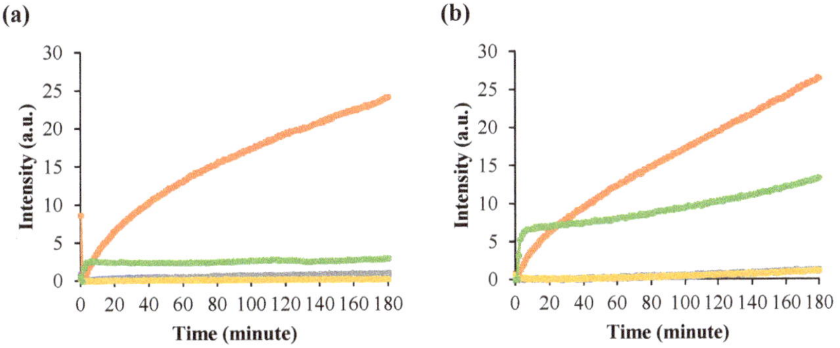

Figure 6. Time-course of RO formation of PDZ3 variants monitored by (**a**) ThT fluorescence at λ_{Em} 480 nm. The dye concentration was 12 µM; (**b**) ANS fluorescence at λ_{Em} 480 nm. The dye concentration was 20 µM. Protein concentration was 1 mg/mL in pH 7.5. Mixture samples were measured at 60 °C for 180 min. PDZ3-F340A/E310L (blue); PDZ3-F340A/N326L (orange); PDZ3-F340A/R309L (gray); PDZ3-F340A (yellow); and PDZ3-wt (green).

The ThT fluorescence intensity of PDZ3-F340A/N326L upon incubation at pH 7.5 and 1 mg/mL at 60 °C and 70 °C for 3 h (Figures 6a and S7a) increased within 5 min and became 5 times higher than that of PDZ3-wt. On the other hand, the ThT fluorescence of PDZ3-F340A/E310L, PDZ3-F340A/R309L, and PDZ3-F340A was essentially negligible. Thus, a single mutation, N326L, not only induced RO but strongly increased amyloidogenicity. For the purpose of discussion, let us note that the simultaneous increase of ANS and ThT fluorescence was also observed for Lysozyme [29].

ANS fluorescence of PDZ3-F340A/N326L slightly increased while PDZ3-wt increased within 10 min when incubated at 60 °C and 70 °C at pH 7.5 (Figures 6b and S7b). After 3 h of incubation, the fluorescent intensity of PDZ3-F340A/N326L was 2.5-fold higher than the PDZ3-wt signal. The ANS fluorescence of the other variants was small. A similar phenomenon appeared at 70 °C incubation, but the fluorescence intensity of PDZ3-F340A/N326L and PDZ3-wt rapidly increased at the beginning and with higher intensity than at 60 °C incubation. This result suggests that PDZ3-F340A/N326L in the RO state has a molten globule-like property.

Finally, we measured the hydrodynamic radii (R_h) of the PDZ3 oligomers at temperatures from 25 °C to 90 °C (Figure S8). First, the R_h of PDZ3 variants were two to three-times larger at high than ambient temperatures (Table 4). In contrast, PDZ3-F340A/N326L showed a hydrodynamic radius (R_h) of 6.74 ± 0.13 nm at 70 °C, and the oligomerization was not fully reversible, as assessed by R_h measured after cooling the sample down to 25 °C. However, at a protein concentration of 0.5 mg/mL, the oligomerization was fully reversible (Figure S9 and Table S4), strongly suggesting the reversibility of the oligomer formed by N326L mutant, which we defined as a basic property of the RO state (R stands for reversible).

Table 4. Hydrodynamic radius (nm, R_h) by DLS.

Name	25 °C	40 °C	60 °C	70 °C	80 °C	90 °C	25 °C (Reverse)
PDZ3-F340A/E310L	1.76 ± 0.04	1.46 ± 0.24	0.99 ± 0.11	0.94 ± 0.55	1.88 ± 0.02	1.63 ± 0.10	1.58 ± 0.07
PDZ3-F340A/N326L	1.15 ± 0.62	1.60 ± 0.03	0.82 ± 0.06	6.74 ± 0.13	7.94 ± 0.22	7.36 ± 0.98	5.68 ± 0.24
PDZ3-F340A/R309L	1.62 ± 0.03	1.62 ± 0.05	1.70 ± 0.01	2.00 ± 0.05	2.08 ± 0.01	2.08 ± 0.02	1.48 ± 0.02
PDZ3-F340A	1.54 ± 0.04	1.56 ± 0.02	1.66 ± 0.00	1.81 ± 0.07	2.23 ± 0.02	2.41 ± 0.13	1.61 ± 0.02
PDZ3-wt	1.65 ± 0.00	1.46 ± 0.02	1.65 ± 0.03	3.61 ± 0.07	3.06 ± 0.03	2.41 ± 0.05	1.66 ± 0.06

DLS was measured at 1 mg/mL, pH 7.5, and 25–90 °C, and after cooling the sample back to 25 °C after heating. R_h values were calculated from size-volume graphs. The errors are the standard deviation of three-times measurements with the same sample.

2.5. The N326L Mutation Induces RO and Amyloidogenesis

Despite being still rare, the number of high-temperature RO observations is gradually increasing, albeit under non-physiological conditions (Cytochrome c) [30], artificially re-designed protein for controlling solubility (tagged BPTI) [19], or globular domains with natural sequences (Dengue4 envelope domain 3 (DEN4ED3) [31]. In addition, we previously showed that the intermediate state (or RO) could be fully inhibited by replacing a single hydrophobic residue at the crystal interface with an alanine [16,21,31], which correlated with the inhibition of amyloidogenicity.

Here, we used a reversed engineering strategy to assess whether increasing the hydrophobicity of the crystal interface induces RO and concurrent amyloidogenicity. We constructed three variants and confirmed that their structure and physicochemical properties were conserved at low and ambient temperatures (Figures S1 and S6, and Table S5). Our experiment unambiguously indicated that one of the variants, PDZ3-F340A/N326L undergoes a three-state denaturation (N↔ 1/4(I_4) ↔D), indicating that a single mutation can induce the formation of RO at high-temperature. In contrast, PDZ3-F340A/E310L and PDZ3-F340A/R309L do not induce RO as shown by a single endothermic peak in DSC thermograms and undergo a two-state denaturation (N↔D) like our template PDZ3-F340A.

It remains unclear why the N326L mutation-induced RO, but not R309L nor E310L, despite all three mutations being located at the crystallographic tetramer interface. Further inspection indicated that N326 is located on the β2 strand of PDZ3-F340A, whereas E310 and R309 are in a loop close to the β1-strand. In addition, the aggregation-prone region of PDZ3-F340A calculated by TANGO [32] indicated that N326 is placed in the 323–328 β2 strand, which has a high amyloidogenicity tendency (Figure 1b). On the other hand, R309 and E310 are in nonamyloidogenic regions. These regions were confirmed in PDZ3-wt by heteronuclear NMR experiments, being β2 engaged in the intermediate arrangement of the unfolding intermediate, whereas the β1 region does not [33]. The F340A and L342A mutations, which abolished RO and amyloidogenicity in the PDZ3-wt, were also located in the 335–343 region crossing the β3 strand amyloidogenic regions (data not shown). Thus, although E310 and R309 are located at the interface, which we assumed would increase the hydrophobic interaction and induce RO formation, they did not induce RO nor amyloidogenicity. In addition, the accessible surface area model created by PyMOL showed that the 323–328 region is buried when proteins arrange into the tetramer, but the 335–343 region is not (Figure 1c).

3. Materials and Methods

3.1. Protein Expression, Purification, and Identification

The proteins were prepared according to our previously reported protocol. In short, single mutations were introduced using a Quikchange protocol and a synthetic gene encoding PDZ3 cloned into a pBAT4 vector as the template. All variants were overexpressed in *Escherichia coli* strain BL21(DE3) with 1 L of LB medium. Protein expression was induced by adding 0.2 mM IPTG when the OD at 590 nm reached 0.6, and the culture was further incubated at 37 °C, 120 rpm for 4 h. The harvested cells were dissolved in 20 mL of 50 mM Tris-HCl (pH 8.7) and lysed by ultrasonication. The supernatant fraction of the cell lysate was then acidified to pH 3 by adding around 1 mL of 1 M HCl and ultracentrifuged. The recombinant proteins were purified from the supernatant by reverse-phase HPLC, lyophilized, and stored at −30 °C until use, as we reported in previous reports [34,35].

The molecular weight of the protein was confirmed by matrix-assisted laser desorption/ionization-time of flight (MALDI-TOF) MS measurements using the plate with Autoflex speed TOF/TOF (Bruker Daltonics, Fremont, CA, USA). The matrix solution was prepared by dissolving 10 mg of sinapic acid in 1 mL of a solution containing 300 μL of acetonitrile, 100 μL of 1% trifluoroacetic acid, and 600 μL of Milli Q water. Protein samples were prepared by mixing 1 μL of protein solution with 9 μL of the matrix solution. 1 μL of 10 μM, 1 μM, and 0.1 μM sample mixtures were spotted and air-dried on the MALDI-TOF MS plate. The molecular weights of all proteins were within 7 Da of the theoretical values (Table S6) computed with ExPASy's ProtParam tool (https://web.expasy.org/protparam/ accessed on 30 May 2020).

Samples were prepared by dissolving lyophilized proteins in Milli Q water, and the protein concentrations of samples were adjusted to 0.2, 0.5, and 1.0 mg/mL in 50 mM potassium phosphate buffer (pH 7.5). The protein concentrations were determined by measuring absorbance at 280 nm (ε = 2980 M^{-1} cm^{-1}) using a Nanodrop (Thermo Fisher Scientific, Waltham, MA, USA). The pH of the samples was confirmed just before performing the experiments. Freshly prepared samples were used for CD, DLS, and fluorescence spectroscopy measurements.

3.2. *Differential Scanning Calorimetry (DSC) Measurements*

Samples were prepared by dissolving lyophilized proteins in Milli Q water and dialyzed for 18 h at 4 °C in 50 mM potassium phosphate buffer (pH 7.5) using a Spectra/Por 3 membrane (MWCO of 3.5 kDa) with one buffer exchange. After dialysis, the protein concentration of samples was adjusted to 1 mg/mL and filtered with Disposable Ultrafiltration Unit, 200 K MWCO (ADVANTEC®, Tokyo, Japan), to remove aggregates. Protein concentrations and pHs of the samples were confirmed just before performing experiments.

DSC measurements were performed using a VP-DSC microcalorimeter (Malvern Panalytical Ltd., Malvern, UK) at a scan rate of +1.0 °C/min in the temperature range of 20 to 100 °C, essentially in line with our previous reports [36,37]. Baselines were recorded before measurements using a 50 mM potassium phosphate buffer (pH 7.5). The reversibility of the thermal unfolding was checked by repeating scans of the same sample. Thermodynamic parameters (T_{mid} and $\Delta H(T_{mid})$) were determined by analyzing the apparent heat capacity curves using a non-linear, least-squares fitting algorithm, DDCL3, and assuming a linear temperature dependence of the heat capacity of the native and denatured states [5,38].

3.3. *Circular Dichroism (CD) Measurements*

CD measurements was performed using a Jasco-J820 spectropolarimeter (Tokyo, Japan). A quartz cuvette with 2-mm optical path length was used. The secondary structure contents from CD spectra were calculated using BeStSel [39]. Thermal stability was measured at a protein concentration of 0.5 mg/mL in 50 mM potassium phosphate buffer (pH 7.5), at a +1.0 °C/min scan rate, and monitored between 25 °C and 90 °C using the CD value at 220 nm. Melting temperatures (T_m) were computed through least-squares fittings of experimental data to a two-state model using Origin 2020b (OriginLab Corp, Northampton, MA, USA) [40].

3.4. *Dynamic Light Scattering (DLS) Measurements*

DLS measurements were performed using a glass cuvette with a Zeta-nanosizer (Nano S, Malvern, UK). The sample was measured at 25–90 °C and reversed to 25 °C. The hydrodynamic radius (R_h) was calculated using the Stokes-Einstein equation from size-number plots [41].

3.5. Fluorescence Spectroscopy Measurement

ANS fluorescence was measured at an excitation wavelength of 380 nm at 60, 70 °C for 53 h. The emission spectra were monitored from 400 to 600 nm. ThT fluorescence was measured with an excitation wavelength of 444 nm, and the emission spectra were observed from 460 to 640 nm. The final concentration of Thioflavin T (ThT) and 8-Anilino-1-naphthalenesulfonate (ANS) was 12 and 20 μM, respectively. The dye was mixed with 300 μL of the samples, from which 60 μL were transferred to a 3.00 mm Hellma® micro cuvette.

3.6. Analytical Ultracentrifugation (AUC) Measurements

Samples were prepared by dissolving lyophilized proteins in Milli Q water and dialyed for 18 h at 4 °C against 50 mM potassium phosphate buffer pH 7.5. The protein concentrations of the samples were adjusted to 1 mg/mL by diluting with the dialyzed buffer. Samples were filtered with a 0.20 μm membrane filter (MilliporeSigma, Burlington, VT, USA) for removing aggregates. Protein concentrations and pH values of the samples were confirmed just before performing the experiments.

Sedimentation velocity experiments were carried out using an Optima XL-I analytical ultracentrifuge (Beckman-Coulter) with An-50 Ti analytical 8-place titanium rotor at 25 °C. Samples were transferred to a 12-mm double-sector epon charcoal-filled centerpiece and centrifuged at a rotor speed of 50,000 rpm, and the absorbance was monitored at 280 nm. Sedimentation velocity data were analyzed using the continuous distribution c(s) analysis module in the SEDFIT software [42]. The range of sedimentation coefficients, where the main peak was present, was integrated to obtain the weighted average sedimentation coefficient. The c(s) distribution was converted into c(M), a molar mass distribution. Solvent density, viscosity, and protein partial specific volumes were calculated using SEDTERP [43].

4. Conclusions

The reverse engineering strategy confirmed that hydrophobic interaction at the interface of the monomeric unit in the crystal tetramer induces the RO formation at a high-temperature. Thermodynamic analysis of the DSC denaturation curves indicated that the molar fraction of RO becomes maximal at 60 to 70 °C and that the secondary structures of PDZ3-F340A/N326L and PDZ3-wt in the intermediate state (RO state) are unfolded. Furthermore, the reversed engineering strategy confirmed the relationship between RO appearance and amyloidogenicity. Kinetic experiments need to establish whether ROs are on or off-pathway for amyloidogenesis, but the present results strongly favor the on-pathway hypothesis.

Supplementary Materials: The following supporting information can be downloaded at: https://www.mdpi.com/article/10.3390/molecules27092813/s1, Figure S1: Accessible surface models of PDZ3 variants and PDZ3-wt by Pymol; Figure S2: Sedimentation velocity analysis of PDZ3 variants by AUC measurements; Figure S3: CD spectra of PDZ3 variants; Figure S4: Concentration dependence of DSC thermograms of PDZ3 variants fitting by DDCL3 analysis; Figure S5: Molar fraction of PDZ3 variants calculated by DDCL3 analysis of DSC thermograms; Figure S6: Fitting curve calculated by CD thermal denaturation superimposed to the molar fraction of I_n + D state calculated from DSC thermograms; Figure S7: Time-course fluorescence spectra of PDZ3 variants at 70 °C by monitoring ThT and ANS fluorescence; Figure S8: Hydrodynamic radii (R_h) of PDZ3 variants using DLS measurements; Figure S9: Bar plot of Hydrodynamic radius (nm, R_h) by DLS measurements; Table S1: Accessible Surface Area values of artificial crystallographic PDZ3-F340A calculated by DSSP; Table S2: Molecular weight of PDZ3 variants determined by MALDI-TOF MS; Table S3: Secondary structure contents of PDZ3 variants calculated by BeStSel; Table S4: Sedimentation velocity analysis of PDZ3 variants at 25°C analyzed using SEDFIT and SEDNTERP; Table S5: Residues of PDZ3 variants between DSC raw data and fitting curves calculated with DDCL3; Table S6: Hydrodynamic radius (nm, R_h) by DLS.

Author Contributions: S.O., T.S. and Y.K. designed the study and wrote the manuscript. S.O. designed and purified the recombinant proteins and performed the spectroscopic measurements. T.S. and S.-i.K. performed DSC measurements and the data analysis. S.O., K.M. and J.R.H.T. carried out AUC analysis. J.C.M. provided the materials and co-wrote the manuscript. All authors have read and agreed to the published version of the manuscript.

Funding: This research was funded by a MEXT scholarship to SO, a JSPS grant-in-aid for scientific research (KAKENHI: 18H02385, 21K05288, and 21K15049) and the TUAT's Institute of Global Innovation Research.

Institutional Review Board Statement: Not applicable.

Informed Consent Statement: Not applicable.

Data Availability Statement: Data is contained within the article or Supplementary Materials.

Acknowledgments: We are grateful to Brindha Subbaian and Kuroda lab members for suggestions with site-directed mutagenesis and technical assistance.

Conflicts of Interest: The authors declare no conflict of interest.

Sample Availability: The PDZ3-wt plasmid is available from JCM.

References

1. Privalov, P.L. Stability of proteins: Small globular proteins. *Adv. Protein Chem.* **1979**, *33*, 167–241. [CrossRef] [PubMed]
2. Pfeil, W. Thermodynamics of alpha-lactalbumin unfolding. *Biophys. Chem.* **1981**, *13*, 181–186. [CrossRef]
3. Privalov, P.L.; Gill, S.J. Stability of protein structure and hydrophobic interaction. *Adv. Protein Chem.* **1988**, *39*, 191–234. [CrossRef]
4. Makhatadze, G.I.; Privalov, P.L. Protein interactions with urea and guanidinium chloride. A calorimetric study. *J. Mol. Biol.* **1992**, *226*, 491–505. [CrossRef]
5. Kidokoro, S.-I.; Wada, A. Determination of thermodynamic functions from scanning calorimetry data. *Biopolymers* **1987**, *26*, 213–229. [CrossRef]
6. Ohgushi, M.; Wada, A. 'Molten-globule state': A compact form of globular proteins with mobile side-chains. *FEBS Lett.* **1983**, *164*, 21–24. [CrossRef]
7. Kuwajima, K. The molten globule state as a clue for understanding the folding and cooperativity of globular-protein structure. *Proteins Struct. Funct. Bioinform.* **1989**, *6*, 87–103. [CrossRef]
8. Kuroda, Y.; Kidokoro, S.-I.; Wada, A. Thermodynamic characterization of cytochrome c at low pH: Observation of the molten globule state and of the cold denaturation process. *J. Mol. Biol.* **1992**, *223*, 1139–1153. [CrossRef]
9. Bychkova, V.E.; Semisotnov, G.V.; Balobanov, V.A.; Finkelstein, A.V. The molten globule concept: 45 years later. *Biochemistry* **2018**, *83*, S33–S47. [CrossRef]
10. Kuroda, Y.; Endo, S.; Nakamura, H. How a novel scientific concept was coined the "molten globule state". *Biomolecules* **2020**, *10*. [CrossRef]
11. Vassilenko, K.S.; Uversky, V.N. Native-like secondary structure of molten globules. *Biochim. Biophys. Acta* **2002**, *1594*, 168–177. [CrossRef]
12. Chen, X.; Levy, J.M.; Hou, A.; Winters, C.; Azzam, R.; Sousa, A.A.; Leapman, R.D.; Nicoll, R.A.; Reese, T.S. PSD-95 family MAGUKs are essential for anchoring AMPA and NMDA receptor complexes at the postsynaptic density. *Proc. Natl. Acad. Sci. USA* **2015**, *112*, E6983. [CrossRef] [PubMed]
13. Murciano-Calles, J.; Marin-Argany, M.; Cobos, E.S.; Villegas, S.; Martinez, J.C. The impact of extra-domain structures and post-translational modifications in the folding/misfolding behaviour of the third PDZ domain of MAGUK neuronal protein PSD-95. *PLoS ONE* **2014**, *9*, e98124. [CrossRef] [PubMed]
14. Savioz, A.; Leuba, G.; Vallet, P.G. A framework to understand the variations of PSD-95 expression in brain aging and in Alzheimer's disease. *Ageing Res. Rev.* **2014**, *18*, 86–94. [CrossRef] [PubMed]
15. Zhang, J.; Petit, C.M.; King, D.S.; Lee, A.L. Phosphorylation of a PDZ domain extension modulates binding affinity and interdomain interactions in postsynaptic density-95 (PSD-95) protein, a membrane-associated guanylate kinase (MAGUK). *J. Biol. Chem.* **2011**, *286*, 41776–41785. [CrossRef] [PubMed]
16. Saotome, T.; Mezaki, T.; Brindha, S.; Unzai, S.; Martinez, J.C.; Kidokoro, S.I.; Kuroda, Y. Thermodynamic analysis of point mutations inhibiting high-temperature reversible oligomerization of PDZ3. *Biophys. J.* **2020**, *119*, 1391–1401. [CrossRef]
17. Murciano-Calles, J.; Cobos, E.S.; Mateo, P.L.; Camara-Artigas, A.; Martinez, J.C. An oligomeric equilibrium intermediate as the precursory nucleus of globular and fibrillar supramacromolecular assemblies in a PDZ domain. *Biophys. J.* **2010**, *99*, 263–272. [CrossRef]
18. Murciano-Calles, J.; Cobos, E.S.; Mateo, P.L.; Camara-Artigas, A.; Martinez, J.C. A comparative analysis of the folding and misfolding pathways of the third PDZ domain of PSD95 investigated under different pH conditions. *Biophys. Chem.* **2011**, *158*, 104–110. [CrossRef]

19. Nakamura, S.; Kibria, M.G.; Unzai, S.; Kuroda, Y.; Kidokoro, S.I. Reversible oligomerization and reverse hydrophobic effect induced by Isoleucine tags attached at the C-terminus of a simplified BPTI variant. *Biochemistry* **2020**, *59*, 3660–3668. [CrossRef]
20. Krissinel, E.; Henrick, K. Inference of macromolecular assemblies from crystalline state. *J. Mol. Biol* **2007**, *372*, 774–797. [CrossRef]
21. Saotome, T.; Onchaiya, S.; Brindha, S.; Mezaki, T.; Unzai, S.; Noguchi, K.; Martinez, J.C.; Kidokoro, S.I.; Kuroda, Y. Blocking PSD95-PDZ3's amyloidogenesis through point mutations that inhibit high-temperature reversible oligomerization (RO). *FEBS J.* **2021**. [CrossRef] [PubMed]
22. Emsley, P.; Lohkamp, B.; Scott, W.G.; Cowtan, K. Features and development of Coot. *Acta Cryst. D Biol. Cryst.* **2010**, *66*, 486–501. [CrossRef] [PubMed]
23. Tien, M.Z.; Meyer, A.G.; Sydykova, D.K.; Spielman, S.J.; Wilke, C.O. Maximum allowed solvent accessibilities of residues in proteins. *PLoS ONE* **2013**, *8*, e80635. [CrossRef] [PubMed]
24. Hawe, A.; Sutter, M.; Jiskoot, W. Extrinsic fluorescent dyes as tools for protein characterization. *Pharm. Res.* **2008**, *25*, 1487–1499. [CrossRef]
25. Dasgupta, M.; Kishore, N. Characterization and analysis of binding of Thioflavin T with partially folded and native states of alpha-lactalbumin protein by calorimetric and spectroscopic techniques. *Int. J. Biol. Macromol.* **2017**, *95*, 376–384. [CrossRef]
26. Krebs, M.R.; Bromley, E.H.; Donald, A.M. The binding of thioflavin-T to amyloid fibrils: Localisation and implications. *J. Struct. Biol.* **2005**, *149*, 30–37. [CrossRef]
27. Guliyeva, A.J.; Gasymov, O.K. ANS fluorescence: Potential to discriminate hydrophobic sites of proteins in solid states. *Biochem. Biophys. Rep.* **2020**, *24*, 100843. [CrossRef]
28. Kibria, M.G.; Akazawa-Ogawa, Y.; Rahman, N.; Hagihara, Y.; Kuroda, Y. The immunogenicity of an anti-EGFR single domain antibody (VHH) is enhanced by misfolded amorphous aggregation but not by heat-induced aggregation. *Eur. J. Pharm. Biopharm.* **2020**, *152*, 164–174. [CrossRef]
29. Sulatsky, M.I.; Sulatskaya, A.I.; Povarova, O.I.; Antifeeva, I.A.; Kuznetsova, I.M.; Turoverov, K.K. Effect of the fluorescent probes ThT and ANS on the mature amyloid fibrils. *Prion* **2020**, *14*, 67–75. [CrossRef]
30. Nakamura, S.; Seki, Y.; Katoh, E.; Kidokoro, S.-I. Thermodynamic and structural properties of the acid molten globule state of horse cytochrome c. *Biochemistry* **2011**, *50*, 3116–3126. [CrossRef]
31. Saotome, T.; Nakamura, S.; Islam, M.M.; Nakazawa, A.; Dellarole, M.; Arisaka, F.; Kidokoro, S.; Kuroda, Y. Unusual reversible oligomerization of unfolded dengue envelope protein domain 3 at high temperatures and its abolition by a point mutation. *Biochemistry* **2016**, *55*, 4469–4475. [CrossRef]
32. Fernandez-Escamilla, A.M.; Rousseau, F.; Schymkowitz, J.; Serrano, L. Prediction of sequence-dependent and mutational effects on the aggregation of peptides and proteins. *Nat. Biotechnol.* **2004**, *22*, 1302–1306. [CrossRef] [PubMed]
33. Marin-Argany, M.; Candel, A.M.; Murciano-Calles, J.; Martinez, J.C.; Villegas, S. The interconversion between a flexible beta-sheet and a fibril beta-arrangement constitutes the main conformational event during misfolding of PSD95-PDZ3 domain. *Biophys. J.* **2012**, *103*, 738–747. [CrossRef] [PubMed]
34. Khan, M.A.; Islam, M.M.; Kuroda, Y. Analysis of protein aggregation kinetics using short amino acid peptide tags. *Biochim. Biophys. Acta (BBA)-Proteins Proteom.* **2013**, *1834*, 2107–2115. [CrossRef] [PubMed]
35. Rathnayaka, T.; Tawa, M.; Nakamura, T.; Sohya, S.; Kuwajima, K.; Yohda, M.; Kuroda, Y. Solubilization and folding of a fully active recombinant Gaussia luciferase with native disulfide bonds by using a SEP-Tag. *Biochim. Biophys. Acta (BBA)-Proteins Proteom.* **2011**, *1814*, 1775–1778. [CrossRef] [PubMed]
36. Kato, A.; Yamada, M.; Nakamura, S.; Kidokoro, S.-I.; Kuroda, Y. Thermodynamic properties of BPTI variants with highly simplified amino acid sequences. *J. Mol. Biol.* **2007**, *372*, 737–746. [CrossRef]
37. Islam, M.M.; Sohya, S.; Noguchi, K.; Kidokoro, S.-I.; Yohda, M.; Kuroda, Y. Thermodynamic and structural analysis of highly stabilized BPTIs by single and double mutations. *Proteins Struct. Funct. Bioinform.* **2009**, *77*, 962–970. [CrossRef]
38. Kidokoro, S.-I.; Uedaira, H.; Wada, A. Determination of thermodynamic functions from scanning calorimetry data. II. For the system that includes self-dissociation / association process. *Biopolymers* **1988**, *27*, 271–297. [CrossRef]
39. Micsonai, A.; Wien, F.; Kernya, L.; Lee, Y.H.; Goto, Y.; Refregiers, M.; Kardos, J. Accurate secondary structure prediction and fold recognition for circular dichroism spectroscopy. *Proc. Natl Acad Sci USA* **2015**, *112*, E3095–E3103. [CrossRef]
40. Greenfield, N.J. Using circular dichroism collected as a function of temperature to determine the thermodynamics of protein unfolding and binding interactions. *Nat. Protoc.* **2006**, *1*, 2527–2535. [CrossRef]
41. Stetefeld, J.; McKenna, S.A.; Patel, T.R. Dynamic light scattering: A practical guide and applications in biomedical sciences. *Biophys Rev.* **2016**, *8*, 409–427. [CrossRef] [PubMed]
42. Schuck, P. Size-distribution analysis of macromolecules by sedimentation velocity ultracentrifugation and lamm equation modeling. *Biophys J.* **2000**, *78*, 1606–1619. [CrossRef]
43. Laue, T.M. *Computer-aided interpretation of analytical sedimentation data for proteins. Analytical Ultracentrifugation in Biochemistry and Polymer Science*; Royal Society of Chemistry: London, UK, 1992; pp. 90–125.

molecules

MDPI

Article

Near-Wall Aggregation of Amyloidogenic Aβ 1-40 Peptide: Direct Observation by the FRET

Natalia Katina, Alisa Mikhaylina, Nelly Ilina, Irina Eliseeva and Vitalii Balobanov *

Institute of Protein Research, Pushchino, 142290 Moscow, Russia; nkatina@phys.protres.ru (N.K.); alisamikhaylina15@gmail.com (A.M.); nelly.ilyina@mail.ru (N.I.); yeliseeva@vega.protres.ru (I.E.)
* Correspondence: balobanov@phys.protres.ru

Abstract: The formation of amyloid fibrils is one of the variants of the self-organization of polypeptide chains. For the amyloid aggregation, the solution must be oversaturated with proteins. The interface of the liquid (solution) and solid (vessel walls) phases can trigger the adsorption of protein molecules, and the resulting oversaturation can initiate conformational transitions in them. In any laboratory experiment, we cannot exclude the presence of surfaces such as the walls of vessels, cuvettes, etc. However, in many works devoted to the study of amyloid formation, this feature is not considered. In our work, we investigated the behavior of the Aβ 1-40 peptide at the water-glass, water–quartz, and water–plastic interface. We carried out a series of simple experiments and showed that the Aβ 1-40 peptide is actively adsorbed on these surfaces, which leads to a significant interaction and aggregation of peptides. This means that the interface can be the place where the first amyloid nucleus appears. We suggest that this effect may also be one of the reasons for the difficulty of reproducing kinetic data when studying the aggregation of the amyloid of the Aβ 1-40 peptide and other amyloidogenic proteins

Keywords: amyloidogenesis; aggregation; adsorption; Aβ 1-40 peptide; boundary of liquid phase

Citation: Katina, N.; Mikhaylina, A.; Ilina, N.; Eliseeva, I.; Balobanov, V. Near-Wall Aggregation of Amyloidogenic Aβ 1-40 Peptide: Direct Observation by the FRET. *Molecules* **2021**, *26*, 7590. https://doi.org/10.3390/molecules26247590

Academic Editors: Kunihiro Kuwajima, Yuko Okamoto, Tuomas Knowles and Michele Vendruscolo

Received: 17 November 2021
Accepted: 13 December 2021
Published: 15 December 2021

1. Introduction

Amyloid aggregation remains one of the most exciting areas in the biophysics of protein molecules. Great efforts have been directed to studying the first stages of aggregation—the formation of the nucleus of amyloid fibrils. Many researchers suggest their spontaneous appearance in solution [1,2]. This interpretation draws an analogy between the process of amyloid aggregation and the process of crystallization. Goto also draws an analogy between amyloid formation and crystallization in his recent works [3]. As is known, under natural conditions, crystallization begins at inhomogeneities—dust grains, surface defects, etc. The rate of crystallization on such inhomogeneities significantly exceeds the rate of spontaneous crystallization from solution [4,5]. Studies devoted to the effect of various surfaces on amyloid formation also indicate similar phenomena during the initiation of the amyloid fibrils growth [6–8]. This behavior is typical for nucleation and elongation systems. The speed of the whole process depends on a random event—the appearance of the first seed of a new phase [9]. This makes the process difficult to experimentally reproduce. Poor reproducibility of the results is one of the complex problems in the experimental study of the kinetics of amyloid fibril formation. The search for a solution to this problem led us to two possible explanations. The first follows from the theory of nucleation and elongation—these are just the nature of the aggregation process. That is, a single random event triggers the aggregation. Second, there is a systematic factor that is not taken into account by researchers. Instead, these two explanations are combined into one: an unaccounted factor leads to the appearance of seeds.

The area of interest of the presented work was the effect of the surface of the vessel, cuvette, or test tube, in which amyloid aggregates are formed. This interest is easy to explain. On the one hand, papers indicate that the surface can modulate the aggregation process [10,11]; on the other hand, in most works, the question of the surface effect influence on the process in bulk is omitted [12]. To investigate the processes occurring on the glass surface, in the conditions of our experiments, studies of the aggregation of Aβ 1-40 peptides were carried out. We labeled Aβ 1-40 peptides by fluorescent dyes Cy5 and Cy3 and used them for observation. We have shown that Aβ 1-40 peptide is adsorbed in significant amounts on the glass surface, which leads to the significant interaction and aggregation of peptides.

2. Results and Discussion

2.1. The Choice of the Method of Observation

A number of spectral and microscopic methods can be used to monitor protein aggregation and amyloid formation [10]. Each of them has its area of application, advantages, and disadvantages. For our research, the possibility of direct observation without the stage of sample preparation is essential. In AFM and TEM, a small portion of the solution is taken as a sample, limiting the field of view. In addition, fixation of the sample to the base sheet can lead to artifacts. The SPR method requires fixing one of the interacting components on a previously prepared substrate. Having analyzed the available research methods in this way, we concluded that fluorescent methods are the most acceptable. Additionally, since there are no tryptophan residues in the Aβ peptide 1-40 sequence, we need an added fluorescent label. The most commonly used label for amylolide structures is the ThioflavinT dye. However, it does not fully satisfy our needs, since it has a low binding to early-stage aggregation species compared to mature fibrils. Namely, these stages are the most interesting and least studied.

So, we chose the fluorescence of the attached dyes as the observation method. In this case, we used two dyes that form a FRET pair—Cy3 and Cy5. When these dyes become closer fluorescence energy of Cy3 can transfer to Cy5. This effect allows to evaluate the interaction between peptides. According to the data of a structural study, in the amyloid fibril of Aβ peptide 1–40, the peptides are located so that their C-terminus is covered from the solvent, and the N-terminus are relatively accessible and close one by one (see Figure 1) [13]. Therefore, we assumed that the peptides marked on the N-terminus would bring the dyes closer together during the formation of amyloid structures, which we will register by the change in the fluorescence spectrum. It was planned in this way to record the kinetics of the initial stages of aggregation.

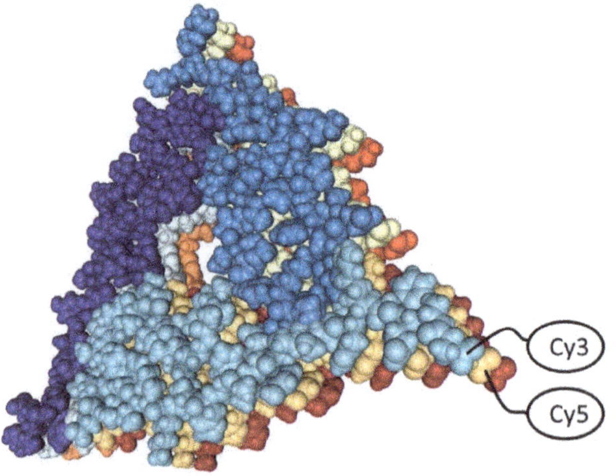

Figure 1. Structure of Aβ 1-40 amyloid fibril (PDB: 2M4J). Fluorescent dyes Cy3 and Cy5 attached to the N terminus of peptides are schematically indicated.

2.2. FRET Measurement in Bulk Solution and on Surfaces

One-half of the peptides were labeled Cy3 dye, which served as a fluorescence donor. The other half was marked with Cy5 dye, which served as an acceptor. The experiments described below were used such peptides in a 1:1 ratio, mixed before or during the experiment. The first experiments carried out in the bulk volume of a solution did not give the desired results. At the initial moments, the FRET signal was not registered. Further research shows that this can be explained by the fact that the objects of our interest were not located in the observed area. We schematically illustrated this idea in Figure 2A. As can be seen, when using the classical design of the fluorescence experiment, only the signal in crossing the excitation light beam and the detector's field of view can be registered. The walls of the cuvette are not into this area; therefore, it is not possible to observe what is happening on them.

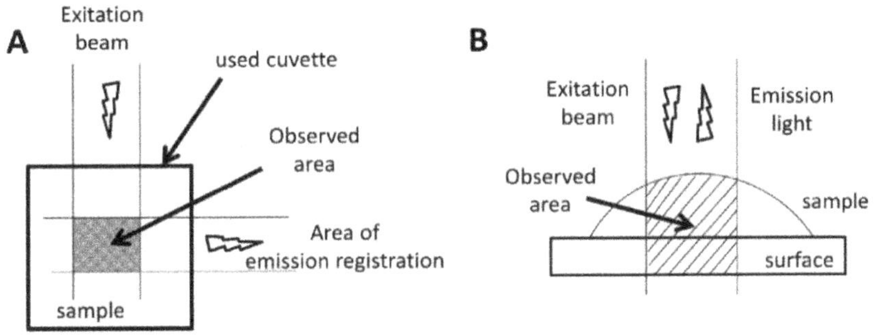

Figure 2. Scheme of fluorescence research in the standard version (**A**) and the fluorescence study on the surface (**B**). Again, the observed areas are shaded.

To test our hypothesis, we changed the design of the experiment. Fluorescence at the interface now also fell into the field of view (Figure 2B). First, the fluorescence of a droplet of a peptide solution labeled with Cy3 dye was measured on the glass surface. Then, a solution of the peptide labeled with the Cy5 dye was added to it. In this case, a fluorescence peak of Cy5 appeared, which indicates the interaction of peptides labeled with different dyes. After that, we gently washed the glass surface and applied a drop of the buffer solution without peptides. The spectrum retained the Cy5 fluorescence peak. The results present in Figure 3 show that the peptides were adsorbed on the surface and were not washed off. This result redirected our experiments towards finding and refining the site of initial aggregation of Aβ 1-40 peptides.

Since the study of amyloid aggregation is carried out not only in glass containers but also in plastic test tubes and in quartz cuvettes, we investigated the influence of these surfaces as well. We obtained similar results. The appearance of FRET is also observed on these surfaces. This signal also remains after flushing the surface.

A reasonable question arises: are the attached dyes the driving force behind the sorption of peptides on the surface? We carried out an additional experiment, which showed that dyes Cy3 and Cy5 by themselves in concentrations comparable to those used in our experiments are not adsorbed in significant amounts on the glass surface. This indicates that they are not the driving force behind peptide–surface interactions. In what follows, we proceeded from the assumption that the observed effects are a consequence of the properties of the peptides.

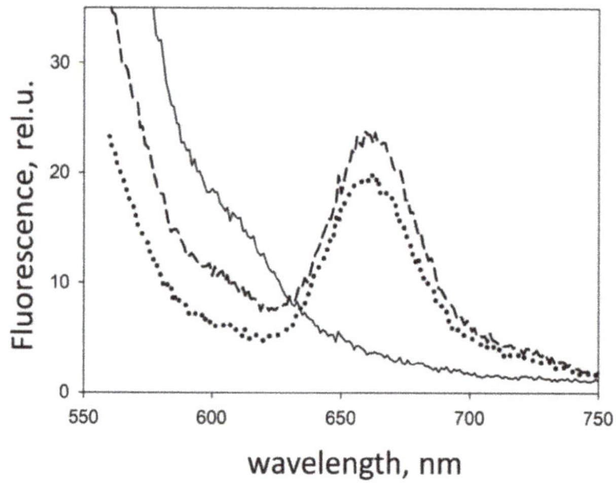

Figure 3. Fluorescence of a droplet of a peptide solution on the glass surface with excitation at 513 nm. A peptide labeled only with Cy3 is a solid line. A mixture of peptides marked Cy3 and Cy5 is a dashed line. The fluorescence of a glass surface gently washed after measuring the mix of peptides is a dotted line.

2.3. *Visualization of the Aggregation Site Using Confocal Fluorescence Microscopy*

To clarify and visualize the localization of interacting peptides, we used a confocal fluorescence microscope. Solution of mixed peptides was applied onto the microscope slides and covered with a coverslip slide. The entire space was examined layer by layer, from the slide to the coverslip, capturing their surfaces. In the channel of general fluorescence (excitation and registration of fluorescence of both dyes), we found that most fluorescent particles are concentrated on glass surfaces (see Figure 4). Moreover, a significant transfer of fluorescent energy (excitation of Cy3 fluorescence and registration of Cy5 fluorescence) is observed precisely on the glass surface. This fact indicates the interface as a possible place for a significant increase in the concentration of peptides and, consequently, acceleration of their interaction and aggregation.

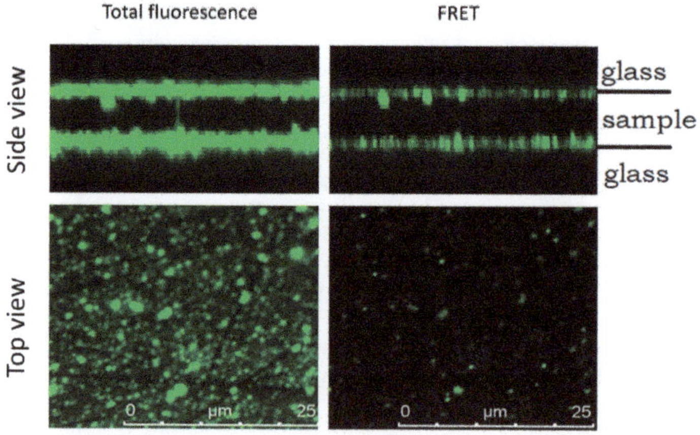

Figure 4. Fluorescent confocal microscopy images. The top view and reconstruction of the side view in the channels of general fluorescence (excitation at 532 nm, emission registration at 533–778 nm) and the FRET fluorescence (excitation at 532 nm emission registration at 650–778 nm) are shown.

2.4. Can We Avoid Surface Effects?

The next logical step in this research is to find ways to inhibit surface effects. To do this, we need to change the properties of the surface and reduce its ability to adsorb protein. Without inventing a new one, we used a well-proven way with preliminary processing of the glass surface with a BSA solution. The experiments described in Section 2.2 were repeated. The results of this experiment are shown in Figure 5. On the surface treated with BSA, the effect of fluorescent energy transfer was almost completely suppressed. That means no close interaction between peptides without accessible glass surfaces. No peptide adsorption was observed on the treated surface. The absence of adsorption indicates the absence of binding of the Aβ to BSA located on the surface. In conjunction with the fact that the BSA does not go from the surface into the solution (we checked this spectrophotometrically), we can conclude that the reason for the change in behavior is precisely in the hiding of the glass surface from peptides.

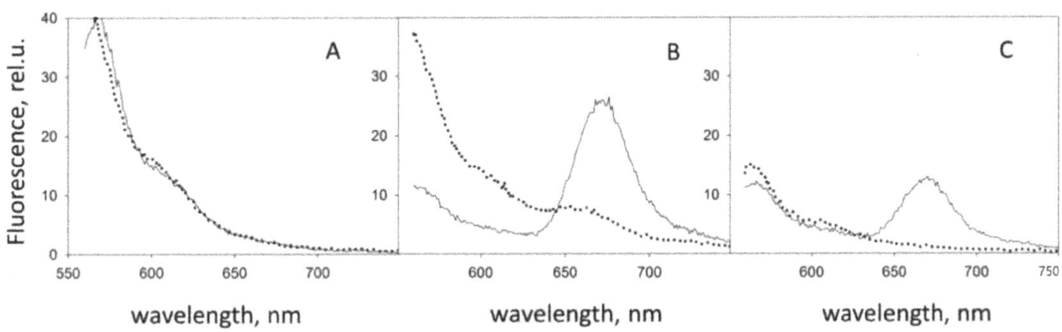

Figure 5. The results of FRET experiments on pure glass (solid lines) and glass surfaces treated by BSA (dotted lines). (**A**) A peptide labeled with Cy3 only; (**B**) a peptide labeled with Cy5 is added; (**C**) fluorescence of the glass surface gently washed with a buffer after measuring the mixture.

This experiment also answered the critical question: does the interaction occur between already sorbed proteins, or does aggregation occur first, and then adsorption? Since no energy transfer is observed when glass surfaces are inaccessible, the interaction between peptides occurs only on the accessible surface already after their adsorption and not in the bulk solution. In addition, since there is no significant increase in FRET during the experiment with a treated surface, the rate of aggregation of peptides in bulk solution is much lower. This fact indicates the surface as a factor of aggregation accelerating.

2.5. Conclusions

The analysis of the presented results concludes that it is necessary to take into account the phenomena occurring with proteins on the surface of the vessels in which the experiment is carried out, especially when it happens by concentration-dependent nucleation and elongation as in the case of aggregation and amyloid formation. Any vessel used in an experiment has walls that can affect protein adsorption and change the protein structure. The critical question: what is the influence of surface effects on the behavior of the protein in bulk? However, answering this question remains outside the scope of this article. This question is especially acute in the presence of solution movement, for example, when pouring or stirring. Undoubtedly, it should be the subject of further research.

The influence of various surfaces on amyloid aggregation has been studied for a long time. However, as the results of this work show, this factor can take place in any experiment without the addition of specific surfaces. The interface between the liquid phase and the vessel wall is sufficient. As the analysis of the literature has shown, this fact falls out of the field of view of researchers almost always when surface effects are not an object of interest. From our point of view, the described effects are pretty suitable for the role of unaccounted systemic factors affecting the reproducibility of experiments on the study of amyloid aggregation. Projecting this thought onto living organisms, we see a variety of different surfaces, each of which can influence the process of amyloid formation in its way. However, it is worth noting that there should not be so many surfaces that provoke amyloid formation; otherwise, amyloidosis would be much more frequent.

3. Materials and Methods

3.1. Gene Expression; Isolation and Purification of Aβ Peptide 1-40

The plasmid vector pET-32 LIC allows one to express the target peptide gene fused with the trx gene of thioredoxin within a hybrid protein. The hybrid protein also carries the 6×His sequence to ensure rapid and efficient purification of the product using metal chelate chromatography. The expression level of hybrid protein in this construct was appreciably high, making it possible to produce more target proteins. The expression of genes was performed in E. coli strain BL-21 DE3. Once expression had been completed, E. coli cells were precipitated by centrifugation at 6000× *g* and 4 °C for 15 min.

The resulting cell biomass was resuspended in 50 mM Tris-HCl, pH 8.0, 1 mM EDTA, 0.02 M β-mercaptoethanol; $MgCl_2$ (to a final concentration of 5 mM) and DNase I (Sigma-Aldrich, St. Louis, MO, USA) (50 µg per g of cells) were added. The cell suspension was homogenized on a French press (Spectronic Instruments, Inc., Irvine, CA, USA). The water-insoluble fraction of the homogenate was precipitated by centrifugation at 30,000 rpm for 30 min; approximately 90% of hybrid protein was present in the water-soluble fraction. The hybrid protein was isolated from the cell lysate by ion-exchange chromatography on a DEAE-cellulose column (Sigma-Aldrich, St. Louis, MO, USA) in 20 mM Tris-HCl, pH 8.0, 100 mM NaCl. The hybrid protein was eluted with 50–500 mM NaCl gradient in the same buffer.

Next, the hybrid protein was purified on a column packed with nickel-nitrilotriacetic acid (Ni-NTA)metal chelate resin (Qiagen, Hilden, Germany) equilibrated with 50 mM Tris-HCl buffer, pH 8.0, containing 300 mM NaCl. The hybrid protein was eluted with 20–250 mM imidazole gradient. The purified hybrid protein was concentrated on the DEAE-cellulose column until a concentration of at least 10 mg/mL. The hybrid protein was subjected to enzymatic cleavage using factor Xa protease (Sigma-Aldrich, St. Louis, MO, USA) in 20 mM Tris-HCl buffer, pH 8.0, containing 100 mM NaCl and 5 mM $CaCl_2$, at room temperature for 24 h at a 1:2000 molar ratio between the enzyme and the protein. The reaction mixture was applied to the column packed with Ni-NTA in the respective buffer; the target Aβ peptide 1-40 was not bound to the sorbent and was eluted from the column immediately after the void volume mark. The purity and homogeneity of the obtained peptide was assessed by electrophoresis in 15% SDS-PAAG.

3.2. Peptide Labeling

To obtain fluorescently labeled peptides, we used fluorescent dyes manufactured by Jena Bioscience Gmbh (Jena, Germany). The peptides were chemically bound to Cy3 Cy5 dyes according to the manufacturer's protocol. The pH of the solution was 7, following the manufacturer's recommendations for selective binding to the N-terminus of peptides. Separation of labeled peptides and unbound dye was performed by SEC on a Sephadex G-25 column. The labeling efficiency was assessed spectrophotometrically in accordance with the manufacturer's guidelines.

3.3. Fluorescence Spectroscopy

The fluorescence measurements were made on a Cary Eclipse fluorescence spectrophotometer (Varian, Palo Alto, CA, USA). Measurements were taken in a quartz cuvette 3 × 3 mm for a standard variant and used a plate reader to measure the fluorescence on the surface. The fluorescence spectra were recorded at 550–750 nm at an excitation wavelength of 513 nm. The final concentration of the peptide was 0.03 mg/mL.

3.4. Confocal Fluorescence Microscopy

A Leica TCS SPE confocal fluorescent microscope was used to clarify and visualize the localization of interacting peptides. The specimens (5 μL) were applied onto the microscope slides and covered with a coverslip slide. Confocal fluorescence microscopy images were recorded with excitation at 532 and emission registration at 553–778 nm for general fluorescence and at 650–778 nm for FRET. The final concentration of the peptide was 0.03 mg/mL.

Author Contributions: N.K. and V.B. planned and designed experiments; A.M. and N.I. participated equally by performing the gene construction and protein purification; V.B. performed the fluorescence studies; I.E. performed the fluorescence microscopy studies; N.K., A.M. and V.B. performed the data analysis; V.B. supervised the project. The manuscript was written in collaboration with all the authors. All authors have read and agreed to the published version of the manuscript.

Funding: The study was supported by the Russian Science Foundation (grant No. 21-14-00268).

Institutional Review Board Statement: Not applicable.

Informed Consent Statement: Not applicable.

Data Availability Statement: Not applicable.

Acknowledgments: The authors are grateful to A.V. Finkelstein for initiating this work and for fruitful discussion of the results obtained.

Conflicts of Interest: The authors declare no conflict of interest.

Sample Availability: Samples of the compounds are not available from the authors.

References

1. Lutter, L.; Serpell, C.J.; Tuite, M.F.; Xue, W.F. The molecular lifecycle of amyloid—Mechanism of assembly, mesoscopic organisation, polymorphism, suprastructures, and biological consequences. *BBA Proteins Proteom.* **2019**, *1867*, 140257. [CrossRef]
2. Ghosh, P.; Vaidya, A.; Kumar, A.; Rangachari, V. Determination of critical nucleation number for a single nucleation amyloid-β aggregation model. *Math. Biosci.* **2016**, *273*, 70–79. [CrossRef]
3. So, M.; Hall, D.; Goto, Y. Revisiting supersaturation as a factor determining amyloid fibrillation. *Curr. Opin. Struct. Biol.* **2016**, *36*, 32–39. [CrossRef] [PubMed]
4. Dessau, M.A.; Modis, Y. Protein crystallization for X-ray crystallography. *J. Vis. Exp.* **2011**, *47*, 2285. [CrossRef] [PubMed]
5. Du, Q.; Dai, B.; Hou, J.; Hu, J.; Zhang, F.; Zhang, Y. A comparative study on the self-assembly of an amyloid-like peptide at water-solid interfaces and in bulk solutions. *Microsc. Res. Tech.* **2015**, *78*, 375–381. [CrossRef] [PubMed]
6. Babinchak, W.M.; Surewicz, W.K. Liquid-Liquid Phase Separation and Its Mechanistic Role in Pathological Protein Aggregation. *J. Mol. Biol.* **2020**, *432*, 1910–1925. [CrossRef] [PubMed]
7. Rabe, M.; Soragni, A.; Reynolds, N.P.; Verdes, D.; Liverani, E.; Riek, R.; Seeger, S. On-surface aggregation of α-synuclein at nanomolar concentrations results in two distinct growth mechanisms. *ACS Chem. Neurosci.* **2013**, *20*, 408–417. [CrossRef]
8. Jiang, D.; Dinh, K.L.; Ruthenburg, T.C.; Zhang, Y.; Su, L.; Land, D.P.; Zhou, F. A kinetic model for beta-amyloid adsorption at the air/solution interface and its implication to the beta-amyloid aggregation process. *J. Phys. Chem. B* **2009**, *113*, 3160–3168. [CrossRef] [PubMed]
9. Burra, G.; Maina, M.B.; Serpell, L.C.; Thakur, A.K. Nucleation-dependent Aggregation Kinetics of Yeast Sup35 Fragment GNNQQNY. *J. Mol. Biol.* **2021**, *5*, 166732. [CrossRef]
10. Grigolato, F.; Arosio, P. The role of surfaces on amyloid formation. *Biophys. Chem.* **2021**, *270*, 106533. [CrossRef]
11. Lin, Y.C.; Petersson, E.J.; Fakhraai, Z. Surface effects mediate self-assembly of amyloid-β peptides. *ACS Nano* **2014**, *8*, 10178–10186. [CrossRef]
12. Wang, W.; Roberts, C.J. Protein aggregation—Mechanisms, detection, and control. *Int. J. Pharm.* **2018**, *25*, 251–268. [CrossRef] [PubMed]
13. Gallardo, R.; Ranson, N.A.; Radford, S.E. Amyloid structures: Much more than just a cross-β fold. *Curr. Opin. Struct. Biol.* **2020**, *60*, 7–16. [CrossRef]

molecules MDPI

Article

Advances in Mixer Design and Detection Methods for Kinetics Studies of Macromolecular Folding and Binding on the Microsecond Time Scale

Takuya Mizukami and Heinrich Roder *

Molecular Therapeutics Program, Fox Chase Cancer Center, Philadelphia, PA 19111, USA; Takuya.Mizukami@fccc.edu
* Correspondence: Heinrich.Roder@fccc.edu

Abstract: Many important biological processes such as protein folding and ligand binding are too fast to be fully resolved using conventional stopped-flow techniques. Although advances in mixer design and detection methods have provided access to the microsecond time regime, there is room for improvement in terms of temporal resolution and sensitivity. To address this need, we developed a continuous-flow mixing instrument with a dead time of 12 to 27 µs (depending on solution viscosity) and enhanced sensitivity, sufficient for monitoring tryptophan or tyrosine fluorescence changes at fluorophore concentrations as low as 1 µM. Relying on commercially available laser microfabrication services, we obtained an integrated mixer/flow-cell assembly on a quartz chip, based on a cross-channel configuration with channel dimensions and geometry designed to minimize backpressure. By gradually increasing the width of the observation channel downstream from the mixing region, we are able to monitor a reaction progress time window ranging from ~10 µs out to ~3 ms. By combining a solid-state UV laser with a Galvano-mirror scanning strategy, we achieved highly efficient and uniform fluorescence excitation along the flow channel. Examples of applications, including refolding of acid-denatured cytochrome c triggered by a pH jump and binding of a peptide ligand to a PDZ domain, demonstrate the capability of the technique to resolve fluorescence changes down to the 10 µs time regime on modest amounts of reagents.

Keywords: turbulent mixing; continuous flow; fluorescence; reaction mechanism; protein folding; protein–ligand interactions

Citation: Mizukami, T.; Roder, H. Advances in Mixer Design and Detection Methods for Kinetics Studies of Macromolecular Folding and Binding on the Microsecond Time Scale. *Molecules* **2022**, *27*, 3392. https://doi.org/10.3390/molecules27113392

Academic Editors: Kunihiro Kuwajima, Yuko Okamoto, Tuomas Knowles and Michele Vendruscolo

Received: 29 April 2022
Accepted: 21 May 2022
Published: 25 May 2022

1. Introduction

Fast time-resolved measurements are essential for gaining mechanistic insight into biological processes such as enzyme catalysis, protein and RNA folding and protein–ligand interactions, which often occur on a time scale extending into the microsecond range. Various techniques have been used to trigger reactions, including temperature and pressure jumps [1–3], flash photolysis [4], electron transfer [5], hydrodynamic focusing under laminar flow conditions [6,7] and turbulent mixing [8–11]. Flash photolysis and temperature jump techniques have been used to trigger protein folding reactions within a few microseconds or less, but they have been limited in scope because of the requirement of a photochemical trigger and cold-denatured initial state, respectively. Elegant hydrodynamic focusing experiments have been developed that can achieve about 90% efficiency for mixing a macromolecule and a small molecule (e.g., denaturant or acid) in less than 10 µs [7,12], but they have not been applied widely due to the high protein concentrations and challenging laser optics needed for monitoring reaction progress in a thin (<1 µm) layer of protein solution. In contrast, rapid mixing of two (or more) reactants by turbulent mixing has proven to be a more generally applicable and versatile approach for initiating biomolecular reactions [11,13–15].

The time resolution of a rapid mixing instrument is limited by the instrumental dead time, i.e., the time delay between the onset of mixing and the first reliable instrumental reading of reaction progress, which in practice is determined empirically by extrapolation of a first-order reaction back in time [16,17]. Factors contributing to the dead time include mixing efficiency, which determines the time required to achieve (nearly) complete mixing of the reactants, flow rate and dead volume between the mixing region and first point of observation. The venerable stopped-flow technique [18–20], coupled with a wide range of detection methods, including absorbance, fluorescence emission, fluorescence lifetime, circular dichroism, light scattering and small-angle X-ray scattering, has long been the main source of kinetic information and played a central role in elucidating kinetic mechanisms of chemical reactions, protein folding, protein–protein interaction and enzyme catalysis. Major strengths of the stopped-flow technique include its versatility, relative sample economy and wide time window, typically ranging from a few milliseconds to minutes (limited by diffusion across the mixer and convection artifacts). However, the time delay and potential artifacts caused by the need to abruptly arrest the flow have made it difficult to achieve dead times shorter than about 1 ms. These complications can be avoided by continuously pumping the reactants through the mixer while monitoring reaction progress along a flow channel downstream from the mixing region [21–23]. More recent advances in mixer technology and detection methods [8–11,24–30] have resulted in major improvements in time resolution and sensitivity of continuous-flow instrumentation. By coupling continuous-flow mixing with a variety of detection methods, including tryptophan or tyrosine fluorescence, fluorescence resonance energy transfer (FRET), fluorescence life time, absorbance, circular dichroism, small-angle X-ray scattering and single-molecule spectroscopy, turbulent mixing devices have yielded a wealth of dynamic and structural information on early stages of protein folding on the microsecond-to-millisecond time scale [14,15,31–52].

2. Methods and Results

2.1. Design Criteria for Turbulent Mixers

At the molecular level, complete mixing of fluids is ultimately limited by diffusion. For example, the diffusion constant of urea used as a denaturant in many protein folding studies is 1.382×10^{-9} m^2/s in water, indicating that the variance in molecular position after 10 μs of diffusion is 1.382×10^{-2} μm^2 = (0.12 μm)^2. Thus, mixing of low molecular weight solutes within 10 μs can be achieved only if the fluid components can be interspersed to within about 0.1 μm, and rapid mixing on this time scale requires submicron scale flow profiles. Turbulent flow mixers achieve nearly complete mixing of reactants by relying on the small fluid eddies generated under highly turbulent flow conditions [23,53]. The size of eddies is inversely related to the Reynold's number and is thus a function of flow rate, viscosity and channel dimensions. Efficient mixing therefore requires high flow velocities, typically ~10 m/s, for micron scale mixers, and large sample volumes (1–10 mL).

Building on earlier design principles [23], Shastry et al. [8] developed a capillary mixer consisting of two concentric quartz tubes with a ~100-μm diameter platinum sphere placed near the exit of the inner capillary (Figure 1A, Mixer 1). To ensure precise positioning of the sphere, three quartz posts with a 10 μm diameter were fused to the inner surface of the outer capillary. The highly turbulent and uniform flow field in the wake of the sphere ensures rapid and complete mixing of the two solutions injected into the inner capillary and the space between the inner and outer capillaries, respectively. By forcing the solutions through a narrow (~10 μm) gap around the perimeter of the sphere, the size of eddies generated downstream is expected to be in the micron-to-submicron size range, resulting in diffusion-limited mixing times in the 10 to 100 μs range. Consistent with this estimate, the dead times we measured by using the quenching of tryptophan by N-bromosuccinimide (NBS) as a test reaction [17] were of the order of 50 μs [8,33], as illustrated in Figure 1A. Compared to earlier designs, which monitored reaction progress point by point on a free-flowing jet [9,23], we were able to greatly improve the reproducibility and sensitivity of continuous-flow measurements by mounting our capillary mixer on top of a quartz cuvette with a 0.25 × 0.25 mm^2 flow channel and using a CCD-based fluorescence imaging setup to generate a continuous profile vs. distance downstream from the mixer. However, construction of these delicate mixers is challenging, and their performance characteristics, including dead time, backpressure and optical properties, can be quite variable.

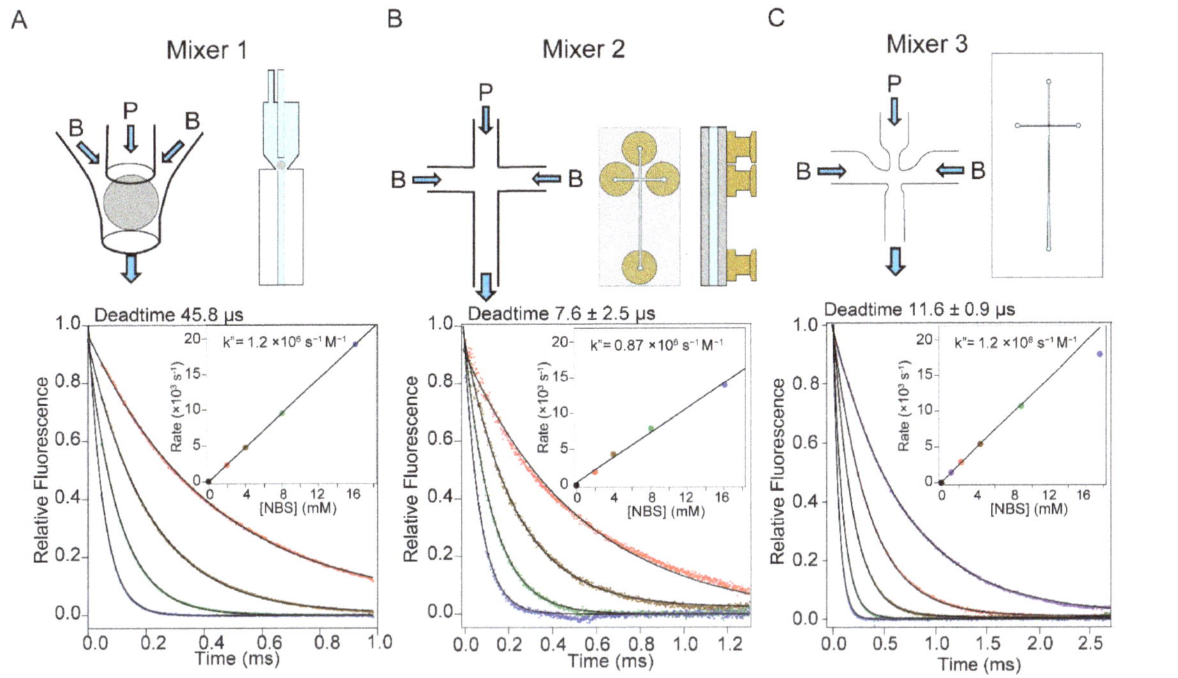

Figure 1. Summary of the mixer design and dead-time calibration of three continuous-flow mixing devices. Upper panels show schematics of the mixing regions and mixer/flow-cell assemblies for: (**A**) Mixer 1, the capillary mixer of Shastry et al. [8], (**B**) Mixer 2, a simple quartz cross-channel mixing chip (Translume, Inc., Ann Arbor, MI, USA) and (**C**) Mixer 3, a customized cross-channel mixing chip designed for reduced backpressure and expanded observation window (see Figure A1 in Appendix A for expanded diagram). The lower panels show representative examples of dead-time calibration using the pseudo-first-order NAT-NBS reaction. Typically, the fluorescent reagent (e.g., protein or NAT) is injected into central channel (*P*), and the second component (e.g., buffer, denaturant, ligand) is injected into the side channels *B*. The total flow rates in the experiments shown in A, B and C were 1 mL/s, 0.4 mL/s and 0.7 mL/s, respectively. The fluorescence decay curves are color-coded according to NBS concentration: purple for 1 mM, red for 2 mM, brown for 4 mM, green for 8 mM and blue for 16 m. Insets: determination of the second-order rate constant of the NAT-NBS reaction, based on the linear NBS concentration dependence of observed rate constants (determined by single-exponential non-linear least-squares fitting; solid lines).

2.2. Design of a Microfabricated Mixer

As a more robust approach to producing mixers, we relied on commercially available laser edging techniques (Translume, Inc., Ann Arbor, MI, USA) for microfabrication of channels with submicron tolerance in fused silica (quartz) [54]. Two related mixing chips based on a cross-channel design are illustrated in panels B and C of Figure 1. The mixers have three inlet ports and one outlet channel. Reactant P (e.g., unfolded protein) is injected into the central port, and reactant B (e.g., refolding buffer) is injected into each of the two side ports. The inlet ports are connected to the mixing chamber via a straight channel for the simple cross-mixer (Mixer 2) while a bottleneck-shaped channel for Mixer 3 is designed to increase the linear velocity of incoming flow to the chamber while keeping the backpressure low. The downstream observation channel of Mixer 3 has a conical shape in order to enhance the time resolution at short times after mixing and to extend the observable time window out to several milliseconds. At the same time, the increase in average channel width substantially lowers the backpressure for Mixer 3 (10–20 bar (1–2 Mpa) at flow velocities of 0.7 to 1.0 mL/s) compared to Mixer 2 (30 bar at 0.4 mL/s). In Figure 1 (lower panels), we compare the performance of these micro-fabricated mixers with that of our original capillary mixer by using the quenching of N-acetyltryptophan (NAT) fluorescence by NBS as a test reaction [17]. At a flow velocity of 1 mL/s, the observable time window of the capillary mixer ranges from the dead time (46 μs under these conditions) out to $t_{max} \approx 1$ ms. The dead time of the simple cross-mixer (Mixer 2) was 8 ± 2.5 μs and $t_{max} \approx 1$ ms at a flow velocity of 0.4 mL/s. Mixer 3 had a similar dead time (11.6 ± 0.9 μs) but an extended time window ($t_{max} = 3$ ms at 0.7 mL/s). A major advantage of the microfabricated mixers is reproducibility. Because of the submicron tolerance of the laser manufacturing process, the two lots of cross-mixers (Mixer 3) we compared showed virtually identical performance characteristics, confirming that the performance of microfabricated mixers is far more consistent compared to our previous hand-made capillary mixers.

2.3. *Solution Delivery and Fluorescence Detection*

To ensure constant, well-defined flow velocities for two reagents at variable mixing ratios, we replaced the pneumatic syringe drive used in our initial setup with a set of motor-driven syringe pumps (CETONI GmbH, Korbussen, Germany). The time axis of continuous-flow data can be accurately calculated from the known flow velocity and dimensions of the mixing channel, as detailed below.

To increase the sensitivity of fluorescence detection, we replaced the arc lamp and monochromator used in our previous setup [8] with a 30 W Q-switched diode-pumped solid-state laser (5 ns pulse duration, 5 μJ of single pulse energy) operating at 266 nm (Shanghai Dream Lasers Technology, Ltd., Shanghai, China). This intense and stable (<5% of power stability over 2 h) UV laser is well suited for excitation of tryptophan and tyrosine fluorescence in proteins at low concentrations. In earlier continuous-flow fluoresce experiments using powerful (350–500 W) Hg or Hg-Xe arc lamps for excitation, including our initial folding experiments on cytochrome c (cyt c) [8,32], we typically worked at protein concentrations of 20–40 μM (after mixing). Our current laser-based setup now yields data of comparable quality at protein concentrations as low as 1 μM, as documented below.

In previous laser-based continuous-flow instruments, the laser was focused on the flow channel and reaction progress was sampled point by point downstream from the mixing region [9,10,24,25]. Generating a complete kinetic trace using this approach requires dozens of separate measurements while maintaining continuous-flow conditions for extended periods of time, consuming large amounts of reagents. In contrast, we used a pair of Galvano mirrors and a cylindrical mirror to focus the laser onto the flow channel and control its position (Figure 2). The mirrors are regulated to scan the length of the flow channel using a triangular wave function, resulting in uniform distribution of light intensity vs. distance over a 23 mm segment of the channel. As in our original setup (Figure 1A), a quartz lens (f = 3.5 mm) and 305 nm high-pass interference filter (Semrock, Rochester, NY, USA) are used to project a magnified image of the flow channel onto a CCD detector (Figure 2). Thus, in a single continuous-flow experiment typically lasting 10 s (including a ~3 s acceleration time prior to activating the CCD camera), we can record a complete trace of fluorescence emission vs. time after mixing over the time range from t_d (~10 μs) to t_{max} (~2.5 ms) with an effective time resolution of ~1 μs in the narrow portion of the observation channel near the mixing region and ~4 μs near the end. We found empirically that a scanning frequencies of 20 Hz is optimal for achieving stable and uniform fluorescence excitation. This scan rate is negligibly slow compared to the pulse repetition rate of the laser but fast compared to the exposure times of ~10 s typically used in continuous-flow experiments. Thus, this novel excitation scheme is equivalent to continuous-wave illumination using a conventional light source but achieves more uniform and intense fluorescence excitation.

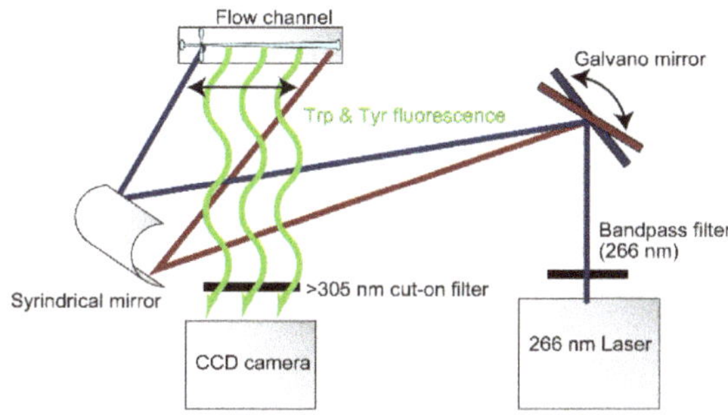

Figure 2. Optical configuration of our new continuous-flow instrument, using a UV laser in conjunction with a pair of Galvano mirrors to achieve uniform fluorescence excitation along the flow channel of the microfluidic mixing chip (Figure 1B,C).

2.4. Data Collection and Analysis

The procedures for data collection and analysis using Mixer 3 are similar to those described by Shastry et al. [8], except for the calculation of the time axis. Because of the conical shape of the flow channel of Mixer 3, the reaction time at constant flow rate is no longer a linear function of the distance from the mixing region. At the flow rates typically used (0.5–1 mL/s), we estimate Reynolds numbers Re of 2300–4500, corresponding to moderately to strongly turbulent flow conditions. In this case, wall effects can be neglected, and the flow velocity profile, $v(\mathbf{r})$, can be approximately described as a 2-dimensional radial function of the position, $\mathbf{r}(r, \theta)$, within the widening flow channel of constant depth (Figure 3):

$$v(\mathbf{r}) = \frac{V}{2\theta_0 d r^2}\mathbf{r}, \tag{1}$$

where V, θ_0, d, r and θ are the volume velocity, opening angle of channel, channel depth and the magnitude and angle of the pointing vector $\mathbf{r}$, respectively (Figure 3A). Using Cartesian coordinate system with origin placed at the exit of the mixing region, the reaction time t_r at each point is represented as follows:

$$t_r = \frac{\theta_0 d}{V}((x-x_0)^2 + y^2) = \frac{\theta_0 d}{V}(1 + tan^2(\theta))(x-x_0)^2, \tag{2}$$

where x_0 is the position of the virtual vertex of the cone and $tan(\theta) = y/(x-x_0)$; thus, $-\theta_0 < \theta < \theta_0$. Given the shallow open angle of the channel, $tan^2(\theta)$ is neglected (~7×10^{-5}), and thus t_r is effectively a function of x only. As a result, the position dependence of t_r simplifies to:

$$t_r(x) \approx \frac{d}{V}\left(\frac{w_L - w_0}{2L}x^2 + w_0 x\right) + \Delta t, \tag{3}$$

where w_L and w_0 are the channel width at $x = L$ and 0, respectively, and Δt is the mixing time. The first term is consistent with the time required to fill the volume between [0, x] at constant flow rate V.

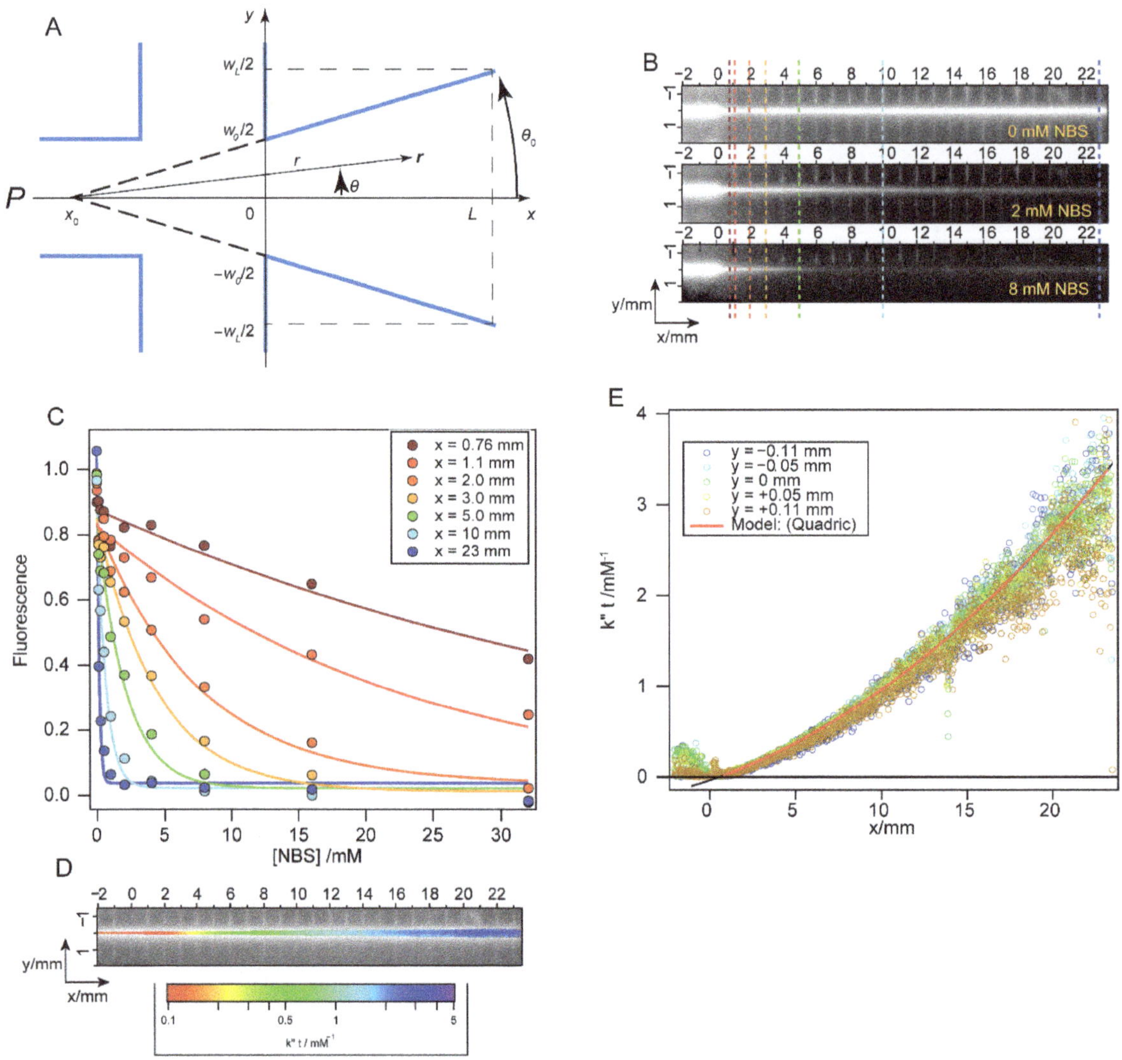

Figure 3. Calibration of the time axis of Mixer 3. (**A**) Schematic of the mixing region and conical flow channel (not to scale). The central inlet port (P in Figure 1) is on the left. The right exit is the observation channel. The x-axis is set along the observation channel with the origin on the entrance of it. x_0 is the virtual vertex of the conical observation channel. The exit of the flow channel is located at $x = L$. The y-axis is set along the cross-section of the channel. The channel widths at the entrance ($x = 0$) and the exit ($x = L$) of the flow channel are w_0 and w_L, respectively. (**B**) Representative profiles of the NAT-NBS quenching reaction. The positions at x = 0.76 mm, 1.1 mm, 2.0 mm, 3.0 mm, 5.0 mm, 10 mm and 23 mm are shown by the lines in brown, red, orange-yellow, green-cyan and blue, respectively. (**C**) The concentration dependence of fluorescence intensity at representative positions. The data are fitted to Equation (4). (**D**) The profile of $k''t$ value obtained by curve fitting in panel C. (**E**) Dependence of distance x on $k''t$.

To confirm the validity of Equation (3) for describing the nonlinear time axis of Mixer 3, we measured the quenching reaction of NAT fluorescence in the presence of excess NBS (Figure 3B). The kinetics of this pseudo first-order reaction is given by:

$$f(t, [NBS]) \approx f_0 exp(-k''[NBS]t), \tag{4}$$

where f_0 is the initial fluorescence intensity, and k'' is the second-order rate constants. Normally, the time course of reaction is analyzed by fitting to a single-exponential function at a fixed NBS concentration to obtain the apparent rate constant, $k''[NBS]$. In contrast, we evaluate Equation (4) at a fixed time (or a fixed pixel position) and variable NBS concentration (125 μM to 32 mM) in order to estimate $k''t$ (Figure 3C,D). The $k''t$ values obtained were a quadratic function of x, as predicted by Equation (3) (Figure 3E), thus confirming the validity of the underlying approximation. On the basis of the scatter of the points in panel E, we estimate that the error of the time-axis calibration is 0.3% at the earliest time points and 4% at long times.

2.5. Optimization of Mixing Conditions

Key considerations in optimizing mixing efficiency and performance are mixing ratio and flow rate. Higher mixing ratios are desirable for achieving large changes in solvent conditions but require more concentrated stock solutions, which may be limited by solubility. First, we used the NAT-NBS reaction to estimate the dead time of Mixer 3 at constant flow rate (0.7 mL/s) and variable mixing ratios (Figure 4A). The shortest dead time, 12 μs, was obtained when mixing 1 part of solution P (center port) with 10 parts of solution B (side ports; see Figure 1). Thus, Mixer 3 shows the best performance when reactions are initiated by a large concentration jump, which was one of our design criteria. Interestingly, the second-best dead time, 17 μs, was obtained for 1:1 mixing. With respect to flow rate optimization, higher flow velocities are expected to produce more turbulent flow in the mixing region, resulting in improved mixing efficiency and shorter dead times, although this comes at the cost of higher consumption of reagents and a shortened observation window. However, further increases in flow rate not only lead to a sharp increase in backpressure but may also give rise to light scattering artifacts due to cavitation, i.e., bubble formation due to the large pressure drop below the mixing region. By measuring the NAT-NBS reaction at different flow rates (1:10 mixing ratio), we found that the dead time of Mixer 3 decreases sharply with increasing flow rate, reaching values of about 20 μs at 0.7 mL/s to 1.0 mL/s. Under these conditions, we expect turbulent flow, based on estimated Reynolds numbers (3200 to 4500). At flow rates of 0.7 mL/s and above, the apparent rate of the NAT-NBS reaction increases linearly with NBS concentration up to 32 mM, as expected for a second-order reaction (Figure 4C). However, at flow rates less than 0.7 mL/s, the apparent rates at high NBS concentrations fall below those expected for a second-order reaction, indicating that the observed kinetics is increasingly limited by the rate of mixing under these sub-optimal flow conditions (Reynolds numbers less than 2400). In summary, optimal performance of Mixer 3, including dead times of 20 μs or less and a time window extending out to 2 to 3 ms, is achieved by using a mixing ratio of 1:10 and flow rates of 0.7 to 1.0 mL/s.

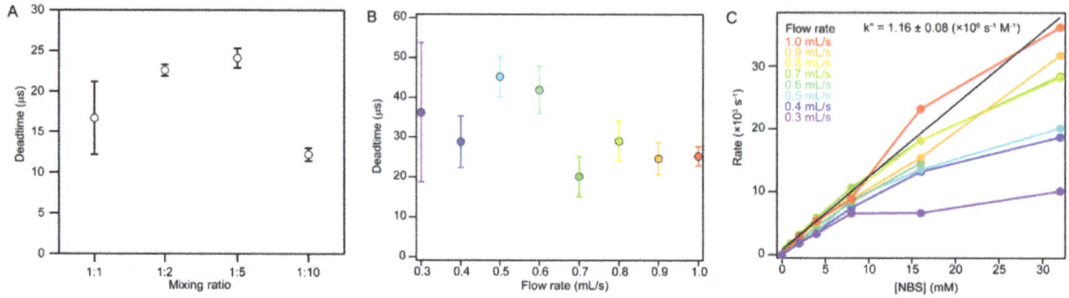

Figure 4. Optimization of mixing ratio and flow rate for Mixer 3. (**A**) Mixing ratio dependence and (**B**) flow rate dependence of dead time. (**C**) The NBS concentration dependence of rate constant obtained at several flow rates. The color code is shown in panel C.

2.6. Mixing Efficiency

To quantify mixing efficiency, we relied on the quenching of NAT fluorescence by potassium iodide (Figure 5A). This collisional quenching process is fast compared to the time scale of mixing, and quenching efficiency is concentration-dependent (90% at 1 mM potassium iodide). Thus, the normalized decrease in NAT fluorescence upon mixing with iodide provides a direct measure of mixing efficiency. At a flow rate of 0.8 mL/s, the fluorescence intensity dropped to 14% of the original value at the first observable data point after the mixing region, indicating that the mixing efficacy is ~96%. At lower flow rates (0.6 and 0.4 mL/s), the residual fluorescence was much higher and decayed more slowly, resulting in substantially longer dead times (≥100 µs). At a flow rate of 1.0 mL/s, the initial fluorescence was slightly higher compared to 0.8 mL/s, most likely due to the onset of cavitation and concomitant light scattering artifacts. We also performed mixing efficiency tests in the presence of 8 M urea, which is commonly used in protein unfolding and refolding experiments (Figure 5B). The iodide quenching profiles obtained were similar to those in the absence of urea, indicating that high concentrations of urea do not impair mixing efficiency. This was confirmed by measuring the dead time, using the NAT-NBS reaction in the presence of 16.2% sucrose with viscosity matching that of an 8 M urea solution (note that NBS reacts with urea). To mimic conditions of a typical protein unfolding experiment, an aqueous solution of NAT was mixed with 10-fold excess of NBS dissolved in 17.9% sucrose, yielding a final concentration of 16.2%. As expected, the increase in viscosity results in a somewhat longer dead time (27 µs), but mixing efficiency remains high and mixing artifacts negligible even when mixing solutions of widely differing density and viscosity (Figure 6).

2.7. Accuracy

To assess the accuracy of the kinetics data obtained by Mixer 3, we measured the second-order rate, k'', of the NAT-NBS reaction by measuring the observed rate constant vs. NBS concentration under pseudo-first-order conditions (Figure 1C). The observed k'' of 1.2×10^6 M^{-1} s^{-1} is in excellent agreement with that measured previously using a capillary mixer (Figure 1A). Taken together, these test results confirm that our microfluidic mixer design allows accurate and reliable measurements of rate constants as fast as ~10^5 s^{-1}.

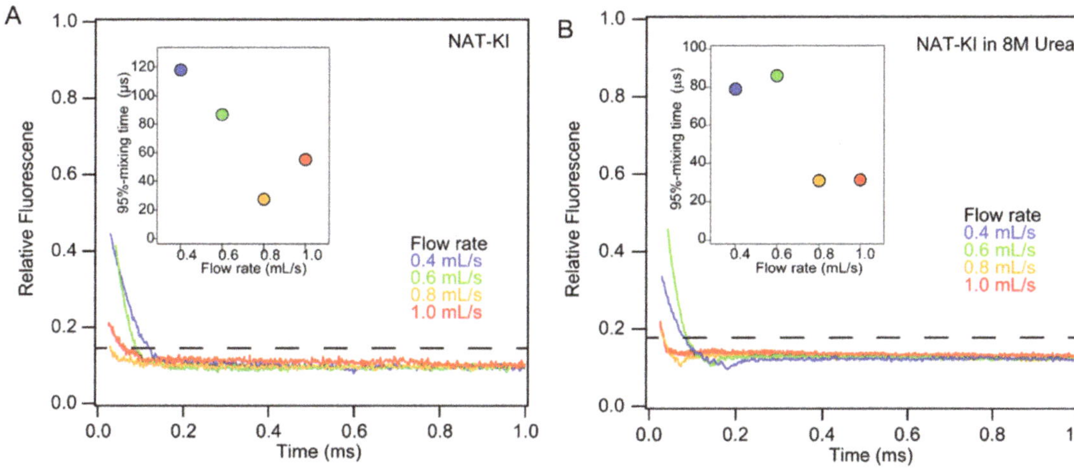

Figure 5. Mixing efficiency of Mixer 3 vs. flow rate assayed using potassium iodine quenching of tryptophan fluorescence. Residual fluorescence of NAT (10 mM) upon mixing with 1 mM potassium iodide in the absence (**A**) and presence (**B**) of 8 M urea. Color code: 0.4 mL/s (blue), 0.6 mL/s (green), 0.8 mL/s (orange) and 1.0 mL (red). The 95% mixing level is shown by the broken lines. The inlet panels show the time to achieve 95% mixing.

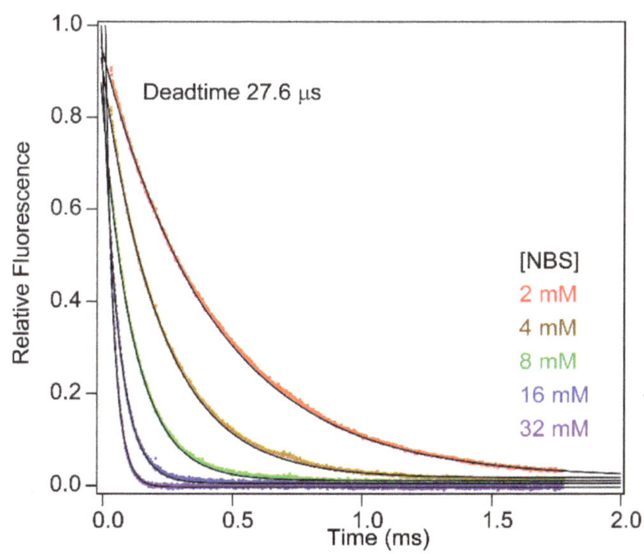

Figure 6. Dead-time calibration performed in the presence of sucrose. NBS is dissolved in 17.9% sucrose, which matches the viscosity of 8.8 M urea (η/η_0 = 1.777). The viscosity of sucrose after 1:10 mixing (16.2%, η/η_0 = 1.663) matches that of 8 M urea.

3. Applications

3.1. Folding of Cytochrome c

As a first application of our improved continuous-flow setup, we used Mixer 3 to monitor early stages of folding of cytochrome c (cyt c), which has long served as a test case for the development of new methodologies [4,5,31,32,55,56]. Horse cyt c is a 104-residue protein with a covalently attached heme group. The fluorescence of its sole tryptophan residue, Trp59, is governed by Förster energy transfer interaction with the heme, resulting in complete quenching in the folded state [57]. Prior studies using commercial stopped-flow instruments with a dead time of 1 ms or longer showed that the fluorescence change upon refolding accounts for at most 15% of the total fluorescence change, indicating that major conformational changes occur on the sub-millisecond time scale [58]. The development of a capillary mixing apparatus with a dead time of ~40 μs [8] made it possible to resolve the entire fluorescence change associated with the refolding of acid-denatured cyt c triggered by a jump in pH from 2.0 to 4.5, including a major phase with time constant τ_1 = 41 μs accounting for 59% of the amplitude and a second phase with time constant τ_2 = 648 μs accounting for 41% of the amplitude (green trace in Figure 7A; c.f. [32]). When we repeated this experiment using Mixer 3 (red trace), we also observed a biphasic decay in fluorescence with identical (within error) time constants (τ_1 = 45 μs, τ_2 = 605 μs) and similar amplitudes (A_1 = 0.54 and A_2 = 0.46). Minor differences in the amplitudes of the kinetic traces obtained with our new setup (Figure 1C) compared to the previous configuration (Figure 1A) can be explained by the fact that the ratio of tryptophan vs. tyrosine fluorescence excitation is lower at the wavelength of the UV laser (266 nm) compared to that used previously (280 nm). In both cases, a double-exponential fit accurately describes the observed fluorescence decay, and the extrapolated value at t = 0 is within ±5% the fluorescence of the acid-unfolded initial state, indicating that we are able to resolve the complete time course of folding associated with changes in intrinsic fluorescence. To document reproducibility, we compared the kinetic traces obtained in four independent experiments. The mean values and standard deviations for the corresponding kinetic parameters were as follows: $<\tau_1>$ = 51 ± 14 μs; $<A_1>$ = 0.55 ± 0.09; $<\tau_2>$ = 602 ± 51 μs; $<A_2>$ = 0.39 ± 0.05.

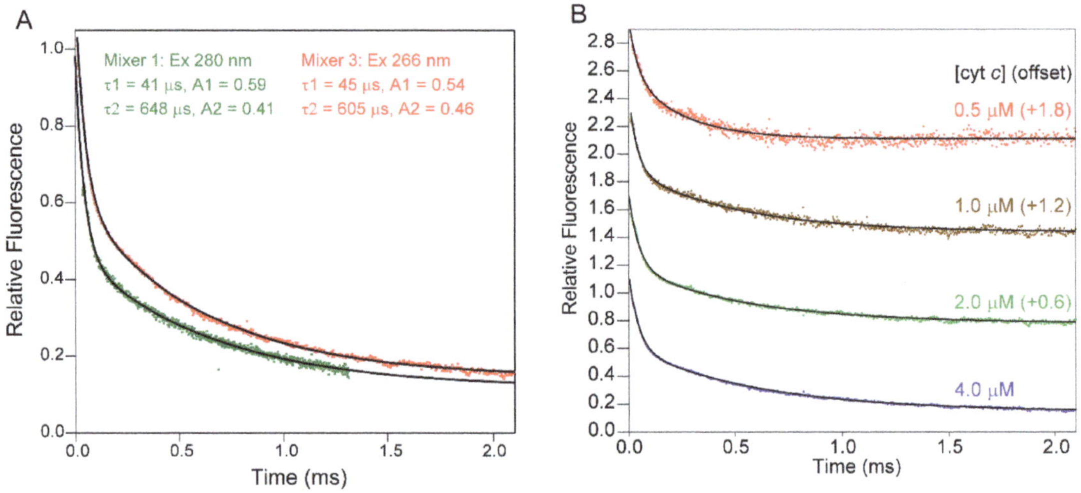

Figure 7. (**A**) Kinetics of refolding of acid-denatured cyt c (pH 2, salt-free) triggered by a jump to pH 4.5 (100 mM sodium acetate) recorded using Mixer 1 (green trace, 280 nm excitation) and Mixer 3 (red trace, 266 nm excitation). (**B**) Kinetics of refolding of cyt c (same conditions as in panel (**A**)) at different protein concentrations, as indicated.

We have previously attributed the fast phase during refolding of cyt c in terms of a barrier-limited collapse of the polypeptide chain, based on its large amplitude consistent with a major decrease in average distance between the covalently attached heme and Trp59, as well as the temperature dependence of its rate constant indicative of a small but significant activation energy [32,33,59]. In a more recent study, using a microfluidic mixer to perform quenched H/D exchange measurements on the sub-millisecond time scale, we found that amide groups in two α-helices in the C-terminal half of cyt c were partially protected already during the fast (40 μs) phase, indicating that the initial compaction of the polypeptide chain involved formation of short-range helix–helix contacts, and is thus inconsistent with a random polymer collapse [60]. Helical structure in the N-terminal region begins to appear only during the second (600 μs) folding phase, suggesting that long-range tertiary contacts are established during the later stages of folding. Mitic et al. recently used an absorbance-detected continuous-flow mixing setup with a 4 μs dead time to monitor the kinetics of refolding of acid-denatured cyt c [11]. In addition to two kinetic phases with time constants similar to those of our two fluorescence-detected phases (83 and 345 μs, respectively), Mitic et al. observed an additional process with a time constant of 4.7 μs they attribute to binding of the His18 side chain to the initially four-coordinated heme iron. Because the heme group is covalently linked to the protein via thioester linkages to Cys 14 and Cys 17, this is a local conformational event not expected to be associated with a change in Trp59 fluorescence.

In addition to documenting the fact that the kinetic data obtained using Mixer 3 are fully consistent with those obtained previously using Mixer 1, Figure 7A also indicates a major improvement in the signal-to-noise ratio (S/N) and data quality. Our new excitation scheme using a 266 nm UV laser in combination with a set of Galvano mirrors (Figure 2) resulted in a high-quality kinetic trace with ~2-fold higher S/N on a 4 μM cyt c solution compared to that measured previously at a fivefold higher protein concentration using a Hg-Xe arc lamp and monochromator tuned to 280 nm. By repeating these experiments at several lower cyt c concentrations, we obtained reproducible kinetic traces with acceptable S/N down to protein concentrations as low as 0.5 μM (Figure 7B). Even though the 266 nm laser is suboptimal in terms of excitation of tryptophan fluorescence, comparison with the data obtained previously using 280 nm excitation (green trace in Figure 7A) indicates that our redesigned optical configuration has resulted in a ~20-fold increase in S/N.

3.2. *Binding of a Peptide Ligand to a PDZ Domain*

Rapid mixing methods have also played a central role in mechanistic studies of protein–ligand interactions. If ligand binding is coupled with a change in protein conformation, one can envision two limiting mechanisms, depending on whether the second-order binding step precedes the conformational change ("induced fit") or occurs after formation of a binding-competent conformation ("conformational selection"). Both scenarios have been documented for different systems [61,62] or even for the same protein–ligand pair under different conditions [63,64]. One approach for distinguishing between these mechanisms is to measure the kinetics of binding as a function of either ligand or protein concentration [62]. The difference in predicted kinetics is especially pronounced at high ligand (or protein) concentrations where rate constants often exceed 1000 s^{-1} and thus require instrumentation for ultra-fast mixing [39]. As an illustrative example, we show in Figure 8 the kinetics of binding of a cognate peptide ligand to the first PDZ domain of NHERF1/EBP50 [65], a signaling adaptor at the membrane–cytoskeleton interface comprising a pair of PDZ domains and a C-terminal ezrin-binding motif. High-affinity binding of the ligand peptides studied here, $CFTR_6$ (Ac-VQDTRL) and $CFTR_{10}$ (Ac-TEEEVQDTRL), derived from the C-terminus of the cystic fibrosis transmembrane conductance regulator (CFTR), is primarily mediated by interactions of the C-terminal DTRL motif with a deep groove on the surface the PDZ domain [66].

To follow the time course of binding as a function of peptide concentration, we monitored the ligand-induced change in fluorescence of a tyrosine residue located in a pocket recognizing the C-terminal leucine of the CFTR peptide (Figure 8A). The rate constants observed under pseudo-first-order conditions increased sharply with increasing peptide concentration, reaching a value of 43,000 ± 3000 s^{-1} (τ = 23 μs) at 3.5 mM (Figure 8B). The results document the ability of our instrument to resolve changes in inherently weak tyrosine fluorescence on the 10 μs time scale. Initially, the observed rate constant increases linearly with peptide concentration but begins to level off somewhat at concentrations above 1 mM. This behavior is intermediate between the strongly hyperbolic concentration dependence reported for PTP-BL PDZ2 and the strictly linear dependence on ligand concentration found for PSD-95 PDZ3 [39], indicating that these structurally closely related protein–ligand pairs experience different degrees of conformational changes and exhibit a surprising range of kinetic properties.

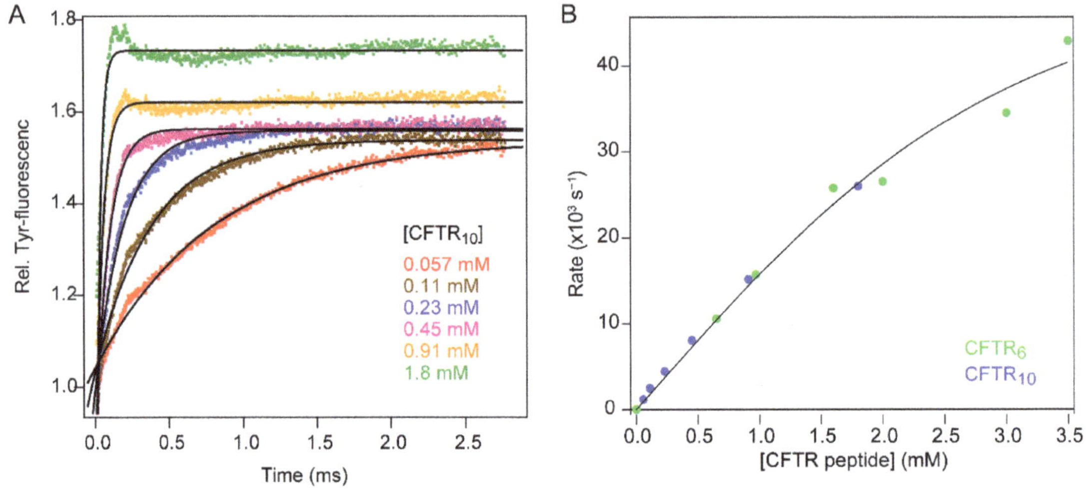

Figure 8. Binding kinetics of peptides from the C-terminus of CFTR to the first PDZ domain of NHERF1. (**A**) Kinetic traces of $CFTR_{10}$ binding to PDZ1 of NHRF1. (**B**) Rate constants of the binding reaction for $CFTR_{10}$ (blue) and $CFTR_6$ (green). The line represents a hyperbolic fit of the rate constant vs. ligand concentration.

4. Conclusions

We have described the design and testing of an instrument for continuous-flow fluorescence measurements on the microsecond-to-millisecond time scale that combines a highly efficient microfluidic mixer/flow cell assembly on a quartz chip with a novel laser-scanning scheme for enhanced fluorescence excitation. The instrument features several performance characteristics that are critical for routine application to the ultrafast kinetic analysis of biomolecular folding and binding reactions. The dead time of 12 μs we measured under optimal flow conditions at low viscosity is at or near the low end of the fluorescence-based rapid mixing devices reported in the literature. Mixing efficiency remains high even under more viscous conditions, which is essential for studies of denaturant-induced protein folding or unfolding reactions. Further reductions in mixer and channel dimensions have been reported to result in somewhat shorter dead times, but this comes at the cost of major increases in backpressure and sample concentration. In contrast, by combining a tapered observation channel with a novel Galvano-mirror laser-scanning strategy, our instrument yields high-quality kinetic traces ranging from tens of microseconds to several milliseconds on protein concentrations as low as 1 μM. Representative kinetic data on early events during folding of cytochrome c and binding of a peptide ligand to a PDZ domain illustrate the power of the method for exploring the dynamics of macromolecular folding and binding reactions.

Author Contributions: Conceptualization, T.M. and H.R.; methodology, T.M. and H.R.; software, T.M.; testing and validation, T.M.; data analysis, T.M.; resources, H.R.; writing—original draft preparation, T.M.; writing—review and editing, H.R. and T.M.; visualization, T.M.; project administration, H.R.; funding acquisition, H.R. All authors have read and agreed to the published version of the manuscript.

Funding: This research was funded by the National Institutes of Health, grant number GM116911 (to H.R.), and a Cancer Center Support Grant from the National Cancer Institute (CA06927).

Institutional Review Board Statement: Not applicable.

Informed Consent Statement: Not applicable.

Data Availability Statement: Not applicable.

Conflicts of Interest: The authors declare no conflict of interest. The funders had no role in the design of the study; in the collection, analyses or interpretation of data; in the writing of the manuscript or in the decision to publish the results.

Appendix A

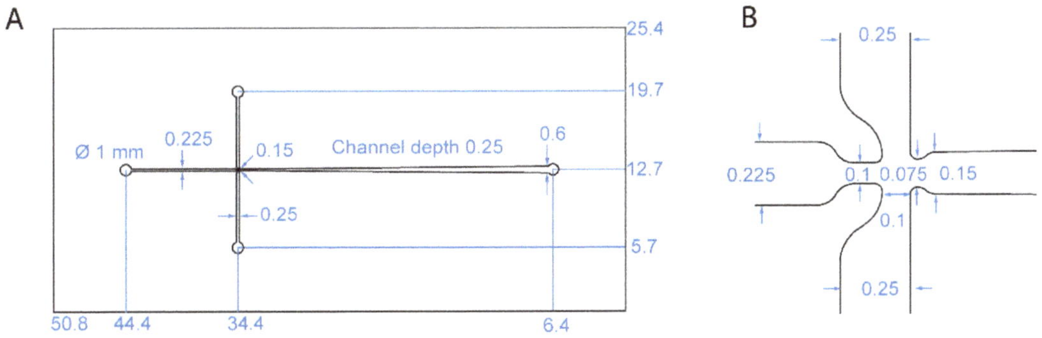

Figure A1. Detailed schematic of the layout of Mixer 3, including dimensions (in mm). (**A**) Overview of the quartz chip. Channel depth is 0.25 mm. (**B**) Expanded diagram of the mixing region at the intersection of the cross-channels.

References

1. Gruebele, M.; Sabelko, J.; Ballew, R.; Ervin, J. Laser temperature jump induced protein folding. *Acc. Chem. Res.* **1998**, *31*, 699–707. [CrossRef]
2. Dyer, R.B.; Brauns, E.B. Laser-induced temperature jump infrared measurements of RNA folding. *Methods Enzymol.* **2009**, *469*, 353–372. [PubMed]
3. Prigozhin, M.B.; Liu, Y.; Wirth, A.J.; Kapoor, S.; Winter, R.; Schulten, K.; Gruebele, M. Misplaced helix slows down ultrafast pressure-jump protein folding. *Proc. Natl. Acad. Sci. USA* **2013**, *110*, 8087–8092. [CrossRef] [PubMed]
4. Jones, C.M.; Henry, E.R.; Hu, Y.; Chan, C.-K.; Luck, S.D.; Bhuyan, A.; Roder, H.; Hofrichter, J.; Eaton, W.A. Fast events in protein folding initiated by nanosecond laser photolysis. *Proc. Natl. Acad. Sci. USA* **1993**, *90*, 11860–11864. [CrossRef]
5. Pascher, T.; Chesick, J.P.; Winkler, J.R.; Gray, H.B. Protein folding triggered by electron transfer. *Science* **1996**, *271*, 1558–1560. [CrossRef] [PubMed]
6. Knight, J.B.; Vishwanath, A.; Brody, J.P.; Austin, R.H. Hydrodynamic focusing on a silicon chip: Mixing nanoliters in microseconds. *Phys. Rev. Lett.* **1998**, *80*, 3863–3866. [CrossRef]
7. Lapidus, L.J.; Yao, S.; McGarrity, K.S.; Hertzog, D.E.; Tubman, E.; Bakajin, O. Protein hydrophobic collapse and early folding steps observed in a microfluidic mixer. *Biophys. J.* **2007**, *93*, 218–224. [CrossRef]
8. Shastry, M.C.R.; Luck, S.D.; Roder, H. A continuous-flow capillary mixer to monitor reactions on the microsecond time scale. *Biophys. J.* **1998**, *74*, 2714–2721. [CrossRef]
9. Chan, C.-K.; Hu, Y.; Takahashi, S.; Rousseau, D.L.; Eaton, W.A.; Hofrichter, J. Submillisecond protein folding kinetics studied by ultrarapid mixing. *Proc. Natl. Acad. Sci. USA* **1997**, *94*, 1779–1784. [CrossRef]
10. Bilsel, O.; Kayatekin, C.; Wallace, L.A.; Matthews, C.R. A microchannel solution mixer for studying microsecond protein folding reactions. *Rev. Sci. Instr.* **2005**, *76*, 014302. [CrossRef]
11. Mitic, S.; Strampraad, M.J.F.; Hagen, W.R.; de Vries, S. Microsecond time-scale kinetics of transient biochemical reactions. *PLoS ONE* **2017**, *12*, e0185888. [CrossRef] [PubMed]
12. DeCamp, S.J.; Naganathan, A.N.; Waldauer, S.A.; Bakajin, O.; Lapidus, L.J. Direct observation of downhill folding of lambda-repressor in a microfluidic mixer. *Biophys. J.* **2009**, *97*, 1772–1777. [CrossRef] [PubMed]
13. Roder, H.; Maki, K.; Cheng, H.; Shastry, M.C. Rapid mixing methods for exploring the kinetics of protein folding. *Methods* **2004**, *34*, 15–27. [CrossRef] [PubMed]

14. Roder, H.; Maki, K.; Cheng, H. Early events in protein folding explored by rapid mixing methods. *Chem. Rev.* **2006**, *106*, 1836–1861. [CrossRef] [PubMed]
15. Kathuria, S.V.; Guo, L.; Graceffa, R.; Barrea, R.; Nobrega, R.P.; Matthews, C.R.; Irving, T.C.; Bilsel, O. Minireview: Structural insights into early folding events using continuous-flow time-resolved small-angle X-ray scattering. *Biopolymers* **2011**, *95*, 550–558. [CrossRef]
16. Tonomura, B.; Nakatani, H.; Ohnishi, M.; Yamaguchi-Ito, J.; Hiromi, K. Test reactions for a stopped-flow apparatus. Reduction of 2,6-dichlorophenolindophenol and potassium ferricyanide by L-ascorbic acid. *Anal. Biochem.* **1978**, *84*, 370–383. [CrossRef]
17. Brissette, P.; Ballou, D.P.; Massey, V. Determination of the dead time of a stopped-flow fluorometer. *Anal. Biochem.* **1989**, *181*, 234–238. [CrossRef]
18. Chance, B. Accelerated and stopped-flow apparatus. II. *J. Franklin Inst.* **1940**, *220*, 455–613. [CrossRef]
19. Gibson, Q.H.; Milnes, L. Apparatus for rapid and sensitive spectrophotometry. *Biochem. J.* **1964**, *91*, 161–171. [CrossRef]
20. Berger, R.L.; Balko, B.; Borcherdt, W.; Friauf, W. High speed optical stopped-flow apparatus. *Rev. Sci. Instrum.* **1968**, *39*, 486–493. [CrossRef]
21. Chance, B. *Rapid Mixing and Sampling Techniques in Biochemistry*; Academic Press: New York, NY, USA, 1964.
22. Moskowitz, G.W.; Bowman, R.L. Muliticaplillary mixer of solutions. *Science* **1966**, *153*, 428–429. [CrossRef] [PubMed]
23. Regenfuss, P.; Clegg, R.M.; Fulwyler, M.J.; Barrantes, F.J.; Jovin, T.M. Mixing liquids in microseconds. *Rev. Sci. Instrum.* **1985**, *56*, 283–290. [CrossRef]
24. Paeng, K.; Paeng, I.; Kincaid, J. Time-resolved resonance raman spectroscopy using a fast mixing device. *Anal. Sci.* **1994**, *10*, 157–159. [CrossRef]
25. Takahashi, S.; Ching, Y.-c.; Wang, J.; Rousseau, D.L. Microsecond generation of oxygen-bound cytochrome c oxidase by rapid solution mixing. *J. Biol. Chem.* **1995**, *270*, 8405–8407. [CrossRef]
26. Akiyama, S.; Takahashi, S.; Ishimori, K.; Morishima, I. Stepwise formation of alpha-helices during cytochrome c folding. *Nat. Struct Biol.* **2000**, *7*, 514–520. [CrossRef]
27. Matsumoto, S.; Yane, A.; Nakashima, S.; Hashida, M.; Fujita, M.; Goto, Y.; Takahashi, S. A rapid flow mixer with 11-mus mixing time microfabricated by a pulsed-laser ablation technique: Observation of a barrier-limited collapse in cytochrome c folding. *J. Am. Chem. Soc.* **2007**, *129*, 3840–3841. [CrossRef]
28. Jiang, L.; Zeng, Y.; Sun, Q.; Sun, Y.; Guo, Z.; Qu, J.Y.; Yao, S. Microsecond protein folding events revealed by time-resolved fluorescence resonance energy transfer in a microfluidic mixer. *Anal. Chem.* **2015**, *87*, 5589–5595. [CrossRef]
29. Inguva, V.; Kathuria, S.V.; Bilsel, O.; Perot, B.J. Computer design of microfluidic mixers for protein/RNA folding studies. *PLoS ONE* **2018**, *13*, e0198534. [CrossRef]
30. Liu, C.; Li, Y.; Liu, B.F. Micromixers and their applications in kinetic analysis of biochemical reactions. *Talanta* **2019**, *205*, 120136. [CrossRef]
31. Takahashi, S.; Yeh, S.-R.; Das, T.K.; Chan, C.-K.; Gottfried, D.S.; Rousseau, D.L. Folding of cytochrome *c* initiated by submillisecond mixing. *Nat. Struct. Biol.* **1997**, *4*, 44–50. [CrossRef]
32. Shastry, M.C.R.; Roder, H. Evidence for barrier-limited protein folding kinetics on the microsecond time scale. *Nat. Struct. Biol.* **1998**, *5*, 385–392. [CrossRef] [PubMed]
33. Shastry, M.C.R.; Sauder, J.M.; Roder, H. Kinetic and structural analysis of submillisecond folding events in cytochrome *c*. *Acc. Chem. Res.* **1998**, *31*, 717–725. [CrossRef]
34. Park, S.-H.; Shastry, M.C.R.; Roder, H. Folding dynamics of the B1 domain of protein G explored by ultrarapid mixing. *Nat. Struct. Biol.* **1999**, *6*, 943–947. [CrossRef] [PubMed]
35. Capaldi, A.P.; Shastry, R.M.C.; Kleanthous, C.; Roder, H.; Radford, S.E. Ultra-rapid mixing experiments reveal that Im7 folds *via* an on-pathway intermediate. *Nat. Struct. Biol.* **2001**, *8*, 68–72. [PubMed]
36. Teilum, K.; Maki, K.; Kragelund, B.B.; Poulsen, F.M.; Roder, H. Early kinetic intermediate in the folding of acyl-CoA binding protein detected by fluorescence labeling and ultrarapid mixing. *Proc. Natl. Acad. Sci. USA* **2002**, *99*, 9807–9812. [CrossRef] [PubMed]
37. Maki, K.; Cheng, H.; Dolgikh, D.A.; Shastry, M.C.; Roder, H. Early Events During Folding of Wild-type Staphylococcal Nuclease and a Single-tryptophan Variant Studied by Ultrarapid Mixing. *J. Mol. Biol.* **2004**, *338*, 383–400. [CrossRef]
38. Uzawa, T.; Akiyama, S.; Kimura, T.; Takahashi, S.; Ishimori, K.; Morishima, I.; Fujisawa, T. Collapse and search dynamics of apomyoglobin folding revealed by submillisecond observations of α-helical content and compactness. *Proc. Natl. Acad. Sci. USA* **2004**, *101*, 1171–1176. [CrossRef]
39. Gianni, S.; Walma, T.; Arcovito, A.; Calosci, N.; Bellelli, A.; Engstrom, A.; Travaglini-Allocatelli, C.; Brunori, M.; Jemth, P.; Vuister, G.W. Demonstration of long-range interactions in a PDZ domain by NMR, kinetics, and protein engineering. *Structure* **2006**, *14*, 1801–1809. [CrossRef]
40. Maki, K.; Cheng, H.; Dolgikh, D.A.; Roder, H. Folding kinetics of staphylococcal nuclease studied by tryptophan engineering and rapid mixing methods. *J. Mol. Biol.* **2007**, *368*, 244–255. [CrossRef] [PubMed]
41. Apetri, A.C.; Maki, K.; Roder, H.; Surewicz, W.K. Early intermediate in human prion protein folding as evidenced by ultrarapid mixing experiments. *J. Am. Chem. Soc.* **2006**, *128*, 11673–11678. [CrossRef]

42. Arai, M.; Kondrashkina, E.; Kayatekin, C.; Matthews, C.R.; Iwakura, M.; Bilsel, O. Microsecond hydrophobic collapse in the folding of Escherichia coli dihydrofolate reductase, an alpha/beta-type protein. *J. Mol. Biol.* **2007**, *368*, 219–229. [CrossRef] [PubMed]
43. Chen, K.C.; Xu, M.; Wedemeyer, W.J.; Roder, H. Microsecond unfolding kinetics of sheep prion protein reveals an intermediate that correlates with susceptibility to classical scrapie. *Biophys. J.* **2011**, *101*, 1221–1230. [CrossRef] [PubMed]
44. Xu, M.; Beresneva, O.; Rosario, R.; Roder, H. Microsecond folding dynamics of apomyoglobin at acidic pH. *J. Phys. Chem. B* **2012**, *116*, 7014–7025. [CrossRef] [PubMed]
45. Mizukami, T.; Xu, M.; Cheng, H.; Roder, H.; Maki, K. Nonuniform chain collapse during early stages of staphylococcal nuclease folding detected by fluorescence resonance energy transfer and ultrarapid mixing methods. *Protein Sci.* **2013**, *22*, 1336–1348. [CrossRef]
46. Honda, R.P.; Xu, M.; Yamaguchi, K.; Roder, H.; Kuwata, K. A Native-like Intermediate Serves as a Branching Point between the Folding and Aggregation Pathways of the Mouse Prion Protein. *Structure* **2015**, *23*, 1735–1742. [CrossRef]
47. Goluguri, R.R.; Udgaonkar, J.B. Microsecond Rearrangements of Hydrophobic Clusters in an Initially Collapsed Globule Prime Structure Formation during the Folding of a Small Protein. *J. Mol. Biol.* **2016**, *428*, 3102–3117. [CrossRef]
48. Sen, S.; Goluguri, R.R.; Udgaonkar, J.B. A Dry Transition State More Compact Than the Native State Is Stabilized by Non-Native Interactions during the Unfolding of a Small Protein. *Biochemistry* **2017**, *56*, 3699–3703. [CrossRef]
49. Srour, B.; Strampraad, M.J.F.; Hagen, W.R.; Hagedoorn, P.L. Refolding kinetics of cytochrome c studied with microsecond timescale continuous-flow UV-vis spectroscopy and rapid freeze-quench EPR. *J. Inorg. Biochem.* **2018**, *184*, 42–49. [CrossRef]
50. Mizukami, T.; Xu, M.; Fazlieva, R.; Bychkova, V.E.; Roder, H. Complex Folding Landscape of Apomyoglobin at Acidic pH Revealed by Ultrafast Kinetic Analysis of Core Mutants. *J. Phys. Chem. B* **2018**, *122*, 11228–11239. [CrossRef]
51. Mizukami, T.; Bedford, J.T.; Liao, S.; Greene, L.H.; Roder, H. Effects of ionic strength on the folding and stability of SAMP1, a ubiquitin-like halophilic protein. *Biophys. J.* **2022**, *121*, 552–564. [CrossRef]
52. Dingfelder, F.; Wunderlich, B.; Benke, S.; Zosel, F.; Zijlstra, N.; Nettels, D.; Schuler, B. Rapid Microfluidic Double-Jump Mixing Device for Single-Molecule Spectroscopy. *J. Am. Chem. Soc.* **2017**, *139*, 6062–6065. [CrossRef] [PubMed]
53. Berger, R.L.; Balko, B.; Chapman, H.F. High resolution mixer for the study of the kinetics of rapid reactions in solution. *Rev. Sci. Instrum.* **1968**, *39*, 493–498. [CrossRef] [PubMed]
54. Bellouard, Y.; Said, A.; Dugan, M.; Bado, P. Fabrication of high-aspect ratio, micro-fluidic channels and tunnels using femtosecond laser pulses and chemical etching. *Opt. Express* **2004**, *12*, 2120–2129. [CrossRef] [PubMed]
55. Roder, H.; Elöve, G.A.; Englander, S.W. Structural characterization of folding intermediates in cytochrome c by H-exchange labelling and proton NMR. *Nature* **1988**, *335*, 700–704. [CrossRef] [PubMed]
56. Pabit, S.A.; Roder, H.; Hagen, S.J. Internal friction controls the speed of protein folding from a compact configuration. *Biochemistry* **2004**, *43*, 12532–12538. [CrossRef]
57. Tsong, T.Y. The Trp-59 fluorescence of ferricytochrome c as a sensitive measure of the over-all protein conformation. *J. Biol. Chem.* **1974**, *249*, 1988–1990. [CrossRef]
58. Elöve, G.A.; Chaffotte, A.F.; Roder, H.; Goldberg, M.E. Early steps in cytochrome c folding probed by time-resolved circular dichroism and fluorescence spectroscopy. *Biochemistry* **1992**, *31*, 6876–6883. [CrossRef]
59. Hagen, S.J.; Eaton, W.A. Two-state expansion and collapse of a polypeptide. *J. Mol. Biol.* **2000**, *297*, 781–789. [CrossRef]
60. Fazelinia, H.; Xu, M.; Cheng, H.; Roder, H. Ultrafast hydrogen exchange reveals specific structural events during the initial stages of folding of cytochrome c. *J. Am. Chem. Soc.* **2014**, *136*, 733–740. [CrossRef]
61. Vogt, A.D.; Di Cera, E. Conformational selection or induced fit? A critical appraisal of the kinetic mechanism. *Biochemistry* **2012**, *51*, 5894–5902. [CrossRef]
62. Gianni, S.; Dogan, J.; Jemth, P. Distinguishing induced fit from conformational selection. *Biophys. Chem.* **2014**, *189*, 33–39. [CrossRef] [PubMed]
63. Hammes, G.G.; Chang, Y.C.; Oas, T.G. Conformational selection or induced fit: A flux description of reaction mechanism. *Proc. Natl. Acad. Sci. USA* **2009**, *106*, 13737–13741. [CrossRef] [PubMed]
64. Mizukami, T.; Furuzawa, S.; Itoh, S.G.; Segawa, S.; Ikura, T.; Ihara, K.; Okumura, H.; Roder, H.; Maki, K. Energetics and kinetics of substrate analog-coupled staphylococcal nuclease folding revealed by a statistical mechanical approach. *Proc. Natl. Acad. Sci. USA* **2020**, *117*, 19953–19962. [CrossRef] [PubMed]
65. Karthikeyan, S.; Leung, T.; Birrane, G.; Webster, G.; Ladias, J.A. Crystal structure of the PDZ1 domain of human Na(+)/H(+) exchanger regulatory factor provides insights into the mechanism of carboxyl-terminal leucine recognition by class I PDZ domains. *J. Mol. Biol.* **2001**, *308*, 963–973. [CrossRef] [PubMed]
66. Cushing, P.R.; Fellows, A.; Villone, D.; Boisguerin, P.; Madden, D.R. The relative binding affinities of PDZ partners for CFTR: A biochemical basis for efficient endocytic recycling. *Biochemistry* **2008**, *47*, 10084–10098. [CrossRef] [PubMed]

molecules MDPI

Article

Crystal Structures of the Plant Phospholipase A1 Proteins Reveal a Unique Dimerization Domain

Yunseok Heo [1], Inhwan Lee [1], Sunjin Moon [1], Ji-Hye Yun [1,2], Eun Yu Kim [3], Sam-Yong Park [4], Jae-Hyun Park [1], Woo Taek Kim [3,*] and Weontae Lee [1,2,*]

[1] Structural Biochemistry & Molecular Biophysics Laboratory, Department of Biochemistry, College of Life Science and Biotechnology, Yonsei University, Seoul 03722, Korea; uoonsek1@yonsei.ac.kr (Y.H.); inhwan@kolmar.co.kr (I.L.); moon@spin.yonsei.ac.kr (S.M.); jihye2@spin.yonsei.ac.kr (J.-H.Y.); jhpark@spin.yonsei.ac.kr (J.-H.P.)

[2] PCG-Biotech, Ltd., 508 KBIZ DMC Tower, Sangam-ro, Seoul 03929, Korea

[3] Department of Systems Biology, College of Life Science and Biotechnology, Yonsei University, Seoul 03722, Korea; eunyukim@cemps.ac.cn

[4] Drug Design Laboratory, Graduate School of Medical Life Science, Yokohama City University, Tsurumi, Yokohama 230-0045, Japan; park@yokohama-cu.ac.jp

* Correspondence: wtkim@yonsei.ac.kr (W.T.K.); wtlee@yonsei.ac.kr (W.L.)

Abstract: Phospholipase is an enzyme that hydrolyzes various phospholipid substrates at specific ester bonds and plays important roles such as membrane remodeling, as digestive enzymes, and the regulation of cellular mechanism. Phospholipase proteins are divided into following the four major groups according to the ester bonds they cleave off: phospholipase A1 (PLA1), phospholipase A2 (PLA2), phospholipase C (PLC), and phospholipase D (PLD). Among the four phospholipase groups, PLA1 has been less studied than the other phospholipases. Here, we report the first molecular structures of plant PLA1s: AtDSEL and CaPLA1 derived from *Arabidopsis thaliana* and *Capsicum annuum*, respectively. AtDSEL and CaPLA1 are novel PLA1s in that they form homodimers since PLAs are generally in the form of a monomer. The dimerization domain at the C-terminal of the AtDSEL and CaPLA1 makes hydrophobic interactions between each monomer, respectively. The C-terminal domain is also present in PLA1s of other plants, but not in PLAs of mammals and fungi. An activity assay of AtDSEL toward various lipid substrates demonstrates that AtDSEL is specialized for the cleavage of *sn*-1 acyl chains. This report reveals a new domain that exists only in plant PLA1s and suggests that the domain is essential for homodimerization.

Keywords: X-ray crystallography; phospholipase A1; homodimer; dimerization domain; catalytic triad; plant protein

Citation: Heo, Y.; Lee, I.; Moon, S.; Yun, J.-H.; Kim, E.Y.; Park, S.-Y.; Park, J.-H.; Kim, W.T.; Lee, W. Crystal Structures of the Plant Phospholipase A1 Proteins Reveal a Unique Dimerization Domain. *Molecules* **2022**, *27*, 2317. https://doi.org/10.3390/molecules27072317

Academic Editors: Kunihiro Kuwajima, Yuko Okamoto, Tuomas Knowles and Michele Vendruscolo

Received: 13 March 2022
Accepted: 31 March 2022
Published: 2 April 2022

1. Introduction

Phospholipase hydrolyzes phospholipids at specific sites. The phospholipases have roles as digestive enzymes, for maintaining and remodeling of a membrane, and for the regulation of cellular mechanisms [1]. The phospholipases are classified into four groups according to the cleavage site; phospholipase A1 (PLA1) cleaves *sn*-1 acyl chain, phospholipase A2 (PLA2) cleaves *sn*-2 acyl chain, phospholipase C (PLC) cleaves before the phosphate, and phospholipase D (PLD) cleaves after the phosphate [2]. Therefore, PLA1 and PLA2 are the acylhydrolase family that catalyzes the substrates containing acyl group, and PLC and PLD belong to phosphodiesterase. Triacylglycerol lipase (TGL), lipoprotein lipase (LPL), and monoacylglycerol lipase (MGL) belong to PLA because they also cleave the *sn*-1 and/or *sn*-2 acyl chain.

PLA1 (EC 3.1.1.32) is present in many different organisms including mammals, fungi, insects, bacteria, metazoans, protozoan parasites, venoms, and plants [1,3]. To date, three-dimensional structures of various PLAs have been elucidated [4–10]. Although these lipases have little sequence identity, they show a typical α/β hydrolase scaffold comprising of a core of seven (or more) β-strands flanked by several α-helices [11]. Other common structural features of lipases are a GXSXG motif, lid domain, and catalytic triad in the α/β hydrolase scaffold [12]. The GXSXG motif is composed of catalytic Ser and two residues before and after the Ser, and the lid domain plays a role for capping the active site; the catalytic triad consists of His, Asp, and the Ser of the GXSXG motif [13–16].

Plant PLAs have important roles in metabolism and lipid biosynthesis, as well as applications in food and biotechnology industries [17,18]. In *Arabidopsis thaliana*, DAD1 (At2g44810) was identified as an *sn*-1 specific acylhydrolase, and several studies have suggested that AtDSEL (*Arabidopsis thaliana* DAD1-like seedling establishment-related lipase), one of the homologues of DAD1, plays significant roles in the regulation of cellular processes such as seed germination, tissue growth, and seedling establishment [19,20]. In the early roots of hot pepper (*Capsicum annuum*), a cDNA encoding PLA1 homolog (CaPLA1) was identified [21]. CaPLA1 was selectively expressed in young roots of the hot pepper and hydrolyzed phospholipids at the *sn*-1 position [21]. CaPLA1 is presumed to be involved in the development of the roots of hot pepper, considering that the expression of CaPLA1 rapidly declined after germination [21].

We have determined the three-dimensional structures of AtDSEL and CaPLA1. They are the first elucidated structures of PLA1 derived from plants. Here, we revealed that these representative plant PLA1s are different from previously reported PLAs; only the plant PLA1s have a domain for homodimerization. Hydrophobic residues in the dimerization domain form a strong dimeric interface. This report broadens our knowledge of PLA by presenting the plant PLA1 structures with novel features.

2. Results and Discussion

2.1. Structural Features of AtDSEL and CaPLA1

Crystal structures of AtDSEL and CaPLA1 were determined at a resolution of 1.8 Å and 2.4 Å, respectively. There are two AtDSEL molecules in the asymmetric unit (Figure 1A), and each monomer chain consists of 16 α-helices (labeled α1–α16) and 11 β-strands (labeled β1–β11) (Figure 1B). AtDSEL shows the typical structure of an α/β hydrolase; core β-strands (β1–β3 and β6–β9) are flanked by several α-helices (Figure 1B). An additional two β-strands (β4 and β5) and surrounding α-helices are also common in PLAs; however, C-terminal α-helices (α14–α16) and β-strands (β10 and β11), colored in blue, are unique features (Figure 1B). No such C-terminal domain has been found in any PLAs. Furthermore, the domain appears to be involved in dimerization; the two monomers interact with each other via the domain (Figure 1A). We used analytical ultracentrifugation (AUC) to confirm whether AtDSEL in solution is homodimerized or not. The theoretical molecular weight of dimeric AtDSEL is 95.7 kDa, and calculated molecular weight from the AUC peak is 95.3 kDa, which strongly suggests that AtDSEL forms homodimers in solution (Figure 1C). In the case of CaPLA1, there are four molecules in the asymmetric unit. Among the four molecules, two molecules make a homodimer (Figure 1D); the other two molecules make homodimers with another symmetry mate, respectively. Each monomer of CaPLA1 has an overall structure similar to that of AtDSEL; α/β hydrolase domain and C-terminal domain are almost the same as each other (Figure 1B,E). However, two additional β-strands between α5 and α7 are not found in CaPLA1 (Figure 1B,E), which is attributed to the disordered residues around the β-strands. The AUC result of CaPLA1 also suggests that CaPLA1 makes homodimers in solution, considering that the molecular weight of the theoretical dimeric CaPLA1 and calculated from the AUC peak are 90.4 kDa and 86.0 kDa, respectively (Figure 1F). Taken together, the homodimerization of AtDSEL and CaPLA1 was demonstrated using AUC, and the C-terminal domain, which is presumed to be present only in plant PLA1s, seems to be related to the dimerization; the two monomers of AtDSEL and CaPLA1 interact with each other via the domain (Figure 1A,D).

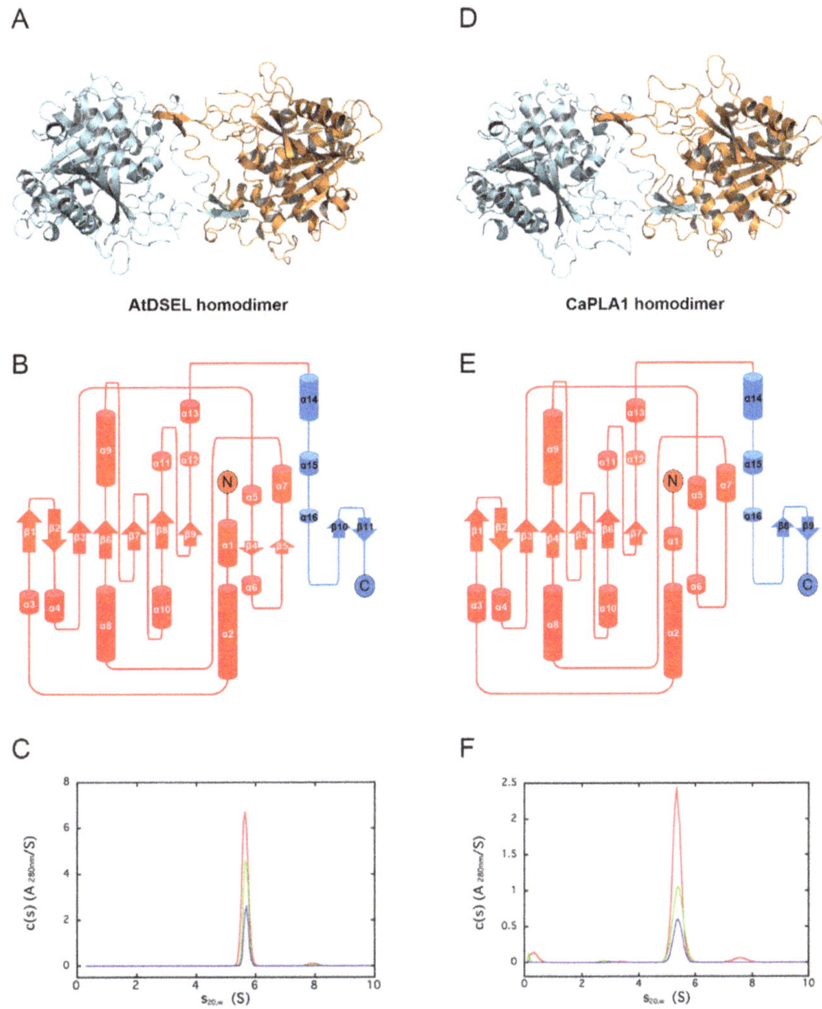

Figure 1. Overall structures of AtDSEL and CaPLA1. (**A**) Crystal structure of full-length AtDSEL. AtDSEL homodimer is shown as a ribbon diagram. Each monomer is colored in cyan and orange, respectively. (**B**) Topology diagram of AtDSEL monomer. α/β hydrolase domain and C-terminal domain are colored in red and blue, respectively. (**C**) Analytical ultracentrifugation (AUC) result of AtDSEL. The experiments were performed at following three protein concentrations: 3.0 mg/mL (colored in red), 1.5 mg/mL (green), and 0.8 mg/mL (blue). Calculated molecular weight of AtDSEL from the AUC peak is 95.3 ± 0.3 kDa, which strongly suggests that AtDSEL forms homodimers in solution, considering that the theoretical molecular weight of dimeric AtDSEL is 95.7 kDa. (**D**) Crystal structure of full-length CaPLA1. CaPLA1 homodimer is shown as a ribbon diagram. (**E**) Topology diagram of CaPLA1 monomer. (**F**) AUC result of CaPLA1. The experiments were performed at following three protein concentrations: 1.1 mg/mL (colored in red), 0.5 mg/mL (green), and 0.3 mg/mL (blue). Calculated molecular weight of CaPLA1 from the AUC peak is 86.0 ± 2.0 kDa, which strongly suggests that CaPLA1 forms homodimers in solution, considering that the theoretical molecular weight of dimeric CaPLA1 is 90.4 kDa.

2.2. Conserved Motifs and Domains of AtDSEL and CaPLA1

AtDSEL and CaPLA1 are divided into the following two domains: the α/β hydrolase domain containing the lid domain, GXSXG motif, and catalytic triad and the C-terminal domain containing the dimerization domain (Figure 2A–D). The α/β hydrolase domains of AtDSEL and CaPLA1 are composed of 340 and 323 residues, respectively, and the their C-terminal domains consist of 79 and 74 residues, respectively (Figure 2B,D). Most PLA structures reported to date have only the α/β hydrolase domain and are composed of approximately 300 residues [4–6]. They do not make any multimers [4–6]. On the other hand, mammal PLAs (human LPL and human pancreatic TGL) have another C-terminal domain in addition to the α/β hydrolase domain; the α/β hydrolase domain is composed of about 340 residues, and the other C-terminal domain consists of about 130 residues [7–9]. The C-terminal domain of mammals shows a typical β-barrel structure [7–9], which is totally different from the C-terminal domains of AtDSEL and CaPLA1. Human MGL is composed of 303 residues and has only the α/β hydrolase domain; however, the lipase makes homodimers [10]. The two monomers of human MGL makes a dimer with an up–down orientation [10]. Taken together, PLAs in fungi and venom have only the α/β hydrolase domain [4–6], and some PLAs in mammals and our plant PLA1s have an additional C-terminal domain as well as the α/β hydrolase domain [7–9] (Figure 2B,D). However, the overall structures and functions of the C-terminal domains of the PLAs in mammals and plants are completely different from each other. The C-terminal domain of AtDSEL and CaPLA1 is composed of three α-helices and one pair of β-strands (Figure 1B,E), while the C-terminal domains of PLAs in mammals are composed of 8–12 β-strands making up the β-barrel structure [7–9]. Moreover, the C-terminal domain of AtDSEL and CaPLA1 is involved in the homodimerization, while the β-barrel of the PLA is related to the interactions with lipids [7]. On the other hand, human MGL can make homodimers though the proteins consisting of only the α/β hydrolase domain [10]. Therefore, only the PLAs in plants have the C-terminal domain for homodimerization.

AtDSEL has two disconnected loops caused by the disordered residues; Arg120 and Glu121 between β1 and β2 and from Pro159 to Leu160 between β3 and α5 could not be assigned due to the poor electron density map of the residues (Figures 1B and 2A). Pro159 and Leu160 belong to the lid domain, which determines the substrate specificities (Figure 2A,B). CaPLA1 also has two disconnected loops resulting from the disordered residues; Pro91 and Asp92 between β1 and β2 and from Asp160 to Phe167 between α6 and α7 could not be refined because of the deficient electron density map of the residues (Figures 1E and 2C). The disordered residues from Asp160 to Phe167 affected the surrounding secondary structures; a pair of β-strands (β4 and β5 in the case of AtDSEL) were not found in CaPLA1 (Figure 1B,E and Figure 2A,C). It has been reported that the length of the lid domain is involved in the type of substrate that the enzyme can catalyze [3]. AtDSEL and CaPLA1 have a lid domain with the proper length for catalyzing the *sn*-1 acyl groups [3] (Figure 2B,D).

The α/β hydrolase domains of AtDSEL and CaPLA1 have a conserved GXSXG motif (Figure 2A,C). In the AtDSEL, the GXSXG motif consists of the residues from Gly234 to Gly238, while the motif is composed of the residues from Gly219 to Gly223 in the case of CaPLA1 (Figure 2B,D). The GXSXG motif is found in various lipases and esterases [22]. The central Ser of the motif functions as a nucleophile [23]. In AtDSEL, the motif exists between β6 and α9 and has the following sequence: GHSLG (Figures 1B and 2A). The motif of CaPLA1 is also located in the same position and has the same sequence as in AtDSEL (Figure 2A,C). It has been reported that the two Gly residues of the GXSXG motif are also essential for the lipase activity [24]. The lipase activities of MGLs in *Arabidopsis thaliana* were completely lost when the researchers applied the following mutagenesis to the lipases: GXAXG, GXSXS, or SXSXG [24], which demonstrates that Gly to the Ser mutant as well as Ser to the Ala mutant causes severe deactivation of the lipases.

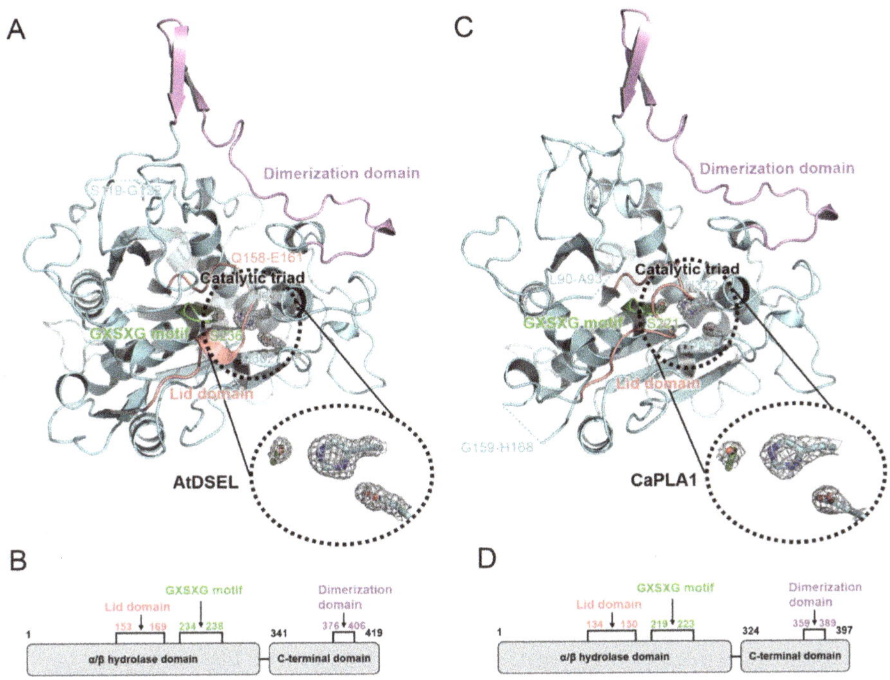

Figure 2. Conserved catalytic triad and domains of AtDSEL and CaPLA1. (**A**) Three residues (Ser236, Asp302, and His339) that compose the catalytic triad of AtDSEL are shown as a stick model. $2mF_O{-}DF_C$ electron density map at a level of 2.0 σ is superimposed to the residues. Lid domain is colored in salmon. Pro159 and Leu160 of the lid domain were disordered, causing the disconnected lid domain in this figure. Conserved GXSXG motif is colored in green, and dimerization domain is colored in purple. (**B**) The construct map of AtDSEL. The lid domain and GXSXG motif belong to α/β hydrolase domain. C-terminal domain containing dimerization domain exists only in plant phospholipase A1 (PLA1). (**C**) Three residues (Ser221, Asp285, and His322), which compose the catalytic triad of CaPLA1, are shown as a stick model. $2mF_O{-}DF_C$ electron density map at a level of 1.8 σ is superimposed to the residues. (**D**) The construct map of CaPLA1.

The catalytic triad of AtDSEL is Ser236, Asp302, and His339, while that of CaPLA1 is Ser221, Asp285, and His322 (Figure 2A,C). The lid domain, which is conserved in lipase proteins, plays the role of capping this active site [25] (Figure 2A,C). All human pancreatic TGL mutants of the catalytic triad (Ser, Asp, and His) showed no detectable activity or highly diminished activity [26]. Therefore, the catalytic triad could not be replaced with other residues.

2.3. *Interactions between Homodimers of AtDSEL and CaPLA1*

The dimerization domains of the AtDSEL homodimer cross each other and stretch in opposite directions (Figure 3A). Met396 and Trp404 of one monomer and Trp388 of the other monomer interact with each other (Figure 3B). The hydrophobic interactions are main driving force behind the formation of the dimerization of AtDSEL. Interactions between the monomers occur throughout the loops (Figure 3B). The length from a sulfur atom of Met396 of one monomer to a sulfur atom of the other Met396 is 23.3 Å (Figure 3B). The dimerization interface of AtDSEL is extremely long to form robust homodimers. The dimerization domain of CaPLA1 is similar to that of AtDSEL; however, more residues are involved in the dimerization (Figure 3C,D). The length from a sulfur atom of one Met379 to that of the other Met379 is 22.1 Å (Figure 3D). In CaPLA1, His376, Met379, and

Trp387 of one monomer and Trp371 and Trp372 of the other monomer make hydrophobic interactions (Figure 3D). Trp372 and His376 in CaPLA1 are replaced by Arg and Asn in AtDSEL, respectively (Figure 3E). Therefore, the dimeric interface of CaPLA1 is more robust than that of AtDSEL (Figure 3B,D).

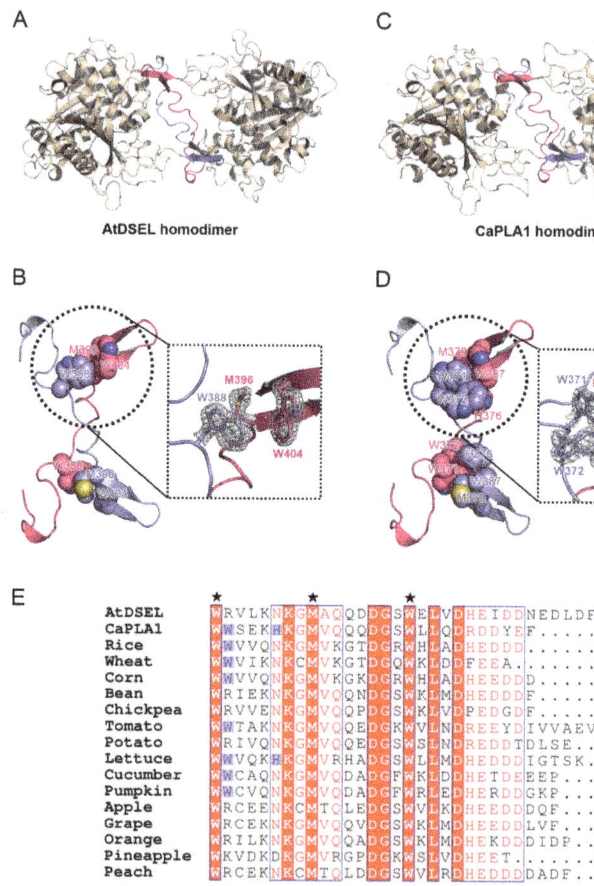

Figure 3. Hydrophobic interactions between homodimers of AtDSEL and CaPLA1. (**A**) AtDSEL homodimer is shown as a ribbon diagram. Each dimerization domain is colored in blue and pink, respectively. The rest is colored in wheat. (**B**) The dimerization domains of AtDSEL are shown as a ribbon diagram, and each domain is colored in blue and pink, respectively. The side-chains of Trp388, Met396, and Trp404 are shown as spheres. The side-chains of the residues are shown as a stick model in the zoomed view, and $2mF_O-DF_C$ electron density map at a level of 2.0 σ is superimposed to the residues. (**C**) CaPLA1 homodimer is shown as a ribbon diagram. (**D**) The dimerization domains of CaPLA1 are shown as a ribbon diagram. The side-chains of Trp371, Trp372, His376, Met379, and Trp387 are shown as spheres. The side-chains of the residues are shown as a stick model in the zoomed view, and $2mF_O-DF_C$ electron density map at a level of 1.8 σ is superimposed to the residues. (**E**) Sequence alignments of plant PLA1s, from Trp388 to C-terminal residue of AtDSEL and Trp371 to C-terminal residue of CaPLA1, were used for the alignments. In the case of other PLA1s, residues from the Trp to C-terminal were used for the alignments. The common three residues of AtDSEL and CaPLA1 involved in the hydrophobic interactions are indicated by black solid pentagrams. The additional two residues involved in the interactions of CaPLA1 are conserved in some plant PLA1s. The residues are highlighted in blue.

The C-terminal residues of AtDSEL, CaPLA1, and 15 PLA1s from various plants are aligned (Figure 3E). The two Trp residues and one Met of AtDSEL are conserved in the 17 PLA1s, indicating that the dimerization is commonplace in plant PLA1s (Figure 3E). Among the 17 PLA1s, eight PLA1s containing CaPLA1 have the two consecutive Trp residues, while the other PLA1s have positively charged residue instead of the second Trp (Figure 3E). His of CaPLA1 is conserved only in lettuce PLA1; the other PLA1s have Asn instead of the His, except for pineapple PLA1 (Figure 3E). The 17 PLA1s have a number of negatively charged residues in the C-terminus (Figure 3E); the reason for this has not yet been clearly elucidated. Taken together, the sequence alignments of the C-terminal regions of the plant PLA1s suggest that they would form homodimers.

2.4. Structural Comparison among PLAs

Data from multiple sequence alignments show that the C-terminal regions of AtDSEL and CaPLA1 are mostly conserved among plant PLA1s (Figure 3E); however, other PLAs with the C-terminal domain have not yet been reported. A structural comparison was performed to confirm the similarities and differences among PLAs (Figure 4A–C). AtDSEL and CaPLA1 are superimposed with an r.m.s. deviation of 1.75 Å for all Cα atoms (Figure 4A). The sequence identity between AtDSEL and CaPLA1 is 52%. The residues comprising the catalytic triad of each protein completely overlap (Figure 4A). Most of the loops as well as α-helices and β-strands are superposed (Figure 4A). In addition, the C-terminal dimerization domains of AtDSEL and CaPLA1 are very well conserved (Figure 4A). Based on the sequence alignments (Figure 3E), PLA1s of other plants are presumed to also have the conserved C-terminal dimerization domain. The substrate preference of AtDSEL toward various lipid substrates was investigated (Figure 4D). AtDSEL specifically hydrolyzed the *sn*-1 acyl chain of 1,3-diacylglycerol (1,3-DAG) and 1-monoacylglycerol (1-MAG) (Figure 4D), suggesting that AtDSEL is an *sn*-1 specific lipase, as previously known [20]. AtDSEL also showed some activity toward 1,2-diacylglycerol (1,2-DAG) and 2-monoacylglycerol (2-MAG) (Figure 4D), which are involved in the *sn*-2 acyl chain.

AtDSEL and human MGL are superposed with an r.m.s. deviation of 6.13 Å for all Cα atoms (Figure 4B). The sequence identity of the two proteins is very low at 13%, and even the secondary structures hardly overlap except for the core β-strands (Figure 4B). However, the residues making up the catalytic triad of human MGL are also Ser, Asp, and His, and they are well overlaid between AtDSEL and human MGL (Figure 4B). On the other hand, the C-terminal dimerization domain of AtDSEL is missing in the human MGL (Figure 4B). The human MGL forms homodimers in an up–down orientation without any additional domain, and human LPL forms homodimers in a head-to-tail orientation with the C-terminal β-barrel domain (Figure 5A,B). Therefore, the dimerization interface of the human PLAs is totally different from that of AtDSEL and CaPLA1. AtDSEL and fungi TGL are overlaid with an r.m.s. deviation of 3.93 Å for all Cα atoms (Figure 4C). Although there is a low sequence identity of 17% between the two proteins, the overall structures of them are well conserved (Figure 4C). The secondary structures of α-helices and β-strands and the catalytic triad residues of fungi TGL are highly similar to those of AtDSEL (Figure 4C). However, fungi TGL also does not have the C-terminal dimerization domain (Figure 4C).

Sequence alignments of PLA1s from various plants and structural comparison based on the structures of AtDSEL, CaPLA1, and lipases from mammals and fungi reveal that the C-terminal dimerization domain of AtDSEL and CaPLA1 is unique to plant PLAs. Thus, we propose that AtDSEL and CaPLA1 are the representative plant PLA1s, which are distinguished by the C-terminal dimerization domain for homodimerization.

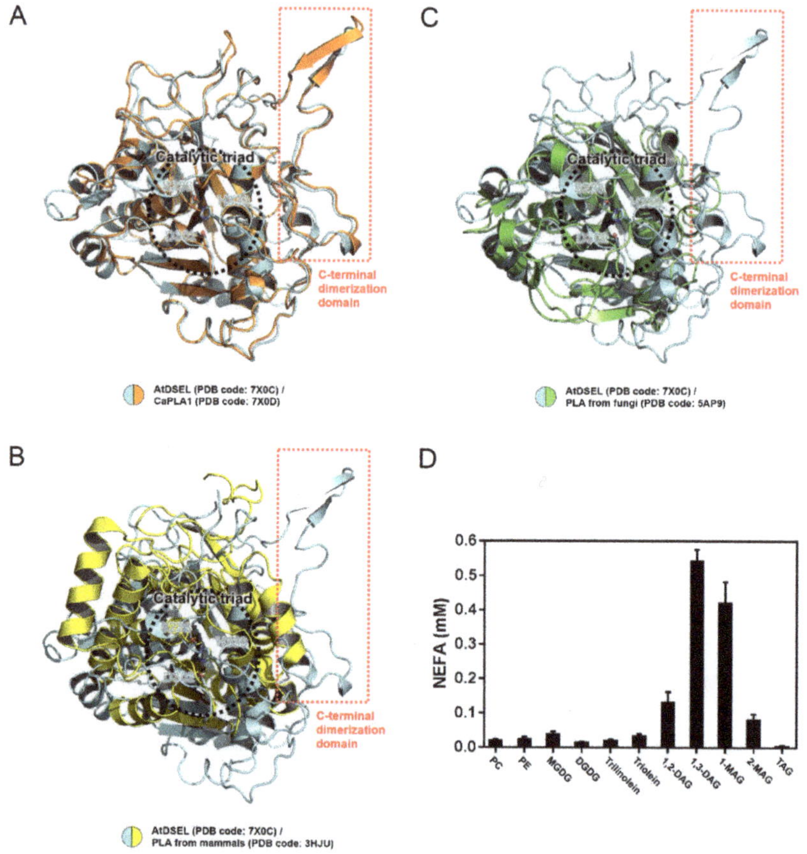

Figure 4. Structural comparison between AtDSEL and other PLAs, and activity assay of AtDSEL toward various lipid substrates. (**A**) Superimposition of AtDSEL and CaPLA1. AtDSEL and CaPLA1 are shown as a ribbon diagram and colored in cyan and orange, respectively. The side-chains of the catalytic triad are shown as a stick model. The additional C-terminal dimerization domains are indicated by a red dotted square. (**B**) Superimposition of AtDSEL and MGL from *Homo sapiens*. AtDSEL and MGL from *Homo sapiens* are colored in cyan and yellow, respectively. (**C**) Superimposition of AtDSEL and TGL from *Thermomyces lanuginosus*. AtDSEL and TGL from *Thermomyces lanuginosus* are colored in cyan and green, respectively. (**D**) Lipolytic activity of AtDSEL toward following substrates: phosphatidylcholine (PC), phosphatidylethanolamine (PE), monogalactosyl diacylglycerol (MGDG), digalactosyl diacylglycerol (DGDG), trilinolein, triolein, 1,2-diacylglycerol (1,2-DAG), 1,3-diacylglycerol (1,3-DAG), 1-monoacylglycerol (1-MAG), 2-monoacylglycerol (2-MAG), triacylglycerol (TAG). The released free fatty acids were measured using a nonesterified fatty acid (NEFA) colorimetric kit. Mean and standard deviation from three independent experiments are presented.

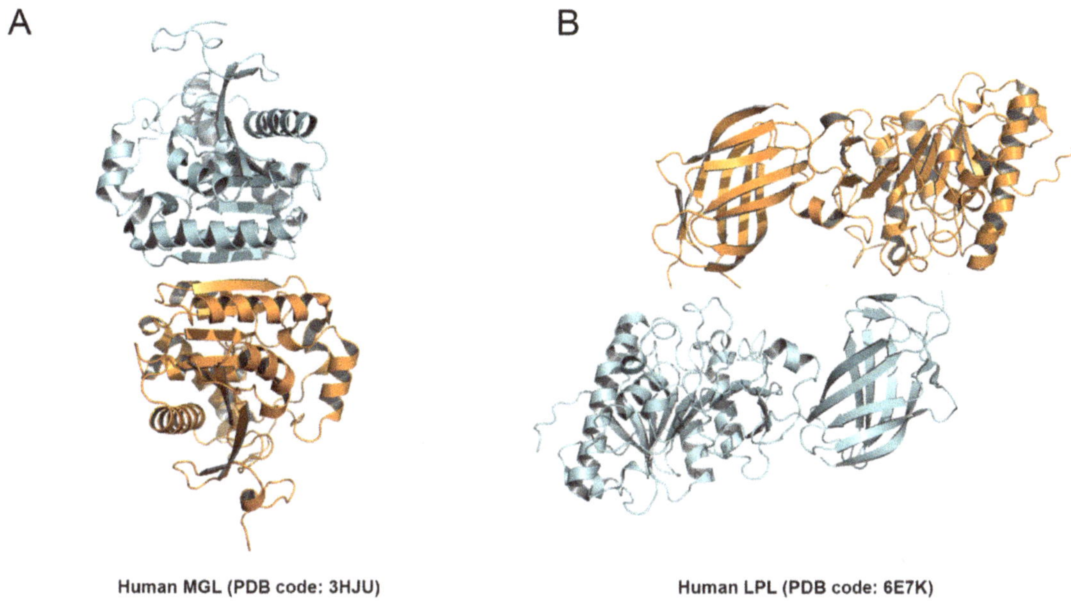

Figure 5. Dimerization interface of human PLAs. (**A**) Dimeric structure of human MGL. Each monomer of human MGL are colored in cyan and orange, respectively. (**B**) Dimeric structure of human LPL.

3. Conclusions

Crystal structures of AtDSEL and CaPLA1, which are the first PLA1 structures derived from plants, reveal the unique features of the plant PLA1s. In addition to the conserved α/β hydrolase scaffold, GXSXG motif, and catalytic triad, plant PLA1s have an extended C-terminal domain composed of approximately 70 residues. In AtDSEL and CaPLA1, these C-terminal regions form a homodimer. Trp388, Met396, and Trp404 in AtDSEL and Trp371, Trp372, His376, Met379, and Trp387 in CaPLA1 play as hydrophobic anchors tethering each monomer of the proteins. A data search by BLAST indicates that the homologous C-terminal regions of AtDSEL and CaPLA1 are only found in plants. Sequence alignments of the C-terminal regions of the plant PLA1s revealed that the two Trp residues and one Met, which are the common residues between the C-terminal domains of AtDSEL and CaPLA1, are conserved. The homodimerizations of AtDSEL and CaPLA1 were supported by our AUC experiments. However, it is still unknown as to why the plant PLA1s form homodimers. The dimerization domain identified by the structures of AtDSEL and CaPLA1 can serve as the structural templates for understanding the plant phospholipase function at the molecular level. Furthermore, we found that some plant PLA1s have two Trp residues in the C-terminal region, while others have one more Trp in the C-terminal region. AtDSEL has two Trp residues and CaPLA1 has three Trp residues in the C-terminal regions, and all Trp residues in the C-terminal region play an important role as the hydrophobic anchors. Whether the difference in the number of Trp in the C-terminal region can characterize the two types of plant PLA1 dimerization domains should be further explained.

4. Materials and Methods

4.1. Cloning, Expression, and Purification

The full-length AtDSEL (Uniprot entry: O49523) and CaPLA1 (Uniprot entry: A5YW95) genes were amplified by polymerase chain reaction (PCR), respectively. The amplified cDNA fragments were subcloned into the pET21b expression vector (Novagen) with a fused 6× His tag and TEV cleavage site (ENLYFQG) at the N-terminus. The resulting plasmid was transformed into *Escherichia coli* (*E. coli*) BL21 (DE3). AtDSEL or CaPLA1 expressing cells were grown in Luria–Bertani (LB) media at 18 °C until they reached an OD_{600} of 0.7, at which time 0.5 mM isopropyl β-D-thiogalactopyranoside (IPTG) was added. The cells were cultured for a further 24 h at 18 °C, and stored at −80 °C after harvesting by centrifugation. Harvested cells were disrupted by sonication in a lysis buffer containing 25 mM sodium phosphate (pH 7.0), 100 mM NaCl, 5 mM β-mercaptoethanol, and protease inhibitor cocktail (Roche). The fusion protein was purified by immobilized metal affinity chromatography (IMAC) on an Ni-NTA column (Amersham Pharmacia Biotech) with the lysis buffer containing 500 mM imidazole. The purified protein was then dialyzed against the lysis buffer (25 mM sodium phosphate (pH 7.0), 100 mM NaCl, 5 mM β-mercaptoethanol). The dialyzed protein was subjected to TEV cleavage for 12 h, and then purified by IMAC on an Ni-NTA column (Amersham Pharmacia Biotech). The protein without His tag was eluted with the lysis buffer containing 20 mM imidazole. The protein containing non-native Gly at the N-terminus after the cleavage was further purified by size exclusion chromatography (SEC) using a HiLoad™ Superdex™ 200 column (GE Healthcare) with 20 mM Tris (pH 8.0), 100 mM NaCl, and 2 mM DTT and concentrated to 20 mg/mL for crystallization.

4.2. Crystallization, Data Collection, and Structure Determination

Crystallization screening of sitting drop was carried out with a mosquito® crystallization robot (TTP Labtech) using commercial screening kits from Hampton Research (Aliso Viejo, CA) and Rigaku (Tokyo, Japan). The initial crystals of AtDSEL were observed in 100 mM sodium acetate (pH 4.6), 30% (*v*/*v*) MPD, and 20 mM calcium chloride by mixing equal volumes (each 300 nL) of protein solution (in 20 mM Tris (pH 8.0), 100 mM NaCl, and 2 mM DTT) and mother liquor. The initial crystals of CaPLA1 were observed in 100 mM HEPES (pH 7.5), 20% (*v*/*v*) PEG 8000, 10% (*v*/*v*) 2-propanol, and 200 mM ammonium sulfate by mixing equal volumes of protein solution and mother liquor. The crystals of both proteins were needle-shaped and approximately 20 μm in size. The crystals were flash-cooled to 100 K in liquid nitrogen. Diffraction data were collected using an EIGER X 16M detector on the BL-17A beamline at the Photon Factory (PF), Tsukuba, Japan. The distance from the detector to the crystal was 210 mm, and a total of 360 images were collected with oscillation widths of 1°, exposing the crystal to the beam for 1 s per image. The data were integrated and scaled using HKL2000 [27]. The diffraction data for AtDSEL and CaPLA1 were collected at a maximum resolution of 1.80 Å and 2.40 Å, respectively. The initial model of AtDSEL and CaPLA1 was determined by molecular replacement (MR) using our AtDSEL structure (PDB code: 2YIJ) with the PHENIX software suite [28]. The model was iteratively refined using COOT and PHENIX [29]. The geometry analysis was performed during the refinement using PHENIX. The final structural coordinates of AtDSEL and CaPLA1 were deposited under the following Protein Data Bank (PDB) accession codes: 7X0C and 7X0D, respectively. Data collection and refinement statistics are summarized in Table 1.

4.3. Analytical Ultracentrifugation (AUC) Experiment

The experiments were conducted at 20 °C using an Optima XL-I analytical ultracentrifuge (Beckman Coulter, Brea, CA, USA) with an An-50 Ti rotor. For sedimentation velocity experiments, cells with a standard Epon two-channel centerpiece and sapphire windows were used. The sample (400 μL) and the reference buffer (420 μL) were loaded into the cells. The rotor temperature was equilibrated at 20 °C in a vacuum chamber for 2 h before the start. The sedimentation velocity experiments for AtDSEL were conducted at the following three protein concentrations: 3.0 mg/mL, 1.5 mg/mL, and 0.8 mg/mL. The experiments for CaPLA1 were conducted at the following three protein concentrations: 1.1 mg/mL, 0.5 mg/mL, and 0.3 mg/mL. Changes in the concentration gradient were monitored with a Rayleigh interference optical system at 10 min intervals during sedimentation at 50 × 103 rpm. The partial specific volume of the protein, solvent density, and solvent viscosity were calculated from standard tables using the SEDNTERP software. The resulting scans were analyzed using the continuous distribution c(s) analysis module in the SEDFIT software. Sedimentation coefficient increments of 100 were used in the appropriate range for each sample. The frictional coefficient was allowed to float during fitting. The weighted average sedimentation coefficient was obtained by integrating the range of sedimentation coefficients in which the peaks were present. The values of the sedimentation coefficient were corrected to 20 °C in pure water (s20,w). The c(s) distribution was converted into c(M), which is the molar mass distribution.

Table 1. Summary of data collection and refinement of AtDSEL and CaPLA1.

Protein	**AtDSEL**			**CaPLA1**		
PDB Code	**7X0C**			**7X0D**		
Data collection						
Space group	$P12_11$			$P12_11$		
Cell dimensions						
a, b, c (Å)	73.977	95.343	79.278	102.313	59.568	147.319
α, β, γ (°)	90.000	105.144	90.000	90.000	90.139	90.000
Resolution (Å)	50–1.80 (1.86–1.80)			50–2.40 (2.44–2.40)		
R_{merge}	0.086 (0.400)			0.119 (0.589)		
I/σ (I)	35.29 (4.70)			13.97 (2.79)		
Completeness (%)	99.8 (100.0)			96.1 (95.5)		
Multiplicity	7.5 (7.5)			3.5 (3.0)		
Refinement						
Resolution (Å)	33.29–1.80			49.11–2.40		
No. reflections	98,085			67,351		
R_{work}/R_{free}	0.1783/0.2117			0.2065/0.2602		
No. atoms						
Protein	6409			12,431		
Water	423			508		
B-factors (Å^2)						
Protein	24.68			26.50		
Water	28.83			28.53		
R.m.s. deviations						
Bond lengths (Å)	0.012			0.002		
Bond angles (°)	1.127			0.545		
Ramachandran plot (%)						
Favored	98.35			97.29		
Allowed	1.65			2.45		
Disallowed	0.00			0.26		

Values in parentheses are for the highest resolution shell.

4.4. In Vitro Lipase Assay

The in vitro lipase activity was measured as previously described with some modifications [30]. To examine phospholipase activity of AtDSEL, the recombinant fusion protein was incubated with 1-palmitoyl-2-^{14}C-palmitoyl-PC (2.22 GBq/mmol; GE Healthcare, Uppsala, Sweden) in 50 mM sodium phosphate buffer (pH 5.8). Lipids were extracted and separated by thin layer chromatography (TLC) (Silica Gel 60; Merck). Radioactive products were detected using the Bio-Imaging Analyzer (BAS2500; Fuji Photo Film). To determine substrate specificity, the MBP-AtDSEL was incubated with 250 μM of various lipid substrates having *sn*-1 or *sn*-2 acyl chains, including phosphatidylcholine (PC), phosphatidylethanolamine (PE), monogalactosyl diacylglycerol (MGDG), digalactosyl diacylglycerol (DGDG), trilinolein, triolein, 1,2-diacylglycerol (1,2-DAG), 1,3-diacylglycerol (1,3-DAG), 1-monoacylglycerol (1-MAG), 2-monoacylglycerol (2-MAG), and triacylglycerol (TAG) in 200 μL of 50 mM sodium phosphate buffer (pH 7.2) for 30 min at 30 °C. The released free fatty acids were extracted and reacted using the nonesterified fatty acid (NEFA) colorimetric kit (Wako Pure Chemicals) and measured colorimetrically at 546 nm.

Author Contributions: Conceptualization, I.L., J.-H.Y. and E.Y.K.; methodology, Y.H., J.-H.Y. and E.Y.K.; software, Y.H. and I.L.; validation, Y.H.; formal analysis, Y.H. and S.-Y.P.; investigation, Y.H.; resources, W.T.K. and W.L.; data curation, Y.H., J.-H.Y. and J.-H.P.; writing—original draft preparation, Y.H., I.L. and S.M.; writing—review and editing, Y.H. and W.L.; visualization, Y.H., I.L. and S.M.; supervision, W.T.K. and W.L.; project administration, W.T.K. and W.L.; funding acquisition, W.T.K. and W.L. All authors have read and agreed to the published version of the manuscript.

Funding: This research was funded by research grants (NRF-2020M3A9G7103934 to W.L. and NRF-2019M3A9F6021810 to W.L.) from the National Research Foundation (NRF) of Korea.

Institutional Review Board Statement: Not applicable.

Informed Consent Statement: Not applicable.

Data Availability Statement: The atomic coordinates for AtDSEL and CaPLA1 have been deposited in the Protein Data Bank under the accession code 7X0C and 7X0D, respectively.

Acknowledgments: We thank the staff at the PAL, Pohang, Korea and BL-17A of the Photon Factory, Tsukuba, Japan.

Conflicts of Interest: The authors declare no conflict of interest.

Sample Availability: Samples of the compounds are not available from the authors.

References

1. Richmond, G.S.; Smith, T.K. Phospholipases A1. *Int. J. Mol. Sci.* **2011**, *12*, 588–612. [CrossRef] [PubMed]
2. Filkin, S.Y.; Lipkin, A.V.; Fedorov, A.N. Phospholipase Superfamily: Structure, Functions, and Biotechnological Applications. *Biochemistry* **2020**, *85*, S177–S195. [CrossRef] [PubMed]
3. Aoki, J.; Inoue, A.; Makide, K.; Saiki, N.; Arai, H. Structure and function of extracellular phospholipase A1 belonging to the pancreatic lipase gene family. *Biochimie* **2007**, *89*, 197–204. [CrossRef] [PubMed]
4. Hou, M.-H.; Chuang, C.-Y.; Ko, T.-P.; Hu, N.-J.; Chou, C.-C.; Shih, Y.-P.; Ho, C.-L.; Wang, A.H.-J. Crystal structure of vespid phospholipase A(1) reveals insights into the mechanism for cause of membrane dysfunction. *Insect Biochem. Mol. Biol.* **2016**, *68*, 79–88. [CrossRef]
5. Skjold-Jørgensen, J.; Vind, J.; Moroz, O.V.; Blagova, E.; Bhatia, V.K.; Swendsen, A.; Wilson, K.S.; Bjerrum, M.J. Controlled lid-opening in Thermomyces lanuginosus lipase- An engineered switch for studying lipase function. *Biochim. Biophys. Acta Proteins Proteom.* **2017**, *1865*, 20–27. [CrossRef]
6. Zhang, M.; Yu, X.-W.; Xu, Y.; Guo, R.-T.; Swapna, G.V.T.; Szyperski, T.; Hunt, J.F.; Montelione, G.T. Structural Basis by Which the N-Terminal Polypeptide Segment of Rhizopus chinensis Lipase Regulates Its Substrate Binding Affinity. *Biochemistry* **2019**, *58*, 3943–3954. [CrossRef]
7. Birrane, G.; Beigneux, A.P.; Dwyer, B.; Strack-Logue, B.; Kristensen, K.K.; Francone, O.L.; Fong, L.G.; Mertens, H.D.T.; Pan, C.Q.; Ploug, M.; et al. Structure of the lipoprotein lipase-GPIHBP1 complex that mediates plasma triglyceride hydrolysis. *Proc. Natl. Acad. Sci. USA* **2019**, *116*, 1723–1732. [CrossRef]
8. Arora, R.; Nimonkar, A.V.; Baird, D.; Wang, C.; Chiu, C.-H.; Horton, P.A.; Hanrahan, S.; Cubbon, R.; Weldon, S.; Tschantz, W.R.; et al. Structure of lipoprotein lipase in complex with GPIHBP1. *Proc. Natl. Acad. Sci. USA* **2019**, *116*, 10360–10365. [CrossRef]
9. Tilbeurgh, H.v.; Sarda, L.; Verger, R.; Cambillau, C. Structure of the pancreatic lipase-procolipase complex. *Nature* **1992**, *359*, 159–162. [CrossRef]
10. Labar, G.; Bauvois, C.; Borel, F.; Ferror, J.-L.; Wouters, J.; Lambert, D.M. Crystal structure of the human monoacylglycerol lipase, a key actor in endocannabinoid signaling. *Chembiochem* **2010**, *11*, 218–227. [CrossRef]
11. Bian, C.; Yuan, C.; Chen, L.; Meehan, E.J.; Jiang, L.; Huang, Z.; Lin, L.; Huang, M. Crystal structure of a triacylglycerol lipase from Penicillium expansum at 1.3 A determined by sulfur SAD. *Proteins* **2010**, *78*, 1601–1605. [PubMed]
12. Noble, M.E.; Cleasby, A.; Johnson, L.N.; Egmond, M.R.; Frenken, L.G. The crystal structure of triacylglycerol lipase from Pseudomonas glumae reveals a partially redundant catalytic aspartate. *FEBS Lett.* **1993**, *331*, 123–128. [CrossRef]
13. Kim, K.K.; Song, H.K.; Shin, D.H.; Hwang, K.Y.; Suh, S.W. The crystal structure of a triacylglycerol lipase from Pseudomonas cepacia reveals a highly open conformation in the absence of a bound inhibitor. *Structure* **1997**, *5*, 173–185. [CrossRef]
14. Derewenda, U.; Swenson, L.; Wei, Y.; Green, R.; Kobos, P.M.; Joerger, R.; Haas, M.J.; Derewenda, Z.S. Conformational lability of lipases observed in the absence of an oil-water interface: Crystallographic studies of enzymes from the fungi Humicola lanuginosa and Rhizopus delemar. *J. Lipid Res.* **1994**, *35*, 524–534. [CrossRef]
15. Yapoudjian, S.; Ivanova, M.G.; Brzozowski, A.M.; Patkar, S.A.; Vind, J.; Svendsen, A.; Verger, R. Binding of Thermomyces (Humicola) lanuginosa lipase to the mixed micelles of cis-parinaric acid/NaTDC. *Eur. J. Biochem.* **2002**, *269*, 1613–1621. [CrossRef]
16. Kohno, M.; Funatsu, J.; Mikami, B.; Kugimiya, W.; Matsuo, T.; Morita, Y. The crystal structure of lipase II from Rhizopus niveus at 2.2 A resolution. *J. Biochem.* **1996**, *120*, 505–510. [CrossRef]
17. Wang, G.; Ryu, S.; Wang, X. Plant phospholipases: An overview. *Methods Mol. Biol.* **2012**, *861*, 123–137.

18. Wang, X. Plant Phospholipases. *Annu. Rev. Plant Physiol. Plant Mol. Biol.* **2001**, *52*, 211–231. [CrossRef]
19. Ishiguro, S.; Kawai-Oda, A.; Ueda, J.; Nishida, I.; Okada, K. The defective in anther dehiscience gene encodes a novel phospholipase A1 catalyzing the initial step of jasmonic acid biosynthesis, which synchronizes pollen maturation, anther dehiscence, and flower opening in Arabidopsis. *Plant Cell* **2001**, *13*, 2191–2209. [CrossRef]
20. Kim, E.Y.; Seo, Y.S.; Kim, W.T. AtDSEL, an Arabidopsis cytosolic DAD1-like acylhydrolase, is involved in negative regulation of storage oil mobilization during seedling establishment. *J. Plant Physiol.* **2011**, *168*, 1705–1709. [CrossRef]
21. Seo, Y.S.; Kim, E.Y.; Mang, H.G.; Kim, W.T. Heterologous expression, and biochemical and cellular characterization of CaPLA1 encoding a hot pepper phospholipase A1 homolog. *Plant J.* **2008**, *53*, 895–908. [CrossRef] [PubMed]
22. Derewenda, Z.S.; Derewenda, U. Relationships among serine hydrolases: Evidence for a common structural motif in triacylglyceride lipases and esterases. *Biochem. Cell Biol.* **1991**, *69*, 842–851. [CrossRef] [PubMed]
23. Ploier, B.; Scharwey, M.; Koch, B.; Schmidt, C.; Schatte, J.; Rechberger, G.; Kollroser, M.; Hermetter, A.; Daum, G. Screening for hydrolytic enzymes reveals Ayr1p as a novel triacylglycerol lipase in Saccharomyces cerevisiae. *J. Biol. Chem.* **2013**, *288*, 36061–36072. [CrossRef] [PubMed]
24. Kim, R.J.; Suh, M.C. The GxSxG motif of Arabidopsis monoacylglycerol lipase (MAGL6 and MAGL8) is essential for their enzyme activities. *Appl. Biol. Chem.* **2016**, *59*, 833–840. [CrossRef]
25. Xu, T.; Liu, L.; Hou, S.; Xu, J.; Yang, B.; Wang, Y.; Liu, J. Crystal structure of a mono- and diacylglycerol lipase from Malassezia globosa reveals a novel lid conformation and insights into the substrate specificity. *J. Struct. Biol.* **2012**, *178*, 363–369. [CrossRef] [PubMed]
26. Lowe, M.E. The catalytic site residues and interfacial binding of human pancreatic lipase. *J. Biol. Chem.* **1992**, *267*, 17069–17073. [CrossRef]
27. Otwinowski, Z.; Minor, W. Processing of X-ray diffraction data collected in oscillation mode. *Methods Enzymol.* **1997**, *276*, 307–326.
28. Liebschner, D.; Afonine, P.V.; Baker, M.L.; Bunkoczi, G.; Chen, V.B.; Croll, T.I.; Hintze, B.; Hung, L.W.; Jain, S.; McCoy, A.J.; et al. Macromolecular structure determination using X-rays, neutrons and electrons: Recent developments in Phenix. *Acta Crystallogr. D Struct. Biol.* **2019**, *75*, 861–877. [CrossRef]
29. Emsley, P.; Cowtan, K. Coot: Model-building tools for molecular graphics. *Acta Crystallogr. D Struct. Biol.* **2004**, *60*, 2126–2132. [CrossRef]
30. Seo, Y.S.; Kim, E.Y.; Kim, J.H.; Kim, W.T. Enzymatic characterization of class I DAD1-like acylhydrolase members targeted to chloroplast in Arabidopsis. *FEBS Lett.* **2009**, *583*, 2301–2307. [CrossRef]

molecules

Article

The Structural Rule Distinguishing a Superfold: A Case Study of Ferredoxin Fold and the Reverse Ferredoxin Fold

Takumi Nishina [1], Megumi Nakajima [1], Masaki Sasai [1,2,3,*] and George Chikenji [1,*]

[1] Department of Applied Physics, Nagoya University, Nagoya 464-8601, Japan; nishina@tbp.ap.pse.nagoya-u.ac.jp (T.N.); nakajima@tbp.cse.nagoya-u.ac.jp (M.N.)
[2] Department of Complex Systems Science, Nagoya University, Nagoya 464-8601, Japan
[3] Fukui Institute for Fundamental Chemistry, Kyoto University, Kyoto 606-8501, Japan
* Correspondence: masakisasai@nagoya-u.jp (M.S.); chikenji@tbp.ap.pse.nagoya-u.ac.jp (G.C.)

Abstract: Superfolds are folds commonly observed among evolutionarily unrelated multiple superfamilies of proteins. Since discovering superfolds almost two decades ago, structural rules distinguishing superfolds from the other ordinary folds have been explored but remained elusive. Here, we analyzed a typical superfold, the ferredoxin fold, and the fold which reverses the N to C terminus direction from the ferredoxin fold as a case study to find the rule to distinguish superfolds from the other folds. Though all the known structural characteristics for superfolds apply to both the ferredoxin fold and the reverse ferredoxin fold, the reverse fold has been found only in a single superfamily. The database analyses in the present study revealed the structural preferences of $\alpha\beta$- and $\beta\alpha$-units; the preferences separate two α-helices in the ferredoxin fold, preventing their collision and stabilizing the fold. In contrast, in the reverse ferredoxin fold, the preferences bring two helices near each other, inducing structural conflict. The Rosetta folding simulations suggested that the ferredoxin fold is physically much more realizable than the reverse ferredoxin fold. Therefore, we propose that minimal structural conflict or minimal frustration among secondary structures is the rule to distinguish a superfold from ordinary folds. Intriguingly, the database analyses revealed that a most stringent structural rule in proteins, the right-handedness of the $\beta\alpha\beta$-unit, is broken in a set of structures to prevent the frustration, suggesting the proposed rule of minimum frustration among secondary structural units is comparably strong as the right-handedness rule of the $\beta\alpha\beta$-unit.

Keywords: protein design; reverse fold; minimum frustration

Citation: Nishina, T.; Nakajima, M.; Sasai, M.; Chikenji, G. The Structural Rule Distinguishing a Superfold: A Case Study of Ferredoxin Fold and the Reverse Ferredoxin Fold. *Molecules* **2022**, *27*, 3547. https://doi.org/10.3390/molecules27113547

Academic Editors: Kunihiro Kuwajima, Yuko Okamoto, Tuomas Knowles and Michele Vendruscolo

Received: 28 March 2022
Accepted: 28 May 2022
Published: 31 May 2022

1. Introduction

A principal goal of protein science is to elucidate the relationship among sequences, structures, and functions [1,2]. Toward such a goal, remarkable progress has been achieved in structure prediction from the knowledge of amino-acid sequences [3,4]. Also, in protein design, which is a reverse problem of structure prediction, elucidation of design principles [5–7] led to an increasing number of successful examples to find amino-acid sequences that can fold into the designed structures [5,6,8–12]. Here, for further advancing the design technology, it is crucial to develop a systematic method to distinguish less designable structures and highly designable ones into each of which a large number of different sequences can fold [13]. Investigating the occurrence of structural folds among natural proteins provides a clue to this problem [14–18]. An ordinary fold appears in only one or a few superfamilies, but a particular fold is shared by a large number of superfamilies; such a particular fold was called a superfold [19]. Here, a superfamily is defined as the largest group of proteins for which common ancestry can be inferred [20]. Superfolds are rare in the entire fold categories but are robust against mutations, suggesting superfolds represent highly designable structures. Each superfold corresponds to many different functions, in sharp contrast to the ordinary folds showing the nearly one-to-one correspondence between fold and function.

Since the discovery of superfolds [19], features distinguishing superfolds from the other ordinary folds have been explored, leading to the several empirical rules that characterize the superfolds, some of which are (1) frequent appearance of super secondary structures [21], (2) avoidance of mixing parallel and anti-parallel β-sheets [14], (3) infrequent jumps between β-strands [16], and (4) high structural symmetry [22]. However, examples of ordinary folds satisfy the rules from (1) through (4), showing the need for further rules to distinguish superfolds. The reverse ferredoxin fold is such an example. The ferredoxin fold, a typical superfold, comprises four β-strands connected in the order and directions as designated in Figure 1A. The reverse ferredoxin fold reverses the N to C terminus direction from the ferredoxin fold (Figure 1B). According to the SCOPe classification [23,24], the ferredoxin fold is found in 62 superfamilies, whereas the reverse ferredoxin fold is found only in one superfamily. Therefore, the reverse ferredoxin fold is not a superfold, but both the ferredoxin fold and the reverse ferredoxin fold satisfy the rules (1) through (4). Other examples show the significant difference between the fold and the reverse fold in the number of occurrences in the spectrum of families [15]. The reason for this difference between folds and reverse folds remains elusive; there have been arguments suggesting physical or functional necessities to avoid the reverse folds [15] and those suggesting the bias occasionally acquired in evolutionary history [25].

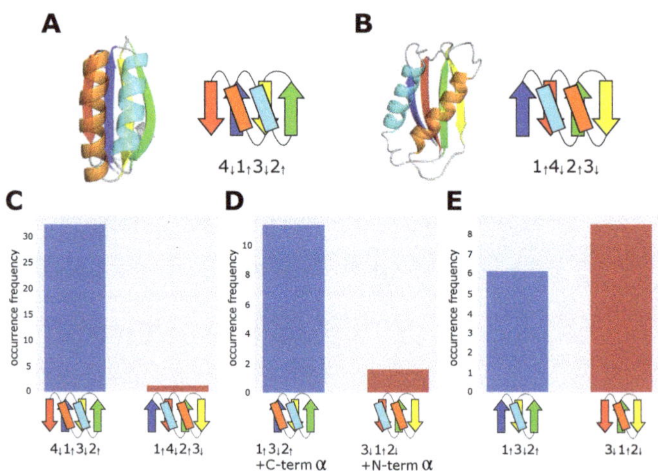

Figure 1. Topology and occurrence frequency of the ferredoxin fold and the reverse ferredoxin fold. (**A**) An example structure (a microcompartment protein, PDB ID: 4QIV) and the topology $4_{\downarrow}1_{\uparrow}3_{\downarrow}2_{\uparrow}$ of the ferredoxin fold. (**B**) An example structure (the catalytic core of human DNA polymerase kappa, PDB ID: 1T94) and the topology $1_{\uparrow}4_{\downarrow}2_{\uparrow}3_{\downarrow}$ of the reverse ferredoxin fold. (**C**) Occurrence frequency of the ferredoxin topology $4_{\downarrow}1_{\uparrow}3_{\downarrow}2_{\uparrow}$ and the reverse ferredoxin topology $1_{\uparrow}4_{\downarrow}2_{\uparrow}3_{\downarrow}$. (**D**) Occurrence frequency of the topology $1_{\uparrow}3_{\downarrow}2_{\uparrow}$ + C-term α and the topology $3_{\downarrow}1_{\uparrow}2_{\downarrow}$ + N-term α. (**E**) Occurrence frequency of the topology $1_{\uparrow}3_{\downarrow}2_{\uparrow}$ and the topology $3_{\downarrow}1_{\uparrow}2_{\downarrow}$. In (**C–E**), the dataset of the 99% sequence identity representatives derived from ECOD was used. Chains are colored from blue (N-terminus) to red (C-terminus). In the topology diagram, β-strands are represented with arrows and α-helices are rectangles.

Here, we explored the factor to distinguish superfolds from the ordinary folds by comparing the ferredoxin fold and the reverse ferredoxin fold as a case study. By analyzing the database, we found the structural tendency shown by the $\alpha\beta$-unit and $\beta\alpha$-unit, suggesting that the structure comprises multiple $\alpha\beta$- and $\beta\alpha$-units should satisfy a rule to minimize the conflict between structural tendencies of these units. We show that the ferredoxin fold satisfies this rule for minimal conflict or frustration, whereas the reverse ferredoxin fold does not. We also performed the Rosetta folding simulations to test the foldability of structures [5]; the test results suggested that the ferredoxin fold is physically much more realizable than the reverse ferredoxin fold. Thus, we propose that the minimum frustration rule to consistently satisfy the structural preference of multiple parts of the protein is a rule to distinguish superfolds from ordinary folds.

2. Results

2.1. Occurrence Frequency of Topologies

Previous analyses showed that the ferredoxin fold is frequently found, whereas the reverse ferredoxin fold is rare among protein families [17,25]. We confirmed this imbalance in the most recent version of a semi-manually curated database, ECOD (version 20210511: develop280), which hierarchically classifies protein domains according to homology, reflecting their evolutionary relationship [26]. ECOD has been frequently updated, suited to estimating the most recent number of homology groups having a topology on which we focus. The ECOD database classifies homologous protein domains according to categories of family and homology. The family (F) group consists of evolutionarily related protein domains with substantial sequence similarity, and the homology (H) group comprises multiple F-groups having functional and structural similarities. The H-group corresponds to the superfamily in the other structural databases, SCOP [27] and CATH [28]. The X-group in ECOD comprises multiple H-groups that share similar features in the structure but lack a convincing evidence for homology. In this study, we used the 99% sequence identity representatives in ECOD as the dataset for the analyses.

We detected secondary structures and hydrogen bonds in protein domains recorded in the dataset using STRIDE [29]. Then, based on the thus found hydrogen-bond pattern among β-strands, we defined the β-sheet topology as in Ref. [15]; we describe the β-sheet topology by representing the strand directions with up and down arrows with the sequential number from the N- to C-termini (4132, for example). Then, topology T of the ferredoxin fold is $T = 4_{\downarrow}1_{\uparrow}3_{\downarrow}2_{\uparrow}$ (Figure 1A) and topology T of the reverse ferredoxin fold is $T = 1_{\uparrow}4_{\downarrow}2_{\uparrow}3_{\downarrow}$ (Figure 1B).

We estimated the occurrence frequency $OF(T)$ of a given topology T by summing the occupation ratio $OR(T,i)$ of protein domains having T in the ith H-group as

$$OF(T) = \sum_{i=1}^{N_{\text{homology}}} OR(T,i), \quad (1)$$

where N_{homology} is the total number of H groups in the dataset, and

$$OR(T,i) = \frac{1}{N_{\text{family}}(i)} \sum_{j=1}^{N_{\text{family}}(i)} \frac{N_{\text{domain}}(T,i,j)}{N_{\text{domain}}(i,j)}. \quad (2)$$

Here, $N_{\text{domain}}(T,i,j)$ is the number of protein domains having topology T in the jth F-group, which belongs to the ith H-group in the dataset. $N_{\text{domain}}(i,j) = \sum_T N_{\text{domain}}(T,j)$ is the total number of protein domains in the jth F-group, and $N_{\text{family}}(i)$ is the number of F-groups in the ith H-group. Figure 1C shows that the occurrence frequency of the ferredoxin topology, $OF(4_{\downarrow}1_{\uparrow}3_{\downarrow}2_{\uparrow})$, is more than 10 times larger than the occurrence frequency of the reverse ferredoxin topology, $OF(1_{\uparrow}4_{\downarrow}2_{\uparrow}3_{\downarrow})$, confirming the previously reported ubiquity of the ferredoxin fold and the rareness of the reverse ferredoxin fold [17,25].

Here, we should note that topology has often been classified with ECOD in terms of X-groups; for example, an X-group called "alpha-beta plaits" has been regarded as the group representing the ferredoxin topology. However, we used STRIDE for a more precise topological classification instead of the X-group classification. Therefore, the $OF(T)$ defined in Equation (1) does not precisely correlate with the number of H-groups in the X-group. Tetracycline resistance protein, tetM (PDB ID: 3J25), for example, belongs to the X-group of alpha-beta plaits, but we did not count tetM as a ferredoxin-topology protein because STRIDE identifies only two β-strands in tetM. Similarly, surface-layer (S-layer) protein

(PDB ID: 3CVZ) belongs to the reverse ferredoxin X-group in ECOD, but we did not count S-layer protein as a protein with the reverse-ferredoxin fold because STRIDE identifies a topology $1_\uparrow 5_\uparrow 4_\downarrow 2_\uparrow 3_\downarrow$ for S-layer protein instead of $1_\uparrow 4_\downarrow 2_\uparrow 3_\downarrow$. See Supplementary Figure S1 for the structure of tetM and S-layer protein.

We examine the minimal structural units that induce the difference between $4_\downarrow 1_\uparrow 3_\downarrow 2_\uparrow$ and $1_\uparrow 4_\downarrow 2_\uparrow 3_\downarrow$. We consider the topology in which the C-terminal strand (β-strand 4) is deleted from the ferredoxin topology by retaining the α-helix connecting β-strands 4 and 3 in the structure, and write the thus obtained topology as $1_\uparrow 3_\downarrow 2_\uparrow +$ C-term α. We also consider the topology in which the C-term α is further deleted from $1_\uparrow 3_\downarrow 2_\uparrow +$ C-term α and write such a topology as $1_\uparrow 3_\downarrow 2_\uparrow$. Similarly, we consider the topology in which the N-terminal strand (β-strand 1) is deleted from the reverse ferredoxin topology by retaining the α-helix connecting β-strands 1 and 2 in the structure. Then, we renumber the strands as $4, 2, 3 \rightarrow 3, 1, 2$, and write the thus-obtained topology as $3_\downarrow 1_\uparrow 2_\downarrow +$ N-term α, which is the reverse of $1_\uparrow 3_\downarrow 2_\uparrow +$ C-term α. We also consider the topology in which the N-term α is further deleted from $3_\downarrow 1_\uparrow 2_\downarrow +$ N-term α and write such a topology as $3_\downarrow 1_\uparrow 2_\downarrow$, which is the reverse of $1_\uparrow 3_\downarrow 2_\uparrow$.

We consider protein domains whose entire (not the partial) structure has the topology $1_\uparrow 3_\downarrow 2_\uparrow +$ C-term α or $3_\downarrow 1_\uparrow 2_\downarrow +$ N-term α, and calculated occurrence frequencies, $OF(1_\uparrow 3_\downarrow 2_\uparrow +$ C-term $\alpha)$ and $OF(3_\downarrow 1_\uparrow 2_\downarrow +$ N-term $\alpha)$ (Figure 1D). We should note that with the topology of $1_\uparrow 3_\downarrow 2_\uparrow +$ C-term α, the C-term α can lie on either side of the β-sheet plane. However, in the ferredoxin fold, this helix is always on the same side of the plane as the α-helix of the $\beta\alpha\beta$-unit consisting of β-strands 1 and 2; therefore, we here calculated $OF(1_\uparrow 3_\downarrow 2_\uparrow +$ C-term $\alpha)$ for the structures in which the C-term α is on the same side of the plane as the α-helix of the $\beta\alpha\beta$-unit. Similarly, we calculated $OF(3_\downarrow 1_\uparrow 2_\downarrow +$ N-term $\alpha)$ for structures in which the N-term α is on the same side of the β-sheet plane as the α-helix of the $\beta\alpha\beta$-unit consisting of β-strands 2 and 3. See the Materials and Methods section for the way to judge which side of the plane the terminal helix lies in a given structure in calculating OFs. Figure 1D shows that $OF(1_\uparrow 3_\downarrow 2_\uparrow +$ C-term $\alpha)$ is significantly larger than $OF(3_\downarrow 1_\uparrow 2_\downarrow +$ N-term $\alpha)$, suggesting that the determining structural factor distinguishing the ferredoxin fold and the reverse ferredoxin fold exists in the difference between $1_\uparrow 3_\downarrow 2_\uparrow +$ C-term α and $3_\downarrow 1_\uparrow 2_\downarrow +$ N-term α. The population of the structures with two helices lying on the opposite side of the β-sheet plane is small in the $1_\uparrow 3_\downarrow 2_\uparrow +$ C-term α topology and in the $3_\downarrow 1_\uparrow 2_\downarrow +$ N-term α topology, and there is no significant difference between occurrence frequencies of two topologies for those structures with helices lying on the opposite side of the plane. The large difference between two topologies only appear for structures in which two helices lie on the same side of the plane (Supplementary Figures S2 and S3).

Similarly, we calculated occurrence frequencies, $OF(1_\uparrow 3_\downarrow 2_\uparrow)$ and $OF(3_\downarrow 1_\uparrow 2_\downarrow)$ (Figure 1E), showing that $OF(3_\downarrow 1_\uparrow 2_\downarrow)$ is mildly larger than $OF(1_\uparrow 3_\downarrow 2_\uparrow)$. These results suggest that the determinant structural factor that induces the difference between $4_\downarrow 1_\uparrow 3_\downarrow 2_\uparrow$ and $1_\uparrow 4_\downarrow 2_\uparrow 3_\downarrow$ is in the difference between $1_\uparrow 3_\downarrow 2_\uparrow +$ C-term α and $3_\downarrow 1_\uparrow 2_\downarrow +$ N-term α. Addition of the C-term α-helix to $1_\uparrow 3_\downarrow 2_\uparrow$ and addition of the N-term α-helix to $3_\downarrow 1_\uparrow 2_\downarrow$ bring about the difference in the occurrence frequency between the ferredoxin topology and the reverse ferredoxin topology. Hereafter, the ferreoxin fold and the $1_\uparrow 3_\downarrow 2_\uparrow +$ C-term α topology are referred to collectively as the ferredoxin-type topology, and the reverse ferredoxin fold and the $3_\downarrow 1_\uparrow 2_\downarrow +$ N-term α topology are referred to collectively as the reverse ferredoxin-type topology.

2.2. Conflict between Structural Preferences of αβ- and βα-Units

Because positions of the $\alpha\beta$- and $\beta\alpha$-units are different in $1_\uparrow 3_\downarrow 2_\uparrow +$ C-term α and $3_\downarrow 1_\uparrow 2_\downarrow +$ N-term α (Figure 1A,B), analyses on these structural units should give critical insights on the difference between $1_\uparrow 3_\downarrow 2_\uparrow +$ C-term α and $3_\downarrow 1_\uparrow 2_\downarrow +$ N-term α. For the structural analyses of these units, we defined the distance x between the plane of the β-pleats in the strand and the α-helix (Figure 2A). See the Materials and Methods section for the precise definition of x. We derived the distribution of x by analyzing the dataset culled from PDB with constraints of the sequence identity < 30%, the finer resolution than 2.0 Å, and the R-factor < 0.25 [30]. For the statistical analyses, we selected typical $\alpha\beta$- and $\beta\alpha$-units following the criterion of Ref. [31]; we used the structural units satisfying the conditions that the linker loop between α-helix and β-strand is shorter than five-residue length and the angle between α-helix and β-strand is less than 60°.

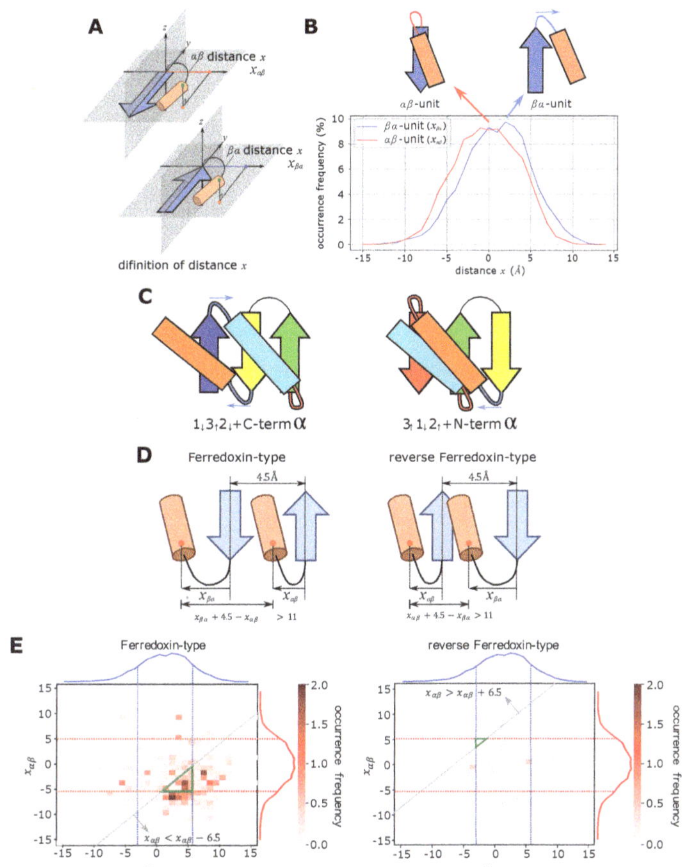

Figure 2. Absence or presence of the structural conflict between α-helices. (**A**) Definition of the distance x between the pleated plane of the β-strand and the α-helix in the $\alpha\beta$-unit (top) and the $\beta\alpha$-unit (bottom). (**B**) Distribution of x in the $\alpha\beta$-unit (red) and the $\beta\alpha$-unit (blue). The distribution was found in the culled PDB dataset with the parameters of the sequence identity < 30%, the finer resolution than 2.0 Å, and the R-factor < 0.25. (**C**) Structural preferences of the the $\alpha\beta$-unit (connected by a red linker) and the $\beta\alpha$-unit (connected by a blue linker) prevent collision between the terminal helix and the helix in the $\beta\alpha\beta$ structure in the $1_\uparrow 3_\downarrow 2_\uparrow$ + C-term α topology (left), while they induce a collision in the $3_\downarrow 1_\uparrow 2_\downarrow$ + N-term α topology (right). Blue arrows show the shift of α-helix induced by the $x > 0$ preference of the $\beta\alpha$-unit. (**D**) The necessary condition to avoid the collision of two helices. $x_{\beta\alpha} - x_{\alpha\beta} + 4.5$ Å > 11Å for the ferredoxin-type topology and $x_{\alpha\beta} - x_{\beta\alpha} + 4.5$ Å > 11Å for the reverse ferredoxin-type topology. (**E**) The realizable area to avoid the collision and the occurrence frequency of $(x_{\beta\alpha}, x_{\alpha\beta})$ in the ECOD database. The realizable area satisfying the three conditions; the necessary condition to avoid the collision, the condition of the frequency > 5% in the $x_{\beta\alpha}$ distribution, and the condition of the frequency > 5% in the $x_{\alpha\beta}$ distribution; is shown with a green triangle on the $(x_{\beta\alpha}, x_{\alpha\beta})$ plane. The occurrence frequency shown with the gray-scale is superposed. Blue and red curves are distributions in (**B**).

Figure 2B shows the distribution of x obtained by the dataset analyses. The distribution of x in the $\beta\alpha$-unit peaked at $2{\sim}4$ Å, whereas the distribution of x in the $\alpha\beta$-unit peaked at $\sim$0 Å, showing a distinct tendency of positive x in the $\beta\alpha$-unit. This positive x distribution implies the tendency of shifting the α-helix toward the direction of blue arrows in Figure 2C. In the $1_{\uparrow}3_{\downarrow}2_{\uparrow}$ + C-term α structure, this shift separates the C-term α-helix from the helix in the $\beta\alpha\beta$ structure, while in the $3_{\downarrow}1_{\uparrow}2_{\downarrow}$ + N-term α structure, the shift induces collision of the N-term α-helix against the helix in the $\beta\alpha\beta$ structure when two helices are on the same side of the β-sheet surface. Therefore, the structural conflict arising between two helices destabilizes the $3_{\downarrow}1_{\uparrow}2_{\downarrow}$ + N-term α structure; and hence, destabilizes the reverse ferredoxin fold.

We can quantitatively assess how the difference in the distribution of the distance x in Figure 2B determines the absence/presence of the structural conflict. We write x in the $\beta\alpha$-unit and the $\alpha\beta$-unit as $x_{\beta\alpha}$ and $x_{\alpha\beta}$, respectively. Considering that a typical distance between two adjacent β-strands in a β-sheet is 4.5 Å [32], the distance between two helices in the ferredoxin-type topology is $x_{\beta\alpha} - x_{\alpha\beta} + 4.5$ Å. Similarly, the distance between two helices in the reverse ferredoxin-type topology is $x_{\alpha\beta} - x_{\beta\alpha} + 4.5$ Å (Figure 2D). Because the helix diameter is approximately 11.0 Å [33], the necessary condition to avoid the collision of two helices is $x_{\beta\alpha} - x_{\alpha\beta} + 4.5\,Å > 11Å$ for the ferredoxin-type topology and $x_{\alpha\beta} - x_{\beta\alpha} + 4.5\,Å > 11Å$ for the reverse ferredoxin-type topology. In Figure 2E, the region satisfying three conditions at the same time is designated by a green triangle on a two-dimensional plane of $x_{\beta\alpha}$ and $x_{\alpha\beta}$: (i) the necessary condition to avoid the collision, (ii) the condition of frequency $> 5\,\%$ in the frequency distribution of $x_{\beta\alpha}$ in Figure 2B, and (iii) the condition of frequency $> 5\,\%$ in the frequency distribution of $x_{\alpha\beta}$ in Figure 2B. The thus-defined green triangle, i.e., the realizable area to avoid the collision, is extremely narrow in the reverse ferredoxin-type topology, whereas it is wide in the ferredoxin-type topology. Figure 2E shows that the occurrence frequency of $(x_{\beta\alpha}, x_{\alpha\beta})$ in the ECOD database is large around the green triangle in the ferredoxin-type fold, while the frequency is small everywhere on the plane of $(x_{\beta\alpha}, x_{\alpha\beta})$ in the reverse ferredoxin-type fold. Thus, the shift of $2{\sim}4$ Å in distributions in Figure 2B is a determining factor for the realizability of the structure. In the reverse ferredoxin-type topology, the structures are realized by breaking at least one of three conditions (i)–(iii). Different ways of breaking the conditions in the reverse ferredoxin-type topology make the distribution scattered on the $(x_{\beta\alpha}, x_{\alpha\beta})$ plane in Figure 2E. Supplementary Figure S4 shows example proteins with the reverse ferredoxin topology showing uncommon configuration of the $\beta\alpha$- or $\alpha\beta$-unit.

We should note that the results shown in Figure 2B,E are the plots for proteins with loops shorter than five-residue length. The longer loops allow the structural variety to obscure the realizability conditions in Figure 2B,E. However, the stability of native structures inversely correlates to the loop length [34,35], making the proteins having the longer loops rare. See Supplementary Figure S5 for the distribution of the loop length found in the ECOD database. Here, it is sufficient to consider non-rare proteins with short enough loops for clarifying how the ferredoxin-type topology is much more realizable than the reverse ferredoxin-type topology.

2.3. Minimum Frustration Rule

The dataset analyses showed that the structural preference of $\alpha\beta$- and $\beta\alpha$-units leads to the structural conflict in the $3_{\downarrow}1_{\uparrow}2_{\downarrow}$ + N-term α structure, while the conflict is avoided in the $1_{\uparrow}3_{\downarrow}2_{\uparrow}$ + C-term α structure. We examined the effect of presence/absence of the structural conflict by performing the Rosetta folding simulations. In these simulations, we substituted all the residues in the model to Valine, and assembled the fragments of one-, three-, or nine-residue length, which have the compatible main-chain dihedral angles with the secondary structures in the blueprints designated in Figure 3. We used the all-Valine sequence to focus on the role of structural consistency among the assembled fragments instead of the effects of the residue-specific interactions. We regard structures generated through the simulations as compatible structures when they have low energy and the same topology as the blueprint. For each blueprint, we performed the fragment-assembly simulation 10,000 times and counted how many compatible structures were obtained through simulations. Koga et al. showed that the topology designated by the blueprint is physically realizable by avoiding the structural conflict when the number of the obtained compatible structures is large, while it is physically unrealizable with the structural inconsistency when the number is small [5]. See the Materials and Methods section for the details of the simulations.

Figure 3A shows the number of structures compatible with the $3_\downarrow 1_\uparrow 2_\downarrow$ + N-term α topology and the number of structures compatible with the $1_\uparrow 3_\downarrow 2_\uparrow$ + C-term α topology. The compatible structures were 229 and 10 for the $1_\uparrow 3_\downarrow 2_\uparrow$ + C-term α topology and the $3_\downarrow 1_\uparrow 2_\downarrow$ + N-term α topology, respectively, showing the $1_\uparrow 3_\downarrow 2_\uparrow$ + C-term α topology is much more realizable than the $3_\downarrow 1_\uparrow 2_\downarrow$ + N-term α topology. We performed the same test for the $1_\uparrow 3_\downarrow 2_\uparrow$ topology and the $3_\downarrow 1_\uparrow 2_\downarrow$ topology. Figure 3B shows that the number of compatible structures for the $1_\uparrow 3_\downarrow 2_\uparrow$ topology is almost same as the number of compatible structures for the $3_\downarrow 1_\uparrow 2_\downarrow$ topology, indicating that there is no significant difference between the realizability of these topologies. Figure 3A,B are qualitatively same as Figure 1D,E, showing that the difference in the realizability of the $1_\uparrow 3_\downarrow 2_\uparrow$ + C-term α topology and the $3_\downarrow 1_\uparrow 2_\downarrow$ + N-term α topology arises from absence/presence of the conflict between local structural units.

Combined analyses of databases and Rosetta folding simulations showed that the structural conflict or frustration is minimized in the largely realizable topology, which characterizes the superfold; therefore, we propose that the minimum frustration among local preferences of secondary structures is the rule to distinguish a superfold from the ordinary folds.

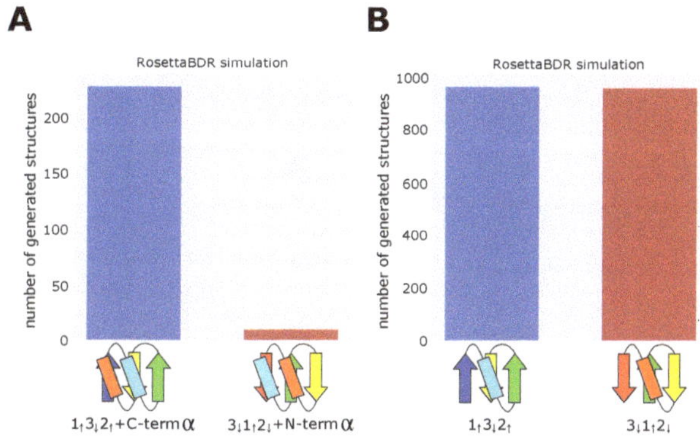

Figure 3. The number of simulated structures compatible with the blueprint. We repeated the Rosetta folding simulations 10,000 times and counted the number of compatible structures generated. (**A**) Comparison between the $1_\uparrow 3_\downarrow 2_\uparrow$ + C-term α topology and the $3_\downarrow 1_\uparrow 2_\downarrow$ + N-term α topology. In simulations, the number of structures in which two helices lie on the same side of the β-sheet surface was counted. (**B**) Comparison between the $1_\uparrow 3_\downarrow 2_\uparrow$ topology and the $3_\downarrow 1_\uparrow 2_\downarrow$ topology.

3. Discussion

In this study, we proposed a rule that the minimum frustration among local structural preferences of secondary structures is the necessary condition for superfolds. In this section, we discuss the meaning of this rule by explaining how the rule predicts occurrence frequency of other structures, the relation of the rule with the other design rule, and the relation with protein function.

3.1. Occurrence Frequency of Other Structures

The present analyses of the ferredoxin fold and the reverse ferredoxin fold showed that the frequently occurring topology is designed to minimize frustration among multiple secondary-structure units that lie near each other on the same side the β-sheet plane. We can examine whether this rule predicts the occurrence frequency of other structures in the dataset. Figure 4A–D are four examples of pairs of topologies; in each pair, one is the topology minimizing frustration, and the other is its reverse topology exhibiting frustration. We should note that pairs in Figure 4B–D have the same arrangement of β-strands but have different connections of terminal α-helices showing different topologies. Our rule of minimum frustration predicts that the topology shown on the left side in each pair in Figure 4 is more realizable than the topology on the right side. We counted the occurrence frequency of these topologies in the dataset and found a significant difference as expected. In particular, we found the zero occurrence frequency of the frustrated topology in Figure 4D. The absence of this topology is reasonable because the frustrated topology of Figure 4D has two positions of structural collisions between helices, whereas the other frustrated topologies in Figure 4A–C show only a single collision in each. These results support our proposal that the minimum frustration among secondary structures is the requirement for the frequently occurring topologies; therefore, the necessary condition for the superfolds.

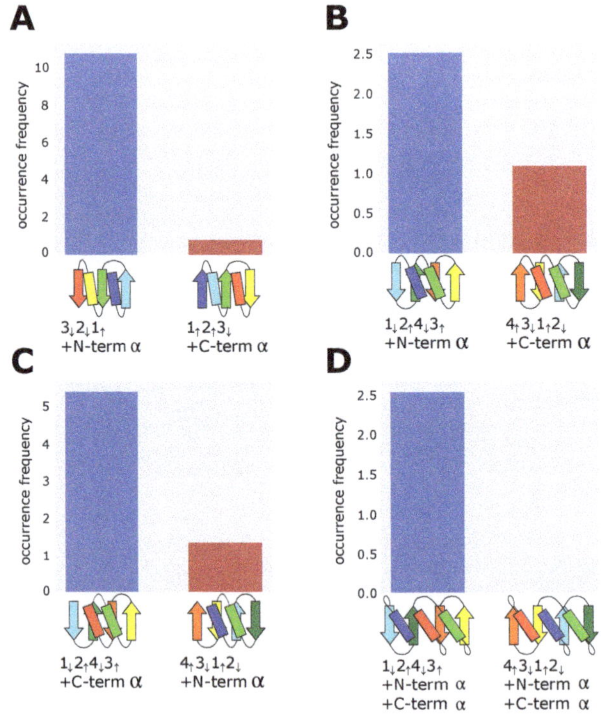

Figure 4. Comparisons of occurrence frequency between topologies minimizing frustration and their reverse topologies exhibiting frustration. **(A)** $3_{\downarrow}2_{\downarrow}1_{\uparrow}$ + N-term α and $1_{\uparrow}2_{\uparrow}3_{\downarrow}$ + C-term α, **(B)** $1_{\downarrow}2_{\uparrow}4_{\downarrow}3_{\uparrow}$ + N-term α and $4_{\uparrow}3_{\downarrow}1_{\uparrow}2_{\downarrow}$ + C-term α, **(C)** $1_{\downarrow}2_{\uparrow}4_{\downarrow}3_{\uparrow}$ + C-term α and $4_{\uparrow}3_{\downarrow}1_{\uparrow}2_{\downarrow}$ + N-term α, and **(D)** $1_{\downarrow}2_{\uparrow}4_{\downarrow}3_{\uparrow}$ + N-term α + C-term α and $4_{\uparrow}3_{\downarrow}1_{\uparrow}2_{\downarrow}$ + N-term α + C-term α. The dataset was the 99% sequence identity representatives derived from the ECOD database.

3.2. The Left-Handed $\beta\alpha\beta$-Unit Is Selectively Found in the $3_{\downarrow}1_{\uparrow}2_{\downarrow}$ + N-termα Structures

We showed that the collision between two helices arising from the structural preference of nearby $\alpha\beta$- and $\beta\alpha$-units decreases the occurrence frequency of the $3_{\downarrow}1_{\uparrow}2_{\downarrow}$ + N-term α topology. However, this collision disappears when the two helices lie on the opposite side of the β-sheet surface. Such configurations are possible in two different ways. One is the configuration that the $\beta\alpha\beta$-unit consisting of β-strands 2 and 3 is right-handed and the terminal helix is on the opposite side; we have a small number of such examples in the dataset as shown in Supplementary Figure S3. The other is the configuration that the $\beta\alpha\beta$-unit is left-handed with the terminal helix in the position similar to that in the reverse ferredoxin fold. Here, we cannot expect the frequent occurrence of the latter structure because more than 98% of the known $\beta\alpha\beta$-unit structures are right-handed [14,36–38]. Indeed, in our dataset derived from ECOD, there is no left-handed $\beta\alpha\beta$-unit in protein domains with the $1_\uparrow 3_\downarrow 2_\uparrow$ + C-term α or the $3_\downarrow 1_\uparrow 2_\downarrow$ + N-term α topology.

However, in the dataset, we found a small number of left-handed $\beta\alpha\beta$-units in protein domains having the extended structures including $1_\uparrow 3_\downarrow 2_\uparrow$ + C-term α or $3_\downarrow 1_\uparrow 2_\downarrow$ + N-term α as a partial structure (Figure 5B,C). See the Materials and Methods section for the method to detect the left-handed $\beta\alpha\beta$-unit in the dataset. Figure 5A shows occurrence frequencies of domains in the dataset having more than four β-strands and include the $1_\uparrow 3_\downarrow 2_\uparrow$ + C-term α or the $3_\downarrow 1_\uparrow 2_\downarrow$ + N-term α topology as their partial structure. For these extended domains, we counted occurrence frequencies separately for those having a left-handed $\beta\alpha\beta$-unit, $OF(\text{Extended-}1_\uparrow 3_\downarrow 2_\uparrow + \text{C-term } \alpha; \text{Left})$ and $OF(\text{Extended-}3_\downarrow 1_\uparrow 2_\downarrow + \text{N-term } \alpha; \text{Left})$, and for those having the right-handed $\beta\alpha\beta$-unit, $OF(\text{Extended-}1_\uparrow 3_\downarrow 2_\uparrow + \text{C-term } \alpha; \text{Right})$ and $OF(\text{Extended-}3_\downarrow 1_\uparrow 2_\downarrow + \text{N-term } \alpha; \text{Right})$. We found $OF(\text{Extended-}1_\uparrow 3_\downarrow 2_\uparrow\text{+C-term } \alpha; \text{Right}) = 73.8$, $OF(\text{Extended-}1_\uparrow 3_\downarrow 2_\uparrow\text{+C-term } \alpha; \text{Left}) = 0.5$, $OF(\text{Extended-}3_\downarrow 1_\uparrow 2_\downarrow\text{+N-term } \alpha; \text{Right}) = 16.0$, and $OF(\text{Extended-}3_\downarrow 1_\uparrow 2_\downarrow\text{+N-term } \alpha; \text{Left}) = 2.5$, leading to the ratios,

$$\frac{OF(\text{Extended-}1_\uparrow 3_\downarrow 2_\uparrow + \text{C-term } \alpha; \text{Left})}{OF(\text{Extended-}1_\uparrow 3_\downarrow 2_\uparrow + \text{C-term } \alpha; \text{Right})} \approx 0.0068,$$
$$\frac{OF(\text{Extended-}3_\downarrow 1_\uparrow 2_\downarrow + \text{N-term } \alpha; \text{Left})}{OF(\text{Extended-}3_\downarrow 1_\uparrow 2_\downarrow + \text{N-term } \alpha; \text{Right})} \approx 0.156, \quad (3)$$

suggesting that some mechanism exists for enhancing the occurrence of the left-handed $\beta\alpha\beta$-unit in the $3_\downarrow 1_\uparrow 2_\downarrow$ + N-term α structure. A plausible explanation is that the left-handed $\beta\alpha\beta$-unit was chosen in these domains to avoid the collision between two helices lying on the same side of the β-sheet in the Extended-$3_\downarrow 1_\uparrow 2_\downarrow$ + N-term α structures. This mechanism suggests that the rule for minimizing frustration between the structural preferences of secondary structures lying nearby on the same side of the β-sheet is comparably strong as the rule of the right-handedness of the $\beta\alpha\beta$-unit.

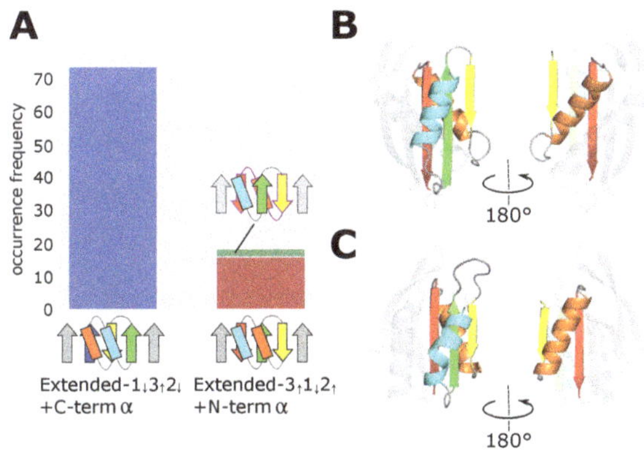

Figure 5. Occurrence of the left-handed and right-handed $\beta\alpha\beta$-units in the extended domains which include the $1_\uparrow 3_\downarrow 2_\uparrow$ + C-term α or $3_\downarrow 1_\uparrow 2_\downarrow$ + N-term α structure. (**A**) Comparison between occurrence frequencies of extended domains that include the $1_\uparrow 3_\downarrow 2_\uparrow$ + C-term α or $3_\downarrow 1_\uparrow 2_\downarrow$ + N-term α as the partial structure. The occurrence frequency of the extended $1_\uparrow 3_\downarrow 2_\uparrow$ + C-term α is 74.3 among which the occurrence frequency of structures having the left-handed $\beta\alpha\beta$-unit is 0.5 (invisible in the figure).

The occurrence frequency of the extended $3_{\downarrow}1_{\uparrow}2_{\downarrow}$ + N-term α structure is 18.5 among which the occurrence frequency of structures having the left-handed $\beta\alpha\beta$-unit is 2.5 (green). (**B**,**C**) Examples of the extended $3_{\downarrow}1_{\uparrow}2_{\downarrow}$ + N-term α domains having the left-handed $\beta\alpha\beta$-unit. (**B**) PDB ID: 2CVE. (**C**) PDB ID: 1RLH.

3.3. Frustration and Function

A remaining critical question is the reason for the existence of protein domains having the reverse ferredoxin topology. Because proteins have evolved not for their stability but their functions, a possible explanation is that frustrated structures are necessary for their functions. Roles of frustration in functions have been discussed with theoretical methods by inferring the local degree of frustration using the coarse-grained energy function of protein conformation [39]. By computationally perturbing the sequence or configuration of a local part of the protein, the local part was regarded as less frustrated when most of the perturbations increase the calculated free energy significantly, while the local part was regarded as frustrated when the free energy change upon perturbations is insignificant [40]. It was shown that the local frustration can guide thermal motions [41] and specific associations [42], suggesting the positive role of frustration in protein functioning.

In this study, we proposed a new definition of frustration as the conflict between structural preferences of local parts of the protein. This definition of frustration should shed further light on the role of frustration. The frustrating interaction between helices in the reverse ferredoxin fold destabilizes the structure. This tendency may be compensated for by a specific residue design to stabilize the fold, or the protein may utilize the tendency to enhance the fluctuation and facilitate the structural change, which is needed for its functioning. An example shown in Figure 1B was the catalytic core of human DNA polymerase kappa. Because the sizeable structural change is necessary for activating a molecular motor motion of DNA polymerase, we can expect that the frustration in this structure helps function DNA polymerase.

The definition of frustration introduced in this study, the structural conflict among the local parts' structural preferences, provides a new perspective to the frustration-function relationship. In particular, the hypothesis proposed in this subsection suggests an intriguing possibility that the designed incorporation of frustration in the structure helps design the protein whose function is related to mobility with the significant structural change. To test this hypothesis, the dynamics and stability of the frustrated proteins and the specific design of sequences to fold the frustrated structures should be examined with further direct and systematic methods.

4. Materials and Methods

4.1. Detecting the Position of the C/N Terminal α-Helix

We explain in Figure 6 the method to judge on which side of the β-sheet plane the C or N-terminal α-helix lies in protein domains. We defined three vectors, **a**, **b**, and **c** in the $1_{\uparrow}3_{\downarrow}2_{\uparrow}$ + C-term α (Figure 6A) and $3_{\downarrow}1_{\uparrow}2_{\downarrow}$ + N-term α (Figure 6B) structures. The terminal α-helix is on the upper side of the β-sheet plane of Figure 6 if $(\mathbf{a} \times \mathbf{b}) \cdot \mathbf{c} > 0$ and the helix is on the lower side of the plane if $(\mathbf{a} \times \mathbf{b}) \cdot \mathbf{c} < 0$.

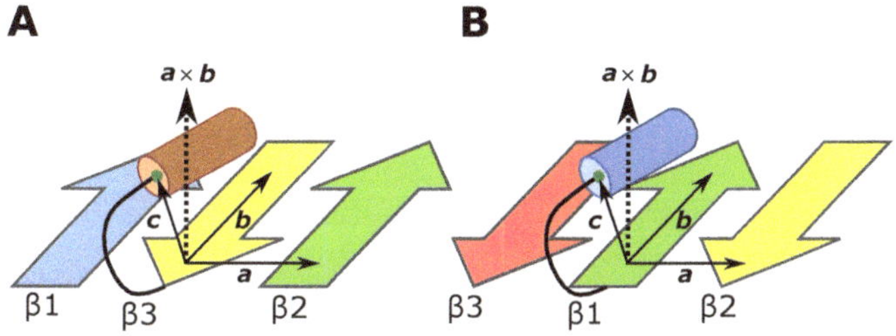

Figure 6. The method to judge on which side of the β-sheet the C or N-terminal α-helix lies. We defined three vectors, **a**, **b**, and **c**. The helix lies on the upper side of the β-sheet plane if $(\mathbf{a} \times \mathbf{b}) \cdot \mathbf{c} > 0$ and the helix lies on the lower side of the plane if $(\mathbf{a} \times \mathbf{b}) \cdot \mathbf{c} < 0$. (**A**) In the $1_{\uparrow}3_{\downarrow}2_{\uparrow}$ + C-term α structure, the vector **a** is a vector extending from the Cα atom of the C-terminal residue of the β-strand 3 (yellow arrow) to the Cα atom of the N-terminal residue of the β-strand 2 (green arrow). The vector **b** is a vector extending from the Cα atom of the C-terminal residue of the β-strand 3 to the Cα atom of the second residue before the C-terminal residue of the β-strand 3. The vector **c** is a vector extending from the Cα atom of the C-terminal residue of the β-strand 3 to the center of mass (green dot) of Cα atoms of four N-terminal residues of the α-helix (orange cylinder). (**B**) In the $3_{\downarrow}1_{\uparrow}2_{\downarrow}$ + N-term α structure, the vector **a** is a vector extending from the Cα atom of the N-terminal residue of the β-strand 1 (green arrow) to the Cα atom of the C-terminal residue of the β-strand 2 (yellow arrow). The vector **b** is a vector extending from the Cα atom of the N-terminal residue of the β-strand 1 to the Cα atom of the second residue after the N-terminal residue of the β-strand 1. The vector **c** is a vector extending from the Cα atom of the N-terminal residue of the β-strand 1 to the center of mass (green dot) of Cα atoms of four C-terminal residues of the α-helix (blue cylinder).

4.2. Definition of the Distance x between the Plane of β-Pleats and the α-Helix in the αβ- or βα-Unit

We measured the distance x between the plane of β-pleats and the α-helix in the $\alpha\beta$- and $\beta\alpha$-units by introducing a xyz-coordinate system in each unit (Figure 7). For defining the coordinate system, we set the direction of the y-axis parallel to the β-strand axis, and set the y-z plane parallel to the plane defined by the terminal three Cα atoms of the β-strand. We set the direction of the z-axis so as to place the helix on the $z > 0$ side. This idea of the coordinate system can be described in a precise way by defining the basis vectors, $\overrightarrow{e_x}$, $\overrightarrow{e_y}$, and $\overrightarrow{e_z}$, of the xyz-coordinate system with $\overrightarrow{e_z}$ being $\overrightarrow{e_z} = \overrightarrow{e_x} \times \overrightarrow{e_y}$.

We defined $\overrightarrow{e_x}$ and $\overrightarrow{e_y}$ as in the following way. Let i be the number of the terminal residue of the β-strand (the C-terminal residue in the $\beta\alpha$-unit and the N-terminal residue in the $\alpha\beta$-unit) and $C\alpha_i$ be the position of the ith Cα atom. We defined $\overrightarrow{e_x}$ by categorizing the $\beta\alpha$- or $\alpha\beta$-unit into two types, the parallel and antiparallel unit (Figure 7A,B). Then, we defined $\overrightarrow{e_x}$ as a normalized vector having the direction, which places both the starting and ending points of the α-helix on the coordinate of $x > 0$;

$$\overrightarrow{e_x} \parallel \begin{cases} \overrightarrow{C\alpha_{i-2}C\alpha_{i-1}} \times \overrightarrow{C\alpha_{i-1}C\alpha_i} & \text{(parallel } \beta\alpha\text{-unit)}, \\ \overrightarrow{C\alpha_i C\alpha_{i-1}} \times \overrightarrow{C\alpha_{i-1}C\alpha_{i-2}} & \text{(antiparallel } \beta\alpha\text{-unit)}, \\ \overrightarrow{C\alpha_i C\alpha_{i+1}} \times \overrightarrow{C\alpha_{i+1}C\alpha_{i+2}} & \text{(parallel } \alpha\beta\text{-unit)}, \\ \overrightarrow{C\alpha_{i+2}C\alpha_{i+1}} \times \overrightarrow{C\alpha_{i+1}C\alpha_i} & \text{(antiparallel } \alpha\beta\text{-unit)}, \end{cases} \quad (4)$$

and $\overrightarrow{e_y}$ is a normalized vector, whose direction is

$$\overrightarrow{e_y} \parallel \begin{cases} \overrightarrow{C\alpha_{i-2}C\alpha_i} & (\beta\alpha\text{-unit}), \\ \overrightarrow{C\alpha_{i+2}C\alpha_i} & (\alpha\beta\text{-unit}). \end{cases} \quad (5)$$

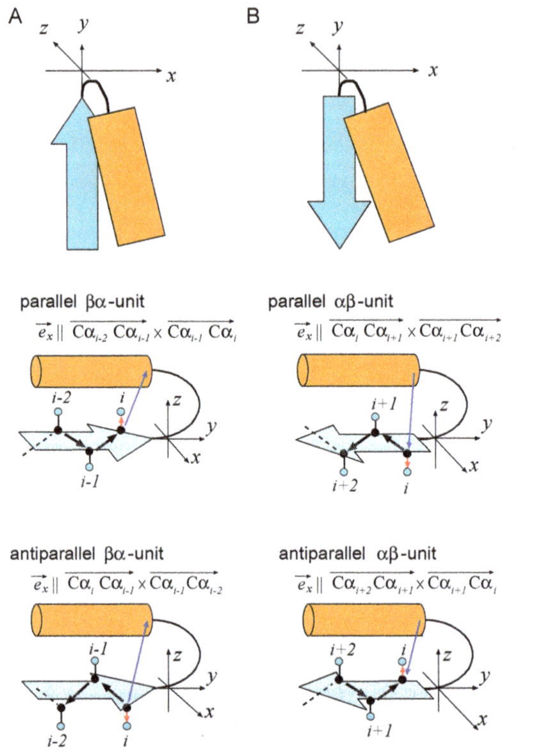

Figure 7. The xyz-coordinate system to define the distance x between the plane of β-pleats and the α-helix. (**A**) The $\beta\alpha$-unit and (**B**) the $\alpha\beta$-unit. These units consist of a β-strand (cyan arrow) and an α-helix (orange rectangle). Top panels represent the rough sketch of the coordinate system. Middle and bottom panels show Cα atoms (black dots), Cβ atoms (cyan dots), a vector spanning from the Cα to the Cβ of the terminal residue of the β-strand (i.e., the residue in the strand nearest to the helix) in each unit (red arrow), and a vector spanning from the Cα of the terminal residue of the β-strand to the center of mass of terminal four residues of the α-helix (i.e., four residues in the helix nearest to the strand) in each unit. Unit is referred to as "parallel" when the inner product of red and blue arrows is positive, and as "antiparallel" when the inner product is negative.

4.3. Rosetta Folding Simulations

We performed the Rosetta folding simulations to test the realizability of the blueprint structures. Here, Rosetta is a software suite that includes algorithms for macromolecular modeling, docking, protein design, etc [43]. Among the many algorithms included in the Rosetta software, we used the Rosetta BluePrintBDR protocol [43] for folding simulations. With this protocol, we performed the folding simulations by assembling one, three, or nine-residue length fragments so as to make the assembled structure compatible with a "blueprint", which describes the length of the secondary structure elements, strand pairings, and backbone torsion ranges for each residue. In thsese simulations, the main chain was represented by N, NH, Cα, C, and CO, and the side chain was represented by a sphere using the centroid model of Rosetta. We used the simulated annealing method to search for low-energy structures, and recorded the last structure of each simulated annealing run as a compatible structure only when the structure met the conditions specified in the blueprint.

As in models of Ref. [44], we represented all the residues as Valine, and used the same energy parameters as in Ref. [44]. The use of the poly-Valine sequence is because our purpose is to determine whether the phenomena observed in the database are explained by backbone properties rather than by the sequence-specific properties. Valine is the smallest and strongest hydrophobic amino acid, which suits this purpose, as shown in Ref. [5]. Figure 8 shows the blueprints we used in the BluePrintBDR protocol. In these blueprints, we used the same length of secondary structures and loops as optimized in Ref. [44]. The purpose of the present Rosetta simulations is to analyze the statistical tendency among different topologies. Because loops in each topology are shorter than five-residue length in most folds, and their distribution is peaked at around the two- to three-residue length (Figure S5), it is sufficient to use the short loops in the blueprints. Here, for the computational efficiency, we restricted ourselves to the loops with two- to three residue length for $\beta\alpha$- and $\alpha\beta$-loops. For β-hairpin loops, we assumed that loop consists of two, four, or five residues in the blueprints because the two or five-residue length is necessary for keeping the chirality rule of the hairpin loop [5] (Figure 8).

In the folding simulations, we did not impose the ABEGO constraint on the loop regions, but we imposed the constraint on the secondary structure regions by making the dihedral angles of the main chain in these regions fall into the ABEGO classes compatible with the secondary structures designated by the bluprint. Here, the ABEGO classification is a coarse-grained representation of the dihedral angles, specifying the regions in a Ramachandran plot with the alphabetic symbols: A, B, E, G, and O denote the right-handed α-helix region, right-handed β-strand region, left-handed β-strand region, left-handed helix region, and the cis peptide conformation, respectively [45].

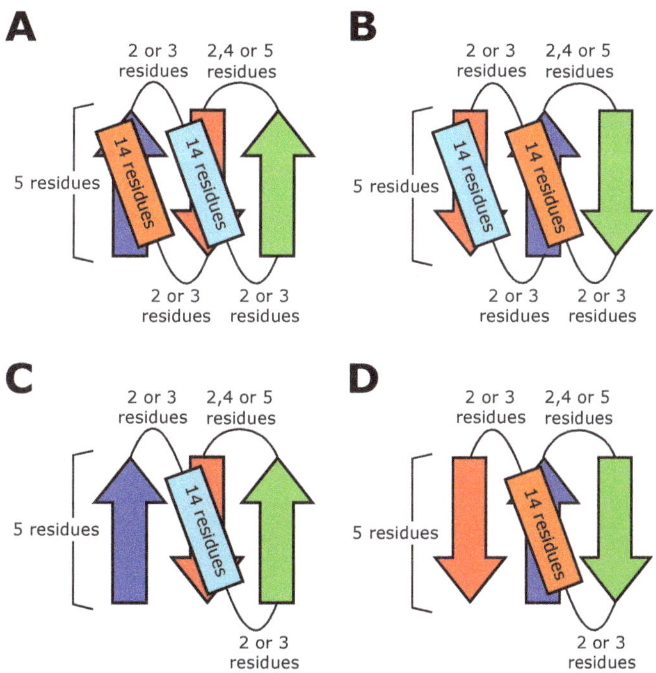

Figure 8. Blueprints used in the Rosetta folding simulations. The blueprints are represented by β-strands (arrows), α-helices (rectangles), and loops (curved lines). Blueprints of (**A**) the $1_{\uparrow}3_{\downarrow}2_{\uparrow}$ + C-term α topology, (**B**) the $3_{\downarrow}1_{\uparrow}2_{\downarrow}$ + N-term α topology, (**C**) the $1_{\uparrow}3_{\downarrow}2_{\uparrow}$ topology, and (**D**) the $3_{\downarrow}1_{\uparrow}2_{\downarrow}$ topology.

4.4. Score to Detect the Left-Handed $\beta\alpha\beta$-Unit

We detected protein domains having the left-handed $\beta\alpha\beta$-unit by calculating the score of the left-handedness (*L-score*) . Here, for defining the *L-score*, we consider a $\beta\alpha\beta$-unit exemplified in Figure 9A. We refer to the N-terminal β-strand in the $\beta\alpha\beta$-unit as $\beta1$, and the C-terminal β-strand as $\beta2$. We should note that the following *L-score* is applicable to evaluating the left-handedness of structures in which $\beta1$ and $\beta2$ are not connected directly to each other by hydrogen bonds, but multiple β-strands intervene between $\beta1$ and $\beta2$. We write the residue length of $\beta1$, $\beta2$, and the linker part connecting $\beta1$ and $\beta2$ as n, m, and l, respectively. We label the residues in those parts as $(N_1, N_2, \cdots, N_n)$, $(C_1, C_2, \cdots, C_m)$, and $(L_1, L_2, \cdots, L_l)$.

We define the residue number $C_{\max}(N_i, N_{i+1})$ so as to maximize the peak angle in Figure 9B when the residues N_i and N_{i+1} are given. Similarly, we define the residue number $N_{\max}(C_j, C_{j+1})$ to maximize the peak angle;

$$\begin{aligned} C_{\max}(N_i, N_{i+1}) &= \arg\max_{C_j}\left[\angle C\alpha_{N_i} C\alpha_{C_j} C\alpha_{N_{i+1}}\right], \\ N_{\max}(C_j, C_{j+1}) &= \arg\max_{N_i}\left[\angle C\alpha_{C_j} C\alpha_{N_i} C\alpha_{C_{j+1}}\right]. \end{aligned} \tag{6}$$

Then, using the Heaviside function, $H[x] = 1$ for $x > 0$ and $H[x] = 0$ for $x \leq 0$, the *L-score* is defined as

$$\begin{aligned} L\text{-}score &= \frac{1}{[(n-1)+(m-1)]\cdot l}\sum_{k=1}^{l}\left[\sum_{i=1}^{n-1} H\left[\left(\overrightarrow{C\alpha_{N_i}C\alpha_{N_{i+1}}} \times \overrightarrow{C\alpha_{N_i}C\alpha_{C_{\max}(N_i,N_{i+1})}}\right)\cdot \overrightarrow{C\alpha_{N_i}C\alpha_{L_k}}\right]\right. \\ &+ \left.\sum_{j=1}^{m-1} H\left[\left(\overrightarrow{C\alpha_{C_{j+1}}C\alpha_{C_j}} \times \overrightarrow{C\alpha_{C_j}C\alpha_{N_{\max}(C_j,C_{j+1})}}\right)\cdot \overrightarrow{C\alpha_{C_j}C\alpha_{L_k}}\right]\right]. \end{aligned} \tag{7}$$

The *L-score* ranges from 0 to 1 (Figure 9C). The higher the score, the more left-handed the $\beta\alpha\beta$-unit becomes. We judged the unit is left-handed when *L-score* ≥ 0.6.

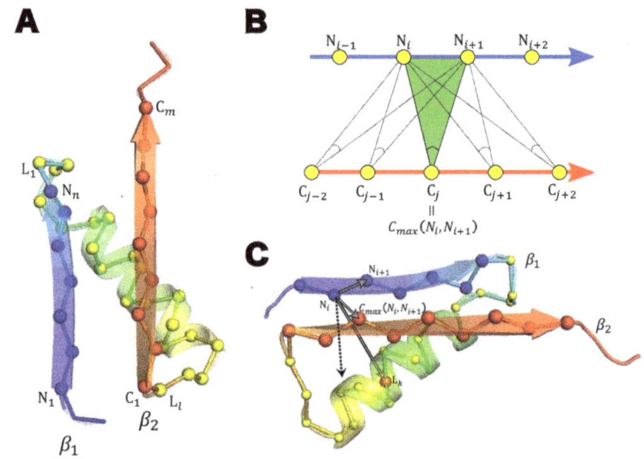

Figure 9. Calculation of the left-handedness score, *L-score*. (**A**) An example left-handed $\beta\alpha\beta$-unit. The cartoon representation and the backbone representation of the main chain are superposed. $C\alpha$ atoms are drawn with spheres in the backbone representation. The first and the last residue numbers of $\beta1$, $\beta2$, and the linker part are labeled on the chain. (**B**) Determination of $C_{\max}(N_i, N_{i+1})$. (**C**) Calculation of a term in *L-score*. The vector connecting $C\alpha_{N_i}C\alpha_{N_{i+1}}$, the one connecting $C\alpha_{N_i}C\alpha_{C_{\max}(N_i,N_{i+1})}$, and the one connecting $C\alpha_{N_i}C\alpha_{L_k}$ in Equation (7) are drawn with gray arrows and the vector product of the first two vectors are drawn with a dashed arrow. The calculated score of this example $\beta\alpha\beta$-unit is *L-score* $= 0.86$.

Supplementary Materials: The following supporting information can be downloaded at: https://www.mdpi.com/article/10.3390/molecules27113547/s1, Supplementary Figures S1–S5.

Author Contributions: Conceptualization, G.C.; methodology, T.N., M.N. and G.C.; software, T.N. and M.N.; investigation, T.N., M.N. and G.C.; writing—original draft preparation, M.S. and G.C.; writing—review and editing, M.S. and G.C. All authors have read and agreed to the published version of the manuscript.

Funding: This research was funded by the KAKENHI Grants, 20H05530, 21H00248, and 22H00406 of Japan Society for the Promotion of Science for M.S. and 19H03166 for G.C. and by Platform Project for Supporting Drug Discovery and Life Science Research (JP21am0101111) from Japan Agency for Medical Research and Development for G.C.

Institutional Review Board Statement: Not applicable.

Informed Consent Statement: Not applicable.

Data Availability Statement: Not applicable.

Conflicts of Interest: The authors declare no conflict of interest.

Abbreviations

The following abbreviations are used in this manuscript:

SCOP	Structural Classification of Proteins
ECOD	Evolutionary Classification of protein Domains
PDB	Protein Data Bank

References

1. Tramontano, A.; Cozzetto, D. *Supramolecular Structure and Function 8*; Springer: Berlin/Heidelberg, Germany, 2004; pp. 15–29.
2. Sadowski, M.I.; Jones, D.T. The sequence-structure relationship and protein function prediction. *Curr. Opin. Str. Biol.* **2019**, *19*, 357–362. [CrossRef] [PubMed]
3. Senior, A.W.; Evans, R.; Jumper, J.; Kirkpatrick, J.; Sifre, L.; Green, T.; Qin, C.; Žídek, A.; Nelson, A.W.R.; Bridgland, A.; et al. Improved protein structure prediction using potentials from deep learning. *Nature* **2020**, *577*, 706–710. [CrossRef] [PubMed]
4. Jumper, J.; Evans, R.; Pritzel, A.; Green, T.; Figurnov, M.; Ronneberger, O.; Tunyasuvunakool, K.; Bates, R.; Žídek, A.; Potapenko, A.; et al. Highly accurate protein structure prediction with AlphaFold. *Nature* **2021**, *596*, 583–589. [CrossRef] [PubMed]
5. Koga, N.; Tatsumi-Koga, R.; Liu, G.; Xiao, R.; Acton, T.B.; Montelione, G.T.; Baker, D. Principles for designing ideal protein structures. *Nature* **2012**, *491*, 222–227. [CrossRef] [PubMed]
6. Marcos, E.; Chidyausiku, T.K.; McShan, A.C.; Evangelidis, T.; Nerli, S.; Carter, L.; Nivón, L.G.G.; Davis, A.; Oberdorfer, G.; Tripsianes, K.; et al. De novo design of a non-local β-sheet protein with high stability and accuracy. *Nat. Struct. Mol. Biol.* **2018**, *25*, 1028–1034. [CrossRef]
7. Murata, H.; Imakawa, H.; Koga, N.; Chikenji, G. The register shift rules for $\beta\alpha\beta$-motifs for de novo protein design. *PLoS ONE* **2021**, *16*, e0256895. [CrossRef]
8. Minami, S.; Kobayashi, N.; Sugiki, T.; Nagashima, T.; Fujiwara, T.; Koga, R.; Chikenji, G.; Koga, N. Exploration of novel $\alpha\beta$-protein folds through de novo design. *bioRxiv* **2021**. [CrossRef]
9. Huang, P.S.; Feldmeier, K.; Parmeggiani, F.; Velasco, D.A.F.; Höcker, B.; Baker, D. De novo design of a four-fold symmetric TIM-barrel protein with atomic-level accuracy. *Nat. Chem. Biol.* **2016**, *12*, 29–34. [CrossRef]
10. Dou, J.; Vorobieva, A.A.; Sheffler, W.; Doyle, L.A.; Park, H.; Bick, M.J.; Mao, B.; Foight, G.W.; Lee, M.Y.; Gagnon, L.A.; et al. De novo design of a fluorescence-activating β-barrel. *Nature* **2018**, *561*, 485–491. [CrossRef]
11. Kuhlman, B.; Dantas, G.; Ireton, G.C.; Varani, G.; Stoddard, B.L.; Baker, D. Design of a novel globular protein fold with atomic-level accuracy. *Science* **2003**, *302*, 1364–1368. [CrossRef]
12. Doyle, L.; Hallinan, J.; Bolduc, J.; Parmeggiani, F.; Baker, D.; Stoddard, B.L.; Bradley, P. Rational design of α-helical tandem repeat proteins with closed architectures. *Nature* **2015**, *528*, 585–588. [CrossRef] [PubMed]
13. Pan, F.; Zhang, Y.; Liu, X.; Zhang, J. Estimating the designability of protein structures. *bioRxiv* **2021**. 11.03.467111. [CrossRef]
14. Richardson, J.S. beta-Sheet topology and the relatedness of proteins. *Nature* **1977**, *268*, 495–500. [CrossRef] [PubMed]
15. Richardson, J.S. The anatomy and taxonomy of protein structure. *Adv. Protein Chem.* **1981**, *34*, 167–339.
16. Ruczinski, I.; Kooperberg, C.; Bonneau, R.; Baker, D. Distributions of beta sheets in proteins with application to structure prediction. *Proteins* **2002**, *48*, 85–97. [CrossRef]
17. Chitturi, B.; Shi, S.; Kinch, L.N.; Grishin, N.V. Compact Structure Patterns in Proteins. *J. Mol. Biol.* **2016**, *428*, 4392–4412. [CrossRef]

18. Minami, S.; Chikenji, G.; Ota, M. Rules for connectivity of secondary structure elements in protein: Two-layer $\alpha\beta$ sandwiches. *Protein Sci.* **2017**, *26*, 2257–2267. [CrossRef]
19. Orengo, C.A.; Jones, D.T.; Thornton, J.M. Protein superfamilles and domain superfolds. *Nature* **1994**, *372*, 631–634. [CrossRef]
20. Murzin, A.G.; Brenner, S.E.; Hubbard, T.; Chothia, C. SCOP: A structural classification of proteins database for the investigation of sequences and structures. *J. Mol. Biol.* **1995**, *247*, 536–540. [CrossRef]
21. Salem, G.M.; Hutchinson, E.G.; Orengo, C.A.; Thornton, J.M. Correlation of observed fold frequency with the occurrence of local structural motifs. *J. Mol. Biol.* **1999**, *287*, 969–981. [CrossRef]
22. Kinoshita, K.; Kidera, A.; Go, N. Diversity of functions of proteins with internal symmetry in spatial arrangement of secondary structural elements. *Protein Sci.* **1999**, *8*, 1210–1217. [CrossRef] [PubMed]
23. Fox, N.K.; Brenner, S.E.; Chandonia, J.M. SCOPe: Structural Classification of Proteins–extended, integrating SCOP and ASTRAL data and classification of new structures. *Nucleic Acids Res.* **2014**, *42*, D304–D309. [CrossRef] [PubMed]
24. Chandonia, J.M.; Fox, N.K.; Brenner, S.E. SCOPe: Classification of large macromolecular structures in the structural classification of proteins-extended database. *Nucleic Acids Res.* **2019**, *47*, D475–D481. [CrossRef] [PubMed]
25. Zhang, C.; Kim, S.H. The anatomy of protein beta-sheet topology. *J. Mol. Biol.* **2000**, *299*, 1075–1089. [CrossRef] [PubMed]
26. Cheng, H.; Schaeffer, R.D.; Liao, Y.; Kinch, L.N.; Pei, J.; Shi, S.; Kim, B.H.; Grishin, N.V. ECOD: An evolutionary classification of protein domains. *PLoS Comput. Biol.* **2014**, *10*, e1003926. [CrossRef]
27. Andreeva, A.; Howorth, D.; Chandonia, J.M.; Brenner, S.E.; Hubbard, T.J.P.; Chothia, C.; Murzin, A.G. Data growth and its impact on the SCOP database: New developments. *Nucleic Acids Res.* **2008**, *36*, D419–D425. [CrossRef]
28. Sillitoe, I.; Bordin, N.; Dawson, N.; Waman, V.P.; Ashford, P.; Scholes, H.M.; Pang, C.S.M.; Woodridge, L.; Rauer, C.; Sen, N.; et al. CATH: increased structural coverage of functional space. *Nucleic Acids Res.* **2021**, *49*, D266–D273. [CrossRef]
29. Frishman, D.; Argos, P. Knowledge-based protein secondary structure assignment. *Proteins* **1995**, *23*, 566–579. [CrossRef]
30. Wang, G.; Dunbrack, R.L., Jr. PISCES: a protein sequence culling server. *Bioinformatics* **2003**, *19*, 1589–1591. [CrossRef]
31. Street, T.O.; Fitzkee, N.C.; Perskie, L.L.; Rose, G.D. Physical-chemical determinants of turn conformations in globular proteins. *Protein Sci.* **2007**, *16*, 1720–1727. [CrossRef]
32. Lesk, A.M.; Brändén, C.I.; Chothia, C. Structural principles of α/β barrel proteins: The packing of the interior of the sheet. *Proteins Str. Funct. Bioinform.* **1989**, *5*, 139–148. [CrossRef] [PubMed]
33. Murzin, A.G.; Finkelstein, A.V. General architecture of the α-helical globule. *J. Mol. Biol.* **1988**, *204*, 749–769. [CrossRef]
34. Nagi, A.D.; Regan, L. An inverse correlation between loop length and stability in a four-helix-bundle protein. *Fold. Des.* **1997**, *2*, 67–75. [CrossRef]
35. Linse, S.; Thulin, E.; Nilsson, H.; Stigler, J. Benefits and constrains of covalency: the role of loop length in protein stability and ligand binding. *Sci. Rep.* **2020**, *10*, 20108. [CrossRef] [PubMed]
36. Richardson, J.S. Handedness of crossover connections in beta sheets. *Proc. Natl. Acad. Sci. USA* **1976**, *73*, 2619–2623. [CrossRef] [PubMed]
37. Sternberg, M.J.E.; Thornton, J.M. On the conformation of proteins: The handedness of the connection between parallel β-strands. *J. Mol. Biol.* **1976**, *110*, 269–283. [CrossRef]
38. Cole, B.J.; Bystroff, C. Alpha helical crossovers favor right-handed supersecondary structures by kinetic trapping: The phone cord effect in protein folding. *Protein Sci.* **2009**, *18*, 1602–1608. [CrossRef]
39. Ferreiro, D.U.; Komives, E.A.; Wolynes, P.G. Frustration, function and folding. *Curr. Opin. Struct. Biol.* **2018**, *48*, 68–73. [CrossRef]
40. Parra, R.G.; Schafer, N.P.; Radusky, L.G.; Tsai, M.Y.; Guzovsky, A.B.; Wolynes, P.G.; Ferreiro, D.U. Protein frustratometer 2: A tool to localize energetic frustration in protein molecules, now with electrostatics. *Nucleic Acids Res.* **2016**, *44*, W356–W360. [CrossRef]
41. Ferreiro, D.U.; Hegler, J.A.; Komives, E.A.; Wolynes, P.G. On the role of frustration in the energy landscapes of allosteric proteins. *Proc. Natl. Acad. Sci. USA* **2011**, *108*, 3499–3503. [CrossRef]
42. Ferreiro, D.U.; Hegler, J.A.; Komives, E.A.; Wolynes, P.G. Localizing frustration in native proteins and protein assemblies. *Proc. Natl. Acad. Sci. USA* **2007**, *104*, 19819–19824. [CrossRef] [PubMed]
43. Fleishman, S.; Leaver-Fay, A.; Corn, J.E.; Strauch, E.M.; Khare, S.D.; Koga, N.; Ashworth, J.; Murphy, P.; Richter, F.; Lemmon, G.; et al. RosettaScripts: A scripting language interface to the Rosetta macromolecular modeling suite. *PLoS ONE* **2011**, *6*, e20161. [CrossRef] [PubMed]
44. Lin, Y.R.; Koga, N.; Tatsumi-Koga, R.; Liu, G.; Clouser, A.F.; Montelione, G.T.; Baker, D. Control over overall shape and size in de novo designed proteins. *Proc. Natl. Acad. Sci. USA* **2015**, *112*, E5478–E5485. [CrossRef] [PubMed]
45. Wintjens, R.T.; Rooman, M.J.; Wodak, S.J. Automatic classification and analysis of alpha alpha-turn motifs in proteins. *J. Mol. Biol.* **1996**, *255*, 235–253. [CrossRef] [PubMed]

molecules

MDPI

Article

CRFalign: A Sequence-Structure Alignment of Proteins Based on a Combination of HMM-HMM Comparison and Conditional Random Fields

Sung Jong Lee [1], Keehyoung Joo [2], Sangjin Sim [3], Juyong Lee [4], In-Ho Lee [5] and Jooyoung Lee [6,*]

1 Basic Science Institute, Changwon National University, Changwon 51140, Korea; yeesj123@gmail.com
2 Center for Advanced Computation, Korea Institute for Advanced Study, Seoul 02455, Korea; newton@kias.re.kr
3 NAVER CLOVA, Seongnam 13561, Korea; dyanos@gmail.com
4 Department of Chemistry, Kangwon National University, Chuncheon 24341, Korea; juyong.lee@kangwon.ac.kr
5 Korea Research Institute of Standards and Science (KRISS), Daejeon 34113, Korea; ihlee@kriss.re.kr
6 School of Computational Sciences, Korea Institute for Advanced Study, Seoul 02455, Korea
* Correspondence: jlee@kias.re.kr

Abstract: Sequence–structure alignment for protein sequences is an important task for the template-based modeling of 3D structures of proteins. Building a reliable sequence–structure alignment is a challenging problem, especially for remote homologue target proteins. We built a method of sequence–structure alignment called CRFalign, which improves upon a base alignment model based on HMM-HMM comparison by employing pairwise conditional random fields in combination with nonlinear scoring functions of structural and sequence features. Nonlinear scoring part is implemented by a set of gradient boosted regression trees. In addition to sequence profile features, various position-dependent structural features are employed including secondary structures and solvent accessibilities. Training is performed on reference alignments at superfamily levels or twilight zone chosen from the SABmark benchmark set. We found that CRFalign method produces relative improvement in terms of average alignment accuracies for validation sets of SABmark benchmark. We also tested CRFalign on 51 sequence–structure pairs involving 15 FM target domains of CASP14, where we could see that CRFalign leads to an improvement in average modeling accuracies in these hard targets (TM-CRFalign $\simeq$ 42.94%) compared with that of HHalign (TM-HHalign $\simeq$ 39.05%) and also that of MRFalign (TM-MRFalign $\simeq$ 36.93%). CRFalign was incorporated to our template search framework called CRFpred and was tested for a random target set of 300 target proteins consisting of Easy, Medium and Hard sets which showed a reasonable template search performance.

Keywords: protein structure prediction; sequence-structure alignment; template-based modeling; conditional random fields; boosted regression trees; CASP

Citation: Lee, S.J.; Joo, K.; Sim, S.; Lee, J.; Lee, I.-H.; Lee, J. CRFalign: A Sequence-Structure Alignment of Proteins Based on a Combination of HMM-HMM Comparison and Conditional Random Fields. *Molecules* **2022**, *27*, 3711. https://doi.org/10.3390/molecules27123711

Academic Editor: Michael Assfalg

Received: 23 March 2022
Accepted: 7 June 2022
Published: 9 June 2022

1. Introduction

Comparing a protein sequence with another sequence or a sequence with a known protein structure is one of the important tasks in bioinformatics, especially in the template-based 3D structure modeling of proteins. In spite of striking new developments in recent years (such as Alphafold [1,2]) on 3D protein structure modeling based on contact predictions via deep learning, the sequence–structure alignment method can be still useful in various stages of protein structure modeling.

In traditional template-based modeling (TBM), the model qualities are highly dependent on finding the best templates and good alignments between the target sequence and the templates. When multiple templates are given, multiple alignments between the sequence and templates [3–6] are utilized. However, multiple alignment is strongly dependent on the alignment accuracies of pair-wise sequence–sequence or sequence–structure alignments. Improving pairwise sequence–structure alignment is also important for finding better templates for protein structure modeling.

Various kinds of profile comparison methods have been developed to improve the alignment quality between sequences. For example, there are several score functions available for calculating the match scores between profiles, such as the dot product score [7], the Jensen–Shannon divergence score [8], the log average score [9] and the Pearson's correlation score [10]. SparksX [11] builds on previous profile–profile (comparison) alignment methods of SP (SP1, SP2, SP3, SP4, SP5) series [12–15] by incorporating additional features incrementally, such as secondary structures and solvent accessibility with a linear combination. The HHpred [16], on the other hand, is based on a comparison of HMM profiles [16,17]. We note here that HHpred was not developed specifically to improve the alignment quality itself but rather mainly for efficient template search even though here in this work we compare our alignment method against HHalign in terms of modeling accuracies. More recently, more discriminative methods of conditional random fields have been applied to pairwise alignment and fold recognition. These include Contralign [18], BoostThreader [19,20] and MRFalign [21]. BoostThreader, in particular, employs a nonlinear scoring function by means of regression trees or neural networks. Recently machine learning methods based on nearest neighbor search was also applied to pairwise alignment of proteins [22]. Among these methods, the HHalign method of the HHpred has been consistently successful with particularly fast performance.

In this work, we built a method for pairwise alignment between a sequence and a structure that combines an HMM-HMM comparison scoring scheme (HHalign, HHblits [23]) and an additional nonlinear scoring function based on pairwise conditional random fields with boosted regression trees [24]. We incorporate boosted regression trees at each stage of the training steps with various features including profile–profile similarity, secondary structure similarity, similarity of the solvent accessibility, as well as environmental features. These nonlinear scoring functions are expected to provide complicated relationships between neighboring features for propensities of match states or gaps. Boosted regression trees in our CRFalign alignment models are trained on a few sets of pairwise alignments selected from SABmark benchmark set [25]. Here in this work, our main focus is on improving the sequence–structure pairwise alignment in terms of the structure modeling accuracies that entails from those alignments. As for the important area of template search [26] based on the present alignment method, we will briefly discuss our test template search results on a set of 300 targets by incorporating CRFalign into our fold recognition framework called CRFpred [27].

One distinctive feature of CRFalign is that it combines (in an additive way) HMM-HMM comparison scores and additional nonlinear scoring scheme implemented in multiple steps of boosted regression trees. That is, the additional nonlinear scoring part is constructed by a sum of residual training steps. Therefore, we expect that the HMM-HMM comparison scores guarantee a reasonable baseline performance and the additional nonlinear scoring part learns the mismatch between the true structural alignment and incorrect alignment (induced from the incorrect HMM profiles) so that it learns better how to align the sequence and structure when the HMM-HMM scores are not reliable enough. These residual learning incorporates comparison of several predicted structural features of the target sequence and the true structural features of the templates including secondary structures and solvent accessibilities such that complex relationships between environmental features can be obtained. These features can be contrasted with MRFalign [21] or ExMachina method [22] which admit only sequence profiles as input. Hence, CRFalign is expected to be relatively effective for hard targets for which close sequence homologues are not available.

We find that improvement of alignment accuracy can be achieved, especially for pairwise alignments between protein sequence and structures with remote homologues. We evaluated the alignment quality of CRFalign by modeling the 3D structures via Modeller for different test sets from SABmark benchmark database and some CASP targets [28,29]. Here, we found that the TM scores and RMSD scores of modeled structures from CRFalign showed improvement over those from base model alignments especially for the case of hard targets. The performance test of CRFalign on 51 sequence–structure pairs involving 15 hard target domains of CASP14 and CRFalign resulted in average modeling accuracies of TM-CRFalign $\simeq$ 42.94% compared with that of HHalign, which shows TM-HHalign $\simeq$ 39.05% and also with that of MRFalign showing TM-MRFalign $\simeq$ 36.93%. As mentioned above, we also performed a test of template search using CRFalign method on a set of 300 targets, which showed a reasonable performance.

2. Materials and Methods

Here, we present the formalism of conditional random fields as applied to pairwise sequence–structure alignment of proteins [19]. For a given pair of protein sequences s (which we denote, in this work, as the sequence for the known *structure*) and t (which we denote as the sequence for the *target* with unknown structure), an arbitrary alignment between the structural sequence s and the target sequence t can be represented as a sequence of match (M), insertion (I) or deletion (D) states. If we suppose that L_s is the sequence length of the *structure*, and that L_t is the sequence length of the *target*, then this sequence of alignment states can also be represented as an alignment path on a rectangular lattice of dimensions $L_s \times L_t$ where a diagonal path corresponds to M, a horizontal one to an I and a vertical one to a D. Here we assume that the target sequence (t) lies along the horizontal with the sequence length L_t and the structure sequence (s) along the vertical with the sequence length L_s (see Figure 1).

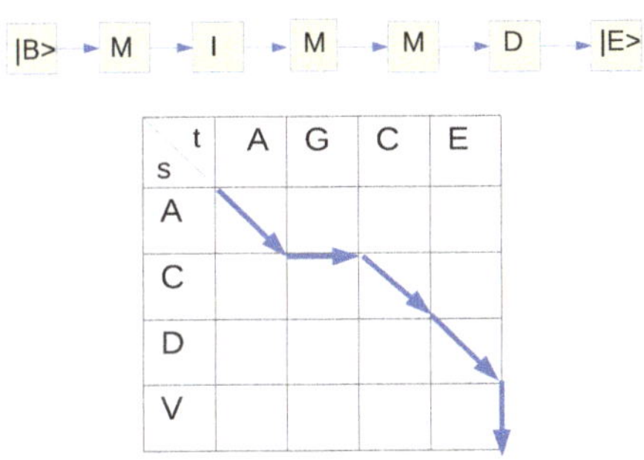

Figure 1. An example of simple pairwise alignment between a target sequence (t) and a structure sequence (s). Note that |B> denotes the BEGIN state and |E> denotes the END state.

Now let us denote the sequence of alignment states (with alignment length L) as $A \equiv (a_0, a_1, a_2, \cdots, a_L, a_{L+1})$ where $a_i \in \{M, I, D\}$ with $i = 1, \cdots, L$ in addition to a_0 which indicates the BEGIN state (|B>) as well as a_{L+1} denoting the END state (|E>). In the formalism of conditional random fields, a probabilistic model for the pairwise alignment is constructed where the probability $P(a|s,t)$ of each of these alignments can be written as,

$$P(a|s,t) = \exp[\sum_{i=1}^{L+1} F(a_{i-1}, a_i|s,t)]/Z(s,t) \tag{1}$$

where the function $F(a_{i-1}, a_i|s,t)$ represents the log-likelihood of the transition from the alignment state a_{i-1} at $i-1$ to the next state a_i at the alignment position i and $Z(s,t)$ is the normalization factor with $Z(s,t) \equiv \sum_a \exp\left[\sum_{i=1}^{L+1} F(a_{i-1}, a_i|s,t)\right]$, which is also called the partition function. The function $F(a_{i-1}, a_i|s,t)$ corresponds to the alignment score at alignment position i in conventional pairwise sequence alignment. Here, however, it depends not just on the present alignment state a_i but also on the previous alignment state a_{i-1} through the local structural or sequence features of residues located around the alignment position $i-1$ and i. This makes it possible to naturally incorporate position dependence or environmental dependence in match scores as well as gap penalties. Important features include similarities of sequence profiles between the sequence and structure, similarities of secondary structures and solvent accessibilities. Originally, conditional random field formalism was based on the function F with linear combination of various scores [18]. J. Xu et al. proposed a method based on nonlinear scoring functions for F such as neural networks or boosted regression trees [19,20]. These nonlinear scoring functions can take nontrivial correlations between different features into account. The optimal choice of the functions F is obtained through training on a set of reference alignments such that the average probability of these reference alignments get maximal values.

In our alignment, F consists of a sum of successive scoring functions as follows,

$$F(a_{i-1}, a_i|s,t) = T_0 + \mu_1 T_1 + \mu_2 T_2 + \cdots \tag{2}$$

where T_0 is a base alignment model of scoring function and T_i's (for $i \geq 1$) are successive nonlinear scoring functions to be determined by optimization of the probabilities of occurrences of some reference alignment set. The constants $\mu_1, \mu_2, \cdots$ are weight parameters that represent the learning rate of the training. Suppose that $P(a_{i-1}, a_i|s,t)$ refers to the net probability that the specific transition from the alignment state a_{i-1} at position $i-1$ to the next state a_i at position i occur (which is also called the posterior probability). Then it is straightforward to show that, for any alignment model defined by $F(a_{i-1}, a_i|s,t)$

$$\frac{\delta \ln P}{\delta F(a_{i-1}, a_i|s,t)} = \delta(a_{i-1}, a_i \in A) - P(a_{i-1}, a_i|s,t), \tag{3}$$

where $\delta(a_{i-1}, a_i \in A)$ is equal to 1 if a_{i-1} at position $i-1$ and the next state a_i at position i actually pass along a reference alignment A and if not, it is equal to zero. A simple way to understand this relation is that, when the alignment model defined by $F(a_{i-1}, a_i|s,t)$ is optimal (i.e., at maximum), then the right hand side of Equation (3) should be zero, in other words, $P(a_{i-1}, a_i|s,t)$ should be equal to 1 (maximal probability) for the states on the alignment path ($\delta = 1$), and equal to zero for the states not on the alignment path ($\delta = 0$). Here, $P(a_{i-1}, a_i|s,t)$ is obtained by summing over all the alignments of s and t with the restriction that a specific transition a_{i-1} to a_i should occur at specific position for the pair of sequences. This can be easily computed by Forward and Backward algorithm [30]. Now, we can see that the successive scoring functions T_k, $k = 1, 2, \cdots$ can be constructed by any machine learning methods for the functional gradient of the $\ln P$ (for the alignment in question to occur) with respect to F, which can be easily sampled from training alignments using the right-hand-side of Equation (3). Here in this work, each T_k is implemented as a boosted regression tree consisting of six decision trees with depth five, which is known to be fast and efficient (Figure 2).

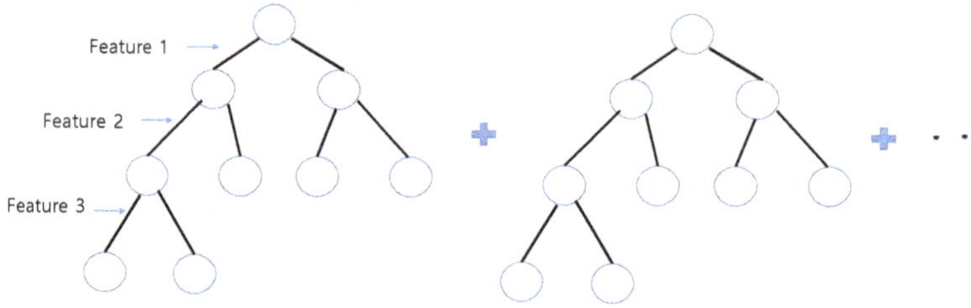

Figure 2. A schematic diagram of a boosted regression tree.

In previous works by Xu et al. [19], the beginning alignment model T_0 for F was chosen as the trivial alignment model with $F_0(a_{i-1}, a_i) = 0$ for all possible transitions at all positions, which roughly corresponds to a random alignment model where all possible pairs of residues have equal probability of alignment as well as equal probabilities for all gaps. Here in our work, instead of beginning with the random alignment model, we chose to begin with some reasonable alignment method which is already available and build our full alignment model by adding nonlinear scorning functions within the framework of conditional random fields. Here, we chose a scoring scheme adapted from HHalign [16] as the base scoring scheme.

HHalign is a pairwise alignment method based on comparison of HMM profiles of protein sequences [16]. In order to construct a pairwise comparison of two HMM's, HHalign introduces five pair-alignment states which are MM, MI, IM, GD and DG where M denotes a Match state in the HMM of a specific residue position for the structure sequence or the target sequence, I an Insertion state, D a deletion state, and finally G denoting a Gap state. For a given alignment between two HMM's, the alignment score of HHalign consists of the HMM-HMM profile match score, transition probability score and secondary structure score as follows.

$$T_0 = \sum_i (S_{aa}(i) + S_{tr}(i) + S_2(i)) \tag{4}$$

where $S_{aa}(i)$ denotes the similarity score between the HMM profiles of the two columns at the alignment position i, S_{tr} represents the propensity of allowed transitions at the alignment position i that are transitions between a pair state and itself and between pair state MM and pair states MI, IM, DG or GD. The last term $S_2(i)$ represents the similarity score between the secondary structure of the template (structure) residue and the predicted secondary structure of the target residue. Usually, for the template structure, the secondary structure information from DSSP [31] is used, while, for the target residue, secondary structure prediction from PSIPRED [32].

Accommodation of the HHalign-type scoring into our CRF alignment model with additional nonlinear scoring function may be implemented in two different ways that are called in this work as the three-state scheme or the five-state scheme. In the three-state scheme of CRF alignment model, we reduce the five states MM, MI, IM, DG, and GD of HHalign to the usual three states M, I, D via the reduction mapping of ($MM \rightarrow$ Match(M), $MI \rightarrow$ Insertion(I), $IM \rightarrow$ Deletion(D), $DG \rightarrow$ Insertion(I), $GD \rightarrow$ Deletion(D)). Note that MI and DG are both reduced to the same alignment state of I, while IM and GD to D. For a given alignment path in this three-state scheme, the zeroth order scoring is obtained by reduction of the HHalign scores such that for an Insertion (I) or a Deletion (D) state, the larger score of the two corresponding states in the original five-state model from HHalign is chosen.

It is also possible to construct a CRF alignment model with a five-state scheme (i.e., without reduction to the three states) incorporating the full HHalign scoring and additional nonlinear scoring functions. Moreover, in this scheme of five state, for the purpose of training from the reference alignments, it is necessary to assign an appropriate five-state label for each of the alignment positions between the target and the template. Since the reference alignments are built by some structure alignments without relation to the HMM profiles of the targets or templates, it is not straightforward to assign five state labels to the alignment states due to the ambiguity between DG vs. MI as well as GD vs. IM. One possible solution to this problem is to choose the unique assignment along the reference alignment path for which the HHalign score becomes maximum. We implemented and tested both the three-state model and the five-state model.

Once the base alignment model for the three-state scheme is fixed, additional nonlinear scoring functions ($T_1, T_2, \dots$) can be constructed via training on a set of reference alignments as follows. One can evaluate the right-hand side of the Equation (3) based on the base alignment model T_0 to get the functional derivative $\delta \ln P/\delta F(a_{i-1}, a_i)|_{F=T_0}$ for any pairwise alignment. Now, we sample transitions from the set of reference training alignments. Both positive samples (i.e., those transitions appearing in the training alignments) as well as negative samples (i.e., those transitions that are not appearing in the training alignments) are taken. For these samples, one can compute the right-hand side of the Equation (3) $1 - \langle P_{a_{i-1},a_i} \rangle$. These target values together with the relevant input features can now be used to train the first additional scoring function T_1 i.e., first correction to the alignment model, where any machine learning methods can be employed. The constant factor μ_1 can be chosen to control the degree of greediness of the training.

For training these gradients, we used so-called gradient boosted regression trees [24]. In addition, the partition function $Z(s,t)$ can be calculated using the standard Forward-Backward algorithm for the given alignment model. Now, when this training is completed for T_1, we are now equipped with a first order corrected alignment model, which can again be used for training the next order regression trees T_2 for further correction, using new samples evaluated at $T_0 + \mu_1 \cdot T_1$. The constants μ_i ($i = 1, 2, 3, \ldots$) are weight parameters that can be adjusted to control the degree of convergence in the training, where we chose $\mu_i = 0.2$ for all i in this work. For each of the three (for the three-state scheme) or five (for the five-state scheme) alignment states, there corresponds a boosted regression tree. Input features for the regression trees include position-dependent structural features such as secondary structures and solvent accessibilities for the known protein structure. In addition, for the sequence with unknown 3D structure, predicted secondary structures and solvent accessibilities are employed instead. In all, there are six features for the Match state and there are another six features for the Gap state. For further details on the input features to the boosted regression trees, refer to Appendix A.

In the CRF alignment model, two kinds of alignments are possible. One is the so-called Viterbi alignment algorithm which selects the one highest scoring alignment (i.e., highest probability). The other method of alignment is the so-called MAP (MAximum Posterior Probability) alignment which first calculates the net probability $P(s_i, t_j)$ for a specific pair of residues s_i (of the template structure) and t_j (of the target) may align in all possible alignments, and then find, through a standard dynamic programming, the alignment that optimizes the sum of these values without gap costs,

$$S(a) \equiv \sum_a P(s_i, t_j) \tag{5}$$

where the sum is performed over all pair matches. This is also called as Maximum Accuracy Alignment (MAC). MAP alignment tends to produce more true matches compared with the Viterbi alignment (see the section on Results).

In contrast to global alignment (where the alignment begins at the first residues of the target and the template), local alignments can be generated by allowing the alignment to begin (and end) at any position of the target and the template without scoring for the end gaps. In the case of MAP alignment, this can be conveniently implemented by introducing a threshold value m_{th} for the match which can range from 0 to 1 and subtract m_{th} from $P(s_i, t_j)$ for all matches as follows,

$$S_{local}(a) \equiv \sum_a (P(s_i, t_j) - m_{th}) \tag{6}$$

where, in the dynamic programming, the alignment can begin at any position of the target and template with no costs for end gaps. As for internal gaps, additional penalties of $-0.5 * m_{th}$ are added to avoid unnatural internal gaps being produced. A special case of alignment mode which is called *glocal* (*glocal* + *local*) alignment is commonly adopted, where the alignment can begin (and end) at internal positions of either the target or the template but not *both*. If we suppose that the sequence of the structure template runs along vertical axis on the left boundary of a rectangular lattice, with that of the target running along horizontal axis on the top boundary, this corresponds (in terms of the alignment path) to beginning the alignment on the upper or left boundaries of the rectangle and ending on the opposite sides (bottom or right boundaries) in the dynamic programming.

3. Results

As for the reference alignment set for training our sequence–structure alignment method, we chose the SABmark (version 1.65) benchmark set [25] which was designed to assess protein sequence alignment algorithms, especially for the case of remote homologous pairs of proteins. SABmark consists of two sets of pairwise and multiple alignment sets with high-resolution X-ray structures derived from the SCOP classification. These sets, Twilight Zone and Superfamilies, are known to cover the entire known fold space with sequences very low to low and low to intermediate similarity, respectively.

The Twilight Zone set consists of 209 sequence groups that each represent a SCOP fold. Sequence similarity is very low with the sequence identities ranging between 0% and 25% and also with the structures being distantly similar. SABmark homepage states that "This set therefore, represents the worst case scenario for sequence alignment, which unfortunately is also the most frequent one, as most related sequences share less than 25% identity" [25]. On the other hand, the Superfamilies set consists of 425 groups, each of which representing a SCOP superfamily. The sequence pairs share at most 50% identity. Even though this set in general consists of less difficult pairs (than the Twilight Zone) they still represent challenging problems for sequence alignments.

We chose three sets of reference alignment, each consisting of 200 pairwise alignments from the Superfamilies set and from the Twilight Zone. These three sets are labeled as NG200, NF200 and TW200, respectively. These are chosen in such a way that the pairs are evenly distributed among different groups of families so that as many groups of families as possible are covered. Among these three sets, two of them (NG200 and NF200 set) are from the Superfamilies set with average sequence identities of 24.2%, 21.2%, respectively. The remaining set of TW200 is derived from the Twilight zone set with average sequence identity of 13.8%. By training our sequence–structure alignment methods on these sets with different levels of sequence similarity, we may be able to compare the modeling capabilities of the resulting alignment methods and choose the most efficient one among those results.

HMM files were generated by using hhmake tools of HMM hhsuite [16,23]. In our training on the three sets (NG200, NF200 and TW200), we employed the three-state scheme. That is, at each step of the training, e.g., T_i with $i = 1, 2, \ldots,$ there are three different boosted regression trees, one for each of the three states at the current position: match (M), insertion (I) and deletion (D) state, respectively (i.e., the three-state scheme).

For each pairwise alignment of the reference training set, we take each of the alignment states along the reference alignment path as a positive sample (for training the scoring function). If it is a match state (M), we add the set of corresponding features together with the target label value $1 - P(a_{i-1}, a_i)$ in Equation (3) to the sample set for the boosted regression tree for the match state. Similarly for insertion (I) or deletion (D) states along the reference alignment path, we add the corresponding features and the label value to the boosted regression trees for I and D respectively. Now, we have to also collect negative samples, that is, those transitions that do not appear on the reference alignments. Suppose that the sequence length of the structure s and the target t are L_s and L_t respectively. If we let the alignment length to be L_a, then we have $L_a \leq L_s + L_t$. Then, we can see that (e.g., for three state alignment model) there are about $\simeq 3(L_s \cdot L_t - L_a)$ negative samples which is usually much larger than the alignment length L_a. We randomly selected some integer (N_f) times the alignment length L_a for the size of the negative samples, where we took $N_f = 16$. That is, we took $16 \cdot L_a$ transitions that are not on the alignment path. These transitions were distributed evenly among the three alignment states M, I and D. We tried other values for N_f, but the present value was found to be most effective in terms of the training accuracy and training time. This resulted in around 2×10^5 samples for each of the three states. At each step of the training, the boosted regression trees consist of six regression trees with each tree having a depth of five. As for the choice of the parameters μ_k, as mentioned above, we simply set $\mu_k = 0.2$ for all steps k. Change of this parameter did not show much difference in the performance of the resulting alignment model.

For each of the above three sets (NG200, NF200 and TW200), in order to perform training of our alignment model and then perform validation test in terms of alignment accuracies, we divided the set into four subsets of 50 pairs each, and then carried out a four-fold training and test with 150 pairs for training and the remaining 50 pairs for test in turn. Alignment accuracies are measured as the ratio of the number of aligned pairs out of the true aligned pairs in the reference structure alignment of the sequence pairs of the SABmark benchmark. For training and test on these sets presented, we employed a three-state scheme.

Figure 3 shows the training and test accuracies (with Viterbi scoring) for the three sets (NG200, NF200 and TW200) as the training step increases. We see that on all three cases, the training and the test accuracies on average improve up to a certain regression steps (about five to seven, depending on the sets), then after that, the average accuracies tend to fluctuate somewhat. Figure 4 shows the relative improvements of the CRFalign test accuracies over the base (zeroth order) model for the three sets. We can clearly see that, for the hard alignment set of NF200 and TW200 as compared with the easier set of NG200, more improvement is achieved especially in terms of the Viterbi alignment accuracies.

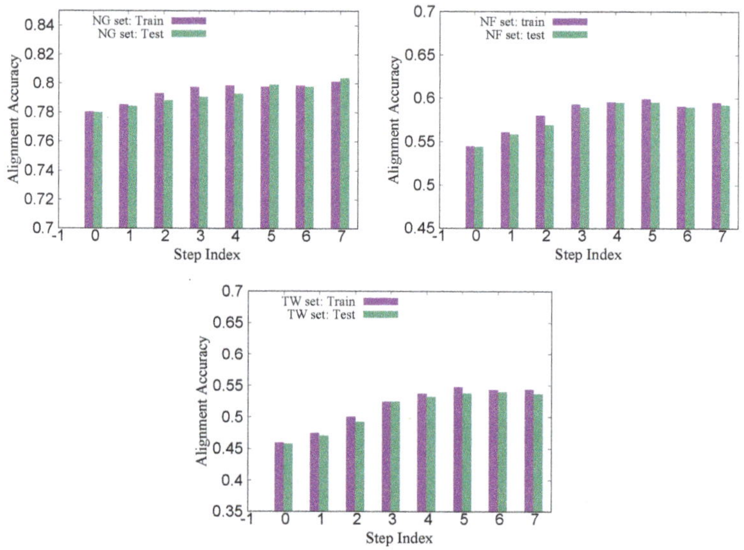

Figure 3. Training and test accuracies for the Viterbi alignment on the three sets NG (**top left**), NF (**top right**) and TW (**bottom**) of 200 reference alignments from SABmark. Note that we are using the three-state scheme here.

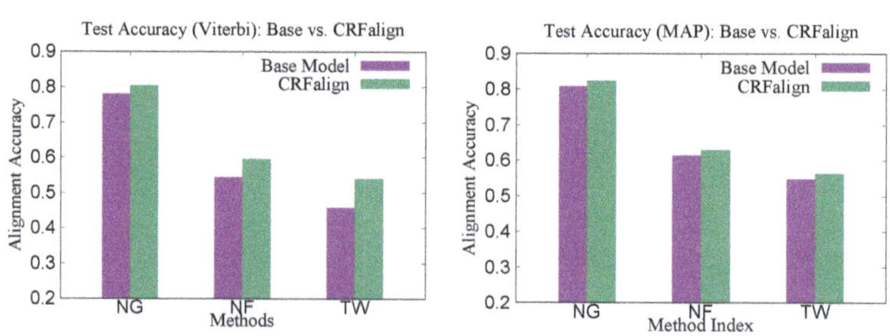

Figure 4. Training and test accuracies at Maximum for (**left**) the Viterbi alignment as well as (**right**) the MAP alignment on the three sets (NG, NF and TW) of 200 reference alignments with the three-state scheme.

In order to assess our alignment method in terms of protein structure modeling for proteins in the SABmark set, we chose the whole 200 pairs of the NF200 set to train our alignment model and then performed sequence–structure alignment with the trained model together with structure modeling on two independent test sets using the Modeller program based on the alignment. One of the two test sets called NG64 consists of 64 pairs chosen from the Superfamilies set of the SABmark. On the other hand, the second test sets called TW55 consists of 55 pairs chosen from the Twilight Zone set of the SABmark, representing more difficult alignment situations. Both sets consist of pairs of sequences that are less than 30% sequence identity against those of the training set NF200 with the average of the sequence identity against the training set being 16.6% (NG64) and 16.3% (TW55), respectively.

Figure 5 shows the modeling results on NG64 and TW55 test sets comparing the CRFalign method with HHalign, where we see that some improvements were made by CRFalign over HHalign results. Table 1 shows the average TM score for the modeling results where again we find that for the hard set of TW55 the improvement was bigger. The average TM score of NG64 set by CRFalign was 71.96%, while that for HHalign was 71.39%. On the other hand, the average TM score of TW55 set by CRFalign was 48.83%, while that for HHalign was 46.32%.

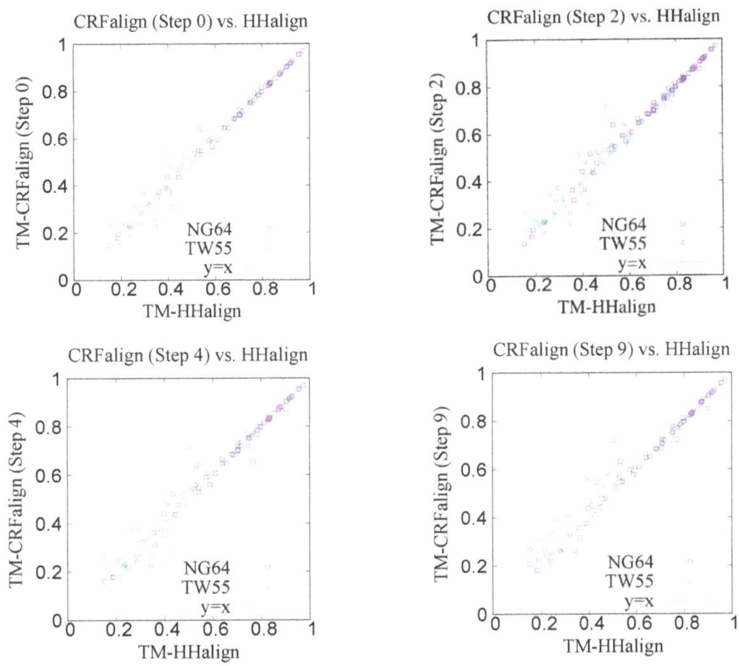

Figure 5. TM scores of structure models of NG64 set and TW55 set obtained by running Modeller on the CRFalign alignments (three-state scheme) at various train steps (**top left**: steps 0, **top right**: step 2, **bottom left**: step 4 and **bottom right**: 9) in comparison with the results of HHalign.

Table 1. Modeling accuracies in TM score of the test sets NG64 and TW55 based on training with NF200.

Set	HHalign	Base Model	CRFalign (3-State Scheme)
NG64	0.7139	0.7176	0.7196
TW55	0.4632	0.4669	0.4883

Figure 6 shows one example where CRFalign result was fed into Modeller with the resulting model exhibiting significant improvement over that of HHalign. Shown is the 3D structure of the chain A of d1nr0a1 (which is the seven-bladed beta propeller domain of C. elegans actin-interacting protein 1) with the template d1fwxa2 (d1nr0a1-d1fwxa2,

$TM_{ref} = 75.38\%$, ID $= 9.4\%$). Note that the sequence ID to the template sequence is only 9.4% but still CRFalign in combination with Modeller could produce a structure of TM score with TM $\simeq 71.8\%$ which is close to the optimal TM score limit of $\simeq 75.38\%$. In contrast HHalign could not properly close the propeller shaped domain with relatively poor value of the TM score of TM-HHalign= 50.36%. In the CRFalign alignment between the template and our target sequence (not shown), we could see that there are a few large gaps in the alignment which would be difficult to correctly align without the help of structure based features and nonlinear scoring model for the alignment.

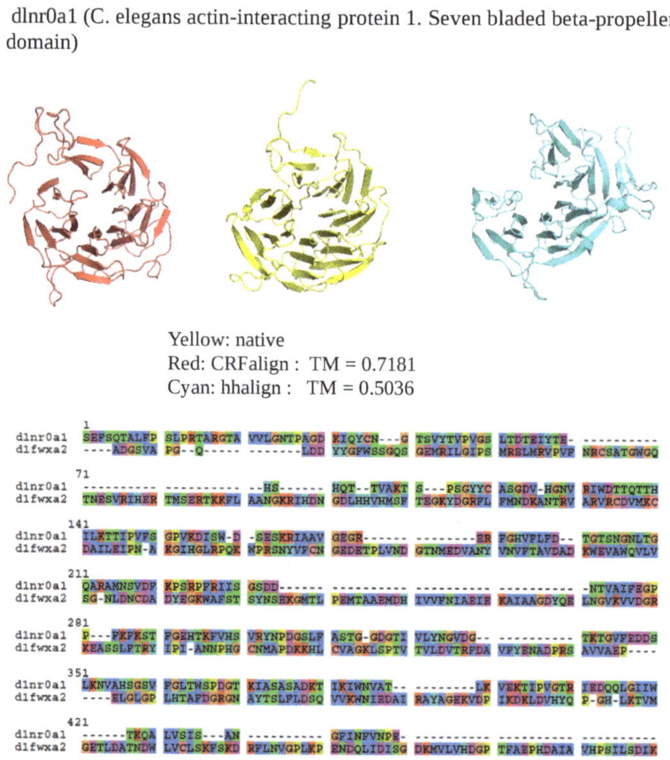

Figure 6. Structure models of d1nr0a1 (C.elegans actin-interacting protein1 Seven-bladed beta-propeller domain) based on d1nr0a1-d1fwxa2 alignment (Reference TM = 0.75381, ID = 9.4%) from CRFalign (**top left**, red) with TM-CRFalign = 0.7181, and HHalign (**top right**, cyan) with TM-HHalign = 0.5036; at the to center (yellow) is the native structure. The **bottom** figure shows the CRFalign pairwise alignment where a few large gaps are recognized for proper alignment.

Another example is shown in Figure 7 where β protein (The Outer Membrane Protein OMPX from E. Coli 1qj8a) is illustrated based on the alignment (d1qj8a-d1i78a $TM_{ref} = 72.65\%$, ID $= 3.4\%$) which could roughly produce the correct fold pattern with TM-CRFalign = 56.38% as compared with HHalign, which failed in producing the correct β patterns on one side with TM-HHalign = 39.02. In this case, the sequence identity is even lower with ID $= 3.4\%$. Here also, the CRFalign alignment to the template (not shown) shows regions of big gaps.

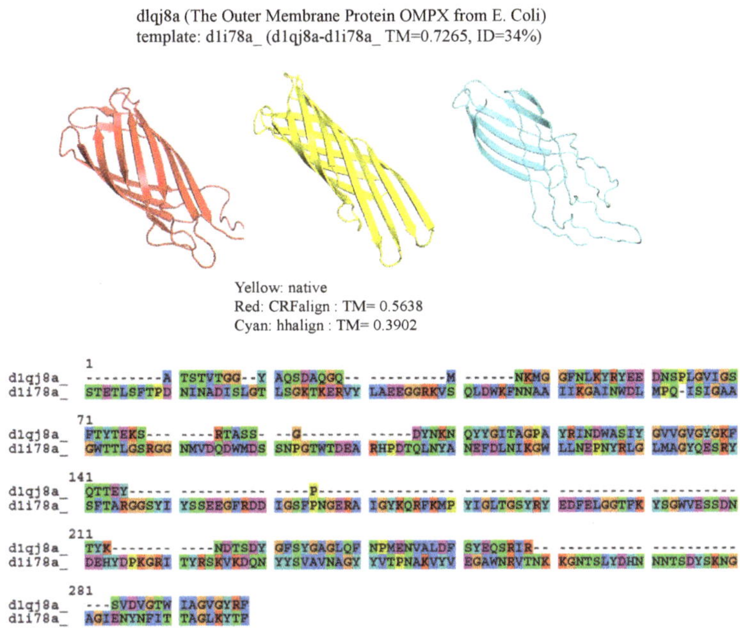

Figure 7. Structure models of d1qj8a (The Outer Membrane Protein OMPX from E. Coli) based on d1qj8a-d1i78a_ alignment (Reference TM = 0.7265, ID = 3.4%) from CRFalign (**top left**, red) with TM-CRFalign = 0.5638, and HHalign (**top right**, cyan) with TM-HHalign = 0.3902; at the top center (yellow) is the native structure. At the **bottom** is shown the CRFalign alignment where also a large gap is recognized.

The final example is shown in Figure 8, which shows the structure of d1hnja2 (Beta-Ketoacyl-acyl carrier protein synthase III) with the alignment d1hnja2-d1hnja1 (TM_{ref} = 59.88%, ID = 10.4%). Here, CRFalign could produce TM-CRFaligna = 57.68% which is quite close to the ideal value of TM_{ref} = 59.88%. In contrast, HHalign could produce the model with TM-HHalign = 48.62% only, failing to reproduce much of the secondary structural elements.

For building working alignment models (targeted for blind structure prediction such as CASP), we trained several hundred different three-state alignment models (i.e., accumulating different sets of boosted regression trees) on the three sets (NG200, NF200 and TW200) using the whole 200 pairwise alignments for each of the three sets. In order to choose optimal alignment models, we need to test these for their modeling capabilities. For this, we tested these on CASP10 targets with appropriate templates by performing alignment and modeling. We chose 58 single-domain targets from CASP10, for which there exist templates. Among these, for 50 of them, we could choose two templates. Hence, in all, we have 108 pairs to align and model.

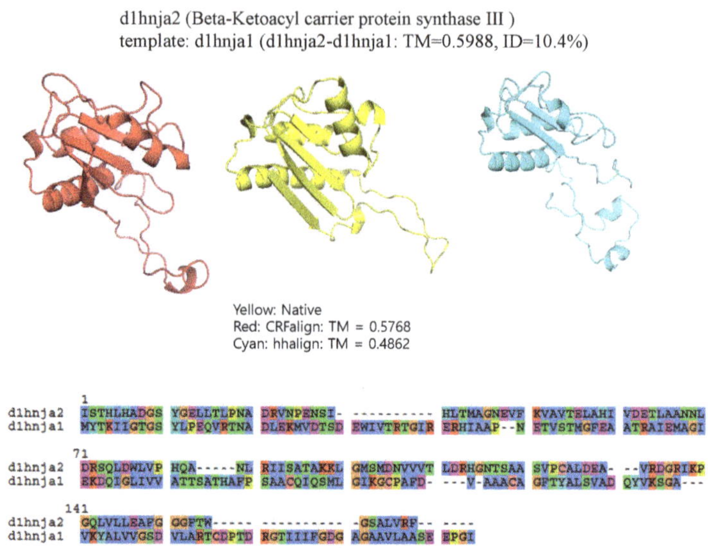

Figure 8. Structure models of d1hnja2 (Beta-Ketoacyl-acyl carrier protein synthase III) based on d1hnja2-d1hnja1 alignment (Reference TM = 0.5988, ID = 10.4%) from CRFalign (**top left**, red) with TM-CRFalign = 0.5768, and HHalign (**top right**, cyan) with TM-HHalign = 0.4862; at the top center (yellow) is the native structure. The **bottom** figure shows the CRFalign alignment.

Figure 9 shows the comparison of the TM scores for the modeled structures with the base alignment model vs. HHalign where we see that the average TM score with the base alignment model ($TM_{base} \simeq 0.5246$) is lower than that of the HHalign model ($TM_{hha} \simeq 0.5286$). However, on the right side, shown is the comparison between the CRFalign (three-state) result and HHalign, where we see various targets for which the base model gave relatively poor result are now showing some improvement with the resulting average TM score of TM-CRFalign $\simeq$ 0.5321. This CRFalign alignment with the three-state scheme was applied successfully to CASP11 [27] and CASP12 [33] .

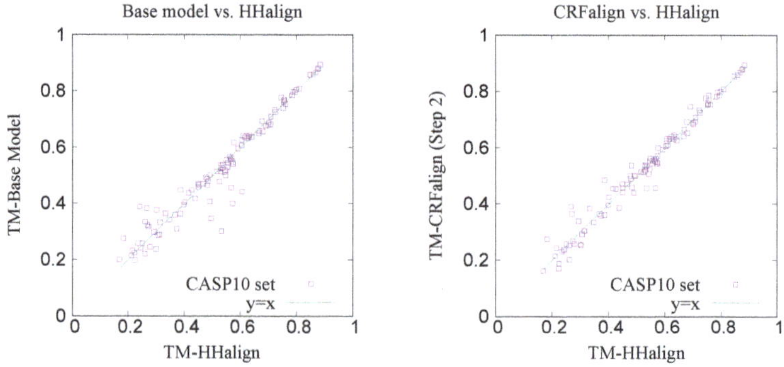

Figure 9. Comparison of TM scores for CASP10 targets by Modeller modeling based on (**left**) base alignment (average TM score = 0.5246) vs. HHalign alignment (average TM score = 0.5286) and (**right**) CRFalign alignment (average TM score = 0.5321) vs. HHalign.

Recently, we constructed a larger training set called TR367 from SABmark for CRFalign with the five-state alignment model. The training set TR367 consists of 367 pairs of proteins from SABmark superfamily (299 pairs) and twilight zone set (68 pairs). These were carefully selected more or less uniformly among different folds and families. In order to assess the pair-wise alignments of the five-state alignment model, we prepared two test sets W200 and S200 from SABmark benchmark set. The W200 set consists of 200 pairs chosen from the twilight zone subset of SABmark set, while those of S200 are 200 pairs from the superfamily set, where all the sequences in the test set have sequence identities less than 20% against those sequences in the training set (TR367). Therefore, the pairs in W200 set should be considered, in general, harder (i.e., remote homologues) than those of S200 set. Figure 10 shows the training and test accuracy of alignments for the TR367 training set and the W200 test set as well as S200 test set. We can see here also that the average alignment accuracy in the W200 set shows larger relative improvement than that in the case of S200 set (Table 2).

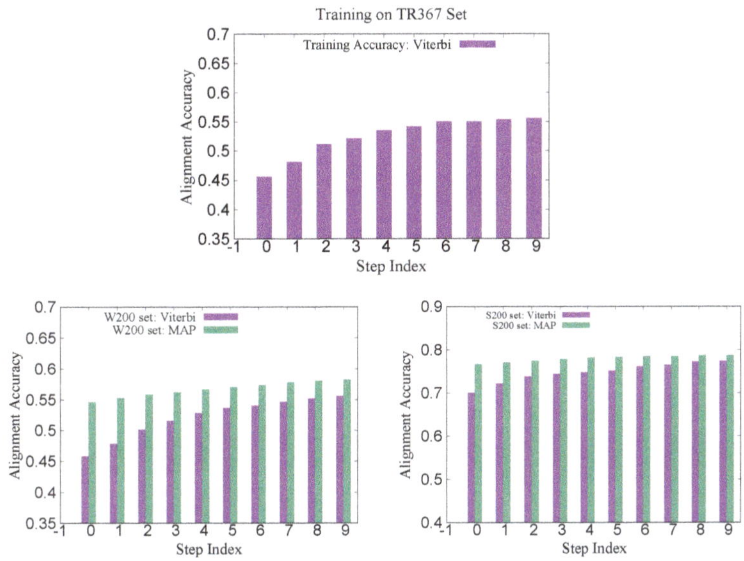

Figure 10. (**Top**) Training alignment accuracy on TR367 set with the five-state model training. (**Bottom left**) Test alignment accuracies on W200 set with the same five-state model; both Viterbi alignment and MAP alignment accuracies are shown. (**Bottom right**) Test alignment accuracies on S200 set with the same five-state model (below).

Table 2. Alignment accuracies in the test sets W200 and S200 based on training with TR367.

Set	Base Model	CRFalign (Five-State Scheme)
W200	0.546	0.583
S200	0.767	0.787

Table 3 shows the modeling accuracies on the two test sets W200 and S200. We can recognize larger relative improvement in the TM score in the case of W200 set compared with that of S200 set which is consistent with the alignment accuracies shown above. Figure 11 shows a comparison of the TM scores of individual targets for W200 set based on CRFalign (at different steps of 1, 4, 7 and 10) vs. the Base model. Here also, one can recognize significant relative improvement in the targets of low TM score region.

Table 3. Modeling accuracies in average TM score of the test sets W200 and S200 based on the five-state training with TR367 set.

Step Number	W200	S200
0	0.4304	0.5776
1	0.4331	0.5792
2	0.4368	0.5812
3	0.4372	0.5808
4	0.4377	0.5809
5	0.4406	0.5826
6	0.4431	0.5827
7	0.4455	0.5820
8	0.4473	0.5833
9	0.4462	0.5849
10	0.4499	0.5870

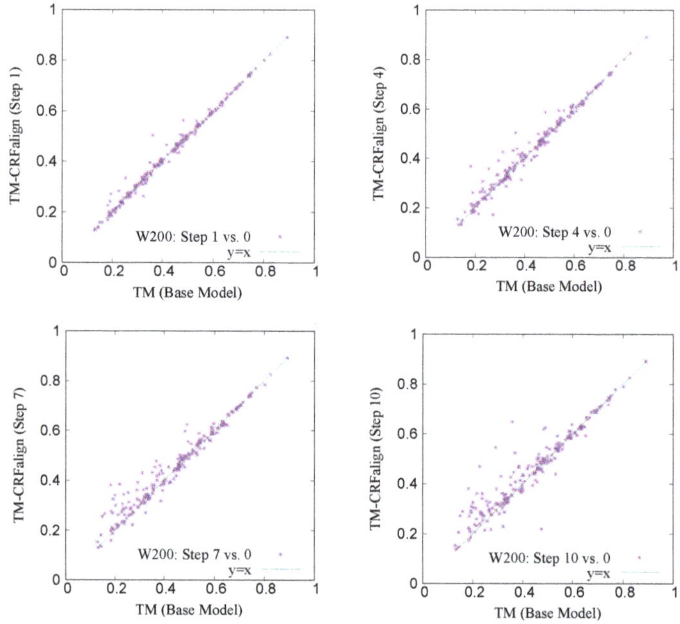

Figure 11. TM scores of structure models on the W200 test set with CRFalign (the five-state model) at steps 1 (**top left**), 4 (**top right**), 7 (**bottom left**), 10 (**bottom right**) vs. the Base model (step 0), respectively.

Figures 12–14 show examples of significantly improved model structures from the W200 set using CRFalign (and Modeller) with the five-state model as compared with those from the base model. The first example domain is d1a1w__(FADD death-effector domain) which consists of mostly alpha-helices. The structure model produced from CRFalign with its template (d1dgna_) exhibits TM score of 0.6477 and $rmsd = 2.318$ with a sizable improvement over that from the Base alignment with TM score = 0.3575 and $rmsd = 12.111$ (Figure 12). The next example is d1gjwa1 (Thermotoga maritima maltosyltransferase) which consists of mostly beta sheets. In this case also, we observe that the CRFalign (of d1gjwa1-d1ktba1) leads to a model structure with TM score = 0.6243 and $rmsd = 2.781$ which is a significant improvement over that based on the base model alignment with TM score = 0.4907 and $rmsd = 4.616$ (Figure 13). The final example is d1mwma2 (ParM from plasmid R1 ADP form), which consists of alpha and beta structures. The CRFalign alignment (of d1mwma2-d1nbwa3) and Modeller produce a model with TM score = 0.6182 and *rmsd* = 3.749 in comparison with that based on the Base alignment with TM score = 0.4829, *rmsd* = 7.159 (Figure 14).

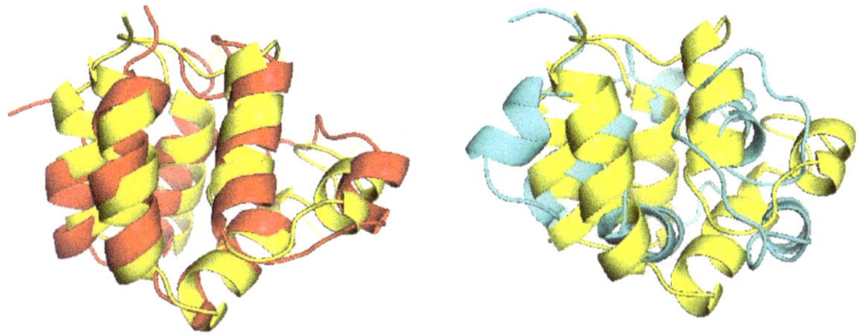

Figure 12. (**Left**) Structure model of d1a1w__(FADD death-effector domain) based on d1a1w__-d1dgna_ CRFalign alignment (red) overlapped with the experimental structure of d1a1w__ (yellow) (with TM score = 0.6477, *rmsd* = 2.318) and (**right**) that based on base model alignment (cyan) overlapped with the experimental structure (yellow) (with TM score = 0.3575, *rmsd* = 12.111).

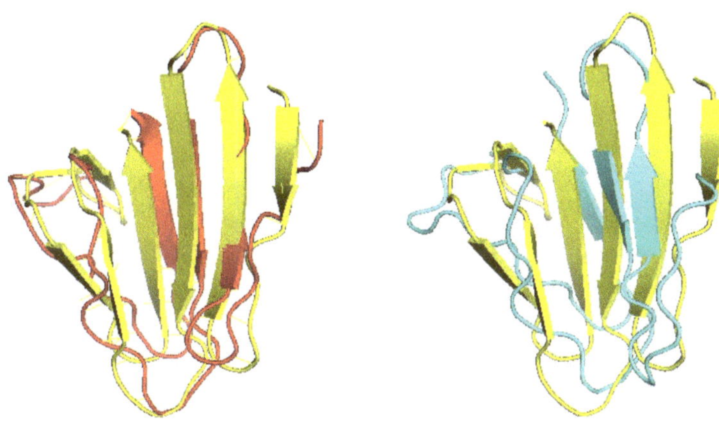

Figure 13. (**Left**) Structure model of d1gjwa1 (Thermotoga maritima maltosyltransferase) based on CRFalign alignment (red) of d1gjwa1-d1ktba1 overlapped with the experimental structure of d1gjwa1 (yellow) (with TM score = 0.6243, *rmsd* = 2.781) and (**right**) that based on the base model alignment (cyan) overlapped with the experimental structure (yellow) (with TM score = 0.4907, *rmsd* = 4.616).

We also tested the performance of CRFalign (five-state scheme) on hard targets of CASP14 [29], where the targets are selected from the 15 protein domains listed as FM (Free Modeling) or FM/TBM (of CASP14). These target domains are T1026-D1, T1030-D1, T1032-D1, T1033-D1, T1038-D2, T1039-D1, T1046s1-D1, T1046s2-D1, T1056-D1, T1067-D1, T1074-D1, T1079-D1, T1080-D1, T1082-D1 and T1099-D1. The structural homologues of these targets (obtained by using LGA structure alignment [34]) can be retrieved from CASP14 homepage (https://predictioncenter.org/download_area/CASP14/templates/LGA/, last accessed on 1 February 2022). For each of these domain targets, 1–6 homologues are available for templates. We ended up with 51 pairs of target and templates involving 15 CASP14 domains for sequence–structure alignment. We compared the TM scores of the structure models obtained via Modeller based on the pairwise alignments using CRFalign and HHalign, respectively. Table 4 shows the average TM score of the structure models from CRFalign at each step number (with both the five-state and three-state models). This can be compared with the TM score result for HHalign TM-HHalign = 0.3905 which is close to the result of base model (in three-state alignment model). For reference and comparion, we tested (baseline) pairwise alignments using BLOSUM62 scores [35] and also recent MRFalign alignment on the above set. These alignments produced (through Modeller) 3D models with average TM scores of TM-Blosum62 = 0.3077 and TM-MRFalign = 0.3693 Table 5. We see that the highest average TM score for CRFalign (five-state model) is around TM = 0.4294 (at step 9) which improves upon HHalign by nearly 3.9% point, and upon MRFalign by about 6.0% respectively as shown in Table 5. The maximal possible average TM score using TM-align (=0.5973) is also shown in Table 5 which shows that there are significant gaps between the maximal TM score and CRFalign. However, it is important to note that these CASP14 targets are classified as hard targets where the corresponding templates are very hard to find from typical template searches, and that these templates are identified from structural alignments such as LGA.

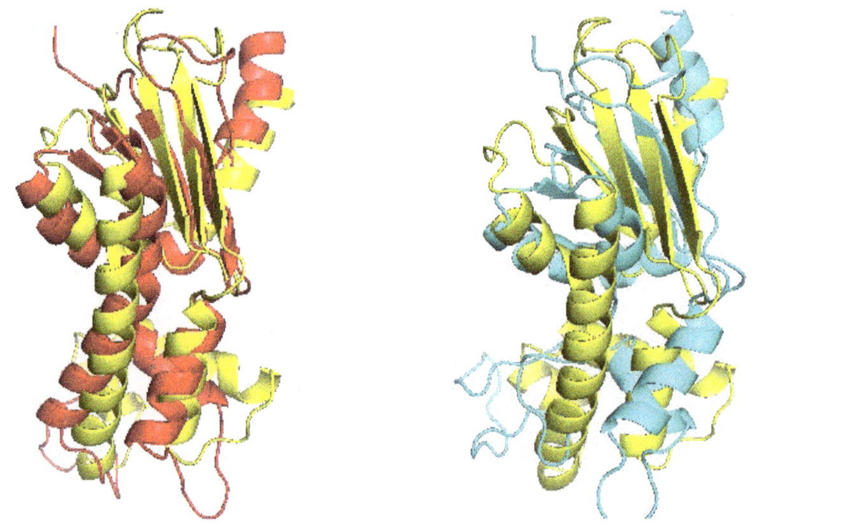

Figure 14. (**Left**) Structure model of d1mwma2 (ParM from plasmid R1 ADP form) based on CRFalign alignment (red) of d1mwma2-d1nbwa3 overlapped with the experimental structure of d1mwma2 (yellow) (with TM score = 0.6182, *rmsd* = 3.749) and (**right**) that based on Base alignment (cyan) overlapped with the experimental structure (yellow) (with TM score = 0.4829, *rmsd* = 7.159).

Table 4. Modeling accuracies in average TM score of 51 pairs involving the CASP14 hard targets based on the five-state model as well as the three-state model.

Step Number	CRFalign (5-State)	CRFalign (3-State)
0	0.3938	0.3902
1	0.3983	0.3958
2	0.4009	0.4029
3	0.4027	0.4041
4	0.4082	0.4045
5	0.4107	0.4071
6	0.4156	0.4074
7	0.4201	0.4011
8	0.4253	0.4034
9	0.4294	0.4129
10	0.4237	0.4198

Table 5. Average TM score of models by various alignment methods on 51 pairs involving the CASP14 hard targets together with the maximal TM score (of TM-align).

Blosum62	MRFalign	HHalign	CRFalign (3-State)	CRFalign (5-State)	TM-Align
0.3077	0.3693	0.3905	0.4198	0.4294	0.5973

Figure 15 shows the x-y comparison plot of the TM scores of the 51 pairs by CRFalign (with the five-state scheme at step 9) against those of HHalign (left) and another comparison of CRFalign against MRFalign [21] (right). We can see that CRFalign improves the modeling accuracy significantly for some of the hard targets in comparison with both HHalign and MRFalign. Figure 16 shows an example of the modeled structures among the CASP14 hard targets which exhibit notable improvement in the modeling accuracies. This target is T1082-D1, which consists of alpha helices where CRFalign results in TM score = 0.5656 that exhibits a large improvement over TM score = 0.2499 of HHalign. We used the above set of 51 pairs of CASP14 target-templates for estimating the running speed of CRFalign alignments. With the average sequence length per target of 178 residues, the average running time of a pairwise alignment was 1.76 seconds on our single CPU of AMD EPYC 7543 (2.80 Ghz).

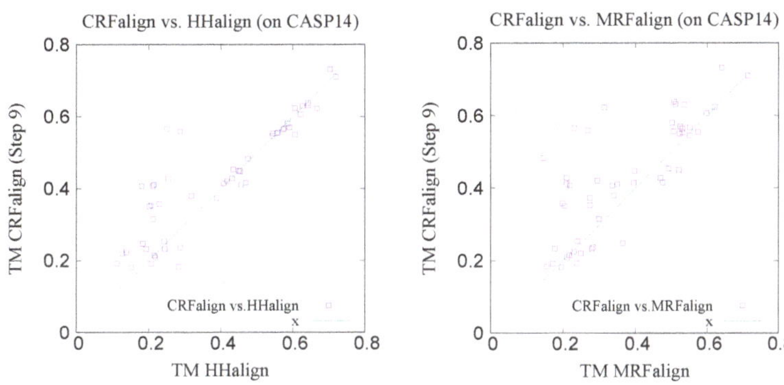

Figure 15. Comparison of the TM scores of structure models on hard targets of CASP14 by CRFalign (five-state model, step 9) against HHalign (**left**) and MRFalign (**right**) respectively.

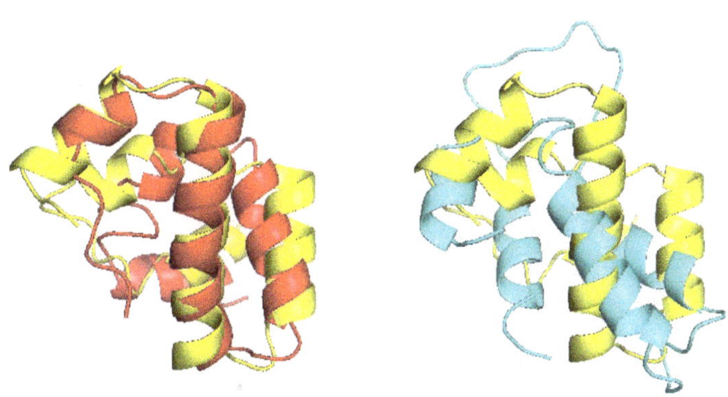

Figure 16. (**Left**) Structure model of the target T1082-D1 based on CRFalign alignment of T1082-D1-6h7bC (red) overlapped with the experimental structure of T1082-D1 (yellow) (with TM score = 0.5656, $rmsd = 3.01$) and (**right**) that based on HHalign alignment (cyan) overlapped with the experimental structure (yellow) (with TM score = 0.2499, $rmsd = 8.83$).

Sequence–structure alignment is useful in fold recognition i.e., template search. In order to test the template search capability of CRFalign alignment, we incorporated CRFalignment (with the five-state scheme) into our fold recognition framework called CRFpred [27,33]. CRFpred utilizes a set of machine learning methods such as random forest, boosted regression tree, support vector machine and linear regression, on features obtained from CRFalignment output. These features include profile scores, secondary structure scores and solvent accessibility scores. Details of the CRFpred will be presented in future publications. The structure database of 35539 proteins was built with 40% sequence identity cutoff. We randomly selected 300 targets from the database with the sequence length ranging from 100 to 500. These targets can be divided into three groups according to the levels of difficulties as measured from TM scores between the targets and best templates (excluding the targets themselves) among the database. These three sets are EASY (170 targets, TM > 80%), MEDIUM (89 targets, 60% < TM < 80%) and HARD (41 targets, TM < 60%) sets.

For each target, search is made from the database (excluding the target itself) and the top five templates are chosen from the result of CRFpred search. Table 6 shows the average TM scores of the targets with the best predicted (CRFpred) template and with the top templates respectively. Figure 17 shows the xy-plot comparison of the TM scores of 300 target proteins with the best template among the top five predicted templates by CRFpred (CRFalign) vs. the TM scores of the same proteins with the top templates from the database. In order to check how well CRFalign and CRFpred can detect a template that is close enough to the top template, we plot in Figure 18, for each of the three taregt sets, the success rate of finding a template wth a TM score that exceeds a given cutoff ratio of the TM score of the top template. We can see that, for the Easy target set, the detection rate at cutoff ratio of 95% reaches 95.9%, while for the case of the Medium target set, at the same cutoff ratio of 95%, the detection rate is down at 84.3%. On the other hand, for the case of the Hard target set, the detection rate at cutoff ratio of 95% is only 48.8%. By lowering the cutoff ratio to 85%, we can see that the detection rate increases to 68.3%. This manifests some measure of the difficulty in detecting a reasonable template for the case of the Hard target set.

Table 6. Summary of Average TM scores of 300 targets.

	Easy	Medium	Hard	All
Top template from database	0.8902	0.7240	0.4882	0.7860
Best in CRFpred top five	0.8837	0.7050	0.4431	0.7704

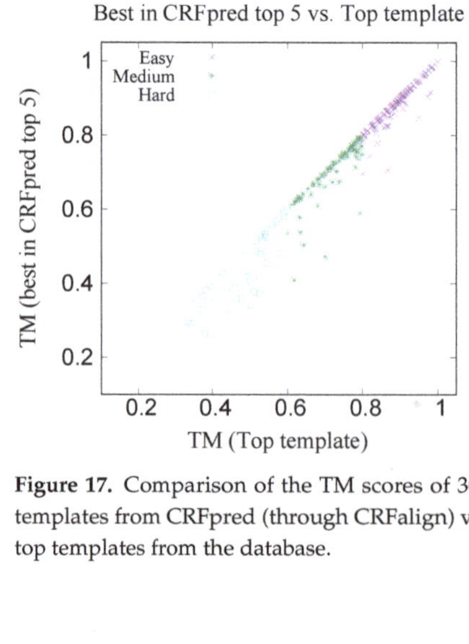

Figure 17. Comparison of the TM scores of 300 proteins with the best among top five predicted templates from CRFpred (through CRFalign) vs. the TM scores of the same proteins with the (true) top templates from the database.

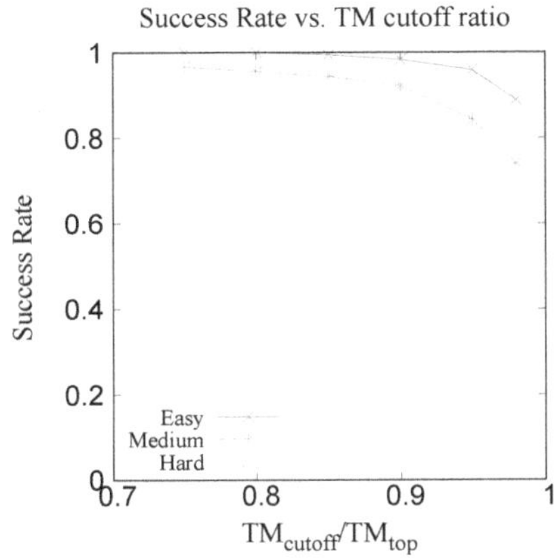

Figure 18. Success rate of finding (among the top five CRFpred templates) a template that is within a certain cutoff ratio of the maximal possible TM score for three subsets of Easy, Medium and Hard targets.

4. Discussion

A sequence–structure alignment method CRFalign is presented that improves upon a reduced three-state or five-state scheme of HMM-HMM profile alignment model by means of conditional random fields with nonlinear scoring on the sequence and structural features implemented with boosted regression trees. CRFalign can extract and exploit complex nonlinear relationships among sequence profiles and structural features, including secondary structures, solvent accessibilities, environment-dependent properties that give rise to position-dependent and environment-dependent match scores and gap penalties. Training of the CRFalign is performed on a chosen set of reference pairwise alignments from the SABmark benchmark set, which consists of Twilight Zone set and Superfamilies set with pairs of sequences very low to low, and low to intermediate sequence similarity, respectively. We found that our alignment method produces relative improvement in terms of average alignment accuracies, especially for the alignment of remote homologous proteins. Comparison of the modeling capabilities of our alignment on independent pairs of SABmark set with those of HHalign showed that our alignment method produced better modeling results especially in the relatively hard targets. This was also confirmed in recent tests on hard targets of CASP14. CRFalign was successfully applied to the initial stages of fold recognition and as an input to the multiple sequence alignment called (MSACSA) in the CASP11 and CASP12 competition on protein structure predictions.

Author Contributions: Conceptualization, J.L. (Jooyoung Lee); methodology, J.L. (Jooyoung Lee), S.J.L., J.L. (Juyong Lee) and I.-H.L.; software, S.J.L., K.J., S.S., J.L. (Juyong Lee) and I.-H.L.; validation, S.J.L. and K.J.; formal analysis, S.J.L.; investigation, S.J.L. and K.J.; data curation, K.J. and S.J.L.; writing—review and editing, S.J.L. and K.J. All authors have read and agreed to the published version of the manuscript.

Funding: This research was funded by the Ministry of Science and ICT, KOREA with grant number NRF-2017R1E1A1A01077717 and NRF-2018R1D1A1B07049312.

Institutional Review Board Statement: Not applicable.

Informed Consent Statement: Not applicable.

Data Availability Statement: Training and test sets are available from S.J.L. (yeesj123@gmail.com).

Acknowledgments: This research was supported by the Basic Science Research Program through the National Research Foundation of Korea (NRF) funded by the Ministry of Science and ICT (NRF-2017R1E1A1A01077717 and NRF-2018R1D1A1B07049312). We thank the KIAS Center for Advanced Computation for providing computing resources.

Conflicts of Interest: The authors declare no conflict of interest.

Appendix A

We consider the following input features for the regression trees of the function F.

(A) Features for match states:

We assume that the i-th position of the template (structure) sequence is aligned (i.e., matched) to the j-th position of the target sequence.

(1) Similarity between the HMM-profile of the structure and that of the template, which is calculated as the logarithm with base 2 of the dot product between the HMM-profiles of s and t. The HMM profiles are obtained from HHblits developed by Söding, et al.

$$\mathrm{HH}(i,j) \equiv \log_2(\sum_k H^s{}_{i,k} \cdot H^t{}_{j,k}) \tag{A1}$$

where $H^s{}_{i,k}$ denotes the k-th component of the hmm profile (position specific scoring matrix) at the residue position i of the structure template, and $H^t{}_{j,k}$ denotes the k-th component of the hmm profile at the residue position j of the target.

(2) Secondary structure similarity : We employ three-state classification for secondary structures, i.e., C (=Coil), H (=α-Helix), and E (=β-strand). For the template structures s, each residue position is assigned by DSSP [31] one of these three states. For the targets, we use the 3-state secondary structure prediction of PSIPRED which gives three values of relative likelihood of secondary structures corresponding to (C, H, E). For a match between the observed secondary structure in 3-state representation at i-th position of the template structure and the predicted secondary structure at j-th position of the target, the secondary structure similarity is determined as the component value of the predicted secondary structure at j-the position of the target corresponding to the observed secondary structure at i-th position of the template structure as follows

$$\mathrm{SS}(i,j) \equiv \mathrm{SSP}(j, k_{i,obs}) \tag{A2}$$

where $k_{i,obs}$ denotes the observed secondary structure (one of C, H, or E) at the residue position i of the template, and $\mathrm{SSP}(i, k_{i,obs})$ denotes the corresponding predicted (PSIPRED) value of the secondary structure at the residue position j of the target.

(3) Solvent accessibility : As for the solvent accessibility feature, we use a three-state classification scheme with labels of 'Buried', 'Intermediate', and 'Exposed' states. These are determined by the values of the relative solvent accessibility (RSA) of a specific residue with ranges 0–9% ('Buried'), 9–36% ('Intermediate'), and 36–100% ('Exposed'). For the template structure, the solvent accessibility state is determined by DSSP. Also for the case of target (query) sequence, we employ the program SANN [36] which predicts RSA propensity with 3-state scheme for a sequence with unknown 3D structure.

For a match between the observed RSA (obtained from DSSP) at i-th position of the template and the predicted RSA (from SANN) at j-th position of the target, the feature value is chosen as

$$\mathrm{SA}(i,j) \equiv \mathrm{SANN}(j, a_{i,obs}) \tag{A3}$$

where $a_{i,obs}$ denotes the observed RSA (one of the three states 'Exposed', 'Buried', or 'Intermediate') at the residue position i of the template (from DSSP), and $\mathrm{SANN}(j, a_{i,obs})$ denotes the corresponding predicted value of the RSA at the residue position j of the target.

(4) BLOSUM matrix, Gonnet matrix, and Kihara matrix : For a match state between s at position i and t at position j, we take as additional features the corresponding matrix

elements of the BLOSUM62 matrix [35] $B(s_i, t_j)$, Gonnet250 matrix [37] $G(s_i, t_j)$, and the Kihara matrix [38] $K(s_i, t_j)$.

(5) Environmental fitness score :

For the residue position i of the template, DSSP provides the secondary structure and the solvent accessibility for the residue. The environmental fitness score for the match between i-th residue position of the template and j-th residue position of target is obtained from the weighted average of the environmental fitness potential of the template's local environment state with the position specific frequency matrix (PSFM) $H^t(j,k)$ derived from the HMM profile of the target as follows

$$\mathrm{Env}(i,j) \equiv \sum_k \phi_{env}(k, ss_i, sa_i) \cdot H^t(j,k) \tag{A4}$$

where $\phi_{env}(k, ss_i, sa_i)$ denotes the environmental fitness potential of amino acid k for the secondary structure state ss_i and solvent accessibility state sa_i of the template at i-th residue position (which was borrowed from the PROSPECT II [39]).

(6) Neighborhood similarity score :

For the match between i-th position of the template and j-th position of the target, this score measures the similarity between the neighboring residues within a fixed window. Suppose we set the window size $n_w \equiv 2 * f + 1$ with $f \geq 1$, then the neighborhood similarity between the template and the target at offset position k from (i,j) is defined as the sum of Pearson correlations of the PSFM, SS and SA at $(i+k)$-th residue position of the template and the $(j+k)$-th residue position of the target with $k = 0, \pm 1, \pm 2, \ldots, \pm f$.

$$\mathrm{Ns}(i,j) \equiv \sum_{k=-f}^{k=+f} (\mathrm{Corr}(H^s(i+k), H^t(j+k)) + \mathrm{Corr}(SS(i+k), SS(j+k)) + \mathrm{Corr}(SA(i+k), SA(j+k))). \tag{A5}$$

Here $\mathrm{Corr}(x,y)$ denotes the Pearson correlation of vectors x and y with the same length of components which is defined as

$$\mathrm{Corr}(x,y) \equiv \frac{\sum_{i=1}^{N}(x_i - \overline{x})(y_i - \overline{y})}{\sqrt{\sum_{i=1}^{N}(x_i - \overline{x})^2}\sqrt{\sum_{i=1}^{N}(y_i - \overline{y})^2}} \tag{A6}$$

where $\overline{x} \equiv \sum_{i=1}^{N} x_i / N$ is the average value of x_i, $i = 1, \ldots, N$. As for the size of the neighboring window, we set $n_w = 9$ (i.e., $f = 4$).

(B) Input features for the Gap states:

(1) Seven-component reduced-profile features derived from the HMM profile based on seven classes of residues:

For a gap state (both at the template and target), the corresponding inserted residue was classified into seven classes as (i) class I of hydrophobic and aliphatic residues including Ala, Ile, Leu, and Val. (ii) class II of hydrophobic and aromatic residues including Phe, Trp, and Tyr. (iii) class III of polar residues including Asn, Cys, Gln, Met, Ser, and Thr. (iv) class IV of Acidic charged residues including Asp, and Glu. (v) class V of Basic charged residues including Arg, His and Lys. (vi) class VI of the residue Gly. (vii) class VII of the residue Pro.

Suppose there is a gap at position i in the template with the corresponding residue at position j of the target, one can construct a simple reduced profile by collecting and summing those component values of the PSFM (derived from the HMM profile) of the target residue at j-th position, such that those components belonging to the same class are summed over. That is,

$$C_{7,s}(i,j,u) \equiv \sum_{k \in u}^{20} H^t(j,k) \tag{A7}$$

where $H^t{}_{(}j,k)$ denotes the k-th component of the HMM profile (position specific scoring matrix) at the residue position i of the target. The seven-class index u ranges from the class I to the class VII as indicated above. Similar method can be applied to the case of a gap in the target.

(2) Secondary structure :

For a gap state at the target (i.e., insertion at the template), the observed secondary structure (from DSSP) of the template at the corresponding inserted residue is used as the secondary structure feature, which is represented in a 3-vector form with the Coil (C) state corresponding to $(1,0,0)$, the Helix (H) state to $(0,1,0)$ and the Extended Beta (E) to $(0,0,1)$. On the other hand, for a gap state at the template (i.e., insertion at the target), the predicted secondary structure propensity with three components (from PSIPRED) of the target at the corresponding inserted residue is used as the secondary structure feature.

(3) Solvent accessibility :

Similarly to the case of secondary structure, for a gap state at the target (i.e., insertion at the template), the observed solvent accessibility in three states (derived from DSSP) of the template at the corresponding inserted residue is used as the input feature. The three states of "Buried" (B), "Intermediate" (I), and "Exposed" (E) are represented as $(1,0,0)$, $(0,1,0)$, and $(0,0,1)$ respectively. On the other hand, for a gap state at the template (i.e., insertion at the target), the predicted solvent accessibility in three states (from SANN) of the target at the corresponding inserted residue is used as the input feature.

(4) Local gap propensity from secondary structure environment:

For a gap state at the target (i.e., insertion at the template), the predicted secondary structure information (from PSIPRED) of the target in the seven neighboring residues (i.e., from -3 to $+3$ separation from the gap position) is combined to give the Coil (or Loop) propensity as

$$SS^g(i,j) \equiv \sum_{k=-3}^{+3} ((SP(j+k,C) - SP(j+k,H)) + (SP(j+k,C) - SP(j+k,E))) \tag{A8}$$

where $SP(j+k,C)$ denotes the component of the predicted secondary structure at the residue position $j+k$ of the target, to be found in a Coil state C, $SP(j+k,H)$ the component of the predicted secondary structure to be found in a Helix state H, and similarly $SP(j+k,E)$ the component for the Beta strand state E.

On the other hand, for a gap state at the template (i.e., insertion at the target), we use similar formula on the neighboring residues for the template except that here, instead of the predicted secondary structure, the observed secondary structure is used, which is represented in a 3-vector form with the Coil (C) state corresponding to $(1,0,0)$, the Helix (H) state to $(0,1,0)$ and the Extended Beta (E) to $(0,0,1)$.

(5) Local gap propensity from solvent accessibility environment:

For a gap state at the target (i.e., insertion at the template), the predicted solvent accessibility information (from SANN) of the target in the seven neighboring residues (i.e., from -3 to $+3$ separation from the gap position) is combined to give the Coil (or Loop) propensity as

$$\mathrm{SA}^g(i,j) \equiv \sum_{k=-3}^{+3} (0.2 * \mathrm{SANN}(j+k,I) + 0.7 * \mathrm{SANN}(j+k,E)) \tag{A9}$$

where $\mathrm{SANN}(j+k,I)$ denotes the component of the predicted solvent accessibility at the residue position $j+k$ of the target, to be found in an "Intermediate" state I, $\mathrm{SANN}(j+k,E)$ the component of the solvent accessibility to be found in an "Exposed" state E. Hence we put more weight on the exposed residues than buried or intermediate residues.

Similarly for a gap state at the template (i.e., insertion at the target), we use similar formula on the neighboring residues for the template except that here, instead of the predicted solvent accessibility, the observed solvent accessibility (in normalized 3-vector form) is used. Therefore, the summand in the above formula becomes 0.2 for the state "Intermediate" and 0.7 for the "Exposed" state.

(6) Additional features for indicators of the end position of the sequences:

In addition to the above input features for the gap states, we also use separate indicators for the end positions of each of the two sequences for special treatment of end gaps. That is, we define

$$\mathrm{End}(i,j) \equiv \delta_i + \delta_j \tag{A10}$$

where δ_i (δ_j) $= 1$ if i (j) is at the beginning (N-terminal) or at the end (C-terminal) of the sequence and 0 otherwise.

References

1. Senior, A.W.; Evans, R.; Jumper, J.; Kirkpatrick, J.; Sifre, L.; Green, T.; Qin, C.; Zidek, A.; Nelson, A.W.R.; Bridgland, A.; et al. Improved protein structure prediction using potentials from deep learning. *Nature* **2020**, *577*, 706–710. [CrossRef] [PubMed]
2. Jumper, J.; Evans, R.; Pritzel, A.; Green, T.; Figurnov, M.; Ronneberger, O.; Tunyasuvunakool, K.; Bates, R.; Žídek, A.; Potapenko, A.; et al. Highly accurate protein structure prediction with AlphaFold. *Nature* **2021**, *596*, 583–589. [CrossRef] [PubMed]
3. Cheng, J.; Baldi, P. A Machine Learning Information Retrieval Approach to Protein Fold Recognition. *Bioinformatics* **2006**, *22*, 1456–1463. doi:10.1093/bioinformatics/btl102. [CrossRef] [PubMed]
4. Cheng, J. A multi-template combination algorithm for protein comparative modeling. *BMC Struct. Biol.* **2008**, *8*, 18. [CrossRef]
5. Zhang, Y. Progress and challenges in protein structure prediction. *Curr. Opin. Struct. Biol.* **2008**, *18*, 342–348. [CrossRef]
6. Joo, K.; Lee, J.; Lee, S.; Seo, J.H.; Lee, S.J.; Lee, J. High accuracy template based modeling by global optimization. *Proteins* **2007**, *69* (Suppl. S8), 83–89. [CrossRef]
7. Rychlewski, L.; Jaroszewski, L.; Li, W.; Godzik, A. Comparison of sequence profiles. Strategies for structural predictions using sequence information. *Protein Sci.* **2000**, *9*, 232–241. doi:10.1110/ps.9.2.232. [CrossRef]
8. Yona, G.; Levitt, M. Within the twilight zone: A sensitive profile-profile comparison tool based on information theory. *J. Mol. Biol.* **2002**, *315*, 1257–1275. [CrossRef]
9. von Ohsen, N.; Zimmer, R. Improving profile-profile alignment via log average scoring. In *Lecture Notes in Computer Science*; Springer: Berlin/Heidelberg, Germany, 2001; Volume 2149, pp. 11–26.
10. Tomii, K.; Akiyama, Y. FORTE: A profile-profile comparison tool for protein fold recognition. *Bioinformatics* **2004**, *20*, 594–595. doi:10.1093/bioinformatics/btg474. [CrossRef]
11. Yang, Y.; Faraggi, E.; Zhao, H.; Zhou, Y. Improving protein fold recognition and template-based modeling by employing probabilistic-based matching between predicted one-dimensional structural properties of query and corresponding native properties of templates. *Bioinformatics* **2011**, *27*, 2076–2082. [CrossRef]
12. Zhou, H.; Zhou, Y. Single-Body Residue-Level Knowledge-Based Energy Score Combined With Sequence-Profile and Secondary Structure Information for Fold Recognition. *Proteins* **2004**, *55*, 1005–1013. [CrossRef] [PubMed]
13. Zhou, H.; Zhou, Y. Fold recognition by combining sequence profiles derived from evolution and from depth-dependent structural alignment of fragments. *Proteins* **2005**, *58*, 321–328. [CrossRef] [PubMed]
14. Liu, S.; Zhang, C.; Liang, S.; Zhou, Y. Fold Recognition by Concurrent Use of Solvent Accessibility and Residue Depth. *Proteins* **2007**, *68*, 636–645. [CrossRef] [PubMed]
15. Zhang, W.; Liu, S.; Zhou, Y. SP5: Improving protein fold recognition by using predicted torsion angles and profile-based gap penalty. *PLoS ONE* **2008**, *3*, e2325. [CrossRef] [PubMed]
16. Söding, J. Protein homology detection by HMM-HMM comparison. *Bioinformatics* **2005**, *21*, 951–960. [CrossRef]
17. Deng, X.; Cheng, J. Enhancing HMM-based protein profile-profile alignment with structural features and evolutionary coupling information. *BMC Bioinform.* **2014**, *15*, 252. [CrossRef]
18. Do, C.B.; Gross, S.S.; Batzoglou, S. CONTRAlign: Discriminative Training for Protein Sequence Alignment. *Res. Comput. Mol. Biol.* **2006**, *3909*, 160–174. [CrossRef]
19. Peng, J.; Xu, J. Boosting Protein Threading Accuracy. *Res. Comput. Mol. Biol.* **2009**, *5541*, 31–45. [CrossRef]
20. Ma, J.; Peng, J.; Wang, S.; Xu, J. A conditional neural fields model for protein threading. *Bioinformatics* **2012**, *28*, i59–i66. [CrossRef]
21. Ma, J.; Wang, S.; Wang, Z.; Xu, J. MRFalign: Protein Homology Detection through Alignment of Markov Random Fields. *PLoS Comput. Biol.* **2014**, *10*, e1003500. [CrossRef]
22. Makigaki, S.; Ishida, T. Sequence alignment using machine learning for accurate template-based protein structure prediction *Bioinformatics* **2020**, *36*, 104–111. [CrossRef] [PubMed]
23. Remmert, M.; Biegert, A.; Hauser, A.; Söding, J. HHblits: Lightning-fast iterative protein sequence searching by HMM-HMM alignment. *Nat. Methods* **2011**, *9*, 173–175. [CrossRef] [PubMed]
24. Tyree, S.; Weinberger, K.Q.; Agrawal, K.; Paykin, J. Parallel boosted regression trees for web search ranking. In Proceedings of the 20th International Conference on World Wide Web, Hyderabad, India, 28 March–1 April 2011; ACM: New York, NY, USA, 2011; pp. 387–396.
25. Walle, I.V.; Lasters, I.; Wyns, L. SABmark-a benchmark for sequence alignment that covers the entire known fold space. *Bioinformatics* **2005**, *21*, 1267–1268. [CrossRef] [PubMed]

26. Yan, R.; Xu, D.; Yang, J.; Waker, S.; Zhang, Y. A comparative assessment and analysis of 20 representative sequence alignment methods for protein structure prediction. *Sci. Rep.* **2013**, *3*, 2619. [CrossRef]
27. Joo, K.; Joung, I.; Lee, S.Y.; Kim, J.Y.; Cheng, Q.; Manavalan, B.; Joung, J.Y.; Heo, S.; Lee, J.; Nam, M.; et al. Template based protein structure modeling by global optimization in CASP11. *Proteins* **2016**, *84* (Suppl. S1), 221–232. [CrossRef]
28. Joo, K.; Lee, J.; Sim, S.; Lee, S.Y.; Lee, K.; Heo, S.; Lee, I.H.; Lee, S.J.; Lee, J. Protein structure modeling for CASP10 by multiple layers of global optimization. *Proteins* **2014**, *82* (Suppl. S2), 188–195. [CrossRef]
29. Moult, J.; Kryshtafovych, A. Special Issue:CASP14: Critical Assessment of methods of protein Structure Prediction, 14th round. *Proteins* **2021**, *89*, 1603–1997.
30. Durbin, R.; Eddy, S. R.; Krogh, A.; Mitchison, G. *Biological Sequence Analysis*, 1st ed.; Cambridge University Press: Cambridge, UK, 1998.
31. Kabsch, W.; Sander, C. Dictionary of protein secondary structure: Pattern recognition of hydrogen-bonded and geometrical features. *Biopolymers* **1983**, *22*, 2577–2637. [CrossRef]
32. Jones, D.T. Protein secondary structure prediction based on position-specific scoring matrices. *J. Mol. Biol.* **1999**, *292*, 195–202. [CrossRef]
33. Hong, S.H.; Joung, I.; Flores-Canales, J.C.; Manavalan, B.; Cheng, Q.; Heo, S.; Kim, J.Y.; Lee, S.Y.; Nam, M.; Joo, K.; et al. Protein structure modeling and refinement by global optimization in CASP12. *Proteins* **2018**, *86*, 122–135. [CrossRef]
34. Zemla, A. LGA: A method for finding 3D similarities in protein structures. *Nucleic Acids Res.* **2003**, *31*, 3370–3374. [CrossRef] [PubMed]
35. Henikoff, S.; Henikoff, J.G. Amino Acid Substitution Matrices from Protein Blocks. *Proc. Natl. Acad. Sci. USA* **1992**, *89*, 10915–10919. [CrossRef] [PubMed]
36. Joo, K.; Lee, S.J.; Lee, J. SANN: Solvent accessibility prediction of proteins by nearest neighbor method. *Proteins* **2012**, *80*, 1791–1797. [CrossRef] [PubMed]
37. Gonnet, G.H.; Cohen, M.A.; Benner, S.A. Exhaustive matching of the entire protein sequence database. *Science* **1992**, *256*, 1443–1445. [CrossRef]
38. Tan, Y.H.; Huang, H.; Kihara, D. Statistical Potential-Based Amino Acid Similarity Matrices for Aligning Distantly Related Protein Sequences. *Proteins* **2006**, *64*, 587–600. [CrossRef]
39. Kim, D.; Xu, D.; Guo, J.t.; Ellrott, K.; Xu, Y. PROSPECT II: Protein structure prediction program for genome-scale applications. *Protein Eng.* **2003**, *16*, 641–650. [CrossRef]

molecules

Article

Modified Protein-Water Interactions in CHARMM36m for Thermodynamics and Kinetics of Proteins in Dilute and Crowded Solutions

Daiki Matsubara [1], Kento Kasahara [1,2], Hisham M. Dokainish [3], Hiraku Oshima [1] and Yuji Sugita [1,3,4,*]

1 Laboratory for Biomolecular Function Simulation, RIKEN Center for Biosystems Dynamics Research, Kobe 650-0047, Hyogo, Japan
2 Division of Chemical Engineering, Graduate School of Engineering Science, Osaka University, Toyonaka 560-8531, Osaka, Japan
3 Theoretical Molecular Science Laboratory, RIKEN Cluster for Pioneering Research, Wako 351-0198, Saitama, Japan
4 Computational Biophysics Research Team, RIKEN Center for Computational Science, Kobe 650-0047, Hyogo, Japan
* Correspondence: sugita@riken.jp; Tel.: +81-48-462-1407

Abstract: Proper balance between protein-protein and protein-water interactions is vital for atomistic molecular dynamics (MD) simulations of globular proteins as well as intrinsically disordered proteins (IDPs). The overestimation of protein-protein interactions tends to make IDPs more compact than those in experiments. Likewise, multiple proteins in crowded solutions are aggregated with each other too strongly. To optimize the balance, Lennard-Jones (LJ) interactions between protein and water are often increased about 10% (with a scaling parameter, $\lambda = 1.1$) from the existing force fields. Here, we explore the optimal scaling parameter of protein-water LJ interactions for CHARMM36m in conjunction with the modified TIP3P water model, by performing enhanced sampling MD simulations of several peptides in dilute solutions and conventional MD simulations of globular proteins in dilute and crowded solutions. In our simulations, 10% increase of protein-water LJ interaction for the CHARMM36m cannot maintain stability of a small helical peptide, $(AAQAA)_3$ in a dilute solution and only a small modification of protein-water LJ interaction up to the 3% increase ($\lambda = 1.03$) is allowed. The modified protein-water interactions are applicable to other peptides and globular proteins in dilute solutions without changing thermodynamic properties from the original CHARMM36m. However, it has a great impact on the diffusive properties of proteins in crowded solutions, avoiding the formation of too sticky protein-protein interactions.

Keywords: molecular dynamics simulation; enhanced sampling method; molecular force fields; van der Waals interaction; CHARMM36m; NBFIX; intrinsically disordered proteins; crowding simulations

Citation: Matsubara, D.; Kasahara, K.; Dokainish, H.M.; Oshima, H.; Sugita, Y. Modified Protein-Water Interactions in CHARMM36m for Thermodynamics and Kinetics of Proteins in Dilute and Crowded Solutions. *Molecules* **2022**, *27*, 5726. https://doi.org/10.3390/molecules27175726

Academic Editors: Kunihiro Kuwajima, Yuko Okamoto, Tuomas Knowles and Michele Vendruscolo

Received: 11 July 2022
Accepted: 30 August 2022
Published: 5 September 2022

1. Introduction

Atomistic descriptions of biomolecules using molecular dynamics (MD) simulation are important to investigate structure-dynamics-function relationships [1,2]. Long-time MD simulations provide us reliable thermodynamic and kinetic properties of proteins and other biomolecules in solution, membrane, and other cellular environments, if accurate molecular force fields are available [3,4]. Atomistic MD simulations used to focus on conformational dynamics of small globular proteins, while large membrane proteins, nucleic acids, and protein-nucleic acid complexes like ribosome have become target systems of the simulations in these days [5–8]. Intrinsically disordered regions/proteins (IDRs/IDPs) and their assemblies including aggregations and liquid droplets in cells have also been examined using atomistic [9–12] and coarse-grained (CG) MD simulations [13–18]. All of these molecules are necessary for modeling highly heterogeneous and crowded cellular environments [19,20].

To investigate protein structure, dynamics, and function in cellular environments, the effect of macromolecular crowding, or the excluded volume effect, is crucial [21]. In addition, weak and non-specific interactions between macromolecules or between macromolecules and metabolites play important roles, as observed in recent experimental [22–24] and computational studies [19,20,25–29]. Such interactions are also used to form liquid-liquid phase separations (LLPSs) in the cytoplasm, membrane, and nucleus, which could be functional platforms for biomolecules [12,30,31]. However, water is the most abundant molecules even in the cellular environments [32]. It suggests that a proper balance between protein-protein and protein-water interactions is necessary for reliable thermodynamic and kinetic properties of proteins predicted in atomistic MD simulations [27,33–35].

Molecular force fields for classical MD simulations are usually described as a sum of bonded (bond, angle, and dihedral) and non-bonded (electrostatic and van der Waals) interactions [36,37]. The reliability of the MD simulations strongly depends on the quality of the molecular force field. Therefore, continuous improvements of the molecular force-field parameters have been carried out [38–40]. Using the latest well-known force fields like AMBER [41,42], CHARMM [43,44], and OPLS [45], conformational stability and dynamics of globular native proteins are well reproduced in atomistic MD simulations, although there are still debates on protein folding pathways or mechanisms predicted in the simulations [46]. In contrast, IDR/IDP structures and intermolecular interactions between them are difficult to be predicted in MD simulations [9,40,47]. The radius of gyration (R_g) for IDR/IDP predicted in the MD simulations tends to be more compact than that observed in experiments [34,47]. Crowded protein solutions, which mimic the cytoplasmic environments in the cells, are also difficult to be investigated in the MD simulations with conventional molecular force fields [19,20,25–29]. In the simulations, proteins could be aggregated more strongly than experiments, and thereby, translational, and rotational diffusive motions of proteins become slower [27]. To overcome these problems, too sticky protein-protein interactions should be avoided from the conventional molecular force fields.

Since water is the most abundant molecules in most of biological simulations, it is reasonable to modify protein-water interactions, without changing protein-protein interactions from the original ones. Based on the idea, Best and his colleagues suggested the optimal scaling parameter of Lennard-Jones (LJ) interactions between protein and water [34]. With the scaling factor, $\lambda = 1.1$ (10% increase of protein-water interaction) for AMBER ff99SB [48] and ff03 [49] in conjunction with TIP3P [50] or TIP4P/2005 water models [51], they successfully simulated structures and dynamics of IDPs including Cold-shock protein and ACTR, consistent with experimental data. As well as in the case of all-atom force fields, the scaling of protein-water interaction in Martini 3 CG model yielded the improved agreement with the small-angle X-ray scattering (SAXS) experiments for IDPs [52]. Recently, Tang et al. further modified the backbone torsion angle parameters of Ser, Thr, and Gln, to improve ff99SBws in conjunction with TIP4P/2005 (ff99SBws-STQ) [35]. Nawrocki et al. also proposed a scaling parameter, $\lambda = 1.09$, for CHARMM36 [43] in conjunction with the modified TIP3P model [53] to avoid aggregations of villin headpiece subdomains (villins) in their high-concentration solution [27]. Using the modified force fields, they could investigate slowing down of translational/rotational diffusions of villins, as the concentration of the proteins increases. The slowdown of diffusions was related to the transient cluster formations in crowded solutions, which was also observed by Hummer and coworkers in their extensive μs MD simulations [28] using Amber99SB*-ILDN-Q [54] in conjunction with TIP4P-D water model [55].

These studies suggest the importance of a proper balance between protein-protein and protein-water interactions, which can be controlled via optimizing the scaling parameter, λ, of protein-water LJ interactions. Although similar scaling parameters were found for AMBER ff99SB/ff03 and CHARMM36, the values are not necessarily applicable to other force fields. Our aim, in this study, is to find an optimal scaling parameter for the simulations of globular and disordered proteins in dilute or crowded solutions, when we employ CHARMM36m [44] in conjunction with the modified TIP3P water [53]. We first examine various scaling parameters for LJ interactions between protein and water in MD simulations of $(AAQAA)_3$ in solution. To confirm the most reasonable scaling parameter, λ, several proteins including TDP-43 in the C-terminal domain (CTD) [31], chignolin [56], c-Src kinase [57], and villin [58] are simulated in dilute or crowded solution. Considering the slow folding/unfolding processes of these peptides/proteins, enhanced sampling algorithms like replica-exchange molecular dynamics (REMD) [59] for $(AAQAA)_3$, Gaussian accelerated MD (GaMD) [60] for TDP-43, and Gaussian accelerated replica-exchange umbrella sampling (GaREUS) [61] for chignolin were employed. We employed different protocols based on the features of each peptide and analysis. In the case of $(AAQAA)_3$, the calculation of the temperature dependence of helicity is required for comparison with the experiments. The REMD method is suitable for evaluating the temperature dependence while enforcing the conformational sampling. Since chignolin is a β-hairpin forming protein and its timescale for folding/unfolding process is longer than that for α-helix, enhanced sampling methods using the collective variables (CVs), such as GaREUS, are needed. TDP-43 is intrinsically disordered, and any CVs are not needed. Thus, GaMD can be used for accelerating the conformational changes without increasing the computational cost from the cMD. Following the work by Nawrocki et al. [27], we test the scaling parameters of protein-water LJ interactions in the range between 1.00 and 1.09. The most reasonable value that we obtain for CHARMM36m in conjunction with the modified TIP3P is largely different from the previously proposed values for AMBER ff99SB/ff03 and CHARMM36, suggesting that a better balance between protein-protein and protein-water interactions is realized in CHARMM36m [44]. However, the small scaling value that we suggest in this study could change the diffusive properties of proteins in crowded solutions.

2. Results

2.1. Folding and Stability of Small Peptides in Solution

2.1.1. $(AAQAA)_3$

We performed REMD simulations [59] of $(AAQAA)_3$ using 58 replicas covering the temperature range between 275.0 and 382.0 K. In the simulations, we aim to examine the temperature dependence of the helicity with different scaling parameters, λ, for protein-water LJ interactions. Since the scaling is introduced in CHARMM using the NBFIX function, the scaling parameter, λ, and the term, NBFIX, are used as the same meaning in this paper. Following to the previous studies [44], we regard that the corresponding residues form an α-helix, when the backbone dihedral angles (ϕ and ψ) for three or more consecutive residues satisfy $-100^\circ < \phi < -30^\circ$ and $-67^\circ < \psi < -7^\circ$, respectively. The fraction of helix is computed from the number of residues forming α-helix, divided by the total number of residues. In Figure 1a, five curves obtained in the REMD simulations with λ = 1.00 (CHARMM36m), 1.03, 1.04, 1.06, and 1.09, are compared to the experimental results [62]. Although none of the simulation curves fit to the temperature dependence observed in the experiment, the fraction of helix at 300 K for λ = 1.00 (CHARMM36m) (0.17 $\pm$ 0.01) and 1.03 (0.15 $\pm$ 0.01) are close to the experimental one (~0.2). Note that at λ = 1.00 in our simulation shows similar value to the previously reported in the CHARMM36m paper [44], despite the difference in water box size. In contrast, using larger scaling values, λ = 1.06 and 1.09, the fraction of helix is drastically reduced to less than 0.1 at all the temperatures. Even at 300 K, it is around 0.05, which is much smaller than the experimental value and those with λ = 1.00 and 1.03. Interestingly, the curve with λ = 1.04 shows medium fractions of helicity between λ = 1.03 and 1.06. In Figure 1b, the faction helix at 300 K versus the scaling parameter, λ, is well fitted with the sigmoidal curve,

$$f(\lambda) = \frac{A}{1+\exp\left(-\left(\frac{\lambda-B}{C}\right)\right)} + D, \tag{1}$$

and the fitted parameters are $(A,B,C,D) = (0.11, 1.04, -0.0049, 0.0055)$. Parameter B corresponds to the scaling factor for the middle point of the helicity curve.

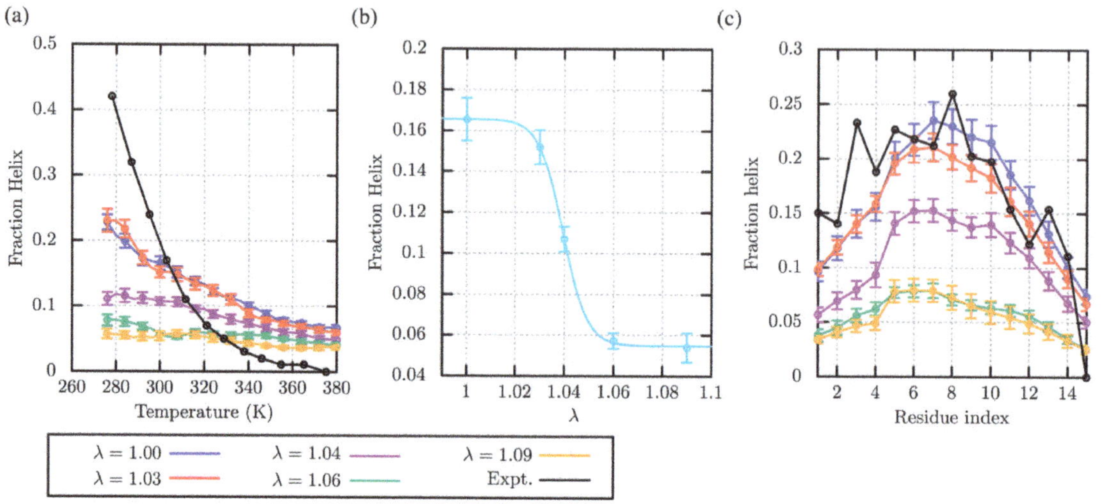

Figure 1. The α-helicity of $(AAQAA)_3$ using a different scaling parameter, λ. (**a**) Temperature dependency of the fraction of helix, (**b**) the λ-dependency of the fraction of helix at 300 K, (**c**) the fraction of α-helix in each residue at 300 K.

In Figure 1c, the fraction of helix in each residue at 300 K is compared between the REMD simulations with the scaling parameter, λ, and the experimental results. The residue profiles with λ = 1.00 and 1.03 are similar to the experimental one, although that in residue 3 (Q_3) is largely underestimated in the simulations. Interestingly, these profiles are also comparable to those predicted in recent AMBER force fields for IDPs (ff99SBws-STQ [35]). In other profiles with λ = 1.04, 1.06, and 1.09, the fraction of helix in each residue is scaled down almost uniformly from those with λ = 1.00 and 1.03.

We next examine the conformational spaces, which were explored in the REMD simulations with different scaling parameters. For this purpose, the two-dimensional potential of mean forces (2D-PMFs) at 300 K are shown along with the end-to-end distance, d, and the Cα root mean square deviation (RMSD) from the ideal α-helix conformation (Figure 2). The folded structures are localized at d~24 Å and RMSD~0.5 Å, while the unfolded structures have a broad region with RMSD > 4 Å. Except for λ = 1.09, these two basins (for the folded and unfolded ones) clearly exist in all the PMFs. The PMFs with λ = 1.03 is close to that with the original CHARMM36m (λ = 1.00), while the unfolded basin is slightly emphasized due to stronger protein-water interactions. At λ = 1.04, 1.06, and 1.09, the unfolded basins show deeper and wider free-energy minima as the protein-water interaction increases via the scaling values. These analyses suggest that the shape of the free-energy landscapes of $(AAQAA)_3$ does not change significantly with different scaling parameters, λ, while the populations of folded and unfolded states are drastically altered.

The REMD simulation results of $(AAQAA)_3$ suggest that the scaling parameter, λ = 1.04, or larger values could underestimate the stability of α-helix, when it is applied to CHARMM36m in conjunction with the modified TIP3P. Unexpectedly, the previously proposed scaling parameter for CHARMM36, λ = 1.09, does not work well for $(AAQAA)_3$. Since the peptide contains only two types of amino acids (Ala and Gln), more through tests are required to examine the applicability of λ = 1.03 in many other protein systems. Hereafter, we mainly discuss two scaling parameters, λ = 1.00 and 1.03 and examine several protein systems in dilute or crowded solution, whereas the results with λ = 1.09 are presented in Supplementary Information.

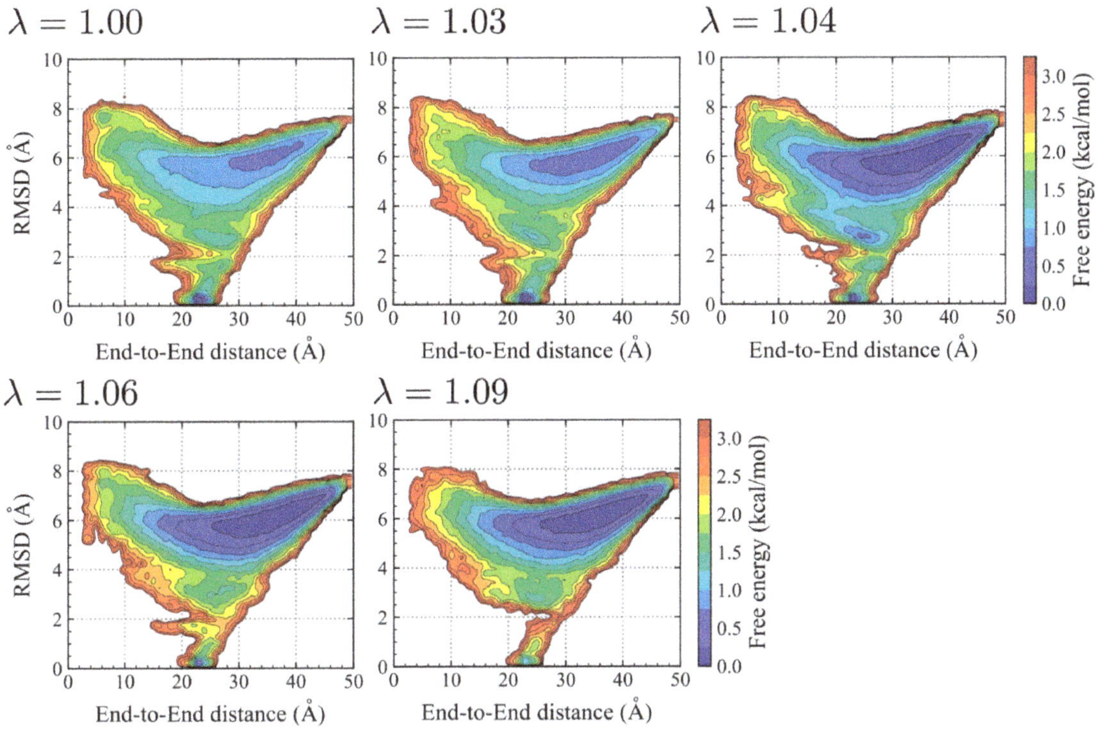

Figure 2. Two-dimensional potential of mean forces (2D-PMFs) of $(AAQAA)_3$ on the end-to-end distance, *d*, and the Cα root mean square deviation (RMSD) from the ideal helix for different λ = 1.00, 1.03, 1.04, 1.06, and 1.09.

2.1.2. Chignolin

The folding of chignolin (PDB ID: 1UAO [56]), a β-hairpin forming protein with 10 amino-acid residues (sequence: GYDPETGTWG), is simulated with GaREUS [61], where replica-exchange umbrella sampling (REUS) [63] and Gaussian accelerated MD (GaMD) [60] are combined. Since a folding process of a β-hairpin takes longer time than α-helical peptide, a powerful sampling method, such as GaREUS, is quite useful to get the converged thermodynamics within reasonable computational times. Experimentally, chignolin is reported to form both a stable β-hairpin and misfolded structures, where the folded population from the NMR measurement is ~60%. We first test two scaling parameters, λ = 1.00, and 1.03 for the GaREUS simulations of chignolin in water and compare them with λ = 1.09.

Figure 3a shows the probability densities along the Cα-RMSD with respect to the NMR structure are shown. In the CHARMM36m paper [44], chignolin, and its double mutant, CLN025 were simulated using REMD, showing the native populations of 2.6% and 41%, respectively. Interestingly, the folded population in the simulation with λ = 1.00 (CHARMM36m) has increased to 27%, suggesting the superiority of sampling efficiency in GaREUS compared to REMD for β-hairpin peptides. As λ increases, the folded conformations (RMSD < 2.2 Å) increases up to 33% for λ = 1.03. Note that the convergence check of the folded population using different trajectory lengths indicates that folded populations are well converged in the present simulation (Supplementary Figure S1). The decrease of the folded population is observed for λ = 1.09 (22%) (Supplementary Figure S2). In comparison, the folded population for λ = 1.09 is lower than those for λ = 1.00 and 1.03. Although the populations for λ = 1.00 and 1.03 are still lower than the NMR measurement [56], the performance of CHARMM36m for the structure prediction of β-hairpin peptides is not too bad as discussed in the original paper. Figure 3b shows the 2D-PMFs spanned with the ASP3-GLY7 and ASP3-THR8 distances, $d(\mathrm{ASP3}-\mathrm{GLY7})$ and $d(\mathrm{ASP3}-\mathrm{THR8})$ which are often used to describe the folded, misfolded, and unfolded states of chignolin [64]. As revealed in the previous study, the regions around ($d(\mathrm{ASP3}-\mathrm{GLY7})$, $d(\mathrm{ASP3}-\mathrm{THR8})$)~(3 Å, 6 Å) and ~(6.5 Å, 3 Å) correspond to the folded and misfolded states. The overall shapes of the 2D-PMFs for λ = 1.00 and λ = 1.03 are similar to each other. However, the misfolded population for λ = 1.03 is slightly increased and the unfolded population is instead reduced compared to that for λ = 1.00. It is interesting that the folded population of the β-hairpin increases as the protein-water LJ interaction is slightly stronger, while the detailed mechanism is still unknown.

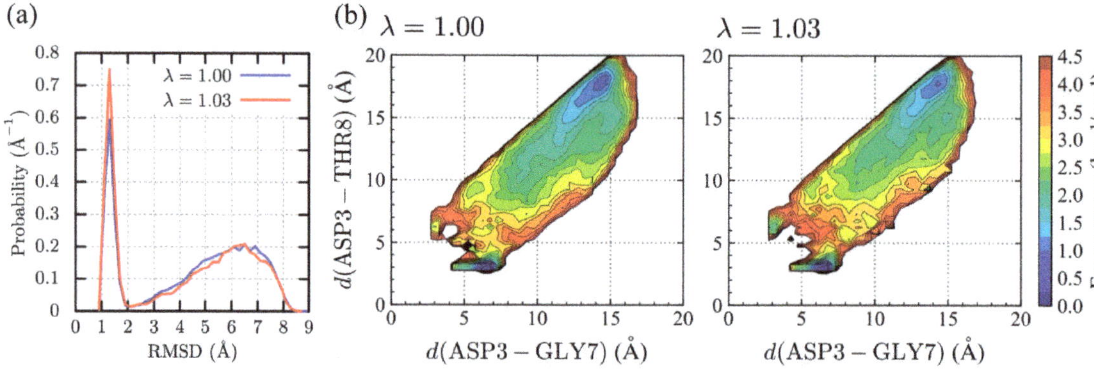

Figure 3. Distributions for the chignolin conformation. (**a**) Probability densities along the Cα-RMSD. (**b**) 2D-PMFs spanned with the ASP3-GLY7 and ASP3-THR8 distances for λ = 1.00 (**left**) and 1.03 (**right**).

2.1.3. TDP-43 in the CTD

Transactive response DNA binding protein 43 (TDP-43) is a versatile nucleic-acid binding protein, playing a central role in amyotrophic lateral sclerosis (ALS) pathogenesis. TDP-43 consists of a well-folded N-terminal domain (NTD), two-highly conserved RNA-recognition motifs (RPMs), and an unstructured prion-like C-terminal domain (CTD). Residues 320–334 in the CTD are found to form a transient α-helix in the previous studies with MD simulations and NMR spectroscopy [31]. We therefore select the residues 310–340 of TDP-43 as the second target system. In the simulations, we focus on the secondary structure formation of TDP-43 in the CTD and use the GaMD method [60] as an enhanced sampling method, which can reduce energy barriers between minimum energy states using a GaMD boost potential. We performed ten replicas of GaMD simulations (each for 1 μs) and took the averages of the fraction of helix in each residue. The effect of GaMD boost potential was removed by the reweighting scheme with the cumulant expansion proposed by Miao et al. [60].

The GaMD simulations were conducted with the scaling parameters of λ = 1.00 and λ = 1.03 to make a comparison with the experiment (Figure 4). The fractions of helices show large statistical errors, as the averaged values are just taken from ten independent GaMD runs. Comparison of λ = 1.00 and 1.03 show a small reduction of helicity in the CTD of TDP-43, as protein-water LJ interaction is stronger. For λ = 1.00, the computed helicity in residues 320–330 (region I, sequence: PAMMAAQAA) is in accord with the experimental values, while the helix formed around 335–345 (region II, sequence: GMMGMLASQQ) is found to be too stabilized. The stability of helix of region II is reduced with λ = 1.03, while the reduction of helicity in region I is also observed. In the case of λ = 1.09, the helicity of region II becomes close to the experimental observation, but the difference from the experiment in region I is further emphasized (Supplementary Figure S3).

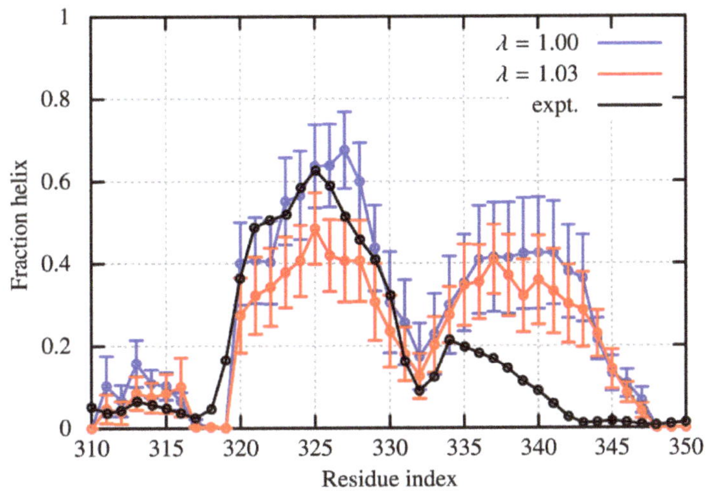

Figure 4. Fraction of helix in the C-terminal domain of TDP-43 for the conditions of λ = 1.00 (blue) and of λ = 1.03 (red), and for the NMR measurement (black). The helicity is evaluated through the DSSP algorithm.

The results suggest that the 1.03 scaling of protein-water LJ interaction keeps major structural features of TDP-43 in the CTD, with some differences from the NMR experimental data. Rg of TDP-43 in the CTD with λ = 1.03 does not change from that with λ = 1.00 (Supplementary Figure S4). It suggests that the scattering function of TDP-43 corresponding to the SAXS profile, which reflects the averaged feature of the protein structures, would be unaltered using the present modification of protein-water LJ interactions. To reduce helicity in the region II, fine tuning of the backbone torsional angles might be necessary like the previous studies of AMBER ff99SBws-STQ [35].

2.2. *Dynamics and Stability of Globular Proteins in Solution*

c-Src Kinase

The c-Src kinase (PDB ID: 1Y57) [57] is one of the essential protein kinases, which consists of the kinase domain (276 residues), SH2, and SH3 domains. Although these two domains are necessary for the activation, we simulate only the kinase domain in water as the previous computational works [65,66]. Five independent MD trajectories for each value of λ (1.00 and 1.03) are analyzed to examine the effect of the scaling parameter, λ = 1.03, on globular protein structures (Figure 5). For comparison, we also performed one MD simulation for λ = 1.09, The distribution of the Cα-RMSDs excluding the residues near the N- and C-terminals have a peak around ~2.5 Å for λ = 1.00 and 1.03 (Figure 5a), suggesting that the kinase domain stability is kept in both cases. Interestingly, the population of the native state (<3.0 Å) with λ = 1.03 slightly increases compared to that with λ = 1.00. The distribution for λ = 1.09 shows a peak spreading around the RMSD of 4.5 Å, which is absent in the cases of λ = 1.00 and 1.03 (Supplementary Figure S5). It indicates the distortion of the kinase domain structure. Referring to Figure 5b, the Cα root mean square fluctuation (Cα-RMSF) analysis shows that the difference is hardly discernible between λ = 1.00 and λ = 1.03, except for the first 40 residues and the residues around 210–220.

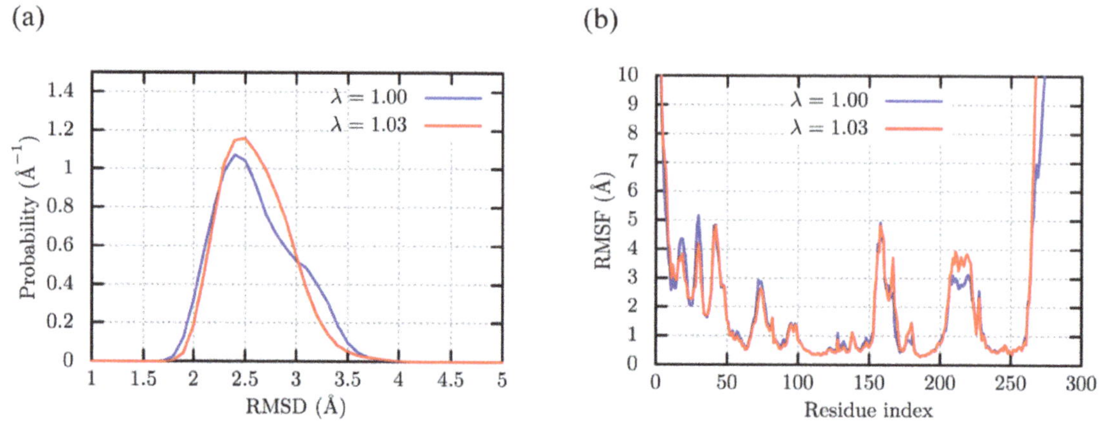

Figure 5. Structural fluctuation of c-Src kinase in water. (**a**) Distribution of Cα-RMSD. The terminal residues are excluded for the analysis. (**b**) The Cα root mean square fluctuation (Cα-RMSF).

The time-series of the Cα-RMSDs with all the 276 residues in Supplementary Figure S6 reveal larger deviations for λ = 1.03. This suggests that the conformational fluctuations near the N- and C-terminals are enhanced as protein-water LJ interaction increases (with λ = 1.03). The results of the Cα-RMSDs with and without N- or C-terminal residues (10 residues for each terminal) are reasonable, since we do not change intra-protein interactions of the CHARMM36m, but slightly emphasize protein-water interactions using the scaling parameter of λ = 1.03. The latter seems to affect the motions of N- and C-terminal residues in water, primarily, while the globular domain stability is almost unaltered. Note that the Cα-RMSD and Cα-RMSF obtained from the MD simulations with the LJ scaling parameters, λ = 1.00 and 1.03, are similar to those in a previous study using the AMBER ff99SB-ILDN [48]. Since the AMBER force field was tuned to reproduce the conformational dynamics and stability of folded native structures, the results obtained here seem to be promising for simulating the folded native protein structures in solution.

2.3. *Structural Stability and Diffusivity of Globular Proteins in the Crowded Solutions*

Finally, a small globular protein, villin headpiece subdomain (villin), was simulated with the conventional MD simulations in dilute solution as well as in crowded solution. Villin contains 35 amino-acid residues, composing of three α-helices [PDB ID: 1VII [58]]. Due to the small size and conformational stability, it was often used to test simulation protocols, folding mechanisms [46], and the effect of macromolecular crowding and weak non-specific interactions [25–27]. Here, we prepared two systems: in one system, a single villin is simulated in water (dilute solution) and in the other, eight villins are simulated in solution (crowded solution). Note that in the crowded solution, both target and crowder proteins are villins. The concentration of the crowded solution is about 32 mM, which is the same as those used in our previous study [27].

The time-series of Cα-RMSDs with respect to the crystal structure in the dilute solution (Supplementary Figure S7a) reveals that villin keeps the native structure (2–3 Å) during the simulations with λ = 1.00. As for λ = 1.03, the RMSDs fluctuate around 2–3 Å for five trajectories, while one trajectory shows larger values of RMSD (~4 Å) as compared to the others due to the orientational change of the N-terminal α-helix (Supplementary Figure S7b). On the other hand, the structural characterization with the DSSP algorithm [67,68] reveals that the native secondary structures in the native structure are well conserved for all the trajectories (Figure 6a). In the crowded solution, most villins keep the structures close to the native state (2–3 Å) with λ = 1.00 (Supplementary Figure S8a), while a partial unfolding is observed in one of eight villins. This is rather consistent with our previous studies, which suggests the partial unfolding due to the weak and non-specific interactions. At λ = 1.03, the number of villins showing the large fluctuation compared with the dilute solution increases, while there was no unfolding trajectory in the eight villins (Supplementary Figure S8b. The secondary structures are conserved in the crowded solution (Figure 6b), which suggests that tertiary structures might be broken due to protein-protein interactions. By further increasing the scaling factor up to λ = 1.09, the breakdown of the native structure is observed both for the dilute and crowded solutions (Supplementary Figures S9 and S10).

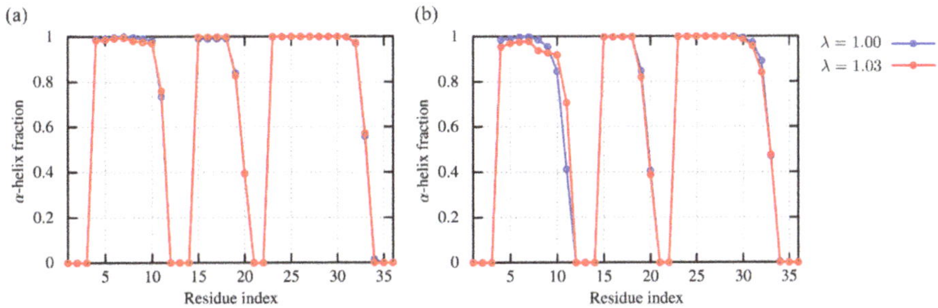

Figure 6. The α-helix fraction of villin in the dilute (**a**) and crowded solutions (**b**). The fraction is computed through DSSP algorithm. As for the crowded solution, the averages of the fraction over eight villins are shown.

The translational diffusions of villin in the dilute and crowder solutions are examined. Note that the diffusion coefficient depends on the system size due to the periodic boundary condition (PBC). In the present study, we employ the correction proposed by Yeh and Hummer [69], represented as

$$D = D_{\mathrm{MSD}} + D_{\mathrm{PBC}}(R_h, \eta, L), \tag{2}$$

where D_{MSD} and D_{PBC} are the diffusion coefficient obtained from the slope of the mean square displacement (MSD), and correction term for PBC defined from the hydrodynamic radius of villin (R_h), solvent viscosity (η), and box length of the system (L). The value of R_h is taken from HYDROPRO [70] estimation as 13.86 Å. The viscosity of the TIP3P water, η_{TIP3P} = 0.35 cP [71], is used for the dilute solution. As for the crowder solution, the Einstein's relationship between the crowder volume fraction (ϕ) and viscosity of the crowder solution (η_c) for suspension is employed. In addition, we utilize the viscosity correction to the diffusion coefficient with the experimental water viscosity, η_{expt} = 0.89 cP [72]. The corrected coefficient is defined as D'. Further description is available in the Supplementary Information.

In dilute solution, the values of D_{MSD} for λ = 1.00 and 1.03 obtained from the linear fitting of the MSD at $30 \leq t/\mathrm{ns} \leq 50$ are also close to each other (Table 1). The corrected coefficients (D') are 0.19 and 0.20 $\mathrm{nm}^2/\mathrm{ns}$ for λ = 1.00 and 1.03, respectively. These values after the PBC and viscosity corrections are in excellent agreement with that predicted from HYDROPRO, 0.18 $\mathrm{nm}^2/\mathrm{ns}$. Note that the HYDROPRO predictions are generally close to the experimental values under the dilute condition, and hence this agreement indicates the reliability of CHARMM36m with λ = 1.03 about the description of the diffusive properties as well as the original CHARMM36m for the dilute solutions.

Table 1. Diffusion coefficients of villin in the dilute and crowder solutions with NBFIX 1.03 scaling factor. The values for 1.00 scaling are shown in parenesis.

	D_{MSD} ($\mathrm{nm}^2/\mathrm{ns}$)	D_{PBC} ($\mathrm{nm}^2/\mathrm{ns}$)	D ($\mathrm{nm}^2/\mathrm{ns}$)	D' ($\mathrm{nm}^2/\mathrm{ns}$)
Dilute	0.33 (0.30)	0.17 (0.17)	0.50 (0.48)	0.20 (0.19)
Crowded	0.048 (0.031)	0.15 (0.15)	0.20 (0.18)	0.077 (0.070)

As for the crowder solution, the diffusion coefficients from the MSDs (D_{MSD}) are 0.031 nm^2/ns for λ = 1.00 and 0.048 nm^2/ns for λ = 1.03, respectively (Figure 7b and Table 1). Also, the diffusion coefficient with the PBC correction (D_{PBC} is 0.20 nm^2/ns) is for λ = 1.03, which is close to that from CHARMM36 with λ = 1.09, 0.22 nm^2/ns [27]. The small modification of protein-water LJ interaction leads to ~1.5 times acceleration of translational motions, judging from MSD results (Figure 7b). These results suggest that the optimal scaling parameter, λ = 1.03, does not change conformational stability of globular proteins significantly, while it avoids from slowdown of translational diffusion for proteins in crowded solution, probably reducing too sticky protein-protein interactions in the MD simulations.

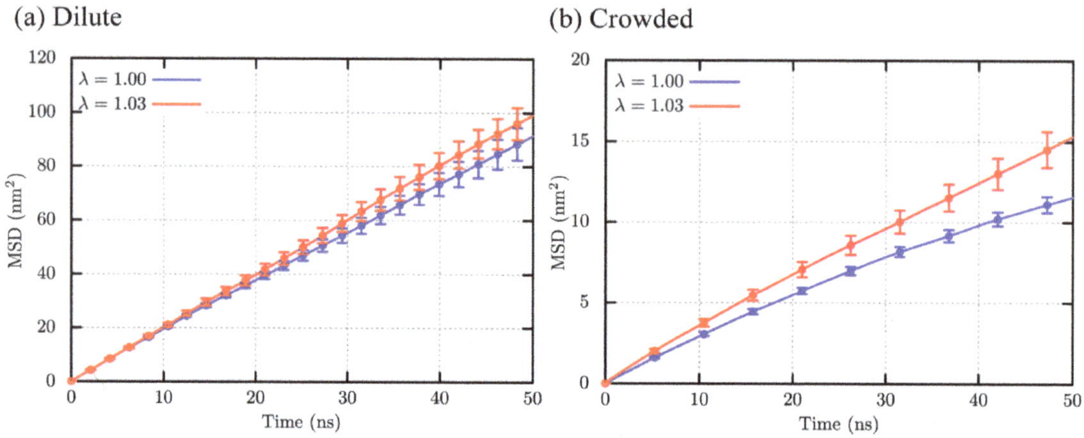

Figure 7. Mean square displacements (MSDs) of villin in (**a**) the dilute and (**b**) crowder solutions.

2.4. Solvation Free Energies of Amino Acid Analogues

So far, we used the single scaling parameter, λ, to modify protein-water LJ interactions in MD simulations of peptides or proteins. To understand how the modification of the CHARMM36m affects molecular interactions between water and each amino acid, we compute the solvation free energies of the 14 amino acid homologs, associated with the solubility of molecules, through free energy perturbation (FEP) method (Figure 8). For all the species, CHARMM36, CHARMM36m (λ = 1.00), and the modified CHARMM36m (λ = 1.03) are compared to the experimental values of ΔG_{solv} [73]. Figure 8 shows that all the calculations with λ = 1.03 become closer to the experimental results compared with others, suggesting that the modification seems to be valid for each amino acid. However, to achieve the quantitative agreement with the experiments, in particular, for the analogues of Asn, Gln, Met, Trp, and Tyr, a larger scaling factor is required. For instance, to achieve the quantitative agreement of solvation free-energies between FEP and experiments, about a 40% increase of protein-water LJ interactions (λ = 1.40) is necessary, which may not be applicable to MD simulations of peptides and proteins in solution. To resolve this inconsistency, furthermore careful tunings in molecular force fields are necessary not only for proteins but also for water molecules.

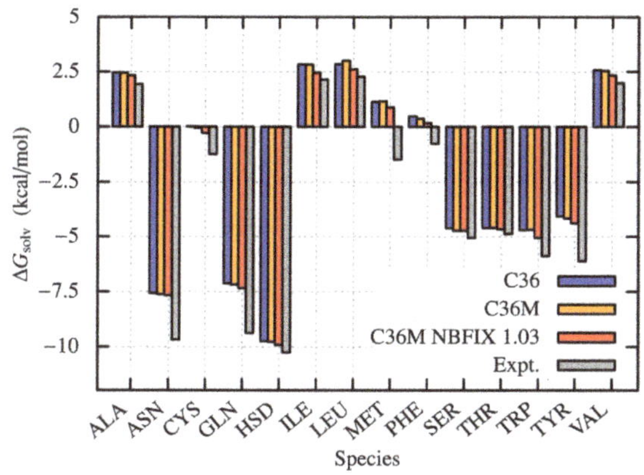

Figure 8. Solvation free energies (ΔG_{solv}) of amino acid analogues computed using free energy perturbation (FEP) method.

3. Discussion and Conclusions

In this study, we performed enhanced sampling MD simulations of small peptides in dilute solution and conventional MD simulations of globular proteins in dilute and crowded solutions using various scaling parameters of protein-water LJ interactions. For CHARMM36m in conjunction with the modified TIP3P water models, about 10% increase of the protein-water interaction affects peptide conformational stability significantly and only small increases (up to 3%) is allowed to keep a good balance between protein-protein and protein-water interactions. The modified force field is applicable not only to α-helical but also β-hairpin peptides and globular proteins in dilute solution. As enhanced sampling methods, we used REMD for $(AAQAA)_3$, GaREUS for chignolin, and GaMD for TDP-43 at the CTD in dilute solution. The GaREUS simulation for chignolin shows higher populations of the folded conformation both for the original (λ = 1.00) and modified (λ = 1.03) CHARMM36m. Note that the original CHARMM36m paper suggested much lower population of the folded state using T-REMD simulation. The GaREUS can enhance conformational sampling with the boost potential, replica-exchange, and umbrella potential for collective variables, which is more suitable to study slow folding/unfolding simulations of a β-hairpin peptide than T-REMD. It suggests the importance of enhanced conformational sampling methods for evaluating the quality of molecular force fields for different peptides or proteins.

In crowded villin simulations, translational and rotational diffusive motions are significantly affected by the small modification in the force field, keeping the conformational stability of villin compared to the simulation with the original CHARMM36m force field. This suggests that 3% increase of protein-water LJ interactions reduces too much sticky protein-protein interaction of the original force fields to avoid aggregation in the high-concentration protein solution. It should be noted that the over sticky protein association could lead to the underestimation of the protein diffusivity, and hence the modification of the protein-water interactions proposed by the present study is useful for elucidating the kinetics in the crowded environments.

For IDRs/IDPs simulations, the scaling of protein-water LJ interactions might not be sufficient to explore disordered conformations in dilute and crowded solutions. Although too much sticky protein-protein or intra-protein interactions are avoided, local conformational properties, such as dihedral angle distributions, should be simulated more carefully. As introduced in AMBER ff99SBws-STQ, a careful high-tuning in the backbone dihedral angle potential energy is useful in the simulations of IDRs/IDPs.

The test systems in this study include an α-helical peptide, $(AAQAA)_3$, a β-hairpin peptide, chignolin, a disordered region with some helical fraction, C-terminal TDP-43, globular proteins, c-Src kinase and villin in dilute and crowded solutions. Ideally, a single molecular force field is applicable to such a variety of peptides and proteins in different conditions, predicting their intrinsic structures, dynamics, and interactions at high precisions. This study provides a minimal set for testing the quality of the existing force fields.

4. Materials and Methods

All the MD simulations were performed using the GENESIS software [74,75]. CHARMM36m [44] with or without NBFIX corrections are used together with TIP3P water model. After modeling each protein/peptide and adding water molecules and ions, energy minimization was carried out and then short MD simulations were performed for equilibration in NPT and NVT ensembles. The system temperature and pressure were controlled using the Bussi thermostat/barostat [76] with a time step of 2 fs. Water molecules and protein/peptide bonds involving hydrogens were constrained with SETTLE and SHAKE [77], respectively. Long-range electrostatic interaction was calculated using particle mesh Ewald (PME) method [78], while LJ interaction was smoothly reduced to zero from 10–12 Å using a switch function.

4.1. $(AAQAA)_3$

Initial structure of $(AAQAA)_3$ was built as an extended conformation using VMD plugin. The N- and C-terminal residues were capped with the acetyl and amide groups, respectively. The peptide was solvated in a periodic box (63 × 63 × 63 Å^3), which contained 7826 water molecules. Five NBFIX scaling values for protein-water interactions, $\lambda = 1.00$, 1.03, 1.04, 1.06, and 1.09, were tested. To explore a wide conformational space of $(AAQAA)_3$, including an α-helical, unfold, and extended structures, temperature replica-exchange MD (T-REMD) [59] with 58 replicas were performed. The temperatures for 58 replicas were 275.00, 276.64, 278.29, 279.95, 281.61, 283.29, 284.97, 286.66, 288.36, 290.07, 291.78, 293.50, 295.23, 296.97, 298.72, 300.48, 302.24, 304.02, 305.80, 307.59, 309.39, 311.20, 313.02, 314.84, 316.68, 318.50, 320.36, 322.22, 324.09, 325.97, 327.86, 329.75, 331.66, 333.58, 335.50, 337.44, 339.38, 341.34, 343.30, 345.27, 347.26, 349.24, 351.25, 353.26, 355.28, 357.31, 359.35, 361.40, 363.46, 365.53, 367.61, 369.71, 371.81, 373.92, 376.04, 378.17, 380.00, and 382.0 K, respectively. The temperatures were determined using the web server, REMD Temperature generator [79].

In production runs, each replica was simulated for 525 ns with NVT ensemble. Replica exchanges were attempted every 1500-time step. The equations of motion were integrated using the r-RESPA multiple time step method [80] with a 3.5 fs time step for fast motions and a 7.0 fs time step for slow motions. To make the simulation trajectories stable, optimized hydrogen repartitioning (HMR) and group-based temperature/pressure approach (group T/P) [81] were employed. Structures and energies were written every 1500 steps. Trajectories from 105 to 525 ns were used for analysis.

4.2. Chignolin

Initial structure of chignolin was taken from the protein data bank (PDB ID: 1UAO [56]) and was solvated in a periodic box (80 × 80 × 80 Å^3). The simulation system consists of a chignolin, 2 Na^+, 17,500 water molecules. Three NBFIX scaling values for protein-water interactions, λ = 1.00, 1.03, and 1.09 were tested. Since folding of chignolin is known as slow processes, we used Gaussian accelerated replica-exchange umbrella sampling (GaREUS) [61] as the enhanced sampling method. In GaREUS, global motions are enhanced with the GaMD biasing potential, while umbrella potentials along the pre-defined collective variables (CVs) were additionally used to enhance conformational sampling along the CVs. In the simulations, we selected the Cα atom distance between Gly1 and Gly10 as a CV in GaREUS. 24 replicas were simulated, each of which was characterized with harmonic restraint forces along the CV. For λ = 1.00, the force constants and equilibrium distances in the harmonic potentials of the 24 replicas are 0.5, 0.5, 0.5, 0.5, 0.5, 0.5, 0.5, 0.5, 0.5, 0.5,

0.5, 0.5, 1.0, 1.0, 1.0, 1.0, 1.0, 1.0, 1.0, 1.0, 1.0, 1.0, 1.0, 1.0 kcal/mol/Å^2 and 29.57, 28.32, 26.89, 25.54, 24.43999, 23.15999, 21.87, 20.53999, 19.33999, 18.11999, 16.91999, 15.64999, 14.51999, 13.66999, 12.69999, 11.80999, 10.87999, 9.86999, 8.85999, 8.06, 7.05, 6.35, 5.26, 4.0 Å, respectively. For λ = 1.03 and 1.09, those are 0.5, 0.5, 0.5, 0.5, 0.5, 0.5, 0.5, 0.5, 0.5, 0.5, 0.5, 0.5, 1.0, 1.0, 1.0, 1.0, 1.0, 1.0, 1.0, 1.0, 1.0, 1.0, 1.0, 1.0 kcal/mol/Å^2 and 29.11998, 27.89998, 26.46998, 25.22998, 23.97998, 22.66998, 21.45998, 20.14998, 18.89998, 17.70998, 16.40998, 15.20999, 14.21999, 13.26999, 12.48999, 11.63999, 10.73999, 9.75999, 8.80, 7.89, 7.05, 6.14, 5.18, 4.0 Å, respectively. The boost potentials in GaMD for λ = 1.00 and 1.03 were initially guessed from an initial 4-ns NVT simulation without boosting. After the initial determination, the boost potentials were updated every 100 ps in a 5-ns NVT simulation while boosting the system. The boost potentials obtained from λ = 1.03 were also used for λ = 1.09. The thresholds in the boosting potentials are set to the lower bound. σ_0 was set to 6 kcal/mol for both potentials.

In production runs, we performed 1-μs GaREUS simulations with NVT ensemble in each replica for λ = 1.00 and 1.03, and 0.425-μs GaREUS simulation for λ = 1.09. To obtain unbiased free-energy landscapes, the two-step reweighting scheme, which consists of the second-order cumulant expansion for removing the GaMD biases and multistate Bennett acceptance ratio (MBAR) [82] for harmonic restraint potentials in REUS, were applied to the simulation trajectories. A 5-fs time step was used to integrate the equation of motion using velocity Verlet integrator with HMR and group T/P. Structures and energies were output every 1000 steps and every 500 steps, respectively. The first 0.1-μs in the GaREUS simulation for each condition were omitted as equilibration.

4.3. TDP-43

The C-terminal domain of TDP-43 (residues 310-340) in water was simulated using dual-boost GaMD [60]. The N- and C-terminal residues were capped with the acetyl and amide groups, respectively. The peptide was solvated in a periodic box (65 × 65 × 65 Å^3), which contain 8395 water molecules, 24 Na^+, and 24 Cl^-. Three NBFIX scaling values, λ = 1.00, 1.03, and 1.09, were tested. Initial structure of TDP-43 was first built as an extended conformation. The structure was collapsed using the short MD simulation in vacuum. In dual-boost GaMD, one boost potential is applied to the dihedral and CMAP energy terms and another to the total potential excluding the dihedral and CMAP energies. To obtain the initial guesses of the both boosting potentials, the thresholds and force constants were calculated from an initial 5-ns simulation without boosting. After the initial determination of the GaMD parameters, the boost potentials were added to the system while continuing the update of parameters every 100 ps in a 15-ns NVT simulation. The finally determined parameters were used for production runs. The thresholds in the boosting potentials are set to the lower bound. σ_0 was set to 6 kcal/mol for both potentials.

In production runs, ten independent GaMD simulations for 1 μs each were performed with NVT ensemble, and the results were averaged. A 5-fs time step was used to integrate the equation of motion using velocity Verlet integrator with HMR and group T/P. To analyze the GaMD simulation trajectories, we avoid the effect of GaMD biasing potential using the reweighting scheme based on the cumulant expansions [60]. Structures and energies were output every 20,000 steps and every 2000 steps, respectively. Trajectories from 0.1 to 1 μs were used for analysis.

4.4. c-Src Kinase

Initial structure of c-Src kinase was taken from PDB (PDB ID: 1Y57 [57]) and was solvated in a periodic box (102 × 102 × 102 Å^3). The system consists of 103,592 atoms, where c-Src kinase, 94 Na^+, 89 Cl^-, and 33,000 water molecules exist. Three NBFIX values (1.00, 1.03 and 1.09) were used. Five independent conventional MD (cMD) simulations for 1 μs for λ = 1.00 and 1.03 were performed with NVT ensemble, while one simulation for 1 μs was done for λ = 1.09. The r-RESPA integrator with a 3.5 fs were employed with HMR and group T/P. Structures and energies were written every 6000 steps and every 3000 steps, respectively. Trajectories from 0.2 to 1 μs were used for analysis.

4.5. Villin Crowding Solution

For the dilute system, one villin was solvated in a periodic box (102 × 102 × 102 Å^3). The system consists of 99,776 atoms, where villin, 89 Na^+, 91 Cl^-, and 33,000 water molecules exist. Three NBFIX values (1.00, 1.03 and 1.09) were used. Five independent cMD simulations for 1 μs each were performed with NVT ensemble for λ = 1.00 and 1.03. As for λ = 1.09, one cMD simulation for 1 μs was performed. The r-RESPA integrator with a 3.5 fs were employed with HMR and group T/P. Structures and energies were output every 6000 steps and every 3000 steps, respectively. The first 0.2 μs trajectory for each was omitted as equilibration.

For the crowded solution, eight villins were solvated in a periodic box (78 × 78 × 78 Å^3). Initial structure of villin was taken from PDB (PDB ID: 1VII [58]). The system consists of 42,262 atoms, where 8 villins, 34 Na^+, 50 Cl^-, 12,466 water molecules exist. Three NBFIX values (1.00, 1.03, and 1.09) were used. The trajectory lengths of cMD simulations with NVT ensemble are 1.5 μs for λ = 1.00 and 1.03, and 1.0 μs for λ = 1.09. The r-RESPA integrator with a 3.5 fs were employed with HMR and group T/P. Structures and energies were output every 15,000 steps and every 5000 steps, respectively. Trajectories from 0.2 to 1.5 μs were used for analysis.

4.6. Amino-Acid Analogues

The 14 species of amino-acid side-chain analogues (Ala, Asn, Cys, Gln, Hsd, Ile, Leu, Met, Phe, Ser, Thr, Trp, Tyr, and Val) were used for calculations of absolute solvation free energies. The initial structure of each analogue was taken from the ChemSpider [83] and PubChem databases. Force field parameters for the beta carbon were changed and one hydrogen atom was added to the beta carbon [84]. 14 NBFIX values (λ = 1.00, 1.01, 1.02, 1.03, 1.04, 1.05, 1.06, 1.07, 1.08, 1.09, 1.1, 1.2, 1.3, and 1.4) were used. Each analogue was solvated within a periodic box (50 × 50 × 50 Å^3), which contains ~4000 water molecules.

For each analogue, the absolute solvation free energy was estimated as $\Delta G_{solv} = \Delta G^{vacuum} - \Delta G^{water}$, where ΔG^{water} and ΔG^{vacuum} are the free-energy changes upon annihilating nonbonded interactions of the analogue in water and in vacuum, respectively. ΔG^{water} and ΔG^{vacuum} were calculated using FEP/λ-REMD. In FEP/λ-REMD calculations, the electrostatic and Lennard-Jones (LJ) interactions of the analogue are scaled by λ_{elec} and λ_{LJ}, respectively. The following 24 windows were used with different coupling parameters: λ_{elec} = 1, 0.9, 0.8, 0.7, 0.6, 0.5, 0.4, 0.3, 0.2, 0.1, 0, 0, 0, 0, 0, 0, 0, 0, 0, 0, 0, 0, 0, 0, and λ_{LJ} = 1, 1, 1, 1, 1, 1, 1, 1, 1, 1, 1, 0.9, 0.8, 0.7, 0.6, 0.5, 0.4, 0.3, 0.2, 0.15, 0.1, 0.05, 0.025, 0. In each calculation, the simulation was run for 3 ns per each window with NPT ensemble, and trajectories from 1 to 3 ns were used for analysis. Replica exchanges were attempted every 800 steps. The free-energy changes are estimated using the Bennett's acceptance ratio (BAR) method. The obtained trajectories are decomposed into three blocks. The mean and the standard error were calculated using the block averages. For the water system, the r-RESPA integrator with a 2.5 fs were employed. For the vacuum system, the velocity Verlet integrator with a 2.0-fs time step were employed, and instead of using PME, non-bonded interactions were truncated at a cutoff distance of 1000 Å.

Supplementary Materials: The following supporting information can be downloaded at: https://www.mdpi.com/article/10.3390/molecules27175726/s1, Scheme of computing diffusion coefficient of villin; Figure S1: Convergence check of folded population of chignolin; Figure S2: Probability densities along the Cα-RMSD of chignolin; Figure S3: Fraction helix of TDP-43 for λ = 1.09; Figure S4: The radius of gyration, R_g, of TDP-43; Figure S5: Distribution of Cα-RMSD of c-Src kinase; Figure S6: The Cα-RMSD of c-Src kinase in dilute solution; Figure S7: The Cα-RMSD of villin in dilute solution; Figure S8: The Cα-RMSD of villin in crowded solution; Figure S9: The Cα-RMSD of villin in the dilute and crowded solutions for λ = 1.09; Figure S10: The helicity of villin in the dilute and crowded solutions.

Author Contributions: Conceptualization, Y.S.; methodology, H.O., K.K. and D.M.; software, K.K., H.O. and Y.S.; validation, K.K., H.O. and Y.S.; formal analysis, D.M., K.K. and H.O.; investigation, D.M., K.K. and H.M.D.; resources, Y.S.; data curation, Y.S.; writing—original draft preparation, K.K., H.O. and Y.S.; writing—review and editing, H.M.D. and Y.S.; visualization, D.M. and K.K.; supervision, Y.S.; project administration, Y.S.; funding acquisition, Y.S. All authors have read and agreed to the published version of the manuscript.

Funding: This research was funded, in part, by MEXT, "Program for Promoting Research on the Supercomputer Fugaku (Biomolecular dynamics in a living cell (JPMXP1020200101)/MD-driven precision medicine (JPMXP1020200201)), MEXT Kakenhi grant number 19H05645 and 21H05249, RIKEN pioneering projects in "Biology of Intracellular Environments", and "Glycolipidologue" (to Y.S.).

Data Availability Statement: The data that support the findings of this study are available from the corresponding author upon reasonable request.

Acknowledgments: We used the computational resources provided by the HPCI system research project (Project ID: hp200129, hp200135, hp210172, hp210177, hp220164, and hp220170) and those in RIKEN Hokusai "BigWaterFall". We are grateful for Michael Feig for the discussion about the optimal scaling parameter of protein-water LJ interactions for CHARMM36m.

Conflicts of Interest: The authors declare no conflict of interest.

Sample Availability: Samples of the compounds are available from the authors.

References

1. Brini, E.; Simmerling, C.; Dill, K. Protein storytelling through physics. *Science* **2020**, *370*, eaaz3041. [CrossRef]
2. Schlick, T.; Portillo-Ledesma, S. Biomolecular modeling thrives in the age of technology. *Nat. Comput. Sci.* **2021**, *1*, 321–331. [CrossRef] [PubMed]
3. Shaw, D.E.; Maragakis, P.; Lindorff-Larsen, K.; Piana, S.; Dror, R.O.; Eastwood, M.P.; Bank, J.A.; Jumper, J.M.; Salmon, J.K.; Shan, Y.; et al. Atomic-level characterization of the structural dynamics of proteins. *Science* **2010**, *330*, 341–346. [CrossRef] [PubMed]
4. Dror, R.O.; Dirks, R.M.; Grossman, J.P.; Xu, H.; Shaw, D.E. Biomolecular simulation: A computational microscope for molecular biology. *Annu. Rev. Biophys.* **2012**, *41*, 429–452. [CrossRef] [PubMed]
5. Munro, J.B.; Altman, R.B.; Tung, C.S.; Cate, J.H.; Sanbonmatsu, K.Y.; Blanchard, S.C. Spontaneous formation of the unlocked state of the ribosome is a multistep process. *Proc. Natl. Acad. Sci. USA* **2010**, *107*, 709–714. [CrossRef]
6. Song, X.; Jensen, M.O.; Jogini, V.; Stein, R.A.; Lee, C.H.; McHaourab, H.S.; Shaw, D.E.; Gouaux, E. Mechanism of NMDA receptor channel block by MK-801 and memantine. *Nature* **2018**, *556*, 515–519. [CrossRef] [PubMed]
7. Mori, T.; Jung, J.; Kobayashi, C.; Dokainish, H.M.; Re, S.; Sugita, Y. Elucidation of interactions regulating conformational stability and dynamics of SARS-CoV-2 S-protein. *Biophys. J.* **2021**, *120*, 1060–1071. [CrossRef] [PubMed]
8. Kobayashi, C.; Matsunaga, Y.; Jung, J.; Sugita, Y. Structural and energetic analysis of metastable intermediate states in the E1P-E2P transition of Ca(2+)-ATPase. *Proc. Natl. Acad. Sci. USA* **2021**, *118*, e2105507118. [CrossRef]
9. Man, V.H.; He, X.; Derreumaux, P.; Ji, B.; Xie, X.Q.; Nguyen, P.H.; Wang, J. Effects of All-Atom Molecular Mechanics Force Fields on Amyloid Peptide Assembly: The Case of Abeta16-22 Dimer. *J. Chem. Theory Comput.* **2019**, *15*, 1440–1452. [CrossRef]
10. Zerze, G.H.; Zheng, W.; Best, R.B.; Mittal, J. Evolution of All-Atom Protein Force Fields to Improve Local and Global Properties. *J. Phys. Chem. Lett.* **2019**, *10*, 2227–2234. [CrossRef]
11. Zheng, W.; Dignon, G.L.; Jovic, N.; Xu, X.; Regy, R.M.; Fawzi, N.L.; Kim, Y.C.; Best, R.B.; Mittal, J. Molecular Details of Protein Condensates Probed by Microsecond Long Atomistic Simulations. *J. Phys. Chem. B* **2020**, *124*, 11671–11679. [CrossRef]
12. Shea, J.E.; Best, R.B.; Mittal, J. Physics-based computational and theoretical approaches to intrinsically disordered proteins. *Curr. Opin. Struct. Biol.* **2021**, *67*, 219–225. [CrossRef]
13. Dignon, G.L.; Zheng, W.; Kim, Y.C.; Best, R.B.; Mittal, J. Sequence determinants of protein phase behavior from a coarse-grained model. *PLoS Comput. Biol.* **2018**, *14*, e1005941. [CrossRef]
14. Joseph, J.A.; Reinhardt, A.; Aguirre, A.; Chew, P.Y.; Russell, K.O.; Espinosa, J.R.; Garaizar, A.; Collepardo-Guevara, R. Physics-driven coarse-grained model for biomolecular phase separation with near-quantitative accuracy. *Nat. Comput. Sci.* **2021**, *1*, 732–743. [CrossRef]
15. Tan, C.; Jung, J.; Kobayashi, C.; Torre, D.U.; Takada, S.; Sugita, Y. Implementation of residue-level coarse-grained models in GENESIS for large-scale molecular dynamics simulations. *PLoS Comput. Biol.* **2022**, *18*, e1009578. [CrossRef]
16. Dannenhoffer-Lafage, T.; Best, R.B. A Data-Driven Hydrophobicity Scale for Predicting Liquid-Liquid Phase Separation of Proteins. *J. Phys. Chem. B* **2021**, *125*, 4046–4056. [CrossRef]
17. Regy, R.M.; Thompson, J.; Kim, Y.C.; Mittal, J. Improved coarse-grained model for studying sequence dependent phase separation of disordered proteins. *Protein Sci.* **2021**, *30*, 1371–1379. [CrossRef]

18. Kumar, A.; Chakraborty, D.; Mugnai, M.L.; Straub, J.E.; Thirumalai, D. Sequence Determines the Switch in the Fibril Forming Regions in the Low-Complexity FUS Protein and Its Variants. *J. Phys. Chem. Let.t* **2021**, *12*, 9026–9032. [CrossRef]
19. Yu, I.; Mori, T.; Ando, T.; Harada, R.; Jung, J.; Sugita, Y.; Feig, M. Biomolecular interactions modulate macromolecular structure and dynamics in atomistic model of a bacterial cytoplasm. *eLife* **2016**, *5*, e19274. [CrossRef]
20. Feig, M.; Yu, I.; Wang, P.H.; Nawrocki, G.; Sugita, Y. Crowding in Cellular Environments at an Atomistic Level from Computer Simulations. *J. Phys. Chem. B* **2017**, *121*, 8009–8025. [CrossRef]
21. Zimmerman, S.B.; Minton, A.P. Macromolecular crowding: Biochemical, biophysical, and physiological consequences. *Annu. Rev. Biophys. Biomol. Struct.* **1993**, *22*, 27–65. [CrossRef] [PubMed]
22. Zhou, H.X.; Rivas, G.; Minton, A.P. Macromolecular crowding and confinement: Biochemical, biophysical, and potential physiological consequences. *Annu. Rev. Biophys.* **2008**, *37*, 375–397. [CrossRef] [PubMed]
23. Miklos, A.C.; Sarkar, M.; Wang, Y.; Pielak, G.J. Protein crowding tunes protein stability. *J. Am. Chem. Soc.* **2011**, *133*, 7116–7120. [CrossRef] [PubMed]
24. Speer, S.L.; Stewart, C.J.; Sapir, L.; Harries, D.; Pielak, G.J. Macromolecular Crowding Is More than Hard-Core Repulsions. *Annu. Rev. Biophys.* **2022**, *51*, 267–300. [CrossRef]
25. Harada, R.; Sugita, Y.; Feig, M. Protein crowding affects hydration structure and dynamics. *J. Am. Chem. Soc.* **2012**, *134*, 4842–4849. [CrossRef]
26. Harada, R.; Tochio, N.; Kigawa, T.; Sugita, Y.; Feig, M. Reduced native state stability in crowded cellular environment due to protein-protein interactions. *J. Am. Chem. Soc.* **2013**, *135*, 3696–3701. [CrossRef]
27. Nawrocki, G.; Wang, P.H.; Yu, I.; Sugita, Y.; Feig, M. Slow-Down in Diffusion in Crowded Protein Solutions Correlates with Transient Cluster Formation. *J. Phys. Chem. B* **2017**, *121*, 11072–11084. [CrossRef]
28. von Bulow, S.; Siggel, M.; Linke, M.; Hummer, G. Dynamic cluster formation determines viscosity and diffusion in dense protein solutions. *Proc. Natl. Acad. Sci. USA* **2019**, *116*, 9843–9852. [CrossRef]
29. Rickard, M.M.; Zhang, Y.; Gruebele, M.; Pogorelov, T.V. In-Cell Protein-Protein Contacts: Transient Interactions in the Crowd. *J. Phys. Chem. Lett.* **2019**, *10*, 5667–5673. [CrossRef]
30. Murthy, A.C.; Dignon, G.L.; Kan, Y.; Zerze, G.H.; Parekh, S.H.; Mittal, J.; Fawzi, N.L. Molecular interactions underlying liquid-liquid phase separation of the FUS low-complexity domain. *Nat. Struct. Mol. Biol.* **2019**, *26*, 637–648. [CrossRef]
31. Conicella, A.E.; Dignon, G.L.; Zerze, G.H.; Schmidt, H.B.; D'Ordine, A.M.; Kim, Y.C.; Rohatgi, R.; Ayala, Y.M.; Mittal, J.; Fawzi, N.L. TDP-43 alpha-helical structure tunes liquid-liquid phase separation and function. *Proc. Natl. Acad. Sci. USA* **2020**, *117*, 5883–5894. [CrossRef]
32. Ball, P. Water is an active matrix of life for cell and molecular biology. *Proc. Natl. Acad. Sci. USA* **2017**, *114*, 13327–13335. [CrossRef]
33. Nerenberg, P.S.; Jo, B.; So, C.; Tripathy, A.; Head-Gordon, T. Optimizing solute-water van der Waals interactions to reproduce solvation free energies. *J. Phys. Chem. B* **2012**, *116*, 4524–4534. [CrossRef]
34. Best, R.B.; Zheng, W.; Mittal, J. Balanced Protein-Water Interactions Improve Properties of Disordered Proteins and Non-Specific Protein Association. *J. Chem. Theory Comput.* **2014**, *10*, 5113–5124. [CrossRef]
35. Tang, W.S.; Fawzi, N.L.; Mittal, J. Refining All-Atom Protein Force Fields for Polar-Rich, Prion-like, Low-Complexity Intrinsically Disordered Proteins. *J. Phys. Chem. B* **2020**, *124*, 9505–9512. [CrossRef]
36. Best, R.B. Atomistic Force Fields for Proteins. *Methods Mol. Biol.* **2019**, *2022*, 3–19.
37. van der Spoel, D. Systematic design of biomolecular force fields. *Curr. Opin. Struct. Biol.* **2021**, *67*, 18–24. [CrossRef]
38. Mackerell, A.D., Jr.; Feig, M.; Brooks, C.L., 3rd. Extending the treatment of backbone energetics in protein force fields: Limitations of gas-phase quantum mechanics in reproducing protein conformational distributions in molecular dynamics simulations. *J. Comput. Chem.* **2004**, *25*, 1400–1415. [CrossRef]
39. Wu, H.N.; Jiang, F.; Wu, Y.D. Significantly Improved Protein Folding Thermodynamics Using a Dispersion-Corrected Water Model and a New Residue-Specific Force Field. *J. Phys. Chem. Lett.* **2017**, *8*, 3199–3205. [CrossRef]
40. Robustelli, P.; Piana, S.; Shaw, D.E. Developing a molecular dynamics force field for both folded and disordered protein states. *Proc. Natl. Acad. Sci. USA* **2018**, *115*, E4758–E4766. [CrossRef] [PubMed]
41. Maier, J.A.; Martinez, C.; Kasavajhala, K.; Wickstrom, L.; Hauser, K.E.; Simmerling, C. ff14SB: Improving the Accuracy of Protein Side Chain and Backbone Parameters from ff99SB. *J. Chem. Theory Comput.* **2015**, *11*, 3696–3713. [CrossRef] [PubMed]
42. Tian, C.; Kasavajhala, K.; Belfon, K.A.A.; Raguette, L.; Huang, H.; Migues, A.N.; Bickel, J.; Wang, Y.; Pincay, J.; Wu, Q.; et al. ff19SB: Amino-Acid-Specific Protein Backbone Parameters Trained against Quantum Mechanics Energy Surfaces in Solution. *J. Chem. Theory Comput.* **2020**, *16*, 528–552. [CrossRef] [PubMed]
43. Best, R.B.; Zhu, X.; Shim, J.; Lopes, P.E.; Mittal, J.; Feig, M.; Mackerell, A.D., Jr. Optimization of the additive CHARMM all-atom protein force field targeting improved sampling of the backbone phi, psi and side-chain chi(1) and chi(2) dihedral angles. *J. Chem. Theory Comput.* **2012**, *8*, 3257–3273. [CrossRef]
44. Huang, J.; Rauscher, S.; Nawrocki, G.; Ran, T.; Feig, M.; de Groot, B.L.; Grubmuller, H.; MacKerell, A.D., Jr. CHARMM36m: An improved force field for folded and intrinsically disordered proteins. *Nat. Methods* **2017**, *14*, 71–73. [CrossRef]
45. Lu, C.; Wu, C.; Ghoreishi, D.; Chen, W.; Wang, L.; Damm, W.; Ross, G.A.; Dahlgren, M.K.; Russell, E.; Von Bargen, C.D.; et al. OPLS4: Improving Force Field Accuracy on Challenging Regimes of Chemical Space. *J. Chem. Theory Comput.* **2021**, *17*, 4291–4300. [CrossRef]

46. Piana, S.; Lindorff-Larsen, K.; Shaw, D.E. How robust are protein folding simulations with respect to force field parameterization? *Biophys. J.* **2011**, *100*, L47–L49. [CrossRef]
47. Piana, S.; Klepeis, J.L.; Shaw, D.E. Assessing the accuracy of physical models used in protein-folding simulations: Quantitative evidence from long molecular dynamics simulations. *Curr. Opin. Struct. Biol.* **2014**, *24*, 98–105. [CrossRef] [PubMed]
48. Hornak, V.; Abel, R.; Okur, A.; Strockbine, B.; Roitberg, A.; Simmerling, C. Comparison of multiple Amber force fields and development of improved protein backbone parameters. *Proteins* **2006**, *65*, 712–725. [CrossRef]
49. Duan, Y.; Wu, C.; Chowdhury, S.; Lee, M.C.; Xiong, G.; Zhang, W.; Yang, R.; Cieplak, P.; Luo, R.; Lee, T.; et al. A point-charge force field for molecular mechanics simulations of proteins based on condensed-phase quantum mechanical calculations. *J. Comput. Chem.* **2003**, *24*, 1999–2012. [CrossRef]
50. Jorgensen, W.L.; Chandrasekhar, J.; Madura, J.D.; Impey, R.W.; Klein, M.L. Comparison of simple potential functions for simulating liquid water. *J. Chem. Phys.* **1983**, *79*, 926–935. [CrossRef]
51. Abascal, J.L.; Vega, C. A general purpose model for the condensed phases of water: TIP4P/2005. *J. Chem. Phys.* **2005**, *123*, 234505. [CrossRef]
52. Thomasen, F.E.; Pesce, F.; Roesgaard, M.A.; Tesei, G.; Lindorff-Larsen, K. Improving Martini 3 for Disordered and Multidomain Proteins. *J. Chem. Theory Comput.* **2022**, *18*, 2033–2041. [CrossRef]
53. MacKerell, A.D.; Bashford, D.; Bellott, M.; Dunbrack, R.L.; Evanseck, J.D.; Field, M.J.; Fischer, S.; Gao, J.; Guo, H.; Ha, S.; et al. All-atom empirical potential for molecular modeling and dynamics studies of proteins. *J. Phys. Chem. B* **1998**, *102*, 3586–3616. [CrossRef]
54. Best, R.B.; de Sancho, D.; Mittal, J. Residue-specific alpha-helix propensities from molecular simulation. *Biophys. J.* **2012**, *102*, 1462–1467. [CrossRef]
55. Piana, S.; Donchev, A.G.; Robustelli, P.; Shaw, D.E. Water dispersion interactions strongly influence simulated structural properties of disordered protein states. *J. Phys. Chem. B* **2015**, *119*, 5113–5123. [CrossRef] [PubMed]
56. Honda, S.; Yamasaki, K.; Sawada, Y.; Morii, H. 10 residue folded peptide designed by segment statistics. *Structure* **2004**, *12*, 1507–1518. [CrossRef]
57. Cowan-Jacob, S.W.; Fendrich, G.; Manley, P.W.; Jahnke, W.; Fabbro, D.; Liebetanz, J.; Meyer, T. The crystal structure of a c-Src complex in an active conformation suggests possible steps in c-Src activation. *Structure* **2005**, *13*, 861–871. [CrossRef]
58. McKnight, C.J.; Matsudaira, P.T.; Kim, P.S. NMR structure of the 35-residue villin headpiece subdomain. *Nat. Struct. Biol.* **1997**, *4*, 180–184. [CrossRef]
59. Sugita, Y.; Okamoto, Y. Replica-exchange molecular dynamics method for protein folding. *Chem. Phys. Lett.* **1999**, *314*, 141–151. [CrossRef]
60. Miao, Y.; Feher, V.A.; McCammon, J.A. Gaussian Accelerated Molecular Dynamics: Unconstrained Enhanced Sampling and Free Energy Calculation. *J. Chem. Theory Comput.* **2015**, *11*, 3584–3595. [CrossRef]
61. Oshima, H.; Re, S.; Sugita, Y. Replica-Exchange Umbrella Sampling Combined with Gaussian Accelerated Molecular Dynamics for Free-Energy Calculation of Biomolecules. *J. Chem. Theory Comput.* **2019**, *15*, 5199–5208. [CrossRef]
62. Shalongo, W.; Dugad, L.; Stellwagen, E. Distribution of Helicity within the Model Peptide Acetyl(AAQAA)3amide. *J. Am. Chem. Soc.* **2002**, *116*, 8288–8293. [CrossRef]
63. Sugita, Y.; Kitao, A.; Okamoto, Y. Multidimensional replica-exchange method for free-energy calculations. *J. Chem. Phys.* **2000**, *113*, 6042–6051. [CrossRef]
64. Satoh, D.; Shimizu, K.; Nakamura, S.; Terada, T. Folding free-energy landscape of a 10-residue mini-protein, chignolin. *FEBS Lett.* **2006**, *580*, 3422–3426. [CrossRef]
65. Re, S.; Oshima, H.; Kasahara, K.; Kamiya, M.; Sugita, Y. Encounter complexes and hidden poses of kinase-inhibitor binding on the free-energy landscape. *Proc. Natl. Acad. Sci. USA* **2019**, *116*, 18404–18409. [CrossRef]
66. Kasahara, K.; Re, S.; Nawrocki, G.; Oshima, H.; Mishima-Tsumagari, C.; Miyata-Yabuki, Y.; Kukimoto-Niino, M.; Yu, I.; Shirouzu, M.; Feig, M.; et al. Reduced efficacy of a Src kinase inhibitor in crowded protein solution. *Nat. Commun.* **2021**, *12*, 4099. [CrossRef]
67. Kabsch, W.; Sander, C. Dictionary of protein secondary structure: Pattern recognition of hydrogen-bonded and geometrical features. *Biopolymers* **1983**, *22*, 2577–2637. [CrossRef]
68. Touw, W.G.; Baakman, C.; Black, J.; te Beek, T.A.; Krieger, E.; Joosten, R.P.; Vriend, G. A series of PDB-related databanks for everyday needs. *Nucleic Acids Res.* **2015**, *43*, D364–D368. [CrossRef]
69. Yeh, I.-C.; Hummer, G. System-Size Dependence of Diffusion Coefficients and Viscosities from Molecular Dynamics Simulations with Periodic Boundary Conditions. *J. Phys. Chem. B* **2004**, *108*, 15873–15879. [CrossRef]
70. Ortega, A.; Amoros, D.; Garcia de la Torre, J. Prediction of hydrodynamic and other solution properties of rigid proteins from atomic- and residue-level models. *Biophys. J.* **2011**, *101*, 892–898. [CrossRef]
71. Feller, S.E.; Pastor, R.W.; Rojnuckarin, A.; Bogusz, S.; Brooks, B.R. Effect of Electrostatic Force Truncation on Interfacial and Transport Properties of Water. *J. Phys. Chem.* **1996**, *100*, 17011–17020. [CrossRef]
72. Korson, L.; Drost-Hansen, W.; Millero, F.J. Viscosity of water at various temperatures. *J. Phys. Chem.* **1969**, *73*, 34–39. [CrossRef]
73. Wolfenden, R.; Andersson, L.; Cullis, P.M.; Southgate, C.C. Affinities of amino acid side chains for solvent water. *Biochemistry* **1981**, *20*, 849–855. [CrossRef]

74. Jung, J.; Mori, T.; Kobayashi, C.; Matsunaga, Y.; Yoda, T.; Feig, M.; Sugita, Y. GENESIS: A hybrid-parallel and multi-scale molecular dynamics simulator with enhanced sampling algorithms for biomolecular and cellular simulations. *Wiley Interdiscip. Rev. Comput. Mol. Sci.* **2015**, *5*, 310–323. [CrossRef]
75. Kobayashi, C.; Jung, J.; Matsunaga, Y.; Mori, T.; Ando, T.; Tamura, K.; Kamiya, M.; Sugita, Y. GENESIS 1.1: A hybrid-parallel molecular dynamics simulator with enhanced sampling algorithms on multiple computational platforms. *J. Comput. Chem.* **2017**, *38*, 2193–2206. [CrossRef]
76. Bussi, G.; Donadio, D.; Parrinello, M. Canonical sampling through velocity rescaling. *J. Chem. Phys.* **2007**, *126*, 014101. [CrossRef]
77. Ryckaert, J.-P.; Ciccotti, G.; Berendsen, H.J.C. Numerical integration of the cartesian equations of motion of a system with constraints: Molecular dynamics of n-alkanes. *J. Comput. Phys.* **1977**, *23*, 327–341. [CrossRef]
78. Essmann, U.; Perera, L.; Berkowitz, M.L.; Darden, T.; Lee, H.; Pedersen, L.G. A smooth particle mesh Ewald method. *J. Chem. Phys.* **1995**, *103*, 8577–8593. [CrossRef]
79. Patriksson, A.; van der Spoel, D. A temperature predictor for parallel tempering simulations. *Phys. Chem. Chem. Phys.* **2008**, *10*, 2073–2077. [CrossRef]
80. Tuckerman, M.; Berne, B.J.; Martyna, G.J. Reversible multiple time scale molecular dynamics. *J. Chem. Phys.* **1992**, *97*, 1990–2001. [CrossRef]
81. Jung, J.; Kasahara, K.; Kobayashi, C.; Oshima, H.; Mori, T.; Sugita, Y. Optimized Hydrogen Mass Repartitioning Scheme Combined with Accurate Temperature/Pressure Evaluations for Thermodynamic and Kinetic Properties of Biological Systems. *J. Chem. Theory Comput.* **2021**, *17*, 5312–5321. [CrossRef]
82. Shirts, M.R.; Chodera, J.D. Statistically optimal analysis of samples from multiple equilibrium states. *J. Chem. Phys.* **2008**, *129*, 124105. [CrossRef]
83. Hettne, K.M.; Williams, A.J.; van Mulligen, E.M.; Kleinjans, J.; Tkachenko, V.; Kors, J.A. Automatic vs. manual curation of a multi-source chemical dictionary: The impact on text mining. *J. Cheminform.* **2010**, *2*, 3. [CrossRef]
84. Shirts, M.R.; Pitera, J.W.; Swope, W.C.; Pande, V.S. Extremely precise free energy calculations of amino acid side chain analogs: Comparison of common molecular mechanics force fields for proteins. *J. Chem. Phys.* **2003**, *119*, 5740–5761. [CrossRef]

molecules

MDPI

Review

Functional Bacterial Amyloids: Understanding Fibrillation, Regulating Biofilm Fibril Formation and Organizing Surface Assemblies

Thorbjørn Vincent Sønderby [1,2], Zahra Najarzadeh [1] and Daniel Erik Otzen [1,*]

[1] Interdisciplinary Nanoscience Center (iNANO), Aarhus University, Gustav Wieds Vej 14, 8000 Aarhus, Denmark; thorbjoern.soenderby@gmail.com (T.V.S.); zahra_najarzadeh@yahoo.com (Z.N.)

[2] Sino-Danish Center (SDC), Eastern Yanqihu Campus, University of Chinese Academy of Sciences, 380 Huaibeizhuang, Huairou District, Beijing 101400, China

* Correspondence: dao@inano.au.dk

Abstract: Functional amyloid is produced by many organisms but is particularly well understood in bacteria, where proteins such as CsgA (*E. coli*) and FapC (*Pseudomonas*) are assembled as functional bacterial amyloid (FuBA) on the cell surface in a carefully optimized process. Besides a host of helper proteins, FuBA formation is aided by multiple imperfect repeats which stabilize amyloid and streamline the aggregation mechanism to a fast-track assembly dominated by primary nucleation. These repeats, which are found in variable numbers in Pseudomonas, are most likely the structural core of the fibrils, though we still lack experimental data to determine whether the repeats give rise to β-helix structures via stacked β-hairpins (highly likely for CsgA) or more complicated arrangements (possibly the case for FapC). The response of FuBA fibrillation to denaturants suggests that nucleation and elongation involve equal amounts of folding, but protein chaperones preferentially target nucleation for effective inhibition. Smart peptides can be designed based on these imperfect repeats and modified with various flanking sequences to divert aggregation to less stable structures, leading to a reduction in biofilm formation. Small molecules such as EGCG can also divert FuBA to less organized structures, such as partially-folded oligomeric species, with the same detrimental effect on biofilm. Finally, the strong tendency of FuBA to self-assemble can lead to the formation of very regular two-dimensional amyloid films on structured surfaces such as graphite, which strongly implies future use in biosensors or other nanobiomaterials. In summary, the properties of functional amyloid are a much-needed corrective to the unfortunate association of amyloid with neurodegenerative disease and a testimony to nature's ability to get the best out of a protein fold.

Keywords: bacterial amyloid; biofilm; curli; FapC; imperfect repeats

Citation: Sønderby, T.V.; Najarzadeh, Z.; Otzen, D.E. Functional Bacterial Amyloids: Understanding Fibrillation, Regulating Biofilm Fibril Formation and Organizing Surface Assemblies. *Molecules* **2022**, *27*, 4080. https://doi.org/10.3390/molecules27134080

Academic Editors: Kunihiro Kuwajima, Yuko Okamoto, Tuomas Knowles and Michele Vendruscolo

Received: 6 June 2022
Accepted: 22 June 2022
Published: 24 June 2022

1. Dedication to Sir Chris Dobson by Daniel Otzen

Chris was a luminary in his field—he lit up the literature on amyloid proteins like a lighthouse, shining bright insight and knowledge into many different aspects of this part of the protein universe. Aside from the prodigious outpouring of original research from him and his group, not to mention his always inspiring and witty lectures (which were never the same, despite his frequent stage appearances), the scientific landscape is dotted with his incisive (and often single-author) reviews that appeared at a breathtaking pace over several decades. For me, one of the real eye-openers was his 2003 review in Nature [1], simply titled "Protein folding and misfolding," which I read as soon as it came out. I leant heavily on it to prepare a lecture for high-school teachers as part of a Danish *"life-long learning"* initiative to keep these teachers informed about the latest developments in their fields. It struck me as very appropriate to consult Chris in such a situation. Although I was reasonably informed about the amyloid field at this stage, Chris' magisterial overview revealed many new aspects for me. One sentence in particular caught my attention: "Nevertheless, there is increasing evidence that the unique properties of amyloid structures have been exploited by some species, including bacteria, fungi and even mammals, for specific (and carefully regulated) purposes.". This brought my hitherto unenlightened attention to the world of functional amyloid, which had important consequences for me. It introduced me to wonderful colleagues such as Matt Chapman (whose seminal work on curli and CsgA was referenced by Chris in the above quote) and inspired me, along with fellow colleagues such as Per Halkjær Nielsen and Morten Dueholm at my host institute (at the time Aalborg University), to embark on a journey into the world of useful aggregation. We started by showing that amyloid was all around us in the bacterial world [2,3] and then went on to focus on some of those systems in more detail. That journey is still ongoing almost two decades later, and we will use this opportunity to tell a little about our research in that area in this paper. Thank you, Chris, for that tip!

2. The Conundrum of Protein Misfolding and Aggregation

Amyloids are highly organized, β-sheet-rich protein aggregates [4]. They can form from a wide variety of different proteins, suggesting—as Chris Dobson vigorously and convincingly emphasized—that proteins have a *generic* ability to form amyloid in addition to their tendency to assume globular folds [4,5]. Indeed, in some cases (again as shown by Chris and colleagues [5]), the amyloid state is more thermodynamically favorable than the native state (Figure 1a). Fortunately, that does not mean that life automatically degenerates to amyloid soup (although it may have started this way) [6]. This is due to several protective mechanisms. One of these is the cooperative nature of protein folding [7]. As a result, only a miniscule fraction of the protein population will be unfolded at any given time, and most aggregation-prone regions remain safely tucked away into the protein interior [7]. Therefore, proteins need to cross a high activation barrier to unfold and reach aggregation-prone states. On top of that, they need to encounter and associate with other similarly conformationally inclined molecules to start the aggregation reaction. These aspects combine to make amyloid formation a kinetic, rather than a thermodynamic, challenge [5]. In addition, cells contain protective mechanisms such as folding chaperones, which aid protein folding, e.g., by screening aggregation-prone regions during folding and sequestering proteins in compartments until they are safely folded [7]. Further, degradation mechanisms quickly remove misfolded proteins in normally functioning cells [7]. Nevertheless, destabilization of the native state (or stabilization of the denatured state) can reduce the kinetic barrier between the folded and denatured state, leading to more frequent traffic between the two states, leading to greater exposure of aggregation-prone regions [4]. The barrier to unfolding is completely removed in intrinsically disordered proteins (IDPs), which are largely disordered and only contain transient structure. For them, the only remaining obstacle is the suite of intermolecular contacts and rearrangements that need to be made to start the process of organized aggregation to the amyloid state [4] (Figure 1b).

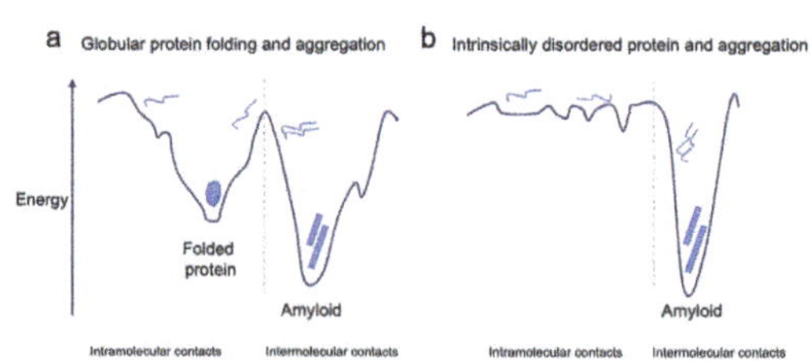

Figure 1. Schematic energy landscapes of globular proteins (**a**) and IDPs (**b**). Both energy landscapes are split into two halves illustrating intramolecular contacts and intermolecular contacts. (**a**) The globular folded protein is the conformation with the lowest energy based on intramolecular contacts. However, aggregation can result in conformations that are even lower in energy, illustrating that amyloid fibril formation is thermodynamically more favorable than the native state. (**b**) IDPs exist in an ensemble of conformations that may have transient intra- and intermolecular contacts. Intermolecular contacts leading to aggregation of IDPs can result in highly stable amyloid aggregates.

3. Amyloids in Sickness and in Health

It follows from the above considerations that unwanted accumulation of amyloid in vivo will be favored by either protein destabilization or the involvement of IDPs. Examples of the former include the protein transthyretin, a tetramer which upon mutagenic destabilization can dissociate and misfold to amyloid structures, leading to cardiac amyloidosis [8]. Stabilization of the tetramer should in principle slow down or even prevent disease progression. Indeed, the drug tafamidis (which binds to two otherwise empty ligand-binding sites on the tetramer) does exactly that [9,10], making it the first and so far only example of a clinically approved drug that specifically prevents amyloid formation. Unfortunately, most amyloid-associated diseases involve proteins or peptides that are IDPs [4], e.g., Parkinson's (α-synuclein), Alzheimer's disease (amyloid-β and tau) and type-II diabetes (Islet amyloid polypeptide) [4,11–13]. They do not lend themselves to such straightforward stabilizing ligand strategies as using tafamidis but require more complicated approaches that somehow target the unwanted associated species without compromising their (more) innocent precursors.

These misfolding diseases should not prevent us from appreciating some innately striking features of amyloids: their stability and simplicity [14]. Their structure only requires copies of a single type of protein which can rapidly stack up on top of each other to form long fibrils [15]. Why not use these properties for something useful? Indeed, many organisms use amyloids for beneficial purposes, and so-called **functional amyloids** (usually based on IDPs) are widespread among organisms and even exist in humans [16–20]. Bacteria in particular have been very active in their exploitation of these self-assemblies [19–21], which amongst other things can be used to strengthen the biofilm in which most bacteria are embedded [19,20,22]. Although useful for the bacteria, biofilms are generally a nuisance for humans, as they protect bacteria against physical, chemical and biological attacks. This leads to increased antibiotic resistance of bacteria on surfaces such as wounds and medical implants [23], but also in more innocuous settings, such as clothes, contact lenses, water flow systems and even oil pipes [24]. Thus, even when an amyloid plays a useful role (for some), there is strong motivation (for others) to prevent it from forming. Nevertheless, their formation and structure are highly instructive and provide an excellent showcase of how to control and benefit from a potentially proteotoxic fold. As we will describe in greater detail below, the best-studied functional amyloids (those in *E. coli* and *Pseudomonas*) are organized as multiple copies of relatively simple repeats, which lack structure in the monomeric state but fold at the multimeric level in a simple but robust framework thanks to a large number of stabilizing inter- and intramolecular contacts, which are built up in a hierarchical manner. This is in stark contrast to pathological amyloid proteins where there are no repeat sequences to provide an orderly framework for assembly; consequently, many different aggregate types (polymorphs) exist which can form in different ways and typically involve cytotoxic partially-folded intermediate states.

4. Functional Bacterial Amyloids

Matt Chapman and colleagues reported the first example of functional bacterial amyloid (FuBA), namely, curli in *Escherichia coli* (*E. coli*) [20], hence giving rise to the term *curliated E. coli*. Since this discovery in 2002, numerous other FuBAs have been discovered, including functional amyloid in *Pseudomonas* (Fap) [19,25]. The curli and Fap systems remain the most well-characterized FuBA systems, and we will briefly introduce these two systems.

5. The Curli System

The curli system is very widespread, occurring in several distinct phyla within the bacterial kingdom [26]. Curli fibers, which extend into the extracellular matrix of *E. coli*, are involved in biofilm formation and play a role in surface colonization and host-tissue contact [20,27]. Curli fibers consist of two proteins, CsgA and CsgB, in vivo [20], of which CsgA constitutes the vast majority [28]. While purified and isolated CsgA very readily and robustly forms fibrils on its own in vitro, the nucleator protein, CsgB, is essential for correct CsgA aggregation in vivo [29]. The biogenesis of curli is tightly regulated and involves five other proteins besides CsgA and CsgB. These are encoded from two operons that are divergently linked, *csgBAC* and *csgDEFG* (Figure 2) [30] but which collaborate in beautiful concord to orchestrate the carefully controlled production of curli projecting from the surface of *E. coli*'s outer membrane. The transcription regulator, CsgD, activates the expression of the curli subunits, CsgA and CsgB, from the *csgBAC* operon, in addition to the periplasmic chaperone, CsgC [30]. The Sec export pathway is responsible for the transport of all curli proteins across the inner membrane to the periplasmic space (except the transcription activator, CsgD) [31]. Once in the periplasmic space, CsgA, CsgB, and CsgF are further transported across the outer membrane through the outer membrane pore-forming protein, CsgG [31]. CsgE forms a complex with the pore-forming protein, CsgG, and functions as a gatekeeper for the secretion of CsgA and CsgB through CsgG by recognizing outer-membrane signal sequences in CsgA and CsgB [32]. The nucleator CsgB is transported to the cell surface, where it is attached to the outer membrane, stabilized by contact with CsgF (Figure 2) [33]. CsgB is homologous to CsgA and presents a binding surface for CsgA, which is secreted from the cell as an IDP, but fibrillates rapidly once it makes contact with CsgB [29,34]. CsgC is a very efficient amyloid inhibitor/chaperone which binds CsgA and CsgB during transport in the periplasmic space and maintains them in the monomeric state, thereby inhibiting premature fibrillation of CsgA and CsgB [35,36].

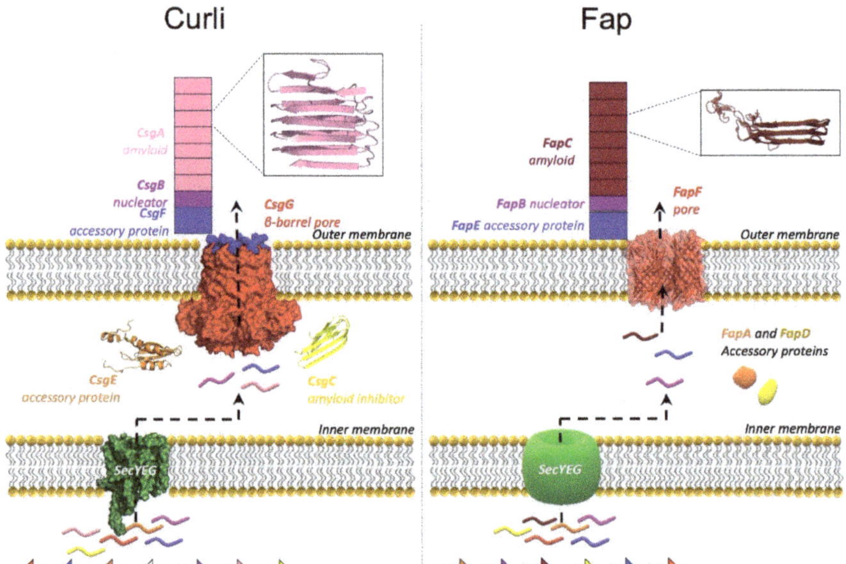

Figure 2. The proteins involved in curli biogenesis are encoded from two divergently transcribed operons, whereas the Fap proteins are encoded from a single operon. The proteins in both systems are transported across the inner membrane through the Sec secretion pathway. The major amyloid components of curli and Fap, CsgA and FapC, respectively, are kept monomeric with the help of chaperone-like accessory proteins, during their transport through the periplasmic space. In each system, the amyloid components are transported to the outer membrane, through their respective pore proteins, CsgG (PDB ID 6SI7) and FapF (PDB ID 5O65). The nucleator proteins CsgB and FapB prime the aggregation of CsgA and FapC, respectively, once the proteins have reached the outer membrane. The insets show structural models of CsgA and FapC. Both CsgA and FapC are predicted to form a β-helix, where the imperfect repeat sequences are stacked on top of each other.

6. The Fap System

The curli and Fap systems are evolutionarily unrelated, yet their biogenesis is remarkably similar [19,20]. The Fap system is especially abundant within the Proteobacteria phylum [37,38] and may contribute to the virulence of the pathological strain *Pseudomonas aeruginosa* (*P. aeruginosa*) [19,39], thanks to its ability to stabilize the biofilm mechanically [40]. Like curli, Fap biogenesis requires several accessory proteins (FapA–F) whose roles appear to be similar to those in the curli system [41]. However, there are differences all the same. Unlike curli, Fap proteins are transcribed from a single operon (Figure 2) [19]. FapC is the main structural component of Fap fibrils, but they also contain smaller amounts of FapB and FapE [19,41]. FapB is predicted to be a nucleator protein analogous to CsgB [19,41]. The role of FapE is less clear but it was found to be essential for Fap secretion [42]. The roles of FapA and FapD are unclear, but they are likely to be chaperones [19,41,42]. Curiously, deletion of FapA leads to Fap fibrils completely made of FapB [41]. Like the curli proteins, the Fap proteins utilize the Sec transport pathway for transport across the inner membrane [42]. Additionally, as in the curli system, a dedicated outer membrane pore protein (FapF) allows transportation of FapB, FapC and FapE across the outer membrane [19,42]. However, FapF and CsgG are completely unrelated. FapF is a trimeric porin, each subunit of which appears to be able to work as an independent channel [42], whereas CsgG is a 9-mer, and the individual subunits each contribute to the formation of a single channel [43].

Insets of amyloid structures are from [38,44]. The solved structures of CsgC and CsgE have been deposited in the PDB with PDB IDs 2Y2Y and 2NA4, respectively.

7. The Major Curli Component, CsgA

The complicated and elegantly regulated biogenesis of curli and Fap illustrates that these systems are evolutionarily top-tuned fibrillation machineries. This becomes even clearer when we inspect the predicted structures of the major amyloid components, CsgA and FapC, both of which contain multiple imperfect repeats.

8. Predicting FuBA Structure Is Easier Than Predicting Pathological Amyloid Structure

The difficulties in solving amyloid structure experimentally (e.g., due to their insoluble and non-crystalline nature), have motivated the use of computational methods to predict the structures of functional amyloids [30,38,44–48]. This has been applied particularly to CsgA [30,44], the curli nucleator protein, CsgB [45,46] and FapC [38]. These amyloid structure predictions have primarily depended on multiple sequence alignment (MSA) to detect patterns of co-evolutionary residue pairs, which often indicates residue-residue contacts in the folded protein [49]. This method has appeared to work well for functional bacterial amyloids, which are functional in their fibril state and are evolutionarily widespread, allowing MSA of many homologous proteins [30,38,44–48]. In addition, MSA has provided insight into the polarity of bactofilin filaments from *Thermus thermophilus*, which complimented cryo-electron microscopy data of the bactofilin filaments [48]. In contrast, MSA has little relevance for pathological amyloids, whose aggregation results from misfolding rather than evolutionary optimization [4] and often results in amyloids with a significant degree of polymorphism and heterogeneity compared to functional amyloids [50]. The newly released AlphaFold2 protein structure predictor is also based on co-evolutionary MSA analysis, coupled with the training of a neural network based on protein structures from the PDB [47,51]. AlphaFold2 is a promising tool for predicting structures of functional amyloid [52]. FuBAs are an interesting test of the capabilities of AlphaFold, because they are IDPs that gain structure upon association, but in contrast to pathological amyloids, their final structure is the result of evolutionary pressure.

Evolution has also ensured that CsgA consists of five 19–23 residue-long imperfect repeat sequences preceded by a 22-residue N22 sequence (Figure 3a,b) [30,44]. The overall length of mature CsgA is 131 residues (after cleavage of a 20-residue inner membrane signal sequence) [20]. All five repeats contain a Ser–Xxx_5–Gln–Xxx_4–Asn–Xxx–Ala–Xxx_3–Gln consensus sequence (Figure 3b) [53,54]. Each repeat sequence of CsgA is predicted to form a hairpin motif consisting of two β-strands linked by a β-turn (Figure 3a) [30,44]. These hairpins are stacked on top of each other, so that each strand engages with the other hairpin strands, forming two parallel β-sheets side by side. The repeats are internally connected by short (4–5 residue) loop sequences [44,55]. This arrangement results in a β-helix structure [30,44–47]. Several studies have predicted similar β-helix structures of CsgA based on in silico covariance analysis and molecular dynamics [30,44–47] which are consistent with (admittedly limited) solid state NMR data [56]. However, the in silico predictions cannot distinguish between a left-handed or right-handed β-helix, since they are equally stable [44]. Frustratingly, CsgA structure predictions deposited in the AlphaFold structure database also include both left- and right-handed β-helix predictions [47,51]. While the AlphaFold predictions of CsgA have not been submitted to peer review and should be interpreted with caution, they are very similar to the previously peer-reviewed predictions [30,44–47]. Despite these ambiguities, all in silico predictions of CsgA structures consist of multiple parallel ladders resulting from the vertical alignment of conserved residues. A hydrophobic core formed from "ladders" of L/I/M/V/A residues is flanked by ladders of conserved Ser, Gln and Asn residues, which are predicted to form hydrogen-bond networks (Figure 3c) [30,44–47].

Figure 3. (**a**) CsgA is predicted to form a β-helix (left- or right-handed) where the five imperfect repeats R1–R5 stack on top of each other. Here a right-handed β-helix is shown that was predicted using AlphaFold. (**b**) The primary sequence of CsgA contains an N22 signal sequence, which is important for transport across the outer membrane through CsgG. The five imperfect repeats R1–R5 are found in residues 23–131. The 7 bars illustrate residue ladders formed in the CsgA β-helix due to stacking of conserved residues. Gatekeeper residues (Asp and Gly) are underlined. (**c**) The top-view of the β-helix shown in (**a**). Ladders formed from conserved Ser, Gln, Ala and varied hydrophobic residues are indicated.

Although the five repeat sequences of CsgA are similar, they differ from each other in important ways. Repeats R2–R4 contain gatekeeper Asp and Gly residues which, given their higher polarity, reduce the aggregation propensity of CsgA (Figure 3b) [57]. They are called gatekeepers because their mutation to the corresponding residues found in R1 and R5 (Asp/Gly → Asn/Ser/Leu/His) accelerates fibrillation markedly compared to wild-type CsgA [57]; i.e., they serve to block (or reduce access to) certain types of interactions. They do so by disrupting a sequence of residues with a certain physical-chemical profile, just as we proposed for the amyloidogenic Aβ peptide [58]. Ironically, the more aggressive aggregation of the gatekeeper-less CsgA mutant compromises bacterial viability and make fibrillation independent of the helper proteins CsgB and CsgF, in vivo [57]. These observations underline an important point: too vigorous fibrillation is to be avoided because it threatens efficient control over the fibrillation process [57]. Perhaps unsurprisingly, the two terminal repeats R1 and R5, which constitute the two flanks of each CsgA monomer, are the most important repeats for driving the fibrillation of CsgA [59]. Exchanging either R1 or R5 with R3, resulting in the mutants R32345 or R12343, drastically reduces curli formation [59]. On the other hand, R1 and R5 are interchangeable, since R12341 and R52345 both form curli fibrils that are indistinguishable from wild-type fibrils [59]. All this is corroborated at the molecular level when the repeat sequences are synthesized as individual peptides and allowed to fibrillate. The peptides fibrillate in the ranking R5 > R1 > R3, whereas R2 and R4 do not fibrillate at all [53,59].

9. The Major Fap Component FapC: A Flexible Number of Imperfect Repeats and Variable Linker Lengths

Like CsgA, FapC contains imperfect repeats [19]. FapC is a considerably longer protein (250–340 residues, depending on the species [19]) than CsgA (131 residues), but only three imperfect repeats are found in FapC, as opposed to five in CsgA [19]. In *Pseudomonas* sp. UK4 (where we originally discovered FapC [19]), the three repeat sequences of FapC are each about 34 residues long, and are connected by two linker sequences of 31 and 39 residues, respectively [19]. These sequences are clearly much longer than the tight turns found in CsgA. The Fap operon is found within three classes of the phylum Proteobacteria, where FapC is found to vary considerably in length due to variations both in linker length (70 in *Pseudomonas* sp. UK4 but 160 in *Pseudomonas aeruginosa*) and in the number of imperfect repeats (three in most species but up to 16 in *Vibrio*) [37]. Nevertheless, the same combination of covariance-derived restraints and molecular dynamics used for CsgA also led to a predicted β-helix structure for FapC in which each repeat forms a β-hairpin (Figure 4a) [38].

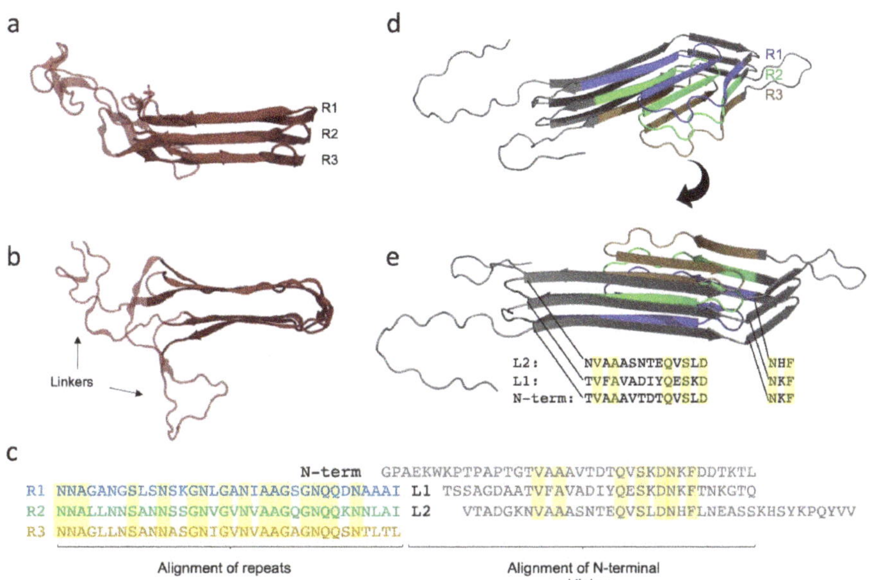

Figure 4. (**a**) FapC is predicted to form a β-helix where the three imperfect repeats R1–R3 stack on top of each other. (**b**) Top-view of the predicted β-helix of FapC. FapC's long linkers (31 and 39 residue lengths) are marked. Panels a and b modified from [38]. (**c**) Alignment of the primary sequence of the repeats R1, R2 and R3 and the N-terminal and linkers, L1 and L2. "Ladders" formed from different kinds of conserved residues are highlighted. (**d**,**e**) Predicted FapC structure from AlphaFold seen at two different angles. The repeat sequences are colored blue (R1), green (R2), orange (R3).

Like CsgA, the imperfect repeats of FapC contain conserved Ser, Ala, Gln and Asn residues (Figure 4c) [19]. The longer repeat sequences and linkers in FapC relative to CsgA [19,20] indicate a less streamlined design could be attributed to the Fap system's evolutionary youthfulness compared to the curli system. That is, FapC has not yet had time to lose some of its unordered parts protruding from its lateral regions (Figure 4b) [37]. However, there are also other possibilities. Structural prediction of FapC by AlphaFold results in a markedly different type of structure which integrates the entire sequence into a highly structured fold (Figure 4d). Here, both the predicted repeat sequences, the linkers and the N-terminal sequence form stacked β-strands in the core of protein. In this predicted structure, the N-terminal and the linkers L1 and L2 stack on top of each other and form "ladders" of conserved residues, just like the repeat sequences (Figure 4e). There is a good deal of evidence that the repeat sequences are the main driving forces of the amyloidogenicity of FapC [19,60,61]. Systematic removal of repeat sequences from FapC has two important consequences: it makes FapC more vulnerable to fragmentation during fibrillation [61] and it reduces the stability of mature FapC fibrils [60]. FuBA stability is here quantified as resistance to dissolution in formic acid [60], one of the few solvents able to reduce FapC to monomers that are visible on SDS-PAGE [19,20]. Formic acid has a pK_a of 3.75 so is not a particularly strong acid [62], but the high concentration needed to dissolve FuBA (~17 M formic acid is needed to dissolve 50% of wildtype FapC [60]) leads to a combination of low pH (<2) and a high concentration of organic solvent, which seems to do the trick [60].

Remarkably, FapC is able to fibrillate even after the complete removal of repeat sequences (although the fibrils are enormously destabilized against formic acid), indicating that the linker regions of FapC play a role in fibrillation (or at least step in as amyloid substituents if the main drivers are removed) [61,63,64]. This is supported by our study on the effect of denaturants on FuBA fibrillation [63], which suggested paradoxically that removal of FapC repeats *increases* the amount of buried surface area per protein molecule. That is, that the number of residues in the fibril backbone *increased* once we removed two or three repeat sequences [63]. The best explanation for this observation is that the linker regions step in to become part of the fibril backbone, once the "proper" amyloid components (repeat sequences) are removed. Currently this remains speculative; however, AlphaFold predictions of a series of different deletions of FapC show quite dramatic changes in structure once two or three repeats are removed (Figure 5), vindicating the observed dramatic changes in stability [61]. Interestingly, in computational disorder predictions of FapC, the repeat sequences obtained higher scores of disorder propensity compared to the linker regions [63]. This is in good agreement with previous systematic analysis which showed a clear relationship between intrinsic disorder and amyloid propensity [61]. The hypothetical structure of a FapC homologue from *Acidothiobacillus* with eight imperfect repeats assigns one "layer" to each repeat in an 8-layer β-helix (Figure 5). Clearly, it will be very exciting to see the experimental structure of FapC once it is hopefully revealed by CryoEM.

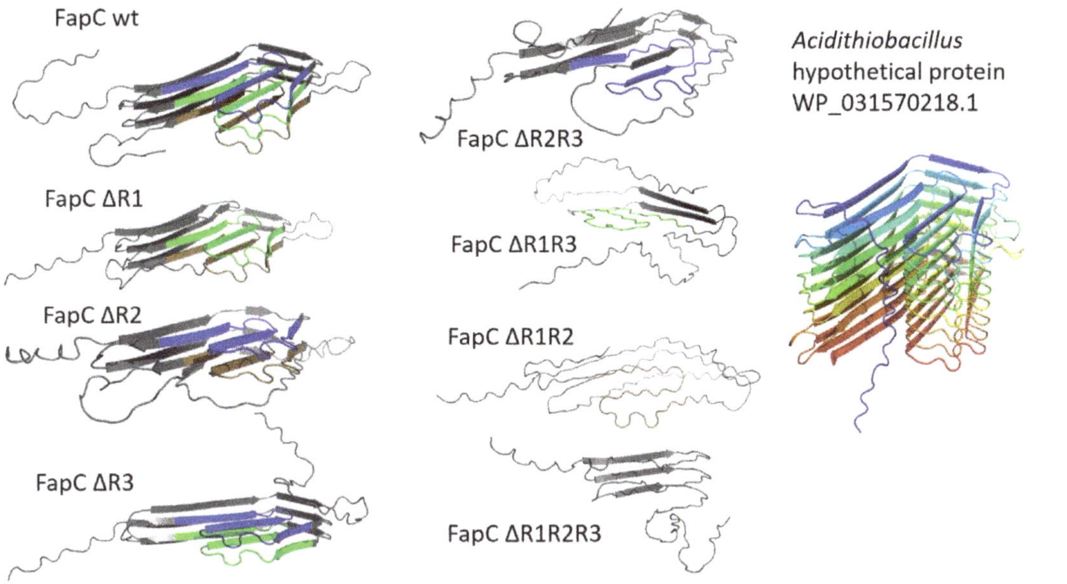

Figure 5. AlphaFold predictions of the structures of FapC imperfect repeats. The two leftmost columns show different FapC UK4 constructs where one or more of the three repeats R1–R3 have been removed (nomenclature and numbering as in [65]), and the structure of the resulting sequence was analyzed by AlphaFold. The rightmost column shows the predicted structure of a FapC homologue from *Acidothiobacillus* which has 8 imperfect repeats, each of which forms a layer in the ensuing β-helix structure. Note the similar cross-sections of FapC wildtype and the *Acidothiobacillus* protein.

There is an important biological lesson in these observations which reinforces the conclusions from gatekeeper residues in CsgA. Uncontrolled aggregation through secondary processes, such as fragmentation or nucleation along the sides of the fibrils, is undesirable because it can lead to runaway processes which accelerate exponentially (i.e., go from baseline to saturation in a very short time). In contrast, primary processes (nucleation followed by elongation) occur at a much more moderate rate, leading to a steady increase in fibril growth over a longer time course which is compatible with the time scales of bacterial growth and biofilm consolidation [66]. Indeed, there is good evidence that primary processes dominate in benign aggregation and secondary processes in pathological ditto (G. Meisl et al. and T.P.J. Knowles, unpublished observations). This may rationalize why FuBAs, under normal conditions, are not cytotoxic to their host organisms [57], and provides more insight into the cytotoxicity of pathological amyloids in human diseases [67,68].

10. Probing FuBA Fibrillation as Folding Steps with Denaturants: Similar Folding during Nucleation and Elongation Activation Steps

The fact that CsgA and FapC are evolutionarily unrelated yet remarkably similar gives us a unique chance to better understand the "design" of amyloid. How do unrelated amyloids converge to relatively similar design blueprints? [63]. Monomeric CsgA is very unstructured according to hydrogen-deuterium exchange [69], light scattering, single fiber AFM-analysis [70], secondary structure determination [71] and NMR studies on the pre-amyloid state [72] and MD simulations of peptide segments of CsgA [73]. However, upon association, CsgA folds into a uniform amyloid state in what appears to be a simple two-state model [69,70]. FapC appears to follow the same model [74]. Thus, the aggregation of both proteins is analogous to the folding of globular proteins, and should therefore be subject to the same type of scrutiny, e.g., using denaturants such as urea and guanidinium chloride. Titration of proteins with increasing concentrations of denaturant typically leads to a sigmoidal equilibrium unfolding curve [75], which is extensively used to study protein stability and folding [76]. Denaturants generally favor unfolded states by preferential binding, so the larger the surface area available for binding, the more the protein is stabilized. The free energy of unfolding depends linearly on denaturant concentration with a slope called the *m*-value, which is proportional to the increase in surface area associated with the unfolding step [77,78]. Thus, the *m*-value directly reports on the extent of folding. This applies both to equilibrium values and kinetics; in the latter case, *m*-values report on the difference in surface area between the ground state (e.g., the native state) and the transition state. Since denaturants decrease folding rates, they would also be expected to reduce aggregation rates—unless aggregation requires prior unfolding from a native state. Thus, denaturants reduce the rate of amyloid formation by the IDPs Aβ40 and Aβ42 (which do not need to unfold before aggregating) [79,80] but increase rates for aggregation of the globular protein insulin [81]. The situation is much less complex with FuBA: Urea reduces both the nucleation rate and the elongation rate of both CsgA and FapC [63], and *m*-values for nucleation and elongation are very similar, suggesting a similar degree of structural consolidation (folding, i.e., burial of surface area) in nucleation and elongation. That is, elongation of FuBA is a natural extension of nucleation. Interestingly, for the FapC mutant lacking all three imperfect repeats, we see an increase in *m*-values, suggesting that more of the protein sequence is integrated into the amyloid core, which again is consistent with the concept of "stand-by" amyloidogenic sequences stepping up to form fibrillar regions once the core amyloidogenic repeats have been removed (Figure 6).

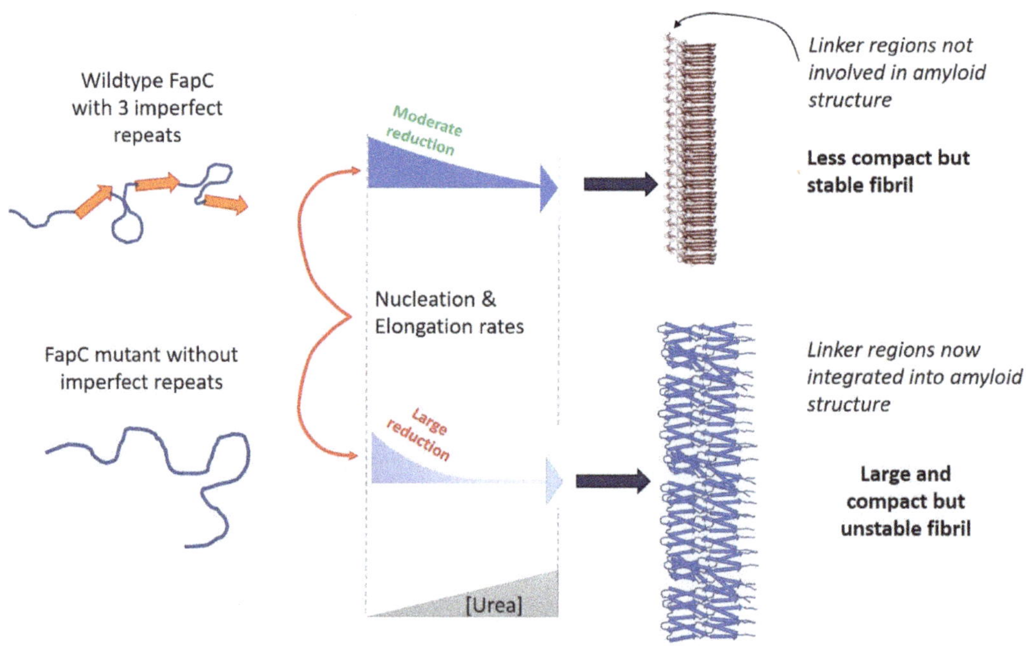

Figure 6. The dependence of FapC fibrillation (nucleation and elongation) on denaturant concentration sheds more light on the folding events accompanying fibrillation. Removal of imperfect repeats leads to a more compact but much less stable fibril, suggesting mobilization of the linker regions to form amyloid. Modified from [63].

The simplicity of FuBA folding is revealed in other ways. In globular protein folding kinetics, a tell-tale sign of folding intermediates is the lack of linearity or roll-over at low denaturant concentrations in chevron plots (log folding rate constants vs. [denaturant]) [77,78]. Conversely, strict linearity implies simple two-state folding. Both CsgA and FapC show a log-linear relationship between elongation rate constants and [urea] between 0 and 8 M urea [63]. This rules out detectable intermediates during FuBA elongation, in contrast to Aβ1-40, which was recently proposed to go through a "frustrated" complex intermediate [82]. This is manifested as in increase in the elongation rate of Aβ up to 1.5 M urea [82], after which there is a decrease in the elongation rate, as described by other groups for both Aβ1-40 and Aβ1-42 [79,80]. For globular proteins, an increase in folding rate with [denaturant] at low [denaturant] is evidence of an off-pathway species which has to unfold back to the denatured state for folding to proceed [83,84]. This usually gives way to a more conventional decline in folding rates at higher [denaturant], so the off-pathway species is destabilized and does not accumulate. Similarly, the increase in elongation rate of Aβ at low [urea] can be seen as a destabilization of the "frustrated" (off-pathway) intermediate which leads to a "smoothing" of the energy landscape and causes less kinetic trapping in the intermediate state [82]. The apparent lack of any frustrated intermediate in FuBA elongation indicates that this is a "cleaner" two-state process with a single harmonious folding step [63].

Nevertheless, there is an important difference between the nucleation and elongation steps: molecular chaperones which inhibit fibrillation of CsgA and FapC primarily target the nucleation step [85] and not the elongation step. We reached this conclusion after analyzing four different human chaperones and the curli chaperone CsgC. Remarkably, the best chaperone was not CsgC but the human protein DNAJB6 which works at such low sub-stoichiometric ratios that it most likely blocks growth of the nucleus itself rather than the isolated monomer; this also explains its preference for nucleation rather than elongation. This common mode of action by completely unrelated chaperones is a strong reflection of the simplicity of bacterial amyloid formation, and primary nucleation is the main step generating new fibrils. At the same time, the fact that so many different chaperones with no evolutionary relationship to each other can target the same (optimized) aggregators suggests a commonality of anti-amyloid activities across nature, perhaps driven by the recognition of amyloidogenic hotspots within the primary structure. In support of this, CsgC can inhibit human αSN fibrillation [35,86] and reduces both CsgA and FapC fibrillation in vitro [36]. The specific molecular basis for chaperone action must remain speculative, though the efficiency of inhibition seems to scale with affinity of binding to the monomeric state [85]. In addition, we note that FapC is more resistant to chaperone action than CsgA, and this may reflect the presence of less amyloidogenic sequences in between the imperfect repeats which sterically hinder effective chaperone binding.

11. Structure-Based Design of Anti-FuBA Peptides

Given that pathological and functional amyloid have different aggregation strategies, do we need to combat their formation in different ways? We can address this question by considering how they respond to specific modulators, such as peptides [87]. Given that amyloid forms through self-recognition, peptides based on the sequence of the target protein are likely to influence aggregation by binding to the protein and thus affect its ability to bind to other copies of the same protein [88–90]. This means that the most important step in peptide design is to identify the most aggregation-prone regions (APR) of a given protein. To appreciate this, we need to consider a strategy that has recently emerged from the Schymkowitz-Rousseau SWITCH laboratory, namely, the ability to induce the aggregation of proteins that are otherwise not amyloidogenic (Figure 7).

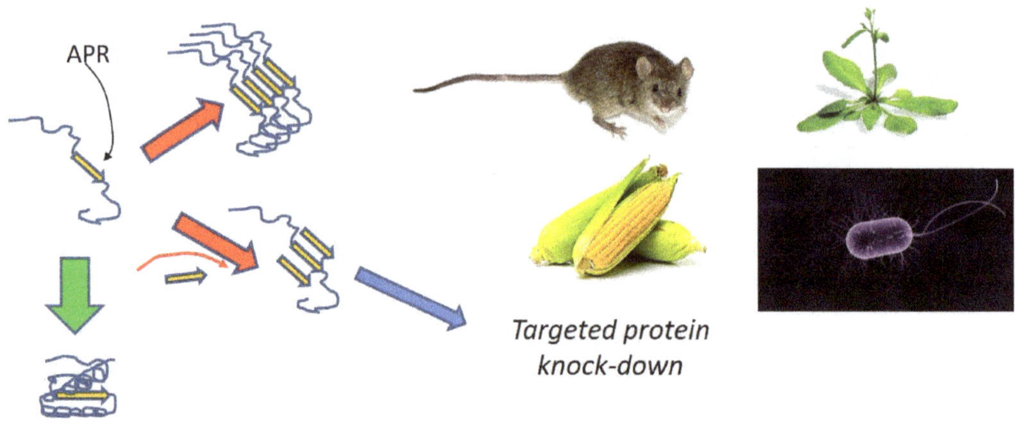

Figure 7. The SWITCH strategy of using APRs (aggregation prone regions) in proteins to knock down protein expression in vivo [91]. APR sequences are usually buried in the protein's hydrophobic core, but during protein biosynthesis, they may be targeted by peptides containing the same APR sequences, leading to aggregation and (presumably) degradation.

APR in globular proteins normally hide in the hydrophobic core of the protein where they avoid contact with other proteins; they only become problematic when they are exposed, e.g., during protein biosynthesis where the nascent polypeptide chain emerges from the ribosome and has not yet folded [92,93]. Thus, if a peptide encoding the APR is present in the cell at this stage, it may trap the protein in a misfolded state which can be maintained and used to accumulate additional aggregates by complexing with other copies of the same trapped protein. This strategy has been exploited to induce simultaneous aggregation of many endogenous proteins in a whole range of hosts, ranging from plants and animals to bacteria [91]. The ability to target vital bacterial proteins makes APR peptides potential antibiotics [92,93]. The strategy involves the prediction of APRs in the target protein using computational algorithms [94], followed by the design of peptides containing the APR of the target protein, e.g., tandem repeats of the APR, flanked by charged gatekeeper residues to increase solubility (Figure 7). Such peptides have proven effective in *Staphylococcus epidermidis* and *E. coli* strains [92,93]. We have used the same method to target CsgA in *E. coli*. This may seem pointless given that CsgA is supposed to aggregate anyway, but the advantage of the APR approach is to alter the nature of the ensuing aggregate. Incorporation of peptides into CsgA leads to non-functional aggregates which are less stable than the original CsgA fibrils and reduce biofilm formation [95].

Using a peptide microarray displaying the whole CsgA sequence as staggered arrays of 14-mer peptides, we measured which peptides were particularly good at binding to full-length CsgA. In this way we identified three segments in CsgA which appeared to be the main drivers of CsgA-CsgA interactions: the R3 and terminal R1 and R5 repeats. Additional computational analysis of the aggregation profile of CsgA showed that R1 and R5 contain some of the most aggregation-prone segments. The terminal repeats of CsgA have previously been experimentally verified to form amyloid-like aggregates [96,97], and we therefore decided to focus primarily on APRs within the terminal repeats of CsgA. This is further supported by the predicted β-helical structure of CsgA, which implies that curli amyloid fibrils are formed through stacking of CsgA monomers by intermolecular R1–R5 contacts [45,98,99]. APRs identified in R1 and R5 were then used as the basis for the design of tandem repeat peptides consisting of two identical hepta-sequences (containing the APR identified in either R1 or R5) linked by a flexible (Gly–Ser) or rigid (Pro–Pro) linker. The peptides pushed CsgA to form destabilized aggregates which were less stable towards formic acid and had a different morphology than untreated CsgA according to transmission electron microscopy. We suspect that the peptides work by interacting with monomeric CsgA to promote aggregation into non-native CsgA aggregates.

APRs can be used not only to promote aggregation, but also to block aggregation by blocking the interaction between fibril ends and incoming monomers [100]. This computational design approach uses the APRs of experimentally solved amyloid structures from the PDB as templates; it measures the effect of peptide docking onto the fibril end of the template structures using the FoldX force field [100]. In addition to using an APR amyloid core template from the PDB, the peptide inhibitor calculations can be based on a computationally predicted APR amyloid core template, e.g., predicted from the Cordax method [101]. This approach was extended to the design of structure-based anti-CsgA peptides [95]. Insertions of R and W residues were predicted in silico to have high potential for fibril end-capping [95,100]. We therefore supplemented our anti-CsgA peptide library by designing peptides aimed to "cap" fibril ends by incorporation of Arg and Trp residues into the APR regions of our peptides [100]. Multiple CsgA-targeting peptides worked rapidly on CsgA, inducing a prompt increase in turbidity of the samples [95]. Remarkably, multiple peptides also modulated the fibrillation of FapC, which is possible by binding to FapC segments which are similar to CsgA in terms of sequence and/or structure.

The final proof is an "application-relevant" test, i.e., the impact of the peptides on FuBA fibrillation in a biofilm. Indeed, several peptides reduced biofilm formation in both *E. coli* and *P. aeruginosa*, indicating that our peptides modulated CsgA and FapC fibrillation, in vivo. It is likely that some peptides predominantly function by an "induction mechanism," where peptides interact with monomeric CsgA and divert the protein into non-native aggregates. This mechanism is supported by the fact that some peptides caused rapid formation of destabilized aggregates with different morphologies, which are much less resistant to formic acid compared to untreated fibrils. Other peptides may work by a "capping mechanism," where peptides are bound to fibril ends, which blocks further elongation through charge repulsion and steric hindrance. The fibril end "capping" mechanism is supported by our observation that some peptides almost completely suppressed the presence of a sigmoidal thioflavin T fibrillation curve. The peptides designed as fibril end "cappers" appeared to work more efficiently in vitro than the peptides designed to induce non-native aggregation, because they resulted in prompter suppression of the CsgA fibrillation and in higher fibril destabilization towards formic acid. However, both types of anti-CsgA peptide strategies appeared to work well in in vivo anti-biofilm assays. A possible explanation for this could be that the peptides are likely taken up by the bacteria and may therefore work intracellularly. Positively charged residues in peptides have been shown to enhance uptake by bacteria [102]. Our most effective peptides contained 4–9 Arg residues, and the reduction in biofilm formation correlated with the number of Arg residues. Clearly there is scope for more development in this area to target different aspects of bacterial growth and biofilm formation.

12. Using Small Molecules and Polyphenols to Target FuBA and Biofilm

Besides smart peptides, another way to target FuBA is to use small molecules acting as aggregation inhibitors. A prime example is the polyphenol epigallocatechin-3-gallate (EGCG) (Figure 8a), which inhibits the formation of both pathological amyloid [103–106] and FuBA [40]. Although EGCG has shown promising pre-clinical results against pathological amyloids, the compound has not successfully passed clinical trials, possibly because of low chemical stability in vivo (it is prone to epimerization and auto-oxidation above pH 6 [107]) and limited penetration of the blood–brain barrier [106,108]. While EGCG also can inhibit the formation of bacterial amyloid and reduce biofilm formation [40,109], stability issues have also hampered its clinical use against infections [110–112]. Nevertheless, EGCG and other (suitably stabilized) small molecules could target biofilm formation in three complementary ways:

(i) Biofilm regulation: Impacts on biochemical processes inside or outside the bacterial cell that regulate biofilm formation. For example, EGCG disrupts quorum-sensing

(QS) signaling by increasing the binding of pyocyanin (a central QS molecule) to FapC fibrils in *P. aeruginosa* [40]. EGCG also activates expression of the small non-coding RNA molecule RybB that binds to initiation codon of the *csgD* mRNA and inhibits expression of the curli transcription factor CsgD, thereby down-regulating production of the two main components of *E. coli* biofilm, namely, curli and pEtN-cellulose (bacterial cellulose where every second glucose group is modified with phosphoethanolamine) [109,113].

(ii) General antibacterial effects: Reduction of biofilm formation at the source, i.e., antimicrobial mechanisms such as disruption of the bacterial cell membrane or inhibition of fatty acid synthesis and enzyme activity [114–116].

(iii) Direct inhibition: EGCG is known to inhibit fibrillation of amyloidogenic monomers secreted from bacteria and remodeling preformed fibrils to amorphous aggregates [40]. EGCG interacts with FapC monomers and redirects it to relatively stable off-pathway oligomers [117], just as is seen for proteins and peptides involved in neurodegenerative diseases such as α-synuclein and Aβ [118]. As with chaperones [85], the main target of inhibition is the nucleation phase, and FapC/CsgA monomers are redirected to off-pathway oligomers [119]. These off-pathway oligomers are SDS-stable and contain a mixture of β-sheets and random coils [120]. This inhibition of FapC fibrillation can take place even in the presence of amyloid inducers such as SDS, rhamnolipids and LPS [121]. By combining a peptide microarray and the EGCG-binding compound Nitro blue tetrazolium, we have shown that EGCG binds to amyloidogenic hot spots containing the sequence "GVNVAA" in repeats R2 and R3 and even linker regions of FapC sequence [117,122]. Small-angle X-ray scattering measurements revealed a core-shell structure for FapC off-pathway oligomers that consist of ~7 monomers with a 25–26 nm short-axis diameter, which is much bigger than would be expected for on-pathway fibril precursors (~ 10 nm) [117].

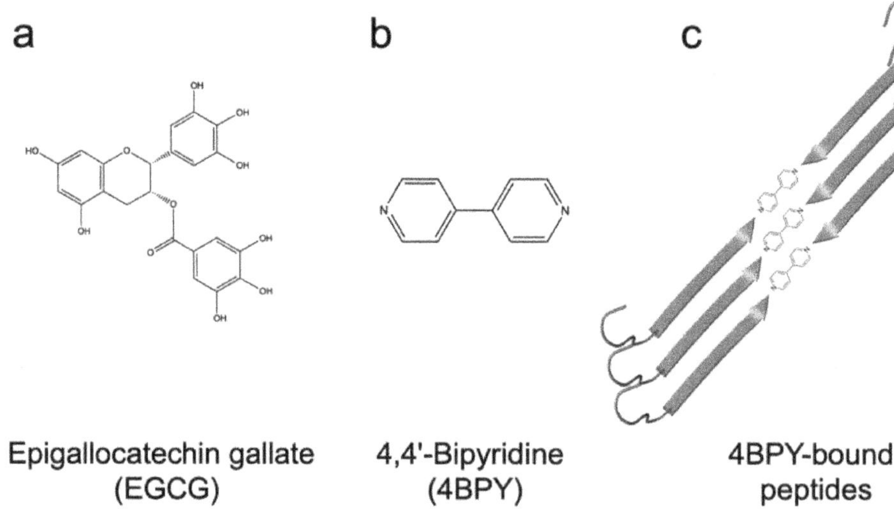

Figure 8. Molecular structures of (**a**) the polyphenol EGCG, (**b**) the chaperone-like small molecule 4BPY. (**c**) 4BPY-bound peptides seen in surface adsorbed assemblies, stabilized by hydrogen bonds between the nitrogen groups of 2BPY and peptide carboxy termini (arrow tips).

There have been many structure–activity studies on EGCG and its sub-scaffolds, such as gallic acid (three hydroxyl groups on a benzoic acid ring), catechin, epicatechin and 3-hydroxytyrosol. However, intact EGCG remains the most effective inhibitor, highlighting the advantages of combining multiple polyphenolic groups. As a logical extension, structures such as penta-*O*-galloyl-β-d-glucose (PGG) that contain five gallic acid groups on a central glucose actually slightly surpass EGCG and show stronger inhibitory effects [117,119,123]. Nevertheless, a fundamental weakness is that the gallic acid groups in both EGCG and PGG are attached to the core of the molecule via ester bonds that are not completely stable above pH 7 [117]. In vitro studies have identified numerous potential alternatives to polyphenols for inhibiting amyloid formation based on in vitro studies [124–126], but their use in vivo is a challenge. Fortunately, there is a growing number of studies highlighting the loading of small molecules into nanoparticles, such as lipid vesicles or polymer-coated albumin aggregates [127–131]. This packaging improves the ability to cross the blood–brain barrier (for therapy) or penetrate biofilm (to combat functional amyloid), allowing the small molecules to target amyloidogenic monomers.

Small molecules may also induce the fibrillation of amyloids, as demonstrated with 4,4′-bipyridine (4BPY) (Figure 8b), which induces Aβ fibrillation in vitro [132]. The nitrogen atoms of 4BPY are capable of hydrogen bonding with the -COOH group of the peptides' C-termini in surface adsorbed peptide assemblies (Figure 8c) [132]. These play a role in our final topic, namely, the ability to assemble bacterial amyloid in an organized fashion on a surface.

13. Functional Bacterial Amyloid as a Bionanomaterial

Functional bacterial amyloids are built to last. The fibrillation of CsgA is remarkably robust and can take place under a wide range of conditions, including extreme values of pH [71] and at high denaturant concentrations [63]. Mature CsgA fibrils are also very stable once formed and can withstand high denaturant concentrations and boiling SDS [20,63,133]. In contrast to pathological amyloids, the fibrils formed from CsgA appear to lack polymorphism, suggesting very precise and steady assembly [134]. In addition, CsgA can tolerate quite substantial modifications, e.g., by fusion to large proteins without losing its ability to fibrillate [135,136]. Together, these abilities have motivated the development of several different CsgA-based bionanomaterials [137–142]. Among the examples are the fusions of polypeptides which provide completely new functionalities to CsgA, such as increased binding to various types of inorganic substrates, including silica, graphene and silver nanoparticles [136,137,139]. Even more flexibility of CsgA-fusion proteins has been demonstrated using the versatile SpyCatcher-SpyTag technology, which allows a "switch-like" covalent coupling of virtually any other SpyCatcher-tagged heterologous protein to Spy-tagged CsgA [136,137]. Interestingly, even unmodified CsgA can effectively bind many different surfaces, including highly hydrophilic or hydrophobic substrates, and the resulting CsgA protein coatings are very robust against harsh conditions [137,143]. The research of protein interactions with abiotic surfaces is important because as new nanomaterials are being developed, it opens new possibilities for developing novel hybrid bionanomaterials. The effective CsgA fibrillation and the robustness of the final assemblies, combined with CsgA's high tolerance for protein engineering, makes it a promising candidate for producing novel bionanomaterial [144,145].

14. Steering FuBA Assemblies Using the Structured Surface of Graphene

One example of such novel bionanomaterials builds on graphene, a carbon nanomaterial (CN) made of sp [2] bonded carbon atoms connected in a planar hexagonal (honeycomb) lattice [146,147]. Graphene consists of a single planar level of the hexagonal lattice and is one atom thick. Graphene can be produced from graphite, which is simply stacked sheets of graphene. Graphite is both produced by the mining of natural sources and synthesized from high-molecular-weight hydrocarbons [148]. The interactions between amyloids and CN are especially interesting. Both the fibrillation kinetics and the final amyloid assemblies can be drastically modulated by CNs [149]. Interestingly, CN can cause either inhibition or induction of amyloid fibrillation [149]. Spherical fullerenes (also known as C_{60}) can bind the hydrophobic motif KLVFF in Aβ and inhibit fibrillation [150]. On the other hand, graphite has been demonstrated to promote the formation of β-strand-dominated assemblies associated with amyloid formation from peptides which are initially dominated by α-helical and random-coil secondary structures [149,151–153]. Additionally, the final assemblies of amyloids can be modulated by CN [149]. There is growing evidence that structural patterns in the graphite lattice can be templates for the assembly of amyloids [149,154–156]. CsgA aggregation is stimulated by graphite (Figure 9a), and it forms very systematic β-strand rich assemblies on highly oriented pyrolytic graphite (HOPG), manifested as lamellar-like structures using scanning tunneling microscopy (STM) analysis (Figure 9b) [157]. The assemblies are large (>10,000 nm [2]) and do not represent individual fibrils, as observed in other studies, but are instead uninterrupted "film-like" assemblies. The distance between β-strands is ~4.8 Å, and β-strand lengths (~4 nm) are very systematic and correlated over longer distances. Further, the directions of the assemblies appeared to be controlled by the underlying graphite lattice (always with an angle of 60° with respect to each other) (Figure 9c), which led to the expected 4.9 Å separation if the peptides were arranged orthogonally to the sides of the hexagon (Figure 9d). MD simulations have indicated that a CsgA-derived peptide initially adsorbs unstructured clusters which gradually mature into lamellar-like structures in directions guided by the graphite lattice [157].

Finally, the orientation of CsgA peptides can be controlled using the small chaperone-like 4BPY (which hydrogen-bonds to the C-terminus of a peptide sequence [158]), allowing precise and tailored assembly of CsgA peptides on graphite. While individual 4BPY tethering molecules are clearly visible on surfaces assembled with short peptides, they are not seen with full-length CsgA. It is possible that full-length CsgA's strong preference for ordered assembly (i.e., the avidity effect of having multiple beta-strand motifs available within one polypeptide chain combined with strong van der Waals interactions with graphene) overrules the need for co-assembly with 4BPY. This nicely underlines the strong innate self-organizing principles inherent in functional amyloid.

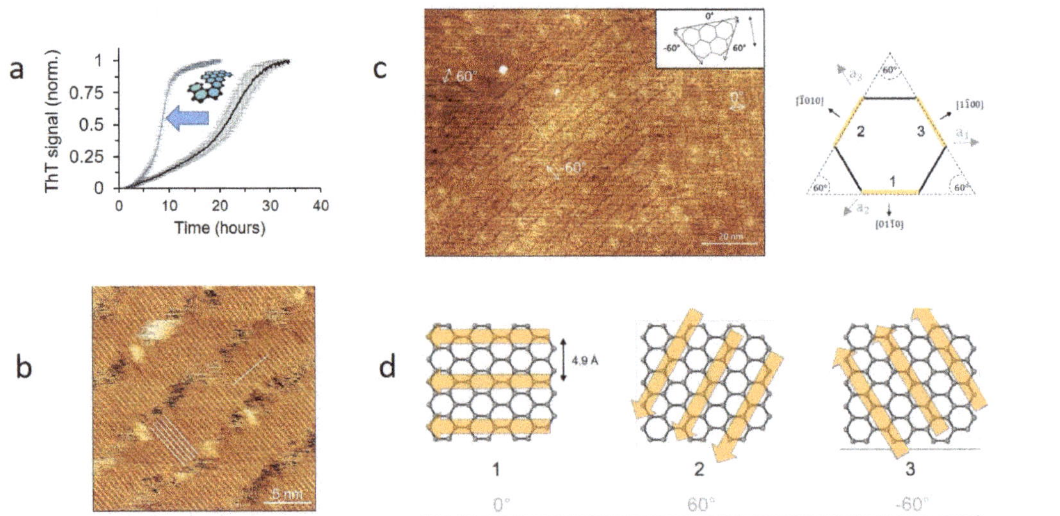

Figure 9. Effect of a structured surface on CsgA self-assembly. (**a**) Graphite nanoparticles promote aggregation of CsgA in vitro. (**b**) STM image of full-length CsgA deposited on HOPG. The five parallel white bars in the lower left corner highlight adjacent β-strands; the orthogonally placed white bar in the upper right corner measures the average distance over 10 strands to be ca. 4.8 Å. (**c**) STM image illustrates the 3 preferred orientations of full-length CsgA on HOPG (angles illustrated in the insert) and their orientation relative to the HOPG orthogonal structure. (**d**) The positions of individual strands in these arrangements are dictated by the orientation of the graphene atoms and lead to a strand distance of 4.9 Å. Adapted from [157].

Author Contributions: Conceptualization, D.E.O. and T.V.S.; writing—original draft preparation, T.V.S., Z.N. and D.E.O.; writing—review and editing, T.V.S. and D.E.O.; visualization, T.V.S. and D.E.O.; supervision, D.E.O.; project administration, D.E.O.; funding acquisition, D.E.O. All authors have read and agreed to the published version of the manuscript.

Funding: D.E.O. and T.V.S. gratefully acknowledge support from the Independent Research Foundation Denmark | Natural Sciences (grant no. 8021-00208B and 8021-00133B) and from the Sino-Danish Center. D.E.O. also acknowledges support from the Independent Research Foundation Denmark | Technical Sciences (grant no. 6111-00241B) and the Lundbeck Foundation (grant no. R276-2018-671).

Conflicts of Interest: The authors declare no conflict of interest.

Abbreviations

APR: aggregation prone region; 4BPY, 4,4′-bipyridine; CN, carbon nanomaterial; EGCG, epigallocatechin-3-gallate; Fap, functional amyloid in Pseudomonas; FuBA, functional bacterial amyloid; HOPG, highly oriented pyrolytic graphite; IDP, intrinsically disordered proteins; MSA, multiple sequence alignment; PGG, penta-O-galloyl-β-d-glucose; STM, scanning tunneling microscopy.

References

1. Dobson, C.M. Protein folding and misfolding. *Nature* **2003**, *426*, 884–890. [CrossRef] [PubMed]
2. Larsen, P.; Dueholm, M.; Christiansen, G.; Nielsen, J.L.; Otzen, D.E.; Nielsen, P.H. Amyloid adhesins are abundant in natural biofilms. *Environ. Microbiol.* **2007**, *9*, 3077–3090. [CrossRef] [PubMed]
3. Larsen, P.; Nielsen, J.L.; Otzen, D.E.; Nielsen, P.H. Amyloid-like adhesins in floc-forming and filamentous bacteria in activated sludge. *Appl. Environ. Microbiol.* **2008**, *74*, 1517–1526. [CrossRef] [PubMed]
4. Chiti, F.; Dobson, C.M. Protein Misfolding, Amyloid Formation, and Human Disease: A Summary of Progress Over the Last Decade. *Annu. Rev. Biochem.* **2017**, *86*, 27–68. [CrossRef]
5. Baldwin, A.J.; Knowles, T.P.J.; Tartaglia, G.G.; Fitzpatrick, A.W.; Devlin, G.L.; Shammas, S.L.; Waudby, C.A.; Mossuto, M.F.; Meehan, S.; Gras, S.L.; et al. Metastability of native proteins and the phenomenon of amyloid formation. *J. Am. Chem. Soc.* **2011**, *133*, 14160–14163. [CrossRef]
6. Greenwald, J.; Riek, R. On the possible amyloid origin of protein folds. *J. Mol. Biol.* **2012**, *421*, 417–426. [CrossRef]
7. Dobson, C.M. Protein misfolding, evolution and disease. *Trends Biochem. Sci.* **1999**, *24*, 329–332. [CrossRef]
8. Sekijima, Y.; Wiseman, R.L.; Matteson, J.; Hammarström, P.; Miller, S.R.; Sawkar, A.R.; Balch, W.E.; Kelly, J.W. The Biological and Chemical Basis for Tissue-Selective Amyloid Disease. *Cell* **2005**, *121*, 73–85. [CrossRef]
9. Lamb, Y.N.; Deeks, E.D. Tafamidis: A Review in Transthyretin Amyloidosis with Polyneuropathy. *Drugs* **2019**, *79*, 863–874. [CrossRef]
10. Bulawa, C.E.; Connelly, S.; Devit, M.; Wang, L.; Weigel, C.; Fleming, J.A.; Packman, J.; Powers, E.T.; Wiseman, R.L.; Foss, T.R.; et al. Tafamidis, a potent and selective transthyretin kinetic stabilizer that inhibits the amyloid cascade. *Proc. Natl. Acad. Sci. USA* **2012**, *109*, 9629–9634. [CrossRef]
11. Fink, A.L. Factors affecting the fibrillation of α-synuclein, a natively unfolded protein. In *Misbehaving Proteins*; Springer: New York, NY, USA, 2006; pp. 265–285.
12. Chiti, F.; Dobson, C.M. Protein Misfolding, Functional Amyloid, and Human Disease. *Annu. Rev. Biochem.* **2006**, *75*, 333–366. [CrossRef] [PubMed]
13. Westermark, P.; Grimelius, L. The pancreatic islet cells in insular amyloidosis in human diabetic and non-diabetic adults. *Acta Pathol. Microbiol. Scand. Sect. A Pathol.* **1973**, *81*, 291–300. [CrossRef] [PubMed]
14. Jahn, T.R.; Makin, O.S.; Morris, K.L.; Marshall, K.E.; Tian, P.; Sikorski, P.; Serpell, L.C. The Common Architecture of Cross-β Amyloid. *J. Mol. Biol.* **2010**, *395*, 717–727. [CrossRef]
15. Knowles, T.P.J.; Buehler, M.J. Nanomechanics of functional and pathological amyloid materials. *Nat. Nanotechnol.* **2011**, *6*, 469–479. [CrossRef] [PubMed]
16. Fowler, D.M.; Koulov, A.V.; Balch, W.E.; Kelly, J.W. Functional amyloid—From bacteria to humans. *Trends Biochem. Sci.* **2007**, *32*, 217–224. [CrossRef]
17. Coustou, V.; Deleu, C.; Saupe, S.; Begueret, J. The protein product of the het-s heterokaryon incompatibility gene of the fungus Podospora anserina behaves as a prion analog. *Proc. Natl. Acad. Sci. USA* **1997**, *94*, 9773–9778. [CrossRef]
18. Claessen, D.; Rink, R.; de Jong, W.; Siebring, J.; de Vreugd, P.; Boersma, F.H.; Dijkhuizen, L.; Wösten, H.A. A novel class of secreted hydrophobic proteins is involved in aerial hyphae formation in Streptomyces coelicolor by forming amyloid-like fibrils. *Genes Dev.* **2003**, *17*, 1714–1726. [CrossRef]
19. Dueholm, M.S.; Petersen, S.V.; Sønderkær, M.; Larsen, P.; Christiansen, G.; Hein, K.L.; Enghild, J.J.; Nielsen, J.L.; Nielsen, K.L.; Nielsen, P.H.; et al. Functional amyloid in *Pseudomonas*. *Mol. Microbiol.* **2010**, *77*, 1009–1020. [CrossRef]
20. Chapman, M.R.; Robinson, L.S.; Pinkner, J.S.; Roth, R.; Heuser, J.; Hammar, M.; Normark, S.; Hultgren, S.J. Role of *Escherichia coli* curli operons in directing amyloid fiber formation. *Science* **2002**, *295*, 851–855. [CrossRef]

21. Schwartz, K.; Syed, A.K.; Stephenson, R.E.; Rickard, A.H.; Boles, B.R. Functional amyloids composed of phenol soluble modulins stabilize *Staphylococcus aureus* biofilms. *PLoS Pathog.* **2012**, *8*, e1002744. [CrossRef]
22. Dueholm, M.S.; Nielsen, P.H.; Chapman, M.; Otzen, D.; Romero, D.; Kolter, R.; Dueholm, M.S.; Nielsen, P.H.; Chapman, M.; Otzen, D. *Functional Amyloids in Bacteria*; Wiley-VCH Verlag GmbH & Co. KGaA: Weinheim, Germany, 2013; pp. 411–438.
23. Hall-Stoodley, L.; Costerton, J.; Stoodley, P. Bacterial biofilms: From the natural environment to infectious diseases. *Nat. Rev. Microbiol.* **2004**, *2*, 95–108. [CrossRef] [PubMed]
24. Hasan, J.; Crawford, R.J.; Ivanova, E.P. Antibacterial surfaces: The quest for a new generation of biomaterials. *Trends Biotechnol.* **2013**, *31*, 295–304. [CrossRef] [PubMed]
25. Christensen, L.F.B.; Schafer, N.; Wolf-Perez, A.; Madsen, D.J.; Otzen, D.E. Bacterial Amyloids: Biogenesis and Biomaterials. In *Advances in Experimental Medicine and Biology*; Springer: New York, NY, USA, 2019; Volume 1174, pp. 113–159.
26. Dueholm, M.S.; Albertsen, M.; Otzen, D.; Nielsen, P.H. Curli Functional Amyloid Systems Are Phylogenetically Widespread and Display Large Diversity in Operon and Protein Structure. *PLoS ONE* **2012**, *7*, e51274. [CrossRef] [PubMed]
27. Pawar, D.M.; Rossman, M.L.; Chen, J. Role of curli fimbriae in mediating the cells of enterohaemorrhagic *Escherichia coli* to attach to abiotic surfaces. *J. Appl. Microbiol.* **2005**, *99*, 418–425. [CrossRef]
28. White, A.P.; Collinson, S.K.; Banser, P.A.; Gibson, D.L.; Paetzel, M.; Strynadka, N.C.J.; Kay, W.W. Structure and characterization of AgfB from Salmonella enteritidis thin aggregative fimbriae. *J. Mol. Biol.* **2001**, *311*, 735–749. [CrossRef]
29. Hammer, N.D.; Schmidt, J.C.; Chapman, M.R. The curli nucleator protein, CsgB, contains an amyloidogenic domain that directs CsgA polymerization. *Proc. Natl. Acad. Sci. USA* **2007**, *104*, 12494–12499. [CrossRef]
30. Barnhart, M.M.; Chapman, M.R. Curli biogenesis and function. *Annu. Rev. Microbiol.* **2006**, *60*, 131–147. [CrossRef]
31. Robinson, L.S.; Ashman, E.M.; Hultgren, S.J.; Chapman, M.R. Secretion of curli fibre subunits is mediated by the outer membrane-localized CsgG protein. *Mol. Microbiol.* **2006**, *59*, 870–881. [CrossRef]
32. Nenninger, A.A.; Robinson, L.S.; Hammer, N.D.; Epstein, E.A.; Badtke, M.P.; Hultgren, S.J.; Chapman, M.R. CsgE is a curli secretion specificity factor that prevents amyloid fibre aggregation. *Mol. Microbiol.* **2011**, *81*, 486–499. [CrossRef]
33. Nenninger, A.A.; Robinson, L.S.; Hultgren, S.J. Localized and efficient curli nucleation requires the chaperone-like amyloid assembly protein CsgF. *Proc. Natl. Acad. Sci. USA* **2009**, *106*, 900–905. [CrossRef]
34. Wang, X.; Chapman, M.R. Sequence Determinants of Bacterial Amyloid Formation. *J. Mol. Biol.* **2008**, *380*, 570–580. [CrossRef] [PubMed]
35. Evans, M.L.; Chorell, E.; Taylor, J.D.; Aden, J.; Gotheson, A.; Li, F.; Koch, M.; Sefer, L.; Matthews, S.J.; Wittung-Stafshede, P.; et al. The bacterial curli system possesses a potent and selective inhibitor of amyloid formation. *Mol. Cell* **2015**, *57*, 445–455. [CrossRef] [PubMed]
36. Taylor, J.D.; Hawthorne, W.J.; Lo, J.; Dear, A.; Jain, N.; Meisl, G.; Andreasen, M.; Fletcher, C.; Koch, M.; Darvill, N.; et al. Electrostatically-guided inhibition of Curli amyloid nucleation by the CsgC-like family of chaperones. *Nat. Publ. Group* **2016**, *6*, 24656. [CrossRef]
37. Dueholm, M.S.; Otzen, D.; Nielsen, P.H. Evolutionary insight into the functional amyloids of the pseudomonads. *PLoS ONE* **2013**, *8*, e76630. [CrossRef]
38. Rouse, S.L.; Matthews, S.J.; Dueholm, M.S. Ecology and Biogenesis of Functional Amyloids in Pseudomonas. *J. Mol. Biol.* **2018**, *430*, 3685–3695. [CrossRef] [PubMed]
39. Zeng, G.; Vad, B.S.; Dueholm, M.S.; Christiansen, G.; Nilsson, M.; Tolker-Nielsen, T.; Nielsen, P.H.; Meyer, R.L.; Otzen, D.E. Functional bacterial amyloid increases Pseudomonas biofilm hydrophobicity and stiffness. *Front. Microbiol.* **2015**, *6*, 1099. [CrossRef]
40. Stenvang, M.; Dueholm, M.S.; Vad, B.S.; Seviour, T.; Zeng, G.; Geifman-Shochat, S.; Søndergaard, M.T.; Christiansen, G.; Meyer, R.L.; Kjelleberg, S.; et al. Epigallocatechin Gallate Remodels *Pseudomonas aeruginosa* Functional Amyloids and Increases Biofilm Susceptibility to Antibiotic Treatment. *J. Biol. Chem.* **2016**, *291*, 26540–26553. [CrossRef]
41. Dueholm, M.S.; Sondergaard, M.T.; Nilsson, M.; Christiansen, G.; Stensballe, A.; Overgaard, M.T.; Givskov, M.; Tolker-Nielsen, T.; Otzen, D.E.; Nielsen, P.H. Expression of Fap amyloids in *Pseudomonas aeruginosa*, *P. fluorescens*, and *P. putida* results in aggregation and increased biofilm formation. *MicrobiologyOpen* **2013**, *2*, 365–382. [CrossRef]
42. Rouse, S.L.; Hawthorne, W.J.; Berry, J.L.; Chorev, D.S.; Ionescu, S.A.; Lambert, S.; Stylianou, F.; Ewert, W.; Mackie, U.; Morgan, R.M.L.; et al. A new class of hybrid secretion system is employed in Pseudomonas amyloid biogenesis. *Nat. Commun.* **2017**, *8*, 263. [CrossRef]
43. Goyal, P.; Krasteva, P.V.; Van Gerven, N.; Gubellini, F.; Van den Broeck, I.; Troupiotis-Tsailaki, A.; Jonckheere, W.; Pehau-Arnaudet, G.; Pinkner, J.S.; Chapman, M.R.; et al. Structural and mechanistic insights into the bacterial amyloid secretion channel CsgG. *Nature* **2014**, *516*, 250–253. [CrossRef]
44. Tian, P.; Boomsma, W.; Wang, Y.; Otzen, D.E.; Jensen, M.H.; Lindorff-Larsen, K.; Lindor, K. Structure of a Functional Amyloid Protein Subunit Computed Using Sequence Variation. *J. Am. Chem. Soc.* **2015**, *137*, 19–22. [CrossRef] [PubMed]
45. Louros, N.N.; Bolas, G.M.P.; Tsiolaki, P.L.; Hamodrakas, S.J.; Iconomidou, V.A. Intrinsic aggregation propensity of the CsgB nucleator protein is crucial for curli fiber formation. *J. Struct. Biol.* **2016**, *195*, 179–189. [CrossRef] [PubMed]
46. Debenedictis, E.P.; Ma, D.; Keten, S. Structural predictions for curli amyloid fibril subunits CsgA and CsgB. *RSC Adv.* **2017**, *7*, 48102–48112. [CrossRef]

47. Jumper, J.; Evans, R.; Pritzel, A.; Green, T.; Figurnov, M.; Ronneberger, O.; Tunyasuvunakool, K.; Bates, R.; Žídek, A.; Potapenko, A. Highly accurate protein structure prediction with AlphaFold. *Nature* **2021**, *596*, 583–589. [CrossRef]
48. Deng, X.; Gonzalez Llamazares, A.; Wagstaff, J.M.; Hale, V.L.; Cannone, G.; McLaughlin, S.H.; Kureisaite-Ciziene, D.; Löwe, J. The structure of bactofilin filaments reveals their mode of membrane binding and lack of polarity. *Nat. Microbiol.* **2019**, *4*, 2357–2368. [CrossRef]
49. Göbel, U.; Sander, C.; Schneider, R.; Valencia, A. Correlated mutations and residue contacts in proteins. *Proteins Struct. Funct. Bioinform.* **1994**, *18*, 309–317. [CrossRef]
50. Gallardo, R.; Ranson, N.A.; Radford, S.E. Amyloid structures: Much more than just a cross-β fold. *Curr. Opin. Struct. Biol.* **2020**, *60*, 7–16. [CrossRef]
51. Varadi, M.; Anyango, S.; Deshpande, M.; Nair, S.; Natassia, C.; Yordanova, G.; Yuan, D.; Stroe, O.; Wood, G.; Laydon, A. AlphaFold Protein Structure Database: Massively expanding the structural coverage of protein-sequence space with high-accuracy models. *Nucleic Acids Res.* **2022**, *50*, D439–D444. [CrossRef]
52. Pinheiro, F.; Santos, J.; Ventura, S. AlphaFold and the amyloid landscape. *J. Mol. Biol.* **2021**, *433*, 167059. [CrossRef]
53. Wang, X.; Smith, D.R.; Jones, J.W.; Chapman, M.R. In vitro polymerization of a functional *Escherichia coli* amyloid protein. *J. Biol. Chem.* **2007**, *282*, 3713–3719. [CrossRef]
54. Cherny, I.; Rockah, L.; Levy-Nissenbaum, O.; Gophna, U.; Ron, E.Z.; Gazit, E. The formation of *Escherichia coli* curli amyloid fibrils is mediated by prion-like peptide repeats. *J. Mol. Biol.* **2005**, *352*, 245–252. [CrossRef] [PubMed]
55. Collinson, S.K.; Parker, J.M.R.; Hodges, R.S.; Kay, W.W. Structural predictions of AgfA, the insoluble fimbrial subunit of Salmonella thin aggregative fimbriae. *J. Mol. Biol.* **1999**, *290*, 741–756. [CrossRef]
56. Shewmaker, F.; McGlinchey, R.P.; Thurber, K.R.; McPhie, P.; Dyda, F.; Tycko, R.; Wickner, R.B. The functional curli amyloid is not based on in-register parallel β-sheet structure. *J. Biol. Chem.* **2009**, *284*, 25065–25076. [CrossRef] [PubMed]
57. Wang, X.; Zhou, Y.; Ren, J.-J.; Hammer, N.D.; Chapman, M.R. Gatekeeper residues in the major curlin subunit modulate bacterial amyloid fi ber biogenesis. *Proc. Natl. Acad. Sci. USA* **2010**, *107*, 163–168. [CrossRef] [PubMed]
58. Otzen, D.E.; Kristensen, P.; Oliveberg, M. Designed protein tetramer zipped together with an Alzheimer sequence: A structural clue to amyloid assembly. *Proc. Natl. Acad. Sci. USA* **2000**, *97*, 9907–9912. [CrossRef]
59. Wang, X.; Hammer, N.D.; Chapman, M.R. The molecular basis of functional bacterial amyloid polymerization and nucleation. *J. Biol. Chem.* **2008**, *283*, 21530–21539. [CrossRef]
60. Christensen, L.F.B.; Nowak, J.S.; Sonderby, T.V.; Frank, S.A.; Otzen, D.E. Quantitating denaturation by formic acid: Imperfect repeats are essential to the stability of the functional amyloid protein FapC. *J. Biol. Chem.* **2020**, *295*, 13031–13046. [CrossRef]
61. Rasmussen, C.B.; Christiansen, G.; Vad, B.S.; Lynggaard, C.; Enghild, J.J.; Andreasen, M.; Otzen, D. Imperfect repeats in the functional amyloid protein FapC reduce the tendency to fragment during fibrillation. *Protein Sci.* **2019**, *28*, 633–642. [CrossRef]
62. Riddick, J.; Bunger, W.; Sakano, T. *Techniques of Chemistry, Organic Solvents*; John Wiley and Sons: New York, NY, USA, 1985; Volume 2.
63. Sonderby, T.V.; Rasmussen, H.O.; Frank, S.A.; Skov Pedersen, J.; Otzen, D.E. Folding Steps in the Fibrillation of Functional Amyloid: Denaturant Sensitivity Reveals Common Features in Nucleation and Elongation. *J. Mol. Biol.* **2021**, *434*, 167337. [CrossRef]
64. Christensen, L.F.B.; Jensen, K.F.; Nielsen, J.; Vad, B.S.; Christiansen, G.; Otzen, D.E. Reducing the Amyloidogenicity of Functional Amyloid Protein FapC Increases Its Ability To Inhibit α-Synuclein Fibrillation. *ACS Omega* **2019**, *4*, 4029–4039. [CrossRef]
65. Jin, F.; Liu, Z. Inherent relationships among different biophysical prediction methods for intrinsically disordered proteins. *Biophys. J.* **2013**, *104*, 488–495. [CrossRef] [PubMed]
66. Andreasen, M.; Meisl, G.; Taylor, J.D.; Michaels, T.C.T.; Otzen, D.E.; Chapman, M.; Dobson, C.M.; Matthews, S.; Knowles, T.P.J. Physical determinants of amyloid assembly in biofilm formation. *mBio* **2019**, *10*, e02279-18. [CrossRef] [PubMed]
67. Caughey, B.; Lansbury Jr, P.T. Protofibrils, pores, fibrils, and neurodegeneration: Separating the responsible protein aggregates from the innocent bystanders. *Annu. Rev. Neurosci.* **2003**, *26*, 267–298. [CrossRef] [PubMed]
68. Verma, M.; Vats, A.; Taneja, V. Toxic species in amyloid disorders: Oligomers or mature fibrils. *Ann. Indian Acad. Neurol.* **2015**, *18*, 138.
69. Wang, H.; Shu, Q.; Rempel, D.L.; Frieden, C.; Gross, M.L. Understanding curli amyloid-protein aggregation by hydrogen–deuterium exchange and mass spectrometry. *Int. J. Mass Spectrom.* **2017**, *420*, 16–23. [CrossRef]
70. Sleutel, M.; Broeck, I.V.D.; Gerven, N.V.; Feuillie, C.; Jonckheere, W.; Valotteau, C.; Dufrêne, Y.F.; Remaut, H. Nucleation and growth of bacterial functional amyloid at single-fiber resolution. *Nat. Chem. Biol.* **2017**, *13*, 902–908. [CrossRef]
71. Dueholm, M.S.; Nielsen, S.B.; Hein, K.L.; Nissen, P.; Chapman, M.; Christiansen, G.; Nielsen, P.H.; Otzen, D.E. Fibrillation of the major curli subunit CsgA under a wide range of conditions implies a robust design of aggregation. *Biochemistry* **2011**, *50*, 8281–8290. [CrossRef] [PubMed]
72. Sewell, L.; Stylianou, F.; Xu, Y.; Taylor, J.; Sefer, L.; Matthews, S. NMR insights into the pre-amyloid ensemble and secretion targeting of the curli subunit CsgA. *Sci. Rep.* **2020**, *10*, 7896. [CrossRef]
73. Tian, P.; Lindorff-Larsen, K.; Boomsma, W.; Jensen, M.H.; Otzen, D.E. A Monte Carlo Study of the Early Steps of Functional Amyloid Formation. *PLoS ONE* **2016**, *11*, e0146096. [CrossRef]
74. Rasmussen, H.Ø.; Otzen, D.E.; Pedersen, J.S. A multimethod approach for analyzing FapC fibrillation and determining mass per length. *Biophys. J.* **2021**, *120*, 2262–2275. [CrossRef]

75. Kauzmann, W. Some factors in the interpretation of protein denaturation. In *Advances in Protein Chemistry*; Elsevier: Amsterdam, The Netherlands, 1959; Volume 14, pp. 1–63.
76. Canchi, D.R.; García, A.E. Cosolvent effects on protein stability. *Annu. Rev. Phys. Chem.* **2013**, *64*, 273–293. [CrossRef] [PubMed]
77. Jackson, S.E.; Fersht, A.R. Folding of chymotrypsin inhibitor 2.1. Evidence for a two-state transition. *Biochemistry* **1991**, *30*, 10428–10435. [CrossRef] [PubMed]
78. Fersht, A. *Structure and Mechanism in Protein Science: A Guide to Enzyme Catalysis and Protein Folding*; Macmillan: New York, NY, USA, 1999.
79. Kim, J.R.; Muresan, A.; Lee, K.Y.C.; Murphy, R.M. Urea modulation of β-amyloid fibril growth: Experimental studies and kinetic models. *Protein Sci.* **2004**, *13*, 2888–2898. [CrossRef] [PubMed]
80. Tiiman, A.; Krishtal, J.; Palumaa, P.; Tõugu, V. In vitro fibrillization of Alzheimer's amyloid-β peptide (1-42). *AIP Adv.* **2015**, *5*, 092401. [CrossRef]
81. Knowles, T.P.; Shu, W.; Devlin, G.L.; Meehan, S.; Auer, S.; Dobson, C.M.; Welland, M.E. Kinetics and thermodynamics of amyloid formation from direct measurements of fluctuations in fibril mass. *Proc. Natl. Acad. Sci. USA* **2007**, *104*, 10016–10021. [CrossRef] [PubMed]
82. Xu, Y.; Knapp, K.; Le, K.N.; Schafer, N.P.; Safari, M.S.; Davtyan, A.; Wolynes, P.G.; Vekilov, P.G. Frustrated peptide chains at the fibril tip control the kinetics of growth of amyloid-beta fibrils. *Proc. Natl. Acad. Sci. USA* **2021**, *118*, e2110995118. [CrossRef] [PubMed]
83. Otzen, D.E.; Oliveberg, M. Salt-induced detour through compact regions of the protein folding landscape. *Proc. Nat. Acad. Sci. USA* **1999**, *96*, 11746–11751. [CrossRef]
84. Baldwin, R. On-pathway versus off-pathway folding intermediates. *Fold. Des.* **1996**, *1*, R1–R8. [CrossRef]
85. Nagaraj, M.; Najarzadeh, Z.; Pansieri, J.; Biverstål, H.; Musteikyte, G.; Smirnovas, V.; Matthews, S.; Emanuelsson, C.; Johansson, J.; Buxbaum, J.N.; et al. Chaperones mainly suppress primary nucleation during formation of functional amyloid required for bacterial biofilm formation. *Chem. Sci.* **2022**, *13*, 536–553. [CrossRef]
86. Chorell, E.; Andersson, E.; Evans, M.L.; Jain, N.; Götheson, A.; Åden, J.; Chapman, M.R.; Almqvist, F.; Wittung-Stafshede, P. Bacterial Chaperones CsgE and CsgC Differentially Modulate Human α-Synuclein Amyloid Formation via Transient Contacts. *PLoS ONE* **2015**, *10*, e0140194. [CrossRef]
87. Armiento, V.; Spanopoulou, A.; Kapurniotu, A. Peptide-Based Molecular Strategies To Interfere with Protein Misfolding, Aggregation, and Cell Degeneration. *Angew. Chem. Int. Ed.* **2020**, *59*, 3372–3384. [CrossRef] [PubMed]
88. El-Agnaf, O.M.A.; Paleologou, K.E.; Greer, B.; Abogrein, A.M.; King, J.E.; Salem, S.A.; Fullwood, N.J.; Benson, F.E.; Hewitt, R.; Ford, K.J.; et al. A strategy for designing inhibitors of α-synuclein aggregation and toxicity as a novel treatment for Parkinson's disease and related disorders. *FASEB J.* **2004**, *18*, 1315–1317. [CrossRef] [PubMed]
89. Tjernberg, L.O.; Näslundt, J.; Lindqvist, F.; Johansson, J.; Karlström, A.R.; Thyberg, J.; Tereniust, L.; Nordstedt, C. Arrest of β-amyloid fibril formation by a pentapeptide ligand. *J. Biol. Chem.* **1996**, *271*, 8545–8548. [CrossRef] [PubMed]
90. Yang, A.; Wang, C.; Song, B.; Zhang, W.; Guo, Y.; Yang, R.; Nie, G.; Yang, Y.; Wang, C. Attenuation of β-Amyloid Toxicity In Vitro and In Vivo by Accelerated Aggregation. *Neurosci. Bull.* **2017**, *33*, 405–412. [CrossRef] [PubMed]
91. Beerten, J.; Schymkowitz, J.; Rousseau, F. Aggregation prone regions and gatekeeping residues in protein sequences. *Curr. Top. Med. Chem.* **2012**, *12*, 2470–2478. [CrossRef]
92. Bednarska, N.G.; van Eldere, J.; Gallardo, R.; Ganesan, A.; Ramakers, M.; Vogel, I.; Baatsen, P.; Staes, A.; Goethals, M.; Hammarström, P.; et al. Protein aggregation as an antibiotic design strategy. *Mol. Microbiol.* **2016**, *99*, 849–865. [CrossRef]
93. Khodaparast, L.; Khodaparast, L.; Gallardo, R.; Louros, N.N.; Michiels, E.; Ramakrishnan, R.; Ramakers, M.; Claes, F.; Young, L.; Shahrooei, M.; et al. Aggregating sequences that occur in many proteins constitute weak spots of bacterial proteostasis. *Nat. Commun.* **2018**, *9*, 866. [CrossRef]
94. Fernandez-Escamilla, A.M.; Rousseau, F.; Schymkowitz, J.; Serrano, L. Prediction of sequence-dependent and mutational effects on the aggregation of peptides and proteins. *Nat. Biotechnol.* **2004**, *22*, 1302–1306. [CrossRef]
95. Sønderby, T.V.; Louros, N.N.; Khodaparast, L.; Khodaparast, L.; Madsen, D.J.; Nagaraj, M.; Strømgaard, K.; Rousseau, F.; Schymkowitz, J.; Otzen, D.E. Sequence-targeted peptides divert Functional Bacterial Amyloid towards destabilized aggregates and reduce biofilm formation. *Biophys. J. in review*.
96. Louros, N.; Konstantoulea, K.; De Vleeschouwer, M.; Ramakers, M.; Schymkowitz, J.; Rousseau, F. WALTZ-DB 2.0: An updated database containing structural information of experimentally determined amyloid-forming peptides. *Nucleic Acids Res.* **2020**, *48*, D389–D393. [CrossRef]
97. Perov, S.; Lidor, O.; Salinas, N.; Golan, N.; Tayeb-Fligelman, E.; Deshmukh, M.; Willbold, D.; Landau, M. Structural Insights into Curli CsgA Cross-beta Fibril Architecture Inspire Repurposing of Anti-amyloid Compounds as Anti-biofilm Agents. *PLoS Pathog.* **2019**, *15*, e1007978. [CrossRef] [PubMed]
98. Dunbar, M.; DeBenedictis, E.; Keten, S. Dimerization energetics of curli fiber subunits CsgA and CsgB. *npj Comput. Mater.* **2019**, *5*, 27. [CrossRef]
99. Sleutel, M.; Pradhan, B.; Remaut, H. Structural analysis of the bacterial amyloid curli. *bioRxiv* **2022**. [CrossRef]
100. Louros, N.; Ramakers, M.; Michiels, E.; Konstantoulea, K.; Morelli, C.; Garcia, T.; Moonen, N.; D'Haeyer, S.; Goossens, V.; Thal, D.R. Mapping the sequence specificity of heterotypic amyloid interactions enables the identification of aggregation modifiers. *Nat. Commun.* **2022**, *13*, 1351. [CrossRef] [PubMed]

101. Louros, N.; Orlando, G.; De Vleeschouwer, M.; Rousseau, F.; Schymkowitz, J. Structure-based machine-guided mapping of amyloid sequence space reveals uncharted sequence clusters with higher solubilities. *Nat. Commun.* **2020**, *11*, 3314. [CrossRef]
102. Hancock, R.E.; Chapple, D.S. Peptide antibiotics. *Antimicrob. Agents Chemother.* **1999**, *43*, 1317–1323. [CrossRef]
103. Mohammad-Beigi, H.; Kjaer, L.; Eskandari, H.; Aliakbari, F.; Christiansen, G.; Ruvo, G.; Ward, J.L.; Otzen, D.E. A possible connection between plant longevity and the absence of protein fibrillation: Basis for identifying aggregation inhibitors in plants. *Front. Plant Sci.* **2019**, *10*, 148. [CrossRef]
104. Mohammad-Beigi, H.; Aliakbari, F.; Sahin, C.; Lomax, C.; Tawfike, A.; Schafer, N.P.; Amiri-Nowdijeh, A.; Eskandari, H.; Møller, I.M.; Hosseini-Mazinani, M.; et al. Oleuropein derivatives from olive fruit extracts reduce—Synuclein fibrillation and oligomer toxicity. *J. Biol. Chem.* **2019**, *294*, 4215–4232. [CrossRef]
105. Bieschke, J.; Russ, J.; Friedrich, R.P.; Ehrnhoefer, D.E.; Wobst, H.; Neugebauer, K.; Wanker, E.E. EGCG remodels mature alpha-synuclein and amyloid-beta fibrils and reduces cellular toxicity. *Proc. Natl. Acad. Sci. USA* **2010**, *107*, 7710–7715. [CrossRef]
106. Youn, K.; Ho, C.-T.; Jun, M. Multifaceted neuroprotective effects of (-)-epigallocatechin-3-gallate (EGCG) in Alzheimer's disease: An overview of pre-clinical studies focused on β-amyloid peptide. *Food Sci. Hum. Wellness* **2022**, *11*, 483–493. [CrossRef]
107. Sahadevan, R.; Singh, S.; Binoy, A.; Sadhukhan, S. Chemico-biological aspects of (-)-epigallocatechin-3-gallate (EGCG) to improve its stability, bioavailability and membrane permeability: Current status and future prospects. *Crit. Rev. Food Sci. Nutr.* **2022**, 1–30. [CrossRef] [PubMed]
108. Fernandes, L.; Cardim-Pires, T.R.; Foguel, D.; Palhano, F.L. Green Tea Polyphenol Epigallocatechin-Gallate in Amyloid Aggregation and Neurodegenerative Diseases. *Front. Neurosci.* **2021**, *15*, 1049. [CrossRef] [PubMed]
109. Serra, D.O.; Mika, F.; Richter, A.M.; Hengge, R. The green tea polyphenol EGCG inhibits E. coli biofilm formation by impairing amyloid curli fibre assembly and downregulating the biofilm regulator CsgD via the σE-dependent sRNA RybB. *Mol. Microbiol.* **2016**, *101*, 136–151. [CrossRef] [PubMed]
110. Widyaningrum, N.; Fudholi, A.; Sudarsono, S.E.; Setyowati, E. Stability of epigallocatechin gallate (EGCG) from green tea (*Camellia sinensis*) and its antibacterial activity against *Staphylococcus epidermidis* ATCC 35984 and *Propionibacterium acnes* ATCC 6919. *Asian J. Biol. Sci* **2015**, *8*, 93–101. [CrossRef]
111. Mereles, D.; Hunstein, W. Epigallocatechin-3-gallate (EGCG) for clinical trials: More pitfalls than promises? *Int. J. Mol. Sci.* **2011**, *12*, 5592–5603. [CrossRef]
112. Hong, J.; Lu, H.; Meng, X.; Ryu, J.-H.; Hara, Y.; Yang, C.S. Stability, cellular uptake, biotransformation, and efflux of tea polyphenol (−)-epigallocatechin-3-gallate in HT-29 human colon adenocarcinoma cells. *Cancer Res.* **2002**, *62*, 7241–7246.
113. Arita-Morioka, K.-i.; Yamanaka, K.; Mizunoe, Y.; Tanaka, Y.; Ogura, T.; Sugimoto, S. Inhibitory effects of Myricetin derivatives on curli-dependent biofilm formation in *Escherichia coli*. *Sci. Rep.* **2018**, *8*, 8452. [CrossRef]
114. Reygaert, W.C. The antimicrobial possibilities of green tea. *Front. Microbiol.* **2014**, *5*, 434. [CrossRef]
115. Zhang, Y.-M.; Rock, C.O. Evaluation of epigallocatechin gallate and related plant polyphenols as inhibitors of the FabG and FabI reductases of bacterial type II fatty-acid synthase. *J. Biol. Chem.* **2004**, *279*, 30994–31001. [CrossRef]
116. Chinnam, N.; Dadi, P.K.; Sabri, S.A.; Ahmad, M.; Kabir, M.A.; Ahmad, Z. Dietary bioflavonoids inhibit *Escherichia coli* ATP synthase in a differential manner. *Int. J. Biol. Macromol.* **2010**, *46*, 478–486. [CrossRef]
117. Najarzadeh, Z.; Mohammad-Beigi, H.; Nedergaard Pedersen, J.; Christiansen, G.; Sønderby, T.V.; Shojaosadati, S.A.; Morshedi, D.; Strømgaard, K.; Meisl, G.; Sutherland, D.; et al. Plant Polyphenols Inhibit Functional Amyloid and Biofilm Formation in *Pseudomonas* Strains by Directing Monomers to Off-Pathway Oligomers. *Biomolecules* **2019**, *9*, 659. [CrossRef] [PubMed]
118. Ehrnhoefer, D.E.; Bieschke, J.; Boeddrich, A.; Herbst, M.; Masino, L.; Lurz, R.; Engemann, S.; Pastore, A.; Wanker, E.E. EGCG redirects amyloidogenic polypeptides into unstructured, off-pathway oligomers. *Nat. Struct. Mol. Biol.* **2008**, *15*, 558–566. [CrossRef] [PubMed]
119. Pruteanu, M.; Hernández Lobato, J.I.; Stach, T.; Hengge, R. Common plant flavonoids prevent the assembly of amyloid curli fibres and can interfere with bacterial biofilm formation. *Environ. Microbiol.* **2020**, *22*, 5280–5299. [CrossRef] [PubMed]
120. Stenvang, M.; Christiansen, G.; Otzen, D.E. Epigallocatechin Gallate Remodels Fibrils of Lattice Corneal Dystrophy Protein, Facilitating Proteolytic Degradation and Preventing Formation of Membrane-Permeabilizing Species. *Biochemistry* **2016**, *55*, 2344–2357. [CrossRef]
121. Najarzadeh, Z.; Pedersen, J.N.; Christiansen, G.; Shojaosadati, S.A.; Pedersen, J.S.; Otzen, D.E. Bacterial amphiphiles as amyloid inducers: Effect of Rhamnolipid and Lipopolysaccharide on FapC fibrillation. *Biochim. Biophys. Acta (BBA)-Proteins Proteom.* **2019**, *1867*, 140263. [CrossRef]
122. Bleem, A.; Christiansen, G.; Madsen, D.J.; Maric, H.; Strømgaard, K.; Bryers, J.D.; Daggett, V.; Meyer, R.L.; Otzen, D.E. Protein engineering reveals mechanisms of functional amyloid formation in Pseudomonas aeruginosa biofilms. *J. Mol. Biol.* **2018**, *430*, 3751–3763. [CrossRef]
123. Hengge, R. Targeting bacterial biofilms by the green tea polyphenol EGCG. *Molecules* **2019**, *24*, 2403. [CrossRef]
124. Eskandari, H.; Ghanadian, M.; Noleto-Dias, C.; Lomax, C.; Tawfike, A.; Christiansen, G.; Sutherland, D.S.; Ward, J.L.; Mohammad-Beigi, H.; Otzen, D.E. Inhibitors of α-synuclein fibrillation and oligomer toxicity in Rosa damascena: The all-pervading powers of flavonoids and phenolic glycosides. *ACS Chem. Neurosci.* **2020**, *11*, 3161–3173. [CrossRef]
125. Young, L.M.; Saunders, J.C.; Mahood, R.A.; Revill, C.H.; Foster, R.J.; Tu, L.-H.; Raleigh, D.P.; Radford, S.E.; Ashcroft, A.E. Screening and classifying small-molecule inhibitors of amyloid formation using ion mobility spectrometry–mass spectrometry. *Nat. Chem.* **2015**, *7*, 73–81. [CrossRef]

126. Doig, A.J.; Derreumaux, P. Inhibition of protein aggregation and amyloid formation by small molecules. *Curr. Opin. Struct. Biol.* **2015**, *30*, 50–56. [CrossRef]
127. Mohammad-Beigi, H.; Morshedi, D.; Shojaosadati, S.A.; Pedersen, J.N.; Marvian, A.T.; Aliakbari, F.; Christiansen, G.; Pedersen, J.S.; Otzen, D.E. Gallic acid loaded onto polyethylenimine-coated human serum albumin nanoparticles (PEI-HSA-GA NPs) stabilizes α-synuclein in the unfolded conformation and inhibits aggregation. *RSC Adv.* **2016**, *6*, 85312–85323. [CrossRef]
128. Aliakbari, F.; Mohammad-Beigi, H.; Abbasi, S.; Rezaei-Ghaleh, N.; Lermyte, F.; Parsafar, S.; Becker, S.; Tafreshi, A.P.; O'Connor, P.B.; Collingwood, J.F. Multiple Protective Roles of Nanoliposome-Incorporated Baicalein against Alpha-Synuclein Aggregates. *Adv. Funct. Mater.* **2021**, *31*, 2007765. [CrossRef]
129. Fernandes, L.; Messias, B.; Pereira-Neves, A.; Azevedo, E.P.; Araújo, J.; Foguel, D.; Palhano, F.L. Green tea polyphenol microparticles based on the oxidative coupling of EGCG inhibit amyloid aggregation/cytotoxicity and serve as a platform for drug delivery. *ACS Biomater. Sci. Eng.* **2020**, *6*, 4414–4423. [CrossRef] [PubMed]
130. Sharma, V.; Ghosh, K.S. Inhibition of amyloid fibrillation by small molecules and nanomaterials: Strategic development of pharmaceuticals against amyloidosis. *Protein Pept. Lett.* **2019**, *26*, 315–323. [CrossRef] [PubMed]
131. Debnath, K.; Sarkar, A.K.; Jana, N.R.; Jana, N.R. Inhibiting Protein Aggregation by Small Molecule-Based Colloidal Nanoparticles. *Acc. Mater. Res.* **2021**, *3*, 54–66. [CrossRef]
132. Liu, L.; Zhang, L.; Mao, X.; Niu, L.; Yang, Y.; Wang, C. Chaperon-Mediated single molecular approach toward modulating aβ peptide aggregation. *Nano Lett.* **2009**, *9*, 4066–4072. [CrossRef]
133. Dorval Courchesne, N.M.; Duraj-Thatte, A.; Tay, P.K.R.; Nguyen, P.Q.; Joshi, N.S. Scalable Production of Genetically Engineered Nanofibrous Macroscopic Materials via Filtration. *ACS Biomater. Sci. Eng.* **2017**, *3*, 733–741. [CrossRef]
134. Deshmukh, M.; Evans, M.L.; Chapman, M.R. *Amyloid by Design: Intrinsic Regulation of Microbial Amyloid Assembly*; Academic Press: London, UK, 2018; Volume 430, pp. 3631–3641.
135. Van Gerven, N.; Goyal, P.; Vandenbussche, G.; De Kerpel, M.; Jonckheere, W.; De Greve, H.; Remaut, H. Secretion and functional display of fusion proteins through the curli biogenesis pathway. *Mol. Microbiol.* **2014**, *91*, 1022–1035. [CrossRef]
136. Nguyen, P.Q.; Botyanszki, Z.; Tay, P.K.R.; Joshi, N.S. Programmable biofilm-based materials from engineered curli nanofibres. *Nat. Commun.* **2014**, *5*, 4945. [CrossRef]
137. Li, Y.; Li, K.; Wang, X.; Cui, M.; Ge, P.; Zhang, J.; Qiu, F.; Zhong, C. Conformable self-assembling amyloid protein coatings with genetically programmable functionality. *Sci. Adv.* **2020**, *6*, eaba1425. [CrossRef]
138. Yang, X.; Li, Z.; Xiao, H.; Wang, N.; Li, Y.; Xu, X.; Chen, Z.; Tan, H.; Li, J. A Universal and Ultrastable Mineralization Coating Bioinspired from Biofilms. *Adv. Funct. Mater.* **2018**, *28*, 1802730. [CrossRef]
139. Zhong, C.; Gurry, T.; Cheng, A.A.; Downey, J.; Deng, Z.; Stultz, C.M.; Lu, T.K. Strong underwater adhesives made by self-assembling multi-protein nanofibres. *Nat. Nanotechnol.* **2014**, *9*, 858–866. [CrossRef] [PubMed]
140. An, B.; Wang, X.; Cui, M.; Gui, X.; Mao, X.; Liu, Y.; Li, K.; Chu, C.; Pu, J.; Ren, S.; et al. Diverse Supramolecular Nanofiber Networks Assembled by Functional Low-Complexity Domains. *ACS Nano* **2017**, *11*, 6985–6995. [CrossRef] [PubMed]
141. Duraj-Thatte, A.M.; Courchesne, N.D.; Praveschotinunt, P.; Rutledge, J.; Lee, Y.; Karp, J.M.; Joshi, N.S. Genetically Programmable Self-Regenerating Bacterial Hydrogels. *Adv. Mater.* **2019**, *31*, e1901826. [CrossRef] [PubMed]
142. Li, Y.; Li, K.; Wang, X.; An, B.; Cui, M.; Pu, J.; Wei, S.; Xue, S.; Ye, H.; Zhao, Y.; et al. Patterned Amyloid Materials Integrating Robustness and Genetically Programmable Functionality. *Nano Lett.* **2019**, *19*, 8399–8408. [CrossRef]
143. DeBenedictis, E.P.; Liu, J.; Keten, S. Adhesion mechanisms of curli subunit CsgA to abiotic surfaces. *Sci. Adv.* **2016**, *2*, e1600998. [CrossRef]
144. So, C.R.; Hayamizu, Y.; Yazici, H.; Gresswell, C.; Khatayevich, D.; Tamerler, C.; Sarikaya, M. Controlling Self-Assembly of Engineered Peptides on Graphite by Rational Mutation. *ACS Nano* **2012**, *6*, 1648–1656. [CrossRef]
145. Walsh, T.R.; Knecht, M.R. Biomolecular Material Recognition in Two Dimensions: Peptide Binding to Graphene, h-BN, and MoS2 Nanosheets as Unique Bioconjugates. *Bioconjug. Chem.* **2019**, *30*, 2727–2750. [CrossRef]
146. Lipson, H.S.; Stokes, A. The structure of graphite. *Proc. R. Soc. Lond. Ser. A Math. Phys. Sci.* **1942**, *181*, 101–105.
147. Oliveira, S.F.; Bisker, G.; Bakh, N.A.; Gibbs, S.L.; Landry, M.P.; Strano, M.S. Protein functionalized carbon nanomaterials for biomedical applications. *Carbon* **2015**, *95*, 767–779. [CrossRef]
148. Surovtseva, D.; Crossin, E.; Pell, R.; Stamford, L. Toward a life cycle inventory for graphite production. *J. Ind. Ecol.* **2022**, *26*, 964–976. [CrossRef]
149. Li, C.; Mezzenga, R. The interplay between carbon nanomaterials and amyloid fibrils in bio-nanotechnology. *Nanoscale* **2013**, *5*, 6207–6218. [CrossRef]
150. Kim, J.E.; Lee, M. Fullerene inhibits β-amyloid peptide aggregation. *Biochem. Biophys. Res. Commun.* **2003**, *303*, 576–579. [CrossRef]
151. Mao, X.; Wang, Y.; Liu, L.; Niu, L.; Yang, Y.; Wang, C. Molecular-level evidence of the surface-induced transformation of peptide structures revealed by scanning tunneling microscopy. *Langmuir* **2009**, *25*, 8849–8853. [CrossRef] [PubMed]
152. Zhang, M.; Mao, X.; Wang, C.; Zeng, W.; Zhang, C.; Li, Z.; Fang, Y.; Yang, Y.; Liang, W.; Wang, C. The effect of graphene oxide on conformation change, aggregation and cytotoxicity of HIV-1 regulatory protein (Vpr). *Biomaterials* **2013**, *34*, 1383–1390. [CrossRef]
153. Mao, X.; Ma, X.; Liu, L.; Niu, L.; Yang, Y.; Wang, C. Structural characteristics of the beta-sheet-like human and rat islet amyloid polypeptides as determined by scanning tunneling microscopy. *J. Struct. Biol.* **2009**, *167*, 209–215. [CrossRef]

154. Brown, C.L.; Aksay, I.A.; Saville, D.A.; Hecht, M.H. Template-Directed Assembly of a de Novo Designed Protein. *J. Am. Chem. Soc.* **2002**, *124*, 6846–6848. [CrossRef]
155. Wang, Z.; Zhou, C.; Wang, C.; Wan, L.; Fang, X.; Bai, C. AFM and STM study of β-amyloid aggregation on graphite. *Ultramicroscopy* **2003**, *97*, 73–79. [CrossRef]
156. Valle-Delgado, J.J.; Liepina, I.; Lapidus, D.; Sabaté, R.; Ventura, S.; Samitier, J.; Fernàndez-Busquets, X. Self-assembly of human amylin-derived peptides studied by atomic force microscopy and single molecule force spectroscopy. *Soft Matter* **2012**, *8*, 1234–1242. [CrossRef]
157. Sønderby, T.V.; Zou, Y.; Wang, P.; Wang, C.; Otzen, D.E. Molecular-level insights into the surface-induced assembly of Functional Bacterial Amyloid. *Biophys. J. in review*.
158. Liu, L.; Zhang, L.; Niu, L.; Xu, M.; Mao, X.; Yang, Y.; Wang, C. Observation of reduced cytotoxicity of aggregated amyloidogenic peptides with chaperone-like molecules. *ACS Nano* **2011**, *5*, 6001–6007. [CrossRef] [PubMed]

molecules

MDPI

Review

Supersaturation-Dependent Formation of Amyloid Fibrils †

Yuji Goto [1,*], Masahiro Noji [2], Kichitaro Nakajima [1] and Keiichi Yamaguchi [1]

1 Global Center for Medical Engineering and Informatics, Osaka University, 2-1 Yamadaoka, Suita 565-0871, Japan
2 Graduate School of Human and Environmental Studies, Kyoto University, Kyoto Sakyo-ku, Kyoto 606-8316, Japan
* Correspondence: gtyj8126@protein.osaka-u.ac.jp

† Dedicated to the memory of Sir Chris Dobson, FRS, who made protein aggregation and amyloid fibril formation one of the most important fields of protein science. On a personal level, YG recalls a series of exciting discussions with Chris on amyloid fibrils, in particular, the role of supersaturation.

Abstract: The supersaturation of a solution refers to a non-equilibrium phase in which the solution is trapped in a soluble state, even though the solute's concentration is greater than its thermodynamic solubility. Upon breaking supersaturation, crystals form and the concentration of the solute decreases to its thermodynamic solubility. Soon after the discovery of the prion phenomena, it was recognized that prion disease transmission and propagation share some similarities with the process of crystallization. Subsequent studies exploring the structural and functional association between amyloid fibrils and amyloidoses solidified this paradigm. However, recent studies have not necessarily focused on supersaturation, possibly because of marked advancements in structural studies clarifying the atomic structures of amyloid fibrils. On the other hand, there is increasing evidence that supersaturation plays a critical role in the formation of amyloid fibrils and the onset of amyloidosis. Here, we review the recent evidence that supersaturation plays a role in linking unfolding/folding and amyloid fibril formation. We also introduce the HANABI (HANdai Amyloid Burst Inducer) system, which enables high-throughput analysis of amyloid fibril formation by the ultrasonication-triggered breakdown of supersaturation. In addition to structural studies, studies based on solubility and supersaturation are essential both to developing a comprehensive understanding of amyloid fibrils and their roles in amyloidosis, and to developing therapeutic strategies.

Keywords: amyloid fibrils; amorphous aggregation; amyloid β; β2-microglobulin; protein misfolding; solubility; supersaturation; ultrasonication

Citation: Goto, Y.; Noji, M.; Nakajima, K.; Yamaguchi, K. Supersaturation-Dependent Formation of Amyloid Fibrils. *Molecules* **2022**, *27*, 4588. https://doi.org/10.3390/molecules27144588

Academic Editor: Angelo Facchiano

Received: 14 May 2022
Accepted: 12 July 2022
Published: 19 July 2022

Publisher's Note: MDPI stays neutral with regard to jurisdictional claims in published maps and institutional affiliations.

1. Supersaturation

Supersaturation is a fundamental natural principle, determining the phases (i.e., vapor, liquid, and solid) of substances [1–6]. Supersaturation of a solution refers to a non-equilibrium phase in which, although the solute's concentration is higher than its thermodynamic solubility, the solute molecules remain soluble for an extended period because of a high free-energy barrier to nucleation. Upon nucleation and crystal formation, the concentration of the solute decreases to its thermodynamic solubility. The supersaturation phenomenon is common in the temperature-dependent liquid–solid or vapor–liquid phase transitions of substances, with one well-known example being the super-cooling of water prior to ice formation. In general, supersaturation is required for the formation of crystals [2]. Supersaturation has been shown to play a role in other ordered protein associations, such as in the fiber formation of hemoglobin S, the molecular basis of sickle cell anemia [7,8]. Supersaturation also plays a role in various types of lithiasis, including urolithiasis, cholelithiasis, and gout, where crystals of calcium oxalate, cholesterol, and uric acid are formed, respectively [9].

A general phase diagram of protein solubility dependent on the precipitant concentration illustrates the role of supersaturation (Figure 1) [3,5,10–12]. The phase diagram consists of a soluble region (Region I), a metastable region (Region II), a labile region (Region III), and an amorphous region (Region IV). Below solubility (Region I), monomers are thermodynamically stable. In the metastable region (Region II), supersaturation is thought to persist in the absence of seeding or agitation. In the labile region (Region III), spontaneous nucleation occurs after a certain lag time. Finally, the amorphous region (Region IV) is dominated by amorphous aggregation occurring without a lag time, possibly by the concomitant formation of many nuclei. It is important to note that, after crystallization or amorphous aggregation, the soluble protein concentration becomes equal to the solubility.

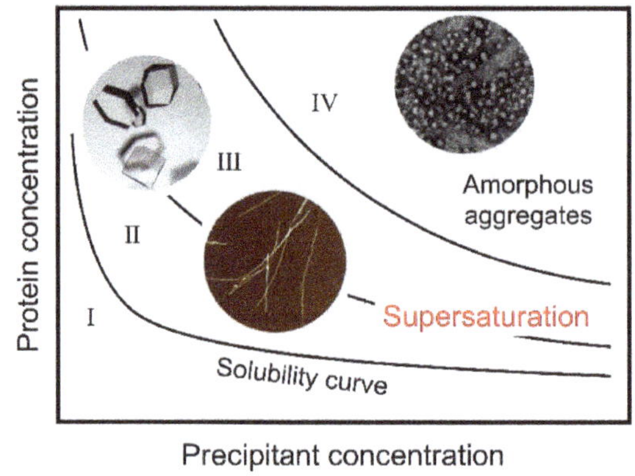

Figure 1. Protein and precipitant concentration-dependent phase diagram common to native crystals and aggregates of denatured proteins. Regions I, II, III, and IV represent undersaturation, the metastable region, the labile region, and the amorphous region, respectively. Crystallization and amyloid fibril formation occur from regions II and III. The figure was modified from So et al. [6], with permission. Copyright 2016 Elsevier.

Supersaturation is quantified in several ways; often used are the supersaturation ratio (S) and the degree of supersaturation (σ) [2,5]:

$$S = [C]/[C]_C \quad (1)$$

$$\sigma = ([C] - [C]_C)/[C]_C \quad (2)$$

where [C] and $[C]_C$ are the initial solute's concentration and thermodynamic solubility, respectively. S and σ are 1 and 0, respectively, at the boundary between Regions I and II, and they increase in Region III with an increase in precipitant concentration (i.e., the driving force of precipitation). In Region IV, amorphous aggregation without a lag time makes evaluations of S or σ difficult. The driving force of nucleation for crystallization is assumed to be proportional to lnS [2]. It is also assumed that the nucleation rate is proportional to the inverse of lag time [13–15].

The physicochemical mechanisms underlying supersaturation have been studied extensively. One of the classical mechanisms underlying supersaturation is the difficulty of nucleation, as modeled for actin polymerization by Oosawa and Kasai [16]; another is classical nucleation theory [17]. However, subsequent studies suggest a more complicated mechanism of supersaturation, in which solutes form a kinetically trapped state, which is located on the distinct pathway(s) to the formation of crystals. Using molecular dynamics simulations, Matsumoto et al. [18,19] showed that the microscopic homochiral domains, which are more energetically stable than average but cannot grow into macroscopic ice structures due to geometrical frustrations, are the major constituent of supercooled liquid water. Studies with small-angle X-ray scattering [20,21], small-angle neutron scattering [22], and dynamic light scattering [23,24] have suggested that equilibrium clusters or networks of solutes with increased stability play a role in maintaining supersaturation. Matsushita et al. [25,26] studied the solution structures of supersaturated sodium acetate and hen egg white lysozyme (HEWL) with diffracted X-ray tracking, and showed that nanoscale networks with increased dynamics and stability are responsible for supersaturation phenomena. These studies generated a new view of supersaturation: it is a kinetically trapped "stable" state, rather than a state before the high barrier of nucleation, as was suggested by the classical actin polymerization model [16].

On other hand, recent work on crystal nucleation focuses on highly concentrated disordered droplets observed before crystallization [4,5,27,28]. More recently, Yamazaki et al. [29] used time-resolved liquid-cell transmission electron microscopy (TEM) to perform an in situ observation of the nucleation of HEWL native crystals. Their TEM images revealed that amorphous solid particles act as heterogeneous nucleation sites. Their findings represent a significant departure from the two-step nucleation and growth mechanism, and suggest that amorphous particle-dependent heterogeneous nucleation is the dominant mechanism of spontaneous crystallization under supersaturation. Amorphous aggregate-assisted amyloid formation was suggested under conditions where oligomer formation is a rare event [30]. We also observed the amorphous oligomeric aggregate-dependent acceleration of HEWL amyloid formation under supersaturation [31]. Taken together, these studies demonstrate that, although supersaturation is an evident phenomenon, there is still uncertainty as to the physicochemical mechanisms that underlie the development, retention, and breakdown of supersaturation, as well as the role of supersaturation in crystal nucleation.

2. Amyloid Fibril Formation

2.1. Similarity with Crystal Formation

As reviewed previously [6], soon after the discovery of prion phenomena such as kuru and scrapie diseases, it was recognized that prion disease transmission and propagation share some similarities with the process of crystallization [32]. In other words, supersaturation was considered to be an important factor underlying prion diseases. Subsequent studies exploring the structural and functional association between amyloid fibrils and amyloidosis solidified this developing paradigm [6,33,34]. The key pieces of evidence supporting this link were that: (i) spontaneous amyloid formation occurs by a nucleation and growth mechanism, requiring a long lag time, (ii) seeding is an efficient approach for accelerating amyloid formation by escaping the high energy barrier associated with nucleation, and (iii) amorphous aggregates predominate under solute conditions well above solubility.

A useful literary analogy that captures the similarities between amyloid fibril formation and crystallization has been [35] drawn from Kurt Vonnegut's novel *Cat's Cradle* [36]. In this fictional work, ice 9, a high temperature ice nucleus (stable at 45.8 °C) is created to help troops escape from mud; however, it is revealed to be an Armageddon device: once a small block of ice 9 is exposed to water, it starts a world-wide water 'freeze' due to seed-dependent propagation.

However, recent studies have not necessarily focused on supersaturation, possibly because of marked advancements in structural studies clarifying the atomic structures of amyloid fibrils based on the X-ray crystallographic, solid-state NMR, and cryoEM approaches [37–42]. Now, the mechanisms of amyloid formation, distinct disease phenotypes, and familial amyloidosis are discussed, based on the atomic structures of amyloid fibrils, their polymorphs, and the effects of mutations on amyloid structures.

On the other hand, there is increasing evidence that supersaturation plays a critical role in the onset of amyloidosis. Investigations into the temporal evolution of major biomarkers of Alzheimer's disease [43,44] in the onset and progression of clinical symptoms have shown that concentrations of amyloid β(1-42) peptide (Aβ(1-42)) in the cerebrospinal fluid (CSF) decline 25 and 15 years before expected symptom onset and Aβ(1-42) deposition, respectively. Similar decreases in the concentration of α-synuclein (αSN) were reported for Parkinson's disease [45–49]. Here, the supersaturation hypothesis provides a simple physical explanation as to why CSF Aβ(1-42) or αSN decreases concomitantly with the deposition of amyloid fibrils [6,50]. The increase in the Aβ(1-40) / Aβ(1–42) ratio for plasma of individuals with Alzheimer's disease [51] might also be explained by the supersaturation hypothesis: the breakdown of supersaturation decreases the soluble concentration of Aβ(1-42) more than that of Aβ(1-40). Moreover, the effects of polyphosphates [52,53], strong acids [54], the isoelectric point [55], or other additives [15,56] indicate that various conditions enhance the driving force of amyloid formation, and that the breakdown of supersaturation is essential to trigger this reaction. Vendruscolo and colleagues addressed the role of supersaturation from a different viewpoint [57,58], stating that the proteins most at risk for aggregation within the cell are those with high expression levels with respect to their solubilities. To achieve a comprehensive understanding of amyloid fibrils and to develop therapeutic strategies, studies based on solubility and supersaturation are essential. This article considers the progress made since the review article published in 2016 [6] on the role of supersaturation in amyloid fibril formation.

2.2. *Supersaturation-Veiled Amyloid Formation Revealed by Heating under Agitation*

β2-microglobulin (β2m), a globular β-barrel protein with 99 amino acid residues and an immunoglobulin fold, is responsible for dialysis-related amyloidosis [59–61]; it is one of the most extensively studied amyloidogenic proteins, because it allows for the detailed study both of protein unfolding/folding and of amyloid fibril formation [39,62–71]. Although amyloid deposits of β2m in dialysis patients are observed at a neutral pH, amyloid formation in vitro has been difficult to detect at a neutral pH because of its resistant native structure [11,62,72]. To further understand the mechanism of amyloid formation in vivo, Noji et al. [12] investigated the association between protein folding/unfolding and misfolding leading to amyloid formation (Figure 2).

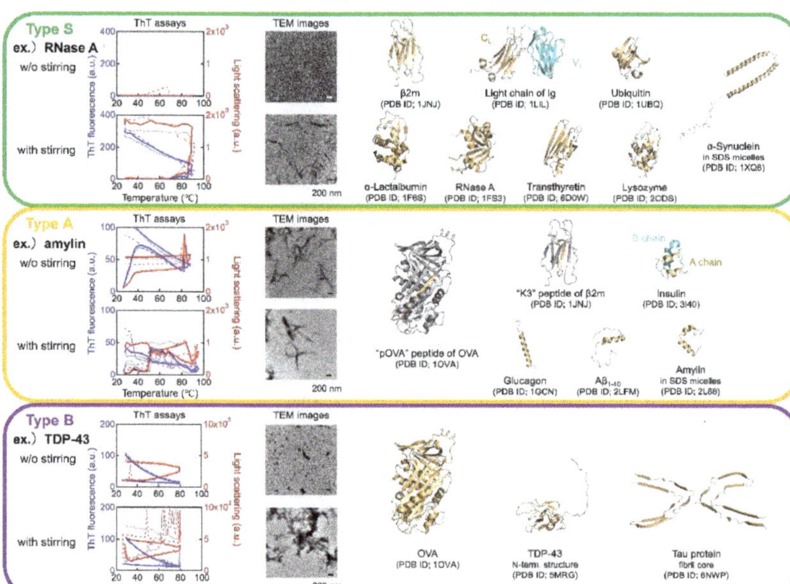

Figure 2. Heating- and agitation-dependent amyloid formation of proteins. According to their aggregation behavior, Type S, A, and B proteins were defined. **Left**: ThT assays upon heating in the presence (upper) or absence (lower) of stirring. The intensities of ThT fluorescence and LS are indicated by blue and red lines, respectively. $n = 3$. Middle: TEM images of the samples after heating in the presence (upper) or absence (lower) of stirring. **Right**: Structures of proteins used with their names and pdb codes. The figure was created based on Noji et al. [2].

The researchers examined the heat denaturation of β2m with or without stirrer agitation; they also monitored amyloid formation via the amyloid-specific thioflavin T (ThT) fluorescence, and the total amount of aggregates via light scattering (LS). They found that β2m efficiently forms amyloid fibrils even at a neutral pH by heating under agitation. They constructed temperature- and NaCl concentration-dependent conformational phase diagrams in the presence or absence of agitation (Figure 3), illustrating how amyloid formation under neutral pH conditions is related to thermal unfolding and the breakdown of supersaturation.

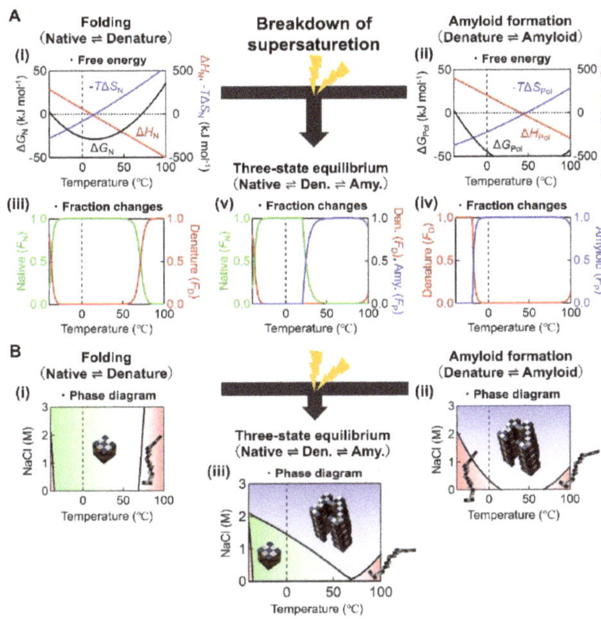

Figure 3. Temperature- and NaCl concentration-dependent conformational phase diagrams of β2m before and after the linkage of folding and misfolding transitions. (**A**) Temperature dependencies of thermodynamic parameters ($\Delta G(T)$, $\Delta H(T)$, and $T\Delta S(T)$) for folding (Mechanism 1, panel (i)) and amyloid formation (Mechanism 2, panel (ii)). Fractions of N, D, and P states for folding (panel (iii)), amyloid formation (panel (iv)), and their linked conditions (panel (v)) are also shown. The plots were made using 0.1 mg/mL β2m, 1.0 M NaCl, and pH 7.0. (**B**) Phase diagrams for folding/unfolding (Mechanism 1, panel (i)), amyloid misfolding (Mechanism 2, panel (ii)), and their linked conditions (panel (iii)). Lines show the simulated phase boundaries at 0.1 mg/mL β2m and pH 7.0. The figure was modified from Noji et al. [12].

Before the breakdown of supersaturation, a "protein concentration-independent" two-state mechanism applies.

$$\mathrm{D} \rightleftharpoons \mathrm{N}\ (\text{Mechanism 1}),$$

The equilibrium constant (K_N) and Gibbs free energy change of folding (ΔG_N) are given by Equations (3) and (4), respectively.

$$K_N = [\mathrm{N}]/[\mathrm{D}] \tag{3}$$

$$\Delta G_N = -RT \ln K_N \tag{4}$$

Here, R and T are the gas constant and the temperature T in Kelvin, respectively. Consideration of the temperature dependences of the thermodynamic parameters (i.e., ΔG_N, enthalpy change (ΔH_N), and entropy change (ΔS_N)) leads to the temperature-dependent changes in fractions of [N] and [D] (Figure 3, see [12] in detail). The combined effects of entropy and enthalpy terms lead to heat- and cold-denaturations [73], although cold-denaturation does not occur for β2m above 0 °C.

Although the detailed mechanisms of amyloid formation remain elusive [14], a simplified model (Mechanism 2) is valid for describing the equilibrium between monomers (D) and polymeric amyloid fibrils (P) [33,34,62,64]:

$$\mathrm{P} + \mathrm{D} \rightleftharpoons \mathrm{P} \ \text{(Mechanism 2)}$$

The elongation of fibrils is defined by the equilibrium association constant (K_{Pol}) as:

$$K_{Pol} = \frac{[P]}{[P][D]} \tag{5}$$

The equilibrium is independent of the molar concentration of amyloid fibrils, [P]. Hence, the equilibrium monomer concentration $[D]_C$ is:

$$[D]_C = \frac{1}{K_{Pol}} \tag{6}$$

$[D]_C$ is referred to as the "critical concentration" [33,34,64,74] because amyloid fibrils form when the concentration of monomers exceeds $[D]_C$. It is noted that $[D]_C$ corresponds to $[C]_C$ in Equations [1] and [2]. By determining $[D]_C$, the apparent free energy change of amyloid formation (ΔG_{Pol}) is obtained by:

$$\Delta G_{Pol} = -RT \ln K_{Pol} = RT \ln[D]_C \tag{7}$$

Assuming that the Gibbs free energy equations (i.e., $\Delta G_{Pol}(T)$, $\Delta H_{Pol}(T)$, and $\Delta S_{Pol}(T)$) also hold true for polymeric amyloid fibrils [12,64,74], and that the heat capacity change upon amyloid formation is the same as that of protein folding, temperature-dependent changes of ΔG_N and fractions of [N], [D], and [P] can be obtained (Figure 3, see [12] for detail). As is the case for the native state, heat- and cold-denaturations of amyloid fibrils are expected.

Upon the breakdown of supersaturation, a three-state mechanism between the native, unfolded, and "protein concentration-dependent" amyloid states determines the overall equilibrium. The transition from the two-state mechanism to the three-state mechanism shifts the overall equilibrium in the direction of amyloid fibrils, apparently destabilizing the native state by the law of mass action (Figure 3B). The results suggest that heating and agitation play important roles in the onset of amyloidosis.

One of the most important findings of a series of studies by Goto and colleagues is that supersaturation-dependent amyloid formation is comprehensively and exactly understood by combining the unfolding/refolding conformational transition under supersaturation, and amyloid formation after the breakdown of supersaturation (Figure 3). As long as we consider monomeric proteins, the former is independent of protein concentration, while the latter is critically dependent on protein concentration (i.e., solubility). Although many studies have addressed the relationship between protein folding and misfolding and schematic folding funnels or energy landscapes [75–77], the exact unification of these two processes, which is in fact simple, has not previously been presented.

2.3. Generality of the Supersaturation-Limited Amyloid Formation

To address the generality of the link between reversible unfolding/refolding under supersaturation and amyloid formation after the breakdown of supersaturation as revealed by β2m, Noji et al. [78] examined the heat denaturation of various proteins with or without stirrer agitation. The study included not only typical amyloidogenic proteins, but also several textbook proteins used previously in folding/unfolding studies (Figure 2).

According to their aggregation behavior, three types of proteins can be defined.

Type S proteins: The first type of protein shows a strict dependence on agitation for amyloid formation at high temperatures. Noji et al. [78] call this transition the "strictly supersaturation-dependent transition" or "S transition". Proteins exhibiting S transitions include those with a native conformation (β2m, variable (V_L) and constant (C_L) domains of antibody light chain, HEWL, and ribonuclease A (RNaseA)) and αSyn. Most interestingly, wild-type RNaseA forms amyloid fibrils upon heating under stirring. The hinge-loop-expanded mutant of RNaseA was reported to generate amyloid-like fibrils via 3D domain swapping, whereas the wild-type RNaseA did not [79,80]. Transthyretin (TTR) at pH 2.0 and 0.1 M NaCl can also be included in Type S, although TTR at a neutral pH tends to form amorphous aggregates [81].

Type A proteins: The second type of protein exhibits spontaneous amyloid formation at high temperatures even without agitation. Noji et al. [78] refer to this type of transition as "autonomous amyloid-forming transition" or "A transition". Type A proteins include insulin, glucagon, islet amyloid polypeptide (IAPP), and Aβ(1-40). In other words, the high amyloidogenicity of these relatively short amyloid peptides does not exhibit intrinsic barriers preventing amyloid formation.

Type B proteins: The third type of protein often produces amorphous aggregates at high temperatures without a lag phase. Noji et al. [78] call this transition the "boiled egg-like transition" or "B transition". Type B proteins (i.e., TDP-43, tau, and ovalbumin (OVA) [82]) are relatively large, and it is possible that their overall amorphous characteristics include amyloid cores (β-spines), producing a "fuzzy coat" morphology [83].

Noji et al. [78] showed that these three types (S, A, and B) of transitions with distinct responses to heating can be located on a general aggregation phase diagram, based on the driving forces of precipitation and protein solubility [12,84,85] (Figure 4A). The S, A, and B transitions are indicated by green, orange, and purple arrows, respectively. This type of diagram is often used to illustrate the crystallization and amorphous precipitation of native proteins and, moreover, for solutes in general (Figure 1) [10].

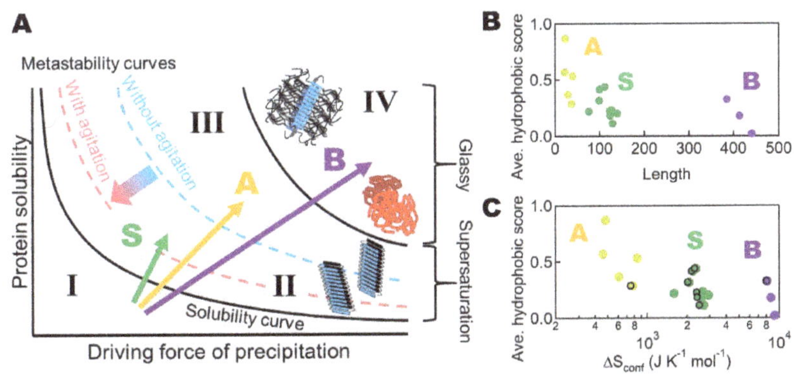

Figure 4. General schematic conformational phase diagram and the three transitions. Three types of amyloidogenic proteins were plotted on a general phase diagram of aggregation (**A**), and diagrams of average hydrophobicity vs. number of amino acid residues (**B**) or ΔS_{conf} (**C**). In c, ΔS_{conf} represents an increase upon denaturation of the main chain with (points within the frame) and without (points outside the frame) the contribution of disulfide bonds. The figure was reproduced from Noji et al. [78].

Thus, the S, A, and B transitions represent those from below solubility (Region I) to the metastable region (Region II), the labile region (Region III), and the amorphous region (Region IV), respectively. These transitions indicate that, with an increase in the driving force of precipitation at high temperatures, the aggregation behavior followed exactly as expected for solutes in general [10].

In terms of the phase diagram of conformational states (Figure 4A), stirring or ultrasonication is a kinetic factor modifying the apparent phase diagram. It is likely that the boundary between the metastable and labile regions is shifted downward upon agitation, decreasing the barrier of supersaturation and inducing spontaneous amyloid formation [3].

To address the mechanism underlying the distinct amyloidogenic transitions, Noji et al. [78] examined the relationship between transition types (i.e., S, A and B types) and various factors which might determine these types (Figure 4B,C). It is evident that the total residue number (abscissa) is the most dominant factor in determining the different types. Then, the hydrophobic score showed a notable correlation with the distinct amyloid types. When viewed from the perspective of size and hydrophobicity (Figure 4B), the S proteins had a moderate size and moderate hydrophobicity, the A proteins had a short length and high hydrophobicity, and the B proteins had a long length and low hydrophobicity.

Another important factor is the disulfide bond [84,86,87]. The reduction in disulfide bonds often reduces amyloidogenicity, as demonstrated for β2m: under acidic conditions, the S-type transition changed to the B-type. These functions of disulfide bonds suggest that a more appropriate scale for evaluating the different types of amyloidogenic proteins is based on the "conformational flexibility of the denatured state".

Although the effects of disulfide bonds in reducing conformational entropy have been addressed [73,88], they are in fact minor in comparison with intrinsic ΔS_{conf} (Figure 4C). More importantly, the disulfide bonds stabilize hydrophobic cores that persist in the denatured state and thus increase amyloidogenicity, as demonstrated for acid-denatured β2m [84,87]. Taken together, the synergetic effects of disulfide bonds (i.e., decreasing the intrinsic conformational entropy and stabilizing the hydrophobic cores) lead to a significant decrease in the flexibility of the denatured states.

3. HANABI, an Ultrasonication-Forced Amyloid Fibril Inducer

3.1. *Ultrasonication-Dependent Breakdown of Supersaturation*

Ultrasonication, conventionally used for amplifying seed amyloid fibrils [62,89,90], is an effective agitation method that triggers the nucleation process [91–95]. By combining an ultrasonicator and a microplate reader, Umemoto et al. [96] developed the HANdai Amyloid Burst Inducer (HANABI) system, which enables high-throughput analysis of amyloid fibril formation [95]. With the HANABI system, ultrasonic irradiation was performed in a water bath; the plate was then moved to the microplate reader, and ThT fluorescence was monitored. These three processes were repeated automatically based on programmed time schedules. Moreover, the plate was moved along the x–y axes in sequence, to ultrasonicate the 96 wells evenly.

Kakuda et al. [97] used the HANABI system to amplify and detect αSN aggregates with seeding activity from CSF, and investigated the correlation between seeding activity and clinical indicators. The seeding activity of CSF correlated with the levels of αSN oligomers measured by an enzyme-linked immunosorbent assay. Moreover, the seeding activity of CSF from patients with Parkinson's disease was higher than that of the control patients. Notably, the lag time of patients with Parkinson's disease was significantly correlated with the ^{123}I-meta-iodobenzylguanidine (MIBG) heart-to-mediastinum (H/M) ratio, one of the most specific radiological features of Parkinson's diseases and dementia with Lewy bodies. These findings showed that the HANABI assay can evaluate the seeding activity of CSF by amplifying misfolded α-synuclein aggregates.

Although the original HANABI system promoted ultrasonication efficacy, several challenges remained. In general, the acoustic field in a sample solution could not remain the same because of changes in temperature, the volume of the water, and the distribution of the dissolved gases in the water bath. Specifically, in the original HANABI system, the fluorescence signal was acquired from the upper surface of the microplate, which was significantly affected by water droplets on the microplate because of the high-power ultrasonication.

Nakajima et al. [98] developed a HANABI-2000 system with an optimized sonoreactor for the amyloid-fibril assay, which improved the reproducibility and controllability of the amyloid fibril formation [95] (Figure 5). First, the water bath was eliminated in order to achieve a reproducible analysis. A single rod-shaped ultrasonic transducer was placed on each sample solution in an assay plate. The resonant frequency of the transducer was 30 kHz, which was optimized for accelerating amyloid fibril formation [99]. Second, a microphone was placed below the assay plate to measure the acoustic intensity of each sample solution. The acoustic intensity measurement allows for the acoustic field in each well to be controlled by individually regulating the voltage and frequency of the driving signal applied to each transducer. Third, a photodetector was placed beneath the microplate to measure the fluorescent signal, which improves the signal-to-noise ratio of the fluorescence measurement because of the absence of the water bath.

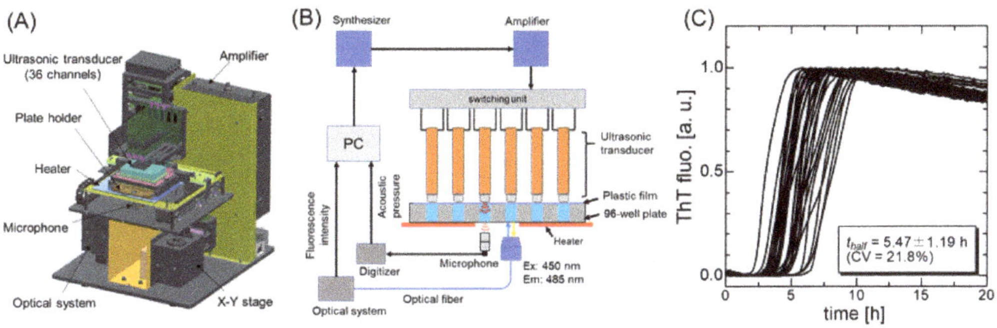

Figure 5. Overview of the HANABI-2000 system. (**A**) A 3D schematic illustration of the optimized sonoreactor for the amyloid-fibril assays, HANABI-2000. The dimensions of the device are 500 × 550 × 550 mm^3. (**B**) A block chart of the control units of HANABI-2000. The figure is reproduced from Nakajima et al. [98] with permission. (**C**) The ThT time–course curves (n = 36), which are irradiated with ultrasound with the compensation procedure. The figure was modified from Nakajima et al. [98] with permission. Copyright 2021 Elsevier.

Using the acid-denatured β2m monomer solution, Nakajima et al. [98] demonstrated that achieving identical acoustic conditions by controlling the oscillation amplitude and frequency of each transducer results in synchronized amyloid fibril formation behavior across 36 solutions with a coefficient of variation (CV) of 22% for a half-time (t_{half}) (Figure 5C).

They then succeeded in detecting 100-fM seeds at an accelerated rate. Moreover, they revealed that acceleration of the amyloid fibril formation reaction with the seeds is achieved by enhancing the primary nucleation and fibril fragmentation. These results suggested the efficacy of HANABI-2000 in the diagnosis of amyloidosis owing to the accelerative seed detection, and the possibility of further early-stage diagnosis even without seeds through the accelerated primary nucleation (i.e., the identification of susceptibility risk biomarkers [49]).

Nakajima et al. [99] studied the molecular mechanism underlying the ultrasonication-dependent acceleration of amyloid fibril formation. They showed that ultrasonic cavitation bubbles behave as catalysts for nucleation: the nucleation reaction is highly dependent on the frequency and pressure of acoustic waves and, under optimal acoustic conditions, the reaction-rate constant for nucleation increased by three orders of magnitude. A theoretical model was proposed to explain the markedly frequency- and pressure-dependent nucleation; in this model, monomers are captured on the bubble's surface during its growth and are highly condensed by the subsequent collapse of the bubble, so that they are transiently exposed to high temperatures [99]. Thus, the dual effects of local condensation and local heating contribute to markedly enhancing the nucleation reaction.

3.2. Comparison of Ultrasonication and Shaking on Breaking Supersaturation

Although both ultrasonication and shaking are effectively used to induce amyloid fibril formation and propagation, the difference between them remained unclear until recently [89,100–103]. Nakajima et al. [104] compared ultrasonication and shaking with respect to the morphology and structure of resultant β2m aggregates, the kinetics of amyloid fibril formation, and seed-detection sensitivity. They focused on t_{half}, the time required for exhibiting half of the maximal ThT fluorescence, and constructed a heat map, which describes the phase diagram of β2m aggregation. The experimental results show that ultrasonication markedly promotes amyloid formation, especially in dilute monomer solutions; it also induces short-dispersed fibrils, and is capable of detecting ultra-trace-concentration seeds with a detection limit of 10 fM.

Most importantly, they indicated that ultrasonication highly alters the energy landscape of an aggregation reaction due to the effects of ultrasonic cavitation. Under shaking (Figure 6B), the metastable region becomes narrower than that under quiescence (Figure 6A), showing that shaking induces a downward shift in the metastable–labile boundary, whereas it has only minimal effects on the labile–amorphous boundary. On the other hand, ultrasonication not only causes a significant downward shift in the metastable–labile boundary but also an upward shift in the labile–amorphous boundary (Figure 6C).

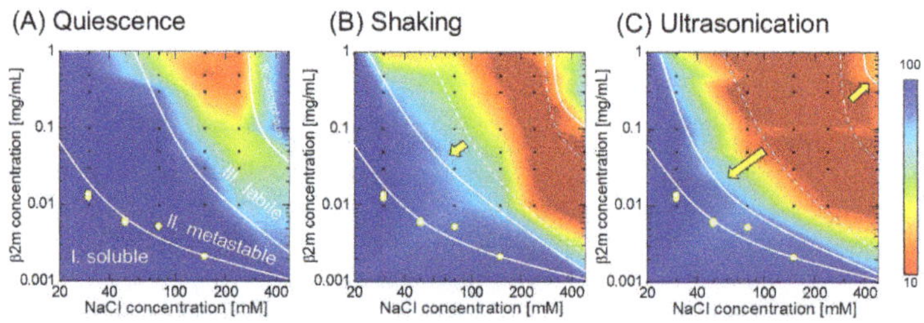

Figure 6. t_{half} heat maps of aggregation reactions. (**A**) Under quiescence, (**B**) under shaking, and (**C**) under ultrasonication. The yellow dots denote the solubility of acidic β2m monomer at each salt concentration, determined by ultracentrifugation and the ELISA assay. The dotted lines in panels (**B**,**C**) indicate the phase boundaries under quiescence, which are varied under agitation as indicated by yellow arrows. The figure is reproduced from Nakajima et al. [104] with permission. Copyright 2021 American Chemical Society.

In the labile region, although the acceleration ability of shaking was similar to that of ultrasonication for solutions with high monomer concentrations, it decreased for solutions with a monomer concentration lower than 0.1 mg/mL. The aggregation acceleration by shaking results from the increase in the apparent mean-free path of monomer movements. Thus, shaking enhances the probability of intermolecular interactions in a condensed solution by increasing the collision frequency among monomers. However, it fails to increase the collision frequency in a dilute solution, diminishing the acceleration effect for nucleation (Figure 6B). In contrast, ultrasonication retains a high acceleration ability, even for dilute monomer solutions, because the cavitation bubble works as a catalyst for the nucleation reaction [99]. On the other hand, if the bubble's surface becomes fully covered with monomers, the acceleration effect saturates. This mechanism consistently explains why the t_{half} value of the ultrasonication-dependent acceleration cannot be decreased below the lower limit of ~10 h for solutions with high monomer concentrations.

4. Liquid–Liquid Phase Separation

One of the most important phenomena related to amyloid formation is liquid–liquid phase separation, which is observed at increasing rates in disordered proteins [17,105,106]. There are cases in which amyloid formation is preceded by the liquid–liquid phase separation. As an example, the low-complexity domain of FUS protein formed phase-separated droplets before the formation of more stable amyloid fibrils [107]. These results are consistent with Ostwald's ripening rule of crystallization, according to which the morphologies of crystals change over time, guided by their kinetic accessibilities and thermodynamic stabilities [108].

Recently, Shimobayashi et al. [17] examined the liquid–liquid phase separation of FUS protein in living cells with a light-controlled oligomerizing system Corelets [109] and constructed the core protein concentration-dependent phase diagram. Their results show that the initial stages of nucleated phase separation can be modelled by classical nucleation theory, which describes how the rate of droplet nucleation depends on the degree of supersaturation. It is noted that, in their phase diagram of polymer science, the "binodal boundary" corresponds to solubility or critical concentration and the "spinodal region" correspond to the amorphous region. We assume that the "macroscopic" phase diagram of conformational states, such as Figures 1, 3, 4 and 6, will be also useful for understanding the liquid–liquid phase separation, where "microscopic" phase diagrams limited by supersaturation might apply to each droplet system.

The relationship between phase-separated droplets and oligomers or relatively small amorphous aggregates is not clear. Amorphous aggregation and amyloid fibrillation have often been considered as separate pathways in direct competition with each other, such that accumulation of amorphous aggregates will always retard fibrillation [3,11,110–112]. On the other hand, Nitani et al. [31] suggest the presence of two types of aggregates, i.e., a frozen amorphous aggregation state, and a labile amorphous aggregate capable of slow fibrillation. The labile amorphous aggregate is consistent with Ostwald's ripening rule of crystallization, in which morphologies of crystals change over time, guided by their kinetic accessibilities and thermodynamic stabilities [6,108]. Moreover, it is possible that spontaneous amyloid formation is assisted by a small amount of amorphous aggregates/amorphous oligomers, as suggested by the new view of crystallization [29].

5. Conclusions

Recent structural studies on amyloid fibrils have advanced remarkably, clarifying their atomic structures and polymorphisms [38–42]. However, amyloid structures do not necessarily fully explain the mechanism of their formation. Physicochemical studies on protein folding and misfolding have established that amyloid fibrils are crystal-like aggregates of denatured proteins, which are formed above solubility by breaking supersaturation [6,33,34,50]. It is evident that no amyloid formation occurs below solubility; moreover, the preformed amyloid fibrils dissolve below solubility [113,114], although the rigidity of amyloid fibrils may prevent rapid dissolution.

Although the validity of Anfinsen's dogma (i.e., reversible unfolding/refolding) is often questioned under high protein concentrations where intermolecular interactions are favored [115], the persistence of supersaturation and the difficulty of amyloid formation have neglected to address the question exactly. Recent studies that focus on solubility and supersaturation have shown that, although protein folding is independent of protein concentration, amyloid formation is critically dependent on protein concentration (i.e., solubility). By combining the concentration-independent folding/unfolding under supersaturation and the concentration-dependent amyloid formation after the breakdown of supersaturation, the exact unification of these two processes has become possible. In other words, the breakdown of supersaturation links Anfinsen's intramolecular folding universe and the intermolecular misfolding universe, extending our understanding of protein folding and misfolding.

The conformational phase diagrams used in this article are similar to a general phase diagram of a solute consisting of soluble (Region I), metastable (Region II), labile (Region III), and amorphous regions (Region IV) (Figures 1, 3, 4 and 6). The phase diagrams based on solubility and supersaturation will be essential for further clarification of the mechanisms of protein folding and misfolding, and their roles in diseases and life functions.

Finally, in relation to supersaturation-dependent amyloid formation, the onset of amyloid deposition is accompanied by a decrease in the soluble concentration of precursor proteins, as discussed in this article [49,50]. In other words, monitoring decreases in precursor protein concentrations might be a promising approach for detecting the onset of amyloidosis.

Author Contributions: Original draft preparation, Y.G. Review and editing, M.N., K.N. and K.Y. All authors have read and agreed to the published version of the manuscript.

Funding: This study was supported by the Japan Society for the Promotion of Science (20K06580, 20K22628, 21K19224, 22H02584, 22K14013, and Core-to-Core Program A: Advanced Research Networks), Ministry of Education, Culture, Sports, Science and Technology (17H06352), and SENTAN from AMED (16809242).

Acknowledgments: This study was performed as part of the Cooperative Research Program for the Institute for Protein Research, Osaka University (CR-21-02).

Conflicts of Interest: The authors declare no conflict of interest.

References

1. Durbin, S.D.; Feher, G. Protein crystallization. *Annu. Rev. Phys. Chem.* **1996**, *47*, 171–204. [CrossRef] [PubMed]
2. Mullin, J.W. *Crystallization*, 4th ed.; Butterworth-Heinemann: Oxford, UK, 2001.
3. Yoshimura, Y.; Lin, Y.X.; Yagi, H.; Lee, Y.H.; Kitayama, H.; Sakurai, K.; So, M.; Ogi, H.; Naiki, H.; Goto, Y. Distinguishing crystal-like amyloid fibrils and glass-like amorphous aggregates from their kinetics of formation. *Proc. Natl. Acad. Sci. USA* **2012**, *109*, 14446–14451. [CrossRef] [PubMed]
4. Wallace, A.F.; Hedges, L.O.; Fernandez-Martinez, A.; Raiteri, P.; Gale, J.D.; Waychunas, G.A.; Whitelam, S.; Banfield, J.F.; De Yoreo, J.J. Microscopic evidence for liquid-liquid separation in supersaturated $CaCO_3$ solutions. *Science* **2013**, *341*, 885–889. [CrossRef]
5. Coquerel, G. Crystallization of molecular systems from solution: Phase diagrams, supersaturation and other basic concepts. *Chem. Soc. Rev.* **2014**, *43*, 2286–2300. [CrossRef] [PubMed]
6. So, M.; Hall, D.; Goto, Y. Revisiting supersaturation as a factor determining amyloid fibrillation. *Curr. Opin. Struct. Biol.* **2016**, *36*, 32–39. [CrossRef] [PubMed]
7. Eaton, W.A.; Hofrichter, J. Hemoglobin S gelation and sickle cell disease. *Blood* **1987**, *70*, 1245–1266. [CrossRef] [PubMed]
8. Hofrichter, J.; Ross, P.D.; Eaton, W.A. Supersaturation in sickle cell hemoglobin solutions. *Proc. Natl. Acad. Sci. USA* **1976**, *73*, 3035–3039. [CrossRef] [PubMed]
9. Gour, N.; Gazit, E. Metabolite assemblies: A surprising extension to the amyloid hypothesis. *Curr. Opin. Chem. Biol.* **2021**, *64*, 154–164. [CrossRef]
10. Bergfors, T. Seeds to crystals. *J. Struct. Biol.* **2003**, *142*, 66–76. [CrossRef]
11. Adachi, M.; Noji, M.; So, M.; Sasahara, K.; Kardos, J.; Naiki, H.; Goto, Y. Aggregation-phase diagrams of β2-microglobulin reveal temperature and salt effects on competitive formation of amyloids versus amorphous aggregates. *J. Biol. Chem.* **2018**, *293*, 14775–14785. [CrossRef]
12. Noji, M.; Sasahara, K.; Yamaguchi, K.; So, M.; Sakurai, K.; Kardos, J.; Naiki, H.; Goto, Y. Heating during agitation of β2-microglobulin reveals that supersaturation breakdown is required for amyloid fibril formation at neutral pH. *J. Biol. Chem.* **2019**, *294*, 15826–15835. [CrossRef] [PubMed]
13. Morris, A.M.; Watzky, M.A.; Agar, J.N.; Finke, R.G. Fitting neurological protein aggregation kinetic data via a 2-step, minimal/"Ockham's razor" model: The Finke-Watzky mechanism of nucleation followed by autocatalytic surface growth. *Biochemistry* **2008**, *47*, 2413–2427. [CrossRef] [PubMed]
14. Morris, A.M.; Watzky, M.A.; Finke, R.G. Protein aggregation kinetics, mechanism, and curve-fitting: A review of the literature. *Biochim. Biophys. Acta* **2009**, *1794*, 375–397. [CrossRef] [PubMed]
15. Sawada, M.; Yamaguchi, K.; Hirano, M.; Noji, M.; So, M.; Otzen, D.; Kawata, Y.; Goto, Y. Amyloid formation of α-synuclein based on the solubility- and supersaturation-dependent mechanism. *Langmuir* **2020**, *36*, 4671–4681. [CrossRef]
16. Oosawa, F.; Kasai, M. A theory of linear and helical aggregations of macromolecules. *J. Mol. Biol.* **1962**, *4*, 10–21. [CrossRef]
17. Shimobayashi, S.F.; Ronceray, P.; Sanders, D.W.; Haataja, M.P.; Brangwynne, C.P. Nucleation landscape of biomolecular condensates. *Nature* **2021**, *599*, 503–506. [CrossRef]
18. Matsumoto, M.; Yagasaki, T.; Tanaka, H. Chiral ordering in supercooled liquid water and amorphous Ice. *Phys. Rev. Lett.* **2015**, *115*, 197801. [CrossRef]
19. Matsumoto, M.; Yagasaki, T.; Tanaka, H. A Bayesian approach for identification of ice Ih, ice Ic, high density, and low density liquid water with a torsional order parameter. *J. Chem. Phys.* **2019**, *150*, 214504. [CrossRef]
20. Narayanan, J.; Liu, X.Y. Protein interactions in undersaturated and supersaturated solutions: A study using light and X-ray scattering. *Biophys. J.* **2003**, *84*, 523–532. [CrossRef]

21. Chattopadhyay, S.; Erdemir, D.; Evans, J.M.B.; Ilavsky, J.; Amenitsch, H.; Segre, C.U.; Myerson, A.S. SAXS study of the nucleation of glycine crystals from a supersaturated solution. *Cryst. Growth Des.* **2005**, *5*, 523–527. [CrossRef]
22. Stradner, A.; Sedgwick, H.; Cardinaux, F.; Poon, W.C.K.; Egelhaaf, S.U.; Schurtenberger, P. Equilibrium cluster formation in concentrated protein solutions and colloids. *Nature* **2004**, *432*, 492–495. [CrossRef] [PubMed]
23. Maes, D.; Vorontsova, M.A.; Potenza, M.A.C.; Sanvito, T.; Sleutel, M.; Giglio, M.; Vekilov, P.G. Do protein crystals nucleate within dense liquid clusters? *Acta Crystallogr. F* **2015**, *71*, 815–822. [CrossRef] [PubMed]
24. Li, Y.; Lubchenko, V.; Vorontsova, M.A.; Filobelo, L.; Vekilov, P.G. Ostwald-Like ripening of the anomalous mesoscopic clusters in protein solutions. *J. Phys. Chem. B* **2012**, *116*, 10657–10664. [CrossRef] [PubMed]
25. Matsushita, Y.; Sekiguchi, H.; Ichiyanagi, K.; Ohta, N.; Ikezaki, K.; Goto, Y.; Sasaki, Y.C. Time-resolved X-ray tracking of expansion and compression dynamics in supersaturating ion-networks. *Sci. Rep.* **2015**, *5*, 17647. [CrossRef]
26. Matsushita, Y.; Sekiguchi, H.; Wong, C.J.; Nishijima, M.; Ikezaki, K.; Hamada, D.; Goto, Y.; Sasaki, Y.C. Nanoscale dynamics of protein assembly networks in supersaturated solutions. *Sci. Rep.* **2017**, *7*, 13883. [CrossRef] [PubMed]
27. Harano, K.; Homma, T.; Niimi, Y.; Koshino, M.; Suenaga, K.; Leibler, L.; Nakamura, E. Heterogeneous nucleation of organic crystals mediated by single-molecule templates. *Nat. Mater.* **2012**, *11*, 877–881. [CrossRef] [PubMed]
28. Vekilov, P.G. Crystal nucleation: Nucleus in a droplet. *Nat. Mater.* **2012**, *11*, 838–840. [CrossRef]
29. Yamazaki, T.; Kimura, Y.; Vekilov, P.G.; Furukawa, E.; Shirai, M.; Matsumoto, H.; Van Driessche, A.E.; Tsukamoto, K. Two types of amorphous protein particles facilitate crystal nucleation. *Proc. Natl. Acad. Sci. USA* **2017**, *114*, 2154–2159. [CrossRef]
30. Auer, S.; Dobson, C.M.; Vendruscolo, M. Characterization of the nucleation barriers for protein aggregation and amyloid formation. *HFSP J.* **2007**, *1*, 137–146. [CrossRef]
31. Nitani, A.; Muta, H.; Adachi, M.; So, M.; Sasahara, K.; Sakurai, K.; Chatani, E.; Naoe, K.; Ogi, H.; Hall, D.; et al. Heparin-dependent aggregation of hen egg white lysozyme reveals two distinct mechanisms of amyloid fibrillation. *J. Biol. Chem.* **2017**, *292*, 21219–21230. [CrossRef]
32. Griffith, J.S. Self-replication and scrapie. *Nature* **1967**, *215*, 1043–1044. [CrossRef] [PubMed]
33. Jarrett, J.T.; Lansbury, P.T., Jr. Seeding "one-dimensional crystallization" of amyloid: A pathogenic mechanism in Alzheimer's disease and scrapie? *Cell* **1993**, *73*, 1055–1058. [CrossRef]
34. Wetzel, R. Kinetics and thermodynamics of amyloid fibril assembly. *Acc. Chem. Res.* **2006**, *39*, 671–679. [CrossRef]
35. Lansbury, P.T., Jr.; Caughey, B. The chemistry of scrapie infection: Implications of the 'ice 9' metaphor. *Chem. Biol.* **1995**, *2*, 1–5. [CrossRef]
36. Vonnegut, K. *Cat's Cradle*; Dell Publishing: New York, NY, USA, 1963.
37. Laganowsky, A.; Liu, C.; Sawaya, M.R.; Whitelegge, J.P.; Park, J.; Zhao, M.; Pensalfini, A.; Soriaga, A.B.; Landau, M.; Teng, P.K.; et al. Atomic view of a toxic amyloid small oligomer. *Science* **2012**, *335*, 1228–1231. [CrossRef] [PubMed]
38. Chiti, F.; Dobson, C.M. Protein misfolding, amyloid formation, and human disease: A summary of progress over the last decade. *Annu. Rev. Biochem.* **2017**, *86*, 27–68. [CrossRef]
39. Iadanza, M.G.; Silvers, R.; Boardman, J.; Smith, H.I.; Karamanos, T.K.; Debelouchina, G.T.; Su, Y.; Griffin, R.G.; Ranson, N.A.; Radford, S.E. The structure of a β2-microglobulin fibril suggests a molecular basis for its amyloid polymorphism. *Nat. Commun.* **2018**, *9*, 4517. [CrossRef]
40. Ke, P.C.; Zhou, R.; Serpell, L.C.; Riek, R.; Knowles, T.P.J.; Lashuel, H.A.; Gazit, E.; Hamley, I.W.; Davis, T.P.; Fandrich, M.; et al. Half a century of amyloids: Past, present and future. *Chem. Soc. Rev.* **2020**, *49*, 5473–5509. [CrossRef]
41. Benson, M.D.; Buxbaum, J.N.; Eisenberg, D.S.; Merlini, G.; Saraiva, M.J.M.; Sekijima, Y.; Sipe, J.D.; Westermark, P. Amyloid nomenclature 2020: Update and recommendations by the International Society of Amyloidosis (ISA) nomenclature committee. *Amyloid* **2020**, *27*, 217–222. [CrossRef]
42. Sawaya, M.R.; Hughes, M.P.; Rodriguez, J.A.; Riek, R.; Eisenberg, D.S. The expanding amyloid family: Structure, stability, function, and pathogenesis. *Cell* **2021**, *184*, 4857–4873. [CrossRef]
43. Bateman, R.J.; Xiong, C.; Benzinger, T.L.; Fagan, A.M.; Goate, A.; Fox, N.C.; Marcus, D.S.; Cairns, N.J.; Xie, X.; Blazey, T.M.; et al. Clinical and biomarker changes in dominantly inherited Alzheimer's disease. *N. Engl. J. Med.* **2012**, *367*, 795–804. [CrossRef] [PubMed]
44. Jack, C.R., Jr.; Knopman, D.S.; Jagust, W.J.; Petersen, R.C.; Weiner, M.W.; Aisen, P.S.; Shaw, L.M.; Vemuri, P.; Wiste, H.J.; Weigand, S.D.; et al. Tracking pathophysiological processes in Alzheimer's disease: An updated hypothetical model of dynamic biomarkers. *Lancet Neurol.* **2013**, *12*, 207–216. [CrossRef]
45. Tokuda, T.; Salem, S.A.; Allsop, D.; Mizuno, T.; Nakagawa, M.; Qureshi, M.M.; Locascio, J.J.; Schlossmacher, M.G.; El-Agnaf, O.M. Decreased α-synuclein in cerebrospinal fluid of aged individuals and subjects with Parkinson's disease. *Biochem. Biophys. Res. Commun.* **2006**, *349*, 162–166. [CrossRef]
46. Hong, Z.; Shi, M.; Chung, K.A.; Quinn, J.F.; Peskind, E.R.; Galasko, D.; Jankovic, J.; Zabetian, C.P.; Leverenz, J.B.; Baird, G.; et al. DJ-1 and α-synuclein in human cerebrospinal fluid as biomarkers of Parkinson's disease. *Brain* **2010**, *133*, 713–726. [CrossRef]
47. Mollenhauer, B.; Locascio, J.J.; Schulz-Schaeffer, W.; Sixel-Doring, F.; Trenkwalder, C.; Schlossmacher, M.G. α-Synuclein and tau concentrations in cerebrospinal fluid of patients presenting with parkinsonism: A cohort study. *Lancet Neurol.* **2011**, *10*, 230–240. [CrossRef]

48. Kang, J.H.; Irwin, D.J.; Chen-Plotkin, A.S.; Siderowf, A.; Caspell, C.; Coffey, C.S.; Waligorska, T.; Taylor, P.; Pan, S.; Frasier, M.; et al. Association of cerebrospinal fluid β-amyloid 1-42, T-tau, P-tau181, and α-synuclein levels with clinical features of drug-naive patients with early Parkinson disease. *JAMA Neurol.* **2013**, *70*, 1277–1287. [CrossRef]
49. Parnetti, L.; Gaetani, L.; Eusebi, P.; Paciotti, S.; Hansson, O.; El-Agnaf, O.; Mollenhauer, B.; Blennow, K.; Calabresi, P. CSF and blood biomarkers for Parkinson's disease. *Lancet Neurol.* **2019**, *18*, 573–586. [CrossRef]
50. Guo, Z. Amyloid hypothesis through the lens of Aβ supersaturation. *Neural. Regen Res.* **2021**, *16*, 1562–1563. [CrossRef]
51. Nakamura, A.; Kaneko, N.; Villemagne, V.L.; Kato, T.; Doecke, J.; Dore, V.; Fowler, C.; Li, Q.X.; Martins, R.; Rowe, C.; et al. High performance plasma amyloid-β biomarkers for Alzheimer's disease. *Nature* **2018**, *554*, 249–254. [CrossRef]
52. Zhang, C.M.; Yamaguchi, K.; So, M.; Sasahara, K.; Ito, T.; Yamamoto, S.; Narita, I.; Kardos, J.; Naiki, H.; Goto, Y. Possible mechanisms of polyphosphate-induced amyloid fibril formation of β2-microglobulin. *Proc. Natl. Acad. Sci. USA* **2019**, *116*, 12833–12838. [CrossRef]
53. Yamaguchi, K.; So, M.; Aguirre, C.; Ikenaka, K.; Mochizuki, H.; Kawata, Y.; Goto, Y. Polyphosphates induce amyloid fibril formation of α-synuclein in concentration-dependent distinct manners. *J. Biol. Chem.* **2021**, *296*, 100510. [CrossRef] [PubMed]
54. Yamaguchi, K.; Hasuo, K.; So, M.; Ikenaka, K.; Mochizuki, H.; Goto, Y. Strong acids induce amyloid fibril formation of β2-microglobulin via an anion-binding mechanism. *J. Biol. Chem.* **2021**, *297*, 101286. [CrossRef] [PubMed]
55. Furukawa, K.; Aguirre, C.; So, M.; Sasahara, K.; Miyanoiri, Y.; Sakurai, K.; Yamaguchi, K.; Ikenaka, K.; Mochizuki, H.; Kardos, J.; et al. Isoelectric point-amyloid formation of α-synuclein extends the generality of the solubility and supersaturation-limited mechanism. *Curr. Res. Struct. Biol.* **2020**, *2*, 35–44. [CrossRef] [PubMed]
56. Muta, H.; Lee, Y.H.; Kardos, J.; Lin, Y.; Yagi, H.; Goto, Y. Supersaturation-limited amyloid fibrillation of insulin revealed by ultrasonication. *J. Biol. Chem.* **2014**, *289*, 18228–18238. [CrossRef]
57. Ciryam, P.; Tartaglia, G.G.; Morimoto, R.I.; Dobson, C.M.; Vendruscolo, M. Widespread aggregation and neurodegenerative diseases are associated with supersaturated proteins. *Cell Rep.* **2013**, *5*, 781–790. [CrossRef]
58. Ciryam, P.; Kundra, R.; Morimoto, R.I.; Dobson, C.M.; Vendruscolo, M. Supersaturation is a major driving force for protein aggregation in neurodegenerative diseases. *Trends Pharmacol. Sci.* **2015**, *36*, 72–77. [CrossRef]
59. Gejyo, F.; Yamada, T.; Odani, S.; Nakagawa, Y.; Arakawa, M.; Kunitomo, T.; Kataoka, H.; Suzuki, M.; Hirasawa, Y.; Shirahama, T.; et al. A new form of amyloid protein associated with chronic hemodialysis was identified as β2-microglobulin. *Biochem. Biophys. Res. Commun.* **1985**, *129*, 701–706. [CrossRef]
60. Gejyo, F.; Homma, N.; Suzuki, Y.; Arakawa, M. Serum levels of β2-microglobulin as a new form of amyloid protein in patients undergoing long-term hemodialysis. *N. Engl. J. Med.* **1986**, *314*, 585–586.
61. Yamamoto, S.; Gejyo, F. Historical background and clinical treatment of dialysis-related amyloidosis. *Biochim. Biophys. Acta* **2005**, *1753*, 4–10. [CrossRef]
62. Naiki, H.; Hashimoto, N.; Suzuki, S.; Kimura, H.; Nakakuki, K.; Gejyo, F. Establishment of a kinetic model of dialysis-related amyloid fibril extension in vitro. *Amyloid* **1997**, *4*, 223–232. [CrossRef]
63. Hoshino, M.; Katou, H.; Hagihara, Y.; Hasegawa, K.; Naiki, H.; Goto, Y. Mapping the core of the β2-microglobulin amyloid fibril by H/D exchange. *Nat. Struct. Biol.* **2002**, *9*, 332–336. [CrossRef]
64. Kardos, J.; Yamamoto, K.; Hasegawa, K.; Naiki, H.; Goto, Y. Direct measurement of the thermodynamic parameters of amyloid formation by isothermal titration calorimetry. *J. Biol. Chem.* **2004**, *279*, 55308–55314. [CrossRef]
65. Gosal, W.S.; Morten, I.J.; Hewitt, E.W.; Smith, D.A.; Thomson, N.H.; Radford, S.E. Competing pathways determine fibril morphology in the self-assembly of β2-microglobulin into amyloid. *J. Mol. Biol.* **2005**, *351*, 850–864. [CrossRef]
66. Valleix, S.; Gillmore, J.D.; Bridoux, F.; Mangione, P.P.; Dogan, A.; Nedelec, B.; Boimard, M.; Touchard, G.; Goujon, J.M.; Lacombe, C.; et al. Hereditary systemic amyloidosis due to Asp76Asn variant β2-microglobulin. *N. Engl. J. Med.* **2012**, *366*, 2276–2283. [CrossRef] [PubMed]
67. Stoppini, M.; Bellotti, V. Systemic amyloidosis: Lessons from β2-microglobulin. *J. Biol. Chem.* **2015**, *290*, 9951–9958. [CrossRef] [PubMed]
68. Pashley, C.L.; Hewitt, E.W.; Radford, S.E. Comparison of the aggregation of homologous β2-microglobulin variants reveals protein solubility as a key determinant of amyloid formation. *J. Mol. Biol.* **2016**, *428*, 631–643. [CrossRef] [PubMed]
69. Benseny-Cases, N.; Karamanos, T.K.; Hoop, C.L.; Baum, J.; Radford, S.E. Extracellular matrix components modulate different stages in β2-microglobulin amyloid formation. *J. Biol. Chem.* **2019**, *294*, 9392–9401. [CrossRef]
70. Hoop, C.L.; Zhu, J.; Bhattacharya, S.; Tobita, C.A.; Radford, S.E.; Baum, J. Collagen I weakly interacts with the β-sheets of β2-microglobulin and enhances conformational exchange to induce amyloid formation. *J. Am. Chem. Soc.* **2020**, *142*, 1321–1331. [CrossRef]
71. Smith, H.I.; Guthertz, N.; Cawood, E.E.; Maya-Martinez, R.; Breeze, A.L.; Radford, S.E. The role of the IT-state in D76N β2-microglobulin amyloid assembly: A crucial intermediate or an innocuous bystander? *J. Biol. Chem.* **2020**, *295*, 12474–12484. [CrossRef]
72. Kihara, M.; Chatani, E.; Sakai, M.; Hasegawa, K.; Naiki, H.; Goto, Y. Seeding-dependent maturation of β2-microglobulin amyloid fibrils at neutral pH. *J. Biol. Chem.* **2005**, *280*, 12012–12018. [CrossRef]
73. Makhatadze, G.I.; Privalov, P.L. Energetics of protein structure. *Adv. Protein Chem.* **1995**, *47*, 307–425.
74. Ikenoue, T.; Lee, Y.H.; Kardos, J.; Yagi, H.; Ikegami, T.; Naiki, H.; Goto, Y. Heat of supersaturation-limited amyloid burst directly monitored by isothermal titration calorimetry. *Proc. Natl. Acad. Sci. USA* **2014**, *111*, 6654–6659. [CrossRef] [PubMed]

75. Clark, P.L. Protein folding in the cell: Reshaping the folding funnel. *Trends Biochem. Sci.* **2004**, *29*, 527–534. [CrossRef]
76. Silva, J.L.; Cordeiro, Y. The "Jekyll and Hyde" Actions of nucleic acids on the prion-like aggregation of proteins. *J. Biol. Chem.* **2016**, *291*, 15482–15490. [CrossRef]
77. Balchin, D.; Hayer-Hartl, M.; Hartl, F.U. In vivo aspects of protein folding and quality control. *Science* **2016**, *353*, aac4354. [CrossRef] [PubMed]
78. Noji, M.; Samejima, T.; Yamaguchi, K.; So, M.; Yuzu, K.; Chatani, E.; Akazawa-Ogawa, Y.; Hagihara, Y.; Kawata, Y.; Ikenaka, K.; et al. Breakdown of supersaturation barrier links protein folding to amyloid formation. *Commun. Biol.* **2021**, *4*, 120. [CrossRef] [PubMed]
79. Liu, Y.; Gotte, G.; Libonati, M.; Eisenberg, D. A domain-swapped RNase A dimer with implications for amyloid formation. *Nat. Struct. Biol.* **2001**, *8*, 211–214. [CrossRef] [PubMed]
80. Sambashivan, S.; Liu, Y.; Sawaya, M.R.; Gingery, M.; Eisenberg, D. Amyloid-like fibrils of ribonuclease A with three-dimensional domain-swapped and native-like structure. *Nature* **2005**, *437*, 266–269. [CrossRef] [PubMed]
81. Saelices, L.; Chung, K.; Lee, J.H.; Cohn, W.; Whitelegge, J.P.; Benson, M.D.; Eisenberg, D.S. Amyloid seeding of transthyretin by ex vivo cardiac fibrils and its inhibition. *Proc. Natl. Acad. Sci. USA* **2018**, *115*, E6741–E6750. [CrossRef]
82. Noji, M.; So, M.; Yamaguchi, K.; Hojo, H.; Onda, M.; Akazawa-Ogawa, Y.; Hagihara, Y.; Goto, Y. Heat-induced aggregation of hen ovalbumin suggests a key factor responsible for serpin polymerization. *Biochemistry* **2018**, *57*, 5415–5426. [CrossRef]
83. Dregni, A.J.; Mandala, V.S.; Wu, H.; Elkins, M.R.; Wang, H.K.; Hung, I.; DeGrado, W.F.; Hong, M. In vitro 0N4R tau fibrils contain a monomorphic β-sheet core enclosed by dynamically heterogeneous fuzzy coat segments. *Proc. Natl. Acad. Sci. USA* **2019**, *116*, 16357–16366. [CrossRef] [PubMed]
84. Yanagi, K.; Sakurai, K.; Yoshimura, Y.; Konuma, T.; Lee, Y.H.; Sugase, K.; Ikegami, T.; Naiki, H.; Goto, Y. The monomer-seed interaction mechanism in the formation of the β2-microglobulin amyloid fibril clarified by solution NMR techniques. *J. Mol. Biol.* **2012**, *422*, 390–402. [CrossRef] [PubMed]
85. Abdolvahabi, A.; Shi, Y.; Rasouli, S.; Croom, C.M.; Chuprin, A.; Shaw, B.F. How do gyrating beads accelerate amyloid fibrillization? *Biophys. J.* **2017**, *112*, 250–264. [CrossRef] [PubMed]
86. Katou, H.; Kanno, T.; Hoshino, M.; Hagihara, Y.; Tanaka, H.; Kawai, T.; Hasegawa, K.; Naiki, H.; Goto, Y. The role of disulfide bond in the amyloidogenic state of β2-microglobulin studied by heteronuclear NMR. *Protein Sci.* **2002**, *11*, 2218–2229. [CrossRef]
87. Muta, H.; So, M.; Sakurai, K.; Kardos, J.; Naiki, H.; Goto, Y. Amyloid formation under complicated conditions in which β2-microglobulin coexists with its proteolytic fragments. *Biochemistry* **2019**, *58*, 4925–4934. [CrossRef]
88. Pace, C.N.; Grimsley, G.R.; Thomson, J.A.; Barnett, B.J. Conformational stability and activity of ribonuclease T1 with zero, one, and two intact disulfide bonds. *J. Biol. Chem.* **1988**, *263*, 11820–11825. [CrossRef]
89. Saborio, G.P.; Permanne, B.; Soto, C. Sensitive detection of pathological prion protein by cyclic amplification of protein misfolding. *Nature* **2001**, *411*, 810–813. [CrossRef]
90. Shahnawaz, M.; Mukherjee, A.; Pritzkow, S.; Mendez, N.; Rabadia, P.; Liu, X.; Hu, B.; Schmeichel, A.; Singer, W.; Wu, G. Discriminating a-synuclein strains in Parkinson's disease and multiple system atrophy. *Nature* **2020**, *578*, 273–277. [CrossRef]
91. Ohhashi, Y.; Kihara, M.; Naiki, H.; Goto, Y. Ultrasonication-induced amyloid fibril formation of β2-microglobulin. *J. Biol. Chem.* **2005**, *280*, 32843–32848. [CrossRef]
92. Chatani, E.; Lee, Y.H.; Yagi, H.; Yoshimura, Y.; Naiki, H.; Goto, Y. Ultrasonication-dependent production and breakdown lead to minimum-sized amyloid fibrils. *Proc. Natl. Acad. Sci. USA* **2009**, *106*, 11119–11124. [CrossRef]
93. So, M.; Yagi, H.; Sakurai, K.; Ogi, H.; Naiki, H.; Goto, Y. Ultrasonication-dependent acceleration of amyloid fibril formation. *J. Mol. Biol.* **2011**, *412*, 568–577. [CrossRef] [PubMed]
94. Kitayama, H.; Yoshimura, Y.; So, M.; Sakurai, K.; Yagi, H.; Goto, Y. A common mechanism underlying amyloid fibrillation and protein crystallization revealed by the effects of ultrasonication. *Biochim. Biophys. Acta* **2013**, *1834*, 2640–2646. [CrossRef] [PubMed]
95. Goto, Y.; Nakajima, K.; Yamaguchi, K.; So, M.; Ikenaka, K.; Mochizuki, H.; Ogi, H. Development of HANABI, an ultrasonication-forced amyloid fibril inducer. *Neurochem. Int.* **2022**, *153*, 105270. [CrossRef]
96. Umemoto, A.; Yagi, H.; So, M.; Goto, Y. High-throughput analysis of the ultrasonication-forced amyloid fibrillation reveals the mechanism underlying the large fluctuation in the lag time. *J. Biol. Chem.* **2014**, *289*, 27290–27299. [CrossRef] [PubMed]
97. Kakuda, K.; Ikenaka, K.; Araki, K.; So, M.; Aguirre, C.; Kajiyama, Y.; Konaka, K.; Noi, K.; Baba, K.; Tsuda, H.; et al. Ultrasonication-based rapid amplification of α-synuclein aggregates in cerebrospinal fluid. *Sci. Rep.* **2019**, *9*, 6001. [CrossRef]
98. Nakajima, K.; Noi, K.; Yamaguchi, K.; So, M.; Ikenaka, K.; Mochizuki, H.; Ogi, H.; Goto, Y. Optimized sonoreactor for accelerative amyloid-fibril assays through enhancement of primary nucleation and fragmentation. *Ultrason. Sonochem.* **2021**, *73*, 105508. [CrossRef]
99. Nakajima, K.; Ogi, H.; Adachi, K.; Noi, K.; Hirao, M.; Yagi, H.; Goto, Y. Nucleus factory on cavitation bubble for amyloid β fibril. *Sci. Rep.* **2016**, *6*, 22015. [CrossRef]
100. Giehm, L.; Otzen, D.E. Strategies to increase the reproducibility of protein fibrillization in plate reader assays. *Anal. Biochem.* **2010**, *400*, 270–281. [CrossRef] [PubMed]
101. Atarashi, R.; Satoh, K.; Sano, K.; Fuse, T.; Yamaguchi, N.; Ishibashi, D.; Matsubara, T.; Nakagaki, T.; Yamanaka, H.; Shirabe, S.; et al. Ultrasensitive human prion detection in cerebrospinal fluid by real-time quaking-induced conversion. *Nat. Med.* **2011**, *17*, 175–178. [CrossRef]

102. Paciotti, S.; Bellomo, G.; Gatticchi, L.; Parnetti, L. Are we ready for detecting α-synuclein prone to aggregation in patients? The case of "protein-misfolding cyclic amplification" and "real-time quaking-induced conversion" as diagnostic tools. *Front. Neurol.* **2018**, *9*, 415. [CrossRef]
103. Wang, Z.J.; Yasuhara, T. An examination of mobile spinal cord stimulators on treating parkinson disease. *Brain Circ.* **2021**, *7*, 8–12. [CrossRef] [PubMed]
104. Nakajima, K.; Toda, H.; Yamaguchi, K.; So, M.; Ikenaka, K.; Mochizuki, H.; Goto, Y.; Ogi, H. Half-time heat map reveals ultrasonic effects on morphology and kinetics of amyloidogenic aggregation reaction. *ACS Chem. Neurosci.* **2021**, *12*, 3456–3466. [CrossRef] [PubMed]
105. Banani, S.F.; Lee, H.O.; Hyman, A.A.; Rosen, M.K. Biomolecular condensates: Organizers of cellular biochemistry. *Nat. Rev. Mol. Cell Biol.* **2017**, *18*, 285–298. [CrossRef] [PubMed]
106. Alberti, S.; Hyman, A.A. Biomolecular condensates at the nexus of cellular stress, protein aggregation disease and ageing. *Nat. Rev. Mol. Cell Biol.* **2021**, *22*, 196–213. [CrossRef]
107. Murray, D.T.; Kato, M.; Lin, Y.; Thurber, K.R.; Hung, I.; McKnight, S.L.; Tycko, R. Structure of FUS protein fibrils and its relevance to self-assembly and phase separation of low-complexity domains. *Cell* **2017**, *171*, 615–627.e16. [CrossRef]
108. Levin, A.; Mason, T.O.; Adler-Abramovich, L.; Buell, A.K.; Meisl, G.; Galvagnion, C.; Bram, Y.; Stratford, S.A.; Dobson, C.M.; Knowles, T.P.; et al. Ostwald's rule of stages governs structural transitions and morphology of dipeptide supramolecular polymers. *Nat. Commun.* **2014**, *5*, 5219. [CrossRef]
109. Bracha, D.; Walls, M.T.; Wei, M.T.; Zhu, L.; Kurian, M.; Avalos, J.L.; Toettcher, J.E.; Brangwynne, C.P. Mapping local and global liquid phase behavior in living cells using photo-oligomerizable seeds. *Cell* **2018**, *175*, 1467–1480.e13. [CrossRef]
110. Bemporad, F.; Chiti, F. Protein misfolded oligomers: Experimental approaches, mechanism of formation, and structure-toxicity relationships. *Chem. Biol.* **2012**, *19*, 315–327. [CrossRef]
111. Miti, T.; Mulaj, M.; Schmit, J.D.; Muschol, M. Stable, metastable, and kinetically trapped amyloid aggregate phases. *Biomacromolecules* **2015**, *16*, 326–335. [CrossRef]
112. Adachi, M.; So, M.; Sakurai, K.; Kardos, J.; Goto, Y. Supersaturation-limited and unlimited phase transitions compete to produce the pathway complexity in amyloid fibrillation. *J. Biol. Chem.* **2015**, *290*, 18134–18145. [CrossRef]
113. Maji, S.K.; Perrin, M.H.; Sawaya, M.R.; Jessberger, S.; Vadodaria, K.; Rissman, R.A.; Singru, P.S.; Nilsson, K.P.; Simon, R.; Schubert, D.; et al. Functional amyloids as natural storage of peptide hormones in pituitary secretory granules. *Science* **2009**, *325*, 328–332. [CrossRef] [PubMed]
114. Yagi, H.; Mizuno, A.; So, M.; Hirano, M.; Adachi, M.; Akazawa-Ogawa, Y.; Hagihara, Y.; Ikenoue, T.; Lee, Y.H.; Kawata, Y.; et al. Ultrasonication-dependent formation and degradation of α-synuclein amyloid fibrils. *Biochim. Biophys. Acta* **2015**, *1854*, 209–217. [CrossRef] [PubMed]
115. Baldwin, A.J.; Knowles, T.P.; Tartaglia, G.G.; Fitzpatrick, A.W.; Devlin, G.L.; Shammas, S.L.; Waudby, C.A.; Mossuto, M.F.; Meehan, S.; Gras, S.L.; et al. Metastability of native proteins and the phenomenon of amyloid formation. *J. Am. Chem. Soc.* **2011**, *133*, 14160–14163. [CrossRef] [PubMed]

molecules

MDPI

Review

Conformational Variability of Amyloid-β and the Morphological Diversity of Its Aggregates †

Maho Yagi-Utsumi [1,2,*] and Koichi Kato [1,2,*]

1 Graduate School of Pharmaceutical Sciences, Nagoya City University, Nagoya 467-8603, Japan
2 Exploratory Research Center on Life and Living Systems and Institute for Molecular Science, National Institutes of Natural Sciences, Okazaki 444-8787, Japan
* Correspondence: mahoyagi@phar.nagoya-cu.ac.jp (M.Y.-U.); kkatonmr@ims.ac.jp (K.K.)

† This article is dedicated to the memory of the late Professor Sir Christopher M. Dobson, who made outstanding contributions to the advancement of studies of protein folding and misfolding.

Abstract: Protein folding is the most fundamental and universal example of biomolecular self-organization and is characterized as an intramolecular process. In contrast, amyloidogenic proteins can interact with one another, leading to protein aggregation. The energy landscape of amyloid fibril formation is characterized by many minima for different competing low-energy structures and, therefore, is much more enigmatic than that of multiple folding pathways. Thus, to understand the entire energy landscape of protein aggregation, it is important to elucidate the full picture of conformational changes and polymorphisms of amyloidogenic proteins. This review provides an overview of the conformational diversity of amyloid-β (Aβ) characterized from experimental and theoretical approaches. Aβ exhibits a high degree of conformational variability upon transiently interacting with various binding molecules in an unstructured conformation in a solution, forming an α-helical intermediate conformation on the membrane and undergoing a structural transition to the β-conformation of amyloid fibrils. This review also outlines the structural polymorphism of Aβ amyloid fibrils depending on environmental factors. A comprehensive understanding of the energy landscape of amyloid formation considering various environmental factors will promote drug discovery and therapeutic strategies by controlling the fibril formation pathway and targeting the consequent morphology of aggregated structures.

Keywords: aggregation; amyloid-β; cryo-electron microscopy; fibril; ganglioside; molecular chaperone; NMR spectroscopy

Citation: Yagi-Utsumi, M.; Kato, K. Conformational Variability of Amyloid-β and the Morphological Diversity of Its Aggregates. *Molecules* **2022**, *27*, 4787. https://doi.org/10.3390/molecules27154787

Academic Editor: Adrian Keller

Received: 4 July 2022
Accepted: 25 July 2022
Published: 26 July 2022

1. Introduction

Protein folding is the most fundamental and universal example of biomolecular self-organization and is characterized as an intramolecular process where nascent unfolded polypeptide chains assemble into their highly ordered native conformations [1]. In the protein folding process, the unfolded protein exhibits various conformations, passes through several folding intermediates, and descends on a potential free-energy surface toward a thermodynamically favorable native state. However, the process of correct protein folding can fail and polypeptide chains can fall into incorrect conformational states. These misfolded proteins can interact with one another, leading to protein aggregation [1–4]. In such misfolding, the amyloid fibrils are in one of the most stable thermodynamic states in the energy landscape [5]. Kinetically trapped misfolded intermediates are assumed to promote specific and nonspecific intermolecular interactions, thereby resulting in the assembly of various forms of aggregates such as oligomers, amorphous aggregates, and amyloid fibrils [5–7]. It is suggested that the energy landscape of amyloid fibril formation is characterized by a large number of minima for different competing low-energy structures [7–9] and, therefore, is much more enigmatic than that of multiple folding pathways.

Furthermore, not only the final amyloid structures but also the aggregation processes are significantly altered by various environmental factors. Thus, to understand the entire energy landscape of protein aggregation, it is important to elucidate the full picture of conformational changes and polymorphisms of amyloidogenic proteins, which can depend on environmental factors.

Amyloid-β (Aβ) is one of the most extensively studied amyloidogenic proteins, primarily because of its pathological significance associated with Alzheimer's disease (AD). Various experimental and theoretical approaches have been employed to characterize the structure and interactions of Aβ, revealing its conformational transformability. Intriguingly, the assembly of Aβ molecules is significantly promoted through an interaction with ganglioside GM1, which is a glycosphingolipid abundant in neuronal cell membranes [10–12]. In this review, we outline our current knowledge on the conformational transitions of Aβ, depending on the surrounding environments and binding molecules, highlighting its conformational variability and the morphological diversity of its aggregates.

2. Transient Interaction of Aβ in a Solution

Aβ is a product of the sequential cleavage of the type-I membrane glycoprotein, amyloid precursor protein (APP), by β- and γ-secretases [13]. The C-terminal region of Aβ is part of the transmembrane domain of APP and originally forms an α-helix conformation in the membrane [14]. After cleavage, the Aβ peptide dissociates from the membrane in the unstructured state and undergoes various structural changes [15]. Molecular dynamics (MD) simulations and nuclear magnetic resonance (NMR) spectroscopy have illustrated that monomeric Aβ conformation is mainly disordered in a solution and rapidly interconverts between many diverse conformational states [16]. Thus, it should be delineated as a conformational ensemble rather than a single dominant folded structure. The dynamic disordered state of Aβ is deeply involved in the process of its aggregation through transient Aβ–Aβ interactions. The dimerization and oligomerization processes of Aβ are well-characterized by MD simulations [17,18], indicating that the C-terminal regions of the Aβ molecules dominantly initiate their interactions.

The transient conformations of monomeric Aβ bound to large fibrils and oligomers have been indirectly observed by NMR techniques such as relaxation dispersion and saturation transfer experiments, which explored the invisible NMR states [19,20]. These data indicated that the central hydrophobic region of monomeric Aβ mainly mediates its interactions with the Aβ(1–40) oligomers whilst the C-terminal hydrophobic regions of both Aβ(1–40) and Aβ(1–42), along with their central hydrophobic regions, are involved in their interactions with the protofibril surface.

Aβ exhibits essential conformational plasticity and adaptability during transient weak interactions with other proteins, lipids, and chemical compounds. For instance, Aβ can interact with the spherical complex, displaying pentasaccharide moieties derived from ganglioside GM1 and enabling the observation of transient glycan–protein interactions [21] (Figure 1). NMR data, along with MD simulations, have indicated that the N-terminal segment of Aβ(1–40), especially the hydrophilic His13-His14-Gln15 segment, is selectively involved in the interaction with the GM1 pentasaccharide cluster whereas the C-terminal segment is scarcely involved in the interaction [21,22]. It has been reported that α-synuclein (αSN), an intrinsically disordered protein involved in Parkinson's disease, also forms weak encounter complexes with ganglioside-embedding small bicelles on an initial membrane-landing process of αSN [23]. Transient interactions of its N-terminal segment were observed for GM1 or GM2, but not GM3, which did not involve any secondary structure formation of αSN. In both Aβ and αSN, the initial encounters are mediated through their N-terminal *ganglioside-philic* segments without any secondary structure formation.

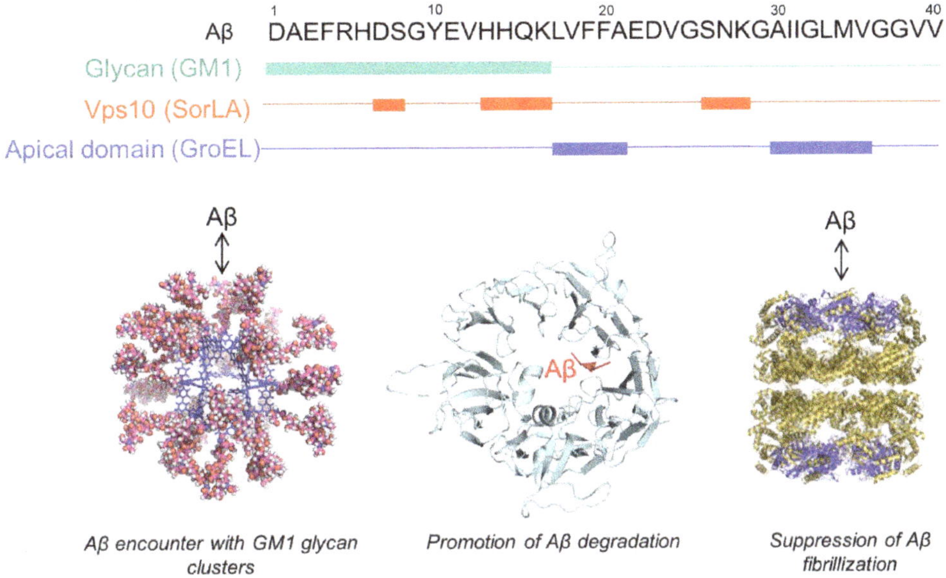

Figure 1. Transient interaction of Aβ with binding molecules. Each binding site on Aβ with the spherical complex displaying GM1 glycans (green), SorLA Vsp10 domain (orange), and apical domain of GroEL (blue) is represented with the primary structure of Aβ. The molecular graphics of GroEL and SorLA Vps10 domain with Aβ are based on PDB: 1KP8 and 3WSZ, respectively. The molecular graphics of the spherical complex displaying GM1 glycans are adopted with permission from reference [21]. 2015, WILEY-VCH Verlag GmbH & Co. KGaA, Weinheim, Germany.

Furthermore, it has been reported that various molecular chaperones such as heat shock proteins, prefoldins, and small heat shock proteins can bind Aβ and thereby inhibit its aggregation and mediate Aβ degradation via the ubiquitin-proteasome system or autophagy [24]. Molecular chaperones assist with the folding of unstructured nascent polypeptide chains into their native conformational state mostly by preventing their off-pathway intermolecular interactions in the energy landscape [5,25]. GroEL, a member of the chaperonine family of molecular chaperons, can suppress Aβ(1–40) amyloid formation by transiently interacting with its two hydrophobic segments, Leu_{17}-Ala_{21} and Ala_{30}-Val_{36} of Aβ(1–40), which contain key residues in fibril formation [26] (Figure 1). Intriguingly, the specific hydrophobic segment of αSN is capable of interacting with the eukaryotic chaperone PDI [27], the bacterial chaperone GroEL [28], and the archaeal chaperone PbaB [29], suggesting that αSN displays a *chaperone-philic* binding motif that can be widely recognized as a mimic of misfolded protein hallmarks. NMR data also indicate that Aβ as well as αSN, when noncovalently tethered to GroEL, remain largely unfolded and highly mobile.

Such dynamic and loose complexes have also been observed for the neuronal sorting receptor SorLA, which captures Aβ inside a tunnel to extend the β-sheet of one of its propeller blades [30] (Figure 1). In conjunction with X-ray crystallography, NMR spectroscopy demonstrated that Aβ can remain attached to SorLA whilst undergoing transitions among different bound states involving multiple capture sequences, suggesting that SorLA binds Aβ monomers through weak interactions and escorts them to lysosomes for degradation.

3. Assembly of the Intermediate Structures of Aβ on Membranes

The aggregation and deposition of Aβ on neuronal cell membranes are deeply involved in the pathogenesis of AD. Aβ can exhibit a free three-dimensional motion in an aqueous solution whilst Aβ molecular motion is restricted at the two-dimensional membrane interface, thereby facilitating Aβ–Aβ interactions [12,31]. Therefore, to understand the molecular mechanisms of Aβ fibrillization, it is necessary to identify the effects of spatial limitations at the membrane interface on the molecular motions and intermolecular interactions of Aβ molecules.

The ganglioside clusters are known to catalyze the self-assembly of amyloidogenic proteins such as Aβ, αSN, and prion protein through their interactions with gangliosides in a nonstoichiometric, but specific, manner [10,11,32–34]. Furthermore, amyloid fibrils on GM1-containing liposomes have been reported to be more toxic than those formed in an aqueous solution [35,36]. To provide structural insights into the conformational transition and molecular assembly of Aβ promoted in membrane environments, a series of NMR studies were carried out to characterize the interactions of Aβ with GM1 clusters by employing various membrane models [37–43] (Figure 2). The NMR data indicated that the GM1 clusters capture Aβ in an α-helical conformation at the hydrophobic/hydrophilic interfaces, restricting its spatial rearrangement: the two helical segments and the C-terminal portion of Aβ are in contact with the hydrophobic interior whilst leaving the remaining regions exposed to the hydrophilic environment of the GM1 cluster [42]. MD simulations have confirmed the topological mode of Aβ at the hydrophobic/hydrophilic interface [31]. The formation of the α-helical structure of Aβ has also been observed in membrane-mimicking micelles [44–46]. However, MD simulations have also indicated that Aβ α-helices are not stable and tend to form a β-hairpin structure because conformational entropy loss on the hairpin formation is smaller at the planar interface than in a free solution [31]. It is conceivable that such entropic effects, along with the higher local concentration of Aβ molecules at the hydrophobic/hydrophilic interface, facilitate their intermolecular interaction coupled with an α-to-β conformational transition on the ganglioside clusters, leading to amyloid fibril formation [31,40]. Indeed, Aβ bound to large, flat vesicles composed of 1,2-dimyristoyl-sn-glycero-3-phosphocholine forms a partially ordered conformation, in which only the C-terminal segments are involved in a parallel β-structure whilst leaving the N-terminal segment disordered [47]. Very recently, a nonfibrillar Aβ assemblage formed on a GM1-containing membrane was identified as a double-layered anti-parallel β-structure [43]. This unique assemblage itself was not transformed into fibrils, but rather provided a solvent-exposed hydrophobic surface that facilitated the conversion of monomeric Aβ into fibrils.

These findings suggest that the GM1 clusters offer a unique platform for binding coupled with the conformational transition of Aβ molecules, thereby restricting their spatial rearrangements to promote specific intermolecular interactions leading to cross-β-sheet formation (Figure 2). This raises a novel medicinal strategy to suppress β-structure formation by stabilizing the α-helical structure of Aβ on the ganglioside clusters. Indeed, it was reported that compounds such as N1-decanoyl-diethylenetriamine that bind and stabilize the α-helical state of Aβ attenuated fibril formation and consequent toxicity in a Drosophila model of AD [48]. On the other hand, hereditary mutations have a potential impact on these on-membrane molecular events [49] as exemplified by the Flemish-type mutation (A21G). This mutation disrupts the first α-helix identified in wild-type Aβ(1–40) bound to lyso-GM1 micelles, rendering the unstructured N-terminal segment tethered to the residual C-terminal helix [37]. Thus, the mutational effects on Aβ conformation depend on the surrounding environments.

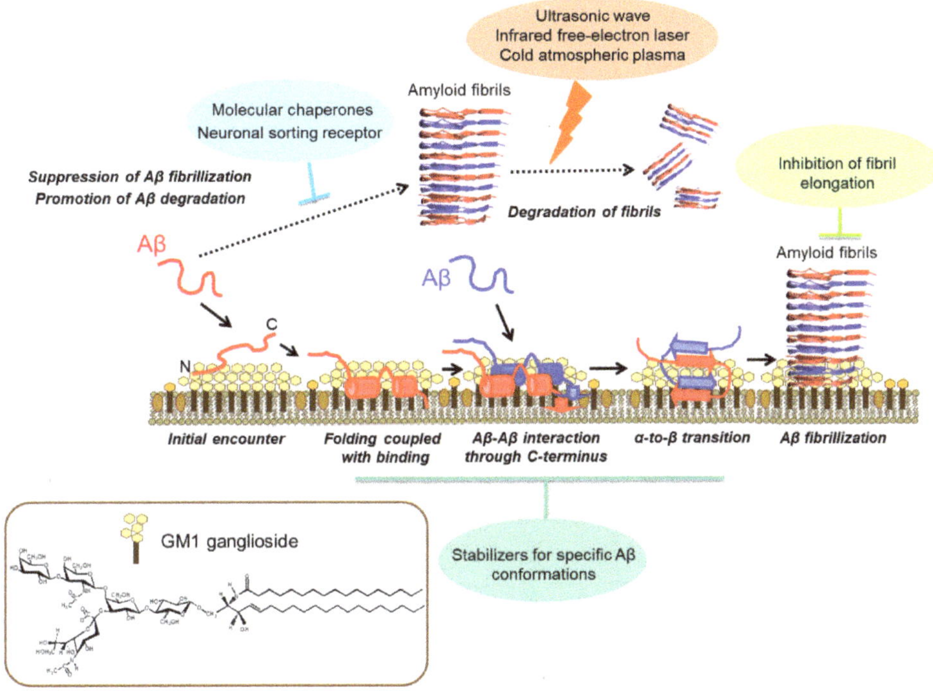

Figure 2. Schematic representation of the structural basis of the conformational transition and molecular assembly of Aβ promoted on GM1 ganglioside clusters on the neuronal cell membrane and the structure-based therapeutic strategies. After the initial encounter, the GM1 cluster captures Aβ at the hydrophobic/hydrophilic interface, which facilitates α-helix formation, thereby restricting the spatial rearrangements of Aβ molecules. Consequently, a specific intermolecular interaction between Aβ molecules is enhanced on the GM1 cluster, leading to their α-to-β conformational transition, resulting in amyloid fibril formation. Several proteins, including molecular chaperones, capture Aβ and thereby suppress its fibrillization. Irradiation with ultrasonic waves, an infrared free-electron laser, and cold atmospheric plasma can break down amyloid fibrils. Adapted with permission from reference [12]. 2019, The Pharmaceutical Society of Japan.

4. Structural Polymorphism of Amyloid Fibrils

Increasing structural data provided by cryo-electron microscopy (cryo-EM) and solid-state NMR spectroscopy demonstrate that the morphology of amyloid fibrils is significantly affected by various solution conditions such as the protein concentration, ionic strength, pH, temperature, and pressure [50–52]. X-ray diffraction studies have shown that amyloid fibrils share similar structural features characterized by a cross-β spine: a double β-sheet with each sheet running parallel to the fibril axis [53]. At the mesoscopic level, however, amyloid fibrils formed under the same conditions show considerable morphological diversity [54,55]. These molecular polymorphisms are assumed to be derived from differences in the number, relative orientation, and internal substructure of the protofilaments. Recent simulation studies have shown that the sequence-specific conformational heterogeneities of monomer ensembles are crucially associated with their aggregation propensities and the fibril polymorphisms can be caused by changes in the population of fibril-like states in the monomeric structures [56,57].

Solid-state NMR-derived high-resolution structural models have visualized that Aβ(1–42) fibrils adopt an S-shaped conformation [58–61] whereas Aβ(1–40) fibrils assume a U-shaped conformation [62,63] (Figure 3). Even in a U-shaped conformation, there is a variation in the interprotofilament interface in Aβ fibrils [64].

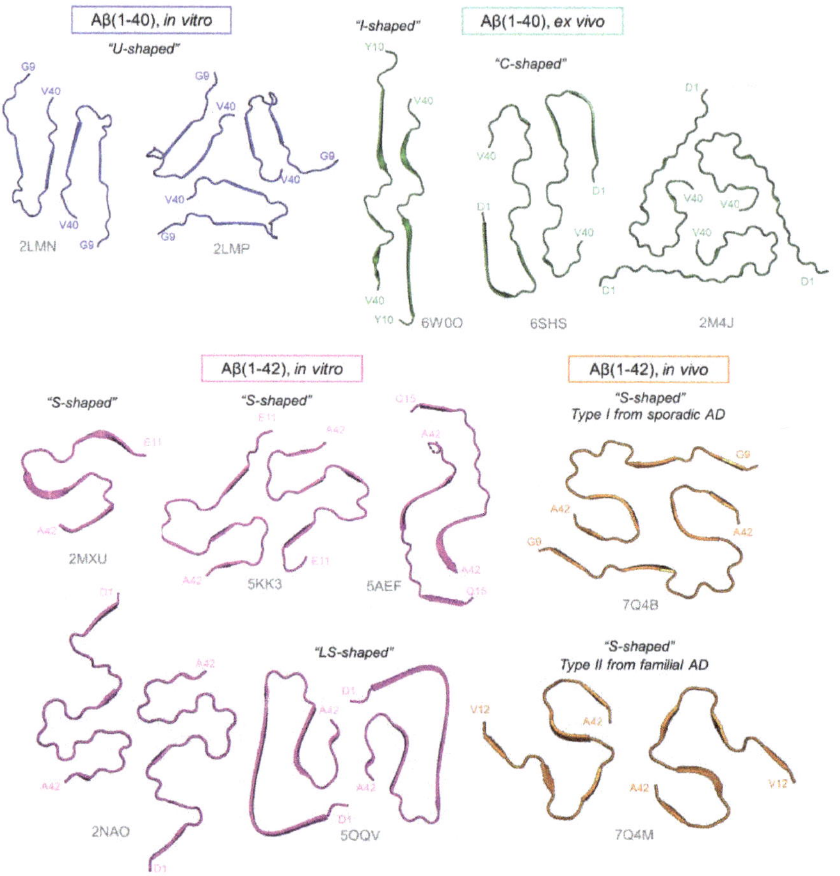

Figure 3. Aβ fibril structures solved by solid-state NMR and cryo-EM. The variety of fibril structures of Aβ(1–40) (blue, PDB: 2LMN, 2LMP) and Aβ(1–42) (magenta, PDB: 2MXU, 5KK3, 5AEF, 2NAO, 5OQV) fibrils prepared in vitro. Ex vivo, Aβ(1–40) seeded fibrils, which were formed by seed aggregation with recombinant Aβ(1–40) and ex vivo fibrils (green, PDB: 6W0O, 6SHS, 2M4J). The Aβ(1–42) fibrils were extracted from human AD brains (orange, PDB: 7Q4B, 7Q4M).

Recent breakthroughs in cryo-EM have yielded the atomic structures of Aβ filaments extracted from AD brains (Figure 3). The structures of the Aβ(1–42) filaments from human AD brains were identified by two types of S-shaped protofilament folds [65] whereas those of the filaments assembled in vitro had an overall LS-shaped topology of individual subunits in the cross-β structure [66]. In the case of Aβ(1–40) fibrils, high-resolution cryo-EM data identified the most prevalent polymorph for fibrils in typical AD patients as I-shaped protofilament folds [67]; another cryo-EM study determined C-shaped folds in brain-derived fibrils [68] where both the N- and C-terminal ends of Aβ were folded back onto the central peptide domain. These morphological differences suggest that Aβ fibrils may adopt disease-specific molecular conformers such as prion and tau strains, depending on the differences in individual brain environments [50,55,69]. Intriguingly, significant differences have also been found in the amyloid formation kinetics and fibril morphology between microgravity-grown and ground-grown Aβ(1–40) amyloids [70]. These data suggest that Aβ fibril formation on the ground is kinetically trapped in a metastable state whereas it proceeds more slowly through a thermodynamic control under microgravity conditions, resulting in the observed morphological differences in Aβ(1–40) fibrils.

The N-terminal regions adjacent to the fibril cores are often invisible or ambiguous in the solid-state NMR- and cryo-EM-derived structures of Aβ fibrils due to structural disorders and/or high mobility. Furthermore, it has been suggested that amyloid fibril cores themselves fluctuate and are heterogeneous, causing morphological diversity in one filament. An MD simulation based on the NMR-derived structural model of an Aβ(1–42) fibril indicated that the protomer at the growing end of an amyloid fibril adopts a β-hairpin conformation with less fluctuation compared with the flexible opposing terminal protomer [71]. These differences in the conformational fluctuation of the two ends of fibrils can explain the experimentally determined unidirectionality of fibril elongation [72,73].

It has been reported that Aβ fibrils can be broken down via irradiation with ultrasonic waves, an infrared free-electron laser, and cold atmospheric plasma by experimental and theoretical approaches [18,74] (Figure 2). Therefore, not only the suppression of the fibril formation but also the degradation of amyloid fibrils can be potential therapeutic strategies for neurodegenerative diseases in the future.

5. Conclusions

Aβ exhibits a high degree of conformational variability upon transiently interacting with binding molecules in an unstructured conformation in a solution, forming an α-helical intermediate conformation on the membrane and undergoing a structural transition to the β-conformation of amyloid fibrils. Despite the cumulative structural data, a comprehensive understanding of the molecular mechanisms behind amyloid polymorphisms remains largely unexplored as a variety of factors can influence the molecular assembly process. Recently, the accuracy of protein structure predictions based on deep learning methods has been dramatically improved and, in the case of natively folded proteins, their three-dimensional structures can now be reliably predicted from amino acid sequences [75,76]. However, it is currently difficult to accurately predict amyloid structures from amino acid sequences because amyloid fibrils are aggregates of many protomers that can form various polymorphic structures despite the same amino acid sequence [9]. Moreover, morphological diversity can be seen for amyloid fibrils grown in the same solution [70,77,78]. In addition, current machine learning methods are not yet capable of predicting protein folding and aggregation pathways.

For a detailed and integrated understanding of the energy landscape of protein aggregation, it is essential to characterize the structures of amyloid fibrils corresponding with the number of heterogeneous minima and to elucidate the amyloid formation processes, including the intermediate structures. Moreover, various environmental factors can significantly influence the amyloid structures and aggregation kinetics. Hence, a comprehensive understanding of the energy landscape of amyloid formation considering such environmental factors will promote drug discovery and therapeutic strategies by controlling the fibril formation pathway and targeting the consequent morphology of the aggregated structures. To address this issue, it is important to further accumulate high-quality data from experimental and computational approaches, to develop informatics-based methods for structure predictions, and to interpret these data from a physicochemical perspective.

Author Contributions: Conceptualization, M.Y.-U. and K.K.; writing—original draft preparation, M.Y.-U.; writing—review and editing, K.K. All authors have read and agreed to the published version of the manuscript.

Funding: This work was supported in part by JSPS KAKENHI (Grant Number JP19K07041 to M.Y.-U.) and by Grant-in-Aid for Research from Nagoya City University (Grant Numbers 2212008 and 2222004 to M.Y.-U).

Institutional Review Board Statement: Not applicable.

Informed Consent Statement: Not applicable.

Acknowledgments: We gratefully thank all collaborators in the research introduced in this review.

Conflicts of Interest: The authors declare no conflict of interest.

References

1. Dobson, C.M. Protein folding and misfolding. *Nature* **2003**, *426*, 884–890. [CrossRef]
2. Chiti, F.; Dobson, C.M. Protein Misfolding, Amyloid Formation, and Human Disease: A Summary of Progress over the Last Decade. *Annu. Rev. Biochem.* **2017**, *86*, 27–68. [CrossRef]
3. Chiti, F.; Dobson, C.M. Protein misfolding, functional amyloid, and human disease. *Annu. Rev. Biochem.* **2006**, *75*, 333–366. [CrossRef]
4. Knowles, T.P.; Vendruscolo, M.; Dobson, C.M. The amyloid state and its association with protein misfolding diseases. *Nat. Rev. Mol. Cell Biol.* **2014**, *15*, 384–396. [CrossRef]
5. Hartl, F.U.; Hayer-Hartl, M. Converging concepts of protein folding in vitro and in vivo. *Nat. Struct. Mol. Biol.* **2009**, *16*, 574–581. [CrossRef]
6. Jahn, T.R.; Radford, S.E. Folding versus aggregation: Polypeptide conformations on competing pathways. *Arch. Biochem. Biophys.* **2008**, *469*, 100–117. [CrossRef]
7. Adamcik, J.; Mezzenga, R. Amyloid Polymorphism in the Protein Folding and Aggregation Energy Landscape. *Angew. Chem. Int. Ed.* **2018**, *57*, 8370–8382. [CrossRef]
8. Eichner, T.; Radford, S.E. A diversity of assembly mechanisms of a generic amyloid fold. *Mol. Cell* **2011**, *43*, 8–18. [CrossRef]
9. Strodel, B. Energy Landscapes of Protein Aggregation and Conformation Switching in Intrinsically Disordered Proteins. *J. Mol. Biol.* **2021**, *433*, 167182. [CrossRef]
10. Matsuzaki, K.; Kato, K.; Yanagisawa, K. Aβ polymerization through interaction with membrane gangliosides. *Biochim. Biophys. Acta* **2010**, *1801*, 868–877. [CrossRef]
11. Matsuzaki, K.; Kato, K.; Yanagisawa, K. Ganglioside-Mediated Assembly of Amyloid β-Protein: Roles in Alzheimer's Disease. *Prog. Mol. Biol. Transl. Sci.* **2018**, *156*, 413–434.
12. Yagi-Utsumi, M. NMR Characterization of Conformational Dynamics and Molecular Assemblies of Proteins. *Biol. Pharm. Bull.* **2019**, *42*, 867–872. [CrossRef]
13. O'Brien, R.J.; Wong, P.C. Amyloid precursor protein processing and Alzheimer's disease. *Annu. Rev. Neurosci.* **2011**, *34*, 185–204. [CrossRef]
14. Song, Y.; Mittendorf, K.F.; Lu, Z.; Sanders, C.R. Impact of bilayer lipid composition on the structure and topology of the transmembrane amyloid precursor C99 protein. *J. Am. Chem. Soc.* **2014**, *136*, 4093–4096. [CrossRef]
15. Chen, Y.C. Impact of a discordant helix on β-amyloid structure, aggregation ability and toxicity. *Eur. Biophys. J.* **2017**, *46*, 681–687. [CrossRef]
16. Newby, F.N.; De Simone, A.; Yagi-Utsumi, M.; Salvatella, X.; Dobson, C.M.; Vendruscolo, M. Structure-Free Validation of Residual Dipolar Coupling and Paramagnetic Relaxation Enhancement Measurements of Disordered Proteins. *Biochemistry* **2015**, *54*, 6876–6886. [CrossRef]
17. Itoh, S.G.; Okumura, H. Dimerization process of amyloid-β(29-42) studied by the Hamiltonian replica-permutation molecular dynamics simulations. *J. Phys. Chem. B* **2014**, *118*, 11428–11436. [CrossRef]
18. Okumura, H.; Itoh, S.G. Molecular Dynamics Simulation Studies on the Aggregation of Amyloid-β Peptides and Their Disaggregation by Ultrasonic Wave and Infrared Laser Irradiation. *Molecules* **2022**, *27*, 2483. [CrossRef]
19. Fawzi, N.L.; Ying, J.; Ghirlando, R.; Torchia, D.A.; Clore, G.M. Atomic-resolution dynamics on the surface of amyloid-β protofibrils probed by solution NMR. *Nature* **2011**, *480*, 268–272. [CrossRef]
20. Fawzi, N.L.; Ying, J.; Torchia, D.A.; Clore, G.M. Kinetics of amyloid β monomer-to-oligomer exchange by NMR relaxation. *J. Am. Chem. Soc.* **2010**, *132*, 9948–9951. [CrossRef]
21. Sato, S.; Yoshimasa, Y.; Fujita, D.; Yagi-Utsumi, M.; Yamaguchi, T.; Kato, K.; Fujita, M. A Self-Assembled Spherical Complex Displaying a Gangliosidic Glycan Cluster Capable of Interacting with Amyloidogenic Proteins. *Angew. Chem. Int. Ed.* **2015**, *54*, 8435–8439. [CrossRef] [PubMed]
22. Tachi, Y.; Okamoto, Y.; Okumura, H. Conformational Change of Amyloid-β 40 in Association with Binding to GM1-Glycan Cluster. *Sci. Rep.* **2019**, *9*, 6853. [CrossRef]
23. Yamaguchi, T.; Uno, T.; Uekusa, Y.; Yagi-Utsumi, M.; Kato, K. Ganglioside-embedding small bicelles for probing membrane-landing processes of intrinsically disordered proteins. *Chem. Commun.* **2013**, *49*, 1235–1237. [CrossRef] [PubMed]
24. Balchin, D.; Hayer-Hartl, M.; Hartl, F.U. In vivo aspects of protein folding and quality control. *Science* **2016**, *353*, aac4354. [CrossRef]
25. Hartl, F.U.; Bracher, A.; Hayer-Hartl, M. Molecular chaperones in protein folding and proteostasis. *Nature* **2011**, *475*, 324–332. [CrossRef]
26. Yagi-Utsumi, M.; Kunihara, T.; Nakamura, T.; Uekusa, Y.; Makabe, K.; Kuwajima, K.; Kato, K. NMR characterization of the interaction of GroEL with amyloid β as a model ligand. *FEBS Lett.* **2013**, *587*, 1605–1609. [CrossRef]

27. Yagi-Utsumi, M.; Satoh, T.; Kato, K. Structural basis of redox-dependent substrate binding of protein disulfide isomerase. *Sci. Rep.* **2015**, *5*, 13909. [CrossRef]
28. Nishida, N.; Yagi-Utsumi, M.; Motojima, F.; Yoshida, M.; Shimada, I.; Kato, K. Nuclear magnetic resonance approaches for characterizing interactions between the bacterial chaperonin GroEL and unstructured proteins. *J. Biosci. Bioeng.* **2013**, *116*, 160–164. [CrossRef] [PubMed]
29. Sugiyama, M.; Yagi, H.; Yamaguchi, T.; Kumoi, K.; Hirai, M.; Oba, Y.; Sato, N.; Porcar, L.; Martel, A.; Kato, K. Conformational characterization of a protein complex involving intrinsically disordered protein by small-angle neutron scattering using the inverse contrast matching method: A case study of interaction between α-synuclein and PbaB tetramer as a model chaperone. *J. Appl. Cryst.* **2014**, *47*, 430–435.
30. Kitago, Y.; Nagae, M.; Nakata, Z.; Yagi-Utsumi, M.; Takagi-Niidome, S.; Mihara, E.; Nogi, T.; Kato, K.; Takagi, J. Structural basis for amyloidogenic peptide recognition by sorLA. *Nat. Struct. Mol. Biol.* **2015**, *22*, 199–206. [CrossRef]
31. Itoh, S.G.; Yagi-Utsumi, M.; Kato, K.; Okumura, H. Effects of a Hydrophilic/Hydrophobic Interface on Amyloid-β Peptides Studied by Molecular Dynamics Simulations and NMR Experiments. *J. Phys. Chem. B* **2018**, *123*, 160–169. [CrossRef]
32. Ariga, T.; McDonald, M.P.; Yu, R.K. Role of ganglioside metabolism in the pathogenesis of Alzheimer's disease—a review. *J. Lipid Res.* **2008**, *49*, 1157–1175. [CrossRef]
33. Fantini, J.; Yahi, N. Molecular insights into amyloid regulation by membrane cholesterol and sphingolipids: Common mechanisms in neurodegenerative diseases. *Expert Rev. Mol. Med.* **2010**, *12*, e27. [CrossRef] [PubMed]
34. Fantini, J.; Yahi, N. The driving force of α-synuclein insertion and amyloid channel formation in the plasma membrane of neural cells: Key role of ganglioside- and cholesterol-binding domains. *Adv. Exp. Med. Biol.* **2013**, *991*, 15–26. [PubMed]
35. Okada, T.; Wakabayashi, M.; Ikeda, K.; Matsuzaki, K. Formation of toxic fibrils of Alzheimer's amyloid β-protein-(1-40) by monosialoganglioside GM1, a neuronal membrane component. *J. Mol. Biol.* **2007**, *371*, 481–489. [CrossRef]
36. Okada, Y.; Okubo, K.; Ikeda, K.; Yano, Y.; Hoshino, M.; Hayashi, Y.; Kiso, Y.; Itoh-Watanabe, H.; Naito, A.; Matsuzaki, K. Toxic Amyloid Tape: A Novel Mixed Antiparallel/Parallel β-Sheet Structure Formed by Amyloid beta-Protein on GM1 Clusters. *ACS Chem. Neurosci.* **2019**, *10*, 563–572. [CrossRef]
37. Yagi-Utsumi, M.; Dobson, C.M. Conformational Effects of the A21G Flemish Mutation on the Aggregation of Amyloid β Peptide. *Biol. Pharm. Bull.* **2015**, *38*, 1668–1672. [CrossRef]
38. Yagi-Utsumi, M.; Kameda, T.; Yamaguchi, Y.; Kato, K. NMR characterization of the interactions between lyso-GM1 aqueous micelles and amyloid β. *FEBS Lett.* **2010**, *584*, 831–836. [CrossRef] [PubMed]
39. Yagi-Utsumi, M.; Kato, K. Structural and dynamic views of GM1 ganglioside. *Glycoconj. J.* **2015**, *32*, 105–112. [CrossRef]
40. Yagi-Utsumi, M.; Matsuo, K.; Yanagisawa, K.; Gekko, K.; Kato, K. Spectroscopic Characterization of Intermolecular Interaction of Amyloid β Promoted on GM1 Micelles. *Int. J. Alzheimer's Dis.* **2010**, *2011*, 925073.
41. Yagi-Utsumi, M.; Yamaguchi, T.; Uekusa, Y.; Kato, K. NMR Characterization of the Conformations, Dynamics, and Interactions of Glycosphingolipids. In *Glycoscience and Glycotechnology*; Kato, K., Peters, T., Eds.; RSC Publishing: Cambridge, UK, 2017; pp. 161–178.
42. Utsumi, M.; Yamaguchi, Y.; Sasakawa, H.; Yamamoto, N.; Yanagisawa, K.; Kato, K. Up-and-down topological mode of amyloid β-peptide lying on hydrophilic/hydrophobic interface of ganglioside clusters. *Glycoconj. J.* **2009**, *26*, 999–1006. [CrossRef]
43. Yagi-Utsumi, M.; Itoh, S.G.; Okumura, H.; Yanagisawa, K.; Kato, K.; Nishimura, K. The double-layered structure of amyloid-β assemblage on GM1-containing membranes catalytically promotes fibrillization. *bioRxiv* **2022**. [CrossRef]
44. Jarvet, J.; Danielsson, J.; Damberg, P.; Oleszczuk, M.; Graslund, A. Positioning of the Alzheimer Aβ(1-40) peptide in SDS micelles using NMR and paramagnetic probes. *J. Biomol. NMR* **2007**, *39*, 63–72. [CrossRef]
45. Wahlstrom, A.; Hugonin, L.; Peralvarez-Marin, A.; Jarvet, J.; Graslund, A. Secondary structure conversions of Alzheimer's Aβ(1-40) peptide induced by membrane-mimicking detergents. *FEBS J.* **2008**, *275*, 5117–5128. [CrossRef] [PubMed]
46. Osterlund, N.; Kulkarni, Y.S.; Misiaszek, A.D.; Wallin, C.; Kruger, D.M.; Liao, Q.; Mashayekhy Rad, F.; Jarvet, J.; Strodel, B.; Warmlander, S.; et al. Amyloid-β Peptide Interactions with Amphiphilic Surfactants: Electrostatic and Hydrophobic Effects. *ACS Chem. Neurosci.* **2018**, *9*, 1680–1692. [CrossRef]
47. Yagi-Utsumi, M.; Kato, K.; Nishimura, K. Membrane-Induced Dichotomous Conformation of Amyloid β with the Disordered N-Terminal Segment Followed by the Stable C-Terminal beta Structure. *PLoS ONE* **2016**, *11*, e0146405. [CrossRef] [PubMed]
48. Nerelius, C.; Sandegren, A.; Sargsyan, H.; Raunak, R.; Leijonmarck, H.; Chatterjee, U.; Fisahn, A.; Imarisio, S.; Lomas, D.A.; Crowther, D.C.; et al. α-helix targeting reduces amyloid-β peptide toxicity. *Proc. Natl. Acad. Sci. USA* **2009**, *106*, 9191–9196. [CrossRef]
49. Yamamoto, N.; Matsuzaki, K.; Yanagisawa, K. Cross-seeding of wild-type and hereditary variant-type amyloid β-proteins in the presence of gangliosides. *J. Neurochem.* **2005**, *95*, 1167–1176. [CrossRef]
50. Creekmore, B.C.; Chang, Y.W.; Lee, E.B. The Cryo-EM Effect: Structural Biology of Neurodegenerative Disease Aggregates. *J. Neuropathol Exp. Neurol.* **2021**, *80*, 514–529. [CrossRef]
51. Tycko, R. Molecular Structure of Aggregated Amyloid-β: Insights from Solid-State Nuclear Magnetic Resonance. *Cold Spring Harb. Perspect Med.* **2016**, *6*, a024083. [CrossRef]
52. Oren, O.; Taube, R.; Papo, N. Amyloid β structural polymorphism, associated toxicity and therapeutic strategies. *Cell. Mol. Life Sci.* **2021**, *78*, 7185–7198. [CrossRef]

53. Stromer, T.; Serpell, L.C. Structure and morphology of the Alzheimer's amyloid fibril. *Microsc. Res. Tech.* **2005**, *67*, 210–217. [CrossRef] [PubMed]
54. Gallardo, R.; Ranson, N.A.; Radford, S.E. Amyloid structures: Much more than just a cross-β fold. *Curr. Opin. Struct. Biol.* **2020**, *60*, 7–16. [CrossRef]
55. Lutter, L.; Aubrey, L.D.; Xue, W.F. On the Structural Diversity and Individuality of Polymorphic Amyloid Protein Assemblies. *J. Mol. Biol.* **2021**, *433*, 167124. [CrossRef] [PubMed]
56. Chakraborty, D.; Straub, J.E.; Thirumalai, D. Differences in the free energies between the excited states of Abeta40 and Abeta42 monomers encode their aggregation propensities. *Proc. Natl. Acad. Sci. USA* **2020**, *117*, 19926–19937. [CrossRef] [PubMed]
57. Kumar, A.; Chakraborty, D.; Mugnai, M.L.; Straub, J.E.; Thirumalai, D. Sequence Determines the Switch in the Fibril Forming Regions in the Low-Complexity FUS Protein and Its Variants. *J. Phys. Chem. Lett.* **2021**, *12*, 9026–9032. [CrossRef]
58. Schmidt, M.; Rohou, A.; Lasker, K.; Yadav, J.K.; Schiene-Fischer, C.; Fandrich, M.; Grigorieff, N. Peptide dimer structure in an Abeta(1-42) fibril visualized with cryo-EM. *Proc. Natl. Acad. Sci. USA* **2015**, *112*, 11858–11863. [CrossRef] [PubMed]
59. Walti, M.A.; Ravotti, F.; Arai, H.; Glabe, C.G.; Wall, J.S.; Bockmann, A.; Guntert, P.; Meier, B.H.; Riek, R. Atomic-resolution structure of a disease-relevant Abeta(1-42) amyloid fibril. *Proc. Natl. Acad. Sci. USA* **2016**, *113*, E4976–E4984. [CrossRef] [PubMed]
60. Xiao, Y.; Ma, B.; McElheny, D.; Parthasarathy, S.; Long, F.; Hoshi, M.; Nussinov, R.; Ishii, Y. Abeta(1-42) fibril structure illuminates self-recognition and replication of amyloid in Alzheimer's disease. *Nat. Struct. Mol. Biol.* **2015**, *22*, 499–505. [CrossRef]
61. Colvin, M.T.; Silvers, R.; Ni, Q.Z.; Can, T.V.; Sergeyev, I.; Rosay, M.; Donovan, K.J.; Michael, B.; Wall, J.; Linse, S.; et al. Atomic Resolution Structure of Monomorphic Aβ42 Amyloid Fibrils. *J. Am. Chem. Soc.* **2016**, *138*, 9663–9674. [CrossRef]
62. Paravastu, A.K.; Leapman, R.D.; Yau, W.M.; Tycko, R. Molecular structural basis for polymorphism in Alzheimer's β-amyloid fibrils. *Proc. Natl. Acad. Sci. USA* **2008**, *105*, 18349–18354. [CrossRef] [PubMed]
63. Lu, J.X.; Qiang, W.; Yau, W.M.; Schwieters, C.D.; Meredith, S.C.; Tycko, R. Molecular structure of β-amyloid fibrils in Alzheimer's disease brain tissue. *Cell* **2013**, *154*, 1257–1268. [CrossRef] [PubMed]
64. Bertini, I.; Gonnelli, L.; Luchinat, C.; Mao, J.; Nesi, A. A new structural model of Aβ40 fibrils. *J. Am. Chem. Soc.* **2011**, *133*, 16013–16022. [CrossRef] [PubMed]
65. Yang, Y.; Arseni, D.; Zhang, W.; Huang, M.; Lovestam, S.; Schweighauser, M.; Kotecha, A.; Murzin, A.G.; Peak-Chew, S.Y.; Macdonald, J.; et al. Cryo-EM structures of amyloid-β 42 filaments from human brains. *Science* **2022**, *375*, 167–172. [CrossRef] [PubMed]
66. Gremer, L.; Scholzel, D.; Schenk, C.; Reinartz, E.; Labahn, J.; Ravelli, R.B.G.; Tusche, M.; Lopez-Iglesias, C.; Hoyer, W.; Heise, H.; et al. Fibril structure of amyloid-β(1-42) by cryo-electron microscopy. *Science* **2017**, *358*, 116–119. [CrossRef]
67. Ghosh, U.; Thurber, K.R.; Yau, W.M.; Tycko, R. Molecular structure of a prevalent amyloid-β fibril polymorph from Alzheimer's disease brain tissue. *Proc. Natl. Acad. Sci. USA* **2021**, *118*, e2023089118. [CrossRef]
68. Kollmer, M.; Close, W.; Funk, L.; Rasmussen, J.; Bsoul, A.; Schierhorn, A.; Schmidt, M.; Sigurdson, C.J.; Jucker, M.; Fandrich, M. Cryo-EM structure and polymorphism of Aβ amyloid fibrils purified from Alzheimer's brain tissue. *Nat. Commun.* **2019**, *10*, 4760. [CrossRef]
69. Chatani, E.; Yuzu, K.; Ohhashi, Y.; Goto, Y. Current Understanding of the Structure, Stability and Dynamic Properties of Amyloid Fibrils. *Int. J. Mol. Sci.* **2021**, *22*, 4349. [CrossRef]
70. Yagi-Utsumi, M.; Yanaka, S.; Song, C.; Satoh, T.; Yamazaki, C.; Kasahara, H.; Shimazu, T.; Murata, K.; Kato, K. Characterization of amyloid β fibril formation under microgravity conditions. *NPJ Microgravity* **2020**, *6*, 17. [CrossRef]
71. Okumura, H.; Itoh, S.G. Structural and fluctuational difference between two ends of Aβ amyloid fibril: MD simulations predict only one end has open conformations. *Sci. Rep.* **2016**, *6*, 38422. [CrossRef]
72. Ban, T.; Hoshino, M.; Takahashi, S.; Hamada, D.; Hasegawa, K.; Naiki, H.; Goto, Y. Direct observation of Aβ amyloid fibril growth and inhibition. *J. Mol. Biol.* **2004**, *344*, 757–767. [CrossRef]
73. Young, L.J.; Kaminski Schierle, G.S.; Kaminski, C.F. Imaging Aβ(1-42) fibril elongation reveals strongly polarised growth and growth incompetent states. *Phys. Chem. Chem. Phys.* **2017**, *19*, 27987–27996. [CrossRef]
74. Yagi-Utsumi, M.; Tanaka, T.; Otsubo, Y.; Yamashita, A.; Yoshimura, S.; Nishida, M.; Kato, K. Cold Atmospheric Plasma Modification of Amyloid β. *Int. J. Mol. Sci.* **2021**, *22*, 3116. [CrossRef]
75. Jumper, J.; Evans, R.; Pritzel, A.; Green, T.; Figurnov, M.; Ronneberger, O.; Tunyasuvunakool, K.; Bates, R.; Zidek, A.; Potapenko, A.; et al. Highly accurate protein structure prediction with AlphaFold. *Nature* **2021**, *596*, 583–589. [CrossRef]
76. Baek, M.; DiMaio, F.; Anishchenko, I.; Dauparas, J.; Ovchinnikov, S.; Lee, G.R.; Wang, J.; Cong, Q.; Kinch, L.N.; Schaeffer, R.D.; et al. Accurate prediction of protein structures and interactions using a three-track neural network. *Science* **2021**, *373*, 871–876. [CrossRef]
77. Fandrich, M.; Meinhardt, J.; Grigorieff, N. Structural polymorphism of Alzheimer Aβ and other amyloid fibrils. *Prion* **2009**, *3*, 89–93. [CrossRef]
78. Fandrich, M.; Schmidt, M.; Grigorieff, N. Recent progress in understanding Alzheimer's b-amyloid structures. *Trends Biochem. Sci.* **2011**, *36*, 338–345. [CrossRef]

molecules

MDPI

Review

DMSO-Quenched H/D-Exchange 2D NMR Spectroscopy and Its Applications in Protein Science †

Kunihiro Kuwajima [1,*], Maho Yagi-Utsumi [2,3,4], Saeko Yanaka [2,3] and Koichi Kato [2,3,4,*]

1 Department of Physics, School of Science, University of Tokyo, 7-3-1 Hongo, Bunkyo-ku, Tokyo 113-0033, Japan
2 Exploratory Research Center on Life and Living Systems and Institute for Molecular Science, National Institutes of Natural Sciences, 5-1 Higashiyama, Myodaiji, Okazaki 444-8787, Aichi, Japan; mahoyagi@ims.ac.jp (M.Y.-U.); saeko-yanaka@ims.ac.jp (S.Y.)
3 Department of Functional Molecular Science, School of Physical Sciences, SOKENDAI (the Graduate University for Advanced Studies), 5-1 Higashiyama, Myodaiji, Okazaki 444-8787, Aichi, Japan
4 Graduate School of Pharmaceutical Sciences, Nagoya City University, 3-1 Tanabe-dori, Mizuho-ku, Nagoya 467-8603, Aichi, Japan
* Correspondence: kuwajima@ims.ac.jp (K.K.); kkatonmr@ims.ac.jp (K.K.)

† This article is dedicated to the memory of the late Professor Sir Christopher M. Dobson, who made outstanding contributions to the advancement of studies of protein folding and the related areas and played an irreplaceable role in the promotion of protein science.

Abstract: Hydrogen/deuterium (H/D) exchange combined with two-dimensional (2D) NMR spectroscopy has been widely used for studying the structure, stability, and dynamics of proteins. When we apply the H/D-exchange method to investigate non-native states of proteins such as equilibrium and kinetic folding intermediates, H/D-exchange quenching techniques are indispensable, because the exchange reaction is usually too fast to follow by 2D NMR. In this article, we will describe the dimethylsulfoxide (DMSO)-quenched H/D-exchange method and its applications in protein science. In this method, the H/D-exchange buffer is replaced by an aprotic DMSO solution, which quenches the exchange reaction. We have improved the DMSO-quenched method by using spin desalting columns, which are used for medium exchange from the H/D-exchange buffer to the DMSO solution. This improvement has allowed us to monitor the H/D exchange of proteins at a high concentration of salts or denaturants. We describe methodological details of the improved DMSO-quenched method and present a case study using the improved method on the H/D-exchange behavior of unfolded human ubiquitin in 6 M guanidinium chloride.

Keywords: hydrogen/deuterium exchange; dimethylsulfoxide; nuclear magnetic resonance

Citation: Kuwajima, K.; Yagi-Utsumi, M.; Yanaka, S.; Kato, K. DMSO-Quenched H/D-Exchange 2D NMR Spectroscopy and Its Applications in Protein Science. *Molecules* **2022**, *27*, 3748. https://doi.org/10.3390/molecules27123748

Academic Editor: Bernhard Loll

Received: 4 May 2022
Accepted: 7 June 2022
Published: 10 June 2022

1. Introduction

Hydrogen/deuterium (H/D) exchange is a powerful tool to investigate the structure, stability, dynamics, and interactions of proteins [1–4]. The advancements in two-dimensional (2D) nuclear magnetic resonance (NMR) techniques have made it possible to monitor the H/D-exchange kinetics of the individually identified peptide amide (NH) protons of proteins at amino-acid residue resolution [5,6]. Recent H/D-exchange studies also employ a mass spectrometric (MS) technique combined with rapid proteolysis and HPLC separation, which allows us to obtain the exchange kinetics at a nearly single-residue resolution [4,7,8]. In H/D-exchange experiments, the exchange reaction of each NH group of a protein to ND takes place in D_2O. The exchange kinetics of the NH proton are determined by a structural opening reaction of the NH group and the intrinsic exchange rate constant, k_{int}, of the NH proton, and are represented by:

$$NH_{cl} \underset{k_{cl}}{\overset{k_{op}}{\rightleftharpoons}} NH_{op} \xrightarrow{k_{int}} ND, \quad (1)$$

where NH_{cl} and NH_{op} denote the closed and open states of the NH group, ND indicates the amide group after H/D exchange, and k_{op} and k_{cl} are the opening and closing rate constants [1–4]. Under the steady-state condition, the observed H/D-exchange rate constant, k_{obs}, is thus given by:

$$k_{obs} = \left(\frac{k_{op}}{k_{op} + k_{cl} + k_{int}}\right) \cdot k_{int}. \quad (2)$$

The k_{int} of each NH proton of a protein depends on solution conditions (pH, temperature, and salt concentration), neighboring residues (i.e., the amino-acid sequence), and isotope effects. Englander and his colleagues [9–12] accurately calibrated these effects on k_{int} and reported the calibration parameters, from which we can quantitatively estimate the k_{int} value of each NH proton of the protein.

The protection factor, P, which represents the degree of protection of the NH group against H/D exchange is given by the ratio of k_{int} to k_{obs} as:

$$P = \frac{k_{int}}{k_{obs}}. \quad (3)$$

The P value of native globular proteins is as large as 10^6–10^9, which corresponds to the structural opening free energy of 8–13 kcal/mol. This free energy change is equivalent to the free energy change of the global unfolding for each protein, indicating that the H/D-exchange reactions of the most stable NH groups are brought about by global unfolding [13,14].

A native-state H/D-exchange method was developed by Bai et al. [15] in 1995, and the H/D-exchange behavior at low concentrations of denaturants was characterized by this method. Temperature or pressure perturbation was also used in the native-state H/D exchange [16–18]. The method is effective for studying different kinds of protein dynamic behavior, ranging from local fluctuations up to sub-global and whole-molecule global unfolding reactions [15,19–21]. The method has been applied to a large number of proteins (reviewed in [4,19–21]).

The H/D-exchange techniques have also been used effectively for studying non-native states, including equilibrium unfolding intermediates and transient intermediates in kinetic refolding reactions of proteins. The molten globule (MG) state, which has a substantial amount of secondary structure but lacks the specific tertiary side-chain packing characteristics of native proteins, is an equilibrium intermediate state under mildly denaturing conditions for numerous globular proteins, many with more than ~100 residues [22–24]. In the 1990s, the structural characterizations of the MG state by H/D-exchange 2D NMR were carried out for a number of globular proteins, including apomyoglobin [25], cytochrome *c* [26,27], α-lactalbumin [28–32], Ca^{2+}-binding milk lysozyme [33,34], and other proteins [35–41]. The P values of slowly exchanging NH protons in the MG state range from 10^2 to 10^3, which is more than three orders of magnitude smaller than the values in the native (N) state. The H/D-exchange rate in the MG state and in other non-native unfolded states [42–52] is usually too fast to follow by 2D NMR spectroscopy. Therefore, the H/D exchange was quenched after the desired period of H/D exchange by rapid refolding, and the NMR spectra were measured in the N state.

The use of a hydrogen-exchange method to detect and characterize a transient folding intermediate of proteins was first reported by Schmid and Baldwin [53] in 1979. They used a tritium-exchange technique and carried out a kinetic competition between folding and hydrogen exchange to investigate the folding kinetics of ribonuclease A. Subsequently, Roder and Wüthrich [54] proposed an extension of this technique by combined use of rapid mixing and NMR analysis. In 1988, two seminal papers, one by Udgaonkar and Baldwin [55] for ribonuclease A and the other by Roder et al. [56] for oxidized cytochrome *c*, appeared and reported the structural characterization of kinetic folding intermediates by hydrogen-exchange labeling and 2D NMR spectroscopy. In both studies, the NH groups of proteins were first fully deuterated in the fully unfolded (U) state in D_2O, and the nonprotected amide ND groups in folding intermediates were proton-labeled in H_2O by a short alkaline pH pulse (pulse-labeling hydrogen exchange), followed by quenching the D/H exchange by rapid refolding to the N state and NMR measurements in the N state. Since then, the pulse-labeling or competition hydrogen-exchange studies combined with 2D NMR to detect and characterize transient folding intermediates have been reported for many globular proteins, including barnase by the Fersht group [57,58], the lysozyme–α-lactalbumin family proteins by the Chris Dobson group [59–63], apomyoglobin and apoleghemoglobin by the Wright group [64–67], and other proteins by other groups [41,68–91]. The early transient folding intermediates thus characterized are very similar in structure and stability to the equilibrium MG state, indicating that the MG state is the equilibrium counterpart of the kinetic folding intermediate formed early during refolding from the U state [23]. However, there are exceptions to this rule. The kinetic folding intermediate of the plant globin apoleghemoglobin under the refolding condition is significantly different in structure from its equilibrium MG state [66]. The early kinetic folding intermediate of ribonuclease H from *Escherichia coli* (*E. coli*) has a well-folded region with a closely packed tertiary structure, which is absent in its equilibrium MG state [92].

In this article, we describe the dimethylsulfoxide (DMSO)-quenched H/D-exchange method, in which the H/D-exchange reaction is quenched by the DMSO solution [93]. As shown above, the hydrogen-exchange reactions in the MG state, other non-native states, and transient kinetic folding intermediates of proteins need to be quenched before measurements of 2D NMR spectra. The rapid refolding was used for quenching the exchange reactions in the above studies, and hence, only NH protons stably protected in the N state are available for analysis. The DMSO-quenched H/D-exchange method is more versatile, because the exchange reactions of all NH groups, including nonprotected NH groups in the N state, are effectively quenched in an aprotic solvent DMSO and are available for the NMR analysis [93]. In the following, we will give a brief historical summary of the DMSO-quenched H/D-exchange method in proteins. In addition, the DMSO-quenched H/D-exchange method has recently been improved by the use of spin desalting columns [94]. This improvement has made it possible to apply the DMSO-quenched method to the exchange reactions of proteins in the presence of a high concentration of salt or denaturant. We thus describe the methodological details of the use of spin desalting columns in the DMSO-quenched H/D-exchange method and present a study on the H/D-exchange behavior of unfolded ubiquitin in 6 M guanidinium chloride (GdmCl), in which the spin desalting column was used in the DMSO-quenched H/D-exchange 2D NMR experiments.

2. DMSO-Quenched H/D-Exchange Method

The DMSO-quenched H/D-exchange method was developed by Zhang et al. [93] in 1995. They investigated various solution conditions to minimize the H/D-exchange rate of NH protons of proteins, and the presence of 95% (v/v) DMSO-d_6 in a DMSO-d_6/D_2O mixture at pH* 5–6 was the best condition, where the H/D-exchange rate was ~100 fold slower than the minimum exchange rate in D_2O. Here, pH* is the uncorrected pH-meter reading, and the pH* was adjusted by dichloroacetic acid-d_2 (DCA-d_2), whose pK_a is 5.72 in the DMSO/D_2O mixture [93]. To use the DMSO-quenched H/D-exchange method to investigate the H/D-exchange reaction of a protein, aliquots of the reaction mixture at various exchange times are first quenched by rapid freezing in liquid nitrogen, and then lyophilized. The lyophilized powder is dissolved in the quenching DMSO solution, and the NMR spectra of the protein are measured. Because proteins are unfolded in the DMSO solution, the NMR peaks may be distributed in a very narrow spectral region. However, the use of isotope (^{15}N and ^{13}C)-enriched proteins and the triple-resonance multi-dimensional NMR techniques [95] have made it feasible to identify most of the cross peaks in the 2D ^{1}H–^{15}N heteronuclear single-quantum coherence (HSQC) spectrum [96] of a small protein and to follow the H/D-exchange kinetics of individually identified NH protons.

2.1. Applications to Folding Intermediates and Amyloid Fibrils

Nishimura et al. [65,66,97] applied the DMSO-quenched H/D-exchange method to elucidate the structure in the equilibrium and transient folding intermediates of apomyoglobin and apoleghemoglobin; they used 99.4% DMSO instead of the 95% DMSO solution as a quenching solvent, however. Using the DMSO-quenched H/D-exchange method, they acquired data for the NH protons of 94 residues for the 153 residues of apomyoglobin, as compared with the 52 residues probed by the conventional pulse-labeling hydrogen-exchange method, in which the exchange was quenched by rapid refolding [97]. The DMSO-quenched method could be applied only for pH-jump refolding experiments because it was difficult to dissolve the lyophilized protein in DMSO in the presence of residual denaturant (urea or GdmCl) after denaturant-jump experiments [65,66]. Sakamoto et al. [98] studied the H/D-exchange kinetics of disulfide-deficient lysozyme in glycerol solution by the DMSO-quenched H/D-exchange method. They removed glycerol from the reaction mixture by reversed-phase HPLC, and the fractionated portion was lyophilized before dissolving in the DMSO solution.

DMSO effectively dissolves amyloid fibrils in vivo and in vitro [99–101]. Therefore, since the early 2000s, the DMSO-quenched H/D-exchange method has been widely used in studies on the H/D-exchange kinetics of amyloid fibrils [102–124] and other protein aggregates, including protein supermolecular complexes [125,126]. These studies have demonstrated the presence of a hydrogen-bonded (H-bonded) core highly resistant to H/D exchange in the amyloid fibrils and the other protein complexes. In the amyloid experiments, insoluble amyloid fibrils were first suspended in D_2O to carry out the H/D-exchange reaction for the desired exchange periods, followed by separation of fibrils by centrifugation and lyophilization. The lyophilized fibrils were dissolved and dissociated into monomers in the DMSO solution, and the unfolded monomeric form was subjected to 2D NMR analysis. To characterize transient kinetic intermediates during the formation of amyloid fibrils, Carulla et al. [127] and Konuma et al. [90] combined the pulse-labeling hydrogen-exchange strategy and the DMSO-quenched method. After a short labeling pH pulse, aliquots of the reaction mixture were frozen and lyophilized, followed by dissolution in the DMSO solution for the subsequent 2D NMR or MS analysis. These studies gave us information about the molecular mechanisms of amyloid formation. Although the DMSO solution based on the standard protocol is composed of 95% DMSO-d_6 at pH* 5–6 adjusted by DCA-d_2, slightly modified compositions, e.g., dry DMSO-d_6 [103,125] and DMSO-d_6/trifluoroacetic acid-d_1 (0.01–1%) mixture [107,110,114–117,119,122,126,128,129] were also used as an H/D-exchange quenching solution. Several excellent review articles on the DMSO-quenched H/D-exchange method have been published and cover more details about the method [130–134].

2.2. Use of Spin Desalting Columns

We improved the DMSO-quenched H/D-exchange NMR method by the use of spin desalting columns for medium exchange from the H/D-exchange buffer to the DMSO solution [94]. As shown above, the conventional DMSO-quenched H/D-exchange experiments had used lyophilization for the medium exchange. Therefore, it was difficult to carry out the DMSO-quenched experiments at a high concentration of salt or denaturant (urea or GdmCl), because the presence of residual salt or denaturant after lyophilization interferes with the 2D NMR analysis of proteins dissolved in DMSO. This is a drawback of the conventional DMSO-quenched method, and it prevents us from using the method to characterize the H/D-exchange behaviors of proteins in the intermediate or unfolded state in denaturant and native proteins under physiological conditions at 0.15 M NaCl (or KCl).

To prepare the quenching DMSO solution, we adjusted the pH* of the 94.5% (v/v) DMSO-d_6/5% (v/v) D_2O/0.5% (v/v) DCA-d_2 solution to between 5 and 6 by adding 10 M NaOD. The addition of NaOD was accompanied by crystalline sodium dichloroacetate-d_1, which was dissolved by stirring with an increase in pH*. Before starting the H/D-exchange reaction of a sample protein, we first prepared a 10-fold concentrated stock solution of the protein in H_2O. The H/D-exchange reaction was started by a 10-fold dilution of the stock solution with a D_2O buffer. At appropriate time points in the H/D-exchange, we collected 1.0 mL aliquots of the reaction solution, which had been pre-dispensed into 2 mL polypropylene cryo-tubes in advance, and froze them by immersing the cryo-tubes in liquid nitrogen to stop the reaction. The frozen aliquots were kept at −85 °C to −80 °C until the medium exchange by a spin desalting column and the subsequent 2D NMR measurements. The frozen aliquots were thawed at an appropriate temperature, at or below room temperature, just before the medium exchange. The dead time of the H/D-exchange measurement depends on how the experiment is conducted. If the experiment is carried out with the cooperation of two experimenters, one quenching the reaction mixture quickly in liquid nitrogen and the other recording the reaction time of the H/D exchange, it is rather easy to realize the dead time of 30 s (see Figure 3).

Figure 1 shows a schematic of the medium exchange procedure from the D_2O buffer of the H/D-exchange reaction mixture to the DMSO solution with the use of spin desalting columns [94]. We used 5 mL columns (Zeba™ Spin Desalting Column 89891, Thermo Fisher Scientific K.K., Tokyo, Japan) and 15 mL polypropylene centrifuge tubes as collection tubes. The procedure consists of the following five steps:

1. Remove the bottom plug of a spin desalting column, and place the column in a collection tube.
2. Centrifuge for 2 min at 1000× *g* to pack the column bed and remove storage buffer.
3. Apply 2.5 mL of the DMSO solution to the column, centrifuge for 2–3 min at 1000× *g*, and discard the flow through. Repeat this process two to three times, but centrifuge for 3 min in the second and third runs, because the DMSO solution is more viscous than water.
4. Apply 1.0 mL of the H/D-exchange sample solution to the column, and centrifuge for 2 min at 20 °C. The application volume should not exceed 1.0 mL, because the application of more than 1.0 mL results in leakage of the H/D-exchange sample solution into the collection tube.
5. Recover the desalted protein sample in the DMSO solution; the protein sample can now be applied to 2D NMR analysis.

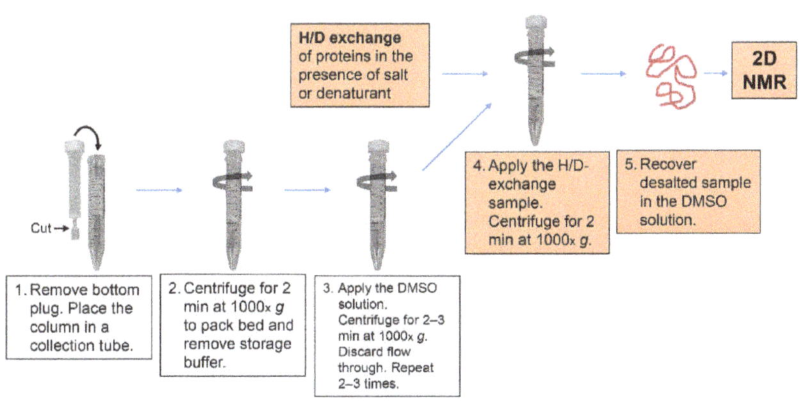

Figure 1. A schematic procedure of the medium exchange by a spin desalting column.

We used 2D ^1H–^{15}N HSQC spectra for the NMR analysis, and hence the sample protein was ^{15}N-labeled, and the assignment of the HSQC cross peaks was carried out by three-dimensional (3D) HN(CA)NNH, HNCA, HN(CO)CA, HNCO, CBCA(CO)NH, and HNCAHA experiments using $^{13}C/^{15}N$-double-labeled proteins. We applied the improved DMSO-quenched 2D NMR method to investigate the H/D-exchange behavior of the *E. coli* co-chaperonin GroES at pH* 6.5 (or 7.5) and 25 °C [135] and unfolded ubiquitin at pH*

3.2 and 15.0 °C in 6 M GdmCl [136], In the following, we will describe the study on the H/D-exchange behavior of unfolded ubiquitin as a case study.

3. A Case Study: Unfolded Ubiquitin in 6 M GdmCl

The characterization of residual structures persistent in unfolded proteins in concentrated denaturant (6 M GdmCl or 8 M urea) is an important issue in studies of protein folding. The problem of protein folding has been described with reference to the Levinthal paradox, in which the initial unfolded state is assumed to be a random coil, and hence, there may exist an astronomically large number of conformations, inaccessible in a reasonable time by a random search, at the beginning of the folding reactions [137–139]. Solving the Levinthal paradox is a fundamental problem in folding studies [140–146]. The presence of the residual structure, if any, in the unfolded state thus invalidates the Levinthal paradox, because such residual structure may form a folding initiation site and guide the subsequent folding reactions. We therefore studied the H/D-exchange behavior of unfolded human ubiquitin in 6 M GdmCl by the DMSO-quenched H/D-exchange 2D NMR method with the use of spin desalting columns [136]. Although the persistence of residual structures in unfolded proteins in concentrated denaturant has been reported for several different proteins [50,147–153], the present method enabled us to estimate the P values of individually identified NH protons, including nonprotected NH protons in the N state [136].

Ubiquitin is a 76-residue α/β protein, composed of a mixed parallel–anti-parallel β-sheet packing against a middle α-helix to form the hydrophobic core (Figure 2) [154]. Ubiquitin is a typical model protein for protein folding studies, and thus its folding reactions have been studied by a variety of biophysical techniques, including stopped-flow [155–158] and continuous-flow [159] kinetic refolding experiments, pulse-labeling hydrogen-exchange experiments combined with 2D NMR spectroscopy [72,160] and electrospray ionization mass spectrometry [161], mutational ϕ-value analysis [162,163], and other techniques [164–166]. These results may be compared with the present results of the H/D-exchange behavior of unfolded ubiquitin in 6 M GdmCl.

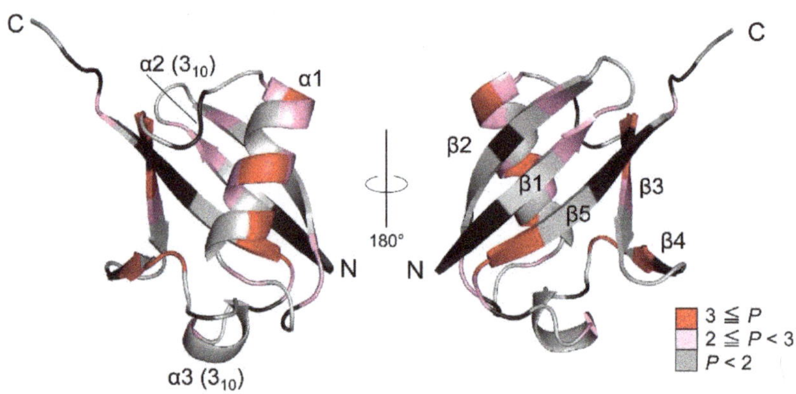

Figure 2. The 3D structure of native ubiquitin (PDB code: 1UBQ). The residues are colored according to the P values of the NH protons, and the red gradient indicates the scale of the P value. The proline residues and the residues that could not be used as probes due to severe broadening or overlapping are shown in black. Adapted with permission from Ref. [136]. Copyright 2020 Biophysical Society.

To analyze the H/D-exchange kinetics of individually identified NH protons of ubiquitin, we first made their spectral assignments in the DMSO solution [136]. Using a combination of 3D spectral measurements, we successfully assigned all the peaks observed in the HSQC spectrum. We then investigated the H/D-exchange kinetics of all the individual NH groups. Excluding NH groups whose residues could not be used as probes due to severe broadening or overlapping, we successfully followed the H/D-exchange kinetics of

60 NH protons [136]. These 60 NH protons include not only the protons stably protected in the native structure but also nonprotected NH protons. The observed kinetic exchange curve, given by the volume, $Y(t)$, of cross peaks in 2D NMR spectra as a function of the H/D-exchange time, t, was a single exponential fitted to the equation:

$$Y(t) = \Delta Y \cdot e^{-k_{\mathrm{obs}} t} + Y(\infty), \tag{4}$$

where ΔY and $Y(\infty)$ are the kinetic amplitude and the final value of the peak volume, respectively. Figure 3 shows the kinetic progress curves of the Val5, Asn25, Gln40, and Glu51 NH protons measured by the DMSO-quenched method. The protection factor P was calculated by Equation (3) for the 60 NH protons, resulting in the protection profile shown in Figure 4, in which the P value is plotted as a function of the residue number.

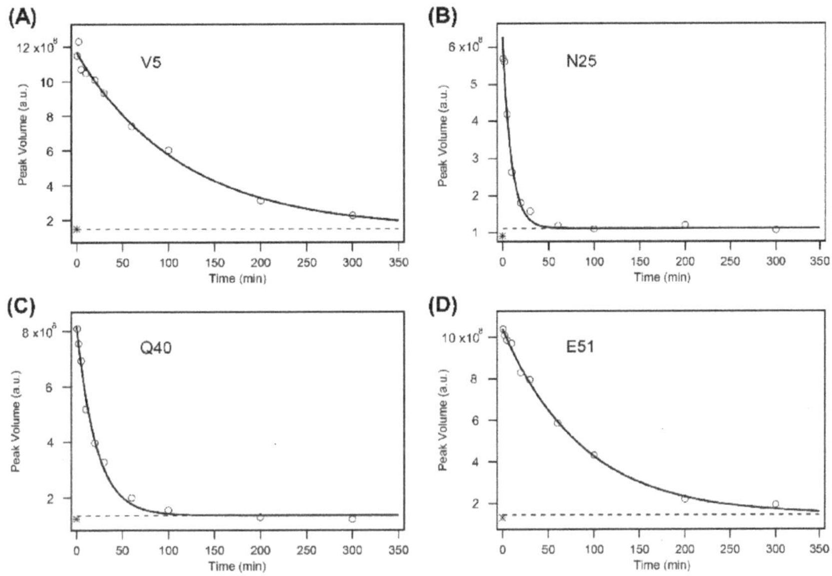

Figure 3. The H/D-exchange curves for Val5 (V5) (**A**), Asn25 (N25) (**B**), Gln40 (Q40) (**C**), and Glu51 (E51) (**D**) of human ubiquitin in 6 M GdmCl at pH* 3.3 and 15 °C. The solid lines are the theoretical curves best fitted to a single-exponential function (Equation (4)). A broken line in each panel indicates the theoretically estimated peak volume after the complete exchange (i.e., $Y(\infty)$ in Equation (4)), and an asterisk "*" in each panel, located between (1–2) $\times 10^8$ of the peak volume, indicates the experimentally observed value after heating the sample at 50 °C for 30 min. Because the reaction mixtures contained 10% H_2O, the final peak volumes did not reach zero. The k_{obs} values for the four residues are: (**A**) $(8.7 \pm 0.8) \times 10^{-3}$ min^{-1}; (**B**) $(10.5 \pm 1.2) \times 10^{-2}$ min^{-1}; (**C**) $(4.6 \pm 0.4) \times 10^{-2}$ min^{-1}; (**D**) $(1.2 \pm 0.1) \times 10^{-2}$ min^{-1}. The first and last time points are at 0.62 and 300.09 min, respectively, for all panels. Adapted with permission from Ref. [136]. Copyright 2020 Biophysical Society.

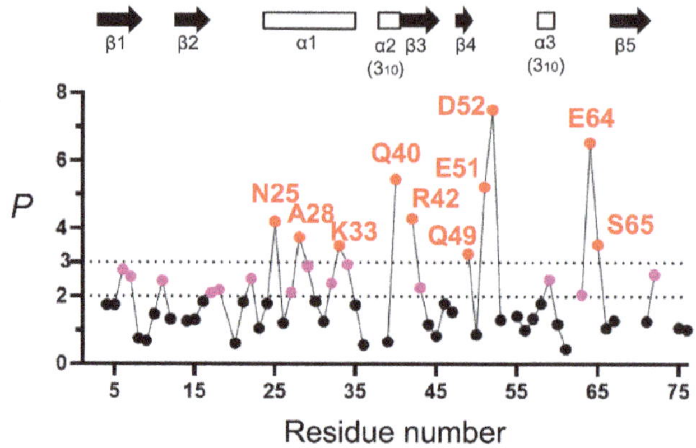

Figure 4. The H/D-exchange protection profile of unfolded ubiquitin in 6 M GdmCl, represented by P as a function of the residue number (pH* 3.3 and 15 °C). The dashed lines indicate the P values of 2 and 3. The amino-acid residues with P values larger than 2 and 3 are indicated in pink and red, respectively, and the other residues in black. The locations of the secondary structures in native ubiquitin (PDB code: 1UBQ) are shown by arrows (β-strands) and open rectangles (helices). The k_{obs} values for the majority of NH protons were obtained by three independent H/D-exchange experiments, and the percent standard error estimate of k_{obs} was ~8%, indicating that the P value of 2.0 can be written as $P = 2.0 \pm 0.16$ (see [136] for the complete list of the standard error estimates of k_{obs} values). Adapted with permission from Ref. [136]. Copyright 2020 Biophysical Society.

From Figure 4, a majority of the NH protons have a P value below 2 and larger than or equal to 0.8, indicating that these peptide NH protons are almost fully exposed to solvent water in 6 M GdmCl. This is consistent with the previous reports that ubiquitin in concentrated denaturant (6 M GdmCl or 8 M urea) at acidic pH is almost fully unfolded [155,167–170]. However, the 10 NH protons of Asn25, Ala28, Lys33, Gln40, Arg42, Gln49, Glu51, Asp52, Glu64, and Ser65 were significantly protected with a P value larger than 3, and the additional 14 NH protons of Lys6, Thr7, Lys11, Val17, Glu18, Thr22, Lys27, Lys29, Asp32, Glu34, Leu43, Tyr59, Lys63, and Arg72 showed P values between 2 and 3. These results thus clearly indicate the presence of residual structures in unfolded ubiquitin. Because the protein was unfolded in 6 M GdmCl, it is most likely that these residues were protected by the formation of an H-bond with a certain acceptor group.

When the H/D-exchange protection is brought about by the formation of the H-bond with a specific acceptor group, we can estimate the fraction of H-bonding (f_{Hbond}) for the protected NH groups. Because only the non-H-bonded form of the NH proton is available for H/D exchange, $(1 - f_{Hbond})$ is equal to k_{obs}/k_{int} (= $1/P$). Therefore, it follows that:

$$f_{Hbond} = 1 - \frac{1}{P} \quad (5)$$

When NH protons have P values larger than 3 and 2, the f_{Hbond} values are larger than 0.67 and 0.50, respectively, from Equation (5). The free energy of the H-bond breakage (0.0–0.41 kcal/mol) estimated from the f_{Hbond} values is thus negligibly small as compared with the unfolding free energy (7.5 kcal/mol) of ubiquitin [155]. Nevertheless, the f_{Hbond} values larger than 0.5 should be significant when we consider the kinetic folding mechanisms of the protein, and some of these weakly protected residues may play an important role in the formation of folding initiation sites at an initial stage of kinetic refolding of the protein from the GdmCl-induced unfolded state.

To understand the relationships between the residual structure in unfolded ubiquitin and the H-bonds formed in native ubiquitin, the H-bonding network in the native structure is shown in Figure 5. From Figures 4 and 5, the NH protons of Asn25, Lys27, Ala28, Lys29,

Asp32, Lys33, and Glu34, which are all significantly protected with a *P* value larger than 2, are involved in the middle α-helix (Ile23–E34) in native ubiquitin, and each NH proton of these residues forms an α-helical H-bond with the peptide CO group of the four amino-acid residues earlier, except for Asn25 (Figure 5B). The Asn25 NH proton forms a more local H-bond with the side-chain O^{γ} atom of Thr22, which acts as an N-cap residue of the helix in native ubiquitin. These results thus clearly demonstrate the presence of a residual structure in this α-helix of ubiquitin in 6 M GdmCl at pH* 3.3 and 15 °C, and Thr22 may also function as a helix stop signal by forming the N-cap conformation as in the native ubiquitin structure [171]. The NH groups of Thr7 and Val17, which form H-bonds with the CO groups of Lys11 and Met1, respectively, in the N-terminal β-hairpin, also have *P* values larger than 2 in the U state (Figure 5C), suggesting that the N-terminal β-hairpin may also be partially preserved in unfolded ubiquitin. The other NH groups having a *P* value larger than 2 in the N-terminal β-hairpin include those of Lys6 and Lys11. Although the NH proton of Lys11 does not form a backbone H-bond in native ubiquitin, it forms a local backbone to the side-chain H-bond with the O^{γ} of Thr7, and hence such a local backbone to the side-chain H-bond may be at least partially preserved in unfolded ubiquitin and may stabilize the residual structure of the N-terminal β-hairpin (Figure 5C). Previous NMR studies on H^N–N^N residual dipolar couplings, chemical shifts, $^3J_{HNHA}$ couplings, relaxation rates, and $^{h3}J_{NC'}$ couplings have also shown that the native-like first β-hairpin conformation was populated to at most 25% in unfolded ubiquitin in 8 M urea [172–175]. From these results, we conclude that there are native-like residual structures in the middle helix and the N-terminal β-hairpin in unfolded ubiquitin in 6 M GdmCl and that these residual structures may play an important role at an initial stage of kinetic refolding from the unfolded state.

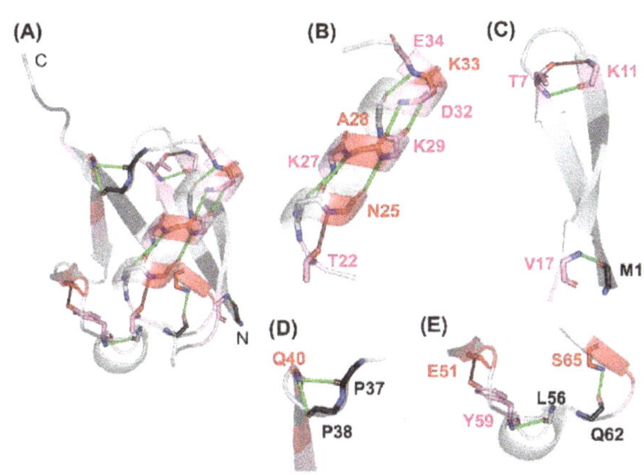

Figure 5. The H-bonding network observed in native ubiquitin (PDB code: 1UBQ). (**A**) A whole view, and (**B–E**) closer views of (**B**) the middle α-helix, (**C**) the N-terminal β-hairpin, (**D**) the one-turn 3_{10} helix (Pro38–Gln40), and (**E**) the Type II β-turn (Gln62–Ser65) and the one-turn 3_{10} helix (Ser57–Tyr59) are shown. The H-bonds of the NH protons of Thr7, Val17, Lys27, Ala28, Lys29, Asp32, Lys33, Glu34, Gln40, Tyr59, and Ser65 with the CO groups of their counterparts are shown as green lines. The local H-bonds formed by the NH protons of Lys11, Asn25, and Glu51 with the side-chain atoms of Thr7, Thr22, and Tyr59, respectively, are shown as brown lines. The red gradient indicates the same *P* value scale as shown in Figure 2. Adapted with permission from Ref. [136]. Copyright 2020 Biophysical Society.

In support of this conclusion, a pulsed H/D-exchange study with rapid mixing methods and 2D NMR analysis by Briggs and Roder [72] has shown that the NH protons in the α-helix and the β-hairpin of the fast-folding species (ca. 80%) of unfolded ubiquitin become protected in an initial 8 ms folding phase from the GdmCl-induced unfolded state of ubiquitin. Went and Jackson [163] performed a comprehensive ϕ-value analysis on the structure of the transition state ensemble of the ubiquitin folding and showed that medium and high ϕ values were found only in the N-terminal β-hairpin and the middle helix, which was also consistent with the above conclusion. The α-helical residual structure as detected by the unfolded-state H/D-exchange NMR spectroscopy in a concentrated denaturant solution was also observed in cytochrome *c* in 6 M urea [91], suggesting that the present observation of the residual structure may not be a rare example.

Three locally stabilized H-bonds, which form two one-turn 3_{10} helices (Gln40–Pro37 and Tyr59–Leu56) and a type II β-turn (Ser65–Gln62) in native ubiquitin, are also partially preserved in 6 M GdmCl. The NH proton of Gln40, which has a P value of 5.5 (f_{Hbond} = 0.81), forms H-bonds with the CO groups of Pro37 and Pro38 (Figure 5D). The NH proton of Tyr59, having a P value of 2.5 (f_{Hbond} = 0.60), forms an H-bond with the CO group of Leu56. The NH proton of Ser65, having a P value of 3.5 (f_{Hbond} = 0.71), forms an H-bond with the CO group of Gln62 (Figure 5E). Therefore, these locally stabilized H-bonds may not be fully disrupted even in 6 M GdmCl. However, it is not yet clear whether these local H-bonding interactions are important for the kinetic folding mechanisms of ubiquitin, because there is a dearth of experimental data concerning the effects of these local H-bonds on the kinetics of the ubiquitin folding.

The NH protons of the three residues of Arg42, Glu64, and Arg72 form native H-bonds with the backbone CO groups of Val70, Gln2, and Gln40, respectively [154], and have P values of 2.7 to 6.5 in unfolded ubiquitin (Figure 4). However, these are nonlocal H-bonds formed between residues at least 28 residues apart from each other, and such nonlocal H-bonds in native ubiquitin may not be stably formed in the U state in 6 M GdmCl. These NH protons may thus be protected by non-native H-bonding interactions. In fact, certain other NH protons, including those of Leu43, Gln49, Asp52, and Lys63, which have P values of 2.1 to 7.5 (Figure 4), are protected by non-native H-bonding interactions in 6 M GdmCl, because these NH protons do in fact lack backbone H-bonds in the native structure. The majority of residues with a P value larger than 3 are either charged or contain a side-chain amide or guanidinium group (Arg42, Gln49, Glu51, and Glu64). It is thus also possible that the protection might be afforded by the H-bonding between the backbone NH and its own side-chain atoms.

4. Conclusions

1. We described the DMSO-quenched H/D-exchange 2D NMR spectroscopy and its applications in protein science. The DMSO-quenched method is superior to the conventionally used refolding-quenched method, because the former allows us to monitor the H/D-exchange kinetics of not only the protected NH protons but also the nonprotected NH protons in the N state.
2. We described the improvement of the DMSO-quenched H/D-exchange method by the use of spin desalting columns and gave the methodological details of the use of spin desalting columns.
3. We presented a case study in which we characterized the H/D-exchange kinetics of 60 peptide NH protons in unfolded ubiquitin in 6 M GdmCl by the improved DMSO-quenched method. The residual structures were preserved in the middle α-helix and the N-terminal β-hairpin in unfolded ubiquitin with a protection factor P larger than 2 (i.e., a fraction of H-bonding f_{Hbond} of larger than 0.5), and these residual structures may play an important role in the folding process as nucleation sites and guide the subsequent folding reactions to the N state.

Author Contributions: Writing—original draft preparation, K.K. (Kunihiro Kuwajima); writing—review and editing, M.Y.-U., S.Y. and K.K. (Koichi Kato). All authors have read and agreed to the published version of the manuscript.

Funding: This work was supported in part by JSPS (Japan Society for the Promotion of Science) KAKENHI Grant Number JP20K06574 and by Joint Research of the Exploratory Research Center on Life and Living Systems (ExCELLS) Program Number 22EXC315. A part of this work was conducted in Institute for Molecular Science, and supported by the Nanotechnology Platform Program <Molecule and Material Synthesis> (JPMXP09S21MS0021) of the Ministry of Education, Culture, Sports, Science and Technology (MEXT) of Japan.

Conflicts of Interest: The authors declare no conflict of interest.

References

1. Hvidt, A.; Nielsen, S.O. Hydrogen exchange in proteins. *Adv. Protein Chem.* **1966**, *21*, 287–386. [CrossRef] [PubMed]
2. Woodward, C.; Carulla, N.; Barany, G. Native state hydrogen-exchange analysis of protein folding and protein motional domains. *Methods Enzymol.* **2004**, *380*, 379–400. [PubMed]
3. Englander, S.W.; Mayne, L.; Krishna, M.M. Protein folding and misfolding: Mechanism and principles. *Q. Rev. Biophys.* **2007**, *40*, 287–326. [CrossRef] [PubMed]
4. Englander, S.W.; Mayne, L.; Kan, Z.Y.; Hu, W. Protein folding—How and why: By hydrogen exchange, fragment separation, and mass spectrometry. *Annu. Rev. Biophys.* **2016**, *45*, 135–152. [CrossRef] [PubMed]
5. Wagner, G.; Wüthrich, K. Amide protein exchange and surface conformation of the basic pancreatic trypsin inhibitor in solution. Studies with two-dimensional nuclear magnetic resonance. *J. Mol. Biol.* **1982**, *160*, 343–361. [CrossRef]
6. Wand, A.J.; Roder, H.; Englander, S.W. Two-dimensional ^{1}H NMR studies of cytochrome *c*: Hydrogen exchange in the N-terminal helix. *Biochemistry* **1986**, *25*, 1107–1114. [CrossRef]
7. Konermann, L.; Simmons, D.A. Protein-folding kinetics and mechanisms studied by pulse-labeling and mass spectrometry. *Mass Spectrom. Rev.* **2003**, *22*, 1–26. [CrossRef]
8. Vadas, O.; Jenkins, M.L.; Dornan, G.L.; Burke, J.E. Using hydrogen-deuterium exchange mass spectrometry to examine protein-membrane interactions. *Methods Enzymol.* **2017**, *583*, 143–172. [CrossRef]
9. Molday, R.S.; Englander, S.W.; Kallen, R.G. Primary structure effects on peptide group hydrogen exchange. *Biochemistry* **1972**, *11*, 150–158. [CrossRef]
10. Bai, Y.; Milne, J.S.; Mayne, L.; Englander, S.W. Primary structure effects on peptide group hydrogen exchange. *Proteins* **1993**, *17*, 75–86. [CrossRef]
11. Connelly, G.P.; Bai, Y.; Jeng, M.F.; Englander, S.W. Isotope effects in peptide group hydrogen exchange. *Proteins* **1993**, *17*, 87–92. [CrossRef] [PubMed]
12. Nguyen, D.; Mayne, L.; Phillips, M.C.; Walter Englander, S. Reference parameters for protein hydrogen exchange rates. *J. Am. Soc. Mass Spectrom.* **2018**, *29*, 1936–1939. [CrossRef] [PubMed]
13. Huyghues-Despointes, B.M.P.; Scholtz, J.M.; Pace, C.N. Protein conformational stabilities can be determined from hydrogen exchange rates. *Nat. Struct. Biol.* **1999**, *6*, 910–912. [PubMed]
14. Huyghues-Despointes, B.M.; Pace, C.N.; Englander, S.W.; Scholtz, J.M. Measuring the conformational stability of a protein by hydrogen exchange. *Methods Mol. Biol.* **2001**, *168*, 69–92.
15. Bai, Y.; Sosnick, T.R.; Mayne, L.; Englander, S.W. Protein folding intermediates: Native-state hydrogen exchange. *Science* **1995**, *269*, 192–197. [CrossRef]
16. Brorsson, A.C.; Kjellson, A.; Aronsson, G.; Sethson, I.; Hambraeus, C.; Jonsson, B.H. The "two-state folder" MerP forms partially unfolded structures that show temperature dependent hydrogen exchange. *J. Mol. Biol.* **2004**, *340*, 333–344. [CrossRef]
17. Rennella, E.; Corazza, A.; Fogolari, F.; Viglino, P.; Giorgetti, S.; Stoppini, M.; Bellotti, V.; Esposito, G. Equilibrium unfolding thermodynamics of β_2-microglobulin analyzed through native-state H/D exchange. *Biophys. J.* **2009**, *96*, 169–179. [CrossRef]
18. Fuentes, E.J.; Wand, A.J. Local stability and dynamics of apocytochrome b_{562} examined by the dependence of hydrogen exchange on hydrostatic pressure. *Biochemistry* **1998**, *37*, 9877–9883. [CrossRef]
19. Bai, Y. Hidden intermediates and Levinthal paradox in the folding of small proteins. *Biochem. Biophys. Res. Commun.* **2003**, *305*, 785–788. [CrossRef]
20. Bai, Y. Protein folding pathways studied by pulsed- and native-state hydrogen exchange. *Chem. Rev.* **2006**, *106*, 1757–1768. [CrossRef]
21. Malhotra, P.; Udgaonkar, J.B. Native state hydrogen exchange-mass spectrometry methods to probe protein folding and unfolding. *Methods Mol. Biol.* **2022**, *2376*, 143–159. [CrossRef] [PubMed]
22. Ptitsyn, O.B. Molten globule and protein folding. *Adv. Protein Chem.* **1995**, *47*, 83–229. [PubMed]
23. Arai, M.; Kuwajima, K. Role of the molten globule state in protein folding. *Adv. Protein Chem.* **2000**, *53*, 209–282. [PubMed]
24. Kuwajima, K. The molten globule, and two-state vs. non-two-state folding of globular proteins. *Biomolecules* **2020**, *10*, 407. [CrossRef]
25. Hughson, F.M.; Wright, P.E.; Baldwin, R.L. Structural characterization of a partly folded apomyoglobin intermediate. *Science* **1990**, *249*, 1544–1548. [CrossRef]
26. Jeng, M.F.; Englander, S.W.; Gülnur, A.; Elöve, G.A.; Wand, A.J.; Roder, H. Structural description of acid-denatured cytochrome c by hydrogen exchange and 2D NMR. *Biochemistry* **1990**, *29*, 10433–10437. [CrossRef]

27. Kuroda, Y.; Endo, S.; Nagayama, K.; Wada, A. Stability of α-helices in a molten globule state of cytochrome c by hydrogen-deuterium exchange and two-dimensional NMR spectroscopy. *J. Mol. Biol.* **1995**, *247*, 682–688. [CrossRef]
28. Alexandrescu, A.T.; Evans, P.A.; Pitkeathly, M.; Baum, J.; Dobson, C.M. Structure and dynamics of the acid-denatured molten globule state of α-lactalbumin: A two-dimensional NMR study. *Biochemistry* **1993**, *32*, 1707–1718. [CrossRef]
29. Chyan, C.-L.; Wormald, C.; Dobson, C.M.; Evans, P.A.; Baum, J. Structure and stability of the molten globule state of guinea-pig α-lactalbumin: A hydrogen exchange study. *Biochemistry* **1993**, *32*, 5681–5691. [CrossRef]
30. Schulman, B.A.; Redfield, C.; Peng, Z.Y.; Dobson, C.M.; Kim, P.S. Different subdomains are most protected from hydrogen exchange in the molten globule and native states of human α-lactalbumin. *J. Mol. Biol.* **1995**, *253*, 651–657. [CrossRef]
31. Engel, M.F.; Visser, A.J.; van Mierlo, C.P. Conformation and orientation of a protein folding intermediate trapped by adsorption. *Proc. Natl. Acad. Sci. USA* **2004**, *101*, 11316–11321. [CrossRef] [PubMed]
32. Nakamura, T.; Makabe, K.; Tomoyori, K.; Maki, K.; Mukaiyama, A.; Kuwajima, K. Different folding pathways taken by highly homologous proteins, goat α-lactalbumin and canine milk lysozyme. *J. Mol. Biol.* **2010**, *396*, 1361–1378. [CrossRef] [PubMed]
33. Morozova-Roche, L.A.; Arico-Muendel, C.C.; Haynie, D.T.; Emelyanenko, V.I.; Van Dael, H.; Dobson, C.M. Structural characterisation and comparison of the native and A-states of equine lysozyme. *J. Mol. Biol.* **1997**, *268*, 903–921. [CrossRef] [PubMed]
34. Kobashigawa, Y.; Demura, M.; Koshiba, T.; Kumaki, Y.; Kuwajima, K.; Nitta, K. Hydrogen exchange study of canine milk lysozyme: Stabilization mechanism of the molten globule. *Proteins* **2000**, *40*, 579–589. [CrossRef]
35. Pan, Y.; Briggs, M.S. Hydrogen exchange in native and alcohol forms of ubiquitin. *Biochemistry* **1992**, *31*, 11405–11412. [CrossRef] [PubMed]
36. Dabora, J.M.; Pelton, J.G.; Marqusee, S. Structure of the acid state of *Escherichia coli* ribonuclease HI. *Biochemistry* **1996**, *35*, 11951–11958. [CrossRef] [PubMed]
37. Chamberlain, A.K.; Marqusee, S. Molten globule unfolding monitored by hydrogen exchange in urea. *Biochemistry* **1998**, *37*, 1736–1742. [CrossRef]
38. Yamasaki, K.; Yamasaki, T.; Kanaya, S.; Oobatake, M. Acid-induced denaturation of *Escherichia coli* ribonuclease HI analyzed by CD and NMR spectroscopies. *Biopolymers* **2003**, *69*, 176–188. [CrossRef]
39. Davis-Searles, P.R.; Morar, A.S.; Saunders, A.J.; Erie, D.A.; Pielak, G.J. Sugar-induced molten-globule model. *Biochemistry* **1998**, *37*, 17048–17053. [CrossRef]
40. Mazon, H.; Marcillat, O.; Forest, E.; Smith, D.L.; Vial, C. Conformational dynamics of the GdmHCl-induced molten globule state of creatine kinase monitored by hydrogen exchange and mass spectrometry. *Biochemistry* **2004**, *43*, 5045–5054. [CrossRef]
41. Nabuurs, S.M.; van Mierlo, C.P.M. Interrupted hydrogen/deuterium exchange reveals the stable core of the remarkably helical molten globule of α-β parallel protein flavodoxin. *J. Biol. Chem.* **2010**, *285*, 4165–4172. [CrossRef] [PubMed]
42. Roder, H.; Wagner, G.; Wüthrich, K. Individual amide proton-exchange rates in thermally unfolded basic pancreatic trypsin-inhibitor. *Biochemistry* **1985**, *24*, 7407–7411. [CrossRef] [PubMed]
43. Robertson, A.D.; Baldwin, R.L. Hydrogen exchange in thermally denatured ribonuclease A. *Biochemistry* **1991**, *30*, 9907–9914. [CrossRef]
44. Radford, S.E.; Buck, M.; Topping, K.D.; Dobson, C.M.; Evans, P.A. Hydrogen exchange in native and denatured states of hen egg-white lysozyme. *Proteins* **1992**, *14*, 237–248. [CrossRef]
45. Buck, M.; Radford, S.E.; Dobson, C.M. A partially folded state of hen egg white lysozyme in trifluoroethanol: Structural characterization and implications for protein folding. *Biochemistry* **1993**, *32*, 669–678. [CrossRef]
46. Fan, P.; Bracken, C.; Baum, J. Structural characterization of monellin in the alcohol-denatured state by NMR: Evidence for β-sheet to α-helix conversion. *Biochemistry* **1993**, *32*, 1573–1582. [CrossRef] [PubMed]
47. Zhang, J.; Peng, X.; Jonas, A.; Jonas, J. NMR study of the cold, heat, and pressure unfolding of ribonuclease A. *Biochemistry* **1995**, *34*, 8631–8641. [CrossRef]
48. Nash, D.P.; Jonas, J. Structure of pressure-assisted cold denatured lysozyme and comparison with lysozyme folding intermediates. *Biochemistry* **1997**, *36*, 14375–14383. [CrossRef]
49. Nash, D.P.; Jonas, J. Structure of the pressure-assisted cold denatured state of ubiquitin. *Biochem. Biophys. Res. Commun.* **1997**, *238*, 289–291. [CrossRef]
50. Wang, H.M.; Yu, C. Investigating the refolding pathway of human acidic fibroblast growth factor (hFGF-1) from the residual structure(s) obtained by denatured-state hydrogen/deuterium exchange. *Biophys. J.* **2011**, *100*, 154–164. [CrossRef]
51. Buck, M.; Radford, S.E.; Dobson, C.M. Amide hydrogen exchange in a highly denatured state. Hen egg-white lysozyme in urea. *J. Mol. Biol.* **1994**, *237*, 247–254. [CrossRef] [PubMed]
52. Katou, H.; Hoshino, M.; Kamikubo, H.; Batt, C.A.; Goto, Y. Native-like β-hairpin retained in the cold-denatured state of bovine β-lactoglobulin. *J. Mol. Biol.* **2001**, *310*, 471–484. [CrossRef] [PubMed]
53. Schmid, F.X.; Baldwin, R.L. Detection of an early intermediate in the folding of ribonuclease A by protection of amide protons against exchange. *J. Mol. Biol.* **1979**, *135*, 199–215. [CrossRef]
54. Roder, H.; Wüthrich, K. Protein folding kinetics by combined use of rapid mixing techniques and NMR observation of individual amide protons. *Proteins* **1986**, *1*, 34–42. [CrossRef] [PubMed]
55. Udgaonkar, J.B.; Baldwin, R.L. NMR evidence for an early framework intermediate on the folding pathway of ribonuclease A. *Nature* **1988**, *335*, 694–699. [CrossRef]

56. Roder, H.; Elöve, G.A.; Englander, S.W. Structural characterization of folding intermediates in cytochrome c by H-exchange labelling and proton NMR. *Nature* **1988**, *335*, 700–704. [CrossRef]
57. Bycroft, M.; Matouschek, A.; Kellis, J.J.; Serrano, L.; Fersht, A.R. Detection and characterization of a folding intermediate in barnase by NMR. *Nature* **1990**, *346*, 488–490. [CrossRef]
58. Matouschek, A.; Serrano, L.; Meiering, E.M.; Bycroft, M.; Fersht, A.R. The folding of an enzyme V. $H/^2H$ exchange–Nuclear magnetic resonance studies on the folding pathway of barnase: Complementarity to and agreement with protein engineering studies. *J. Mol. Biol.* **1992**, *224*, 837–845. [CrossRef]
59. Miranker, A.; Radford, S.E.; Karplus, M.; Dobson, C.M. Demonstration by NMR of folding domains in lysozyme. *Nature* **1991**, *349*, 633–636. [CrossRef]
60. Hooke, S.D.; Radford, S.E.; Dobson, C.M. The refolding of human lysozyme: A comparison with the structurally homologous hen lysozyme. *Biochemistry* **1994**, *33*, 5867–5876. [CrossRef]
61. Eyles, S.J.; Radford, S.E.; Robinson, C.V.; Dobson, C.M. Kinetic consequences of the removal of a disulfide bridge on the folding of hen lysozyme. *Biochemistry* **1994**, *33*, 13038–13048. [CrossRef] [PubMed]
62. Forge, V.; Wijesinha, R.T.; Balbach, J.; Brew, K.; Robinson, C.V.; Redfield, C.; Dobson, C.M. Rapid collapse and slow structural reorganisation during the refolding of bovine α-lactalbumin. *J. Mol. Biol.* **1999**, *288*, 673–688. [CrossRef] [PubMed]
63. Morozova-Roche, L.A.; Jones, J.A.; Noppe, W.; Dobson, C.M. Independent nucleation and heterogeneous assembly of structure during folding of equine lysozyme. *J. Mol. Biol.* **1999**, *289*, 1055–1073. [CrossRef] [PubMed]
64. Jennings, P.A.; Wright, P.E. Formation of a molten globule intermediate early in the kinetic folding pathway of apomyoglobin. *Science* **1993**, *262*, 892–896. [CrossRef] [PubMed]
65. Nishimura, C.; Dyson, H.J.; Wright, P.E. Enhanced picture of protein-folding intermediates using organic solvents in H/D exchange and quench-flow experiments. *Proc. Natl. Acad. Sci. USA* **2005**, *102*, 4765–4770. [CrossRef] [PubMed]
66. Nishimura, C.; Dyson, H.J.; Wright, P.E. The kinetic and equilibrium molten globule intermediates of apoleghemoglobin differ in structure. *J. Mol. Biol.* **2008**, *378*, 715–725. [CrossRef]
67. Uzawa, T.; Nishimura, C.; Akiyama, S.; Ishimori, K.; Takahashi, S.; Dyson, H.J.; Wright, P.E. Hierarchical folding mechanism of apomyoglobin revealed by ultra-fast H/D exchange coupled with 2D NMR. *Proc. Natl. Acad. Sci. USA* **2008**, *105*, 13859–13864. [CrossRef]
68. Udgaonkar, J.B.; Baldwin, R.L. Nature of the early folding intermediate of ribonuclease A. *Biochemistry* **1995**, *34*, 4088–4096. [CrossRef]
69. Houry, W.A.; Scheraga, H.A. Structure of a hydrophobically collapsed intermediate on the conformational folding pathway of ribonuclease A probed by hydrogen-deuterium exchange. *Biochemistry* **1996**, *35*, 11734–11746. [CrossRef]
70. Krishna, M.M.; Lin, Y.; Mayne, L.; Englander, S.W. Intimate view of a kinetic protein folding intermediate: Residue-resolved structure, interactions, stability, folding and unfolding rates, homogeneity. *J. Mol. Biol.* **2003**, *334*, 501–513. [CrossRef]
71. Lu, J.; Dahlquist, F.W. Detection and characterization of an early folding intermediate of T4 lysozyme using pulsed hydrogen exchange and two-dimensional NMR. *Biochemistry* **1992**, *31*, 4749–4756. [CrossRef] [PubMed]
72. Briggs, M.S.; Roder, H. Early hydrogen-bonding events in the folding reaction of ubiquitin. *Proc. Natl. Acad. Sci. USA* **1992**, *89*, 2017–2021. [CrossRef] [PubMed]
73. Mullins, L.S.; Pace, C.N.; Raushel, F.M. Investigation of ribonuclease T1 folding intermediates by hydrogen-deuterium amide exchange-two-dimensional NMR spectroscopy. *Biochemistry* **1993**, *32*, 6152–6156. [CrossRef]
74. Koide, S.; Dyson, H.J.; Wright, P.E. Characterization of a folding intermediate of apoplastocyanin trapped by proline isomerization. *Biochemistry* **1993**, *32*, 12299–12310. [CrossRef] [PubMed]
75. Varley, P.; Gronenborn, A.M.; Christensen, H.; Wingfield, P.T.; Pain, R.H.; Clore, G.M. Kinetics of folding of the all β-sheet protein interleukin-1β. *Science* **1993**, *260*, 1110–1113. [CrossRef] [PubMed]
76. Kuszewski, J.; Clore, G.M.; Gronenborn, A.M. Fast folding of a prototypic polypeptide: The immunoglobulin binding domain of streptococcal protein G. *Protein Sci.* **1994**, *3*, 1945–1952. [CrossRef]
77. Heidary, D.K.; Gross, L.A.; Roy, M.; Jennings, P.A. Evidence for an obligatory intermediate in the folding of interleukin-1β. *Nat. Struct. Biol.* **1997**, *4*, 725–731. [CrossRef]
78. Jacobs, M.D.; Fox, R.O. Staphylococcal nuclease folding intermediate characterized by hydrogen exchange and NMR spectroscopy. *Proc. Natl. Acad. Sci. USA* **1994**, *91*, 449–453. [CrossRef]
79. Walkenhorst, W.F.; Edwards, J.A.; Markley, J.L.; Roder, H. Early formation of a beta hairpin during folding of staphylococcal nuclease H124L as detected by pulsed hydrogen exchange. *Protein Sci.* **2002**, *11*, 82–91. [CrossRef]
80. Jones, B.E.; Matthews, C.R. Early intermediates in the folding of dihydrofolate reductase from *Escherichia coli* detected by hydrogen exchange and NMR. *Protein Sci.* **1995**, *4*, 167–177. [CrossRef]
81. Raschke, T.M.; Marqusee, S. The kinetic folding intermediate of ribonuclease H resembles the acid molten globule and partially unfolded molecules detected under native conditions. *Nat. Struct. Biol.* **1997**, *4*, 298–304. [CrossRef] [PubMed]
82. Parker, M.J.; Dempsey, C.E.; Lorch, M.; Clarke, A.R. Acquisition of native β-strand topology during the rapid collapse phase of protein folding. *Biochemistry* **1997**, *36*, 13396–13405. [CrossRef] [PubMed]
83. Hosszu, L.L.P.; Craven, C.J.; Parker, M.J.; Lorch, M.; Spencer, J.; Clarke, A.R.; Waltho, J.P. Structure of a kinetic protein folding intermediate by equilibrium amide exchange. *Nat. Struct. Biol.* **1997**, *4*, 801–804. [CrossRef] [PubMed]

84. Freund, C.; Gehrig, P.; Holak, T.A.; Plückthun, A. Comparison of the amide proton exchange behavior of the rapidly formed folding intermediate and the native state of an antibody scFv fragment. *FEBS Lett.* **1997**, *407*, 42–46. [CrossRef]
85. Teilum, K.; Kragelund, B.B.; Knudsen, J.; Poulsen, F.M. Formation of hydrogen bonds precedes the rate-limiting formation of persistent structure in the folding of ACBP. *J. Mol. Biol.* **2000**, *301*, 1307–1314. [CrossRef]
86. Forge, V.; Hoshino, M.; Kuwata, K.; Arai, M.; Kuwajima, K.; Batt, C.A.; Goto, Y. Is folding of β-Lactoglobulin non-hierarchic? Intermediate with native-like β-sheet and non-native α-helix. *J. Mol. Biol.* **2000**, *296*, 1039–1051. [CrossRef]
87. Kuwata, K.; Shastry, R.; Cheng, H.; Hoshino, M.; Batt, C.A.; Goto, Y.; Roder, H. Structural and kinetic characterization of early folding events in β-lactoglobulin. *Nat. Struct. Biol.* **2001**, *8*, 151–155. [CrossRef]
88. Schulenburg, C.; Löw, C.; Weininger, U.; Mrestani-Klaus, C.; Hofmann, H.; Balbach, J.; Ulbrich-Hofmann, R.; Arnold, U. The folding pathway of onconase is directed by a conserved intermediate. *Biochemistry* **2009**, *48*, 8449–8457. [CrossRef]
89. Di Paolo, A.; Balbeur, D.; De Pauw, E.; Redfield, C.; Matagne, A. Rapid collapse into a molten globule is followed by simple two-state kinetics in the folding of lysozyme from bacteriophage λ. *Biochemistry* **2010**, *49*, 8646–8657. [CrossRef]
90. Konuma, T.; Chatani, E.; Yagi, M.; Sakurai, K.; Ikegami, T.; Naiki, H.; Goto, Y. Kinetic intermediates of β_2-microglobulin fibril elongation probed by pulse-labeling H/D exchange combined with NMR analysis. *J. Mol. Biol.* **2011**, *405*, 851–862. [CrossRef]
91. Fazelinia, H.; Xu, M.; Cheng, H.; Roder, H. Ultrafast hydrogen exchange reveals specific structural events during the initial stages of folding of cytochrome c. *J. Am. Chem. Soc.* **2014**, *136*, 733–740. [CrossRef] [PubMed]
92. Rosen, L.E.; Connell, K.B.; Marqusee, S. Evidence for close side-chain packing in an early protein folding intermediate previously assumed to be a molten globule. *Proc. Natl. Acad. Sci. USA* **2014**, *111*, 14746–14751. [CrossRef] [PubMed]
93. Zhang, Y.Z.; Paterson, Y.; Roder, H. Rapid amide proton exchange rates in peptides and proteins measured by solvent quenching and two-dimensional NMR. *Protein Sci.* **1995**, *4*, 804–814. [CrossRef] [PubMed]
94. Chandak, M.S.; Nakamura, T.; Takenaka, T.; Chaudhuri, T.K.; Yagi-Utsumi, M.; Chen, J.; Kato, K.; Kuwajima, K. The use of spin desalting columns in DMSO-quenched H/D-exchange NMR experiments. *Protein Sci.* **2013**, *22*, 486–491. [CrossRef]
95. Bax, A.; Grzesiek, S. Methodological advances in protein NMR. *Acc. Chem. Res.* **1993**, *26*, 131–138. [CrossRef]
96. Bax, A.; Ikura, M.; Kay, L.E.; Torchia, D.A.; Tschudin, R. Comparison of different modes of two-dimensional reverse-correlation NMR for the study of proteins. *J. Magn. Reson.* **1990**, *86*, 304–318. [CrossRef]
97. Nishimura, C. Folding of apomyoglobin: Analysis of transient intermediate structure during refolding using quick hydrogen deuterium exchange and NMR. *Proc. Jpn. Acad. Ser. B Phys. Biol. Sci.* **2017**, *93*, 10–27. [CrossRef]
98. Sakamoto, K.; Hirai, K.; Kitamura, Y.; Yamazaki, K.; Yusa, M.; Tokunaga, N.; Doi, G.; Noda, Y.; Tachibana, H.; Segawa, S. Glycerol-induced folding of unstructured disulfide-deficient lysozyme into a native-like conformation. *Biopolymers* **2009**, *91*, 665–675. [CrossRef]
99. van Rijswijk, M.H.; Donker, A.J.; Ruinen, L. Dimethylsulphoxide in amyloidosis. *Lancet* **1979**, *313*, 207–208. [CrossRef]
100. Regelson, W.; Harkins, S.W. "Amyloid is not a tombstone"—A summation. The primary role for cerebrovascular and CSF dynamics as factors in Alzheimer's disease (AD): DMSO, fluorocarbon oxygen carriers, thyroid hormonal, and other suggested therapeutic measures. *Ann. N. Y. Acad. Sci.* **1997**, *826*, 348–374. [CrossRef]
101. Hirota-Nakaoka, N.; Hasegawa, K.; Naiki, H.; Goto, Y. Dissolution of β_2-microglobulin amyloid fibrils by dimethylsulfoxide. *J. Biochem.* **2003**, *134*, 159–164. [CrossRef] [PubMed]
102. Alexandrescu, A.T. An NMR-based quenched hydrogen exchange investigation of model amyloid fibrils formed by cold shock protein A. *Pac. Symp. Biocomput.* **2001**, *2000*, 67–78.
103. Ippel, J.H.; Olofsson, A.; Schleucher, J.; Lundgren, E.; Wijmenga, S.S. Probing solvent accessibility of amyloid fibrils by solution NMR spectroscopy. *Proc. Natl. Acad. Sci. USA* **2002**, *99*, 8648–8653. [CrossRef]
104. Hoshino, M.; Katou, H.; Hagihara, Y.; Hasegawa, K.; Naiki, H.; Goto, Y. Mapping the core of the β_2-microglobulin amyloid fibril by H/D exchange. *Nat. Struct. Biol.* **2002**, *9*, 332–336. [CrossRef] [PubMed]
105. Kuwata, K.; Matumoto, T.; Cheng, H.; Nagayama, K.; James, T.L.; Roder, H. NMR-detected hydrogen exchange and molecular dynamics simulations provide structural insight into fibril formation of prion protein fragment 106–126. *Proc. Natl. Acad. Sci. USA* **2003**, *100*, 14790–14795. [CrossRef] [PubMed]
106. Yamaguchi, K.I.; Katou, H.; Hoshino, M.; Hasegawa, K.; Naiki, H.; Goto, Y. Core and heterogeneity of β_2-microglobulin amyloid fibrils as revealed by H/D exchange. *J. Mol. Biol.* **2004**, *338*, 559–571. [CrossRef] [PubMed]
107. Ritter, C.; Maddelein, M.L.; Siemer, A.B.; Lührs, T.; Ernst, M.; Meier, B.H.; Saupe, S.J.; Riek, R. Correlation of structural elements and infectivity of the HET-s prion. *Nature* **2005**, *435*, 844–848. [CrossRef]
108. Carulla, N.; Caddy, G.L.; Hall, D.R.; Zurdo, J.; Gairi, M.; Feliz, M.; Giralt, E.; Robinson, C.V.; Dobson, C.M. Molecular recycling within amyloid fibrils. *Nature* **2005**, *436*, 554–558. [CrossRef]
109. Whittemore, N.A.; Mishra, R.; Kheterpal, I.; Williams, A.D.; Wetzel, R.; Serpersu, E.H. Hydrogen-deuterium (H/D) exchange mapping of $A\beta_{1-40}$ amyloid fibril secondary structure using nuclear magnetic resonance spectroscopy. *Biochemistry* **2005**, *44*, 4434–4441. [CrossRef]
110. Lührs, T.; Ritter, C.; Adrian, M.; Riek-Loher, D.; Bohrmann, B.; Döbeli, H.; Schubert, D.; Riek, R. 3D structure of Alzheimer's amyloid-β(1–42) fibrils. *Proc. Natl. Acad. Sci. USA* **2005**, *102*, 17342–17347. [CrossRef]
111. Wilson, L.M.; Mok, Y.F.; Binger, K.J.; Griffin, M.D.; Mertens, H.D.; Lin, F.; Wade, J.D.; Gooley, P.R.; Howlett, G.J. A structural core within apolipoprotein C-II amyloid fibrils identified using hydrogen exchange and proteolysis. *J. Mol. Biol.* **2007**, *366*, 1639–1651. [CrossRef] [PubMed]

112. Toyama, B.H.; Kelly, M.J.; Gross, J.D.; Weissman, J.S. The structural basis of yeast prion strain variants. *Nature* **2007**, *449*, 233–237. [CrossRef] [PubMed]
113. Morgan, G.J.; Giannini, S.; Hounslow, A.M.; Craven, C.J.; Zerovnik, E.; Turk, V.; Waltho, J.P.; Staniforth, R.A. Exclusion of the native α-helix from the amyloid fibrils of a mixed α/β protein. *J. Mol. Biol.* **2008**, *375*, 487–498. [CrossRef] [PubMed]
114. Wang, L.; Maji, S.K.; Sawaya, M.R.; Eisenberg, D.; Riek, R. Bacterial inclusion bodies contain amyloid-like structure. *PLoS Biol.* **2008**, *6*, e195. [CrossRef]
115. Vilar, M.; Chou, H.T.; Lührs, T.; Maji, S.K.; Riek-Loher, D.; Verel, R.; Manning, G.; Stahlberg, H.; Riek, R. The fold of α-synuclein fibrils. *Proc. Natl. Acad. Sci. USA* **2008**, *105*, 8637–8642. [CrossRef]
116. Wasmer, C.; Zimmer, A.; Sabaté, R.; Soragni, A.; Saupe, S.J.; Ritter, C.; Meier, B.H. Structural similarity between the prion domain of HET-s and a homologue can explain amyloid cross-seeding in spite of limited sequence identity. *J. Mol. Biol.* **2010**, *402*, 311–325. [CrossRef]
117. Kurouski, D.; Washington, J.; Ozbil, M.; Prabhakar, R.; Shekhtman, A.; Lednev, I.K. Disulfide bridges remain intact while native insulin converts into amyloid fibrils. *PLoS ONE* **2012**, *7*, e36989. [CrossRef]
118. Alexandrescu, A.T. Amide proton solvent protection in amylin fibrils probed by quenched hydrogen exchange NMR. *PLoS ONE* **2013**, *8*, e56467. [CrossRef]
119. Garvey, M.; Baumann, M.; Wulff, M.; Kumar, S.T.; Markx, D.; Morgado, I.; Knüpfer, U.; Horn, U.; Mawrin, C.; Fändrich, M.; et al. Molecular architecture of Aβ fibrils grown in cerebrospinal fluid solution and in a cell culture model of Aβ plaque formation. *Amyloid* **2016**, *23*, 76–85. [CrossRef]
120. Rezaei-Ghaleh, N.; Kumar, S.; Walter, J.; Zweckstetter, M. Phosphorylation interferes with maturation of amyloid-β fibrillar structure in the N Terminus. *J. Biol. Chem.* **2016**, *291*, 16059–16067. [CrossRef]
121. Wälti, M.A.; Orts, J.; Riek, R. Quenched hydrogen-deuterium exchange NMR of a disease-relevant Aβ(1-42) amyloid polymorph. *PLoS ONE* **2017**, *12*, e0172862. [CrossRef] [PubMed]
122. Morimoto, D.; Nishizawa, R.; Walinda, E.; Takashima, S.; Sugase, K.; Shirakawa, M. Hydrogen-deuterium exchange profiles of polyubiquitin fibrils. *Polymers* **2018**, *10*, 240. [CrossRef] [PubMed]
123. Zlatic, C.O.; Mao, Y.; Todorova, N.; Mok, Y.F.; Howlett, G.J.; Yarovsky, I.; Gooley, P.R.; Griffin, M.D.W. Polymorphism in disease-related apolipoprotein C-II amyloid fibrils: A structural model for rod-like fibrils. *FEBS J.* **2018**, *285*, 2799–2812. [CrossRef] [PubMed]
124. Aucoin, D.; Xia, Y.; Theint, T.; Nadaud, P.S.; Surewicz, K.; Surewicz, W.K.; Jaroniec, C.P. Protein-solvent interfaces in human Y145Stop prion protein amyloid fibrils probed by paramagnetic solid-state NMR spectroscopy. *J. Struct. Biol.* **2019**, *206*, 36–42. [CrossRef] [PubMed]
125. Dyson, H.J.; Kostic, M.; Liu, J.; Martinez-Yamout, M.A. Hydrogen-deuterium exchange strategy for delineation of contact sites in protein complexes. *FEBS Lett.* **2008**, *582*, 1495–1500. [CrossRef]
126. Chakraborty, S.; Hosur, R.V. NMR insights into the core of GED assembly by H/D exchange coupled with DMSO dissociation and analysis of the denatured state. *J. Mol. Biol.* **2011**, *405*, 1202–1214. [CrossRef] [PubMed]
127. Carulla, N.; Zhou, M.; Arimon, M.; Gairí, M.; Giralt, E.; Robinson, C.V.; Dobson, C.M. Experimental characterization of disordered and ordered aggregates populated during the process of amyloid fibril formation. *Proc. Natl. Acad. Sci. USA* **2009**, *106*, 7828–7833. [CrossRef]
128. Kulminskaya, N.V.; Yoshimura, Y.; Runager, K.; Sørensen, C.S.; Bjerring, M.; Andreasen, M.; Otzen, D.E.; Enghild, J.J.; Nielsen, N.C.; Mulder, F.A. Near-complete ^{1}H, ^{13}C, ^{15}N resonance assignments of dimethylsulfoxide-denatured TGFBIp FAS1-4 A546T. *Biomol. NMR Assign.* **2016**, *10*, 25–29. [CrossRef]
129. Karsisiotis, A.I.; Deacon, O.M.; Macdonald, C.; Blumenschein, T.M.A.; Moore, G.R.; Worrall, J.A.R. Near-complete backbone resonance assignments of acid-denatured human cytochrome *c* in dimethylsulfoxide: A prelude to studying interactions with phospholipids. *Biomol. NMR Assign.* **2017**, *11*, 165–168. [CrossRef]
130. Hoshino, M.; Katou, H.; Yamaguchi, K.I.; Goto, Y. Dimethylsulfoxide-quenched hydrogen/deuterium exchange method to study amyloid fibril structure. *Biochim. Biophys. Acta* **2007**, *1768*, 1886–1899. [CrossRef]
131. Carulla, N.; Zhou, M.; Giralt, E.; Robinson, C.V.; Dobson, C.M. Structure and intermolecular dynamics of aggregates populated during amyloid fibril formation studied by hydrogen/deuterium exchange. *Acc. Chem. Res.* **2010**, *43*, 1072–1079. [CrossRef]
132. Vilar, M.; Wang, L.; Riek, R. Structural studies of amyloids by quenched hydrogen-deuterium exchange by NMR. *Methods Mol. Biol.* **2012**, *849*, 185–198. [CrossRef] [PubMed]
133. Lee, Y.H.; Goto, Y. Kinetic intermediates of amyloid fibrillation studied by hydrogen exchange methods with nuclear magnetic resonance. *Biochim. Biophys. Acta* **2012**, *1824*, 1307–1323. [CrossRef] [PubMed]
134. Alexandrescu, A.T. Quenched hydrogen exchange NMR of amyloid fibrils. *Methods Mol. Biol.* **2016**, *1345*, 211–222. [PubMed]
135. Chandak, M.S.; Nakamura, T.; Makabe, K.; Takenaka, T.; Mukaiyama, A.; Chaudhuri, T.K.; Kato, K.; Kuwajima, K. The H/D-exchange kinetics of the *Escherichia coli* co-chaperonin GroES studied by 2D NMR and DMSO-quenched exchange methods. *J. Mol. Biol.* **2013**, *425*, 2541–2560. [CrossRef]
136. Yagi-Utsumi, M.; Chandak, M.S.; Yanaka, S.; Hiranyakorn, M.; Nakamura, T.; Kato, K.; Kuwajima, K. Residual structure of unfolded ubiquitin as revealed by hydrogen/deuterium-exchange 2D NMR. *Biophys. J.* **2020**, *119*, 2029–2038. [CrossRef]
137. Levinthal, C. Are there pathways for protein folding? *J. Chim. Phys.* **1968**, *65*, 44–45. [CrossRef]
138. Levinthal, C. How to fold graciously. *Mössbauer Spectrosc. Biol. Syst.* **1969**, *67*, 22–24.

139. Wetlaufer, D.B. Nucleation, rapid folding, and globular intrachain regions in proteins. *Proc. Natl. Acad. Sci. USA* **1973**, *70*, 697–701. [CrossRef]
140. Zwanzig, R.; Szabo, A.; Bagchi, B. Levinthal's paradox. *Proc. Natl. Acad. Sci. USA* **1992**, *89*, 20–22. [CrossRef]
141. Bryngelson, J.D.; Onuchic, J.N.; Socci, N.D.; Wolynes, P.G. Funnels, pathways, and the energy landscape of protein folding: A synthesis. *Proteins* **1995**, *21*, 167–195. [CrossRef] [PubMed]
142. Dill, K.A.; Chan, H.S. From Levinthal to pathways to funnels. *Nat. Struct. Biol.* **1997**, *4*, 10–19. [CrossRef] [PubMed]
143. Dobson, C.M.; Sali, A.; Karplus, M. Protein folding: A perspective from theory and experiment. *Angew. Chem. Int. Ed. Engl.* **1998**, *37*, 868–893. [CrossRef]
144. Englander, S.W.; Mayne, L. The nature of protein folding pathways. *Proc. Natl. Acad. Sci. USA* **2014**, *111*, 15873–15880. [CrossRef] [PubMed]
145. Ivankov, D.N.; Finkelstein, A.V. Solution of Levinthal's paradox and a physical theory of protein folding times. *Biomolecules* **2020**, *10*, 250. [CrossRef]
146. Rose, G.D. Protein folding—Seeing is deceiving. *Protein Sci.* **2021**, *30*, 1606–1616. [CrossRef]
147. Neri, D.; Billeter, M.; Wider, G.; Wüthrich, K. NMR determination of residual structure in a urea-denatured protein, the 434-repressor. *Science* **1992**, *257*, 1559–1563. [CrossRef]
148. Shortle, D.; Ackerman, M.S. Persistence of native-like topology in a denatured protein in 8 M urea. *Science* **2001**, *293*, 487–489. [CrossRef]
149. Klein-Seetharaman, J.; Oikawa, M.; Grimshaw, S.B.; Wirmer, J.; Duchardt, E.; Ueda, T.; Imoto, T.; Smith, L.J.; Dobson, C.M.; Schwalbe, H. Long-range interactions within a nonnative protein. *Science* **2002**, *295*, 1719–1722. [CrossRef]
150. Wirmer, J.; Berk, H.; Ugolini, R.; Redfield, C.; Schwalbe, H. Characterization of the unfolded state of bovine α-lactalbumin and comparison with unfolded states of homologous proteins. *Protein Sci.* **2006**, *15*, 1397–1407. [CrossRef]
151. Danielson, T.A.; Stine, J.M.; Dar, T.A.; Briknarova, K.; Bowler, B.E. Effect of an imposed contact on secondary structure in the denatured state of yeast Iso-1-cytochrome c. *Biochemistry* **2017**, *56*, 6662–6676. [CrossRef] [PubMed]
152. Leavens, M.J.; Spang, L.E.; Cherney, M.M.; Bowler, B.E. Denatured state conformational biases in three-helix bundles containing divergent sequences localize near turns and helix capping residues. *Biochemistry* **2021**, *60*, 3071–3085. [CrossRef] [PubMed]
153. Bhatia, S.; Udgaonkar, J.B. Heterogeneity in protein folding and unfolding reactions. *Chem. Rev.* **2022**, *122*, 8911–8935. [CrossRef] [PubMed]
154. Vijay-Kumar, S.; Bugg, C.E.; Cook, W.J. Structure of ubiquitin refined at 1.8 Å resolution. *J. Mol. Biol.* **1987**, *194*, 531–544. [CrossRef]
155. Khorasanizadeh, S.; Peters, I.D.; Butt, T.R.; Roder, H. Folding and stability of a tryptophan-containing mutant of ubiquitin. *Biochemistry* **1993**, *32*, 7054–7063. [CrossRef] [PubMed]
156. Khorasanizadeh, S.; Peters, I.D.; Roder, H. Evidence for a three-state model of protein folding from kinetic analysis of ubiquitin variants with altered core residues. *Nat. Struct. Biol.* **1996**, *3*, 193–205. [CrossRef]
157. Went, H.M.; Benitez-Cardoza, C.G.; Jackson, S.E. Is an intermediate state populated on the folding pathway of ubiquitin? *FEBS Lett.* **2004**, *567*, 333–338. [CrossRef]
158. Larios, E.; Li, J.S.; Schulten, K.; Kihara, H.; Gruebele, M. Multiple probes reveal a native-like intermediate during low-temperature refolding of ubiquitin. *J. Mol. Biol.* **2004**, *340*, 115–125. [CrossRef]
159. Roder, H.; Maki, K.; Cheng, H. Early events in protein folding explored by rapid mixing methods. *Chem. Rev.* **2006**, *106*, 1836–1861. [CrossRef]
160. Gladwin, S.T.; Evans, P.A. Structure of very early protein folding intermediates: New insights through a variant of hydrogen exchange labelling. *Fold. Des.* **1996**, *1*, 407–417. [CrossRef]
161. Pan, J.; Wilson, D.J.; Konermann, L. Pulsed hydrogen exchange and electrospray charge-state distribution as complementary probes of protein structure in kinetic experiments: Implications for ubiquitin folding. *Biochemistry* **2005**, *44*, 8627–8633. [CrossRef] [PubMed]
162. Sosnick, T.R.; Dothager, R.S.; Krantz, B.A. Differences in the folding transition state of ubiquitin indicated by ϕ and ψ analyses. *Proc. Natl. Acad. Sci. USA* **2004**, *101*, 17377–17382. [CrossRef]
163. Went, H.M.; Jackson, S.E. Ubiquitin folds through a highly polarized transition state. *Protein Eng. Des. Sel.* **2005**, *18*, 229–237. [CrossRef]
164. Kitahara, R.; Akasaka, K. Close identity of a pressure-stabilized intermediate with a kinetic intermediate in protein folding. *Proc. Natl. Acad. Sci. USA* **2003**, *100*, 3167–3172. [CrossRef]
165. Jackson, S.E. Ubiquitin: A small protein folding paradigm. *Org. Biomol. Chem.* **2006**, *4*, 1845–1853. [CrossRef]
166. Wirtz, H.; Schäfer, S.; Hoberg, C.; Reid, K.M.; Leitner, D.M.; Havenith, M. Hydrophobic collapse of ubiquitin generates rapid protein-water motions. *Biochemistry* **2018**, *57*, 3650–3657. [CrossRef]
167. Ibarra-Molero, B.; Loladze, V.V.; Makhatadze, G.I.; Sanchez-Ruiz, J.M. Thermal versus guanidine-induced unfolding of ubiquitin. An analysis in terms of the contributions from charge-charge interactions to protein stability. *Biochemistry* **1999**, *38*, 8138–8149. [CrossRef]
168. Peti, W.; Hennig, M.; Smith, L.J.; Schwalbe, H. NMR spectroscopic investigation of ψ torsion angle distribution in unfolded ubiquitin from analysis of $^3J(C_\alpha, C_\alpha)$ coupling constants and cross-correlated $\Gamma^{CHNN,C\alpha H\alpha}$ relaxation rates. *J. Am. Chem. Soc.* **2000**, *122*, 12017–12018. [CrossRef]

169. Peti, W.; Smith, L.J.; Redfield, C.; Schwalbe, H. Chemical shifts in denatured proteins: Resonance assignments for denatured ubiquitin and comparisons with other denatured proteins. *J. Biomol. NMR* **2001**, *19*, 153–165. [CrossRef]
170. Wirmer, J.; Peti, W.; Schwalbe, H. Motional properties of unfolded ubiquitin: A model for a random coil protein. *J. Biomol. NMR* **2006**, *35*, 175–186. [CrossRef]
171. Harper, E.T.; Rose, G.D. Helix stop signals in proteins and peptides: The capping box. *Biochemistry* **1993**, *32*, 7605–7609. [CrossRef] [PubMed]
172. Meier, S.; Strohmeier, M.; Blackledge, M.; Grzesiek, S. Direct observation of dipolar couplings and hydrogen bonds across a β-hairpin in 8 M urea. *J. Am. Chem. Soc.* **2007**, *129*, 754–755. [CrossRef] [PubMed]
173. Huang, J.R.; Grzesiek, S. Ensemble calculations of unstructured proteins constrained by RDC and PRE data: A case study of urea-denatured ubiquitin. *J. Am. Chem. Soc.* **2010**, *132*, 694–705. [CrossRef]
174. Candotti, M.; Esteban-Martin, S.; Salvatella, X.; Orozco, M. Toward an atomistic description of the urea-denatured state of proteins. *Proc. Natl. Acad. Sci. USA* **2013**, *110*, 5933–5938. [CrossRef]
175. Aznauryan, M.; Delgado, L.; Soranno, A.; Nettels, D.; Huang, J.R.; Labhardt, A.M.; Grzesiek, S.; Schuler, B. Comprehensive structural and dynamical view of an unfolded protein from the combination of single-molecule FRET, NMR, and SAXS. *Proc. Natl. Acad. Sci. USA* **2016**, *113*, E5389–E5398. [CrossRef]

molecules MDPI

Review

The Molten Globule State of a Globular Protein in a Cell Is More or Less Frequent Case Rather than an Exception

Valentina E. Bychkova [1], Dmitry A. Dolgikh [2], Vitalii A. Balobanov [1,*] and Alexei V. Finkelstein [1]

[1] Institute of Protein Research, Russian Academy of Sciences, 142290 Pushchino, Moscow Region, Russia; vbychkova@mail.ru (V.E.B.); afinkel@vega.protres.ru (A.V.F.)
[2] Shemyakin-Ovchinnikov Institute of Bioorganic Chemistry, Russian Academy of Sciences, 117871 Moscow, Russia; dolgikh@nmr.ru
* Correspondence: balobanov@phys.protres.ru

Abstract: Quite a long time ago, Oleg B. Ptitsyn put forward a hypothesis about the possible functional significance of the molten globule (MG) state for the functioning of proteins. MG is an intermediate between the unfolded and the native state of a protein. Its experimental detection and investigation in a cell are extremely difficult. In the last decades, intensive studies have demonstrated that the MG-like state of some globular proteins arises from either their modifications or interactions with protein partners or other cell components. This review summarizes such reports. In many cases, MG was evidenced to be functionally important. Thus, the MG state is quite common for functional cellular proteins. This supports Ptitsyn's hypothesis that some globular proteins may switch between two active states, rigid (N) and soft (MG), to work in solution or interact with partners.

Keywords: globular protein; rigid native state; molten globule; intrinsically disordered; functional state; unfolded state; coil; post-translational modifications; membrane; chaperone

Citation: Bychkova, V.E.; Dolgikh, D.A.; Balobanov, V.A.; Finkelstein, A.V. The Molten Globule State of a Globular Protein in a Cell Is More or Less Frequent Case Rather than an Exception. *Molecules* **2022**, *27*, 4361. https://doi.org/10.3390/molecules27144361

Academic Editors: Kunihiro Kuwajima, Yuko Okamoto, Tuomas Knowles and Michele Vendruscolo

Received: 27 May 2022
Accepted: 3 July 2022
Published: 7 July 2022

1. Introduction

The compact state of a protein molecule, characterized by a pronounced secondary and fluctuating tertiary structure, was theoretically predicted by O.B. Ptitsyn [1] and experimentally proved by his team [2–4]. Later, this state was called a "molten globule" [5–7], and was in vitro observed for many proteins under moderately denaturing conditions in vitro (see reviews [8–10]).

The last three decades demonstrated a significant progress in the theory of protein folding [11–22] and intensive studies of a wide range of proteins [23–33]. The development of experimental approaches and the use of new techniques, especially such as nuclear magnetic resonance NMR (in some modifications) and fluorescence, made it possible to follow changes in the protein structure under cell-like conditions [24–33]. Many studies directly or indirectly imply the presence of molten protein globules under these conditions.

The discovery of intrinsically disordered (or natively unfolded) proteins (IDPs) comprehensively described by Tompa [34] (see reviews [35–37]) revealed that some of them are in the MG state, while the structure of others is closer to the unfolded state. The latter are sometimes referred to as "pre-MGs". It has become clear that many protein functions require the rigid N state of the protein, while the others require their more or less disordered states. The latter have several properties similar to the MG state, but some properties distinguish typical IDPs from typical MGs. (see Table 1 and [38–43]).

It is possible to distinguish the MG state of a protein (or, by the original definition, "protein with fluctuating tertiary structure") from IDP due to the differences in changes of their properties under different impacts.

Here, we do not discuss natively unfolded proteins but focus on the MG of "normal" globular proteins.

Table 1. Differences in the properties of the MG state of proteins and IDP.

Impact Type and Conditions	MG State, Globular Proteins	IDP State, Natively Unfolded Proteins
Unfolding by strong denaturants	Global unfolding to the coil state	Usually, no further global unfolding, but disruption of local structures is possible
Heat effect	Decrease in secondary structure content	Structuring, heat resistance
Different behavior of normalized SAXS curves (Kratky plots) [38,39]	Bell-shaped curves with a pronounced maximum	Monotonic curve rising (no maximum)
H/D exchange [40,41]	Slightly elevated exchange as compared to the N state	The exchange is orders of magnitude higher than that for MG
Gel filtration, electrophoresis [42,43]	Increase in hydrodynamic volume by 20–50% as compared to the N state	Hydrodynamic volume is 400–600% larger when compared to the N state of globular proteins with the same molecular weight

There are three important issues to be addressed:

(1) Why proteins adopt the MG state;
(2) What impacts cellular proteins causing a change in their stability and transition to MG;
(3) What functions the molten protein globule performs in the cell.

For convenience, we divide issue (2) into subsections by the impact type:

- Post-translational protein modifications: acetylation, phosphorylation, ubiquitination, glycation, and others.
- Protein–protein interactions: substrate–receptor, multimers, modular proteins.
- Protein–membrane interactions.
- Protein–chaperone interactions.
- Protein interactions with specific adapter proteins.

2. Physics of the MG

Depending on ambient conditions, the most stable state of a protein chain may be neither rigid (solid) nor completely unfolded (coil) but "molten".

In the majority (but not all) of proteins, the MG in vitro arises either from the N state under the effect of a moderate denaturant, with the further transition to the coil as the denaturant concentration grows, or from the coil due to denaturant dilution [9]. The MG-like state also results from heat denaturation (melting) of a rigid globule. The N-to-MG transition is of the "all-or-none" type. Typically, MG does not undergo further "all-or-none" melting or swelling; rather, its unfolding looks like a cooperative though broad S-shaped transition observed by optical methods such as CD and fluorescence [9]. However, some rigid proteins (especially small ones) unfold directly into coils without any intermediate state [44].

MG is a "soft" [9] state that shows similarity to a rigid protein in many aspects. Reinforced by hydrogen bonds, its secondary structure is a well-developed and stable until the globule is "dissolved" by a solvent. However, the MG's side chains lose their dense packing but acquire freedom of movement (that is, they lose energy but gain entropy). The liberation of the side chain rotational isomerization is the main driving force of protein melting [45].

Since most of the protein chain degrees of freedom relate to the small-scale side chain movements, it is their liberation that can make the MG thermodynamically advantageous. The liberation of small-scale side chain rotational isomerization does not require the complete unfolding of the globule; slight swelling would be enough. This swelling, however, leads to a significant decrease in the van der Waals attraction, which strongly depends on the distance, and even a slight increase in the globule's volume is enough to reduce it greatly. Generally, all of this is like the melting of a crystal, where a slight increase in volume reduces van der Waals interactions and liberates the motion of the molecules.

Unlike common polymer globules, the protein chain covered with different side chains cannot unfold by gradual swelling, because these side chains cannot change their positions independently; the rigid protein chain controls the positions of many side chains sitting at it, and this entire "forest" of side chains has to move as a whole.

Before the discovery of the MG state, protein denaturation was considered as complete unfolding of the protein structure; that is, as the transition to the coil. After this discovery [46–48], it became clear that the denatured protein can be either dense or loose, depending on the solvent's strength and the hydrophobicity of the protein chain [2,45–49].

The pores in the MG (i.e., the vacant space necessary for side chain movements) are usually "wet"; that is, they are occupied by the solvent [45,50] because a water molecule inside the protein is still better than the vacuum. Experimentally, the "wetness" of the MG is proven by the absence of a visible increase in the protein partial volume [30] after denaturation of almost any type. A "dry" state of the pores is thermodynamically less stable [51,52], yet it is observed in some kinetic processes [26,51].

The MG compactness is maintained by residual hydrophobic interactions that are at least three times weaker than those within the native protein [31]; the fact that even these residual contacts are absent for some side chains emphasizes the heterogeneity of MG [9].

3. Cellular Events Causing Changes in Protein Structure Stability and Leading to the Transition to the MG

3.1. Post-Translational Modifications

In terms of protein structure, post-translational modifications can be equated to mutations. Most of them occur on the surface of a protein globule and do not have a significant effect on the protein structure. The result of some other modifications is the loss of the dense packing and transition to the MG-like state. Modifications of certain protein activities are known to require a change in ambient conditions [52] or local unfolding of the sites of the modifications [53–55], or (sometimes) partial denaturation [56], i.e., transition to a "softer" state.

For example, the tumor suppressor protein p53 is susceptible to a variety of modifications that change its functions in response to cellular stress, including acetylation, methylation, phosphorylation, and ubiquitination [56]. Specifically, p53 is the prime example of a protein whose acetylation requires partial denaturation. As another example: the peptidyl prolyl isomerase Pin1 can be modified by phosphorylation, ubiquitination, sumoylation, or oxidation, depending on the function that it will perform in the cell [57].

3.1.1. Acetylation

Acetylation (mainly at Gln or Lys) is a reversible post-translational protein modification crucial for the regulation of gene expression [53,56,58]. It mainly affects large macromolecular complexes, such as chromatin remodeling, cell cycle, splicing, nuclear transport, actin nucleation, and others [53,54,57,59,60]. For histones, acetylation is critical because it triggers DNA transcription [61,62]. Moreover, acetylation can reduce interactions dependent on phosphorylation [61]. The acetylated N-terminus of a protein chain can create a specific signal for chain degradation [63].

Acyl groups that recognize elements of protein–protein interactions can vary from simple acetate to modified long-chain fatty acids required for the interaction with membranes and affecting signal transduction. They attach to various amino acids (Lys, Cys, and Ser/Thr), which can change the hydrophobicity of a protein and modify its functions. Myristoyl- and palmitate-induced modifications of Cys residues increase the protein affinity for membranes [64]. Thus, acylation is one of the key regulators of cellular pathways.

3.1.2. Phosphorylation

Phosphorylation is the main mode of external signal transduction. ATP was shown to induce a conformational transition in proteins [65]. Signal transduction is associated with the modification of proteins that receive a signal from outside. These proteins interact with a large class of molecules known as adapter proteins and organizing centers (hubs) (SH2, SH3, 14-3-3, AKAPs), where the bound proteins undergo modification by phosphatases and kinases, and move within the cell and elsewhere [54,55,66].

3.1.3. Ubiquitination

Ubiquitin (Ub) participates in many cellular events [67–79] including cell division, cell differentiation, signal transduction, movement of proteins within the cell, quality control, signaling, and endocytosis. Additionally, ubiquitin controls protein degradation and participates in DNA repair, endocytosis, autophagy, transcription, and immune system support. Ubiquitin binds to partners either as a single protein or in the form of Ub chains and interacts with Ub-receptors. The attachment of one to three Ub molecules affects the movement of proteins within the cell, and the addition of four or more (up to 16) Ub molecules directs the protein to degradation [69–71].

3.1.4. Glycation

Glycation is another modification that affects the protein structure [80]. The addition of sugars manifests as early as during biosynthesis. Attachment of any type of sugar to in-protein residue results in the surface localization of this residue where it stays during the folding process. When the folding is completed, a specific enzyme removes the sugar molecule, but the labeled amino acid residue remains on the surface of the protein and participates in further events.

3.2. Protein–Protein Interactions

The interaction of protein ligands with protein receptors is a common event in the cell; one of the most studied interacting pairs is insulin and its receptor [81–90]. Using this system, it was demonstrated that during the interaction changes occur in the structures of both the hormone and the receptor. Both insulin and its receptor are water-soluble proteins. Independent experiments showed that in appropriate conditions, the denaturant-affected structures of these proteins undergo the N-to-MG transition. Surprisingly, when in a complex, the structures of both insulin and its receptor remain similar to their MG in solution. This is direct evidence that these proteins change their rigid structure to a softer one capable of interaction.

Similar changes were observed for the relaxin protein family related to insulin [91].

Importantly, during protein–protein interactions, a part of the surface that was in contact with water appears in a hydrophobic environment, which can cause rearrangement of the structure [81–90]. During the formation of multimeric and modular proteins, changes in their structures also result in more stable complexes [92,93].

3.3. Protein–Membrane Interactions

In 1988 we suggested that the membrane surface can influence the structure of proteins. Because of the proximity of a poorly (compared to water) polarized membrane, our experiments were performed at low dielectric permittivity and low pH. They evidenced the actual membrane impact on the protein structure (see reviews [25,94–96]). The study of apomyoglobin showed that in the presence of phospholipid vesicles at neutral pH, both its native and unfolded forms bind to the vesicle surface, showing properties similar to those of MG, i.e., both forms undergo conformational changes upon interaction with the membrane. For myoglobin the effect of membrane proximity is not so pronounced, but it is still significant. This influence is of decisive importance for the myoglobin-induced exchange of oxygen and carbon dioxide near the mitochondrial membrane.

3.4. Protein–Chaperone Interactions

Chaperones (specific assisting proteins) play a very important role in cell life. They bind to both native and unfolded proteins, providing their transition to the MG state and protecting them from random interactions, aggregation, and degradation [97–102].

Chaperones bind kinetic intermediates during protein folding, thereby shielding them from aggregation and side interactions. They ensure the correct folding of both monomeric proteins and oligomeric subunits, and the formation of oligomeric proteins [103]. Chaperones can transport newly synthesized proteins from the place of biosynthesis to the place of functioning. Their main role is to prevent non-specific interactions between proteins and keep them in a state competent for various processes in the cell. The binding of proteins to chaperones and their release consumes ATP energy [103]. Chaperones are also involved in protein quality control [72,104] and protein degradation [105].

3.5. Protein Interactions with Specific Adapter Proteins and Organizing Centers

Signal transduction within the cell requires precise work of the cellular regulatory system; to meet this requirement, specific protein domains interact with each other and different components of the cell. The modular nature of these domains allows their concurrent interaction with a variety of proteins [106,107]. They are known to recognize post-translational modifications [108–113], but because many of them are natively unfolded proteins they are not considered here.

However, among the adapters, there is at least one compact system that includes proteins of the 14-3-3 family [106,107,114–116]. The system received the strange name "14-3-3" due to the technique of isolation on chromatographic columns and testing by electrophoresis. For its main isoform (ζ), the crystal structure was deciphered back in 1995 [117,118]. It is a dimer consisting of two chains, each of which contains nine antiparallel helices; they form a horseshoe-like region between two structures where the substrate binds. There are nine isoforms of this protein, the combination of which allows the binding of multiple proteins. These isoforms recognize substrate proteins phosphorylated at Ser or Thr and can bind them by one or by two, depending on the cellular process. Such binding changes the conformation of the partner and brings the substrate proteins closer to one another, thus facilitating their interaction. It can also open an active site of the bound substrate, while other regions of this substrate can be shielded from currently unnecessary and even dangerous interactions with the main part of the protein [119–121].

The involvement of an adapter protein in the stabilization of the active site structure increases both the substrate binding and the yield of the reaction product, and this directly indicates the regulation of the catalytic activity of the bound substrate protein [117,122].

The structural basis for the interaction of 14-3-3 proteins with their substrates is described in [119–121]. As discovered, even disordered regions of the bound substrate protein acquire a well-structured shape upon interaction with a 14-3-3 protein [120,121,123]. Thus, it can be assumed that binding/release with 14-3-3 proteins can lead to the transition of the protein structure from rigid to soft and vice versa.

4. Functions of the Molten Protein Globule in the Cell

The development of new, more sensitive, and accurate research methods leads to the revision of some of the kinetic and equilibrium data, so the interpretation of these data may change [124,125]. For example, several proteins whose folding was previously believed to be a 2-state process were found to have compact folding intermediates with properties like those of MG [123,124]. This was observed for some cold shock proteins [126], the B1 domain of the G protein [127], single-chain monellin [128], RNase A [129], and the bacterial immune protein Im7 [130].

The experimental discovery of a "dry" MG as an unstable kinetic folding intermediate has presented some new data [26,52]; initially, the "dry" MG was predicted by Shakhnovich and Finkelstein [45,50] along with the much better known "wet" MG which is a stable folding intermediate. The presence of both the "dry" and "wet" MG was detected in monellin [26], the villin headpiece subdomain [131], *E. coli* DHFR [132], and RNase A [133].

Because some steps of enzymatic catalysis require structural flexibility, the biologically active conformational states other than fully folded structures may be more frequent than it was thought previously. For example, according to [134], the "non-native" state of acylphosphatase from *Sulfolobus solfataricus* shows enzymatic activity.

It was found that sometimes MGs occur outside the folding pathway in proteins with substituted amino acids, as exemplified by apoflavodoxin. Since MG are generally prone to aggregation, their presence outside the folding pathway increases the risk of protein accumulation, which can adversely affect the organism [134,135]. In 2003, Dobson comprehensively investigated how MG formation and aggregation can cause protein misfolding and aggregation, which results in numerous pathologies [136]. Later, he detected the other associations between the observed abnormalities in protein folding and diseases [136,137].

Surprisingly, the complex of human milk protein α-lactalbumin (αLA) with oleic acid (HAMLET) can kill cancer cells. In this complex, αLA is neither active nor "native" but preserves the MG state [138,139].

A similar complex with lysozyme also shows bactericidal activity, causing DNA fragmentation [139,140].

To date, ample evidence for the functional significance of many MG-like proteins has been reported [141]. This is true for the protein p53 [98], ferredoxin [142], alpha-mannosidase [143], melanogaster crammer [144], glycated serum albumin [80], and others. Some of them were reviewed by Bychkova et al., 2018 [10]. Moreover, the functional significance of such protein states is shown for the monomeric form of chorismate mutase [145], dihydrofolate reductase [146], ubiquitin [147], periplasmic binding proteins [148], staphylococcal nuclease [148], and α-galactosidase (cicer α-galactosidase) [149] (of course, we do not claim that a functional MG-like state is mandatory for all proteins).

An interesting observation concerning the metmyoglobin (MetMb) structure was reported back in 1998, although its altering conformation was not the focus of this study. The conformation changed during the transition of non-active MetMb to its active form capable of O_2 binding. The transition pathway included an intermediate allowing further relaxation of the protein to form active deoxyMb, as was observed by the changing Soret band. This transition was catalyzed by MetMb reductase [150], which suggests that enzymes interacted with proteins in an intermediate state facilitating the reaction.

5. Conclusions

Summing up this review, we can state the fact that in the cell, some active proteins can have both the rigid structure typical of enzymes and that of an MG. In one of his re-views, Oleg B. Ptitsyn put forward a hypothesis that there could be two types of the "native" protein state, hard and soft. Since the "native state of a protein" usually implies the state that allows protein functioning, this prediction is apparently true for many proteins that display activity when their structure is MG-like. The experimental data obtained for different proteins speak for the proposed hypothesis that the transition of a protein to the state of an MG in a cell is not something extremely rare and exceptional. At least for some proteins, the state of the MG is necessary for the performance of their functions. This situation suggests the need to consider the conformational state of the protein when studying its activity both in the cell and in vitro, especially when projecting research results from one condition to another. The cell is a storehouse for unexpected phenomena and discoveries. Hopefully, the development of new and more sensitive methods or the improvement of those already in use will lead to the discovery of novel proteins functioning in the MG state.

Author Contributions: V.E.B., D.A.D., and V.A.B. performed literature search, A.V.F. and V.E.B. performed editing. All authors have read and agreed to the published version of the manuscript.

Funding: The work was supported by the Russian Foundation for Basic Research grant № 21-14-00268.

Acknowledgments: We are grateful to O.A. Eremeeva for assistance and E.V. Serebrova for editing the English text.

Conflicts of Interest: The authors declare the absence of any conflict of interest.

References

1. Ptitsyn, O.B. Step-wise mechanism of self-organization of protein molecules. *Dokl. Acad. Nauk USSR* **1973**, *210*, 1213–1215.
2. Dolgikh, D.A.; Gilmanshin, R.I.; Brazhnikov, E.V.; Bychkova, V.E.; Semisotnov, G.V.; Venyaminov, S.Y.; Ptitsyn, O.B. Alpha-lactalbumin: Compact state with fluctuating tertiary structure? *FEBS Lett.* **1981**, *136*, 311–315. [CrossRef]
3. Gilmanshin, R.I.; Dolgikh, D.A.; Ptitsyn, O.B.; Finkelshtein, A.V.; Shakhnovich, E.I. Protein globules without the unique tertiary structure: Experimental data for alpha-lactalbumins and the general model. *Biophysics* **1982**, *27*, 1005–1016.
4. Ptitsyn, O.B.; Dolgikh, D.A.; Gilmanshin, R.I.; Shakhnovich, E.I.; Finkelshtein, A.V. Fluctuating state of a protein globule. *Mol. Biol.* **1983**, *17*, 569–576.
5. Ohgushi, M.; Wada, A. 'Molten-globule state': A compact form of globular proteins with mobile side-chains. *FEBS Lett.* **1983**, *164*, 21–24. [CrossRef]
6. Kuroda, Y.; Kidokoro, S.; Wada, A. Thermodynamic characterization of cytochrome c at low pH. Observation of the molten globule state and of the cold denaturation process. *J. Mol. Biol.* **1992**, *223*, 1139–1153. [CrossRef]
7. Creighton, T.E. *Proteins: Structures and Molecular Properties*, 2nd ed.; W. H. Freeman & Co: New York, NY, USA, 1993; Chapters 2 and 7.
8. Bychkova, V.E.; Ptitsyn, O.B. The molten globule in vitro and in vivo. *Chemtracts* **1993**, *4*, 133–163.
9. Ptitsyn, O.B. Molten globule and protein folding. *Adv. Prot. Chem.* **1995**, *47*, 83–229.
10. Bychkova, V.E.; Semisotnov, G.V.; Balobanov, V.A.; Finkelstein, A.V. The molten globule concept: 45 years later. *Biochemistry* **2018**, *83*, S33–S47. [CrossRef]
11. Arai, M.; Kuwajima, K. Role of the molten globule state in protein folding. *Adv. Prot. Chem.* **2000**, *53*, 209–282.
12. Dobson, C.M. Protein folding. Solid evidence for molten globules. *Curr. Biol.* **1994**, *4*, 636–640. [CrossRef]
13. Pande, V.S.; Rokhsar, D.S. Is the molten globule a third phase of proteins? *Proc. Natl. Acad. Sci. USA* **1998**, *95*, 1490–1494. [CrossRef] [PubMed]
14. Finkelstein, A.V.; Ptitsyn, O.B. *Protein Physics. A Course of Lectures*, 2nd ed.; Academic Press: New York, NY, USA, 2016; Chapters 17–21.
15. Finkelstein, A.V. 50+ Years of protein folding. *Biochemistry* **2018**, *83*, S3–S18. [CrossRef] [PubMed]
16. Finkelstein, A.V.; Garbuzynskiy, S.O. Reduction of the search space for the folding of proteins at the level of formation and assembly of secondary structures: A new view on the solution of levinthal's paradox. *Chemphyschem* **2015**, *16*, 3375–3378. [CrossRef]
17. Camilloni, C.; Bonetti, D.; Morrone, A.; Giri, R.; Dobson, C.M.; Brunori, M.; Gianni, S.; Vendruscolo, M. Towards a structural biology of the hydrophobic effect in protein folding. *Sci. Rep.* **2016**, *6*, 28285. [CrossRef]
18. Baldwin, R.L.; Rose, G.D. Molten globules, entropy-driven conformational change and protein folding. *Curr. Opin. Struct. Biol.* **2013**, *23*, 4–10. [CrossRef]
19. Garbuzynskiy, S.O.; Ivankov, D.N.; Bogatyreva, N.S.; Finkelstein, A.V. Golden triangle for folding rates of globular proteins. *Proc. Natl. Acad. Sci. USA* **2013**, *110*, 147–150. [CrossRef]
20. Ivankov, D.N.; Garbuzynskiy, S.O.; Alm, E.; Plaxco, K.W.; Baker, D.; Finkelstein, A.V. Contact order revisited: Influence of protein size on the folding rate. *Prot. Sci.* **2003**, *12*, 2057–2062. [CrossRef]
21. Galzitskaya, O.V.; Garbuzynskiy, S.O.; Ivankov, D.N.; Finkelstein, A.V. Chain length is the main determinant of the folding rate for proteins with three-state folding kinetics. *Proteins-Struct. Funct. Bioinform.* **2003**, *51*, 162–166. [CrossRef]
22. Uversky, V.N.; Finkelstein, A.V. Life in phases: Intra-and inter-molecular phase transitions in protein solutions. *Biomolecules* **2019**, *9*, 842. [CrossRef]
23. Fersht, A. *Structure and Mechanism in Protein Science: A Guide to Enzyme Catalysis and Protein Folding*; W. H. Freeman & Co.: New York, NY, USA, 1999; Chapters 15, 18, 19.
24. Palmer, A.G. Enzyme dynamics from NMR spectroscopy. *Acc. Chem. Res.* **2015**, *48*, 457–465. [CrossRef] [PubMed]
25. Bychkova, V.E.; Basova, L.V.; Balobanov, V.A. How membrane surface affects protein structure. *Biochemistry* **2014**, *79*, 1483–1514. [CrossRef] [PubMed]
26. Jha, S.K.; Udgaonkar, J.B. Direct evidence for a dry molten globule intermediate during the unfolding of a small protein. *Proc. Natl. Acad. Sci. USA* **2009**, *106*, 12289–12294. [CrossRef]
27. Baldwin, R.L.; Frieden, C.; Rose, G.D. Dry molten globule intermediates and the mechanism of protein unfolding. *Proteins-Struct. Funct. Bioinform.* **2010**, *78*, 2725–2737. [CrossRef] [PubMed]
28. Sarkar, S.S.; Udgaonkar, J.B.; Krishnamoorthy, G. Unfolding of a small protein proceeds via dry and wet globules and a solvated transition state. *Biophys. J.* **2013**, *105*, 2392–2402. [CrossRef] [PubMed]
29. Balobanov, V.A.; Katina, N.S.; Finkelstein, A.V.; Bychkova, V.E. Intermediate states of apomyoglobin: Are they parts of the same area of conformations diagram? *Biochemistry* **2017**, *82*, 625–631. [CrossRef]

30. Kharakoz, D.P.; Bychkova, V.E. Molten globule of human alpha-lactalbumin: Hydration, density, and compressibility of the interior. *Biochemistry* **1997**, *36*, 1882–1890. [CrossRef]
31. Samatova, E.N.; Katina, N.S.; Balobanov, V.A.; Melnik, B.S.; Dolgikh, D.A.; Bychkova, V.E.; Finkelstein, A.V. How strong are side chain interactions in the folding intermediate? *Prot. Sci.* **2009**, *18*, 2152–2159. [CrossRef]
32. Li, L.Y.; Shi, X.S.; Guo, X.D.; Li, H.; Xu, C.Q. Ionic protein lipid interaction at the plasma membrane: What can the charge do? *Trends Biochem. Sci.* **2014**, *39*, 130–140. [CrossRef]
33. Mizukami, T.; Xu, M.; Fazlieva, R.; Bychkova, V.E.; Roder, H. Complex folding landscape of apomyoglobin at acidic pH revealed by ultrafast kinetic analysis of core mutants. *J. Phys. Chem. B* **2018**, *122*, 11228–11239. [CrossRef]
34. Tompa, P.; Fersht, A. *Structure and Function of Intrinsically Desordered Proteins*, 1st ed.; Chapman and Hall/CRC: New York, NY, USA, 2009; Chapters 1–3, 8, 10–12, 15.
35. Wright, P.E.; Dyson, H.J. Intrinsically unstructured proteins: Re-assessing the protein structure-function paradigm. *J. Mol. Biol.* **1999**, *293*, 321–331. [CrossRef] [PubMed]
36. Dyson, H.J.; Wright, P.E. Coupling of folding and binding for unstructured proteins. *Curr. Opin. Struct. Biol.* **2002**, *12*, 54–60. [CrossRef]
37. Uversky, V.N. Natively unfolded proteins: A point where biology waits for physics. *Prot. Sci.* **2002**, *11*, 739–756. [CrossRef] [PubMed]
38. Uversky, V.N. What does it mean to be natively unfolded? *Eur. J. Biochem.* **2002**, *269*, 2–12. [CrossRef] [PubMed]
39. Receveur-Brechot, V.; Durand, D. How random are intrinsically disordered proteins? a small angle scattering perspective. *Curr. Protein Pept. Sci.* **2012**, *13*, 55–75. [CrossRef] [PubMed]
40. Permyakov, S.E.; Millett, I.S.; Doniach, S.; Permyakov, E.A.; Uversky, V.N. Natively unfolded C-terminal domain of caldesmon remains substantially unstructured after the effective binding to calmodulin. *Proteins-Struct. Funct. Bioinform.* **2003**, *53*, 855–862. [CrossRef]
41. Bracken, C. NMR spin relaxation methods for characterization of disorder and folding in proteins. *J. Mol. Graph. Model.* **2001**, *19*, 3–12. [CrossRef]
42. Bracken, C.; Iakoucheva, L.M.; Rorner, P.R.; Dunker, A.K. Combining prediction, computation and experiment for the characterization of protein disorder. *Curr. Opin. Struct. Biol.* **2004**, *14*, 570–576. [CrossRef]
43. Csizmok, V.; Szollosi, E.; Friedrich, P.; Tompa, P. A novel two-dimensional electrophoresis technique for the identification of intrinsically unstructured proteins. *Mol. Cell. Proteom.* **2006**, *5*, 265–273. [CrossRef]
44. Jackson, S.E. How do small single-domain proteins fold? *Fold Des.* **1998**, *3*, R81–R91. [CrossRef]
45. Shakhnovich, E.I.; Finkelstein, A.V. Theory of cooperative transitions in protein molecules. I. Why denaturation of globular proteins is a first-order phase transition. *Biopolymers* **1989**, *28*, 1667–1680. [CrossRef] [PubMed]
46. Wong, K.P.; Tanford, C. Denaturation of bovine carbonic anhydrase B by guanidine hydrochloride. A process involving separable sequential conformational transitions. *J. Biol. Chem.* **1973**, *248*, 8518–8523. [CrossRef]
47. Robson, B.; Pain, R.H. The mechanism of folding of globular proteins. Equilibria and kinetics of conformational transitions of penicillinase from Staphylococcus aureus involving a state of intermediate conformation. *Biochem. J.* **1976**, *155*, 331–344. [CrossRef] [PubMed]
48. Kuwajima, K.; Nitta, K.; Yoneyama, M.; Sugai, S. Three-state denaturation of alpha-lactalbumin by guanidine hydrochloride. *J. Mol. Biol.* **1976**, *106*, 359–373. [CrossRef]
49. Ikeguchi, M.; Kuwajima, K.; Sugai, S. Ca^{2+}-induced alteration in the unfolding behavior of alpha-lactalbumin. *J. Biochem.* **1986**, *99*, 1191–1201. [CrossRef]
50. Finkelstein, A.V.; Shakhnovich, E.I. Theory of cooperative transitions in protein molecules. II. Phase diagram for a protein molecule in solution. *Biopolymers* **1989**, *28*, 1681–1694. [CrossRef]
51. Muñoz, V.; Sadqi, M.; Naganathan, A.N.; de Sancho, D. Exploiting the downhill folding regime via experiment. *HFSP J.* **2008**, *2*, 342–353. [CrossRef]
52. Prasad, M.; Thomas, J.L.; Whittal, R.M.; Bose, H.S. Mitochondrial 3 beta-hydroxysteroid dehydrogenase enzyme activity requires reversible ph-dependent conformational change at the intermembrane space. *J. Biol. Chem.* **2012**, *287*, 9534–9546. [CrossRef]
53. Ceskova, P.; Chichgeri, H.; Wallace, M.; Vojtesek, B.; Hupp, T.R. On the mechanism of sequence-specific DNA-dependent acetylation of p53: The acetylation motif is exposed upon DNA binding. *J. Mol. Biol.* **2006**, *357*, 442–456. [CrossRef]
54. Armstrong, B.D.; Choi, J.; Lopez, C.; Wesener, D.A.; Hubbell, W.; Cavagnero, S.; Han, S. Site-specific hydration dynamics in the nonpolar core of a molten globule by dynamic nuclear polarization of water. *J. Am. Chem. Soc.* **2011**, *133*, 5987–5995. [CrossRef]
55. Spolaore, B.; Raboni, S.; Molina, A.R.; Satwekar, A.; Damiano, N.; Fontana, A. Local unfolding is required for the site-specific protein modification by transglutaminase. *Biochemistry* **2012**, *51*, 8679–8689. [CrossRef] [PubMed]
56. Taira, N.; Yoshida, K. Post-translational modifications of p53 tumor suppressor: Determinants of its functional targets. *Histol. Histopathol.* **2012**, *27*, 437–443. [PubMed]
57. Chen, D.M.; Wang, L.; Lee, T.H. Post-translational modifications of the peptidyl-prolyl isomerase pin1. *Front. Cell Dev. Biol.* **2020**, *8*, 129. [CrossRef] [PubMed]
58. Soond, S.M.; Savvateeva, L.V.; Makarov, V.A.; Gorokhovets, N.V.; Townsend, P.A.; Zamyatnin, A.A. Making connections: P53 and the cathepsin proteases as co-regulators of cancer and apoptosis. *Cancers* **2020**, *12*, 3476. [CrossRef]

59. Teixeira, C.S.S.; Cerqueira, N.; Gomes, P.; Sousa, S.E. A Molecular perspective on sirtuin activity. *Int. J. Mol. Sci.* **2020**, *21*, 8609. [CrossRef]
60. Choudhary, C.; Kumar, C.; Gnad, F.; Nielsen, M.L.; Rehman, M.; Walther, T.C.; Olsen, J.V.; Mann, M. Lysine acetylation targets protein complexes and co-regulates major cellular functions. *Science* **2009**, *325*, 834–840. [CrossRef] [PubMed]
61. Barlev, N.A.; Liu, L.; Chehab, N.H.; Mansfield, K.; Harris, K.G.; Halazonetis, T.D.; Berger, S.L. Acetylation of p53 activates transcription through recruitment of coactivators/histone acetyltransferases. *Mol. Cell* **2001**, *8*, 1243–1254. [CrossRef]
62. Sarg, B.; Helliger, W.; Talasz, H.; Koutzamani, E.; Lindner, H.H. Histone H4 hyperacetylation precludes histone H4 lysine 20 trimethylation. *J. Biol. Chem.* **2004**, *279*, 53458–53464. [CrossRef]
63. Hwang, C.S.; Shemorry, A.; Varshavsky, A. N-terminal acetylation of cellular proteins creates specific degradation signals. *Science* **2010**, *327*, 973–977. [CrossRef]
64. Thinon, E.; Hang, H.C. Chemical reporters for exploring protein acylation. *Biochem. Soc. Trans.* **2015**, *43*, 253–261. [CrossRef]
65. Goto, Y.; Okamura, N.; Aimoto, S. ATP-induced conformational transition of denatured proteins. *J. Biochem.* **1991**, *109*, 746–750. [CrossRef] [PubMed]
66. Liu, X.R.; Zhang, Y.P.; Wang, Y.J.; Yang, M.W.; Hong, F.F.; Yang, S.L. Protein phosphorylation in cancer: Role of nitric oxide Signaling Pathway. *Biomolecules* **2021**, *11*, 1009. [CrossRef] [PubMed]
67. Gavrilov, Y.; Hagai, T.; Levy, Y. Nonspecific yet decisive: Ubiquitination can affect the native-state dynamics of the modified protein. *Prot. Sci.* **2015**, *24*, 1580–1592. [CrossRef] [PubMed]
68. Grabbe, C.; Husnjak, K.; Dikic, I. The spatial and temporal organization of ubiquitin networks. *Nat. Rev. Mol. Cell Biol.* **2011**, *12*, 295–307. [CrossRef] [PubMed]
69. Oikawa, D.; Sato, Y.; Ito, H.; Tokunaga, F. Linear ubiquitin code: Its writer, erasers, decoders, inhibitors, and implications in disorders. *Int. J. Mol. Sci.* **2020**, *21*, 3381. [CrossRef] [PubMed]
70. Haglund, K.; Dikic, I. Ubiquitylation and cell signaling. *EMBO J.* **2005**, *24*, 3353–3359. [CrossRef]
71. Thrower, J.S.; Hoffman, L.; Rechsteiner, M.; Pickart, C.M. Recognition of the polyubiquitin proteolytic signal. *EMBO J.* **2000**, *19*, 94–102. [CrossRef]
72. Ciechanover, A.; Kwon, Y.T. Protein Quality control by molecular chaperones in neurodegeneration. *Front. Neurosci.* **2017**, *11*, 185. [CrossRef]
73. Krappmann, D.; Scheidereit, C. A pervasive role of ubiquitin conjugation in activation and termination of IkB kinase pathways. *EMBO Rep.* **2005**, *6*, 321–326. [CrossRef]
74. Santonico, E. Old and new concepts in ubiquitin and NEDD8 recognition. *Biomolecules* **2020**, *10*, 566. [CrossRef]
75. Kliza, K.; Husnjak, K. Resolving the complexity of ubiquitin networks. *Front. Mol. Biosci.* **2020**, *7*, 21. [CrossRef] [PubMed]
76. Wilkinson, K.D. DUBs at a glance. *J. Cell Sci.* **2009**, *122*, 2325–2329. [CrossRef] [PubMed]
77. Dikic, I.; Wakatsuki, S.; Walters, K.J. Ubiquitin binding domains–from structures to functions. *Nat. Rev. Mol. Cell Biol.* **2009**, *10*, 659–671. [CrossRef]
78. Tyrrell, A.; Flick, K.; Kleiger, G.; Zhang, H.W.; Deshaies, R.J.; Kaiser, P. Physiologically relevant and portable tandem ubiquitin-binding domain stabilizes polyubiquitylated proteins. *Proc. Natl. Acad. Sci. USA* **2010**, *107*, 19796–19801. [CrossRef]
79. Iwai, K.; Tokunaga, F. Linear polyubiquitination: A new regulator of NF-kappa B activation. *EMBO Rep.* **2009**, *10*, 706–713. [CrossRef]
80. Indurthi, V.S.K.; Leclerc, E.; Vetter, S.W. Interaction between glycated serum albumin and AGE-receptors depends on structural changes and the glycation reagent. *Arch. Biochem. Biophys.* **2012**, *528*, 185–196. [CrossRef]
81. Nakagawa, S.H.; Tager, H.S. Importance of main-chain flexibility and the insulin fold in insulin-receptor interactions. *Biochemistry* **1993**, *32*, 7237–7243. [CrossRef] [PubMed]
82. Hua, Q.X.; Ladbury, J.E.; Weiss, M.A. Dynamics of a monomeric insulin analogue: Testing the molten-globule hypothesis. *Biochemistry* **1993**, *32*, 1433–1442. [CrossRef]
83. Ward, C.W.; Lawrence, M.C. Ligand-induced activation of the insulin receptor: A multi-step process involving structural changes in both the ligand and the receptor. *Bioessays* **2009**, *31*, 422–434. [CrossRef]
84. Hua, Q.X.; Xu, B.; Huang, K.; Hu, S.Q.; Nakagawa, S.; Jia, W.H.; Wang, S.; Whittaker, J.; Katsoyannis, P.G.; Weiss, M.A. Enhancing the activity of a protein by stereospecific unfolding: Conformational life cycle of insulin and its evolutionary origins. *J. Biol. Chem.* **2009**, *284*, 14586–14596. [CrossRef]
85. Whitten, A.E.; Smith, B.J.; Menting, J.G.; Margetts, M.B.; McKern, N.M.; Lovrecz, G.O.; Adams, T.E.; Richards, K.; Bentley, J.D.; Trewhella, J.; et al. Solution structure of ectodomains of the insulin receptor family: The ectodomain of the type 1 insulin-like growth factor receptor displays asymmetry of ligand binding accompanied by limited conformational change. *J. Mol. Biol.* **2009**, *394*, 878–892. [CrossRef] [PubMed]
86. Smock, R.G.; Gierasch, L.M. Sending signals dynamically. *Science* **2009**, *324*, 198–203. [CrossRef]
87. Hua, Q. Insulin: A small protein with a long journey. *Prot. Cell* **2010**, *1*, 537–551. [CrossRef] [PubMed]
88. Menting, J.G.; Whittaker, J.; Margetts, M.B.; Whittaker, L.J.; Kong, G.K.W.; Smith, B.J.; Watson, C.J.; Zakova, L.; Kletvikova, E.; Jiracek, J.; et al. How insulin engages its primary binding site on the insulin receptor. *Nature* **2013**, *493*, 241–245. [CrossRef] [PubMed]
89. Chen, L.; Balabanidou, V.; Remeta, D.P.; Minetti, C.; Portaliou, A.G.; Economou, A.; Kalodimos, C.G. Structural instability tuning as a regulatory mechanism in protein-protein interactions. *Mol. Cell* **2011**, *44*, 734–744. [CrossRef]

90. Guo, Y.; Tang, C.Y.; Man, X.F.; Tang, H.N.; Tang, J.; Wang, F.; Zhou, C.L.; Tan, S.W.; Feng, Y.Z.; Zhou, H.D. Insulin receptor substrate-1 time-dependently regulates bone formation by controlling collagen I alpha 2 expression via miR-342. *FASEB J.* **2016**, *30*, 4214–4226. [CrossRef]
91. Svendsen, A.M.; Vrecl, M.; Knudsen, L.; Heding, A.; Wade, J.D.; Bathgate, R.A.D.; De Meyts, P.; Nohr, J. Dimerization and negative cooperativity in the relaxin family peptide receptors. *Relaxin Relat. Pept. Fifth Int. Conf.* **2009**, *1160*, 54–59. [CrossRef]
92. Jean-Baptiste, G.; Yang, Z.; Greenwood, M.T. Regulatory mechanisms involved in modulating RGS function. *Cell. Mol. Life Sci.* **2006**, *63*, 1969–1985. [CrossRef]
93. Abramow-Newerly, M.; Roy, A.A.; Nunn, C.; Chidiac, P. RGS proteins have a signalling complex: Interactions between RGS proteins and GPCRs, effectors, and auxiliary proteins. *Cell. Signal.* **2006**, *18*, 579–591. [CrossRef]
94. Bychkova, V.E.; Pain, R.H.; Ptitsyn, O.B. The "molten globule" state is involved in translocation of proteins across membranes? *FEBS Lett.* **1988**, *238*, 231–234. [CrossRef]
95. Bucciantini, M.; Cecchi, C. Biological membranes as protein aggregation matrices and targets of amyloid toxicity. *Protein Misfold. Cell. Stress Dis. Aging Concepts Protoc.* **2010**, *648*, 231–243.
96. Bokvist, M.; Grobner, G. Misfolding of amyloidogenic proteins at membrane surfaces: The impact of macromolecular crowding. *J. Am. Chem. Soc.* **2007**, *129*, 14848–14849. [CrossRef] [PubMed]
97. Hiller, S. Molecular chaperones and their denaturing effect on client proteins. *J. Biomol. NMR* **2021**, *75*, 1–8. [CrossRef] [PubMed]
98. Park, S.J.; Borin, B.N.; Martinez-Yamout, M.A.; Dyson, H.J. The client protein p53 adopts a molten globule-like state in the presence of Hsp90. *Nat. Struct. Mol. Biol.* **2011**, *18*, 537–541. [CrossRef] [PubMed]
99. Oh, J.H.; Hyun, J.Y.; Varshavsky, A. Control of Hsp90 chaperone and its clients by N-terminal acetylation and the N-end rule pathway. *Proc. Natl. Acad. Sci. USA* **2017**, *114*, E4370–E4379. [CrossRef]
100. Park, S.J.; Kostic, M.; Dyson, H.J. Dynamic interaction of hsp90 with its client protein p53. *J. Mol. Biol.* **2011**, *411*, 158–173. [CrossRef]
101. Ortiz-Munoz, A.; Medina-Abarca, H.F.; Fontana, W. Combinatorial protein-protein interactions on a polymerizing scaffold. *Proc. Natl. Acad. Sci. USA* **2020**, *117*, 2930–2937. [CrossRef]
102. Pey, A.L.; Salido, E.; Sanchez-Ruiz, J.M. Role of low native state kinetic stability and interaction of partially unfolded states with molecular chaperones in the mitochondrial protein mistargeting associated with primary hyperoxaluria. *Amino Acids* **2011**, *41*, 1233–1245. [CrossRef]
103. Xu, H.F. ATP-driven nonequilibrium activation of kinase clients by the molecular chaperone Hsp90. *Biophys. J.* **2020**, *119*, 1538–1549. [CrossRef]
104. Bershtein, S.; Mu, W.M.; Serohijos, A.W.R.; Zhou, J.W.; Shakhnovich, E.I. Protein quality control acts on folding intermediates to shape the effects of mutations on organismal fitness. *Mol. Cell* **2013**, *49*, 133–144. [CrossRef]
105. Shiber, A.; Ravid, T. Chaperoning proteins for destruction: Diverse roles of Hsp70 chaperones and their co-chaperones in targeting misfolded proteins to the proteasome. *Biomolecules* **2014**, *4*, 704–724. [CrossRef] [PubMed]
106. Pawson, T.; Nash, P. Assembly of cell regulatory systems through protein interaction domains. *Science* **2003**, *300*, 445–452. [CrossRef] [PubMed]
107. Scott, J.D.; Pawson, T. Cell signaling in space and time: Where proteins come together and when they're apart. *Science* **2009**, *326*, 1220–1224. [CrossRef] [PubMed]
108. Liu, B.A.; Nash, P.D. Evolution of SH2 domains and phosphotyrosine signalling networks. *Philos. Trans. R. Soc. B-Biol. Sci.* **2012**, *367*, 2556–2573. [CrossRef]
109. Bhattacharyya, R.P.; Remenyi, A.; Yeh, B.J.; Lim, W.A. Domains, motifs, and scaffolds: The role of modular interactions in the evolution and wiring of cell signaling circuits. *Annu. Rev. Biochem.* **2006**, *75*, 655–680. [CrossRef]
110. Zeke, A.; Lukacs, M.; Lim, W.A.; Remenyi, A. Scaffolds: Interaction platforms for cellular signaling circuits. *Trends Cell Biol.* **2009**, *19*, 364–374. [CrossRef]
111. Good, M.C.; Zalatan, J.G.; Lim, W.A. Scaffold proteins: Hubs for controlling the flow of cellular information. *Science* **2011**, *332*, 680–686. [CrossRef]
112. Langeberg, L.K.; Scott, J.D. Signalling scaffolds and local organization of cellular behaviour. *Nat. Rev. Mol. Cell Biol.* **2015**, *16*, 232–244. [CrossRef]
113. Schultz, J.E.; Natarajan, J. Regulated unfolding: A basic principle of intraprotein signaling in modular proteins. *Trends Biochem. Sci.* **2013**, *38*, 538–545. [CrossRef]
114. Liu, D.; Bienkowska, J.; Petosa, C.; Collier, R.J.; Fu, H.; Liddington, R. Crystal structure of the zeta isoform of the 14-3-3 protein. *Nature* **1995**, *376*, 191–194. [CrossRef]
115. Yaffe, M.B.; Rittinger, K.; Volinia, S.; Caron, P.R.; Aitken, A.; Leffers, H.; Gamblin, S.J.; Smerdon, S.J.; Cantley, L.C. The structural basis for 14-3-3:phosphopeptide binding specifity. *Cell* **1997**, *91*, 961–971. [CrossRef]
116. Fu, H.; Subramanian, R.R.; Masters, S.C. 14-3-3 Proteins: Structure, function, and regulation. *Annu. Rev. Pharmacol. Toxicol.* **2000**, *40*, 617–647. [CrossRef] [PubMed]
117. Pawson, T. Specificity in signal transduction: From phosphotyrosine-SH2 domain interactions to complex cellular systems. *Cell* **2004**, *116*, 191–203. [CrossRef]
118. JiJin, J.; Pawson, T. Modular evolution of phosphorylation-based signalling systems. *Philos. Trans. R. Soc. B-Biol. Sci.* **2012**, *367*, 2540–2555. [CrossRef] [PubMed]

119. Obsilova, V.; Silhan, J.; Boura, E.; Teisinger, J.; Obsil, T. 14-3-3 Proteins: A family of versatile molecular regulators. *Physiol. Res.* **2008**, *57* (Suppl. S3), S11–S21. [CrossRef] [PubMed]
120. Mitrea, D.M.; Kriwacki, R.W. Regulated unfolding of proteins in signaling. *FEBS Lett.* **2013**, *587*, 1081–1088. [CrossRef] [PubMed]
121. Obsilova, V.; Kopecka, M.; Kosek, D.; Kacirova, M.; Kylarova, S.; Rezabkova, L.; Obsil, T. Mechanism of the 14-3-3 protein function: Regulation of protein function through conformational modulation. *Physiol. Res.* **2014**, *63* (Suppl. S1), S155–S164. [CrossRef]
122. Yaffe, M.B. How do 14-3-3 proteins work?-Gatekeeper phosphorylation and the molecular anvil hypothesis. *FEBS Lett.* **2002**, *513*, 53–57. [CrossRef]
123. Ikeguchi, M.; Kuwajima, K.; Mitani, M.; Sugai, S. Evidence for identity between the equilibrium unfolding intermediate and a transient folding intermediate: A comparative study of the folding reactions of alpha-lactalbumin and lysozyme. *Biochemistry* **1986**, *25*, 6965–6972. [CrossRef]
124. Radford, S.E.; Dobson, C.M.; Evans, P.A. The folding of hen lysozyme involves partially structured intermediates and multiple pathways. *Nature* **1992**, *358*, 302–307. [CrossRef]
125. Sluchanko, N.N.; Beelen, S.; Kulikova, A.A.; Weeks, S.D.; Antson, A.A.; Gusev, N.B.; Strelkov, S.V. Structural basis for the interaction of a human small heat shock protein with the 14-3-3 universal signaling regulator. *Structure* **2017**, *25*, 305–316. [CrossRef] [PubMed]
126. Magg, C.; Kubelka, J.; Holtermann, G.; Haas, E.; Schmid, F.X. Specifity of the initial collapse in the folding of the cold shock protein. *J. Mol. Biol.* **2006**, *360*, 1067–1080. [CrossRef] [PubMed]
127. Park, S.H.; Shastry, M.C.; Roder, H. Folding dynamics of the B1 domain of protein G explored by ultra-rapid mixing. *Nat. Struct. Biol.* **1999**, *6*, 943–947. [CrossRef]
128. Kimura, T.; Uzawa, T.; Ishimori, K.; Morishima, I.; Takahashi, S.; Konno, T.; Akiyama, S.; Fujisawa, T. Specific collapse followed by slow hydrogen-bond formation of beta-sheet in the folding of single-chain monellin. *Proc. Natl. Acad. Sci. USA* **2005**, *102*, 2748–2753. [CrossRef] [PubMed]
129. Juneja, J.; Udgaonkar, J.B. Characterization of the unfolding of Ribonuclease A by a pulsed hydrogen exchange study: Evidence for competing pathways for unfolding. *Biochemistry* **2002**, *41*, 2641–2654. [CrossRef] [PubMed]
130. Capaldi, A.P.; Shastry, M.C.R.; Kleanthous, C.; Roder, H.; Radford, S.E. Ultrarapid mixing experiments reveal that Im7 folds via an on-pathway intermediate. *Nat. Struc. Biol.* **2001**, *8*, 68–72.
131. Reiner, A.; Henklein, P.; Kiefhaber, T. An unlocking/relocking barrier in conformational fluctuations of villin headpiece subdomain. *Proc. Natl. Acad. Sci. USA* **2010**, *107*, 4955–4960. [CrossRef]
132. Hoeltzli, S.D.; Frieden, C. Stopped-flow NMR spectroscopy: Real-time unfolding studies of 6-19F-tryptophan-labeled *Escherichia coli* dihydrofolate reductase DHFR. *Proc. Natl. Acad. Sci. USA* **1995**, *92*, 9318–9322. [CrossRef]
133. Kiefhaber, T.; Labhardt, A.M.; Baldwin, R.L. Direct NMR evidence for an intermediate preceding the rate-limiting step in the unfolding of ribonuclease A. *Nature* **1995**, *375*, 513–515. [CrossRef]
134. Bemporad, F.; Gsponer, J.; Hopearuoho, H.I.; Plakoutsi, G.; Stati, G.; Stefani, M.; Taddei, N.; Vendruscolo, M.; Chiti, F. Biological function in a non-native partially folded state of a protein. *EMBO J.* **2008**, *27*, 1525–1535. [CrossRef]
135. Chiti, F.; Dobson, C.M. Protein misfolding, functional amyloid, and human disease. *Annu. Rev. Biochem.* **2006**, *75*, 333–366. [CrossRef] [PubMed]
136. Dobson, C.M. Protein folding and misfolding. *Nature* **2003**, *426*, 884–890. [CrossRef] [PubMed]
137. Luheshi, L.M.; Dobson, C.M. Bridging the gap: From protein misfolding to protein misfolding diseases. *FEBS Lett.* **2009**, *583*, 2581–2586. [CrossRef] [PubMed]
138. Pettersson-Kastberg, J.; Mossberg, A.K.; Trulsson, M.; Yong, Y.J.; Min, S.; Lim, Y.; O'Brien, J.E.; Svanborg, C.; Mok, K.H. Alpha-Lactalbumin, engineered to be nonnative and inactive, kills tumor cells when in complex with oleic acid: A new biological function resulting from partial unfolding. *J. Mol. Biol.* **2009**, *394*, 994–1010. [CrossRef] [PubMed]
139. Nadeem, A.; Sanborn, J.; Gettel, D.L.; Ho, C.S.J.; Rydstrom, A.; Ngassam, V.N.; Klausen, T.K.; Pedersen, S.F.; Lam, M.; Parikh, A.N.; et al. Protein receptor-independent plasma membrane remodeling by HAMLET: A tumoricidal protein-lipid complex. *Sci. Rep.* **2015**, *5*, 16432. [CrossRef]
140. Clementi, E.A.; Wilhelm, K.R.; Schleucher, J.; Morozova-Roche, L.A.; Hakansson, A.P. A complex of equine lysozyme and oleic acid with bactericidal activity against *Streptococcus pneumoniae*. *PLoS ONE* **2013**, *8*, e80649. [CrossRef]
141. Zhou, B.; Tian, K.G.; Jing, G.Z. An in vitro peptide folding model suggests the presence of the molten globule state during nascent peptide folding. *Prot. Eng.* **2000**, *13*, 35–39. [CrossRef]
142. Leal, S.S.; Gomes, C.M. Studies of the molten globule state of ferredoxin: Structural characterization and implications on protein folding and iron-sulfur center assembly. *Proteins-Struct. Funct. Bioinform.* **2007**, *68*, 606–616. [CrossRef]
143. Kumar, A.; Gaikwad, S.M. Multistate unfolding of alpha-mannosidase from Canavalia ensiformis (Jack Bean): Evidence for the thermostable molten globule. *Biochem. Biophys. Res. Commun.* **2010**, *403*, 391–397. [CrossRef]
144. Tseng, T.S.; Cheng, C.S.; Chen, D.J.; Shih, M.F.; Liu, Y.N.; Hsu, S.T.D.; Lyu, P.C. A molten globule-to-ordered structure transition of Drosophila melanogaster crammer is required for its ability to inhibit cathepsin. *Biochem. J.* **2012**, *442*, 563–572. [CrossRef]
145. Vamvaca, K.; Jelesarov, I.; Hilvert, D. Kinetics and thermodynamics of ligand binding to a molten globular enzyme and its native counterpart. *J. Mol. Biol.* **2008**, *382*, 971–977. [CrossRef] [PubMed]

146. Kasper, J.R.; Park, C. Ligand binding to a high-energy partially unfolded protein. *Prot. Sci.* **2015**, *24*, 129–137. [CrossRef] [PubMed]
147. Kitazawa, S.; Kameda, T.; Yagi-Utsumi, M.; Sugase, K.; Baxter, N.J.; Kato, K.; Williamson, M.P.; Kitahara, R. Solution structure of the q41n variant of ubiquitin as a model for the alternatively folded N-2 state of ubiquitin. *Biochemistry* **2013**, *52*, 1874–1885. [CrossRef] [PubMed]
148. Prajapati, R.S.; Indu, S.; Varadarajan, R. Identification and thermodynamic characterization of molten globule states of periplasmic binding proteins. *Biochemistry* **2007**, *46*, 10339–10352. [CrossRef]
149. Singh, N.; Kumar, R.; Jagannadham, M.V.; Kayastha, A.M. Evidence for a molten globule state in cicer alpha-galactosidase induced by pH, temperature, and guanidine hydrochloride. *Appl. Biochem. Biotechnol.* **2013**, *169*, 2315–2325. [CrossRef]
150. Lamb, D.C.; Ostermann, A.; Prusakov, V.E.; Parak, F.G. From metmyoglobin to deoxymyoglobin: Relaxations of an intermediate state. *Eur. Biophys. J. Biophys. Lett.* **1998**, *27*, 113–125.

molecules MDPI

Review

Molecular Dynamics Simulation Studies on the Aggregation of Amyloid-β Peptides and Their Disaggregation by Ultrasonic Wave and Infrared Laser Irradiation

Hisashi Okumura [1,2,3,*] and **Satoru G. Itoh** [1,2,3]

1 Exploratory Research Center on Life and Living Systems, National Institutes of Natural Sciences, Okazaki 444-8787, Aichi, Japan; itoh@ims.ac.jp
2 Institute for Molecular Science, National Institutes of Natural Sciences, Okazaki 444-8787, Aichi, Japan
3 Department of Structural Molecular Science, SOKENDAI (The Graduate University for Advanced Studies), Okazaki 444-8787, Aichi, Japan
* Correspondence: hokumura@ims.ac.jp

Abstract: Alzheimer's disease is understood to be caused by amyloid fibrils and oligomers formed by aggregated amyloid-β (Aβ) peptides. This review article presents molecular dynamics (MD) simulation studies of Aβ peptides and Aβ fragments on their aggregation, aggregation inhibition, amyloid fibril conformations in equilibrium, and disruption of the amyloid fibril by ultrasonic wave and infrared laser irradiation. In the aggregation of Aβ, a β-hairpin structure promotes the formation of intermolecular β-sheet structures. Aβ peptides tend to exist at hydrophilic/hydrophobic interfaces and form more β-hairpin structures than in bulk water. These facts are the reasons why the aggregation is accelerated at the interface. We also explain how polyphenols, which are attracting attention as aggregation inhibitors of Aβ peptides, interact with Aβ. An MD simulation study of the Aβ amyloid fibrils in equilibrium is also presented: the Aβ amyloid fibril has a different structure at one end from that at the other end. The amyloid fibrils can be destroyed by ultrasonic wave and infrared laser irradiation. The molecular mechanisms of these amyloid fibril disruptions are also explained, particularly focusing on the function of water molecules. Finally, we discuss the prospects for developing treatments for Alzheimer's disease using MD simulations.

Keywords: molecular dynamics simulation; replica permutation method; amyloid-β; aggregation; disaggregation; β-sheet; α-helix; interface; inhibitor; polyphenol

Citation: Okumura, H.; Itoh, S.G. Molecular Dynamics Simulation Studies on the Aggregation of Amyloid-β Peptides and Their Disaggregation by Ultrasonic Wave and Infrared Laser Irradiation. *Molecules* **2022**, *27*, 2483. https://doi.org/10.3390/molecules27082483

Academic Editors: Yuko Okamoto, Kunihiro Kuwajima, Tuomas Knowles and Michele Vendruscolo

Received: 9 March 2022
Accepted: 7 April 2022
Published: 12 April 2022

1. Introduction

Proteins are normally folded correctly in vivo to maintain their functions. However, when their concentration increases due to, for example, aging, they aggregate to form oligomers, spherical aggregates, and amyloid fibrils, needle-like aggregates. These protein aggregates are associated with about 40 human neurodegenerative diseases [1–3]. For instance, amyloid-β (Aβ) peptide is related to Alzheimer's disease. Huntington's disease is caused by polyglutamine. Parkinson's disease is associated with α-synuclein. Dialysis-related amyloidosis is caused by β2-microglobulin.

Alzheimer's disease is one of dementia and is characterized by brain atrophy and senile plaques in the cerebral cortex [4,5]. The senile plaques are caused by the deposition of Aβ peptides on the brain cells [6,7]. Aβ is produced by proteolytic cleavage of the amyloid precursor protein and consists of 39–43 amino acid residues [8]. It usually consists of 40 or 42 amino acid residues. Aβ peptide with 40 residues is referred to as Aβ40, and that with 42 residues is referred to as Aβ42. The amino acid sequence of Aβ40 is DAEFRHDSGYEVHHQKLVFFAEDVGSNKGAIIGLMVGGVV, and that of Aβ42 is DAEFRHDSGYEVHHQKLVFFAEDVGSNKGAIIGLMVGGVVIA.

The structure of the Aβ amyloid fibril has been revealed by several experiments [9–14]. The main secondary structure of the Aβ amyloid fibril is the cross-β-sheet structure [9].

Aβ peptides form two intermolecular β-sheet structures, β1 and β2 [10,11]. The β1 and β2 regions consist of residues 12–24 and residues 30–40, respectively, in Aβ40 [10], while the β1 and β2 regions consist of residues 18–26 and residues 31–42, respectively, in Aβ42 [11]. The structures of individual Aβ peptides in the amyloid fibril models reported in Refs. [10,11] seem to be U-shaped. Other structural models have been also reported because Aβ peptides form polymorphic amyloid fibrils with various molecular structures depending on experimental conditions. For example, Lu et al. reported a three-fold symmetric amyloid fibril model consisting of three Aβ40 peptides [12]. The structure of Aβ42 in an amyloid fibril revealed by Xiao et al. is S-shaped [13]. Gremer et al. reported that the N-terminus of Aβ42 is L-shaped, and the C-terminus is S-shaped, giving the overall Aβ42 peptide an LS-shaped structure in their amyloid fibril model [14].

The typical time course of the amyloid fibril formation is shown in Figure 1. First, several Aβ monomers aggregate to form an oligomer. The oligomer then grows to an amyloid fibril. Aβ peptides are attached to the ends of the amyloid fibril, making the amyloid fibril elongate. When almost all Aβ peptides in the solution aggregate, the system reaches thermal equilibrium, and the amyloid fibril stops the elongation. The amyloid fibril can be destroyed by ultrasonic wave irradiation or infrared laser irradiation.

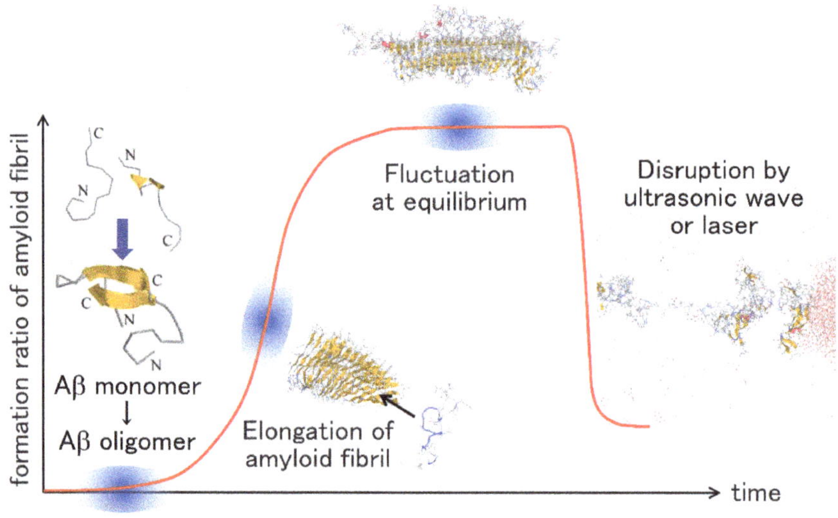

Figure 1. Schematic illustration of oligomerization of Aβ peptides, elongation of the Aβ amyloid fibril, the Aβ amyloid fibril in equilibrium, and disruption of the Aβ amyloid fibril.

The structural changes in the aggregation and disaggregation process have been investigated by molecular dynamics (MD) simulation. Numerous simulation studies have been performed so far on the monomeric state [15–32], dimerization [33–46], oligomerization [47–54], amyloid fibril elongation [55–68], amyloid fibril stability [69–79], and destruction of amyloid fibrils [80–84]. Most of these studies are well summarized in the review articles [85–88]. In this review, we explain the MD simulation studies on the aggregation and disaggregation of Aβ peptides that we have performed. These studies have elucidated the process from aggregation to disaggregation of the Aβ peptides at the atomic level. In Section 2, we present an MD simulation study on the aggregation process of Aβ fragments that revealed that the β-hairpin structure promotes the formation of the intermolecular β-sheet structure [36]. In Section 3, we explain that Aβ peptides at hydrophilic/hydrophobic interfaces form more β-hairpin structures than in the bulk water [30]. This is one of the reasons why aggregation at the interface is promoted. Research on the inhibition of aggregation of Aβ peptides has been ongoing, as well as their aggregation.

Polyphenols have attracted attention as aggregation inhibitors for Aβ peptides. In Section 4, we introduce an MD simulation study on the interaction between an Aβ fragment and polyphenols [31]. When almost all the Aβ peptides form amyloid fibrils in an aqueous solution, the system reaches equilibrium. An MD simulation study has recently revealed that the structures of the two ends of the Aβ amyloid fibril are different in equilibrium [72]. We describe this simulation study in Section 5. Amyloid fibrils can be destroyed by ultrasonic wave irradiation or infrared laser irradiation. In Section 6, we explain an MD simulation study revealing that the cavitation induced by the ultrasonic wave destroys the amyloid fibrils [80]. In Section 7, we introduce an MD simulation study that clarified the function of water molecules in laser-induced amyloid fibril destruction [84]. Section 8 is devoted to the conclusions.

2. Aggregation of Aβ Fragments

To identify the important regions and amino acids in the amyloid fibril and oligomer formation of Aβ peptides, several experiments have been performed using the full-length Aβ peptides and Aβ fragments [89–93]. These studies revealed that the C-terminal region of the Aβ peptide, Aβ(29–42), consisting of the 29th to 42nd amino acid residues, promotes the amyloid fibril formation of the Aβ peptides [89]. Aβ(29–42) was also known to form amyloid fibrils by itself [90–92]. In the early stages of amyloid fibril formation, oligomers are formed. Recent studies have shown that oligomers are more neurotoxic than amyloid fibrils [94,95]. To develop a remedy for Alzheimer's disease, it is necessary to understand the details of the oligomer structure and formation process of the Aβ peptides, but these are not clear. We recently investigated the oligomer formation process of the Aβ(29–42) peptides by MD simulation [36,50,96]. We introduce in this section the MD simulation study on the Aβ(29–42) dimerization [36].

2.1. Hamiltonian Replica-Permutation Molecular Dynamics Simulation of Aβ(29–42) Peptides

We performed Hamiltonian replica-permutation MD simulations of two Aβ(29–42) peptides in explicit water solvent [36]. The replica-permutation method [97] is one of the generalized-ensemble algorithms [98–101] developed by the authors. This method is an improved alternative to the replica-exchange method [102,103]. In the replica-exchange and replica-permutation methods, several copies of the system, referred to as replicas, are prepared, and each replica is assigned a different temperature. The temperatures are exchanged between two replicas during the simulation in the replica-exchange method, as shown in Figure 2a. In the replica-permutation method, on the other hand, the temperatures are permuted between three or more replicas, as shown in Figure 2b. In addition, the Suwa–Todo algorithm [104] is used instead of the Metropolis algorithm [105] for the replica-permutation trials. The Suwa–Todo algorithm is the most efficient Monte Carlo method and is utilized in several generalized-ensemble algorithms [22,97,106–110]. The replica-permutation method is known to provide statistically more reliable data on biomolecular structures than the replica-exchange method [97,106].

There are several variations of the replica-permutation method [22,106–108], such as the Hamiltonian replica-permutation method [22], the isobaric-isothermal replica-permutation method [107], the replica sub-permutation method [106], and the replica-permutation with solute tempering [110]. In the Hamiltonian replica-permutation method, an artificial parameter is introduced in the potential energy, and each replica is assigned a different value for this parameter. Instead of the temperatures, the parameter values are permuted between three or more replicas during the MD simulations. The method used here is the Coulomb replica-permutation method [23], which is a kind of the Hamiltonian replica-permutation method [22]. In this method, a parameter is introduced in the electrostatic potential energy, and the values of this parameter are permuted.

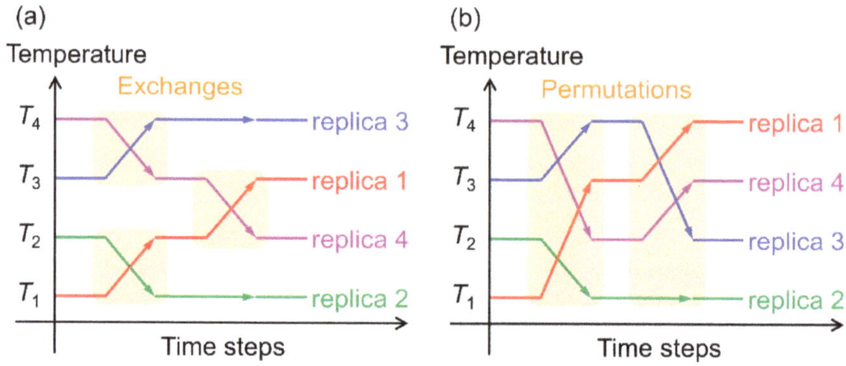

Figure 2. (**a**) Schematic illustration of time series of temperatures in the replica-exchange method. Orange squares mean replica exchange trials with the Metroplis algorithm. (**b**) Schematic illustration of time series of temperatures in the replica-permutation method. Orange rectangles mean replica permutation trials with the Suwa–Todo algorithm.

The MD simulations were performed as follows. Two Aβ(29–42) molecules with explicit water molecules were first prepared in a cubic simulation box. The N-terminus and C-terminus of Aβ(29–42) were blocked by the acetyl group and the N-methyl group, respectively. The amino acid sequence was Ace-GAIIGLMVGGVVIA-Nme. The AMBER parm99SB force field [111] and TIP3P rigid-body model [112] were used for the Aβ(29–42) peptides and water molecules, respectively. Temperature was controlled at 300 K by the Nosé–Hoover thermostat [113–115]. Coulomb replica-permutation MD simulations were performed with eight replicas from three different initial conditions. The simulation time was 200 ns, including 10 ns equilibration, for each replica. The total time length of the production runs of the Coulomb replica-permutation MD simulations was 4.56 μs. Other simulation details can be found in Ref. [36].

2.2. Dimerization of Aβ(29–42) Peptides

The dimerization of the Aβ(29–42) peptides was observed in the Coulomb replica-permutation MD simulations. The MD simulations showed that the dimer formation process proceeds in two steps. First, the β-hairpin structure increases when the two Aβ(29–42) molecules approach each other, as shown in Figure 3a, followed by the formation of a dimer with an intermolecular β-sheet structure. The reason for the increase in the β-hairpin structure in the first step is that a structure like Figure 3b becomes stable. In Figure 3b, Aβ(29–42) shown in yellow forms the β-hairpin structure, which is stabilized by the intermolecular hydrophobic side-chain contact between the amino acid residues shown by the yellow and green dots.

In the second step, it was found that the intermolecular β-sheet structures are readily formed at the amino acid residues with the intramolecular β-sheet structures. In other words, when the other Aβ(29–42) approaches the stable β-hairpin structure, the intermolecular β-sheet structure is easily formed between the β-hairpin and Aβ(29–42). In this way, the β-hairpin structure accelerates the formation of an oligomer with the intermolecular β-sheet structure. Not only our MD simulation study [36] but also some recent experimental and computational studies reported that the β-hairpin structure plays an essential role in the oligomer formation [35,116,117].

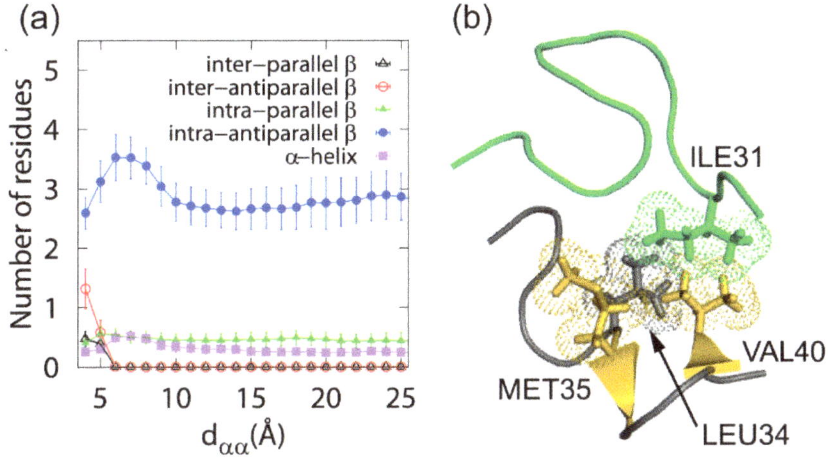

Figure 3. (**a**) The number of amino acid residues forming each secondary structure as a function of the intermolecular C_α–C_α distance $d_{\alpha\alpha}$ between the two Aβ(29–42) peptides. (**b**) A typical β-hairpin structure of Aβ(29–42). Reprinted with permission from Ref. [36]. Copyright 2014 American Chemical Society.

3. Structure of an Aβ Peptide at an Air–Water Interface

Aggregation of Aβ peptides is accelerated at hydrophilic/hydrophobic interfaces, such as air–water interfaces [118,119] and cell membrane surfaces [120,121]. One reason why the Aβ aggregation is accelerated there is that the concentration of Aβ peptides is higher at the interfaces because they have both hydrophobic and hydrophilic residues and tend to exist there. In addition, we recently performed MD simulations of Aβ40 at the air–water interface and found that it takes the β-hairpin structure more than in the bulk water [30]. As shown in the previous section, the β-hairpin structure promotes the intermolecular β-sheet formation. That is, the aggregation of Aβ peptides is enhanced not only by the high concentration but also by the conformation of the Aβ peptide. In this section, we explain the MD simulation study that revealed the structure of the full-length Aβ peptide, Aβ40, at the air–water interface [30].

3.1. Molecular Dynamics Simulation of Aβ40 at the Air–Water Interface

We performed MD simulations of an Aβ40 peptide in a system with air–water interfaces. The air–water interface was prepared by removing half the water molecules in a cubic simulation box. The side length of the box was set to 108.0 Å. For statistical analysis, nine different initial conditions were employed using the combination of three different coordinates and three different velocities. The initial structure of the Aβ40 peptide was fully extended with all dihedral angles φ and ψ of 180° for all the three initial coordinates. The MD simulation was performed from each initial condition for 240 ns including the equilibration period of 10 ns. Temperature was controlled at 350 K using the Nosé–Hoover thermostat [113–115].

For comparison, we also performed MD simulations of the Aβ40 peptide in the bulk water. The initial structure of the Aβ40 peptide in the bulk the water was also fully extended. Nine different initial conditions were prepared as well, with nine different initial velocities. The side length of the cubic unit cell was 91.1 Å. The MD simulation was performed from each initial condition for 240 ns including the equilibration period of 10 ns, again. For other simulation details, please refer to Ref. [30].

3.2. Molecular Structure of Aβ40 at the Air–Water Interface

We observed that Aβ40 existed at the air–water interface in all MD simulations with the interface starting from nine different initial conditions. Figure 4a shows a typical conformation at the air–water interface. The β1 and β2 regions are bound at the interface, and the N-terminal region and the linker region between β1 and β2 are in the aqueous solution. These results mean that Aβ40 tends to exist at the air–water interface because the hydrophobic residues of Aβ40 tend to exist in the hydrophobic region (air), and the hydrophilic residues tend to exist in the hydrophilic region (water). That is, the Aβ peptide can be regarded as an amphiphilic molecule, such as a surfactant, and tends to exist at a hydrophilic/hydrophobic interface.

In order to clarify the Aβ40 structure at the interface, the average distance between the C_α atoms of each residue and the interface was calculated, as shown in Figure 4b. The positive value indicates that the C_α atom of that residue is in the water, and the negative value indicates that it is in the air. We can see that Aβ40 has an up-and-down shape at the air–water interface. This result agrees well with the NMR experiments for the Aβ40 structure on lyso-GM1 micelles [122], in which Val12–Gly25, Ile31–Val36, and Val39–Val40 of Aβ40 (red lines in Figure 4b) were found to bind to lyso-GM1 micelles. In addition, these results also agree with the Aβ40 conformation on GM1 micelles [123]. Thus, we can infer that the up-and-down shape of Aβ40 at the interface may hold for other hydrophilic/hydrophobic interfaces in general.

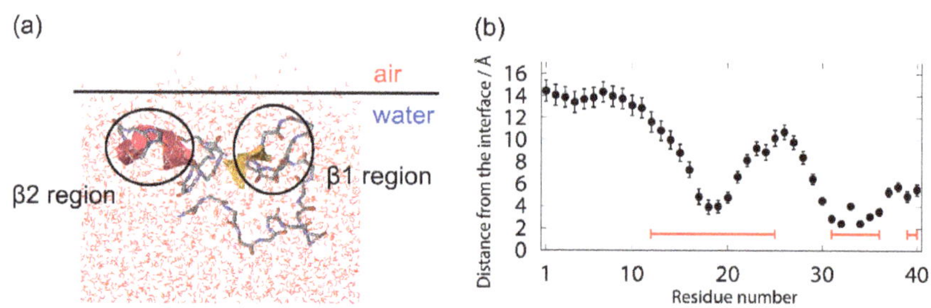

Figure 4. (**a**) A typical snapshot of the Aβ40 peptide at the interface. (**b**) The average distance between the C_α atoms of each amino acid residue of the Aβ40 peptide and the interface. The red lines represent the residues that were bound to the lyso-GM1 micelle in the experiment [122]. Reprinted with permission from Ref. [30]. Copyright 2019 American Chemical Society.

We calculated the intramolecular contact probabilities of the C_α atoms of Aβ40 to reveal the effect of the interface on the Aβ40 conformation. Figure 5a,b show the probabilities at the air–water interface and those in the bulk water, respectively. The β1 and β2 regions form helix structures at the air–water interface. This result agrees well with the experimental results on the lyso-GM1 micelles [122]. A β-hairpin structure is also formed between the β1 and β2 regions. These secondary structures were formed during the MD simulations as follows. The β1 and β2 regions first formed helix structures at the interface. The helix structure of the β1 region was then destroyed, and the extended β1 region approached the β2 region, forming a β-bridge. The helix structure in the β2 region was destroyed, and the β-hairpin structure was finally formed.

In the bulk water, on the other hand, helix structures are formed in the β1 and β2 regions, whereas the β-hairpin structure is hardly formed, as shown in Figure 5b. The difference between the β-hairpin formation probability at the interface and that in the bulk water causes a difference in the oligomer formation ability since the β-hairpin structure accelerates the intermolecular β-sheet formation with other Aβ peptides, as reviewed in the previous section [36,50]. This fact is also pointed out by experimental studies [116,117].

Thus, we can infer that there are two reasons why the aggregation of Aβ peptides is enhanced at the hydrophilic/hydrophobic interfaces. One reason is that the concentration of Aβ peptides increases at the interfaces since they have both hydrophilic and hydrophobic residues and tend to exist there. The other reason is that Aβ peptides take the β-hairpin structure, promoting aggregation.

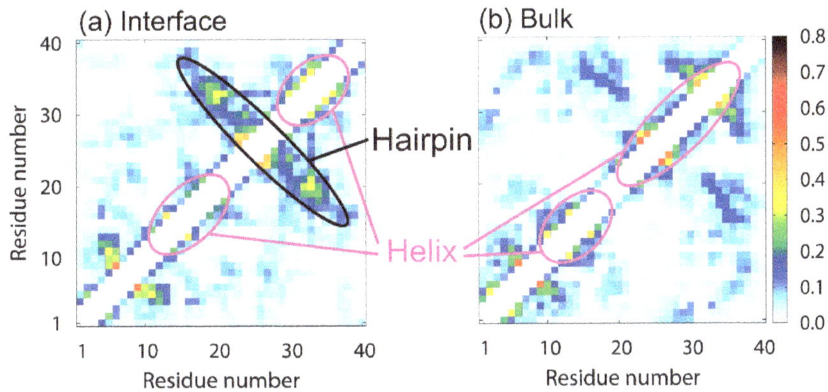

Figure 5. Intramolecular contact probabilities of the C_α atoms of Aβ40 (**a**) at the air–water interface and (**b**) in the bulk water. Reprinted with permission from Ref. [30]. Copyright 2019 American Chemical Society.

Next, we explain why the β-hairpin structure is stabilized at the hydrophilic/hydrophobic interface. Since the β1 and β2 regions tend to exist at the interface, as shown in Figure 4, these regions' motion is restricted at the interface, that is, in two dimensions (Figure 6). In the bulk water, on the other hand, the β1 and β2 regions can move relatively freely in three dimensions. Entropy increases in the bulk water because the β1 and β2 regions can take more conformations. However, the entropy increase is suppressed due to the two-dimensional motion at the interface. To reduce the free energy, it is necessary to reduce enthalpy at the interface. Therefore, hydrogen bonds are formed between the β1 and β2 regions to reduce the enthalpy under this restriction. As a result, the β-hairpin structure is formed more at the interface.

We described here the structure of an Aβ peptide at the air–water interface. Several MD simulations have been performed to investigate the structure of an Aβ peptide at interfaces such as cell membrane surfaces, too [124–131]. An important membrane surface in the body is monosialotetrahexosylganglioside (GM1) clusters on neuronal cell membranes, because it is reported by experiments that Aβ peptide aggregation is promoted there [120,121]. MD simulation studies on the GM1 glycan cluster have also been performed [29,132]. The GM1 glycan cluster in these studies consists of a self-assembled supramolecule and GM1 glycans transplanted on it [133]. The HHQ region (residues 13–15) was found to bind well to the GM1 glycan cluster [29]. This fact is in good agreement with our results at the air–water interface, where the β1 region (residues 10–22) is present at the air–water interface. However, on the GM1 glycan cluster, Aβ formed an α-helix structure in the C-terminal region, but did not form the β-hairpin structure between the β1 and β2 regions. The reasons for this may be considered as follows. The GM1 glycan moiety on the self-assembled supramolecule has lower fluidity than the GM1 clusters on the neural cell membrane. Aβ, therefore, can reach only the GM1 glycan moiety that corresponds to the headgroup of the GM1 cluster on the membrane. The interface between the GM1 glycan region and the aqueous solution is not as different in hydrophilicity and hydrophobicity as the air–water interface because the GM1 glycan moiety is relatively hydrophilic. The reason for the β-hairpin formation is that the β1 and β2 regions are constrained at the hydrophilic/hydrophobic interface, as shown in Figure 6. Thus, we can consider that the β1 and β2 regions were not constrained on the GM1 glycan moieties of the GM1 glycan cluster as much as the air–water interface, and the β-hairpin structure was not formed on the GM1 glycan cluster. We expect that Aβ peptides can reach the interface between the GM1 glycan moiety and the lipid ceramide moiety and form the β-hairpin structure by performing MD simulations of Aβ with the GM1 clusters on the neural cell membrane in the future.

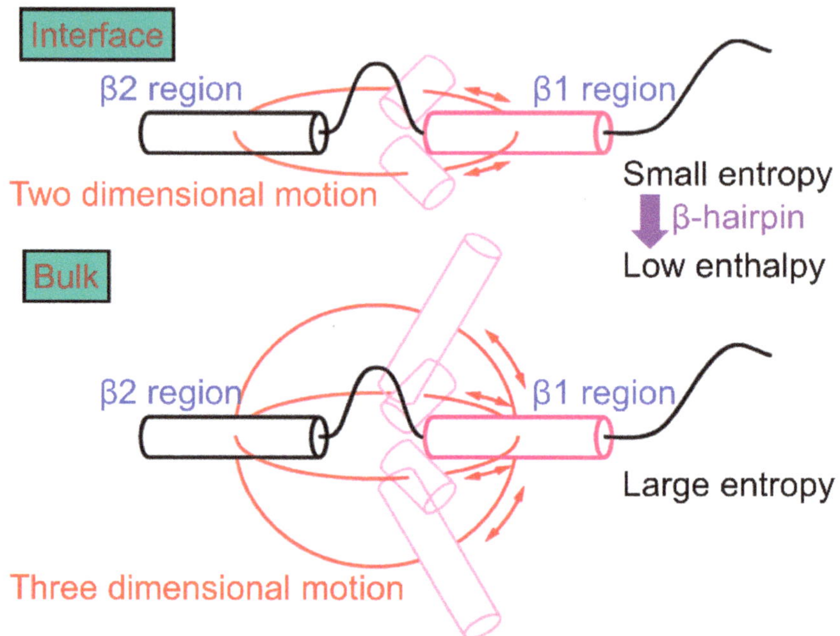

Figure 6. Schematic representation of the conformation of Aβ40 at the air–water interface and that in the bulk water. Reprinted with permission from Ref. [30]. Copyright 2019 American Chemical Society.

4. Inhibitor against Aggregation of Aβ Peptides: Polyphenol

Not only the aggregation of Aβ peptides but also the inhibition of the Aβ aggregation have been studied experimentally [134,135] and computationally [31]. It is known that the aggregation of Aβ peptides is inhibited by polyphenols [135]. The polyphenols thus have attracted attention as drug candidate molecules against Alzheimer's disease. The efficiency in inhibiting the Aβ aggregation has been investigated for several polyphenols [135]. According to recent experiments, myricetin (Myr) and rosmarinic acid (RA) (Figure S1) are most effective in inhibiting the Aβ aggregation [135]. However, the molecular mechanism of these polyphenols inhibiting the Aβ aggregation is not revealed. We recently performed MD simulations of an Aβ(16–22) peptide and these polyphenols to gain insight into this problem [31]. The Aβ(16–22) peptides are known to form amyloid fibrils by experiments [93]. It is relatively easy to reproduce the intermolecular β-sheet formation by MD simulation [136–140]. We present the MD simulation study on the interaction between the Aβ(16–22) peptide and these polyphenols [31] in this section.

4.1. Replica-Permutation MD Simulation of an Aβ(16–22) Peptide and Polyphenols

We performed all-atom replica-permutation MD simulations of an Aβ(16–22) peptide and polyphenols [31]. Each system consists of one Aβ(16–22) peptide, one polyphenol molecule (Myr or RA), and water molecules. For the RA system, we added a Na^+ ion as a counter ion. The N-terminus of the Aβ(16–22) peptide was blocked by the acetyl group, and the C-terminus by the N-methyl group to reduce the effect of the N- and C-terminal electric charges. The amino acid sequence is thus Ace-KLVFFAE-Nme. We used the AMBER parm14SB [141] and generalized AMBER force fields [142] for the Aβ(16–22) peptide and polyphenol molecules, respectively. The TIP3P rigid-body model [112] was used for the water molecules. To control the temperatures, the Nosé–Hoover thermostat [113–115] was used. We employed 14 replicas in the replica-permutation simulations. The temperatures of the replicas were ranged from 300.0 to 500.0 K. The Generalized-Ensemble Molecular Biophysics (GEMB) program was used to perform the MD simulations. This program was developed by one of the authors (H. Okumura) and has been applied to several protein and peptide systems [106–108,110,143–155]. We can perform MD simulations with the generalized-ensemble algorithms [98–100,156], such as the replica-exchange [102,103], replica-permutation [22,97,157], multicanonical [158–161], and multibaric-multithermal [162–165] methods, using this program. Here, a replica-permutation MD simulation was performed for 120 ns for each replica, including the first 20 ns as the equilibration. We then observed how these polyphenols bound to the Aβ(16–22) peptide. Other simulation details can be found in Ref. [31].

4.2. Structure of the Complexes of an Aβ(16–22) Peptide and Polyphenols

As a result of the MD simulations, we observed that polyphenols were bound to the Aβ(16–22) peptide, as shown in Figure 7. Hydrogen bonds were formed, as indicated by the cyan ovals in Figure 7, between the polyphenols and Aβ(16–22) peptide. In the Myr system, the carboxyl group (-COO) of Glu22 often formed a hydrogen bond with a hydroxy group (-OH) of Myr, as shown in Figure 7a. In the RA system, the amine group (-NH3) of Lys16 often bound to the carboxyl group of RA, and the carboxyl group of Glu22 frequently formed a hydrogen bond with a hydroxy group of RA, as shown in Figure 7b.

The contact probability of each amino acid residue of the Aβ(16–22) peptide with these polyphenols was also calculated, as in Figure 8. Myr binds to Glu22 with the probability of 30%, as shown in Figure 8a. However, the other residues of the Aβ(16–22) peptide have much lower contact probabilities with Myr. High contact probabilities in the RA system are found at two residues, Glu22 with 71% and Lys16 with 17%, as shown in Figure 8b. On the other hand, the hydrophobic residues (Leu, Val, Phe, and Ala) have low contact probabilities in both systems. It is known that the Aβ(16–22) peptides form anti-parallel β-sheets because of the electrostatic interaction between the carboxyl group of Glu22, which has a negative charge, and the amine group of Lys16, which has a positive charge [53,137]. We can thus expect that the aggregation of the Aβ(16–22) peptides is inhibited by Myr and RA because they bind to the side chains of Glu22 and Lys16, as shown in Figure 7.

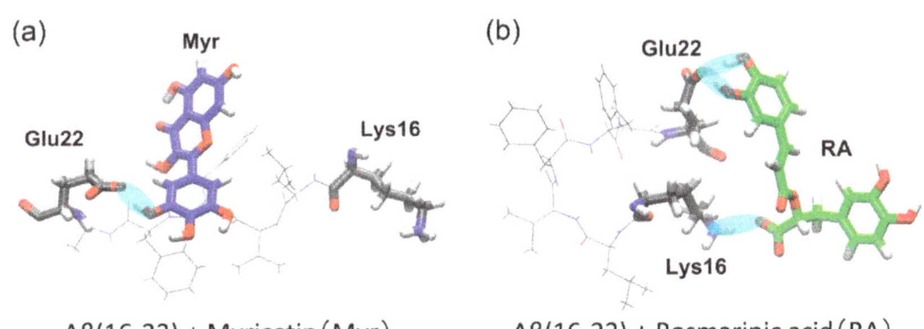

Figure 7. Typical snapshots obtained from the replica-permutation MD simulations of the (**a**) Myr system and (**b**) RA system. The hydrogen bonds between the polyphenols and the Aβ(16–22) peptide are indicated by the cyan ovals. Reprinted with permission from Ref. [31]. Copyright 2020 Elsevier.

The contact probability of each atom of polyphenols was also calculated to specify which atoms of polyphenols contribute to the interaction with the Aβ(16–22) peptide, as shown in Figure 9. As a result, multiple adjacent hydroxy groups around six-membered rings were found to have high contact probabilities with the Aβ(16–22) peptide in both Myr and RA systems. The carboxyl group in RA also contacts the Aβ(16–22) peptide.

Thus, we can expect that these atoms in polyphenols play an essential role in inhibiting the Aβ(16–22) aggregation.

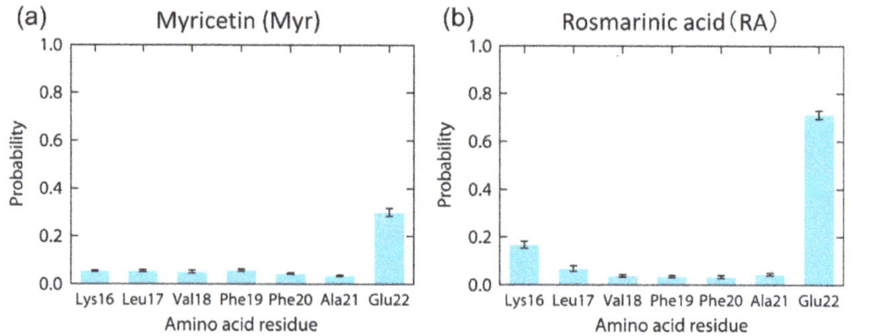

Figure 8. Contact probability of each residue in the Aβ(16–22) peptide with (**a**) Myr and (**b**) RA at 300 K. Reprinted with permission from Ref. [31]. Copyright 2020 Elsevier.

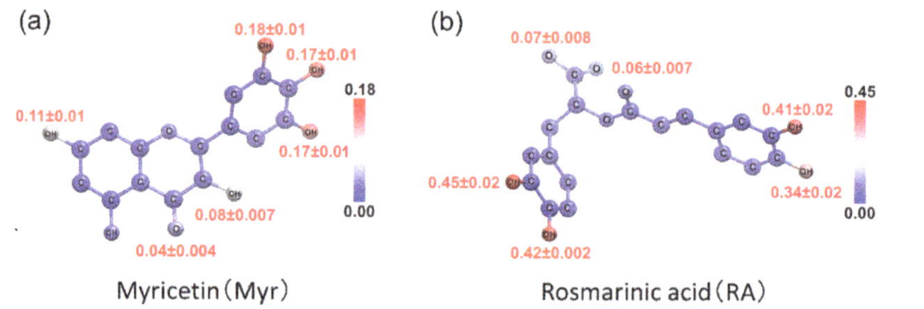

Figure 9. Color mapping to show contact probability of the (**a**) Myr and (**b**) RA atoms with the Aβ(16–22) peptide at 300 K. Reprinted with permission from Ref. [31]. Copyright 2020 Elsevier.

5. Structures of the Two Ends of the Aβ Amyloid Fibril

The structures of Aβ amyloid fibrils have been clarified by X-ray diffraction and solid-state NMR experiments [120,166,167]: the amyloid fibril has a cross-β structure comprising two β-sheets, β1 and β2, as shown in Figure 10a. Here, the β1 and β2 regions correspond to residues 18–26 and 31–42, respectively. However, it is generally known that the structure in the bulk region and that at the interface are different in many materials, known as the surface reconstruction of crystals [168] and polarization on water surface [169,170]. In the case of the amyloid fibril, the bulk region corresponds to the central part of the amyloid fibril, and the interface corresponds to the end of the amyloid fibril. The amyloid fibril structure revealed by the experiments is that in the central region. The structures at the ends of the amyloid fibril have not been revealed because only one or two Aβ peptides constitute the end of the amyloid fibril, which cannot be measured by experimental techniques such as X-rays and NMR. In addition, the amyloid fibril elongates by binding one Aβ peptide to the end of the fibril. It is thus important to clarify the structure of the Aβ peptide at the ends of the amyloid fibril to understand the elongation mechanism of the fibril.

We, therefore, performed MD simulations to investigate the structure of the amyloid fibril ends [72]. As a result, not only the difference in Aβ structure between the ends and the central region but also that between two ends were discovered. The two ends of the Aβ amyloid fibril are referred to as the odd and even ends because C=O and N–H of the odd-numbered (even-numbered) residues in the β1 region are exposed at the odd (even) end [11]. Different molecular conformations between the odd and even ends had not been reported before our MD simulations [10]. In this section, we introduce the MD simulation study to reveal the structural differences at the odd end, in the central region, and at the even end of the Aβ amyloid fibril [72].

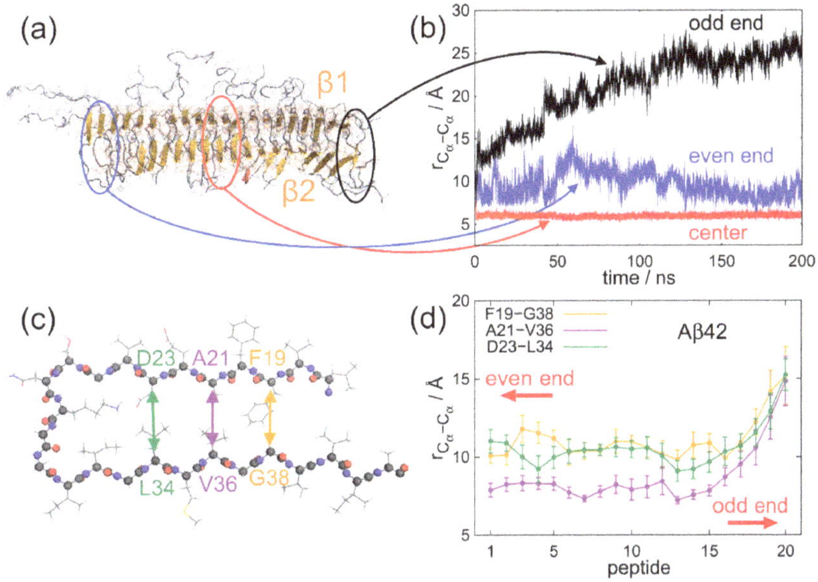

Figure 10. (**a**) A snapshot of the Aβ42 amyloid fibril in the MD simulation. (**b**) Time series of C_α–C_α distance between A21 and V36 at the odd end, in the center region, and at the even end. (**c**) Side view of chain C of model 1 of the PDB conformation (PDB ID: 2BEG) of the Aβ42 amyloid fibril. (**d**) The average C_α–C_α distances of the Aβ42 amyloid fibril between F19 and G38 (orange), A21 and V36 (purple), and D23 and L34 (green). Panels (**a**,**c**) were created using PyMOL [171]. Reprinted with permission from Ref. [72]. Copyright 2016 Springer Nature.

5.1. Molecular Dynamics Simulation of the Aβ Amyloid Fibril

We prepared an amyloid fibril consisting of 20 Aβ42 peptides with explicit water molecules. Because the central structure of the Aβ amyloid fibril is known by solid-state NMR experiments (PDB: 2BEG) [11], the initial structures of the Aβ amyloid fibrils in the MD simulations were modeled using this structure. The AMBER parm99SB was used for the Aβ peptide force field [111], and the TIP3P rigid-body model was used for the water molecules [112]. The electrostatic interaction was calculated using the particle mesh Ewald method [172], and the time step width for the Aβ peptide was set to 0.5 fs and that for the water molecules to 4 fs. The water molecules were treated as rigid-body molecules [144]. The temperature was set to 298 K using the Nosé–Hoover thermostat [113–115], and the pressure was set to 0.1 MPa using the Andersen barostat [173]. Then, 200 ns simulations were performed from nine different initial conditions. We used the GEMB program [148] here again. For other simulation details, please refer to Ref. [72].

5.2. Structure of Aβ Peptides at the Ends of the Aβ Amyloid Fibril

We unexpectedly observed that the N- and C-termini gradually opened at the odd end, whereas these termini remained closed at the even end. In all simulations, the odd end often opened, whereas the even end never opened. Figure 10b shows the time series of the C_α–C_α distance between A21 and V36 at the odd end, in the central region, and at the even end. The pair of C_α atoms of A21 and V36 is illustrated in Figure 10c. The C_α–C_α distance between these residues clearly increased at the odd end. On the other hand, at the even end, this C_α–C_α distance fluctuated, but did not increase so much. In the central region, it was almost constant. Figure 10d shows the averages of three C_α–C_α distances between F19 and G38, A21 and V36, and D23 and L34. The averages were taken over the nine initial conditions at a time ranging from 100 to 200 ns. The differences in the three

C_α–C_α distances between the odd end and even ends are statistically significant. It means that not only from one MD trajectory but also after taking averages of nine MD trajectories, we can see that the β-sheets were well separated at the odd end, whereas the two β-sheets were closely spaced with some fluctuation at the even end. To illustrate this structural difference at both ends clearly, Figure 11 shows the Aβ amyloid fibril and the side views of the Aβ peptides at both ends.

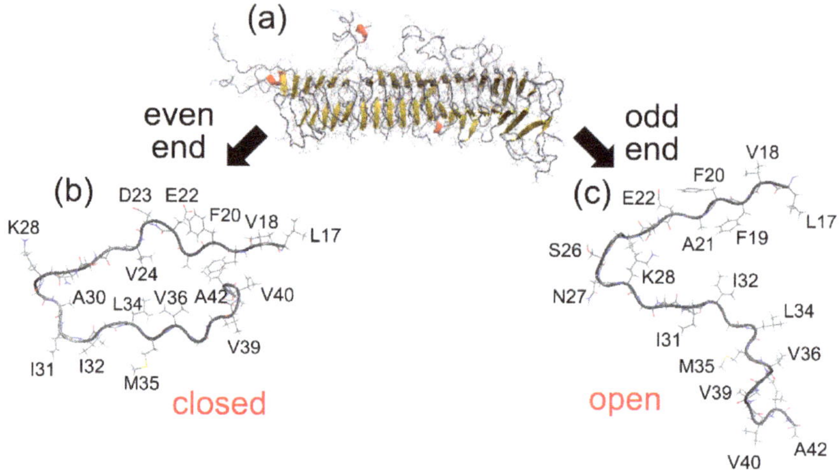

Figure 11. (**a**) A snapshot of the Aβ amyloid fibril. (**b**) Side view of the Aβ peptide at the even end. (**c**) Side view of the Aβ peptide at the odd end. The figures were created using PyMOL [171]. Reprinted with permission from Ref. [72]. Copyright 2016 Springer Nature.

In order to explain why the structures and fluctuations differ between the two ends, we calculated the probability that each amino acid residue forms an intermolecular parallel β-sheet structure, as shown in Figure 12a. Since 20 Aβ42 peptides were used, the horizontal axis represents the peptide number (1–20) and the vertical axis represents the amino acid residue number (1–42). The β2 region has a high formation probability of the intermolecular parallel β-sheet structure, and the β1 region has a much higher probability than the β2 region. The reason for this difference is that β2 contains the glycine residues, which tend to move easily. This result explains the large fluctuation at the odd end as follows. We can see from the PDB structure that β2 does not exist directly below β1 because each Aβ peptide is slightly tilted, as shown in Figure 12b. Therefore, β1 is more exposed to the solvent at the even end, whereas β2 is more exposed to the solvent at the odd end, as indicated by the dashed ellipses in Figure 12b. These two β-strands, β1 at the even end and β2 at the odd end, are both exposed to the solvent and therefore tend to fluctuate. However, as shown in Figure 12a, β1 forms a more stable intermolecular β-sheet structure with the neighboring Aβ peptide, whereas β2 does not form such a stable intermolecular β-sheet structure with the neighboring Aβ peptide. Therefore, the odd end, where β2 is exposed, tends to fluctuate more and to take open and closed conformations.

It was experimentally known that the Aβ fibrils extend only in one direction [174,175]. This unidirectionality of the fibril extension implies that the odd and even ends take different conformations, but it was not clear what exactly the structures of both ends were. Our simulation study is the first work to reveal the difference in the structure and fluctuation between the two ends of an amyloid fibril.

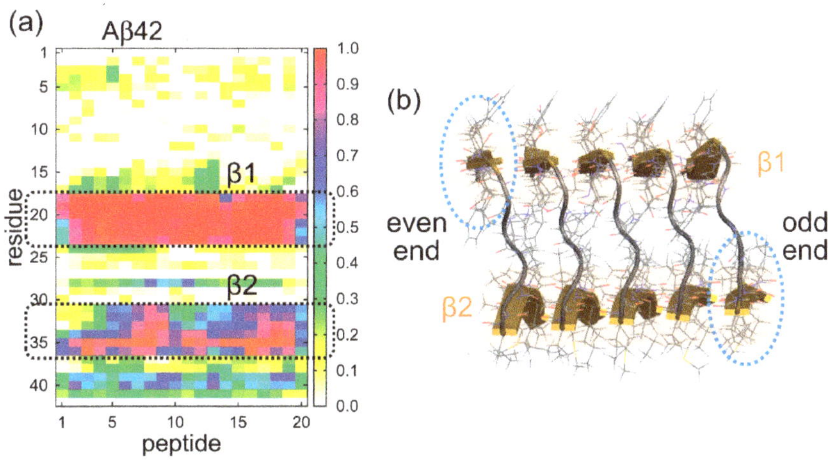

Figure 12. (**a**) The probability that each amino acid residue in each Aβ peptide has a parallel intermolecular β-sheet structure. (**b**) The Aβ amyloid fibril structure revealed by NMR experiments (PDB: 2BEG) [11]. The solvent-exposed β-strands at the even and odd ends are indicated by ellipses. Reprinted with permission from Ref. [72]. Copyright 2016 Springer Nature.

After we performed the MD simulations, the structure of a single amyloid fibril of yeast prion protein sup35 was observed by high-speed atomic force microscopy [176]. This experiment showed that the fluctuation was large at one end and small at the other end, as we predicted from the MD simulations. In other words, the difference in the structures and fluctuations of the two ends of the amyloid fibril that we predicted was confirmed by the experiment.

6. Amyloid Fibril Disruption by Ultrasonic Waves

Amyloid fibrils can be destroyed by ultrasonic wave irradiation or infrared laser irradiation. It has been suggested that the destruction mechanism by the ultrasonic wave is due to cavitation (bubble formation), but the atomic-level details of how the bubbles in water destroy the amyloid fibrils have not been understood experimentally. MD simulation studies on cavitation had been performed mainly for simple liquids such as Lennard-Jones liquids [177–180], but not for biomolecular systems. We recently performed nonequilibrium MD simulations of the destruction of the Aβ amyloid fibril by applying ultrasonic waves [80]. In this section, we review the MD simulations of the amyloid fibril disruption by the ultrasonic waves.

6.1. Molecular Dynamics Simulation to Mimic Ultrasonic Waves

We prepared amyloid fibrils consisting of dodecamer, hexamer, and trimer of Aβ peptides with explicit water molecules. The numbers of water molecules are 10,168, 11,112, and 11,591 for the dodecamer, hexamer, and trimer systems, respectively. Twelve, six, and three sodium ions were also included as counter ions in the dodecamer, hexamer, and trimer systems, respectively. After equilibration MD simulations, nonequilibrium MD simulations were performed with time-dependent pressure to mimic the ultrasonic waves. This pressure is expressed by a sinusoidal curve, which is given by

$$P(t) = P_0 + \Delta P \sin\left(\frac{2\pi t}{\tau}\right), \quad (1)$$

where average pressure P_0, pressure amplitude ΔP, and period τ were set as $P_0 = 100$ MPa, $\Delta P = 200$ MPa, and $\tau = 1$ ns, as illustrated in Figure S2. The temperature was controlled at 298 K with the Nosé–Hoover thermostat [113–115]. The pressure was controlled with the Andersen barostat [173]. We used the AMBER parm99SB force field [111] for the Aβ peptides and the TIP3P rigid-body model [112] for the water molecules. The symplectic [181] quaternion scheme [144,182] was used for the water molecules. The same MD simulations were performed for 10 ns (=10τ) from 20 different initial conditions for statistical analysis. These MD simulations were performed with the GEMB program [148], again. For other simulation details, see Ref. [80].

6.2. Disruption of Aβ Amyloid Fibril by Ultrasonic Waves

The disruption process of the Aβ amyloid fibril by the ultrasonic wave is shown in Figure 13. When the pressure was positive, there was no significant change in the amyloid fibril and water structure. However, when the pressure became negative, a bubble was generated around the amyloid fibril, often near the hydrophobic residues in the β2 region. When the pressure became positive again, the bubble collapsed and a water droplet attacked the amyloid fibrils as a jet flow, resulting in the disruption of the Aβ amyloid fibril.

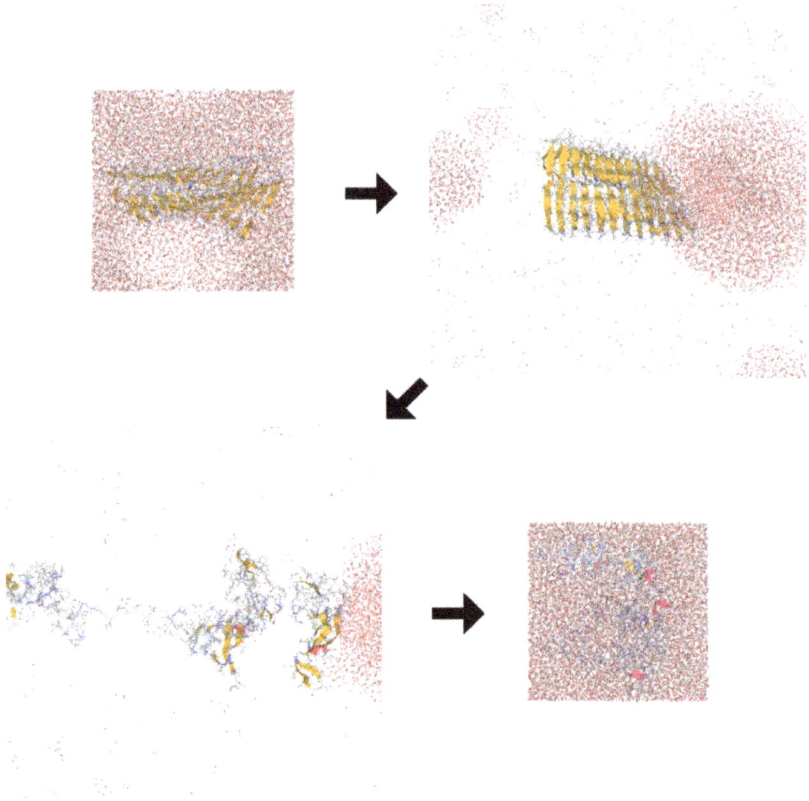

Figure 13. Snapshots of the destruction of the Aβ amyloid fibril by ultrasonic waves. The amyloid fibril was destroyed by the jet flow generated when the bubble collapsed. Reprinted with permission from Ref. [80]. Copyright 2014 American Chemical Society.

Once the amyloid fibril was destroyed, the bubble formation was not observed again. This result suggests that the hydrophobic residues in the β2 region serve as a nucleus for the bubble formation. Even if the same number of hydrophobic residues exist in the water, they cannot function as a nucleus unless assembled as the amyloid fibril. Therefore, we also performed nonequilibrium MD simulations of amyloid fibrils consisting of six and three Aβ peptides. Figure 14 shows how many times the pressure had been negative before the bubbles were formed and the amyloid fibrils were disrupted in twenty MD simulations for each system. In the dodecamer system, a bubble was formed at the first negative pressure in fourteen MD simulations. In four MD simulations, a bubble was formed at the second negative pressure. In two MD simulations, a bubble was formed at the third negative pressure. However, it takes longer for shorter amyloid fibrils to be destroyed. In the trimer system, in particular, a bubble was formed only in one MD simulation out of twenty simulations. These results mean that it takes longer for a shorter amyloid to be a nucleus for the bubble formation. Because the β2 region mainly consists of the hydrophobic residues, these residues can be the nucleus for the bubble formation. The hydrophobic residues in the short amyloid fibrils are not enough to function as a nucleus. This is why it takes time for the bubble formation in the case of short amyloid fibrils.

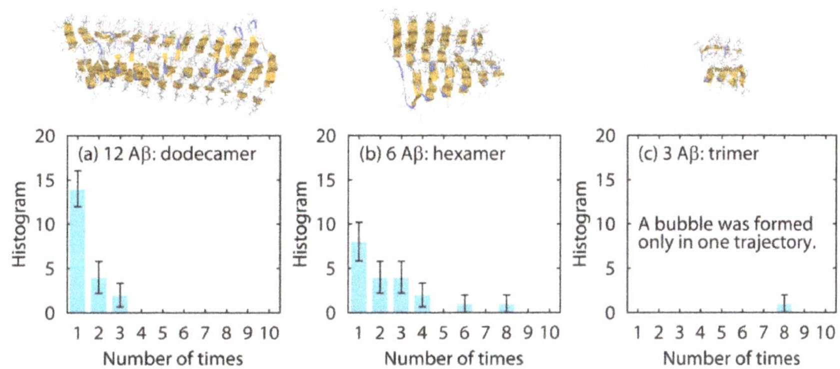

Figure 14. Histograms that show how many times the pressure had been negative before the amyloid fibril was destroyed for the (**a**) dodecamer, (**b**) hexamer, and (**c**) trimer systems. Reprinted with permission from Ref. [80]. Copyright 2014 American Chemical Society.

It was found in experiments that after amyloid fibrils were broken down into shorter fibrils by ultrasonication, the lengths of the short amyloid fibrils were almost the same [183]. This experimental result can be explained from our MD simulations as follows. If the amyloid fibril is longer than some critical length, the region with the hydrophobic residues can be large enough as the nucleus for the bubble formation, and the bubble breaks down the fibrils. On the other hand, if the amyloid fibril is not long enough, the hydrophobic region is not enough, and the amyloid fibrils are not disrupted. This is why ultrasonication makes the length of the amyloid fibril be almost the same.

7. Laser-Induced Disruption of the Aβ Amyloid Fibril

It is also known that amyloid fibrils can be broken down via infrared free-electron laser (IR-FEL) irradiation. The destruction of amyloid fibrils via laser irradiation has been studied using both experimental [184–186] and theoretical techniques [82]. Amyloid fibrils form intermolecular hydrogen bonds between backbone C=O and N–H. Therefore, it was assumed that when a laser that matches the frequency of the C=O stretching vibration is irradiated, the C=O stretching vibration resonates and is amplified, which breaks the hydrogen bonds and results in the disruption of the amyloid fibrils [82]. However, recent experiments showed that Aβ amyloid fibrils under dry conditions are not destroyed by the same laser irradiation; they are only destroyed in the presence of water [185]. This fact suggests that water molecules play an essential role in amyloid fibril destruction. However, the role of the water molecules had not been known.

As the last topic of this review, we introduce our recent MD simulations for the disruption of an Aβ amyloid fibril via laser irradiation in an aqueous solution [84]. In this study, we revealed a new role of water molecules in breaking hydrogen bonds in biomolecules; this mechanism is different from water penetration under high pressure [100,148,187–189] and water jets when ultrasonic waves are applied [80]. In addition, we succeeded in reproducing an experimental observation [185], in which more α-helix structures are formed after the laser irradiation, and explaining the reason for this phenomenon.

7.1. Molecular Dynamics Simulation to Mimic Laser Irradiation

In IR-FEL experiments, a sample is irradiated with an infrared laser that corresponds to the backbone C=O stretching vibration (amide I band). To determine the resonance wavenumber of the C=O stretching vibration of the model Aβ amyloid fibril, we first performed equilibrium MD simulations of an amyloid fibril consisting of twelve Aβ42 peptides in an explicit water solvent. We used the GEMB program [148] again to perform the MD simulations. The initial amyloid fibril conformations were prepared using model 1 of the 2BEG PDB conformation [11]. A total of 1 Aβ amyloid fibril, 36 sodium ions, and 25,480 water molecules were placed in a cubic simulation box with a side length of 96.324 Å. The total number of atoms was 84,000. Six different initial conditions were prepared for the statistical analysis.

We applied the AMBER parm14SB force field [141] to the Aβ peptides and counter ions. We used the TIP3P rigid-body model [112] for the water molecules by adopting the symplectic [181] quaternion scheme [144,182]. The MD simulations were performed at 310 K and 0.1 MPa for 50 ns from the six initial conditions. The temperature was controlled using the Nosé–Hoover thermostat [113–115]. The pressure was controlled using the Andersen barostat [173]. The first 10 ns of the simulations were regarded as the equilibration, and the following 40 ns were used for the analysis. We used the amino acid residues V18–D23 in the β1 region and I31–V36 in the β2 region to calculate the infrared absorption spectrum of the C=O stretching vibration. We then determined the resonance wavenumber in this model fibril as 1676 cm^{-1}. For comparison, we also performed equilibrium MD simulations of an Aβ peptide for α-helix and random coil structures and calculated the infrared absorption spectra of these structures.

After the resonance wavenumber was determined, we performed nonequilibrium MD simulations of the Aβ amyloid fibril, applying a time-varying electric field with the resonance wavenumber to simulate the IR-FEL irradiation. To mimic the IR-FEL irradiation, an electric field was applied as a series of Gaussian-distributed pulses [82] with an interval of 35 ps. Each pulse is expressed as

$$E(t) = E_0 \exp\left(-\frac{(t-t_0)^2}{2\sigma^2}\right)\cos\left(\omega(t-t_0)^2\right) \quad (2)$$

where E_0 is the maximum intensity of the electric field, t is time, t_0 is the time at $E = E_0$, σ is the standard deviation of the Gaussian distribution, and ω is the angular frequency related to the wavenumber ν such that $\omega = 2\pi c\nu$, where c is the speed of light. The wavenumber ν was set to the resonance wavenumber 1676 cm^{-1}, and E_0 was set to 1×10^8 V/cm. The value of σ was set to 1 ps to match that used in the IR-FEL experiments [185]. The final conformations and velocities in the previous equilibrium MD simulations were used as the initial conformations and velocities for the nonequilibrium MD simulations. Constant-temperature MD simulations were then performed at 310 K for 1000 pulses, that is, for 35 ns. Other simulation details can be found in Ref. [84].

7.2. Amyloid Fibril Disruption by Laser Irradiation

We observed that the amyloid fibril was gradually destroyed, as shown in Figure 15. To quantify this result, we calculated the ratio of the amino acid residues that formed the intermolecular parallel β-sheet structure according to the DSSP criteria [190], as shown in Figure 16a. Almost all the intermolecular β-sheet structures were destroyed after 1000 pulses. In Figure 15, we see that many helix structures (red ribbons) formed after the amyloid fibril was disrupted. We calculated the ratio of the amino acid residues in the helix structures, as shown in Figure 16b. Here, the α-, 3_{10}-, and π-helices [190] were included in the helix structures. This figure shows that the helix structures increased as the intermolecular β-sheet structure was destroyed in the MD simulations. These results are consistent with the IR-FEL experiments [185].

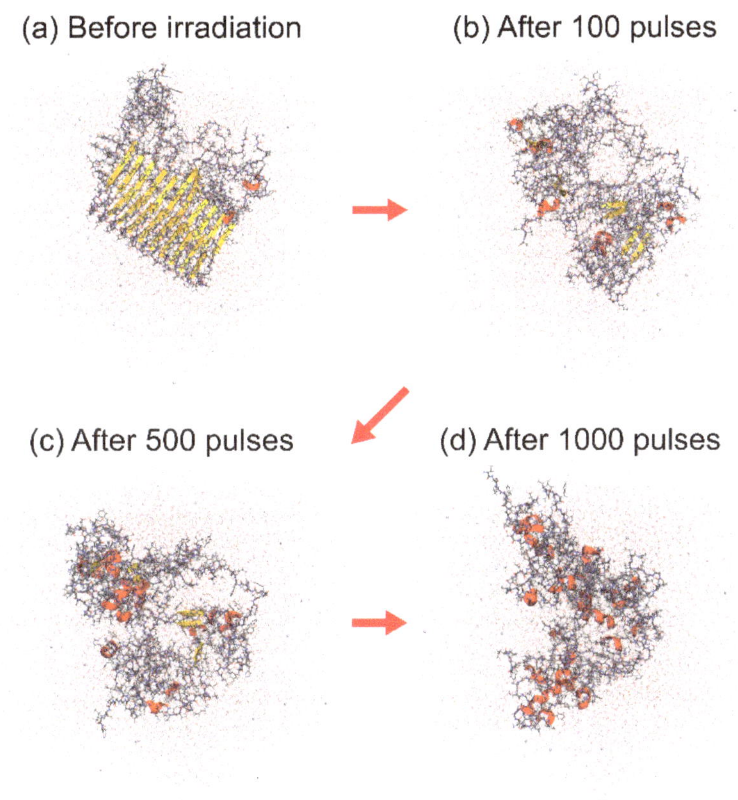

Figure 15. Snapshots during the laser-induced disruption process of the Aβ amyloid fibril in the nonequilibrium MD simulation (**a**) before IR-FEL irradiation, (**b**) after 100 pulses, (**c**) after 500 pulses, and (**d**) after 1000 pulses. The images were created using PyMOL [171].

To examine the role of water molecules in the amyloid fibril disruption, we focused on the intermolecular β-bridges between two Aβ peptides. Figure 17 shows enlarged snapshots of the Aβ amyloid fibril in a typical MD simulation run and the electric field pulse intensity at the same time (red circles). Six intermolecular hydrogen bonds existed between the two β-strands (inside the purple dashed line) in Figure 17a. These intermolecular hydrogen bonds were broken by an electric field pulse in Figure 17b, and most of them were broken by the end of the pulse (Figure 17c). However, these hydrogen bonds re-formed after the pulse. During this re-formation, water molecules sometimes formed hydrogen bonds with the Aβ peptides (Figure 17d), but they soon separated from the peptides. The two β-strands were eventually completely repaired (Figure 17e). Before this pulse, the intermolecular hydrogen bonds between the Aβ peptides were repeatedly broken and repaired after each electric field irradiation in the same way. Immediately after the hydrogen bonds between the Aβ peptides were broken by the next pulse (Figure 17f), however, a water molecule (the pink-highlighted water molecule) entered the space between C=O and N–H, where the intermolecular hydrogen bond had been previously formed (Figure 17g). This water molecule formed hydrogen bonds with the Aβ peptides and prevented the hydrogen bond re-formation between C=O and N–H of the Aβ peptides in Figure 17h). Another water molecule (the blue-highlighted water molecule) also entered the space between the Aβ peptides and formed hydrogen bonds with the Aβ peptides. Some hydrogen bonds of the red-highlighted water molecule were broken in Figure 17i, but the blue-highlighted water molecule still formed some hydrogen bonds with the Aβ peptides. Even after the red-highlighted water molecule separated from the peptides, the blue-highlighted water molecule stayed in this location (Figure 17j). Other water molecules then entered the gap between the Aβ peptides (Figure 17k). Because the hydrogen bonds between the Aβ peptides were replaced by those between the Aβ peptides and the water molecules, the intermolecular hydrogen bonds between the Aβ peptides could not be re-formed before the next laser pulse. As a result, the intermolecular β-sheet of the Aβ amyloid fibril was destroyed (Figure 17l). This phenomenon occurred throughout the amyloid fibril, and the entire fibril was finally disrupted.

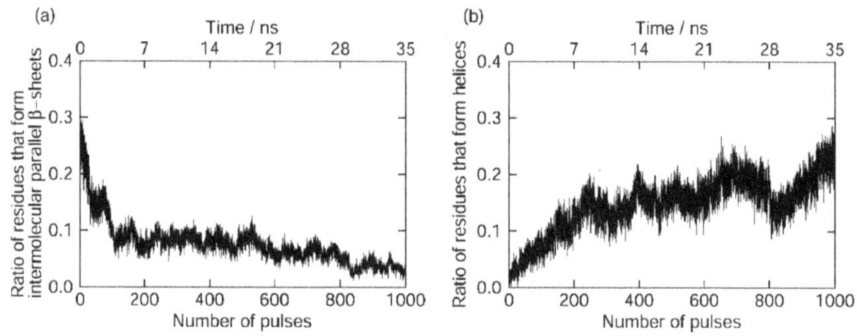

Figure 16. Time series of the ratio of the residues that form (**a**) intermolecular β-sheets and (**b**) helices in one of the nonequilibrium MD simulations.

To understand why helix structures increased after the laser irradiation, we performed additional equilibrium MD simulations for α-helix and random coil structures of an Aβ peptide. We then calculated the infrared absorption spectrum of the C=O stretching vibration, as shown in Figure 18. We found that the resonance wavenumber for the random coil structure was 1675 cm^{-1}, which is close to that for the intermolecular β-sheet structure and the laser wavenumber of this study, while the resonance wavenumber for the α-helix structure was 1697 cm^{-1}, which is far from these wavenumbers. These results mean that helix structures can exist stably without breaking the hydrogen bonds between C=O and N–H because their resonance frequency is different from the laser frequency used to destroy the intermolecular β-sheet structure.

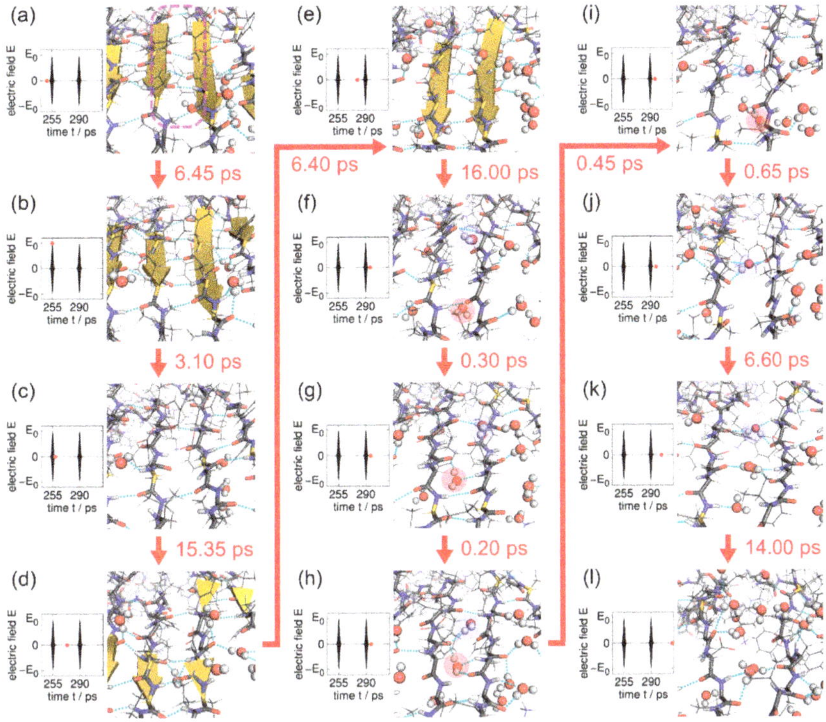

Figure 17. (**a–l**) Disruption process of the hydrogen bonds between the Aβ peptides and the electric field pulse. Two water molecules that disrupted the hydrogen bond re-formation between the Aβ peptides are highlighted with pink and blue circles. The images were created using PyMOL [171]. Reprinted with permission from Ref. [84]. Copyright 2021 American Chemical Society.

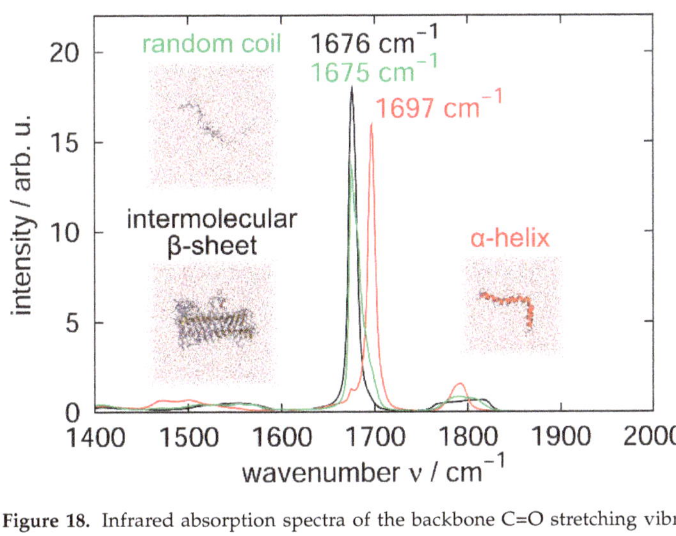

Figure 18. Infrared absorption spectra of the backbone C=O stretching vibration that forms the amyloid fibril (black), α-helix (red), and random coil (green). The snapshot images were created using PyMOL [171]. Reprinted in part with permission from Ref. [84]. Copyright 2021 American Chemical Society.

8. Conclusions

In this review, we presented the molecular dynamics (MD) simulation studies of full-length Aβ peptides and Aβ fragments that revealed the mechanism of their aggregation, the inhibition of the aggregation, the amyloid fibril in equilibrium, and the disruption of the amyloid fibril at the atomic level. We first explained that a β-hairpin structure enhances the formation of an intermolecular β-sheet structure. The β-hairpin structure is more formed at hydrophilic/hydrophobic interfaces. This is one of the reasons that the aggregation of the peptides is accelerated at the interfaces. The other reason is that the Aβ peptide has both hydrophilic and hydrophobic residues and tends to exist at the interfaces.

We also explained how polyphenols such as myricetin and rosmarinic acid interact with an Aβ(16–22) peptide. Because the aggregation of Aβ(16–22) peptides is caused by the electrostatic interaction between charged amino acid residues, Lys16 and Glu22, these polyphenols are expected to inhibit the aggregation by forming hydrogen bonds between these charged residues and the hydroxy and carboxyl groups of the polyphenols.

When almost all of the Aβ peptides in solution form amyloid fibrils, the system reaches equilibrium. The MD simulations of the Aβ amyloid fibril in equilibrium showed that Aβ always takes a closed form at the even end, whereas Aβ fluctuates more and takes an open form at the odd end. The reason for this phenomenon was also clarified. This finding is useful for understanding the mechanism of the amyloid fibril elongation and for designing drugs that inhibit its elongation.

It is possible to destroy the Aβ amyloid fibril by applying ultrasonic waves or infrared laser. The MD simulations also revealed the mechanisms of the Aβ amyloid fibril destruction for the ultrasonic wave and infrared laser irradiations. When the ultrasonic waves are applied, the Aβ amyloid fibril is disrupted by cavitation: a bubble is formed when the pressure is negative, and a water droplet then attacks and disrupts the amyloid fibril after the pressure becomes positive again. When the infrared laser is irradiated, hydrogen bonds between C=O and N−H are broken, but most of them re-form after the laser pulse. However, a water molecule nearby sometimes happens to enter the gap between C=O and N−H. It inhibits the re-formation of the hydrogen bonds, leading to the disruption of the amyloid fibril. In both cases of the ultrasonic wave and infrared laser irradiations, water molecules play an essential role in disrupting the amyloid fibril.

All simulation studies described here are based on the all-atom model with the explicit water solvent. There are several all-atom force fields, such as AMBER14SB [141], CHARMM36m [191], and GROMOS54A7 [192]. Some studies examined the structure of Aβ peptides using these force fields to find the optimal force field [193–195]. Because the all-atom force fields have been improved over the years, it is desirable to use the best force field available at the time. While MD simulation based on the all-atom models has the advantage of analyzing phenomena at the atomic level, it is computationally time-consuming. Since protein aggregation simulations are particularly computationally demanding, simulation studies using implicit solvent models, such as the GB/SA model [196–198], and coarse-grained models, such as the AWSEM [199], MARTINI [200,201], and UNRES force fields [202,203], are also being conducted [204,205]. The implicit solvent models used to be employed often [33,34,51,96] but now are not often used for all-atom simulations because it is known that the interaction between water and solutes plays an important role in the aggregation [206,207] and disaggregation of the amyloid fibrils [80,84]. Although these coarse-grained models do not provide atomic-level details, they can save much computation time. As more simplified models, lattice models have also been used to simulate protein aggregation [208–211]. The lattice models are primarily used to elucidate more general physical principles rather than to examine individual protein aggregates biologically. Depending on the purpose, these models would also continue to be used to study protein aggregation.

There is no established therapy to destroy amyloid fibrils at this time by irradiating the brain of Alzheimer's patients with ultrasonic waves or infrared laser. However, we believe that there is a possibility that such a therapy can be realized in the future. In fact, animal experiments have been conducted to remove Aβ aggregates by irradiating the brain with ultrasonic waves, although its purpose is not to disrupt the Aβ aggregates [212–215]. Delivering therapeutic agents, such as anti-Aβ antibodies, to the brain is a possible approach for Alzheimer's disease. However, the penetration of the therapeutic agents to the brain is hampered by the blood–brain barrier. To make it possible, focused ultrasound is utilized. Focused ultrasound opens the blood–brain barrier and promotes the therapeutic agent delivery to the brain. It was reported that the Aβ aggregates were reduced in Alzheimer's disease model mice, and their behavior was improved [214].

We believe that the destruction of amyloid fibrils by the infrared laser irradiation may also have therapeutic potential in the future. In particular, it is noteworthy that the α-helix structure is formed more after the amyloid fibrils are destroyed by the infrared laser irradiation. This is because, unlike the β-hairpin structure, the α-helix structure can be maintained in the monomeric state and relatively easily excreted from the human body. Techniques to destroy amyloid fibrils may also be useful in developing treatments for other diseases caused by other amyloid fibrils.

As we reviewed here, MD simulation can identify which residues or atoms are important for the aggregation and aggregation inhibition and can be used to design a useful drug molecule for the treatment of Alzheimer's disease and other neurodegenerative diseases. MD simulation can also elucidate the molecular mechanism of amyloid fibril destruction. We hope that MD simulation will become a new tool for developing treatments for these diseases in the future.

Supplementary Materials: The following supporting information can be downloaded at: https://www.mdpi.com/article/10.3390/molecules27082483/s1, Figure S1: The chemical structures of (a) myricetin (Myr) and (b) rosmarinic acid (RA); Figure S2: Time series of the set pressure, which varies sinusoidally. References [31,80] are cited in the supplementary materials.

Author Contributions: Conceptualization, H.O. and S.G.I.; writing—original draft preparation, H.O. and S.G.I.; writing—review and editing, H.O. and S.G.I. All authors have read and agreed to the published version of the manuscript.

Funding: This review received no external funding.

Acknowledgments: We gratefully thank all collaborators of the research works explained in this review. The research presented in this review used supercomputers at the Research Center for Computational Science, Okazaki Research Facilities, National Institutes of Natural Sciences, and the Supercomputer Center, the Institute for Solid State Physics, the University of Tokyo.

Conflicts of Interest: The authors declare no conflict of interest.

References

1. Sipe, J.D.; Cohen, A.S. Review: History of the Amyloid Fibril. *J. Struct. Biol.* **2000**, *130*, 88–98. [CrossRef] [PubMed]
2. Chiti, F.; Dobson, C.M. Protein misfolding, functional amyloid, and human disease. *Annu. Rev. Biochem.* **2006**, *75*, 333–366. [CrossRef] [PubMed]
3. Tycko, R. Amyloid Polymorphism: Structural Basis and Neurobiological Relevance. *Neuron* **2015**, *86*, 632–645. [CrossRef]
4. Post, F. *The Clinical Psychiatry of Late Life*; Pergamon Press: Oxford, UK, 1965.
5. Tomlinson, B.; Blessed, G.; Roth, M. Observations on the brains of non-demented old people. *J. Neurol. Sci.* **1968**, *7*, 331–356. [CrossRef]
6. Glenner, G.G.; Wong, C.W. Alzheimer's disease: Initial report of the purification and characterization of a novel cerebrovascular amyloid protein. *Biochem. Biophys. Res. Commun.* **1984**, *120*, 885–890. [CrossRef]
7. Masters, C.L.; Simms, G.; Weinman, N.A.; Multhaup, G.; McDonald, B.L.; Beyreuther, K. Amyloid plaque core protein in Alzheimer disease and Down syndrome. *Proc. Natl. Acad. Sci. USA* **1985**, *82*, 4245–4249. [CrossRef]
8. Kang, J.; Lemaire, H.-G.; Unterbeck, A.; Salbaum, J.; Masters, C.; Grzeschik, K.-H.; Multhaup, G.; Beyreuther, K.; Müller-Hill, B. The precursor of Alzheimer's disease amyloid A4 protein resembles a cell-surface receptor. *Alzheimer Dis. Assoc. Disord.* **1987**, *1*, 206–207. [CrossRef]
9. Wetzel, R. Ideas of Order for Amyloid Fibril Structure. *Structure* **2002**, *10*, 1031–1036. [CrossRef]
10. Petkova, A.T.; Ishii, Y.; Balbach, J.J.; Antzutkin, O.N.; Leapman, R.D.; Delaglio, F.; Tycko, R. A structural model for Alzheimer's β-amyloid fibrils based on experimental constraints from solid state NMR. *Proc. Natl. Acad. Sci. USA* **2002**, *99*, 16742–16747. [CrossRef]

11. Lührs, T.; Ritter, C.; Adrian, M.; Riek-Loher, D.; Bohrmann, B.; Dobeli, H.; Schubert, D.; Riek, R. 3D structure of Alzheimer's amyloid-β(1-42) fibrils. *Proc. Natl. Acad. Sci. USA* **2005**, *102*, 17342–17347. [CrossRef]
12. Lu, J.X.; Qiang, W.; Yau, W.M.; Schwieters, C.D.; Meredith, S.C.; Tycko, R. Molecular Structure of β-Amyloid Fibrils in Alzheimer's Disease Brain Tissue. *Cell* **2013**, *154*, 1257–1268. [CrossRef] [PubMed]
13. Xiao, Y.; Ma, B.; McElheny, D.; Parthasarathy, S.; Long, F.; Hoshi, M.; Nussinov, R.; Ishii, Y. Aβ(1-42) fibril structure illuminates self-recognition and replication of amyloid in Alzheimer's disease. *Nat. Struct. Mol. Biol.* **2015**, *22*, 499–505. [CrossRef] [PubMed]
14. Gremer, L.; Schölzel, D.; Schenk, C.; Reinartz, E.; Labahn, J.; Ravelli, R.B.G.; Tusche, M.; Lopez-Iglesias, C.; Hoyer, W.; Heise, H.; et al. Fibril Structure of Amyloid-β(1–42) by Cryoelectron Microscopy. *Science* **2017**, *358*, 116. [CrossRef] [PubMed]
15. Sgourakis, N.G.; Yan, Y.; McCallum, S.A.; Wang, C.; Garcia, A.E. The Alzheimer's Peptides Aβ40 and 42 Adopt Distinct Conformations in Water: A Combined MD / NMR Study. *J. Mol. Biol.* **2007**, *368*, 1448–1457. [CrossRef]
16. Allison, J.R.; Varnai, P.; Dobson, C.M.; Vendruscolo, M. Determination of the Free Energy Landscape of α-Synuclein Using Spin Label Nuclear Magnetic Resonance Measurements. *J. Am. Chem. Soc.* **2009**, *131*, 18314–18326. [CrossRef]
17. Vitalis, A.; Caflisch, A. Micelle-Like Architecture of the Monomer Ensemble of Alzheimer's Amyloid-β Peptide in Aqueous Solution and Its Implications for Aβ Aggregation. *J. Mol. Biol.* **2010**, *403*, 148–165. [CrossRef]
18. Sgourakis, N.G.; Merced-Serrano, M.; Boutsidis, C.; Drineas, P.; Du, Z.; Wang, C.; Garcia, A.E. Atomic-Level Characterization of the Ensemble of the Aβ(1–42) Monomer in Water Using Unbiased Molecular Dynamics Simulations and Spectral Algorithms. *J. Mol. Biol.* **2011**, *405*, 570. [CrossRef]
19. Velez-Vega, C.; Escobedo, F.A. Characterizing the structural behavior of selected Aβ-42 monomers with different solubilities. *J. Phys. Chem. B* **2011**, *115*, 4900–4910. [CrossRef]
20. Olubiyi, O.O.; Strodel, B. Structures of the amyloid β-peptides Aβ1–40 and Aβ1–42 as influenced by pH and a D-Peptide. *J. Phys. Chem. B* **2012**, *116*, 3280. [CrossRef]
21. Ball, K.A.; Phillips, A.H.; Wemmer, D.E.; Head-Gordon, T. Differences in β-strand Populations of Monomeric Aβ40 and Aβ42. *Biophys. J.* **2013**, *104*, 2714–2724. [CrossRef]
22. Itoh, S.G.; Okumura, H. Hamiltonian replica-permutation method and its applications to an alanine dipeptide and amyloid-β(29-42) peptides. *J. Comput. Chem.* **2013**, *34*, 2493–2497. [CrossRef] [PubMed]
23. Itoh, S.G.; Okumura, H. Coulomb replica-exchange method: Handling electrostatic attractive and repulsive forces for biomolecules. *J. Comput. Chem.* **2013**, *34*, 622–639. [CrossRef] [PubMed]
24. Rosenman, D.J.; Connors, C.R.; Chen, W.; Wang, C.; Garcia, A.E. Aβ Monomers Transiently Sample Oligomer and Fibril-Like Configurations: Ensemble Characterization Using a Combined MD/NMR Approach. *J. Mol. Biol.* **2013**, *425*, 3338. [CrossRef] [PubMed]
25. Rosenman, D.J.; Wang, C.; García, A.E. Characterization of Aβ Monomers through the Convergence of Ensemble Properties among Simulations with Multiple Force Fields. *J. Phys. Chem. B* **2015**, *120*, 259–277. [CrossRef]
26. Ilie, I.M.; Nayar, D.; den Otter, W.K.; van der Vegt, N.F.A.; Briels, W.J. Intrinsic Conformational Preferences and Interactions in α-Synuclein Fibrils: Insights from Molecular Dynamics Simulations. *J. Chem. Theory Comput.* **2018**, *14*, 3298–3310. [CrossRef]
27. Mudedla, S.K.; Murugan, N.A.; Agren, H. Free Energy Landscape for α-Helix to β-sheet Interconversion in Small Amyloid Forming Peptide under Nanoconfinement. *J. Phys. Chem. B* **2018**, *122*, 9654–9664. [CrossRef]
28. Meng, F.; Bellaiche, M.M.; Kim, J.-Y.; Zerze, G.H.; Best, R.B.; Chung, H.S. Highly Disordered Amyloid-β Monomer Probed by Single-Molecule FRET and MD Simulation. *Biophys. J.* **2018**, *114*, 870–884. [CrossRef]
29. Tachi, Y.; Okamoto, Y.; Okumura, H. Conformational Change of Amyloid-β 40 in Association with Binding to GM1-Glycan Cluster. *Sci. Rep.* **2019**, *9*, 6853. [CrossRef]
30. Itoh, S.G.; Yagi-Utsumi, M.; Kato, K.; Okumura, H. Effects of a Hydrophilic/Hydrophobic Interface on Amyloid-β Peptides Studied by Molecular Dynamics Simulations and NMR Experiments. *J. Phys. Chem. B* **2018**, *123*, 160–169. [CrossRef]
31. Ngoc, L.L.N.; Itoh, S.G.; Sompornpisut, P.; Okumura, H. Replica-permutation molecular dynamics simulations of an amyloid-β(16–22) peptide and polyphenols. *Chem. Phys. Lett.* **2020**, *758*, 137913. [CrossRef]
32. Tavanti, F.; Pedone, A.; Menziani, M.C. Disclosing the Interaction of Gold Nanoparticles with Aβ(1–40) Monomers through Replica Exchange Molecular Dynamics Simulations. *Int. J. Mol. Sci.* **2020**, *22*, 26. [CrossRef] [PubMed]
33. Chebaro, Y.; Mousseau, N.; Derreumaux, P. Structures and thermodynamics of Alzheimer's amyloid-β Aβ(16-35) monomer and dimer by replica exchange molecular dynamics simulations: Implication for full-length Aβ fibrillation. *J. Phys. Chem. B* **2009**, *113*, 7668–7675. [CrossRef] [PubMed]
34. Cote, S.; Laghaei, R.; Derreumaux, P.; Mousseau, N. Distinct dimerization for various alloforms of the amyloid-β protein: Aβ(1-40), Aβ(1-42), and Aβ(1-40)(D23N). *J. Phys. Chem. B* **2012**, *116*, 4043–4055. [CrossRef] [PubMed]
35. Chiang, H.L.; Chen, C.J.; Okumura, H.; Hu, C.K. Transformation between α-helix and β-sheet structures of one and two polyglutamine peptides in explicit water molecules by replica-exchange molecular dynamics simulations. *J. Comput. Chem.* **2014**, *35*, 1430–1437. [CrossRef]
36. Itoh, S.G.; Okumura, H. Dimerization process of amyloid-β(29-42) studied by the Hamiltonian replica-permutation molecular dynamics simulations. *J. Phys. Chem. B* **2014**, *118*, 11428–11436. [CrossRef]
37. Nguyen, P.H.; Sterpone, F.; Campanera, J.M.; Nasica-Labouze, J.; Derreumaux, P. Impact of the A2V Mutation on the Heterozygous and Homozygous Aβ1–40 Dimer Structures from Atomistic Simulations. *ACS Chem. Neurosci.* **2016**, *7*, 823–832. [CrossRef]

38. Tarus, B.; Tran, T.T.; Nasica-Labouze, J.; Sterpone, F.; Nguyen, P.H.; Derreumaux, P. Structures of the Alzheimer's Wild-Type Aβ1-40 Dimer from Atomistic Simulations. *J. Phys. Chem. B* **2015**, *119*, 10478–10487. [CrossRef]
39. Nguyen, P.H.; Sterpone, F.; Pouplana, R.; Derreumaux, P.; Campanera, J.M. Dimerization Mechanism of Alzheimer Aβ40 Peptides: The High Content of Intrapeptide-Stabilized Conformations in A2V and A2T Heterozygous Dimers Retards Amyloid Fibril Formation. *J. Phys. Chem. B* **2016**, *120*, 12111–12126. [CrossRef]
40. Das, P.; Chacko, A.R.; Belfort, G. Alzheimer's Protective Cross-Interaction between Wild-Type and A2T Variants Alters Aβ42 Dimer Structure. *ACS Chem. Neurosci.* **2016**, *8*, 606–618. [CrossRef]
41. Man, V.H.; Nguyen, P.H.; Derreumaux, P. Conformational Ensembles of the Wild-Type and S8C Aβ1-42 Dimers. *J. Phys. Chem. B* **2017**, *121*, 2434–2442. [CrossRef]
42. Man, V.H.; Nguyen, P.H.; Derreumaux, P. High-Resolution Structures of the Amyloid-β 1–42 Dimers from the Comparison of Four Atomistic Force Fields. *J. Phys. Chem. B* **2017**, *121*, 5977–5987. [CrossRef]
43. Sharma, B.; Ranganathan, S.V.; Belfort, G. Weaker N-Terminal Interactions for the Protective over the Causative Aβ Peptide Dimer Mutants. *ACS Chem. Neurosci.* **2018**, *9*, 1247–1253. [CrossRef]
44. Nishizawa, H.; Okumura, H. Classical Molecular Dynamics Simulation to Understand Role of a Zinc Ion for Aggregation of Amyloid-β Peptides. *J. Comput. Chem. Jpn.* **2018**, *17*, 76–79. [CrossRef]
45. Kargar, F.; Emadi, S.; Fazli, H. Dimerization of Aβ40 inside dipalmitoylphosphatidylcholine bilayer and its effect on bilayer integrity: Atomistic simulation at three temperatures. *Proteins Struct. Funct. Bioinform.* **2020**, *88*, 1540–1552. [CrossRef] [PubMed]
46. Yamauchi, M.; Okumura, H. Dimerization of α-Synuclein Fragments Studied by Isothermal–Isobaric Replica-Permutation Molecular Dynamics Simulation. *J. Chem. Inf. Model.* **2021**, *61*, 1307–1321. [CrossRef] [PubMed]
47. Gsponer, J.; Haberthür, U.; Caflisch, A. The role of side-chain interactions in the early steps of aggregation: Molecular dynamics simulations of an amyloid-forming peptide from the yeast prion Sup35. *Proc. Natl. Acad. Sci. USA* **2003**, *100*, 5154–5159. [CrossRef] [PubMed]
48. Urbanc, B.; Betnel, M.; Cruz, L.; Bitan, G.; Teplow, D.B. Elucidation of Amyloid β-Protein Oligomerization Mechanisms: Discrete Molecular Dynamics Study. *J. Am. Chem. Soc.* **2010**, *132*, 4266–4280. [CrossRef]
49. Pacheco, M.C.; Ismail, A.E.; Strodel, B. Oligomer Formation of Toxic and Functional Amyloid Peptides Studied with Atomistic Simulations. *J. Phys. Chem. B* **2015**, *119*, 9696–9705. [CrossRef]
50. Itoh, S.G.; Okumura, H. Oligomer Formation of Amyloid-β(29-42) from Its Monomers Using the Hamiltonian Replica-Permutation Molecular Dynamics Simulation. *J. Phys. Chem. B* **2016**, *120*, 6555–6561. [CrossRef]
51. Barz, B.; Liao, Q.; Strodel, B. Pathways of Amyloid-β Aggregation Depend on Oligomer Shape. *J. Am. Chem. Soc.* **2017**, *140*, 319–327. [CrossRef]
52. Sun, Y.; Ge, X.; Xing, Y.; Wang, B.; Ding, F. β-barrel Oligomers as Common Intermediates of Peptides Self-Assembling into Cross-β Aggregates. *Sci. Rep.* **2018**, *8*, 10353. [CrossRef]
53. Okumura, H.; Itoh, S.G. Molecular dynamics simulations of amyloid-β(16–22) peptide aggregation at air–water interfaces. *J. Chem. Phys.* **2020**, *152*, 95101. [CrossRef] [PubMed]
54. Chrobak, W.; Pacut, D.W.; Blomgren, F.; Rodin, A.; Swenson, J.; Ermilova, I. Component of Cannabis, Cannabidiol, as a Possible Drug against the Cytotoxicity of Aβ(31–35) and Aβ(25–35) Peptides: An Investigation by Molecular Dynamics and Well-Tempered Metadynamics Simulations. *ACS Chem. Neurosci.* **2021**, *12*, 660–674. [CrossRef] [PubMed]
55. Nguyen, P.H.; Li, M.S.; Stock, G.; Straub, J.E.; Thirumalai, D. Monomer adds to preformed structured oligomers of Aβ-peptides by a two-stage dock–lock mechanism. *Proc. Natl. Acad. Sci. USA* **2007**, *104*, 111–116. [CrossRef] [PubMed]
56. O'Brien, E.P.; Okamoto, Y.; Straub, J.E.; Brooks, B.R.; Thirumalai, D. Thermodynamic Perspective on the Dock—Lock Growth Mechanism of Amyloid Fibrils. *J. Phys. Chem. B* **2009**, *113*, 14421–14430. [CrossRef] [PubMed]
57. Takeda, T.; Klimov, D.K. Probing Energetics of Aβ Fibril Elongation by Molecular Dynamics Simulations. *Biophys. J.* **2009**, *96*, 4428–4437. [CrossRef]
58. Takeda, T.; Klimov, D.K. Replica Exchange Simulations of the Thermodynamics of Aβ Fibril Growth. *Biophys. J.* **2009**, *96*, 442–452. [CrossRef] [PubMed]
59. Reddy, A.S.; Wang, L.; Singh, S.; Ling, Y.L.; Buchanan, L.; Zanni, M.T.; Skinner, J.L.; de Pablo, J.J. Stable and Metastable States of Human Amylin in Solution. *Biophys. J.* **2010**, *99*, 2208–2216. [CrossRef]
60. Han, M.; Hansmann, U.H. Replica exchange molecular dynamics of the thermodynamics of fibril growth of Alzheimer's Aβ42 peptide. *J. Chem. Phys.* **2011**, *135*, 65101. [CrossRef]
61. Straub, J.E.; Thirumalai, D. Toward a Molecular Theory of Early and Late Events in Monomer to Amyloid Fibril Formation. *Annu. Rev. Phys. Chem.* **2011**, *62*, 437–463. [CrossRef]
62. Gurry, T.; Stultz, C.M. Mechanism of Amyloid-β Fibril Elongation. *Biochemistry* **2014**, *53*, 6981–6991. [CrossRef] [PubMed]
63. Han, W.; Schulten, K. Fibril Elongation by Aβ17–42: Kinetic Network Analysis of Hybrid-Resolution Molecular Dynamics Simulations. *J. Am. Chem. Soc.* **2014**, *136*, 12450–12460. [CrossRef] [PubMed]
64. Schwierz, N.; Frost, C.V.; Geissler, P.L.; Zacharias, M. Dynamics of Seeded Aβ40-Fibril Growth from Atomistic Molecular Dynamics Simulations: Kinetic Trapping and Reduced Water Mobility in the Locking Step. *J. Am. Chem. Soc.* **2016**, *138*, 527–539. [CrossRef]
65. Sasmal, S.; Schwierz, N.; Head-Gordon, T. Mechanism of Nucleation and Growth of Aβ40 Fibrils from All-Atom and Coarse-Grained Simulations. *J. Phys. Chem. B* **2016**, *120*, 12088–12097. [CrossRef] [PubMed]

66. Bacci, M.; Vymětal, J.; Mihajlovic, M.; Caflisch, A.; Vitalis, A. Amyloid β Fibril Elongation by Monomers Involves Disorder at the Tip. *J. Chem. Theory Comput.* **2017**, *13*, 5117–5130. [CrossRef] [PubMed]
67. Ilie, I.M.; Otter, W.K.D.; Briels, W.J. The attachment of α-synuclein to a fiber: A coarse-grain approach. *J. Chem. Phys.* **2017**, *146*, 115102. [CrossRef]
68. Sun, Y.; Ding, F. αB-Crystallin Chaperone Inhibits Aβ Aggregation by Capping the β-sheet-Rich Oligomers and Fibrils. *J. Phys. Chem. B* **2020**, *124*, 10138–10146. [CrossRef]
69. Buchete, N.V.; Tycko, R.; Hummer, G. Molecular dynamics simulations of Alzheimer's β-amyloid protofilaments. *J. Mol. Biol.* **2005**, *353*, 804–821. [CrossRef]
70. Baumketner, A.; Krone, M.G.; Shea, J.-E. Role of the familial Dutch mutation E22Q in the folding and aggregation of the 15–28 fragment of the Alzheimer amyloid-β protein. *Proc. Natl. Acad. Sci. USA* **2008**, *105*, 6027–6032. [CrossRef]
71. Lemkul, J.A.; Bevan, D.R. Assessing the Stability of Alzheimer's Amyloid Protofibrils Using Molecular Dynamics. *J. Phys. Chem. B* **2010**, *114*, 1652–1660. [CrossRef]
72. Okumura, H.; Itoh, S.G. Structural and fluctuational difference between two ends of Aβ amyloid fibril: MD simulations predict only one end has open conformations. *Sci. Rep.* **2016**, *6*, 38422. [CrossRef] [PubMed]
73. Rodriguez, R.A.; Chen, L.Y.; Plascencia-Villa, G.; Perry, G. Thermodynamics of Amyloid-β Fibril Elongation: Atomistic Details of the Transition State. *ACS Chem. Neurosci.* **2017**, *9*, 783–789. [CrossRef] [PubMed]
74. Davidson, D.S.; Brown, A.M.; Lemkul, J.A. Insights into Stabilizing Forces in Amyloid Fibrils of Differing Sizes from Polarizable Molecular Dynamics Simulations. *J. Mol. Biol.* **2018**, *430*, 3819. [CrossRef] [PubMed]
75. Ilie, I.-M.; Caflisch, A. Disorder at the Tips of a Disease-Relevant Aβ42 Amyloid Fibril: A Molecular Dynamics Study. *J. Phys. Chem. B* **2018**, *122*, 11072–11082. [CrossRef] [PubMed]
76. Nirmalraj, P.N.; List, J.; Battacharya, S.; Howe, G.; Xu, L.; Thompson, D.; Mayer, M. Complete aggregation pathway of amyloid β (1-40) and (1-42) resolved on an atomically clean interface. *Sci. Adv.* **2020**, *6*, eaaz6014. [CrossRef]
77. Gupta, S.; Dasmahapatra, A.K. Destabilization potential of phenolics on Aβ fibrils: Mechanistic insights from molecular dynamics simulation. *Phys. Chem. Chem. Phys.* **2020**, *22*, 19643–19658. [CrossRef]
78. Poma, A.B.; Thu, T.T.M.; Tri, L.T.M.; Nguyen, H.L.; Li, M.S. Nanomechanical Stability of Aβ Tetramers and Fibril-like Structures: Molecular Dynamics Simulations. *J. Phys. Chem. B* **2021**, *125*, 7628–7637. [CrossRef]
79. Jahan, I.; Nayeem, S.M. Destabilization of Alzheimer's Aβ42 protofibrils with acyclovir, carmustine, curcumin, and tetracycline: Insights from molecular dynamics simulations. *New J. Chem.* **2021**, *45*, 21031–21048. [CrossRef]
80. Okumura, H.; Itoh, S.G. Amyloid Fibril Disruption by Ultrasonic Cavitation: Nonequilibrium Molecular Dynamics Simulations. *J. Am. Chem. Soc.* **2014**, *136*, 10549–10552. [CrossRef]
81. Viet, M.; Derreumaux, P.; Nguyen, P.H. Nonequilibrium all-atom molecular dynamics simulation of the bubble cavitation and application to dissociate amyloid fibrils. *J. Chem. Phys.* **2016**, *145*, 174113. [CrossRef]
82. Viet, M.; Derreumaux, P.; Li, M.S.; Roland, C.; Sagui, C.; Nguyen, P.H. Picosecond dissociation of amyloid fibrils with infrared laser: A nonequilibrium simulation study. *J. Chem. Phys.* **2015**, *143*, 155101. [CrossRef] [PubMed]
83. Brogi, S.; Sirous, H.; Calderone, V.; Chemi, G. Amyloid β fibril disruption by oleuropein aglycone: Long-time molecular dynamics simulation to gain insight into the mechanism of action of this polyphenol from extra virgin olive oil. *Food Funct.* **2020**, *11*, 8122–8132. [CrossRef] [PubMed]
84. Okumura, H.; Itoh, S.G.; Nakamura, K.; Kawasaki, T. Role of Water Molecules and Helix Structure Stabilization in the Laser-Induced Disruption of Amyloid Fibrils Observed by Nonequilibrium Molecular Dynamics Simulations. *J. Phys. Chem. B* **2021**, *125*, 4964–4976. [CrossRef] [PubMed]
85. Nasica-Labouze, J.; Nguyen, P.H.; Sterpone, F.; Berthoumieu, O.; Buchete, N.-V.; Coté, S.; De Simone, A.; Doig, A.J.; Faller, P.; Garcia, A.; et al. Amyloid β Protein and Alzheimer's Disease: When Computer Simulations Complement Experimental Studies. *Chem. Rev.* **2015**, *115*, 3518–3563. [CrossRef] [PubMed]
86. Ilie, I.M.; Caflisch, A. Simulation Studies of Amyloidogenic Polypeptides and Their Aggregates. *Chem. Rev.* **2019**, *119*, 6956–6993. [CrossRef] [PubMed]
87. Nguyen, P.H.; Derreumaux, P. Structures of the intrinsically disordered Aβ, tau and α-synuclein proteins in aqueous solution from computer simulations. *Biophys. Chem.* **2020**, *264*, 106421. [CrossRef]
88. Strodel, B. Amyloid aggregation simulations: Challenges, advances and perspectives. *Curr. Opin. Struct. Biol.* **2020**, *67*, 145–152. [CrossRef]
89. Jarrett, J.T.; Berger, E.P.; Lansbury, P.T. The carboxy terminus of the β amyloid protein is critical for the seeding of amyloid formation: Implications for the pathogenesis of Alzheimer's disease. *Biochemistry* **1993**, *32*, 4693–4697. [CrossRef]
90. Hilbich, C.; Kisterswoike, B.; Reed, J.; Masters, C.L.; Beyreuther, K. Aggregation and secondary structure of synthetic amyloid βA4 peptides of Alzheimer's disease. *J. Mol. Biol.* **1991**, *218*, 149–163. [CrossRef]
91. Barrow, C.J.; Yasuda, A.; Kenny, P.T.M.; Zagorski, M.G. Solution conformations and aggregational properties of synthetic amyloid β-peptides of Alzheimer's disease: Analysis of circular dichroism spectra. *J. Mol. Biol.* **1992**, *225*, 1075–1093. [CrossRef]
92. Serpell, L.C. Alzheimer's amyloid fibrils: Structure and assembly. *Biochim. Biophys. Acta* **2000**, *1502*, 16–30. [CrossRef]
93. Balbach, J.J.; Ishii, Y.; Antzutkin, O.N.; Leapman, R.D.; Rizzo, N.W.; Dyda, F.; Reed, J.; Tycko, R. Amyloid Fibril Formation by Aβ16-22, a Seven-Residue Fragment of the Alzheimer's β-Amyloid Peptide, and Structural Characterization by Solid State NMR. *Biochemistry* **2000**, *39*, 13748–13759. [CrossRef] [PubMed]

94. Haass, C.; Selkoe, D.J. Soluble protein oligomers in neurodegeneration: Lessons from the Alzheimer's amyloid β-peptide. *Nat. Rev. Mol. Cell Biol.* **2007**, *8*, 101–112. [CrossRef] [PubMed]
95. Shankar, G.M.; Li, S.M.; Mehta, T.H.; Garcia-Munoz, A.; Shepardson, N.E.; Smith, I.; Brett, F.M.; Farrell, M.A.; Rowan, M.J.; Lemere, C.A.; et al. Amyloid-β protein dimers isolated directly from Alzheimer's brains impair synaptic plasticity and memory. *Nat. Med.* **2008**, *14*, 837–842. [CrossRef]
96. Itoh, S.G.; Okamoto, Y. Amyloid-β(29−42) Dimer Formations Studied by a Multicanonical−Multioverlap Molecular Dynamics Simulation. *J. Phys. Chem. B* **2008**, *112*, 2767–2770. [CrossRef]
97. Itoh, S.G.; Okumura, H. Replica-Permutation Method with the Suwa–Todo Algorithm beyond the Replica-Exchange Method. *J. Chem. Theory Comput.* **2012**, *9*, 570–581. [CrossRef]
98. Mitsutake, A.; Sugita, Y.; Okamoto, Y. Generalized-ensemble algorithms for molecular simulations of biopolymers. *Biopolymers* **2001**, *60*, 96–123. [CrossRef]
99. Itoh, S.G.; Okumura, H.; Okamoto, Y. Generalized-ensemble algorithms for molecular dynamics simulations. *Mol. Simul.* **2007**, *33*, 47–56. [CrossRef]
100. Yamauchi, M.; Mori, Y.; Okumura, H. Molecular simulations by generalized-ensemble algorithms in isothermal–isobaric ensemble. *Biophys. Rev.* **2019**, *11*, 457–469. [CrossRef]
101. Itoh, S.G.; Okumura, H. All-Atom Molecular Dynamics Simulation Methods for the Aggregation of Protein and Peptides: Replica Exchange/Permutation and Nonequilibrium Simulations. In *Computer Simulations of Aggregation of Proteins and Peptides*; Li, M.S., Kloczkowski, A., Cieplak, M., Kouza, M., Eds.; Humana: New York, NY, USA, 2022; pp. 197–220.
102. Hukushima, K.; Nemoto, K. Exchange Monte Carlo Method and Application to Spin Glass Simulations. *J. Phys. Soc. Jpn.* **1996**, *65*, 1604–1608. [CrossRef]
103. Sugita, Y.; Okamoto, Y. Replica-exchange molecular dynamics method for protein folding. *Chem. Phys. Lett.* **1999**, *314*, 141–151. [CrossRef]
104. Suwa, H.; Todo, S. Markov Chain Monte Carlo Method without Detailed Balance. *Phys. Rev. Lett.* **2010**, *105*, 120603. [CrossRef] [PubMed]
105. Metropolis, N.; Rosenbluth, A.W.; Rosenbluth, M.N.; Teller, A.H.; Teller, E. Equation of State Calculations by Fast Computing Machines. *J. Chem. Phys.* **1953**, *21*, 1087–1092. [CrossRef]
106. Yamauchi, M.; Okumura, H. Replica sub-permutation method for molecular dynamics and Monte Carlo simulations. *J. Comput. Chem.* **2019**, *40*, 2694–2711. [CrossRef] [PubMed]
107. Yamauchi, M.; Okumura, H. Development of isothermal-isobaric replica-permutation method for molecular dynamics and Monte Carlo simulations and its application to reveal temperature and pressure dependence of folded, misfolded, and unfolded states of chignolin. *J. Chem. Phys.* **2017**, *147*, 184107. [CrossRef] [PubMed]
108. Nishizawa, H.; Okumura, H. Comparison of Replica-Permutation Molecular Dynamics Simulations with and without Detailed Balance Condition. *J. Phys. Soc. Jpn.* **2015**, *84*, 74801. [CrossRef]
109. Mori, Y.; Okumura, H. Simulated tempering based on global balance or detailed balance conditions: Suwa-Todo, heat bath, and Metropolis algorithms. *J. Comput. Chem.* **2015**, *36*, 2344–2349. [CrossRef]
110. Fukuhara, D.; Itoh, S.G.; Okumura, H. Replica permutation with solute tempering for molecular dynamics simulation and its application to the dimerization of amyloid-β fragments. *J. Chem. Phys.* **2022**, *156*, 84109. [CrossRef]
111. Hornak, V.; Abel, R.; Okur, A.; Strockbine, B.; Roitberg, A.; Simmerling, C. Comparison of multiple Amber force fields and development of improved protein backbone parameters. *Proteins Struct. Funct. Bioinform.* **2006**, *65*, 712–725. [CrossRef]
112. Jorgensen, W.L.; Chandrasekhar, J.; Madura, J.D.; Impey, R.W.; Klein, M.L. Comparison of simple potential functions for simulating liquid water. *J. Chem. Phys.* **1983**, *79*, 926–935. [CrossRef]
113. Nosé, S. A unified formulation of the constant temperature molecular dynamics methods. *J. Chem. Phys.* **1984**, *81*, 511–519. [CrossRef]
114. Nosé, S. A molecular dynamics method for simulations in the canonical ensemble. *Mol. Phys.* **1984**, *52*, 255–268. [CrossRef]
115. Hoover, W.G. Canonical dynamics: Equilibrium phase-space distributions. *Phys. Rev. A* **1985**, *31*, 1695–1697. [CrossRef] [PubMed]
116. Abelein, A.; Abrahams, J.P.; Danielsson, J.; Gräslund, A.; Jarvet, J.; Luo, J.; Tiiman, A.; Wärmländer, S.K.T.S. The hairpin conformation of the amyloid β peptide is an important structural motif along the aggregation pathway. *JBIC J. Biol. Inorg. Chem.* **2014**, *19*, 623–634. [CrossRef]
117. Maity, S.; Hashemi, M.; Lyubchenko, Y.L. Nano-assembly of amyloid β peptide: Role of the hairpin fold. *Sci. Rep.* **2017**, *7*, 2344. [CrossRef]
118. Morinaga, A.; Hasegawa, K.; Nomura, R.; Ookoshi, T.; Ozawa, D.; Goto, Y.; Yamada, M.; Naiki, H. Critical role of interfaces and agitation on the nucleation of Aβ amyloid fibrils at low concentrations of Aβ monomers. *Biochim. Biophys. Acta* **2010**, *1804*, 986–995. [CrossRef]
119. Jean, L.; Lee, C.F.; Vaux, D.J. Enrichment of Amyloidogenesis at an Air-Water Interface. *Biophys. J.* **2012**, *102*, 1154–1162. [CrossRef]
120. Yagi-Utsumi, M.; Kato, K.; Nishimura, K. Membrane-Induced Dichotomous Conformation of Amyloid β with the Disordered N-Terminal Segment Followed by the Stable C-Terminal β Structure. *PLoS ONE* **2016**, *11*, e0146405. [CrossRef]
121. Fantini, J.; Yahi, N. Molecular insights into amyloid regulation by membrane cholesterol and sphingolipids: Common mechanisms in neurodegenerative diseases. *Expert Rev. Mol. Med.* **2010**, *12*, e27. [CrossRef]

122. Yagi-Utsumi, M.; Yamaguchi, Y.; Sasakawa, H.; Yamamoto, N.; Yanagisawa, K.; Kato, K. Up-and-down topological mode of amyloid β-peptide lying on hydrophilic/hydrophobic interface of ganglioside clusters. *Glycoconj. J.* **2008**, *26*, 999–1006. [CrossRef]
123. Yagi-Utsumi, M.; Matsuo, K.; Yanagisawa, K.; Gekko, K.; Kato, K. Spectroscopic Characterization of Intermolecular Interaction of Amyloid β Promoted on GM1 Micelles. *Int. J. Alzheimers Dis.* **2010**, *2011*, 925073. [PubMed]
124. Lemkul, J.A.; Bevan, D.R. Perturbation of membranes by the amyloid β-peptide—A molecular dynamics study. *FEBS J.* **2009**, *276*, 3060–3075. [CrossRef] [PubMed]
125. Na Zhao, L.; Chiu, S.-W.; Benoit, J.; Chew, L.Y.; Mu, Y. Amyloid β Peptides Aggregation in a Mixed Membrane Bilayer: A Molecular Dynamics Study. *J. Phys. Chem. B* **2011**, *115*, 12247–12256. [CrossRef] [PubMed]
126. Ikeda, K.; Yamaguchi, T.; Fukunaga, S.; Hoshino, M.; Matsuzaki, K. Mechanism of amyloid beta-protein aggregation mediated by GM1 ganglioside clusters. *Biochemistry* **2011**, *50*, 6433–6440. [CrossRef] [PubMed]
127. Poojari, C.; Kukol, A.; Strodel, B. How the amyloid-β peptide and membranes affect each other: An extensive simulation study. *Biochim. Biophys. Acta-Biomembr.* **2013**, *1828*, 327–339. [CrossRef] [PubMed]
128. Hoshino, T.; Mahmood, I.; Mori, K.; Matsuzaki, K. Binding and Aggregation Mechanism of Amyloid β-Peptides onto the GM1 Ganglioside-Containing Lipid Membrane. *J. Phys. Chem. B* **2013**, *117*, 8085–8094. [CrossRef]
129. Brown, A.M.; Bevan, D.R. Molecular Dynamics Simulations of Amyloid β-Peptide (1-42): Tetramer Formation and Membrane Interactions. *Biophys. J.* **2016**, *111*, 937–949. [CrossRef]
130. Khondker, A.; Alsop, R.J.; Rheinstädter, M.C. Membrane-Accelerated Amyloid-β Aggregation and Formation of Cross-β Sheets. *Membranes* **2017**, *7*, 49. [CrossRef]
131. Niu, Z.; Zhang, Z.; Zhao, W.; Yang, J. Interactions between amyloid β peptide and lipid membranes. *Biochim. Biophys. Acta-Biomembr.* **2018**, *1860*, 1663–1669. [CrossRef]
132. Tachi, Y.; Okamoto, Y.; Okumura, H. Conformational properties of an artificial GM1 glycan cluster based on a metal-ligand complex. *J. Chem. Phys.* **2018**, *149*, 135101. [CrossRef]
133. Sato, S.; Yoshimasa, Y.; Fujita, D.; Yagi-Utsumi, M.; Yamaguchi, T.; Kato, K.; Fujita, M. A Self-Assembled Spherical Com-plex Displaying a Gangliosidic Glycan Cluster Capable of Interacting with Amyloidogenic Proteins. *Angew. Chem. Int. Ed.* **2015**, *54*, 8435–8439. [CrossRef] [PubMed]
134. Ono, K.; Hasegawa, K.; Naiki, H.; Yamada, M. Curcumin has potent anti-amyloidogenic effects for Alzheimer's β-amyloid fibrils in vitro. *J. Neurosci. Res.* **2004**, *75*, 742–750. [CrossRef] [PubMed]
135. Ono, K.; Li, L.; Takamura, Y.; Yoshiike, Y.; Zhu, L.; Han, F.; Mao, X.; Ikeda, T.; Takasaki, J.-I.; Nishijo, H.; et al. Phenolic Compounds Prevent Amyloid β-Protein Oligomerization and Synaptic Dysfunction by Site-specific Binding. *J. Biol. Chem.* **2012**, *287*, 14631–14643. [CrossRef] [PubMed]
136. Klimov, D.K.; Straub, J.E.; Thirumalai, D. Aqueous urea solution destabilizes Aβ(16-22) oligomers. *Proc. Natl. Acad. Sci. USA* **2004**, *101*, 14760–14765. [CrossRef] [PubMed]
137. Nguyen, P.H.; Li, M.S.; Derreumaux, P. Effects of all-atom force fields on amyloid oligomerization: Replica exchange molecular dynamics simulations of the Aβ(16-22) dimer and trimer. *Phys. Chem. Chem. Phys.* **2011**, *13*, 9778–9788. [CrossRef] [PubMed]
138. Riccardi, L.; Nguyen, P.H.; Stock, G. Construction of the Free Energy Landscape of Peptide Aggregation from Molecular Dynamics Simulations. *J. Chem. Theory Comput.* **2012**, *8*, 1471–1479. [CrossRef]
139. Nguyen, P.H.; Okamoto, Y.; Derreumaux, P. Communication: Simulated tempering with fast on-the-fly weight determination. *J. Chem. Phys.* **2013**, *138*, 61102. [CrossRef]
140. Barz, B.; Wales, D.J.; Strodel, B. A Kinetic Approach to the Sequence–Aggregation Relationship in Disease-Related Protein Assembly. *J. Phys. Chem. B* **2014**, *118*, 1003–1011. [CrossRef]
141. Maier, J.A.; Martinez, C.; Kasavajhala, K.; Wickstrom, L.; Hauser, K.E.; Simmerling, C. ff14SB: Improving the accuracy of protein side chain and backbone parameters from ff99SB. *J. Chem. Theory Comput.* **2015**, *11*, 3696–3713. [CrossRef]
142. Wang, J.; Wolf, R.M.; Caldwell, J.W.; Kollman, P.A.; Case, D.A. Development and testing of a general amber force field. *J. Comput. Chem.* **2004**, *25*, 1157–1174. [CrossRef]
143. Okumura, H.; Okamoto, Y. Multibaric–Multithermal Molecular Dynamics Simulation of Alanine Dipeptide in Explicit Water. *Bull. Chem. Soc. Jpn.* **2007**, *80*, 1114–1123. [CrossRef]
144. Okumura, H.; Itoh, S.G.; Okamoto, Y. Explicit symplectic integrators of molecular dynamics algorithms for rigid-body molecules in the canonical, isobaric-isothermal, and related ensembles. *J. Chem. Phys.* **2007**, *126*, 84103. [CrossRef] [PubMed]
145. Okumura, H.; Okamoto, Y. Temperature and Pressure Dependence of Alanine Dipeptide Studied by Multibaric-Multithermal Molecular Dynamics Simulations. *J. Phys. Chem. B* **2008**, *112*, 12038–12049. [CrossRef]
146. Okumura, H. Partial multicanonical algorithm for molecular dynamics and Monte Carlo simulations. *J. Chem. Phys.* **2008**, *129*, 124116. [CrossRef] [PubMed]
147. Okumura, H. Optimization of partial multicanonical molecular dynamics simulations applied to an alanine dipeptide in explicit water solvent. *Phys. Chem. Chem. Phys.* **2011**, *13*, 114–126. [CrossRef] [PubMed]
148. Okumura, H. Temperature and pressure denaturation of chignolin: Folding and unfolding simulation by multibaric-multithermal molecular dynamics method. *Proteins Struct. Funct. Bioinf.* **2012**, *80*, 2397–2416. [CrossRef] [PubMed]
149. Okumura, H.; Itoh, S.G. Transformation of a design peptide between the α-helix and β-hairpin structures using a helix-strand replica-exchange molecular dynamics simulation. *Phys. Chem. Chem. Phys.* **2013**, *15*, 13852–13861. [CrossRef] [PubMed]

150. Mizukami, T.; Furuzawa, S.; Itoh, S.G.; Segawa, S.; Ikura, T.; Ihara, K.; Okumura, H.; Roder, H.; Maki, K. Energetics and kinetics of substrate analog-coupled staphylococcal nuclease folding revealed by a statistical mechanical approach. *Proc. Natl. Acad. Sci. USA* **2020**, *117*, 19953–19962. [CrossRef]
151. Nguyen, T.H.D.; Itoh, S.G.; Okumura, H.; Tominaga, M. Structural basis for promiscuous action of monoterpenes on TRP channels. *Commun. Biol.* **2021**, *4*, 293. [CrossRef]
152. Tanimoto, S.; Itoh, S.G.; Okumura, H. "Bucket brigade" using lysine residues in RNA-dependent RNA polymerase of SARS-CoV-2. *Biophys. J.* **2021**, *120*, 3615–3627. [CrossRef]
153. Miyazawa, K.; Itoh, S.G.; Watanabe, H.; Uchihashi, T.; Yanaka, S.; Yagi-Utsumi, M.; Kato, K.; Arakawa, K.; Okumura, H. Tardigrade Secretory-Abundant Heat-Soluble Protein Has a Flexible β-Barrel Structure in Solution and Keeps This Structure in Dehydration. *J. Phys. Chem. B* **2021**, *125*, 9145–9154. [CrossRef] [PubMed]
154. Itoh, S.G.; Tanimoto, S.; Okumura, H. Dynamic properties of SARS-CoV and SARS-CoV-2 RNA-dependent RNA polymerases studied by molecular dynamics simulations. *Chem. Phys. Lett.* **2021**, *778*, 138819. [CrossRef] [PubMed]
155. Miyazawa, K.; Itoh, S.G.; Yoshida, Y.; Arakawa, K.; Okumura, H. Tardigrade Secretory-Abundant Heat-Soluble Protein Varies Entrance Propensity Depending on the Amino-Acid Sequence. *J. Phys. Chem. B* **2022**, *126*, 2361–2368. [CrossRef] [PubMed]
156. Okumura, H.; Itoh, S.G.; Okamoto, Y. Generalized-Ensemble Algorithms for Simulations of Complex Molecular Systems. In *Practical Aspects of Computational Chemistry II: An Overview of the Last Two Decades and Current Trends*; Leszczynski, J., Shukla, M.K., Eds.; Springer: Dordrecht, The Netherlands, 2012; pp. 69–101.
157. Yamauchi, M.; La Penna, G.; Itoh, S.G.; Okumura, H. Implementations of replica-permutation and replica sub-permutation methods into LAMMPS. *Comput. Phys. Commun.* **2022**, *276*, 108362. [CrossRef]
158. Berg, B.A.; Neuhaus, T. Multicanonical algorithms for 1st order phase-transitions. *Phys. Lett. B* **1991**, *267*, 249–253. [CrossRef]
159. Berg, B.A.; Neuhaus, T. Multicanonical ensemble: A new approach to simulate first-order phase transitions. *Phys. Rev. Lett.* **1992**, *68*, 9–12. [CrossRef]
160. Hansmann, U.H.E.; Okamoto, Y.; Eisenmenger, F. Molecular dynamics, Langevin and hybrid Monte Carlo simulations in a multicanonical ensemble. *Chem. Phys. Lett.* **1996**, *259*, 321–330. [CrossRef]
161. Nakajima, N.; Nakamura, H.; Kidera, A. Multicanonical ensemble generated by molecular dynamics simulation for enhanced conformational sampling of peptides. *J. Phys. Chem. B* **1997**, *101*, 817–824. [CrossRef]
162. Okumura, H.; Okamoto, Y. Monte Carlo simulations in multibaric–multithermal ensemble. *Chem. Phys. Lett.* **2004**, *383*, 391–396. [CrossRef]
163. Okumura, H.; Okamoto, Y. Monte Carlo simulations in generalized isobaric-isothermal ensembles. *Phys. Rev. E* **2004**, *70*, 026702. [CrossRef]
164. Okumura, H.; Okamoto, Y. Molecular dynamics simulations in the multibaric–multithermal ensemble. *Chem. Phys. Lett.* **2004**, *391*, 248–253. [CrossRef]
165. Okumura, H.; Okamoto, Y. Multibaric-multithermal ensemble molecular dynamics simulations. *J. Comput. Chem.* **2005**, *27*, 379–395. [CrossRef] [PubMed]
166. Sunde, M.; Serpell, L.; Bartlam, M.; Fraser, P.E.; Pepys, M.B.; Blake, C.C. Common core structure of amyloid fibrils by synchrotron X-ray diffraction. *J. Mol. Biol.* **1997**, *273*, 729–739. [CrossRef] [PubMed]
167. Petkova, A.T.; Yau, W.M.; Tycko, R. Experimental constraints on quaternary structure in Alzheimer's β-amyloid fibrils. *Biochemistry* **2006**, *45*, 498–512. [CrossRef]
168. Kittel, C. Surface and interface physics. In *Introduction to Solid State Physics*; Wiley: New York, NY, USA, 2004; pp. 487–514.
169. Buch, V.; Milet, A.; Vácha, R.; Jungwirth, P.; Devlin, J.P. Water surface is acidic. *Proc. Natl. Acad. Sci. USA* **2007**, *104*, 7342–7347. [CrossRef]
170. Beattie, J.K.; Djerdjev, A.M.; Warr, G.G. The surface of neat water is basic. *Faraday Discuss.* **2008**, *141*, 31–39. [CrossRef]
171. *The PyMOL Molecular Graphics System*; Schrödinger, LLC.: New York, NY, USA, 2015.
172. Essmann, U.; Perera, L.; Berkowitz, M.L.; Darden, T.; Lee, H.; Pedersen, L.G. A smooth particle mesh Ewald method. *J. Chem. Phys.* **1995**, *103*, 8577–8593. [CrossRef]
173. Andersen, H.C. Molecular dynamics simulations at constant pressure and/or temperature. *J. Chem. Phys.* **1980**, *72*, 2384–2393. [CrossRef]
174. Ban, T.; Hamada, D.; Hasegawa, K.; Naiki, H.; Goto, Y. Direct Observation of Amyloid Fibril Growth Monitored by Thioflavin T Fluorescence. *J. Biol. Chem.* **2003**, *278*, 16462–16465. [CrossRef]
175. Ban, T.; Hoshino, M.; Takahashi, S.; Hamada, D.; Hasegawa, K.; Naiki, H.; Goto, Y. Direct Observation of Aβ Amyloid Fibril Growth and Inhibition. *J. Mol. Biol.* **2004**, *344*, 757–767. [CrossRef]
176. Uchihashi, T.; Konno, H. Self-Assembly Dynamics of Biological Molecules Observed by High-Speed Atomic Force Microscopy. In Proceedings of the 96th Annual Meeting of the Chemical Society of Japan, No. 1S5-13. Kyotanabe, Japan, 24–27 March 2016.
177. Kinjo, T.; Ohguchi, K.; Yasuoka, K.; Matsumoto, M. Computer simulation of fluid phase change: Vapor nucleation and bubble formation dynamics. *Comput. Mater. Sci.* **1999**, *14*, 138–141. [CrossRef]
178. Xiao, C.; Heyes, D.M.; Powles, J.G. The collapsing bubble in a liquid by molecular dynamics simulations. *Mol. Phys.* **2002**, *100*, 3451–3468. [CrossRef]
179. Okumura, H.; Ito, N. Nonequilibrium molecular dynamics simulations of a bubble. *Phys. Rev. E* **2003**, *67*, 45301. [CrossRef] [PubMed]

180. Watanabe, H.; Inaoka, H.; Ito, N. Ripening kinetics of bubbles: A molecular dynamics study. *J. Chem. Phys.* **2016**, *145*, 124707. [CrossRef]
181. Yoshida, H. Construction of higher order symplectic integrators. *Phys. Lett. A* **1990**, *150*, 262–268. [CrossRef]
182. Miller, T.F.; Eleftheriou, M.; Pattnaik, P.; Ndirango, A.; Newns, D.; Martyna, G.J. Symplectic quaternion scheme for biophysical molecular dynamics. *J. Chem. Phys.* **2002**, *116*, 8649–8659. [CrossRef]
183. Chatani, E.; Lee, Y.-H.; Yagi, H.; Yoshimura, Y.; Naiki, H.; Goto, Y. Ultrasonication-dependent production and breakdown lead to minimum-sized amyloid fibrils. *Proc. Natl. Acad. Sci. USA* **2009**, *106*, 11119–11124. [CrossRef]
184. Kawasaki, T.; Ohori, G.; Chiba, T.; Tsukiyama, K.; Nakamura, K. Picosecond pulsed infrared laser tuned to amide I band dissociates polyglutamine fibrils in cells. *Lasers Med. Sci.* **2016**, *31*, 1425–1431. [CrossRef]
185. Kawasaki, T.; Yaji, T.; Ohta, T.; Tsukiyama, K.; Nakamura, K. Dissociation of β-sheet Stacking of Amyloid β Fibrils by Irradiation of Intense, Short-Pulsed Mid-infrared Laser. *Cell. Mol. Neurobiol.* **2018**, *38*, 1039–1049. [CrossRef]
186. Kawasaki, T.; Tsukiyama, K.; Irizawa, A. Dissolution of a fibrous peptide by terahertz free electron laser. *Sci. Rep.* **2019**, *9*, 10636. [CrossRef]
187. Mori, Y.; Okumura, H. Pressure-Induced Helical Structure of a Peptide Studied by Simulated Tempering Molecular Dynamics Simulations. *J. Phys. Chem. Lett.* **2013**, *4*, 2079–2083. [CrossRef]
188. Mori, Y.; Okumura, H. Molecular dynamics of the structural changes of helical peptides induced by pressure. *Proteins Struct. Funct. Bioinform.* **2014**, *82*, 2970–2981. [CrossRef] [PubMed]
189. Mori, Y.; Okumura, H. Molecular dynamics simulation study on the high-pressure behaviour of an AK16 peptide. *Mol. Simul.* **2015**, *41*, 1035–1040. [CrossRef]
190. Kabsch, W.; Sander, C. Dictionary of protein secondary structure: Pattern recognition of hydrogen-bonded and geometrical features. *Biopolymers* **1983**, *22*, 2577–2637. [CrossRef] [PubMed]
191. Huang, J.; Rauscher, S.; Nawrocki, G.; Ran, T.; Feig, M.; de Groot, B.L.; Grubmüller, H.; MacKerell, A.D. CHARMM36m: An improved force field for folded and intrinsically disordered proteins. *Nat. Methods* **2017**, *14*, 71–73. [CrossRef]
192. Schmid, N.; Eichenberger, A.P.; Choutko, A.; Riniker, S.; Winger, M.; Mark, A.; Van Gunsteren, W.F. Definition and testing of the GROMOS force-field versions 54A7 and 54B7. *Eur. Biophys. J.* **2011**, *40*, 843–856. [CrossRef]
193. Watts, C.R.; Gregory, A.; Frisbie, C.; Lovas, S. Effects of force fields on the conformational and dynamic properties of amyloid β(1-40) dimer explored by replica exchange molecular dynamics simulations. *Proteins Struct. Funct. Bioinform.* **2018**, *86*, 279–300. [CrossRef]
194. Man, V.H.; He, X.; Gao, J.; Wang, J. Effects of All-Atom Molecular Mechanics Force Fields on Amyloid Peptide Assembly: The Case of PHF6 Peptide of Tau Protein. *J. Chem. Theory Comput.* **2021**, *17*, 6458–6471. [CrossRef]
195. Caliskan, M.; Mandaci, S.Y.; Uversky, V.N.; Coskuner-Weber, O. Secondary structure dependence of amyloid-β(1–40) on simulation techniques and force field parameters. *Chem. Biol. Drug Des.* **2021**, *97*, 1100–1108. [CrossRef]
196. Still, W.C.; Tempczyk, A.; Hawley, R.C.; Hendrickson, T. Semianalytical treatment of solvation for molecular mechanics and dynamics. *J. Am. Chem. Soc.* **1990**, *112*, 6127–6129. [CrossRef]
197. Dominy, B.N.; Brooks, C.L. Development of a Generalized Born Model Parametrization for Proteins and Nucleic Acids. *J. Phys. Chem. B* **1999**, *103*, 3765–3773. [CrossRef]
198. Feig, M.; Im, W.; Brooks, C.L. Implicit solvation based on generalized Born theory in different dielectric environments. *J. Chem. Phys.* **2004**, *120*, 903–911. [CrossRef] [PubMed]
199. Davtyan, A.; Schafer, N.P.; Zheng, W.; Clementi, C.; Wolynes, P.G.; Papoian, G.A. AWSEM-MD: Protein Structure Prediction Using Coarse-Grained Physical Potentials and Bioinformatically Based Local Structure Biasing. *J. Phys. Chem. B* **2012**, *116*, 8494–8503. [CrossRef]
200. Marrink, S.J.; Risselada, H.J.; Yefimov, S.; Tieleman, D.P.; de Vries, A.H. The MARTINI Force Field: Coarse Grained Model for Biomolecular Simulations. *J. Phys. Chem. B* **2007**, *111*, 7812–7824. [CrossRef]
201. Monticelli, L.; Kandasamy, S.K.; Periole, X.; Larson, R.G.; Tieleman, D.P.; Marrink, S. The MARTINI Coarse-Grained Force Field: Extension to Proteins. *J. Chem. Theory Comput.* **2008**, *4*, 819–834. [CrossRef] [PubMed]
202. Liwo, A.; Oldziej, S.; Pincus, M.R.; Wawak, R.J.; Rackovsky, S.; Scheraga, H.A. A united-residue force field for off-lattice protein-structure simulations. I. Functional forms and parameters of long-range side-chain interaction potentials from protein crystal data. *J. Comput. Chem.* **1997**, *18*, 849–873. [CrossRef]
203. Liwo, A.; Pincus, M.R.; Wawak, R.J.; Rackovsky, S.; Ołdziej, S.; Scheraga, H.A. A united-residue force field for off-lattice protein-structure simulations. II. Parameterization of short-range interactions and determination of weights of energy terms by Z-score optimization. *J. Comput. Chem.* **1997**, *18*, 874–887. [CrossRef]
204. Co, N.T.; Li, M.S.; Krupa, P. Computational Models for the Study of Protein Aggregation. In *Computer Simulations of Aggregation of Proteins and Peptides*; Li, M.S., Kloczkowski, A., Cieplak, M., Kouza, M., Eds.; Springer: New York, NY, USA, 2022; pp. 51–78.
205. Rojas, A.V.; Maisuradze, G.G.; Scheraga, H.A.; Liwo, A. Probing Protein Aggregation Using the Coarse-Grained UNRES Force Field. In *Computer Simulations of Aggregation of Proteins and Peptides*; Li, M.S., Kloczkowski, A., Cieplak, M., Kouza, M., Eds.; Springer: New York, NY, USA, 2022; pp. 79–104.
206. Chong, S.-H.; Ham, S. Atomic-level investigations on the amyloid-β dimerization process and its driving forces in water. *Phys. Chem. Chem. Phys.* **2011**, *14*, 1573–1575. [CrossRef]

207. Masutani, K.; Yamamori, Y.; Kim, K.; Matubayasi, N. Free-energy analysis of the hydration and cosolvent effects on the β-sheet aggregation through all-atom molecular dynamics simulation. *J. Chem. Phys.* **2019**, *150*, 145101. [CrossRef]
208. Li, M.S.; Klimov, D.K.; Straub, J.E.; Thirumalai, D. Probing the mechanisms of fibril formation using lattice models. *J. Chem. Phys.* **2008**, *129*, 175101. [CrossRef]
209. Co, N.T.; Hu, C.-K.; Li, M.S. Dual effect of crowders on fibrillation kinetics of polypeptide chains revealed by lattice models. *J. Chem. Phys.* **2013**, *138*, 185101. [CrossRef] [PubMed]
210. Irbäck, A.; Jónsson, S.; Linnemann, N.; Linse, B.; Wallin, S. Aggregate Geometry in Amyloid Fibril Nucleation. *Phys. Rev. Lett.* **2013**, *110*, 58101. [CrossRef] [PubMed]
211. Abeln, S.; Vendruscolo, M.; Dobson, C.M.; Frenkel, D. A Simple Lattice Model That Captures Protein Folding, Aggregation and Amyloid Formation. *PLoS ONE* **2014**, *9*, e85185. [CrossRef] [PubMed]
212. Hynynen, K.; McDannold, N.; Vykhodtseva, N.; Jolesz, F.A. Noninvasive MR Imaging–guided Focal Opening of the Blood-Brain Barrier in Rabbits. *Radiology* **2001**, *220*, 640–646. [CrossRef]
213. Jordão, J.F.; Ayala-Grosso, C.A.; Markham, K.; Huang, Y.; Chopra, R.; McLaurin, J.; Hynynen, K.; Aubert, I. Antibodies Targeted to the Brain with Image-Guided Focused Ultrasound Reduces Amyloid-β Plaque Load in the TgCRND8 Mouse Model of Alzheimer's Disease. *PLoS ONE* **2010**, *5*, e10549. [CrossRef]
214. Burgess, A.; Dubey, S.; Yeung, S.; Hough, O.; Eterman, N.; Aubert, I.; Hynynen, K. Alzheimer Disease in a Mouse Model: MR Imaging–guided Focused Ultrasound Targeted to the Hippocampus Opens the Blood-Brain Barrier and Improves Pathologic Abnormalities and Behavior. *Radiology* **2014**, *273*, 736–745. [CrossRef]
215. Hsu, P.-H.; Lin, Y.-T.; Chung, Y.-H.; Lin, K.-J.; Yang, L.-Y.; Yen, T.-C.; Liu, H.-L. Focused Ultrasound-Induced Blood-Brain Barrier Opening Enhances GSK-3 Inhibitor Delivery for Amyloid-β Plaque Reduction. *Sci. Rep.* **2018**, *8*, 12882. [CrossRef]

molecules

MDPI

Review

The Wako-Saitô-Muñoz-Eaton Model for Predicting Protein Folding and Dynamics

Koji Ooka [1,2], Runjing Liu [3] and Munehito Arai [1,3,*]

[1] Department of Physics, Graduate School of Science, The University of Tokyo, 3-8-1 Komaba, Meguro, Tokyo 153-8902, Japan; ooka-koji063@g.ecc.u-tokyo.ac.jp
[2] Komaba Organization for Educational Excellence, College of Arts and Sciences, The University of Tokyo, 3-8-1 Komaba, Meguro, Tokyo 153-8902, Japan
[3] Department of Life Sciences, Graduate School of Arts and Sciences, The University of Tokyo, 3-8-1 Komaba, Meguro, Tokyo 153-8902, Japan; liurunjing@g.ecc.u-tokyo.ac.jp
* Correspondence: arai@bio.c.u-tokyo.ac.jp

Abstract: Despite the recent advances in the prediction of protein structures by deep neutral networks, the elucidation of protein-folding mechanisms remains challenging. A promising theory for describing protein folding is a coarse-grained statistical mechanical model called the Wako-Saitô-Muñoz-Eaton (WSME) model. The model can calculate the free-energy landscapes of proteins based on a three-dimensional structure with low computational complexity, thereby providing a comprehensive understanding of the folding pathways and the structure and stability of the intermediates and transition states involved in the folding reaction. In this review, we summarize previous and recent studies on protein folding and dynamics performed using the WSME model and discuss future challenges and prospects. The WSME model successfully predicted the folding mechanisms of small single-domain proteins and the effects of amino-acid substitutions on protein stability and folding in a manner that was consistent with experimental results. Furthermore, extended versions of the WSME model were applied to predict the folding mechanisms of multi-domain proteins and the conformational changes associated with protein function. Thus, the WSME model may contribute significantly to solving the protein-folding problem and is expected to be useful for predicting protein folding, stability, and dynamics in basic research and in industrial and medical applications.

Keywords: protein folding; statistical mechanical model; WSME model; folding kinetics; folding intermediates; protein dynamics

Citation: Ooka, K.; Liu, R.; Arai, M. The Wako-Saitô-Muñoz-Eaton Model for Predicting Protein Folding and Dynamics. *Molecules* **2022**, *27*, 4460. https://doi.org/10.3390/molecules27144460

Academic Editor: Michael Assfalg

Received: 20 June 2022
Accepted: 8 July 2022
Published: 12 July 2022

1. Introduction

Many proteins fold into specific three-dimensional (3D) structures to perform their functions and drive various biological processes. Therefore, elucidating how proteins fold is essential for understanding the fundamental processes of life. Since the existence of protein-folding pathways was first proposed [1], the detection and characterization of intermediates and transition states during folding reactions have been extensively performed using a variety of experimental techniques [2–5]. Theoretical studies of protein-folding reactions, including molecular-dynamics (MD) simulations of all-atom models and Monte Carlo simulations of coarse-grained models, have made significant progress in explaining experimental observations [6–12]. In particular, statistical mechanical approaches have shown that protein-folding processes can be comprehensively described by the free-energy landscapes of proteins [7–9,13–16].

The Wako–Saitô–Muñoz–Eaton (WSME) model is promising for describing protein-folding reactions [17]. The WSME model is a coarse-grained model of proteins based on a simple and elementary statistical mechanical theory and can readily calculate free-energy landscapes using the 3D native structures of proteins [13,18–20]. The free-energy landscapes obtained by the WSME model comprehensively predict both the thermodynamic stability of proteins under equilibrium conditions and their kinetic folding processes under non-equilibrium conditions, including folding pathways, folding-rate constants, and the structures of the intermediates and transition states. The predictions were consistent with the experimental results in many cases, especially for the folding of small single-domain proteins [13]. To date, many extensions and modifications have been implemented in the model to accommodate a variety of experimental conditions, contributing significantly to our understanding of protein-folding mechanisms. Furthermore, theoretical predictions by the WSME model play an important role in complementing MD simulations, resolving discrepancies between simulations and experiments, and bridging the gap between them. The WSME model has also been used to estimate the effects of amino-acid substitutions on proteins and to explain conformational changes accompanied by protein function, making it a promising protein-engineering tool for industrial and medical applications [21,22]. Because protein folding and dynamics have been extensively studied using the WSME model, it would be useful to summarize the previous progress and discuss the issues that remain to be resolved.

In this review, we first describe the details of the basic WSME model and outline how to calculate the free-energy landscapes of proteins in Section 2. The subsequent sections summarize the applications of the WSME model for predicting the folding processes of small single-domain proteins (Section 3) and of multi-domain proteins with complex folding mechanisms (Section 4). Section 5 presents extended versions of the WSME model for analyzing the protein dynamics observed in intrinsically disordered proteins, functional motions, and amyloid fibril formation. Finally, the future challenges and prospects of the WSME model are discussed in Section 6.

2. WSME Model

2.1. Description of the Model

The WSME model is a coarse-grained statistical-mechanical model based on the 3D structures of proteins. In 1978, Wako and Saitô originally proposed the basic ideas and calculation methods of this model (called the "island model") [18,19]. Approximately 20 years later, Muñoz and Eaton rediscovered it in 1999 [13]. Since then, this model has been termed the Wako–Saitô–Muñoz–Eaton (WSME) model. The WSME model is a Gō-type model that considers only the interactions formed in the native state of proteins without considering non-native interactions [23]. Gō proposed the consistency principle, which holds for ideal proteins and states that the most stable structure of a local fragment taken from a protein is consistent with the native structure of the full-length protein. In other words, the interactions that stabilize the local structure of a protein are consistent with the interactions that stabilize the overall structure of a protein [24]. Such ideal proteins can be virtually constructed by considering only the interactions that stabilize the native structure; this type of potential is called the Gō potential [25]. The WSME model uses the Gō potential, assuming the consistency principle, and describes the folding and dynamics of ideal proteins. The consistency principle is considered equivalent to the principle of minimal frustration, which states that frustration in energy arising from stabilization by non-native contacts is minimized in foldable proteins [7,16]. Therefore, the WSME model also assumes the principle of minimal frustration. Due to its simplicity, the WSME model can readily calculate the free-energy landscape of protein folding.

The basic WSME model is as follows (Figures 1 and 2). First, an Ising-like two-state variable m_k is assigned to each residue of a protein. The index k represents the residue number. m_k is 1 when the residue is in the native-like conformation and 0 when the residue is in other conformations. The protein state $\{m\}$ is defined as a set of residue states (m_1, m_2,

$\ldots, m_N$), where N is the total number of residues. Next, the protein state has 2^N possible conformations. The Hamiltonian of the WSME model is defined as:

$$H(\{m\}) = \sum_{i=1}^{N-1} \sum_{j=i+1}^{N} \varepsilon_{i,j} \Delta_{i,j} m_{i,j}, \tag{1}$$

where $\Delta_{i,j}$ represents the native contact between residues; $\Delta_{i,j} = 1$ when i- and j-th residues are in contact with each other in the native state, otherwise $\Delta_{i,j} = 0$ (Figure 1). $\varepsilon_{i,j}$ is the contact energy between i- and j-th residues in the native state and takes negative values when a stable interaction is formed in the native state (Figure 2). $m_{i,j}$ is defined as:

$$m_{i,j} = m_i m_{i+1} \cdots m_j = \prod_{k=i}^{j} m_k, \tag{2}$$

and $m_{i,j} = 1$ only when all residues between i- and j-th residues are in native-like conformations. This implies that native interactions between i- and j-th residues are established only when all intervening residues are cooperatively folded into their native conformations (Figure 1). Therefore, this model assumes that folding is initiated by local interactions between neighboring residues and spreads to distal regions via the growth and docking of native segments (Figure 1). The number of states W is defined as follows:

$$W(\{m\}) = \exp\left[\left(S_0 + \sum_{i=1}^{N} S_i m_i\right) / k_B\right], \tag{3}$$

where k_B is the Boltzmann constant, S_0 is the conformational entropy of the fully unfolded state, and S_i (<0) is the entropic reduction attributed to the formation of the native conformation. Then, the partition function is described as:

$$\begin{aligned} Z &= \sum_{\text{All states}} W(\{m\}) \exp\left[-\frac{H(\{m\})}{k_B T}\right] \\ &= \sum_{\text{All states}} \exp\left[-\frac{1}{k_B T}\left(\sum_{i=1}^{N-1} \sum_{j=i+1}^{N} \varepsilon_{i,j} \Delta_{i,j} m_{i,j} - T \sum_{i=1}^{N} S_i m_i\right)\right]. \end{aligned} \tag{4}$$

S_0 is neglected in Equation (4) because it is a constant and does not affect the results of the free-energy calculation. Thus, the effective free energy of a native stretch from i- to j-th residue can be described as:

$$F_{i,j} = \sum_{k=i}^{j-1} \sum_{l=k+1}^{j} \varepsilon_{k,l} - T \sum_{k=i}^{j} S_k, \tag{5}$$

where the first and second terms are enthalpy and entropy, respectively. Equation (5) shows that the progress of the folding reaction is enthalpically favorable due to the formation of native interactions, but is entropically unfavorable due to the reduction in the number of possible states. Such enthalpy–entropy compensation is directly reflected in the WSME model, and the balance between them results in free-energy barriers. The extent to which protein folding proceeds is often used as a reaction coordinate, such as the fraction of residues in the native state, $n = \sum_{i=1}^{N} m_i / N$, and the fraction of native contacts formed, $Q = \sum_{i<j} \Delta_{i,j} m_{i,j} / \sum_{i<j} \Delta_{i,j}$. The WSME model can calculate the free energy from the partition function restricted to the value of the reaction coordinate (Figure 2).

The WSME model was originally developed by Wako and Saitô in 1978 to study the statistical mechanical properties of ideal biopolymers, including proteins [18,19]. Subsequently, the model was applied to predict pathways and intermediates in the folding of several proteins by calculating the free-energy landscapes and residue-specific structure formation along reaction coordinates [27]. In the 1980s, the idea underlying the WSME model, that native contacts are formed in local segments (islands) and that native islands grow into entire proteins, was applied to two-dimensional (2D) and 3D lattice models of protein folding [23,24,28] and to a Potts-like model with three states (α-helix, β-strand, and coil) [29], showing that this idea is useful for describing the nature of protein-folding transitions. In the 1980s and 1990s, experimental data characterizing detailed protein-folding reactions increased, especially through the use of Φ-value analysis to investigate structure formation in the transition states [30–32], and through the use of the pulsed-hydrogen exchange nuclear magnetic resonance (NMR) technique to examine the structures of kinetic intermediates [33–36]. This prompted the need for theoretical models to explain the experimental results. In 1999, Muñoz and Eaton rediscovered the WSME model and succeeded in predicting the free-energy landscapes and folding rates for 18 small proteins that folded in a two-state manner; the findings were in good agreement with the experimental data [13]. These results indicated that the WSME model is promising for explaining the experimental results of protein folding. Furthermore, the results suggest that real small proteins can be approximated by ideal proteins that satisfy the consistency principle and principle of minimal frustration.

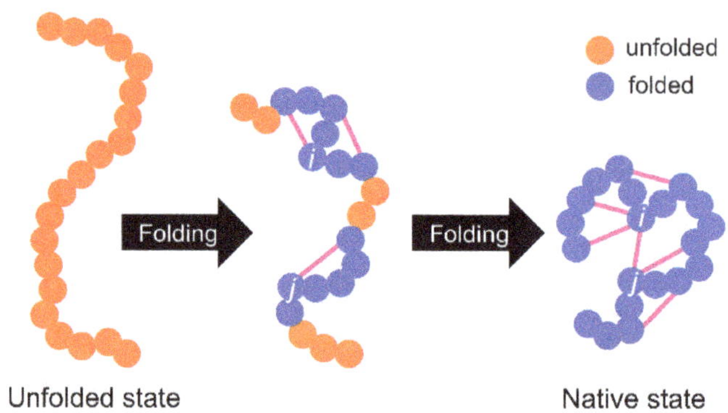

Figure 1. Schematic representation of a protein-folding process in the Wako–Saitô–Muñoz–Eaton (WSME) model. Residues in folded or unfolded conformations are indicated by blue and orange circles, respectively. Native contacts indicated by magenta lines are formed only when all intervening residues are cooperatively folded into the native-like conformations.

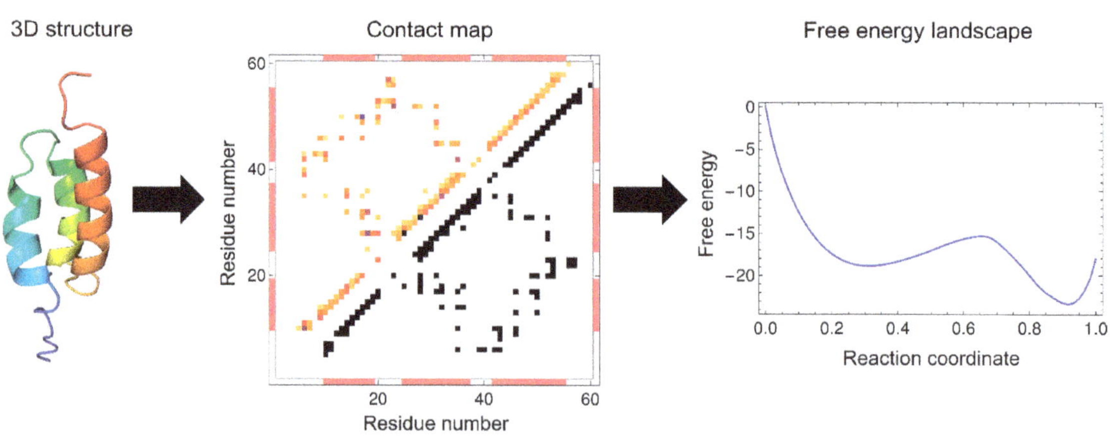

Figure 2. Basic procedure for calculating free energy with the WSME model. Based on the protein three-dimensional structure (left panel: the B domain of protein A (BdpA); PDB ID: 1BDD), a residue–residue contact map of the protein is calculated (middle panel). A pair of *i*- and *j*-th residues

($j > i + 2$) is defined as being in contact when at least one of the distances between the atoms in the i-th residue and those in the j-th residue is less than 4 Å in the native state. The triangle in the lower right half (with black squares) is a binary contact map with a uniform contact energy, and the triangle in the upper left half (with colored squares) is a non-binary contact map weighted by the number of atoms in the native contacts. The partition function is calculated from the Hamiltonian based on the contact map, and the free energy is obtained as a function of the reaction coordinate (right panel). The one-dimensional free-energy landscape calculated using the uniform contact energy is shown. Adapted with permission from Ref. [26]. Copyright (2006) National Academy of Sciences, USA.

2.2. Calculation of the Partition Function

There are several methods for calculating the partition function of the basic WSME model and its variants. Since the number of states for an N-residue protein is 2^N in the WSME model, and the computational complexity increases exponentially with increasing protein size, it is impossible to numerically calculate the partition function using Equation (4), even for a protein with ~50 residues. Thus, approximations that consider only specific protein states along the folding reaction coordinates are sometimes used. For example, single, double, and triple sequence approximations (SSA, DSA, and TSA) assume that up to one, two, and three independently folded segments, respectively, are allowed during the folding process (Figure 3) [13,37]. These approximations reduce the number of states to a polynomial quantity, enabling the calculation of the partition function, even for proteins with ~100 residues. DSA with loops (DSA/L), a variant of DSA, was developed, which involves non-local interactions between two folded segments [38–44]. DSA/L predicts fewer cooperative folding transitions than the original model because of the presence of non-local interactions. Non-local interactions play an essential role in the folding of large proteins because interactions between distant regions can affect their folding processes. In line with this, the introduction of non-local interactions between the N- and C-termini by a virtual linker in a multi-domain protein successfully explained folding processes that were not explained by the original model [45]. In addition, mean-field approximation was discussed for computing the partition function of the WSME model [46,47].

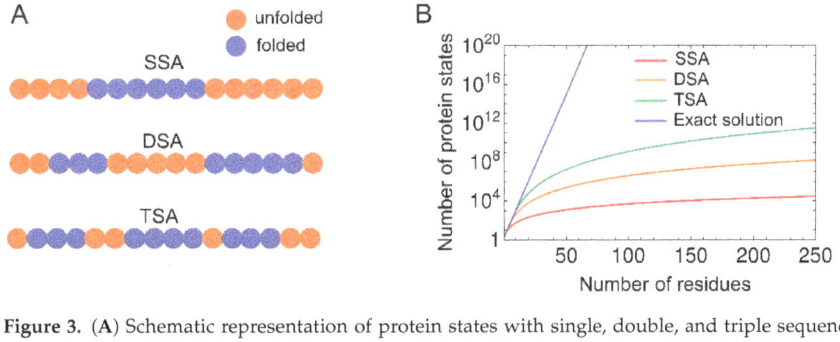

Figure 3. (**A**) Schematic representation of protein states with single, double, and triple sequence approximations (SSA, DSA, and TSA, respectively). Residues in folded or unfolded conformations are indicated by blue and orange circles, respectively. In the SSA, only the protein states with a single native segment are considered. In DSA and TSA, the protein states with up to two and three native segments are considered, respectively. (**B**) Number of microscopic protein states, which is considered in the calculation of the partition function using the SSA, DSA, TSA, and exact solution, is plotted against the number of residues.

An exact solution to the WSME model using transfer matrices was reported by Bruscolini and Pelizzola in 2002 [20]. The exact solution enables efficient calculation of the partition function by summing all 2^N states (Figure 3), and the calculation, even for large proteins with ~200 residues, is completed instantaneously.

Comparison of the approximate solutions, such as SSA, DSA, DSA/L, and TSA, with the exact solution of the WSME model was performed by calculating the free-energy profiles and folding rates for many proteins [37]. SSA did not describe reasonable free-energy barriers on one-dimensional (1D) free-energy landscapes because it is a coarse sampling method and considers only a single folded segment [37]. By contrast, all calculation methods other than SSA were able to predict folding rates, which was consistent with experimental results. The calculated free-energy profiles were almost unchanged, irrespective of the reaction coordinates used (n or Q). The finding that the calculations with DSA and TSA yielded similar results to the exact calculation suggests that the number of native segments formed during the folding process is 2 or 3 for small proteins. This was also shown in long-duration all-atom MD simulations, confirming that simple assumptions, such as DSA/L, are sufficient to explain the folding mechanisms of small proteins [42]. Note that for small proteins with fewer than ~50 residues, calculations with the exact solution overestimated the cooperativity of folding transitions, whereas those with DSA/L were the most consistent with the experiments [42,43].

2.3. Contact Energy

Calculation of the free energy using the WSME model requires setting an entropic cost associated with the folding reaction and preparing a residue–residue contact map and the energy for each contact based on the native structure of a protein (Figure 2). A residue pair is defined as being in contact when the distance between two residues in the native state is within a specific cutoff (typically 4 Å). The simplest way to assign contact energies is to use the same energy value for all the native contacts. Surprisingly, even with this simple treatment, the calculated free-energy landscapes effectively explained the folding processes experimentally observed for many small proteins [26,37,39–43,48,49]. Another way of assigning contact energies is to weigh the contact energy, depending on the number of atoms involved in the contact [13,20,45,50–53]. The use of weighted contact energies may yield more accurate results than the use of uniform contact energies.

Interestingly, sequence-dependent weighting of contact energies is not always necessary to describe the folding pathway for some small proteins [43]. Using DSA/L, Kubelka et al. investigated the effects of contact energy weighting on the free-energy landscapes for two proteins with a helix-turn-helix motif (P22 subdomain and $\alpha t\alpha$). They compared three contact energies: those weighted by the structural ensemble determined by NMR, those statistically determined by Miyazawa and Jernigan, and sequence-independent uniform contact energies. The results showed that the use of uniform contact energies was sufficient to explain the folding processes and to predict different folding pathways for two proteins with the same fold, reflecting subtle differences in local stability [43]. This suggests that the main-chain structure (i.e., protein topology or fold) itself contains sufficient information about the folding processes, and that the assumptions in the WSME (DSA/L) model successfully decode the folding mechanisms encoded in the contact map of the native structure.

Because homologous proteins with similar topologies have similar contact maps, uniform contact energies predict similar folding mechanisms for homologous proteins. However, even proteins with the same fold may have different folding processes, depending on the amino-acid sequence [54–60]. Such differences in the folding mechanisms cannot be distinguished by calculations using uniform contact energies. Therefore, obtaining suitable contact energies to describe the details of folding mechanisms is a challenge in optimizing the WSME model. Several studies have attempted to evaluate residue–residue contacts by considering the contribution of non-covalent interactions that drive protein folding, including electrostatic interactions, hydrogen bonds, van der Waals interactions, and hydrophobic interactions [61–64]. In addition, the introduction of temperature-dependent enthalpy and entropy terms [65,66] and the calculation of contact energies using AMBER force fields [67], which are typically used for MD simulations, have also been proposed. Such rigorous contact evaluations based on physical chemistry allow the calculation of contact energies for all residue pairs, rather than selecting residue–residue contacts according to an inter-residue distance cutoff. These approaches may require the determination of additional parameters, such as scaling constants, in order for predictions to agree with experimental results, including the temperature-dependent denaturation curves monitored by circular dichroism or NMR spectroscopy [43,53,68,69], the temperature dependence of specific heat capacity [40,41,63,65,69–73], and the denaturant dependence of folding/unfolding rate constants (called a chevron plot) [38,74].

3. Prediction of Folding Mechanisms

3.1. One-Dimensional Free-Energy Landscape: Two-State Versus Downhill Folding

The 1D free-energy landscape obtained using the WSME model is a powerful tool for analyzing protein stability under equilibrium conditions. Once the free-energy landscape is calculated, metastable intermediates and free-energy barriers can be clearly visualized, and folding mechanisms can be directly analyzed from the free-energy surface. Furthermore, the WSME model provides an effective analytical method for investigating the temperature and denaturant dependence of folding pathways [67,75].

Although two small proteins, gpW and the SH3 domain (Figure 4A,B), have comparable thermodynamic stability, experiments revealed that gpW folds ~1000-fold faster than SH3 [65]. Consistent with these observations, the WSME model with improved contact-energy calculations predict that the free-energy landscape of gpW has a marginal barrier, whereas that of SH3 has a clear barrier and exhibits cooperative two-state folding [65]. Similarly, the free-energy landscapes were compared between the WW domain of PIN1, a two-state folder, and BBL, which was experimentally shown to be a downhill folder without a clear free-energy barrier [51]. The WSME model predicts that PIN1 has a free-energy landscape with a distinct barrier, whereas BBL has an overall downhill landscape at low temperatures. Thus, the WSME model successfully explains the folding mechanisms of small proteins.

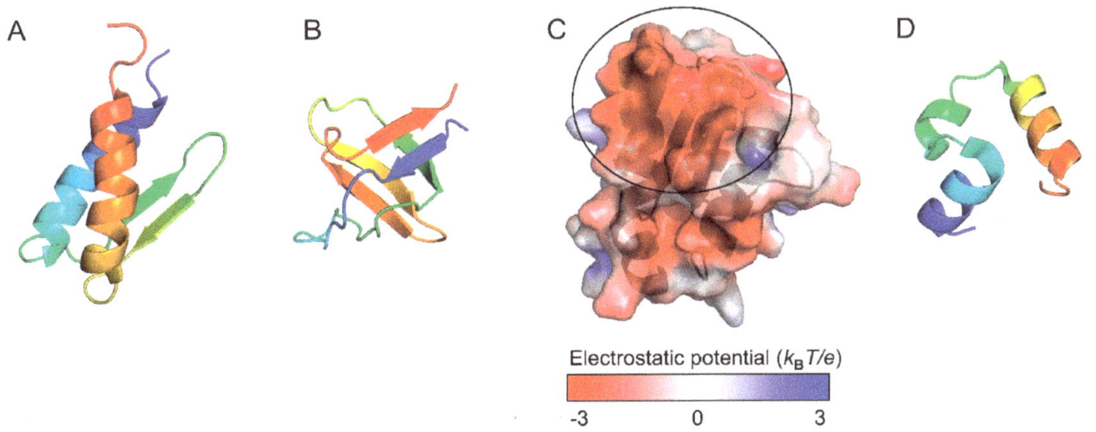

Figure 4. Native structures of gpW (PDB ID: 1HYW) (**A**) and the SH3 domain (PDB ID: 1SHG) (**B**). (**C**) Electrostatic potential energy surface of barstar (PDB ID: 1BTA). The black circle shows the barnase-binding site with large negative potentials. The electrostatic potential was calculated by APBS [76]. (**D**) Native structure of the 35-residue subdomain from the villin headpiece (PDB ID: 1YRF). Figures were drawn using PyMOL Molecular Graphics System, Version 2.4.0 Schrödinger, LLC.

Remarkably, the WSME model proposes that the folding mechanism of BBL is temperature-dependent, involving a downhill folding in a biologically relevant temperature range and a barrier-limited cooperative folding with a slight free-energy barrier at the transition midpoint temperature (T_m) [51,61]. This indicates that downhill and two-state folding mechanisms are continuously connected along temperatures and belong to the same folding class. The WSME model also successfully quantifies the free-energy barrier of PDD, a protein homologous to BBL [69,70,77]. The free-energy landscapes of PDD show downhill folding at low temperatures, but show two-state folding with a small free-energy barrier around the T_m. Therefore, the WSME model can quantitatively characterize the temperature dependence of folding mechanisms, even for proteins with small free-energy barriers.

3.2. Two-Dimensional Free-Energy Landscape: Multiple Folding Pathways

Multi-dimensional representations of free-energy landscapes can be achieved using multiple reaction coordinates corresponding to the structural formation of multiple regions of a protein [78]. Such multi-dimensional free-energy landscapes allow the visualization of detailed folding pathways. Moreover, the WSME model can predict the degree of structure formation of each residue along a folding pathway [26,79–81]. Using the WSME model with uniform contact energies, Sasai et al. calculated the free-energy landscapes of the B domain of protein A (BdpA), consisting of three helices with a symmetric topology (Figure 2) [26]. The 1D free-energy profile indicates that BdpA folds in a two-state manner (Figure 2), and the 2D free-energy landscape identifies two major folding pathways (Figure 5A). These pathways were revealed for the first time by describing the multi-dimensional free-energy landscape. Previous experimental studies of BdpA folding using Φ-value analysis showed that the second helix is the most structured in the transition state [82]. However, MD simulations could not reproduce these observations. By contrast, the WSME model provides folding processes that are in agreement with experimental results. The model suggests that proteins with symmetrical structures, such as BdpA, have two nearly symmetrical folding pathways (Figure 5A) [26]. In the transition state of one pathway (TS1), the first and second helices of BdpA are partially formed, whereas in the transition state of the other pathway (TS2), the second and third helices are partially formed. When these pathways are averaged, the second helix is the most completely formed in the transition state of the BdpA, which is consistent with experimental results [26].

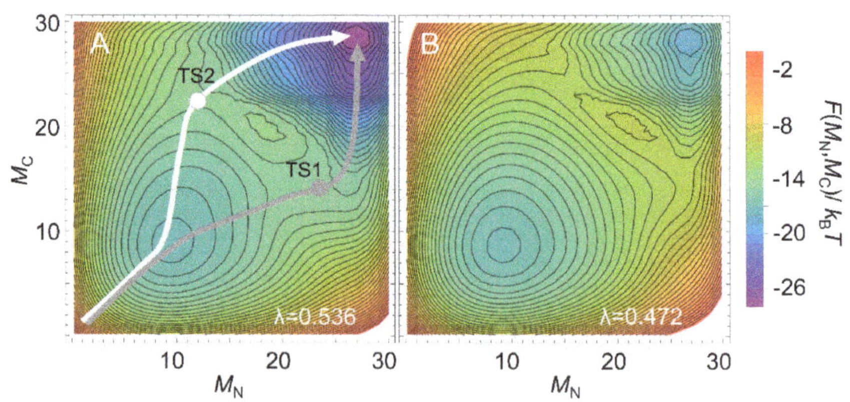

Figure 5. Two-dimensional free-energy landscapes of BdpA under folding conditions (**A**) and at the transition midpoint (**B**). The native structure of BdpA is shown in Figure 2 (left panel). M_N and M_C are the number of folded residues in the N-terminal half of BdpA (involving the first helix and the first half of the second helix) and C-terminal half of BdpA (involving the second half of the second helix and the third helix), respectively. Gray and white arrows denote the dominant folding pathways passing through the saddle points corresponding to the transition state 1 (TS1) and transition state 2 (TS2), respectively. $\lambda = \varepsilon/(k_B T)$ is a parameter related to the uniform contact energy ε and temperature T. Adapted with permission from Ref. [26]. Copyright (2006) National Academy of Sciences, U.S.A.

The above prediction for BdpA used uniform contact energies for all the native contacts, emphasizing the importance of a symmetric topology. Interestingly, the calculations predict that the two contrasting pathways will occur almost equally near room temperature (Figure 5A), whereas at higher temperatures, the symmetry is broken, and the folding is biased toward one pathway (Figure 5B). However, experimental Φ-values at a high temperature (near T_m) did not verify this prediction [83]. Zamparo and Pelizzola examined the temperature dependence of the folding pathways of four proteins (BdpA, albumin-binding domain (ABD), designed α3D protein, and engrailed homeodomain) with similar folds consisting of three helices using contact energies weighted according to the number of atoms involved in the contact rather than uniform contact energies [52]. The results suggest that even for proteins with symmetric structures, the folding abilities of the N- and C-terminal regions depend on subtle differences in the native contacts involved, and the transition-state structure is almost independent of temperature, which is in agreement with the results of experiments. The results also highlight the importance of accurate contact energies for the reliable prediction of protein-folding pathways [52].

3.3. *Effects of Amino-Acid Substitutions on Stability and Folding*

Predicting protein stability is difficult because the 3D structures of proteins are only marginally stabilized by networks of weak non-covalent interactions. Thus, amino-acid substitutions in proteins can have complex effects on the free-energy landscapes, changing the free energies of the native and unfolded states, as well as the number and nature of folding intermediates. Nevertheless, as shown above, the WSME model has the potential to predict the effects of amino-acid substitutions on protein stability and folding by calculating the free-energy landscapes of wild-type and mutant proteins. Such calculations have provided useful insights for protein engineering and medical applications [22,44,64,84–86]. Naganathan et al. proposed a framework for calculating the stability of mutants using the WSME model and developed two programs, pStab [21] and pPerturb [22], which are available online. These methods may be useful as the first steps in screening protein mutants with the desired stability.

The relationship between folding and function has been examined by comparing the free-energy landscapes of homologous proteins with those of proteins with amino-acid substitutions or chemical modifications [62,63,74,80,87–92]. The charge distribution on the protein surface is one of the key factors controlling ligand binding and can also affect protein stability and folding [62,63,74]. For example, it has been suggested that barstar (Figure 4C) maintains its ability to bind to barnase by acquiring a large binding surface with negative electrostatic potential during evolution, resulting in a complex free-energy landscape with multiple folding intermediates [88]. Theoretical calculations predict that amino-acid substitutions to neutralize the charges at the barnase-binding site would improve the stability of barstar and simplify its folding mechanism to "frustration-free" two-state folding [88]. Thus, the WSME model may be useful for evaluating the effects of amino-acid substitutions and clarifying the role of each residue in the stability, folding, and function of proteins.

3.4. *Effects of External Forces on Protein (un)Folding*

An extended WSME model with external forces was constructed as a theoretical model of mechanical unfolding experiments on a single-protein molecule using atomic-force microscopy (AFM) [93]. The model calculates the equilibrium force-extension curves and free-energy landscapes as a function of the end-to-end length of a protein to characterize mechanical unfolding [93–95]. The kinetics of the response to time-dependent external forces (force clamp and dynamic loading) can also be evaluated by combining Monte Carlo simulations. Such analyses of the mechanical unfolding of ubiquitin predict the order of secondary structure formation and the presence of kinetic intermediates, which are consistent with the results from experiments and all-atom MD simulations [96]. In addition, this extended WSME model predicts the major and minor unfolding pathways of green fluorescent protein observed experimentally [97] and was further applied to characterize the equilibrium properties and kinetic unfolding pathways of RNAs, such as an RNA hairpin and the *Tetrahymena thermophila* ribozyme [98,99].

Single-molecule experiments with AFM have also shown that glycerol, a protective osmolyte, stabilizes the native state of globular proteins against mechanical unfolding without changing the position of the transition state on the reaction coordinate [100]. To simulate the mechanical unfolding of a protein in the presence of osmolytes, extended versions of the WSME model that consider the effects of osmolytes were developed [100,101]. The model successfully reproduces the experimental results of mechanical unfolding, in which the position of the transition state along the reaction coordinate is unchanged by osmolytes for the immunoglobulin-binding B1-domain of *Streptococcal* protein G (GB1) and the I27 module of human cardiac titin [100,101]. Thus, the WSME model with external forces is useful for understanding the single-molecule behavior of proteins during mechanical unfolding.

The external force term introduced in the above models has also been used to evaluate the effects of crowded environments, such as inside cells, on protein stability and folding. The predictions for ABD, GB1, and the β-hairpin of GB1 indicate that as the cage size confining a protein gradually decreases, the protein molecule will be stabilized up to a certain threshold cage size, and then destabilized below the threshold [102]. Furthermore, a general relationship between cage size and folding rate has been observed for various proteins [102]. A model for non-equilibrium diffusion dynamics was also developed using an external force term to describe the intracellular translocation of proteins [103]. Thus, the WSME model with external forces is also useful for theoretically evaluating protein stability and folding in various situations in which mechanical forces act on proteins.

3.5. Folding Kinetics and Transition State

The macroscopic kinetic behavior of protein states during folding can be predicted with kinetic models, such as master equations, using the free energies of the unfolded, intermediate, transition, and native states obtained from the WSME model. The theoretical folding rates thus obtained were shown to depend on the protein topology, which was consistent with experimental observations [13,37,50,51,104,105]. For example, the predicted folding rates of the 35-residue subdomain from the villin headpiece, which has three short α-helices (Figure 4D) and exhibits ultrafast folding, are consistent with those measured experimentally [38–41]. Thus, the WSME model is a powerful tool for studying subtle differences in folding rates [38,73–75,106]. Because virtual amino-acid substitutions can be introduced by perturbing specific contact energies, the WSME model with such perturbations can be used to calculate the theoretical Φ-values along the folding pathway [13,26,78,105,107–110]. Sasai et al. calculated theoretical Φ-values for the transition state in the folding of BdpA by averaging the transition-state structures on both of the major folding pathways (Figure 5A) and succeeded in obtaining Φ-values consistent with experiments [26]. Thus, theoretical Φ-value analysis using the WSME model can describe folding reactions at the resolution of individual residues.

The free-energy landscapes obtained by the WSME model can be combined with Monte Carlo simulations using the Metropolis algorithm to simulate single-molecule trajectories and examine microscopic protein-folding kinetics [42,45,53,65,71,72,87,88,106,111,112]. Since the WSME model is a Gō-type coarse-grained model with a limited number of possible conformations, simulations of protein-folding reactions can be performed with low computational complexity. An ensemble average of many single-molecule trajectories reproduces the macroscopic folding behaviors [51,61,65]. The use of such simulations for several proteins suggests that even proteins that exhibit simple two-state folding have a variety of folding pathways with different transition-state structures, and that the experimentally observed transition state is the average of these structures [72,110]. This method is expected to resolve the possible discrepancies between the experimental results and a small number of MD trajectories for protein folding/unfolding reactions, as it provides a rich dataset of single-molecule folding trajectories that cannot be obtained from MD simulations.

Although the spatial resolution of the WSME model is lower than that of all-atom MD simulations, it has been suggested that folding/unfolding simulations with the WSME model reproduce all-atom MD simulations [42]. A comparison of the folding/unfolding trajectories for the villin headpiece based on the WSME model using DSA/L with those of the long-time all-atom MD simulations using explicit solvent, performed by Shaw et al., shows that the folding behaviors are very similar in both simulations, including the rate of transition between relevant conformations and the order of helix formation [42]. Since the WSME model only considers the residue–residue interactions occurring in the native structure, these results highlight the importance of native contacts in determining protein-folding mechanisms.

4. Folding Mechanisms of Multi-Domain Proteins

In the previous sections, we showed that the folding mechanisms of small single-domain proteins are described well by the WSME model. By contrast, the folding mechanisms of multi-domain proteins have been less frequently studied because they have complex, multiple folding pathways and intermediates, making it difficult to theoretically predict the folding processes [53,75,78,81,91,106,108,112–116]. However, multi-domain proteins comprise the majority of the proteome, and more than 70% of eukaryotic proteins contain multiple domains [117,118]. Therefore, the elucidation of the folding mechanisms of multi-domain proteins is an important issue in the life sciences. The two major ways of connecting two globular domains are (1) the tandem connection of two domains by a linker and (2) the insertion of one domain into another domain.

4.1. Tandem Connection of Multiple Domains

The WSME model assumes that folding starts at local segments and then spreads throughout the molecule via the extensions and connections of the folded segments. Thus, the model is suitable for multi-domain proteins consisting of tandemly connected small globular domains, each of which folds in a two-state manner (Figure 6A). Typical examples are repeat proteins, and predictions of their folding processes are in agreement with experiments in terms of the structures of folding intermediates and the order of domain formation [53,75,78,113,114].

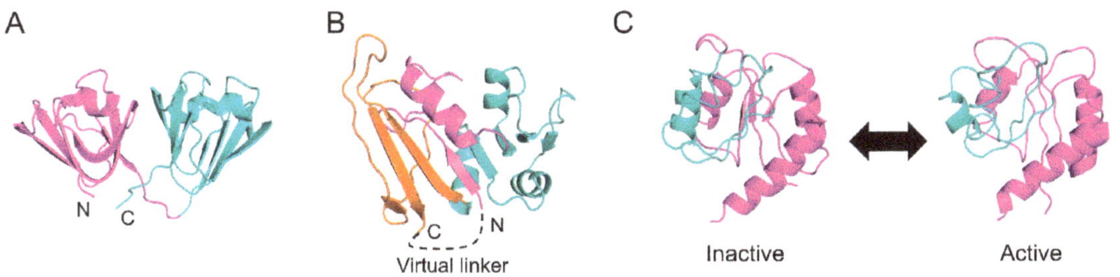

Figure 6. (**A**) Native structure of human γD-crystallin (PDB ID: 1HK0). Domains 1 and 2 are shown in magenta and cyan, respectively. (**B**) Native structure of *Escherichia coli* dihydrofolate reductase (DHFR) (PDB ID: 1RX1). The N- and C-terminal parts of the discontinuous loop subdomain (DLD) are shown in magenta and orange, respectively, and the adenosine-binding subdomain (ABD) is shown in cyan. In the extended WSME (eWSME) model, a virtual linker (dashed line) was implemented to virtually connect the N- and C- termini, both of which are included in the DLD. (**C**) Inactive and active conformations of nitrogen regulatory protein C (NtrC) (PDB ID: 1DC7 and 1DC8, respectively). Phosphorylation of NtrC induces allosteric conformational changes to the residues shown in cyan. Figures were drawn using PyMOL Molecular Graphics System, Version 2.4.0 Schrödinger, LLC.

Sasai et al. applied the WSME model to multi-domain proteins with two globular domains connected in tandem, including γD-crystallin (Figure 6A), spore coat protein S, and R16-R17 spectrin domain, and investigated the effects of domain–domain interactions on folding reactions [108]. The computational results consistently explained the folding pathways and transition-state structures obtained by Φ-value analysis and suggested that the connectivity and interaction between the two domains determine the equilibrium and kinetic folding mechanisms. Furthermore, high-dimensional free-energy landscapes are effective in analyzing complex folding mechanisms and reveal hidden folding pathways, intermediates, and transition states for barnase, nitrogen regulatory protein C (NtrC), and an ankyrin repeat protein [78]. Although the computational complexity increases as the protein size increases, an efficient method to reduce computational complexity has been reported that considers short segments as blocks [114]. Note that when domain–domain interactions are strong in multi-domain proteins, the folding mechanisms may become more complex, making the prediction of folding processes more challenging, even for multi-domain proteins with tandem connections.

4.2. Domain Insertions

Another mode of domain connection is domain insertion. There are many multi-domain proteins in which one domain is inserted into another [119]. Many folding experiments have been performed on multi-domain proteins with domain insertions, including dihydrofolate reductase (DHFR), apomyoglobin, barnase, α-lactalbumin from bovine, human, and goat sources, and lysozyme from hen-egg-white, human, equine, and canine sources [2,4,5,9,30,36,54,55,57,59,60,120–131]. Interestingly, these proteins accumulate molten globule-like folding intermediates in which the discontinuous domain is more organized than the inserted continuous domain [5]. Such intermediates may be formed via a hydrophobic collapse mechanism driven by non-local hydrophobic interactions between distant residues in the amino-acid sequence [5,8,130]. The original WSME model cannot provide free-energy landscapes consistent with experiments for these proteins because it assumes that all the intervening domains must be folded before the discontinuous domain starts to fold.

DHFR is one of the most closely studied proteins in terms of its kinetic folding mechanism [5,120,122,123,127–131]. DHFR consists of two domains, with one globular domain (adenosine-binding subdomain, ABD) inserted into the other globular domain (discontinuous loop subdomain, DLD) (Figure 6B). We previously showed that the folding reaction of DHFR involves at least seven phases and six intermediates [131]. In brief, DHFR first forms a compact intermediate within 35 μs after the initiation of the folding reaction, and then DLD and ABD fold independently with time constants of 550 μs and 200 ms, respectively, accumulating an intermediate in which DLD is more organized than ABD. Finally, both domains dock to form the native structure. We also revealed that after a few milliseconds of folding, the folding behavior of "circular DHFR" with a disulfide bond introduced between the N- and C-termini is almost identical to that of "linear DHFR" with the disulfide bond reduced [128]. This suggests that the interactions between the N- and C-termini involved in DLD are already formed in the early stages of folding. However, these folding processes cannot be explained using the original WSME model.

To facilitate the folding of a discontinuous domain, Sasai et al. developed an extended WSME (eWSME) model, in which a virtual linker was introduced at the N- and C-termini of DHFR (Figure 6B) [45]. In this model, even when the inserted continuous domain (ABD) is not folded, non-local interactions can be formed between the N- and C-terminal regions involved in DLD via the virtual linker. The free-energy landscape calculated by the eWSME model successfully predicts the two of the six folding intermediates reported in the experiments [45,132]. Furthermore, Sasai et al. proposed that the introduction of multiple virtual linkers into a protein molecule may enable the prediction of the folding processes of multi-domain proteins with more than two domains [17]. Thus, the WSME model may be applicable for predicting the free-energy landscapes of a variety of multi-domain proteins after sufficient modifications. However, such modifications may not be easily implemented because it is not clear where and how many virtual linkers should be introduced in a protein molecule. Furthermore, as the number of virtual linkers increases, the mathematics describing them may become more complex. Nevertheless, the development of a modified version of the eWSME model that can be applied to any protein, including both small single-domain proteins and large multi-domain proteins, would represent significant progress toward solving the folding-process component of the "protein-folding problem" [12].

5. Applications beyond Protein Folding

5.1. Intrinsically Disordered Proteins

In addition to protein folding, the WSME model is also applicable to the conformational changes associated with protein function. Intrinsically disordered proteins (IDPs) have disordered structures in isolation but fold into specific structures upon binding to their partners [5,133–135]. For example, the intrinsically disordered region (IDR) of a neuron-restrictive silencer factor (NRSF) takes various β-hairpin-like structures in isolation but forms an α-helical structure when bound to its target protein, Sin3 [136–138]. Disordered structures of NRSF are theoretically created in the absence of Sin3 by introducing interactions favoring the β-hairpin structure into the WSME model; such interactions are different from those stabilizing the NRSF-Sin3 complex [136,137]. Furthermore, the free-energy landscape for the binding of NRSF to Sin3 obtained from this model reproduces the coupled folding and binding behaviors [136,138] commonly observed in many IDPs [5,133–135].

The free-energy landscape of an intrinsically disordered DNA-binding domain of the transcriptional regulator CytR, calculated using the WSME model, suggests that the conformational ensemble of the disordered state involves competition for several specific conformations [68]. By introducing the interaction between CytR and its partner DNA, the model successfully describes how, as the partner DNA approaches CytR, the free-energy landscape of CytR in the disordered state with multiple local minima changes into a landscape with a global minimum corresponding to the DNA-bound form of CytR [111,139]. Furthermore, the free-energy landscapes of CytR in the presence of a polymeric crowder, polyethylene glycol (PEG), mimicking crowded intracellular environments, provide a PEG concentration–temperature phase diagram showing that CytR is more folded both at lower temperatures and at higher PEG concentrations, which is in agreement with experimental results [140].

Thus, the WSME model comprehensively explains both the folding of globular proteins and the structures of IDPs in free and bound forms based on free-energy landscapes. In addition, the model can predict the effects of temperature, osmolytes, and amino-acid substitutions on IDP structures and may be useful for controlling the conformations of IDPs [141,142]. It may also be possible to predict the effects of ion valence and ionic strength on the free-energy landscapes of IDPs by incorporating them into the contact energies. The next target of IDP studies using the WSME model would be to predict the mechanisms of the coupled folding and binding reactions of IDPs [5,135].

5.2. Conformational Changes Associated with Protein Function

Many proteins dynamically change their conformations and exert their functions by binding to specific targets or through post-translational modifications. Free-energy calculations using the WSME model have also been applied to the theoretical analysis of the conformational changes associated with protein functions, such as photocycles and allosteric transitions [17,143–146].

Photoactive yellow protein (PYP) is a model protein for photoreceptors that has a photocycle consisting of three states [147]. The cycle involves coordinated motion on different time scales, from isomerization of chromophores, occurring in nanoseconds, to the partial denaturation of proteins, occurring in milliseconds or more [147]. Sasai et al. constructed an extended model describing motions over a wide range of time scales by adding an energy term to the WSME model that depends on local packing changes. The calculations assuming the ground state of the photocycle to be the native state yield a free-energy landscape that reasonably reproduces the photocycle and predicts the detailed structure of each state involved in the cycle [143,144]. Thus, the WSME model successfully explains the mechanism through which local structural fluctuations induce large-scale conformational changes, suggesting that the close interplay of motions at different time scales plays a crucial role in regulating protein function.

Sasai et al. further modified the WSME model to allow multiple native states and developed an allosteric WSME (aWSME) model that can calculate a free-energy landscape reflecting protein allostery [17,145,146]. The bacterial enhancer-binding protein, NtrC, undergoes an allosteric transition from the inactive to the active state through phosphorylation (Figure 6C). The application of the aWSME model to NtrC yields free-energy landscapes that predict large conformational fluctuations between the inactive and active states, as well as allosteric conformational changes upon phosphorylation [145]. The aWSME model also predicts that the GTP-binding protein Ras is stabilized by binding to GDP, whereas the structure of Ras in the GTP-bound state fluctuates significantly, suggesting that the difference in conformational fluctuations between the GDP- and GTP-bound states regulates signal transduction [146]. Thus, the aWSME model, which allows multiple native states, provides a mechanistic explanation for the transitions between multiple stable conformations and allosteric conformational changes upon effector binding.

Cnu is a transcriptional co-repressor that regulates gene expression upon temperature changes and has also been proposed to be involved in pH-dependent gene expression [148–150]. Using the WSME model with rigorous contact energy-calculations, including electrostatic interactions, Naganathan et al. showed that the distribution of conformations in the native-state ensemble of Cnu is sensitive to changes in both temperature and pH, suggesting that Cnu can serve as both a temperature and a pH sensor [148–150].

5.3. Other Applications

Amyloids are insoluble fibrous aggregates of proteins stabilized primarily by hydrogen bonds and hydrophobic interactions and have a cross β-sheet structure with parallel β-strands aligned perpendicular to the fibril axis [151,152]. Because amyloids are implicated in neurodegenerative diseases, such as Alzheimer's disease, Parkinson's disease, and bovine spongiform encephalopathy, understanding the mechanisms of amyloid fibril formation is an important issue in drug discovery [151,152]. The assumption of the WSME model, in which an entire protein molecule folds through the elongation and docking of local native segments, has also been utilized as a model to describe amyloid formation [153,154]. By introducing both the interactions stabilizing the monomeric form of a protein and those stabilizing the amyloid form, the WSME model was able to qualitatively reproduce a sharp phase transition to amyloid fibrils, which is characteristic of the nucleation-growth model and is consistent with experiments on amyloid formation [153,154].

Since the WSME model can be regarded as a simple 1D lattice model, the exact solution of the model can be calculated using the transfer matrix method, even for systems with non-uniform interactions. Taking advantage of this feature, the WSME model has also been used to describe the growth of strained epitaxy [155,156]. Furthermore, the WSME model itself has been a subject of research in theoretical statistical mechanics, and efforts have been made to develop kinetic analyses by applying the cluster-variation method, which is one of the most precise methods for solving the Ising model when an exact solution is not available [50,157–159]. Other studies have examined the relationship between protein structures and folding mechanisms through the partition function zeros of the WSME model [160–163]; partition functions for various secondary-structure elements and two small proteins (BBL and chymotrypsin inhibitor 2) have shown that the distribution of partition function zeros distinguishes folding mechanisms, such as downhill and two-state folding [161].

6. Summary and Future Perspectives

In this review, we summarized how the WSME model and its extended versions describe protein folding and dynamics. The WSME model can calculate the free-energy landscapes of proteins, which predict the thermodynamic quantities involved in equilibrium-unfolding transitions and the pathways and structures involved in kinetic folding processes. These calculations are consistent with the experimental results of protein folding, especially for small single-domain proteins, suggesting that the WSME model enables the prediction of detailed protein-folding processes that are difficult to measure experimentally, and contributes to our understanding of protein-folding mechanisms. Surprisingly, although the WSME model is a simple coarse-grained model, it can reproduce various aspects of protein folding obtained by all-atom MD simulations. This agreement strongly supports the hypothesis that folding reactions are primarily driven by native interactions and that the free-energy landscape is globally biased toward the native state. This also indicates that the WSME model adequately captures and deciphers the bias encoded in protein conformation. Therefore, the WSME model, when combined with rigorous contact-energy calculations, provides theoretical predictions that are in good agreement with the experimental results for small proteins.

The WSME model has also been applied to predict the folding mechanisms of multi-domain proteins, especially those consisting of tandemly connected small globular domains. Although it is difficult to compute entire folding reactions of large multi-domain proteins using all-atom MD simulations, the WSME model can calculate the free-energy landscapes of such proteins with low computational complexity. Therefore, the WSME model and MD simulations are expected to be important tools for predicting protein-folding mechanisms.

Nevertheless, it is still challenging to predict the folding mechanisms of proteins with complex structures, such as multi-domain proteins with domain insertions and those with strong interactions between domains. Although non-local interactions between distant segments in the amino-acid sequence may be formed early in folding reactions by the hydrophobic collapse mechanism [5], they cannot be considered in the original WSME model. One promising approach to solving this problem is to introduce virtual linkers at non-local contacts that can be formed early in the folding reaction. Indeed, introducing a single virtual linker between the N- and C-termini is effective in predicting the folding processes of DHFR [17,45]. The next challenge would be to introduce multiple virtual linkers at arbitrary positions in a single protein to enable the prediction of the folding mechanisms of any protein, including small single-domain proteins and large multi-domain proteins with complex main-chain topologies.

Another type of interaction that complicates the protein structure is a disulfide bond. The WSME model has never explicitly considered the folding reactions of disulfide-intact proteins or those involving oxidative formation of disulfide bonds. The prediction of such folding reactions is also challenging, but it may be achieved by replacing the virtual linkers introduced above as non-local interactions with covalent linkers.

Because of its simplicity and versatility, the WSME model can be used to analyze various biological events other than protein folding under equilibrium and non-equilibrium conditions by calculating free-energy landscapes using exact or approximate solutions and, subsequently, performing Monte Carlo simulations. Due to this utility, the extended version of the WSME model provides reasonable predictions for protein-conformation changes in IDPs and allosteric conformational changes accompanied by protein functions, such as protein–protein interactions and ligand binding. Furthermore, the model may be applicable to multimer formation, domain swapping, and the coupled folding and binding reactions of IDPs.

Recently, protein-structure prediction has made great advances, through deep-learning approaches, towards solving the structure-prediction component of the "protein-folding problem" [12,164]. However, even state-of-the-art structure prediction methods do not provide an understanding of how proteins fold into specific structures [165]. Therefore, the theoretical prediction of protein-folding processes remains a challenge. Since the WSME model can predict protein folding and dynamics with low computational complexity, the WSME model and its modifications will play an important role in solving the folding-process component of the "protein-folding problem" in the near future.

Author Contributions: Writing—original draft preparation, K.O., R.L. and M.A.; writing—review and editing, K.O., R.L. and M.A. All authors have read and agreed to the published version of the manuscript.

Funding: This work was supported by JSPS KAKENHI, grant numbers JP16H02217, JP19H02521, and JP21K18841 (M.A.), and a Grant-in-Aid for JSPS Fellows, grant number JP20J11762 (K.O.).

Institutional Review Board Statement: Not applicable.

Informed Consent Statement: Not applicable.

Data Availability Statement: Not applicable.

Conflicts of Interest: The authors declare no conflict of interest.

References

1. Levinthal, C. How to fold graciously. In *Mössbauer Spectroscopy in Biological Systems, Proceedings of the Meeting Held at Allerton House, Monticello, IL, USA, 17–18 March 1969*; Department of Chemistry and Physics, University of Illinois: Champaign, IL, USA, 1969; Volume 67, pp. 22–24.
2. Dobson, C.M.; Evans, P.A.; Radford, S.E. Understanding how proteins fold: The lysozyme story so far. *Trends Biochem. Sci.* **1994**, *19*, 31–37. [CrossRef]
3. Jackson, S.E. How do small single-domain proteins fold? *Fold. Des.* **1998**, *3*, R81–R91. [CrossRef]
4. Arai, M.; Kuwajima, K. Role of the molten globule state in protein folding. *Adv. Protein Chem.* **2000**, *53*, 209–282. [CrossRef]
5. Arai, M. Unified understanding of folding and binding mechanisms of globular and intrinsically disordered proteins. *Biophys. Rev.* **2018**, *10*, 163–181. [CrossRef] [PubMed]
6. Šali, A.; Shakhnovich, E.; Karplus, M. How does a protein fold? *Nature* **1994**, *369*, 248–251. [CrossRef]
7. Bryngelson, J.D.; Onuchic, J.N.; Socci, N.D.; Wolynes, P.G. Funnels, pathways, and the energy landscape of protein folding: A synthesis. *Proteins* **1995**, *21*, 167–195. [CrossRef]
8. Dill, K.A.; Bromberg, S.; Yue, K.; Fiebig, K.M.; Yee, D.P.; Thomas, P.D.; Chan, H.S. Principles of protein folding—A perspective from simple exact models. *Protein Sci.* **1995**, *4*, 561–602. [CrossRef]
9. Dobson, C.M.; Šali, A.; Karplus, M. Protein folding: A perspective from theory and experiment. *Angew. Chem. Int. Ed.* **1998**, *37*, 868–893. [CrossRef]
10. Mayor, U.; Guydosh, N.R.; Johnson, C.M.; Grossmann, J.G.; Sato, S.; Jas, G.S.; Freund, S.M.; Alonso, D.O.; Daggett, V.; Fersht, A.R. The complete folding pathway of a protein from nanoseconds to microseconds. *Nature* **2003**, *421*, 863–867. [CrossRef]
11. Lindorff-Larsen, K.; Piana, S.; Dror, R.O.; Shaw, D.E. How fast-folding proteins fold. *Science* **2011**, *334*, 517–520. [CrossRef]
12. Dill, K.A.; Ozkan, S.B.; Shell, M.S.; Weikl, T.R. The protein folding problem. *Annu. Rev. Biophys.* **2008**, *37*, 289–316. [CrossRef]
13. Muñoz, V.; Eaton, W.A. A simple model for calculating the kinetics of protein folding from three-dimensional structures. *Proc. Natl. Acad. Sci. USA* **1999**, *96*, 11311–11316. [CrossRef] [PubMed]
14. Alm, E.; Baker, D. Prediction of protein-folding mechanisms from free-energy landscapes derived from native structures. *Proc. Natl. Acad. Sci. USA* **1999**, *96*, 11305–11310. [CrossRef] [PubMed]
15. Galzitskaya, O.V.; Finkelstein, A.V. A theoretical search for folding/unfolding nuclei in three-dimensional protein structures. *Proc. Natl. Acad. Sci. USA* **1999**, *96*, 11299–11304. [CrossRef]
16. Onuchic, J.N.; Wolynes, P.G. Theory of protein folding. *Curr. Opin. Struct. Biol.* **2004**, *14*, 70–75. [CrossRef] [PubMed]
17. Sasai, M.; Chikenji, G.; Terada, T.P. Cooperativity and modularity in protein folding. *Biophys. Phys.* **2016**, *13*, 281–293. [CrossRef]
18. Wako, H.; Saitô, N. Statistical mechanical theory of the protein conformation. I. General considerations and the application to homopolymers. *J. Phys. Soc. Jpn.* **1978**, *44*, 1931–1938. [CrossRef]
19. Wako, H.; Saitô, N. Statistical mechanical theory of the protein conformation. II. Folding pathway for protein. *J. Phys. Soc. Jpn.* **1978**, *44*, 1939–1945. [CrossRef]
20. Bruscolini, P.; Pelizzola, A. Exact solution of the Muñoz-Eaton model for protein folding. *Phys. Rev. Lett.* **2002**, *88*, 258101. [CrossRef]
21. Gopi, S.; Devanshu, D.; Krishna, P.; Naganathan, A.N. pStab: Prediction of stable mutants, unfolding curves, stability maps and protein electrostatic frustration. *Bioinformatics* **2018**, *34*, 875–877. [CrossRef]
22. Gopi, S.; Devanshu, D.; Rajasekaran, N.; Anantakrishnan, S.; Naganathan, A.N. pPerturb: A server for predicting long-distance energetic couplings and mutation-induced stability changes in proteins via perturbations. *ACS Omega* **2020**, *5*, 1142–1146. [CrossRef] [PubMed]
23. Gō, N.; Abe, H. Non-interacting local-structure model of folding and unfolding transition in globular-proteins. 1. Formulation. *Biopolymers* **1981**, *20*, 991–1011. [CrossRef] [PubMed]
24. Gō, N. Theoretical studies of protein folding. *Annu. Rev. Biophys. Bioeng.* **1983**, *12*, 183–210. [CrossRef]
25. Takada, S. Gō-ing for the prediction of protein folding mechanisms. *Proc. Natl. Acad. Sci. USA* **1999**, *96*, 11698–11700. [CrossRef] [PubMed]

26. Itoh, K.; Sasai, M. Flexibly varying folding mechanism of a nearly symmetrical protein: B domain of protein A. *Proc. Natl. Acad. Sci. USA* **2006**, *103*, 7298–7303. [CrossRef] [PubMed]
27. Miyazawa, S.; Jernigan, R.L. Most probable intermediates in protein folding-unfolding with a noninteracting globule-coil model. *Biochemistry* **1982**, *21*, 5203–5213. [CrossRef] [PubMed]
28. Abe, H.; Gō, N. Non-interacting local-structure model of folding and unfolding transition in globular-proteins. 2. Application to two-dimensional lattice proteins. *Biopolymers* **1981**, *20*, 1013–1031. [CrossRef] [PubMed]
29. Saitô, N. Statistical-mechanics of DNA and protein suitable for computer calculation. *Cell Biophys.* **1987**, *11*, 321–329. [CrossRef]
30. Matouschek, A.; Kellis, J.T., Jr.; Serrano, L.; Fersht, A.R. Mapping the transition state and pathway of protein folding by protein engineering. *Nature* **1989**, *340*, 122–126. [CrossRef]
31. Fersht, A.R.; Matouschek, A.; Serrano, L. The folding of an enzyme. I. Theory of protein engineering analysis of stability and pathway of protein folding. *J. Mol. Biol.* **1992**, *224*, 771–782. [CrossRef]
32. Fersht, A.R.; Sato, S. Φ-value analysis and the nature of protein-folding transition states. *Proc. Natl. Acad. Sci. USA* **2004**, *101*, 7976–7981. [CrossRef] [PubMed]
33. Udgaonkar, J.B.; Baldwin, R.L. NMR evidence for an early framework intermediate on the folding pathway of ribonuclease A. *Nature* **1988**, *335*, 694–699. [CrossRef]
34. Roder, H.; Elöve, G.A.; Englander, S.W. Structural characterization of folding intermediates in cytochrome c by H-exchange labelling and proton NMR. *Nature* **1988**, *335*, 700–704. [CrossRef] [PubMed]
35. Radford, S.E.; Dobson, C.M.; Evans, P.A. The folding of hen lysozyme involves partially structured intermediates and multiple pathways. *Nature* **1992**, *358*, 302–307. [CrossRef] [PubMed]
36. Jennings, P.A.; Wright, P.E. Formation of a molten globule intermediate early in the kinetic folding pathway of apomyoglobin. *Science* **1993**, *262*, 892–896. [CrossRef] [PubMed]
37. Henry, E.R.; Eaton, W.A. Combinatorial modeling of protein folding kinetics: Free energy profiles and rates. *Chem. Phys.* **2004**, *307*, 163–185. [CrossRef]
38. Cellmer, T.; Henry, E.R.; Kubelka, J.; Hofrichter, J.; Eaton, W.A. Relaxation rate for an ultrafast folding protein is independent of chemical denaturant concentration. *J. Am. Chem. Soc.* **2007**, *129*, 14564–14565. [CrossRef]
39. Cellmer, T.; Henry, E.R.; Hofrichter, J.; Eaton, W.A. Measuring internal friction of an ultrafast-folding protein. *Proc. Natl. Acad. Sci. USA* **2008**, *105*, 18320–18325. [CrossRef]
40. Godoy-Ruiz, R.; Henry, E.R.; Kubelka, J.; Hofrichter, J.; Muñoz, V.; Sanchez-Ruiz, J.M.; Eaton, W.A. Estimating free-energy barrier heights for an ultrafast folding protein from calorimetric and kinetic data. *J. Phys. Chem. B* **2008**, *112*, 5938–5949. [CrossRef]
41. Kubelka, J.; Henry, E.R.; Cellmer, T.; Hofrichter, J.; Eaton, W.A. Chemical, physical, and theoretical kinetics of an ultrafast folding protein. *Proc. Natl. Acad. Sci. USA* **2008**, *105*, 18655–18662. [CrossRef]
42. Henry, E.R.; Best, R.B.; Eaton, W.A. Comparing a simple theoretical model for protein folding with all-atom molecular dynamics simulations. *Proc. Natl. Acad. Sci. USA* **2013**, *110*, 17880–17885. [CrossRef] [PubMed]
43. Lai, J.K.; Kubelka, G.S.; Kubelka, J. Sequence, structure, and cooperativity in folding of elementary protein structural motifs. *Proc. Natl. Acad. Sci. USA* **2015**, *112*, 9890–9895. [CrossRef]
44. Lai, J.K.; Kubelka, G.S.; Kubelka, J. Effect of mutations on the global and site-specific stability and folding of an elementary protein structural motif. *J. Phys. Chem. B* **2018**, *122*, 11083–11094. [CrossRef] [PubMed]
45. Inanami, T.; Terada, T.P.; Sasai, M. Folding pathway of a multidomain protein depends on its topology of domain connectivity. *Proc. Natl. Acad. Sci. USA* **2014**, *111*, 15969–15974. [CrossRef] [PubMed]
46. Flammini, A.; Banavar, J.R.; Maritan, A. Energy landscape and native-state structure of proteins—A simplified model. *Eur. Lett.* **2002**, *58*, 623–629. [CrossRef]
47. Liang, K.K.; Hayashi, M.; Shiu, Y.J.; Mo, Y.; Shao, J.S.; Yan, Y.J.; Lin, S.H. Thermodynamics and kinetics of protein folding: A mean field theory. *Phys. Chem. Chem. Phys.* **2003**, *5*, 5300–5308. [CrossRef]
48. Garcia-Mira, M.M.; Sadqi, M.; Fischer, N.; Sanchez-Ruiz, J.M.; Muñoz, V. Experimental identification of downhill protein folding. *Science* **2002**, *298*, 2191–2195. [CrossRef]
49. Lee, E.; Jung, Y. Calculations of free energy surfaces for small proteins and a protein-RNA complex using a lattice model approach. *Bull. Korean Chem. Soc.* **2011**, *32*, 3051–3056. [CrossRef]
50. Zamparo, M.; Pelizzola, A. Kinetics of the Wako-Saitô-Muñoz-Eaton model of protein folding. *Phys. Rev. Lett.* **2006**, *97*, 068106. [CrossRef]
51. Bruscolini, P.; Pelizzola, A.; Zamparo, M. Downhill versus two-state protein folding in a statistical mechanical model. *J. Chem. Phys.* **2007**, *126*, 215103. [CrossRef]
52. Zamparo, M.; Pelizzola, A. Nearly symmetrical proteins: Folding pathways and transition states. *J. Chem. Phys.* **2009**, *131*, 035101. [CrossRef] [PubMed]
53. Faccin, M.; Bruscolini, P.; Pelizzola, A. Analysis of the equilibrium and kinetics of the ankyrin repeat protein myotrophin. *J. Chem. Phys.* **2011**, *134*, 075102. [CrossRef] [PubMed]
54. Radford, S.E.; Dobson, C.M.; Dobson, C.M.; Fersht, A.R. Insights into protein folding using physical techniques: Studies of lysozyme and α-lactalbumin. *Philos. Trans. R. Soc. Lond. B Biol. Sci.* **1995**, *348*, 17–25. [CrossRef]
55. Mizuguchi, M.; Arai, M.; Ke, Y.; Nitta, K.; Kuwajima, K. Equilibrium and kinetics of the folding of equine lysozyme studied by circular dichroism spectroscopy. *J. Mol. Biol.* **1998**, *283*, 265–277. [CrossRef] [PubMed]

56. McCallister, E.L.; Alm, E.; Baker, D. Critical role of β-hairpin formation in protein G folding. *Nat. Struct. Biol.* **2000**, *7*, 669–673. [CrossRef] [PubMed]
57. Chaudhuri, T.K.; Arai, M.; Terada, T.P.; Ikura, T.; Kuwajima, K. Equilibrium and kinetic studies on folding of the authentic and recombinant forms of human α-lactalbumin by circular dichroism spectroscopy. *Biochemistry* **2000**, *39*, 15643–15651. [CrossRef] [PubMed]
58. Karanicolas, J.; Brooks, C.L., 3rd. The origins of asymmetry in the folding transition states of protein L and protein G. *Protein Sci.* **2002**, *11*, 2351–2361. [CrossRef]
59. Schlepckow, K.; Wirmer, J.; Bachmann, A.; Kiefhaber, T.; Schwalbe, H. Conserved folding pathways of α-lactalbumin and lysozyme revealed by kinetic CD, fluorescence, NMR, and interrupted refolding experiments. *J. Mol. Biol.* **2008**, *378*, 686–698. [CrossRef]
60. Nakamura, T.; Makabe, K.; Tomoyori, K.; Maki, K.; Mukaiyama, A.; Kuwajima, K. Different folding pathways taken by highly homologous proteins, goat α-lactalbumin and canine milk lysozyme. *J. Mol. Biol.* **2010**, *396*, 1361–1378. [CrossRef]
61. Yu, W.; Chung, K.; Cheon, M.; Heo, M.; Han, K.H.; Ham, S.; Chang, I. Cooperative folding kinetics of BBL protein and peripheral subunit-binding domain homologues. *Proc. Natl. Acad. Sci. USA* **2008**, *105*, 2397–2402. [CrossRef]
62. Halskau, O.; Perez-Jimenez, R.; Ibarra-Molero, B.; Underhaug, J.; Muñoz, V.; Martinez, A.; Sanchez-Ruiz, J.M. Large-scale modulation of thermodynamic protein folding barriers linked to electrostatics. *Proc. Natl. Acad. Sci. USA* **2008**, *105*, 8625–8630. [CrossRef] [PubMed]
63. Naganathan, A.N. Predictions from an Ising-like statistical mechanical model on the dynamic and thermodynamic effects of protein surface electrostatics. *J. Chem. Theory Comput.* **2012**, *8*, 4646–4656. [CrossRef] [PubMed]
64. Park, S.; Shin, Y.-K.; Yoon, J.-Y.; Nam, K.-H.; Munashingha, P.R.; Park, S.; Park, S.-Y.; Kim, S.; Lee, J.; Seo, M.J.; et al. Computational design of a thermolabile uracil-DNA glycosylase of Escherichia coli. *Biophys. J.* **2022**, *121*, 1276–1288. [CrossRef]
65. Bruscolini, P.; Naganathan, A.N. Quantitative prediction of protein folding behaviors from a simple statistical model. *J. Am. Chem. Soc.* **2011**, *133*, 5372–5379. [CrossRef]
66. Tsai, M.Y.; Yuan, J.M.; Teranishi, Y.; Lin, S.H. Thermodynamics of protein folding using a modified Wako-Saitô-Muñoz-Eaton model. *J. Biol. Phys.* **2012**, *38*, 543–571. [CrossRef]
67. Kim, H.; Kim, S.; Jung, Y.; Han, J.; Yun, J.H.; Chang, I.; Lee, W. Probing the folding-unfolding transition of a thermophilic protein, MTH1880. *PLoS ONE* **2016**, *11*, e0145853. [CrossRef] [PubMed]
68. Naganathan, A.N.; Orozco, M. The conformational landscape of an intrinsically disordered DNA-binding domain of a transcription regulator. *J. Phys. Chem. B* **2013**, *117*, 13842–13850. [CrossRef]
69. Naganathan, A.N.; Muñoz, V. Thermodynamics of downhill folding: Multi-probe analysis of PDD, a protein that folds over a marginal free energy barrier. *J. Phys. Chem. B* **2014**, *118*, 8982–8994. [CrossRef]
70. Naganathan, A.N.; Perez-Jimenez, R.; Muñoz, V.; Sanchez-Ruiz, J.M. Estimation of protein folding free energy barriers from calorimetric data by multi-model Bayesian analysis. *Phys. Chem. Chem. Phys.* **2011**, *13*, 17064–17076. [CrossRef]
71. Rajasekaran, N.; Gopi, S.; Narayan, A.; Naganathan, A.N. Quantifying protein disorder through measures of excess conformational entropy. *J. Phys. Chem. B* **2016**, *120*, 4341–4350. [CrossRef]
72. Gopi, S.; Singh, A.; Suresh, S.; Paul, S.; Ranu, S.; Naganathan, A.N. Toward a quantitative description of microscopic pathway heterogeneity in protein folding. *Phys. Chem. Chem. Phys.* **2017**, *19*, 20891–20903. [CrossRef] [PubMed]
73. Subramanian, S.; Golla, H.; Divakar, K.; Kannan, A.; de Sancho, D.; Naganathan, A.N. Slow folding of a helical protein: Large barriers, strong internal friction, or a shallow, bumpy landscape? *J. Phys. Chem. B* **2020**, *124*, 8973–8983. [CrossRef] [PubMed]
74. Munshi, S.; Naganathan, A.N. Imprints of function on the folding landscape: Functional role for an intermediate in a conserved eukaryotic binding protein. *Phys. Chem. Chem. Phys.* **2015**, *17*, 11042–11052. [CrossRef] [PubMed]
75. Sivanandan, S.; Naganathan, A.N. A disorder-induced domino-like destabilization mechanism governs the folding and functional dynamics of the repeat protein IκBα. *PLoS Comput. Biol.* **2013**, *9*, e1003403. [CrossRef]
76. Jurrus, E.; Engel, D.; Star, K.; Monson, K.; Brandi, J.; Felberg, L.E.; Brookes, D.H.; Wilson, L.; Chen, J.; Liles, K.; et al. Improvements to the APBS biomolecular solvation software suite. *Protein Sci.* **2018**, *27*, 112–128. [CrossRef]
77. Bhattacharjee, K.; Gopi, S.; Naganathan, A.N. A disordered loop mediates heterogeneous unfolding of an ordered protein by altering the native ensemble. *J. Phys. Chem. Lett.* **2020**, *11*, 6749–6756. [CrossRef]
78. Itoh, K.; Sasai, M. Multidimensional theory of protein folding. *J. Chem. Phys.* **2009**, *130*, 145104. [CrossRef]
79. Chung, H.S.; Tokmakoff, A. Temperature-de pendent downhill unfolding of ubiquitin. II. Modeling the free energy surface. *Proteins* **2008**, *72*, 488–497. [CrossRef]
80. Garniz-Arco, G.; Risso, V.A.; Gaucher, E.A.; Gavira, J.A.; Naganathan, A.N.; Ibarra-Molero, B.; Sanchez-Ruiz, J.M. Combining ancestral reconstruction with folding-landscape simulations to engineer heterologous protein expression. *J. Mol. Biol.* **2021**, *433*, 167321. [CrossRef]
81. Naganathan, A.N.; Dani, R.; Gopi, S.; Aranganathan, A.; Narayan, A. Folding intermediates, heterogeneous native ensembles and protein function. *J. Mol. Biol.* **2021**, *433*, 167325. [CrossRef]
82. Sato, S.; Religa, T.L.; Daggett, V.; Fersht, A.R. Testing protein-folding simulations by experiment: B domain of protein A. *Proc. Natl. Acad. Sci. USA* **2004**, *101*, 6952–6956. [CrossRef] [PubMed]
83. Sato, S.; Fersht, A.R. Searching for multiple folding pathways of a nearly symmetrical protein: Temperature dependent Φ-value analysis of the B domain of protein A. *J. Mol. Biol.* **2007**, *372*, 254–267. [CrossRef] [PubMed]

84. Rajasekaran, N.; Naganathan, A.N. A self-consistent structural perturbation approach for determining the magnitude and extent of allosteric coupling in proteins. *Biochem. J.* **2017**, *474*, 2379–2388. [CrossRef] [PubMed]
85. Rajasekaran, N.; Suresh, S.; Gopi, S.; Raman, K.; Naganathan, A.N. A general mechanism for the propagation of mutational effects in proteins. *Biochemistry* **2017**, *56*, 294–305. [CrossRef]
86. Garg, S.; Sagar, A.; Singaraju, G.S.; Dani, R.; Bari, N.K.; Naganathan, A.N.; Rakshit, S. Weakening of interaction networks with aging in tip-link protein induces hearing loss. *Biochem. J.* **2021**, *478*, 121–134. [CrossRef]
87. Gopi, S.; Rajasekaran, N.; Singh, A.; Ranu, S.; Naganathan, A.N. Energetic and topological determinants of a phosphorylation-induced disorder-to-order protein conformational switch. *Phys. Chem. Chem. Phys.* **2015**, *17*, 27264–27269. [CrossRef]
88. Naganathan, A.N.; Sanchez-Ruiz, J.M.; Munshi, S.; Suresh, S. Are protein folding intermediates the evolutionary consequence of functional constraints? *J. Phys. Chem. B* **2015**, *119*, 1323–1333. [CrossRef]
89. Naganathan, A.N. Molecular origins of folding rate differences in the thioredoxin family. *Biochem. J.* **2020**, *477*, 1083–1087. [CrossRef]
90. Narayan, A.; Gopi, S.; Lukose, B.; Naganathan, A.N. Electrostatic frustration shapes folding mechanistic differences in paralogous bacterial stress response proteins. *J. Mol. Biol.* **2020**, *432*, 4830–4839. [CrossRef]
91. Naganathan, A.N.; Kannan, A. A hierarchy of coupling free energies underlie the thermodynamic and functional architecture of protein structures. *Curr. Res. Struct. Biol.* **2021**, *3*, 257–267. [CrossRef]
92. Golla, H.; Kannan, A.; Gopi, S.; Murugan, S.; Perumalsamy, L.R.; Naganathan, A.N. Structural–energetic basis for coupling between equilibrium fluctuations and phosphorylation in a protein native ensemble. *ACS Cent. Sci.* **2022**, *8*, 282–293. [CrossRef]
93. Imparato, A.; Pelizzola, A.; Zamparo, M. Ising-like model for protein mechanical unfolding. *Phys. Rev. Lett.* **2007**, *98*, 148102. [CrossRef] [PubMed]
94. Imparato, A.; Pelizzola, A.; Zamparo, M. Protein mechanical unfolding: A model with binary variables. *J. Chem. Phys.* **2007**, *127*, 145105. [CrossRef] [PubMed]
95. Caraglio, M.; Imparato, A.; Pelizzola, A. Pathways of mechanical unfolding of FnIII(10): Low force intermediates. *J. Chem. Phys.* **2010**, *133*, 065101. [CrossRef] [PubMed]
96. Imparato, A.; Pelizzola, A. Mechanical unfolding and refolding pathways of ubiquitin. *Phys. Rev. Lett.* **2008**, *100*, 158104. [CrossRef] [PubMed]
97. Caraglio, M.; Imparato, A.; Pelizzola, A. Direction-dependent mechanical unfolding and green fluorescent protein as a force sensor. *Phys. Rev. E* **2011**, *84*, 021918. [CrossRef] [PubMed]
98. Imparato, A.; Pelizzola, A.; Zamparo, M. Equilibrium properties and force-driven unfolding pathways of RNA molecules. *Phys. Rev. Lett.* **2009**, *103*, 188102. [CrossRef]
99. Ancona, M.; Bentivoglio, A.; Caraglio, M.; Gonnella, G.; Pelizzola, A. Emergence of effective temperatures in an out-of-equilibrium model of biopolymer folding. *Phys. Rev. E* **2021**, *103*, 062415. [CrossRef]
100. Aioanei, D.; Tessari, I.; Bubacco, L.; Samori, B.; Brucale, M. Observing the osmophobic effect in action at the single molecule level. *Proteins* **2011**, *79*, 2214–2223. [CrossRef]
101. Aioanei, D.; Brucale, M.; Tessari, I.; Bubacco, L.; Samori, B. Worm-like Ising model for protein mechanical unfolding under the effect of osmolytes. *Biophys. J.* **2012**, *102*, 342–350. [CrossRef]
102. Caraglio, M.; Pelizzola, A. Effects of confinement on thermal stability and folding kinetics in a simple Ising-like model. *Phys. Biol.* **2012**, *9*, 016006. [CrossRef] [PubMed]
103. Pelizzola, A.; Zamparo, M. Nonequilibrium dynamics of an exactly solvable Ising-like model and protein translocation. *Eur. Lett.* **2013**, *102*, 10001. [CrossRef]
104. Bruscolini, P.; Pelizzola, A.; Zamparo, M. Rate determining factors in protein model structures. *Phys. Rev. Lett.* **2007**, *99*, 038103. [CrossRef] [PubMed]
105. Niewieczerzal, S.; Cieplak, M. The folding transition state theory in simple model systems. *J. Phys. Condens. Matter* **2008**, *20*, 244134. [CrossRef]
106. Narayan, A.; Naganathan, A.N. Evidence for the sequential folding mechanism in RNase H from an ensemble-based model. *J. Phys. Chem. B* **2014**, *118*, 5050–5058. [CrossRef]
107. Chang, I.; Cieplak, M.; Banavar, J.R.; Maritan, A. What can one learn from experiments about the elusive transition state? *Protein Sci.* **2004**, *13*, 2446–2457. [CrossRef]
108. Itoh, K.; Sasai, M. Cooperativity, connectivity, and folding pathways of multidomain proteins. *Proc. Natl. Acad. Sci. USA* **2008**, *105*, 13865–13870. [CrossRef]
109. Wako, H.; Abe, H. Characterization of protein folding by a Φ-value calculation with a statistical-mechanical model. *Biophys. Phys.* **2016**, *13*, 263–279. [CrossRef]
110. Gopi, S.; Paul, S.; Ranu, S.; Naganathan, A.N. Extracting the hidden distributions underlying the mean transition state structures in protein folding. *J. Phys. Chem. Lett.* **2018**, *9*, 1771–1777. [CrossRef]
111. Gopi, S.; Naganathan, A.N. Non-specific DNA-driven quinary interactions promote structural transitions in proteins. *Phys. Chem. Chem. Phys.* **2020**, *22*, 12671–12677. [CrossRef]
112. Gopi, S.; Lukose, B.; Naganathan, A.N. Diverse native ensembles dictate the differential functional responses of nuclear receptor ligand-binding domains. *J. Phys. Chem. B* **2021**, *125*, 3546–3555. [CrossRef] [PubMed]

113. Hutton, R.D.; Wilkinson, J.; Faccin, M.; Sivertsson, D.M.; Pelizzola, A.; Lowe, A.R.; Bruscolini, P.; Itzhaki, L.S. Mapping the topography of a protein energy landscape. *J. Am. Chem. Soc.* **2015**, *137*, 14610–14625. [CrossRef] [PubMed]
114. Gopi, S.; Aranganathan, A.; Naganathan, A.N. Thermodynamics and folding landscapes of large proteins from a statistical mechanical model. *Curr. Res. Struct. Biol.* **2019**, *1*, 6–12. [CrossRef]
115. Pacheco-Garcia, J.L.; Anoz-Carbonell, E.; Vankova, P.; Kannan, A.; Palomino-Morales, R.; Mesa-Torres, N.; Salido, E.; Man, P.; Medina, M.; Naganathan, A.N.; et al. Structural basis of the pleiotropic and specific phenotypic consequences of missense mutations in the multifunctional NAD(P)H:quinone oxidoreductase 1 and their pharmacological rescue. *Redox Biol.* **2021**, *46*, 102112. [CrossRef] [PubMed]
116. Sannigrahi, A.; Chowdhury, S.; Das, B.; Banerjee, A.; Halder, A.; Kumar, A.; Saleem, M.; Naganathan, A.N.; Karmakar, S.; Chattopadhyay, K. The metal cofactor zinc and interacting membranes modulate SOD1 conformation-aggregation landscape in an in vitro ALS model. *eLife* **2021**, *10*, e61453. [CrossRef] [PubMed]
117. Apic, G.; Gough, J.; Teichmann, S.A. Domain combinations in archaeal, eubacterial and eukaryotic proteomes. *J. Mol. Biol.* **2001**, *310*, 311–325. [CrossRef]
118. Han, J.H.; Batey, S.; Nickson, A.A.; Teichmann, S.A.; Clarke, J. The folding and evolution of multidomain proteins. *Nat. Rev. Mol. Cell Biol.* **2007**, *8*, 319–330. [CrossRef]
119. Aroul-Selvam, R.; Hubbard, T.; Sasidharan, R. Domain insertions in protein structures. *J. Mol. Biol.* **2004**, *338*, 633–641. [CrossRef]
120. Touchette, N.A.; Perry, K.M.; Matthews, C.R. Folding of dihydrofolate reductase from Escherichia coli. *Biochemistry* **1986**, *25*, 5445–5452. [CrossRef]
121. Fersht, A.R. The sixth Datta Lecture. Protein folding and stability: The pathway of folding of barnase. *FEBS Lett.* **1993**, *325*, 5–16. [CrossRef]
122. Jennings, P.A.; Finn, B.E.; Jones, B.E.; Matthews, C.R. A reexamination of the folding mechanism of dihydrofolate reductase from Escherichia coli: Verification and refinement of a four-channel model. *Biochemistry* **1993**, *32*, 3783–3789. [CrossRef] [PubMed]
123. Jones, B.E.; Matthews, C.R. Early intermediates in the folding of dihydrofolate reductase from Escherichia coli detected by hydrogen exchange and NMR. *Protein Sci.* **1995**, *4*, 167–177. [CrossRef]
124. Arai, M.; Kuwajima, K. Rapid formation of a molten globule intermediate in refolding of α-lactalbumin. *Fold. Des.* **1996**, *1*, 275–287. [CrossRef]
125. Arai, M.; Hamel, P.; Kanaya, E.; Inaka, K.; Miki, K.; Kikuchi, M.; Kuwajima, K. Effect of an alternative disulfide bond on the structure, stability, and folding of human lysozyme. *Biochemistry* **2000**, *39*, 3472–3479. [CrossRef] [PubMed]
126. Arai, M.; Ito, K.; Inobe, T.; Nakao, M.; Maki, K.; Kamagata, K.; Kihara, H.; Amemiya, Y.; Kuwajima, K. Fast compaction of α-lactalbumin during folding studied by stopped-flow X-ray scattering. *J. Mol. Biol.* **2002**, *321*, 121–132. [CrossRef]
127. Arai, M.; Maki, K.; Takahashi, H.; Iwakura, M. Testing the relationship between foldability and the early folding events of dihydrofolate reductase from Escherichia coli. *J. Mol. Biol.* **2003**, *328*, 273–288. [CrossRef]
128. Arai, M.; Kataoka, M.; Kuwajima, K.; Matthews, C.R.; Iwakura, M. Effects of the difference in the unfolded-state ensemble on the folding of Escherichia coli dihydrofolate reductase. *J. Mol. Biol.* **2003**, *329*, 779–791. [CrossRef]
129. Arai, M.; Iwakura, M. Probing the interactions between the folding elements early in the folding of Escherichia coli dihydrofolate reductase by systematic sequence perturbation analysis. *J. Mol. Biol.* **2005**, *347*, 337–353. [CrossRef]
130. Arai, M.; Kondrashkina, E.; Kayatekin, C.; Matthews, C.R.; Iwakura, M.; Bilsel, O. Microsecond hydrophobic collapse in the folding of Escherichia coli dihydrofolate reductase, an α/β-type protein. *J. Mol. Biol.* **2007**, *368*, 219–229. [CrossRef]
131. Arai, M.; Iwakura, M.; Matthews, C.R.; Bilsel, O. Microsecond subdomain folding in dihydrofolate reductase. *J. Mol. Biol.* **2011**, *410*, 329–342. [CrossRef]
132. Muñoz, V. A simple theoretical model goes a long way in explaining complex behavior in protein folding. *Proc. Natl. Acad. Sci. USA* **2014**, *111*, 15863–15864. [CrossRef]
133. Wright, P.E.; Dyson, H.J. Intrinsically unstructured proteins: Re-assessing the protein structure-function paradigm. *J. Mol. Biol.* **1999**, *293*, 321–331. [CrossRef] [PubMed]
134. Dunker, A.; Lawson, J.; Brown, C.J.; Williams, R.M.; Romero, P.; Oh, J.S.; Oldfield, C.J.; Campen, A.M.; Ratliff, C.M.; Hipps, K.W.; et al. Intrinsically disordered protein. *J. Mol. Graph. Model.* **2001**, *19*, 26–59. [CrossRef]
135. Arai, M.; Sugase, K.; Dyson, H.J.; Wright, P.E. Conformational propensities of intrinsically disordered proteins influence the mechanism of binding and folding. *Proc. Natl. Acad. Sci. USA* **2015**, *112*, 9614–9619. [CrossRef]
136. Matsushita, K.; Kikuchi, M. Frustration-induced protein intrinsic disorder. *J. Chem. Phys.* **2013**, *138*, 105101. [CrossRef]
137. Matsushita, K.; Kikuchi, M. Short polypeptide with metastable structures. *Interdiscip. Sci. Rev.* **2013**, *19*, 29–34. [CrossRef]
138. Matsushita, K.; Kikuchi, M. Multicanonical simulation of coupled folding and binding of intrinsically disordered protein using an Ising-like protein model. *J. Phys. Conf. Ser.* **2013**, *454*, 012034. [CrossRef]
139. Munshi, S.; Gopi, S.; Asampille, G.; Subramanian, S.; Campos, L.A.; Atreya, H.S.; Naganathan, A.N. Tunable order-disorder continuum in protein-DNA interactions. *Nucleic Acids Res.* **2018**, *46*, 8700–8709. [CrossRef]
140. Rajendran, D.; Mitra, S.; Oikawa, H.; Madhurima, K.; Sekhar, A.; Takahashi, S.; Naganathan, A.N. Quantification of entropic excluded volume effects driving crowding-induced collapse and folding of a disordered protein. *J. Phys. Chem. Lett.* **2022**, *13*, 3112–3120. [CrossRef]
141. Munshi, S.; Rajendran, D.; Naganathan, A.N. Entropic control of an excited folded-like conformation in a disordered protein ensemble. *J. Mol. Biol.* **2018**, *430*, 2688–2694. [CrossRef]

142. Munshi, S.; Rajendran, D.; Ramesh, S.; Subramanian, S.; Bhattacharjee, K.; Kumar, M.R.; Naganathan, A.N. Controlling structure and dimensions of a disordered protein via mutations. *Biochemistry* **2020**, *59*, 171–174. [CrossRef] [PubMed]
143. Itoh, K.; Sasai, M. Coupling of functioning and folding: Photoactive yellow protein as an example system. *Chem. Phys.* **2004**, *307*, 121–127. [CrossRef]
144. Itoh, K.; Sasai, M. Dynamical transition and proteinquake in photoactive yellow protein. *Proc. Natl. Acad. Sci. USA* **2004**, *101*, 14736–14741. [CrossRef]
145. Itoh, K.; Sasai, M. Entropic mechanism of large fluctuation in allosteric transition. *Proc. Natl. Acad. Sci. USA* **2010**, *107*, 7775–7780. [CrossRef]
146. Itoh, K.; Sasai, M. Statistical mechanics of protein allostery: Roles of backbone and side-chain structural fluctuations. *J. Chem. Phys.* **2011**, *134*, 125102. [CrossRef] [PubMed]
147. Hellingwerf, K.J.; Hendriks, J.; Gensch, T. Photoactive yellow protein, a new type of photoreceptor protein: Will this "yellow lab" bring us where we want to go? *J. Phys. Chem. A* **2003**, *107*, 1082–1094. [CrossRef]
148. Narayan, A.; Campos, L.A.; Bhatia, S.; Fushman, D.; Naganathan, A.N. Graded structural polymorphism in a bacterial thermosensor protein. *J. Am. Chem. Soc.* **2017**, *139*, 792–802. [CrossRef] [PubMed]
149. Narayan, A.; Naganathan, A.N. Tuning the continuum of structural states in the native ensemble of a regulatory protein. *J. Phys. Chem. Lett.* **2017**, *8*, 1683–1687. [CrossRef]
150. Narayan, A.; Naganathan, A.N. Switching protein conformational substates by protonation and mutation. *J. Phys. Chem. B* **2018**, *122*, 11039–11047. [CrossRef]
151. Chiti, F.; Dobson, C.M. Protein misfolding, functional amyloid, and human disease. *Annu. Rev. Biochem.* **2006**, *75*, 333–366. [CrossRef]
152. Iadanza, M.G.; Jackson, M.P.; Hewitt, E.W.; Ranson, N.A.; Radford, S.E. A new era for understanding amyloid structures and disease. *Nat. Rev. Mol. Cell Biol.* **2018**, *19*, 755–773. [CrossRef]
153. Zamparo, M.; Trovato, A.; Maritan, A. Simplified exactly solvable model for β-amyloid aggregation. *Phys. Rev. Lett.* **2010**, *105*, 108102. [CrossRef] [PubMed]
154. Xiao, X.H.; Wu, M.C. Simplified lattice model for polypeptide fibrillar transitions. *Phys. Rev. E* **2014**, *90*, 042701. [CrossRef] [PubMed]
155. Tokar, V.I.; Dreysse, H. A model of strained epitaxy on an alloyed substrate. *J. Phys. Condens. Matter* **2004**, *16*, S2203–S2210. [CrossRef]
156. Tokar, V.I.; Dreysse, H. Transfer matrix solution of the Wako-Saitô-Muñoz-Eaton model augmented by arbitrary short range interactions. *J. Stat. Mech.* **2010**, *2010*, P08028. [CrossRef]
157. Pelizzola, A. Exactness of the cluster variation method and factorization of the equilibrium probability for the Wako-Saitô-Muñoz-Eaton model of protein folding. *J. Stat. Mech.* **2005**, *2005*, P11010. [CrossRef]
158. Pelizzola, A. Cluster variation method in statistical physics and probabilistic graphical models. *J. Phys. A Math. Theory* **2005**, *38*, R309–R339. [CrossRef]
159. Zamparo, M.; Pelizzola, A. Rigorous results on the local equilibrium kinetics of a protein folding model. *J. Stat. Mech.* **2006**, *2006*, P12009. [CrossRef]
160. Lee, J. Exact partition function zeros of the Wako-Saitô-Muñoz-Eaton β hairpin model. *Phys. Rev. E* **2013**, *88*, 022710. [CrossRef]
161. Lee, J. Exact partition function zeros of the Wako-Saitô-Muñoz-Eaton protein model. *Phys. Rev. Lett.* **2013**, *110*, 248101. [CrossRef]
162. Lee, J. Phase diagram of the Wako-Saitô-Muñoz-Eaton β hairpin model obtained from the partition-function zeros. *J. Korean Phys. Soc.* **2014**, *65*, 671–675. [CrossRef]
163. Lee, J. Finite-size phase diagram of the Wako-Saitô-Muñoz-Eaton α-helix model. *J. Korean Phys. Soc.* **2015**, *67*, 1187–1191. [CrossRef]
164. Jumper, J.; Evans, R.; Pritzel, A.; Green, T.; Figurnov, M.; Ronneberger, O.; Tunyasuvunakool, K.; Bates, R.; Žídek, A.; Potapenko, A.; et al. Highly accurate protein structure prediction with AlphaFold. *Nature* **2021**, *596*, 583–589. [CrossRef] [PubMed]
165. Outeiral, C.; Nissley, D.A.; Deane, C.M. Current structure predictors are not learning the physics of protein folding. *Bioinformatics* **2022**, *38*, 1881–1887. [CrossRef]

MDPI
St. Alban-Anlage 66
4052 Basel
Switzerland
Tel. +41 61 683 77 34
Fax +41 61 302 89 18
www.mdpi.com

Molecules Editorial Office
E-mail: molecules@mdpi.com
www.mdpi.com/journal/molecules

www.ingramcontent.com/pod-product-compliance
Lightning Source LLC
Chambersburg PA
CBHW041938260926
51850CB00018B/250